Molecular Design and Applications of Photofunctional Polymers and Materials

RSC Polymer Chemistry Series

Series Editors:
Professor Ben Zhong Tang (Editor-in-Chief), *The Hong Kong University of Science and Technology, Hong Kong, China*
Professor Alaa S. Abd-El-Aziz, *University of Prince Edward Island, Canada*
Professor Stephen L. Craig, *Duke University, USA*
Professor Jianhua Dong, *National Natural Science Foundation of China, China*
Professor Toshio Masuda, *Fukui University of Technology, Japan*
Professor Christoph Weder, *University of Fribourg, Switzerland*

How to obtain future titles on publication:
A standing order plan is available for this series. A standing order will bring delivery of each new volume immediately on publication.

For further information please contact:
Book Sales Department, Royal Society of Chemistry, Thomas Graham House, Science Park, Milton Road, Cambridge, CB4 0WF, UK
Telephone: +44 (0)1223 420066, Fax: +44 (0)1223 420247,
Email: booksales@rsc.org
Visit our website at http://www.rsc.org/Shop/Books/

Molecular Design and Applications of Photofunctional Polymers and Materials

Edited by

Wai-Yeung Wong
*Department of Chemistry and Centre for Advanced Luminescence Materials,
Hong Kong Baptist University, Hong Kong*
Email: rwywong@hkbu.edu.hk

Alaa S Abd-El-Aziz
University of Prince Edward Island, Canada

RSC Publishing

RSC Polymer Chemistry Series No. 2

ISBN: 978-1-84973-575-9
ISSN: 2044-0790

A catalogue record for this book is available from the British Library

Published by The Royal Society of Chemistry,
Thomas Graham House, Science Park, Milton Road,
Cambridge CB4 0WF, UK

Registered Charity Number 207890

For further information see our web site at www.rsc.org

Printed in the United Kingdom by Henry Ling Limited, at the Dorset Press, Dorchester, DT1 1HD

Preface

Photofunctional organic and organometallic polymers and materials have become a field of intense activity in optoelectronic/photonic and energy research since they have unique advantages in device fabrication – they are low cost and light weight, offer ease of device fabrication and have potential use in flexible devices. They hold great promise as versatile functional materials for use in energy interconversions. New synthetic methods need to be developed to produce technologically useful materials with specific functional roles. Such research clearly presents a promising way out of the worldwide energy problem. The transformations of light into electricity (*i.e.* electrical energy generation in photovoltaic cells) and electricity into light (*i.e.* light generation in organic light-emitting diodes) are two important interrelated but complementary areas that have attracted considerable research interest in recent years. Work in other optoelectronic, electronic and nanotechnological applications are also emerging rapidly and attract considerable academic and industrial attention. This edited book aims to survey recent research at the frontiers of the topic, with emphasis on fundamental concepts and current applications. Here, a group of experts in the field have made significant contributions to this volume. Other topics such as photochemically degradable polymers, electrochromic and photochromic materials, biosensing and bioimaging materials, and low- and high-refractive index materials are also included. This contribution is very timely and will attract much attention in the scientific community.

Chapter 1, written by Ho and Wong, presents the development of some yellow- and orange-emitting metallophosphors. These phosphorescent dyes find key applications in fabricating monochromatic and white organic light-emitting devices (OLEDs). Chapter 2, by Brady and Tyler, describes some photochemically degradable polymers containing metal-metal bonds and their

RSC Polymer Chemistry Series No. 2
Molecular Design and Applications of Photofunctional Polymers and Materials
Edited by Wai-Yeung Wong and Alaa S Abd-El-Aziz
© The Royal Society of Chemistry 2012
Published by the Royal Society of Chemistry, www.rsc.org

synthetic and photochemical aspects. The applications of metal acetylide complexes and polymers as photofunctional materials and molecular wires are summarized by Raithby and Long in Chapters 3 and 4, respectively. Chapter 5, by Lo and co-workers, covers the growing field of luminescent transition-metal complexes as biomolecular labels and probes. In Chapters 6 and 7, Nishihara *et al.* first report the exciting area of photoactive metallopolymers whereas Wong and Yam survey the interesting field of photofunctional metal-containing complexes in the latter case. All of these molecules can function effectively as photochromic and optoelectronic materials. In Chapter 8, Mak and co-workers highlight the synthesis and applications of some rhenium(I)-containing polymers. Chapter 9, by Abd-El-Aziz and Strohm, discusses the recent advances in the chemistry of azo-conjugated organometallic and coordination compounds, ranging from small molecules to polymers. Tang *et al.* cover in Chapter 10 the synthesis, characterization and potential photonic applications of some hyperbranched aryleneethynylene polymers. In Chapter 11, Tian and co-workers also nicely present the molecular design of various organic photosensitizing dyes with various donor-π-acceptor-type structures which can be applied in dye-sensitized solar cells, whereas Jin also discusses the related work in Chapter 15 using functional polymers instead of small molecules. A good overview of the recent advances in conjugated polymer nanoparticles is given by Liu *et al.* in Chapter 12, in which they review their different synthetic methods and applications ranging from optoelectronics, cellular imaging to biosensors. In Chapters 13 and 14, Wu and Xie describe some recent studies on polymer-based white light-emitting devices and organic photovoltaic cells, respectively. Within the realm of energy science, both of these topics have important implications towards energy-saving and energy-producing applications.

Clearly, photofunctional materials chemistry represents an intriguing interdisciplinary subject that deserves further research attention in the forthcoming years. Many emerging future challenges as well as opportunities exist for both exploratory and application-oriented research. We are very optimistic to the continuing developments of these light-driven functional materials in the scientific community.

Wai-Yeung Wong
Alaa S. Abd-El-Aziz

Contents

RSC Polymer Chemistry Series No. 2
Molecular Design and Applications of Photofunctional Polymers and Materials
Edited by Wai-Yeung Wong and Alaa S Abd-El-Aziz
© The Royal Society of Chemistry 2012
Published by the Royal Society of Chemistry, www.rsc.org

**Chapter 3   Metal Acetylide Complexes, Oligomers and Polymers in
Photofunctional Materials Chemistry   56**
Marek Jura, Paul R. Raithby and Paul J. Wilson

**Chapter 4   Metal σ-Alkynyl Complexes as Molecular Wires and Devices:
A Comparative Study of Electron Density and Delocalisation   85**
Michael S. Inkpen and Nicholas J. Long

CHAPTER 1

Heavy-Metal Organometallic Complexes as Yellow and Orange Triplet Emitters for Organic Light-Emitting Diodes

CHEUK-LAM HO AND WAI-YEUNG WONG*

Institute of Molecular Functional Materials[†] and Department of Chemistry, Hong Kong Baptist University, Waterloo Road, Kowloon Tong, Hong Kong, P. R. China
*E-mail: rwywong@hkbu.edu.hk
[†]Areas of Excellence Scheme, University Grants Committee, Hong Kong

1.1 Introduction

While lighting applications account for about 19% of the electricity consumption of the world,[1] the need to reduce energy consumption associated with the low efficiency of conventional lighting systems (*e.g.* the incandescent bulbs) has prompted researchers to pay considerable research attention to developing new energy-saving technologies such as organic light-emitting devices (OLEDs). Actually, incandescent bulbs that have long been the most common lighting sources are very inefficient (converting only 5–10% of this energy into light) and dissipate the main part of the electrical energy absorbed as heat. Even the energy-saving compact fluorescent lamps are only about 20% energy efficient with typical power efficiency of 40–70 lm W^{-1}. Moreover, fluorescent lamps contain a small but significant amount of toxic mercury in the tube,[2] which

RSC Polymer Chemistry Series No. 2
Molecular Design and Applications of Photofunctional Polymers and Materials
Edited by Wai-Yeung Wong and Alaa S Abd-El-Aziz
© The Royal Society of Chemistry 2012
Published by the Royal Society of Chemistry, www.rsc.org

complicates their disposal and causes an important environmental impact. Recently, the efficiencies of white organic light-emitting devices (WOLEDs) have been shown to approach or surpass those of the fluorescent lamps due to recent advances in novel material synthesis and optimisation of device structures in the past few years.[3,4] The key advantages of OLEDs for flat-panel display applications are their self-emitting property, high luminous efficiency, full colour capability, wide viewing angle, high contrast, low power consumption, low weight, potentially large-area colour displays and flexibility.[5] In particular, recent developments in using phosphorescent materials have led to significant improvements in OLED performance up to 100 lm W^{-1},[6] thus providing organic semiconducting lighting with a very bright future and allowing WOLEDs to become the next generation of light illumination systems.

White-light emission can be obtained based on the principle of additive colour mixing. In practice, this is mostly done by mixing the three primary colours (red, green and blue, RGB). Besides R-G-B phosphorescent emitters, phosphors showing complementary colours, such as blue (B) and orange or yellow (O or Y), can also be utilised to produce white-light emission in the devices. This approach can eliminate the necessity for excessive emissive dopants in a device, hence reducing structural heterogeneities and the device fabrication process can generally be simplified. Building on these attractive properties, research on two-colour WOLEDs still remains scarce and is driving many researchers to investigate high-efficiency yellow or orange triplet emitters.

Of the various heavy-metal ions that could be envisaged for promoting radiative emission of triplet states in OLEDs, iridum(III), platinum(II) and other transition metals have attracted most attention to date. Flourishing studies over the past decade have revealed how the excited-state energies and hence emission colours in several classes of their complexes can be controlled through rational ligand design.[7] Here, we summarise a number of triplet emitters that show yellow or orange electroluminescence (EL) that may serve as good candidates for WOLED applications.

1.2   Iridium(III) Complexes

Iridium(III) complexes are considered to be the seminal generation of phosphorescent emitters. As a general approach, the emission peak wavelength was found to be greatly dependent on the molecular design of the cyclometallating ligand chelates.

1.2.1   Homoleptic and Heteroleptic Iridium(III) Complexes with Modified 2-Phenylpyridyl Moieties

Recently, for OLEDs based on heavy-metal Ir(III) complexes, the benchmark green emitters *fac*-[Ir(ppy)₃] and [Ir(ppy)₂(acac)][8] (Hppy = 2-phenylpyridine, Hacac = acetylacetone) have drawn great attention due to their ease of synthesis and high efficiency. To generate yellow or orange colour, a very versatile avenue can be adopted towards emission colour tuning of Ir(III) complexes *via* the facile derivatisation of the phenyl or pyridyl moiety of ppy with various substituents or functionalities. Attachment of the main-group moieties SO_2 and PO to ppy in **Ir-1** and **Ir-2** was reported by Wong *et al.* in which the approach successfully shifts the charge-transfer character from the pyridyl group in some ppy-type complexes to the electron-withdrawing main-group moieties.[9] This kind of complex improves their electron-injection (EI) and electron-transporting (ET) features. The strongly electron-withdrawing inductive influence of the polar PO group in **Ir-2** was shown to lower significantly the lowest unoccupied molecular orbital (LUMO) energy, thus redshifting the emission peak as compared to [Ir(ppy)$_2$(acac)]. **Ir-2** shows a shorter emission wavelength at 541 nm than **Ir-1** (550 nm). The vacuum-deposited yellow-emitting device based on **Ir-1** turned on at 3.7 V, with a maximum luminance (L_{max}) of 48 567 cd m^{-2} at 19.9 V and maximum external quantum efficiency (η_{ext}) of 10.7%, luminance efficiency (η_L) of 35.1 cd A^{-1}, and power efficiency (η_P) of 23.1 lm W^{-1}, while a higher turn-on voltage (12.1 V) was found in the solution-processed yellow-orange device based on **Ir-2** with L_{max} of 7103 cd m^{-2} at 27.9 V and maximum efficiencies of 3.49%, 11.89 cd A^{-1} and 2.40 lm W^{-1}.

By extending the π-electron delocalisation of the aromatic ligand, the energy gap between the ground and lowest excited states can be effectively reduced to provide a redshifted emission. Complexes **Ir-3** to **Ir-7** consist of additional fused aromatic rings either on the phenyl or pyridyl ring of the ppy ligand. Yellow phosphorescence from **Ir-3** exhibits a very broad and featureless peak at *ca.* 550 nm with a wide full spectral width at half-maximum (FWHM) of 87 nm without the formation of excimer emission.[10] It gives high peak forward-viewing η_L, η_P and η_{ext} of 33.6 cd A^{-1}, 21.8 lm W^{-1} and 10.5% along with Commission international de l'Eclairage (CIE) coordinates of (0.45, 0.54). This broad emission and nonexcimer formation effectively overcome the problem of inevitably low colour rendering indexes (CRI) that are common in many WOLEDs. Highly efficient WOLEDs with maximum efficiencies as high as 46 cd A^{-1} and 41 lm W^{-1} under forward voltage bias can be achieved.

3-Phenylisoquinolinyl (**Ir-4** to **Ir-6**)[11] and phenylbenzoquinoline-based (**Ir-7**)[12] Ir(III) complexes also achieve yellow or orange electrophosphorescence.

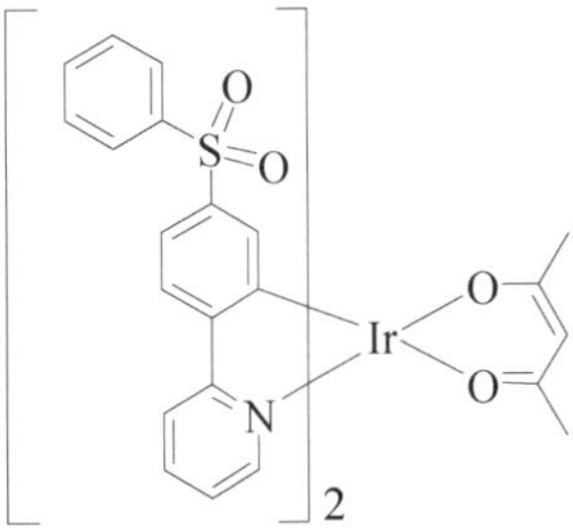

Ir-1

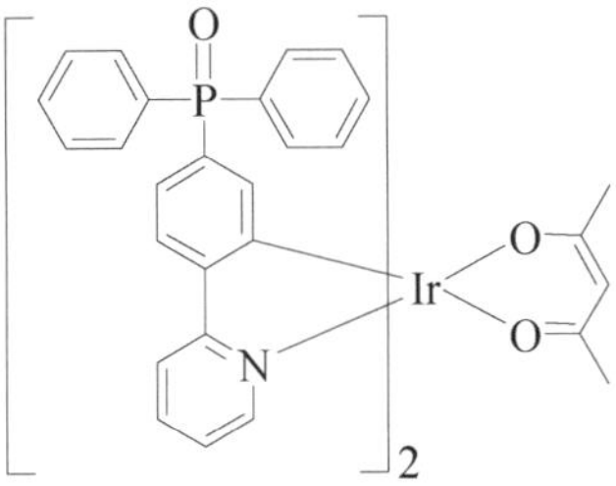

Ir-2

Ir-3 X = H Ir-4
X = CF$_3$ Ir-5 Ir-6 Ir-7

Their corresponding CIE coordinates are (0.49, 0.51), (0.46, 0.53), (0.41, 0.54) and (0.55, 0.44) for **Ir-4** to **Ir-7**, respectively. The better EL performance observed in **Ir-4** can be ascribed to its shorter phosphorescence lifetime. The triplet–triplet (T–T) annihilation becomes more serious for the fluoro derivatives **Ir-5** and **Ir-6**.

Substitution of benzothiazole for pyridine leads to a redshift in the phosphorescence emission. The EL peaks of benzothiazole-based complexes **Ir-8** and **Ir-9** [7e,13,14] appear at around 560 nm in the yellow region and the replacement of S by O in the chromophore causes a blueshift in the emission wavelength to the green region for **Ir-10** due to the lower polarisability and basicity of oxygen relative to sulfur. However, by introducing a CF$_3$ substituent to **Ir-10**,[15] the phosphor dye emits yellow light at 544 nm with luminance up to 9200 cd m^{-2} at 100 mA cm^{-2}. Orange-emitting OLEDs were also fabricated by using crosslinkable compound **Ir-11**,[16] which can be crosslinked with a hole-conducting matrix *via* cationic ring-opening polymerisation to yield an insoluble emitting layer by using the spin-coating method. The optimal device has a maximum η_L of 18.4 cd A^{-1} at 100 cd m^{-2} and η_P of 11.7 lm W^{-1}. At a luminous density of 1000 cd m^{-2}, the efficiency was still 15 cd A^{-1} and 11.5 lm W^{-1}, respectively. The results gave a performance that compares favourably with other multilayered devices fabricated by thermal evaporation.

Kwon and coworkers designed a Ir(III) complex **Ir-12**[17] with high device efficiency owing to its regulated energy levels and high stability stemming from

Ir-8 **Ir-9** **Ir-10**

Ir-11   **Ir-12**

the introduction of the imide group as well as the suppressed T–T annihilation in the EL devices in the presence of a bulky isopropyl unit. All the devices with different doping concentrations emit intense orange EL at 566–570 nm with the CIE chromaticity coordinates of (0.51, 0.49) to (0.52, 0.47). The orange-emitting device (λ = 570 nm) with 4 wt.% dopant level exhibited a low operation voltage of 5.3 V at 1000 cd m^{-2}, a maximum η_{ext} of 13.8%, and a maximum η_P of 32.7 lm W^{-1}. Adoption of a double layer-structure enhanced the η_{ext} further to 14.4%.

The use of styrylbenzoimidazole derivatives instead of the traditional pyridine derivatives enhances the ET ability and OLED efficiency at high brightness levels. Devices based on **Ir-13** and **Ir-14** gave yellow EL at 570 and 585 nm, respectively.[18] The performances of the devices based on **Ir-14** are inferior to those of the corresponding devices based on **Ir-13**. The device based on **Ir-13** exhibited a L_{max} of 56 162 cd m^{-2}, accompanied by high η_L of 25.7 cd A^{-1} at high brightness of 1000 cd m^{-2} and 20.7 cd A^{-1} at 10 000 cd m^{-2}.

To investigate the structure–property relationship and develop durable, high-efficiency devices, the groups of Guo and Zou synthesised a series of

R = H   **Ir-13**
CH$_3$ **Ir-14**

Ir-15

Ir-16

Ir-17

Ir(III) complexes **Ir-15** to **Ir-17** bearing the pyrazine ligands. High yellow colour purity with a featureless emission peak at 580 nm was obtained using **Ir-15**.[19] A systematic study on colour tuning *via* variation of the cyclometallating and ancillary ligands was then carried out. Their phosphorescence peak wavelength can be fine tuned in the yellow range. The emission shifts to the blue in **Ir-16**[20] (λ_{EL} = 561 nm; CIE = (0.45, 0.54)) by adding some fluoro groups to the ligand. The use of picolinic acid instead of acetylacetonate as the ancillary ligand in **Ir-17** also causes a hypsochromic shift in the EL wavelength to 569 nm with CIE of (0.48, 0.50). This result suggests that the strategies for tuning the emission colour and colour purity by changing the ligand substituents can also be applied to the pyrazine system.

The use of a sterically hindered amidinate as the chelating ancillary ligand in **Ir-18** effectively relieves the self-quenching problem in the EL emission and such a group also acts as a good ambipolar charge-transporting material with high hole and electron mobilities.[21] The highest occupied molecular orbital (HOMO) level of this complex is higher than that with acac ligand and therefore the energy gap of **Ir-18** is significantly smaller than that with acac and this can be ascribed in part to the higher π-bonding ability of the amidinate ligand. The better hole-transporting ability of **Ir-18** resulted in excellent EL performance for the orange-emitting device with peak η_{ext} of 18.4%, η_L of 21.4 cd A^{-1} and η_P of 18.7 lm W^{-1}. A WOLED with the structure of ITO/NPB/**Ir-18**:Bepp$_2$/Bepp$_2$/LiF/Al was fabricated (Bepp$_2$ = bis(2-(2-hydroxyphenyl)-pyridine)beryllium, a deep blue-emitting fluorescent complex). The CIE coordinates remain almost unchanged (0.35 $\pm$ 0.02, 0.33 $\pm$ 0.02) upon varying the luminance and so the device possesses a high colour reproduci-

Ir-18

bility. A reasonably low turn-on voltage (2.7 V) was recorded with remarkable EL efficiency at a peak η_{ext} of 27.8% (η_L of 60.8 cd A^{-1}) at a luminance of 300 cd m^{-2} (4.4 V) and the highest η_P of 48.8 lm W^{-1} at 80 cd m^{-2} (3.8 V). Such high performance of the WOLED should be attributed to the carrier direct-injection mechanism as well as efficient energy transfer from Bepp$_2$ to **Ir-18** in the emitting layer.

Devices based on the heteroleptic Ir(III) cyclometallates **Ir-19** and **Ir-20** containing both azolate and diphenylphosphinoaryl chelates displayed bright orange emission.[22] Their photophysical properties are almost identical, regardless of the identity of X and their phosphorescence quantum efficiency can reach near unity for good device performance. The device structure employed was ITO/PEDOT:PSS/NPD/TCTA/host:dopant (10%)/TPBI/LiF/Al (PEDOT:PSS = poly(ethylenedioxythiophene)-poly(styrenesulfonic acid); NPD = *N,N'*-di-1-naphthyl-*N,N'*-diphenyl-1,1'-biphenyl-4,4'diamine; TCTA = 4,4',4"-tris(carbazol-9-yl)-triphenylamine; TPBI = 1,3,5-tris[*N*-(phenyl)ben-zimidazole]benzene). Two different host materials, namely, 4,5-diaza-2',7'-bis(carbazol-9-yl)-9,9'-spirobifluorene and 4,4'-*N,N'*-dicarbazolebiphenyl (CBP), were used for comparison. The results identified that effective energy confinement and appropriate guest–host combinations were observed for the devices using the spirobifluorene-based host that showed higher EL efficiencies with L_{max} of 19 300 cd m^{-2}, η_{ext} = 17.1%, η_P = 49.3 lm W^{-1}, CIE at (0.51, 0.48) for **Ir-19**, and L_{max} of 21 100 cd m^{-2}, η_{ext} = 15%, η_P = 37 lm W^{-1}, CIE at (0.51, 0.49) for **Ir-20**.

Besides modification of the ligand structures, colour tuning was also realised by incorporation of ligands with different electrochemical properties in some heteroleptic coumarin-based structures. Complexes **Ir-21** to **Ir-24** were reported by Ren and coworkers.[23] Due to the fact that coumarin is strongly electron withdrawing and very electron deficient, any replacement of ppy ligands in Ir(ppy)$_3$ with coumarin derivatives can significantly decrease the HOMO energy but it has a less pronounced effect on the LUMO energy, consequently leading to an increased HOMO–LUMO gap and a blueshift in the emission energy. Therefore, the λ_{PL} of **Ir-21** shifts from 550 to 536 nm in **Ir-22**, whereas **Ir-23** shifts from 570 to 544 nm in **Ir-24**. The OLEDs fabricated

X = CH **Ir-19**

N **Ir-20**

using these coumarin-based Ir(III) complexes as emissive dopants are highly efficient and stable with η_{ext} as high as 20–21% and η_P above 45 lm W^{-1} at 1 mA cm^{-2}. Improved charge balance in the emissive layer was observed in **Ir-21**; thanks to the fact that **Ir-21** is already a good charge-transporting material. Its η_{ext} showed a steady increase as the doping concentration was increased.

Dendrimers are found to be useful in high-performance OLEDs since they can get rid of the lower-purity problem in polymers but can still be processed by solution-based techniques and possess higher thermal properties and glass-transition temperatures than small molecules. To generate yellow EL, Bolink *et al.* synthesised a planar nondelocalised cyclic phosphazene **Ir-25** where the central core is inert, and its optical and electronic properties are dependent on the nature of the dendron, *i.e.* the Ir(ppy)$_3$ fragment in **Ir-25**.[24] This approach is synthetically facile and these rigid spheres can also improve the amorphous properties of the resulting complexes. The EL colour of the device is yellow (CIE coordinates at $x = 0.45$, $y = 0.54$) and becomes redshifted with respect to [Ir(ppy)$_2$(acac)] (0.31, 0.57), because of elongation of the effective conjugation length *via* the biphenyl unit. The device also provides better results by over 30–35% than that with [Ir(ppy)$_2$(acac)] fabricated using similar device architectures showing η_L of 24 cd A^{-1} and luminance of 3362 cd m^{-2}.

Single-layer devices were fabricated using charged Ir(III) complex **Ir-26** and it was found to emit yellow light with a brightness that exceeds 300 cd m^{-2} and a luminous power efficiency that exceeds 10 lm W^{-1} at just 3.0 V.[25] Their EL peak maximum varied with the applied bias, and it appeared at 560 nm under forward bias and showed a redshift of 20 nm under reverse bias. The CIE coordinates are $x = 0.425$, $y = 0.549$ for the forward bias and $x = 0.491$, $y = 0.496$ for the reverse one. The PF$_6$$^-$ space charge was found to dominate the device characteristics.

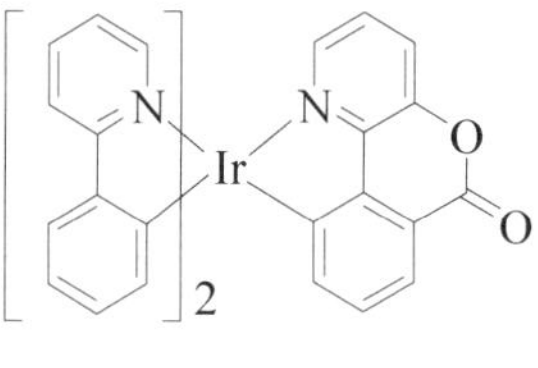

Ir-21

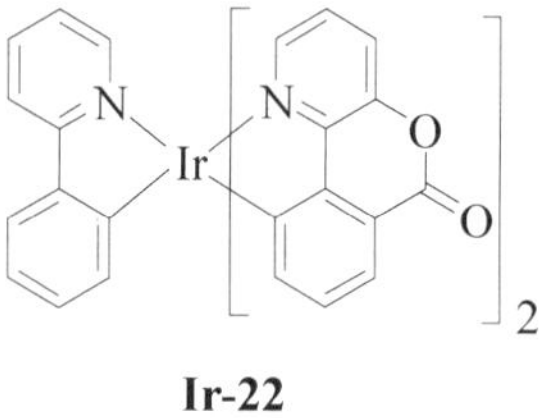

Ir-22

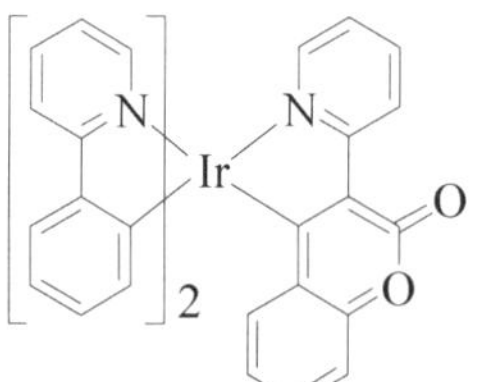

Ir-23

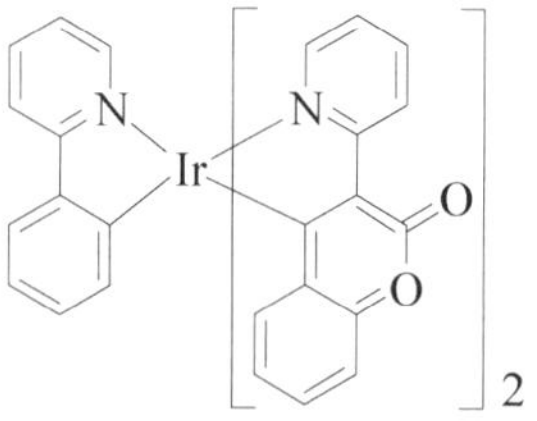

Ir-24

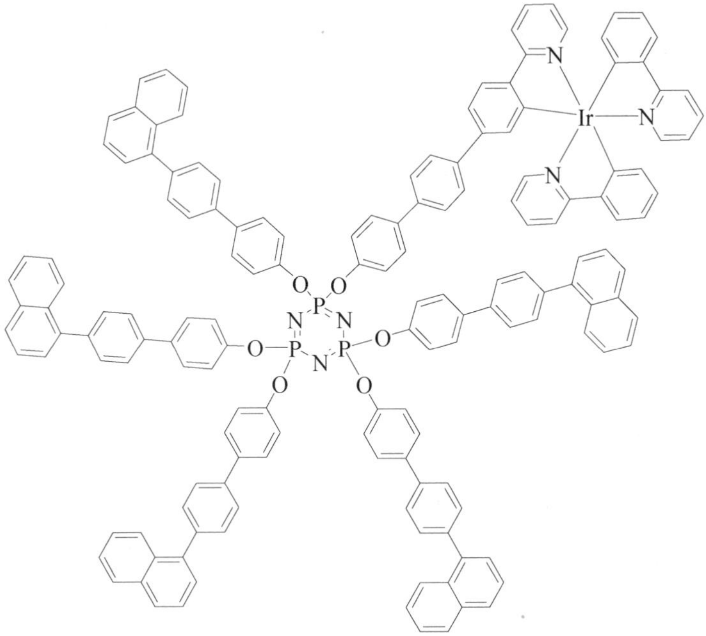

Ir-25

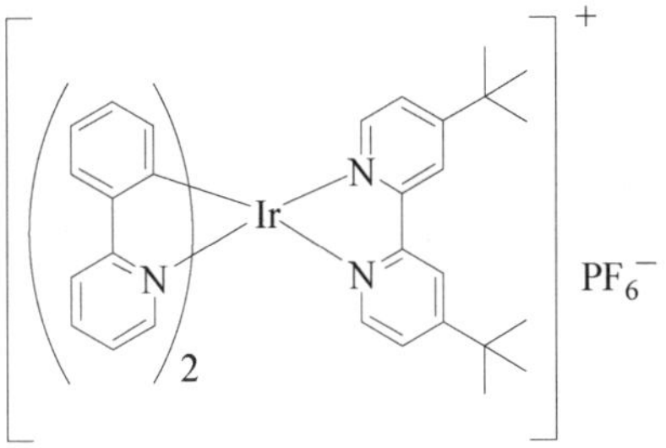

Ir-26

1.2.2 Iridium(III) Complexes Containing Fluorene, Carbazole or Triphenylamine

Recent research endeavours for Ir(III) complexes in OLEDs have involved the use of fluorene-based chromophores, which possess great promise as highly stable and efficient emissive cores in the synthesis of useful Ir(III) complexes. Fluorene-bridged materials possess the advantages of ease of functionalisation at their 9-position of the fluorene ring.[26] By replacing the phenyl ring in [Ir(ppy)$_3$] with fluorene as for **Ir-27**,[27] the PL peak maximum was largely shifted to 548 nm with a minor shoulder peak at around 588 nm and this yellow phosphorescence did not show a strong dependence on the dopant level. The apparently yellow emission results from the mixing of green and orange colours in the device. The device at the 2 wt.% doping concentration shows the highest η_{ext} of 10.3%, which corresponds to a peak η_L of 36.7 cd A^{-1} and a η_P of 17.7 lm W^{-1}. The corresponding brightness reached $\sim$ 8190 cd m^{-2} at 12 V and the CIE colour coordinates of 2 wt.% **Ir-27**-doped OLED is (0.44, 0.55), falling into the yellow region of the chromaticity diagram. This promising phosphor was then used to fabricate colour-stable WOLEDs with a four-emission-layer structure. With the optimised device structure and conditions, the highest efficiencies reached 23 cd A^{-1} and 13%, corresponding to a maximum total η_P of $\sim$ 21 lm W^{-1}.

Extension of the π-conjugation through incorporation of electron-rich and hole-transporting carbazole unit in **Ir-28–Ir-31**[28] and triphenylamine unit in **Ir-32–Ir-33**[29] to the fluorene fragment shows an obvious redshift in their emission profile and destabilises the ground state of the complex by electron donation. This approach increases the HOMO levels and improves the charge balance in the complexes. Better hole injection (HI) and hole transport (HT) were observed in carbazole- or triphenylamine-capped metallophosphors. The HOMO levels are higher for the 3-substituted 9-phenylcarbazole end group than for the *N*-coordinated carbazole unit (*i.e.* **Ir-30** > **Ir-28** and **Ir-31** > **Ir-29**), making the PL peak shift more to the red in the former. They all give intense yellow or orange electrophosphorescene with high EL efficiencies. OLEDs fabricated from multicomponent complexes **Ir-28–Ir-31** as the solution-processed emissive layer have been fabricated which showed very high peak efficiencies (9.6%, 29.8 cd A^{-1} and 13.4 lm W^{-1}). Enhanced thermal stability and more morphologically stable amorphous thin-film formation were observed in **Ir-32–Ir-33** as compared to **Ir-27**. For optimised orange devices at 5 wt.% of **Ir-33**, the L_{max} reached $\sim$ 45 530 cd m^{-2} at 12 V and the highest η_L is 34.8 cd A^{-1}, corresponding to a peak η_P of 18.2 lm W^{-1} and an η_{ext} of 10.5%.

Work was extended to the ionic metallophosphors as well. **Ir-34** is expected to possess many merits for solid-state lighting and display applications and can show good charge-transfer properties as compared to the neutral species.[30] In spite of the impressive scope that charged complexes can offer in the OLED

Ir-27

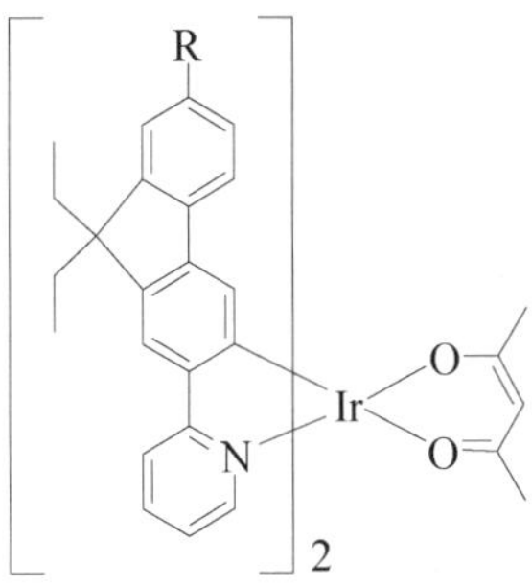

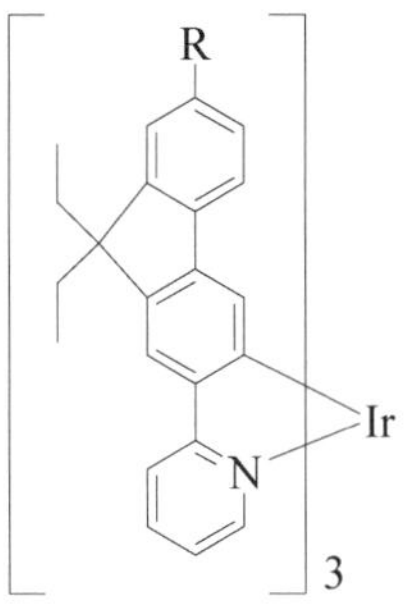

R 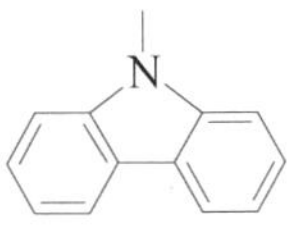 = **Ir-28**

R = **Ir-29**

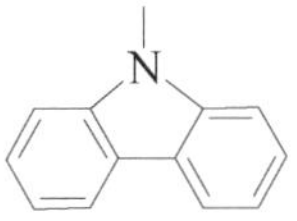

R = **Ir-30**

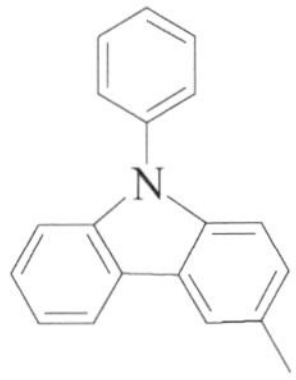

R = **Ir-31**

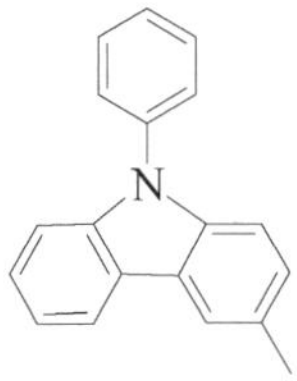

industry, the poor sublimability of charged Ir(III) dopant and its inferior compatibility with common hydrophobic polymer host tends to hinder their widespread practical applications. Interestingly, cationic **Ir-34** coordinated by the HT groups can be vacuum sublimed without significant decomposition to make evaporated orange-emitting OLEDs successfully. Efficient yellow OLEDs doped with 5 wt.% **Ir-34** can be fabricated, showing peak efficiencies of 6.5%, 19.7 cd A^{-1} and 18.4 lm W^{-1} with EL peak maximum at 565 nm. The

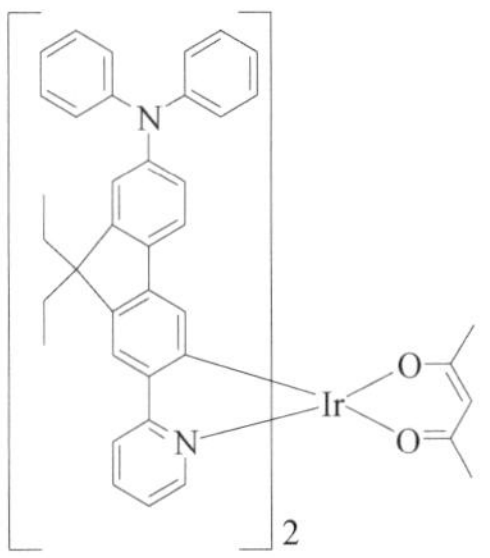

Ir-32 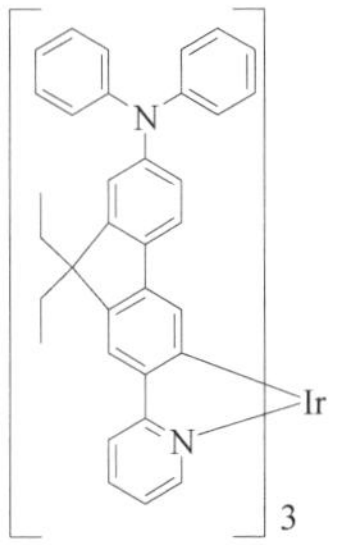   **Ir-33** 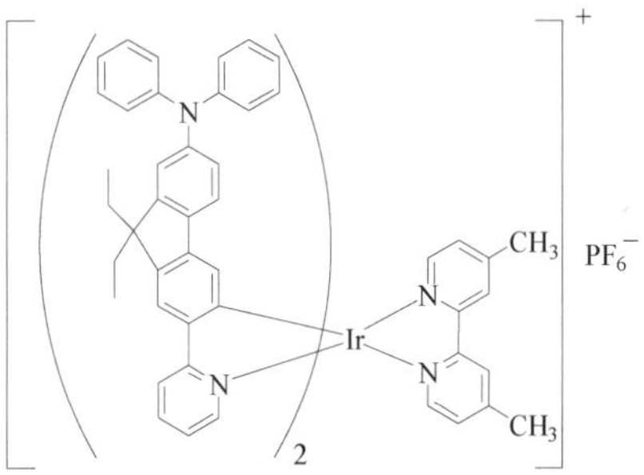   **Ir-34**

Ir-35

Ir-36

work provides a good platform for developing vacuum-sublimable charged metallophosphors for the design of highly efficient OLEDs and light-emitting electrochemical cells (LECs).

New yellow or yellow-orange Ir(III) triplet emitters with minimised T–T annihilation were reported by Wong and coworkers by functionalisation at the 9-position of fluorene ring. Complexes **Ir-35** and **Ir-36** with two sterically bulky triphenylamine moieties have sufficiently high glass-transition temperature to avoid crystallisation and also prevent close contact between dopant molecules.[31] Both amorphous molecules have the potential of facilitating HI, alleviating the T–T annihilation problem at high current density and hence fabricating simple solution-processible OLEDs even with the use of a small-molecule host only, which makes them good multifunctional phosphorescent materials. Remarkably, efficient and stable OLEDs can be obtained using the configuration of ITO/PEDOT:PSS/x% **Ir-35** or **Ir-36**:CBP/BCP/Alq$_3$/LiF/Al without the need for a HT layer (BCP: 2,9-dimethyl-4,7-diphenyl-1,10-phenanthroline; Alq$_3$: tris(8-hydroxyquinolinato)aluminium). The best orange-emitting devices made from **Ir-35** (or **Ir-36**) achieved L_{max} of 25 660 (20 100) cd m^{-2} at 24 (25) V, maximum η_{ext} of 4.63 (6.43)%, η_L of 15.07 (20.42) cd A^{-1} and η_P of 2.36 (3.05) lm W^{-1}. Thanks to the sterically bulky nature of the structure, the roll-off of η_{ext} is only gentle with increasing current density even at the higher doping level. Another way to suppress T–T annihilation and concentration quenching and to improve the film quality is to use a 3-dimensional spirobifluorene core as the cyclometallating ligands, as reported by Chen and his coworkers.[32] Three yellow OLEDs derived from **Ir-37** to **Ir-39** were fabricated by doping the complexes at the concentration of 10 wt.% into the polymeric host material consisting of poly(*N*-vinylcarbazole) (PVK) and 2-(4-biphenylyl)-5(4-*tert*-butyl-phenyl)-1,3,4-oxadiazole (PBD), showing satisfactory device performance. Complex **Ir-37** achieved a peak efficiency of 36.4

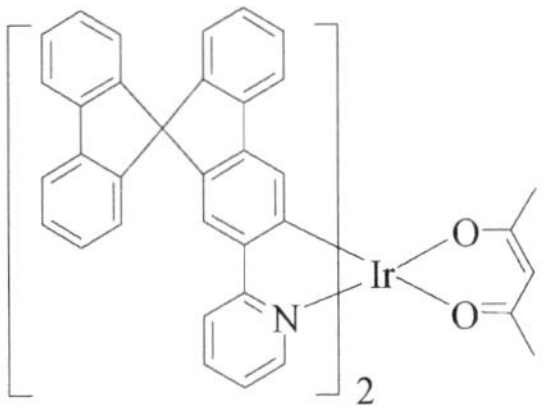

Ir-37 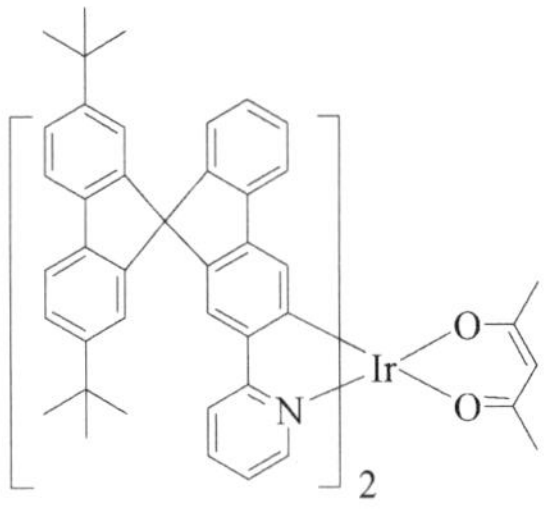   **Ir-38** 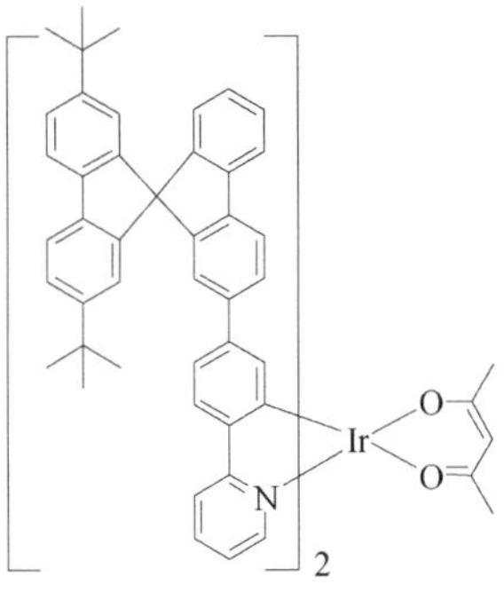   **Ir-39**

cd A^{-1} (η_{ext} of 10.1%) at 198 cd m^{-2} with λ_{EL} = 555 nm. Extending the electronic conjugation in **Ir-39** shifts the λ_{EL} to 560 nm. The η_{ext} did not decrease obviously with increasing current density, which confirms that the T–T annihilation was largely suppressed by the complexes.

Ir(III) complexes bearing polyphenylene dendritic ligands **Ir-40** to **Ir-42** through simple synthesis and purification procedures are another way to prevent molecular aggregation or $\pi-\pi$ stacking that can show highly efficient EL using solution-processing device fabrication.[33] The emission wavelengths of the materials could be effectively tuned from 549 nm to 582 nm by changing the conjugation of the ligands through incorporating additional aromatic segment (*e.g.* phenyl or fluorenyl group) onto the basic dendritic ligand. The best device performance was achieved by using **Ir-41** as the dopant with the configuration of ITO/PEDOT:PSS (50 nm)/PVK:PBD (40 wt.%):Ir complex (6 wt.%) (70 nm)/ BCP (12 nm)/ Alq$_3$ (20 nm) / Mg:Ag (150 nm) that gave peak η_L of 34.0 cd A^{-1} and η_{ext} of 10.3%. The efficiencies can be further improved to 46.3 cd A^{-1} and 13.9% if TPBI was used as the charge-blocking layer.

Another approach to achieve yellow EL with functional features such as electron transport (ET) and hole blocking (HB) was achieved using ligand consisting of benzoimidazole unit in the Ir(III) complex (**Ir-43**). The complex emits at λ_{EL} = 568 nm, L_{max} of 21 105 cd m^{-2}, η_{ext} of 7.3%, η_L of 21 cd A^{-1}

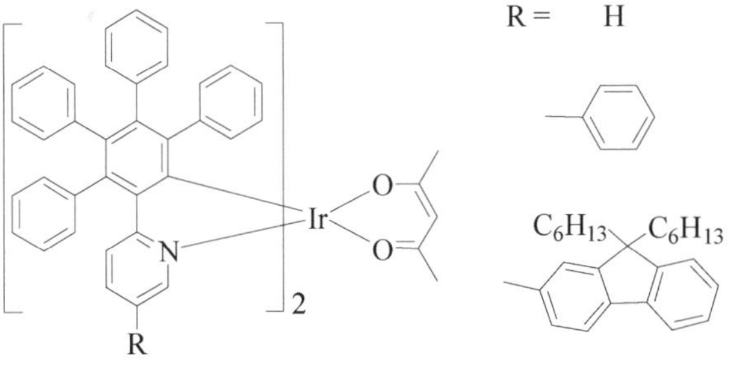

R =   H   **Ir-40**

          Ir-41

C_6H_{13} C_6H_{13} **Ir-42**

Ir-43

and η_P of 4.7 lm W^{-1} at a current density of 100 mA cm^{-2}, which compare favourably well with those of the green-emitting [(ppy)$_2$Ir(acac)] (λ_{EL} = 525 nm).[7e]

Many research groups have extensively investigated the use of carbazole in the development of new organic electronics materials. Many carbazole-based organic oligomers are high-mobility HT materials that are characterised by tunable and high-energy triplet levels, and they are widely used as the host materials for different metal-organic guests emitting a wide spectrum of visible colours.[34] The robust Ir-carbazolyl complexes have been recognised for their prominent role in ensuring high EL efficiency by elevating the HOMO levels and hence enhancing the HI/HT or hole-trapping features, enhancing the morphological stability as well as EL efficiencies as compared to the ppy and 1-phenylisoquinolinato (piq) congeners.[35] With the bulky, inductively electron-withdrawing CF$_3$ group on the pyridyl ring in **Ir-44**, an interesting orange electrophosphor (λ_{EL} = 556 nm) was produced that can afford well-performed devices with the highest efficiencies at 40.2 cd A^{-1}, corresponding to η_{ext} of 12.4% and η_P of 24.0 lm W^{-1}. On the other hand, by simply changing the carbazole substitution position from C-3 to C-2 in **Ir-45**,[36] marked changes can be seen in the EL properties, including a redshift of 85 nm in the emission wavelength and a change of the CIE coordinates from green (0.22, 0.60) for C-3 to orange (0.58, 0.39) for C-2. If the carbazole unit is not directly bonded to the metal centre (**Ir-46**), blueshifts of PL and EL spectra were observed that gave rise to yellow emission only.[37]

A multifunctional carbazole-based Ir(III) complex **Ir-47** was reported by Tang *et al.*[38] that contains both a HT carbazole group and an ET oxadiazole group as the main components. This approach was expected to be beneficial to improve the EL performance. Polymer light-emitting diodes (PLEDs) based on a structure of ITO/PEDOT:PSS/PVK:PBD:**Ir-47**/TPBI/CsF/Al were fabricated by solution-processed technology and the resulting EL spectra showed the emission of both monomers and exciplexes. The PLED doped with **Ir-47** at 4 wt.% doping concentration shows the optimal data at L_{max} of 7746 cd cm^{-2},

Ir-44   **Ir-45**   **Ir-46**

maximum η_L of 14.0 cd A^{-1} and η_{ext} of 5.8% with the CIE coordinates of (0.48, 0.50).

If arylamine groups are linked to 2-phenylpyridine backbone in the Ir complexes (**Ir-48 to Ir-49**),[39] their HOMO levels are shown to be higher compared to [Ir(ppy)$_2$(acac)], therefore an improved HT stability was confirmed. Their EL lies in the orange region of CIE diagram. Due to the increased π-conjugated system of the ligands, the peak emission wavelength of **Ir-49** (λ_{EL} = 589 nm; CIE = (0.58, 0.45)) shows a slight redshift as compared to that of **Ir-48** (λ_{EL} = 581 nm; CIE = (0.54, 0.45)).

The highly amorphous and soluble heteroleptic Ir(III) complex **Ir-50** also experiences a redshift in emission wavelength relative to [Ir(ppy)$_3$] because of the longer conjugation length.[40] The authors believe that there is an efficient interligand energy transfer within the Ir complex. However, a decreased phosphorescence quantum yield (Φ_P) was observed, probably due to the

Ir-47

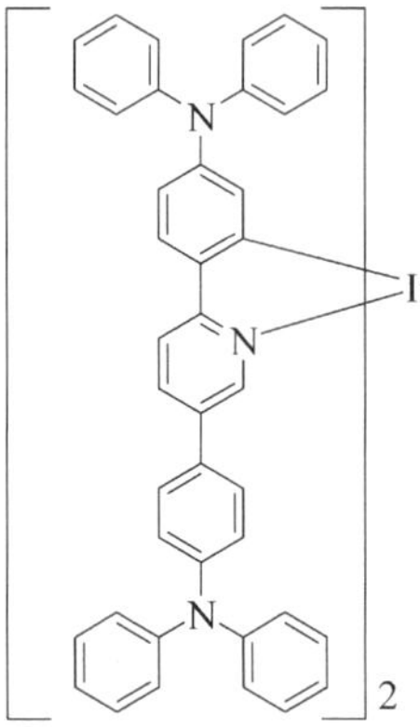

Ir-48

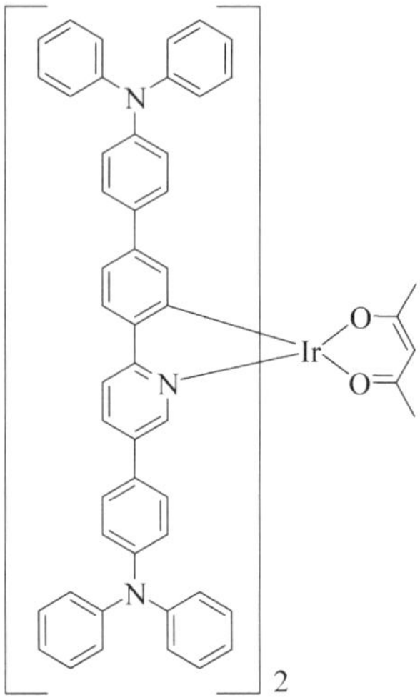

Ir-49

reduced triplet energy of the ligand. The best device performance was obtained with **Ir-50** doped in PVK, which was found to have a L_{max} above 15 000 cd m^{-2} and a peak η_L of 21 cd A^{-1}.

Three phosphorescent dendrimers (**Ir-51** to **Ir-53**) with an Ir(III) complex core and oligocarbazole- or oligofluorene-substituted ligands were reported by Lu and coworkers.[41] Hyperbranched oligocarbazoles or oligofluorenes were attached to the Ir(III) complex core to minimise the interaction between the phosphorescent cores that also functioned as the charge-transporting host for the Ir complex core. The structures of the oligocarbazole were designed to maintain high triplet energy of the ligands so that phosphorescence quenching in the resulting dendrimers can be avoided, while the oligofluorene in **Ir-53**

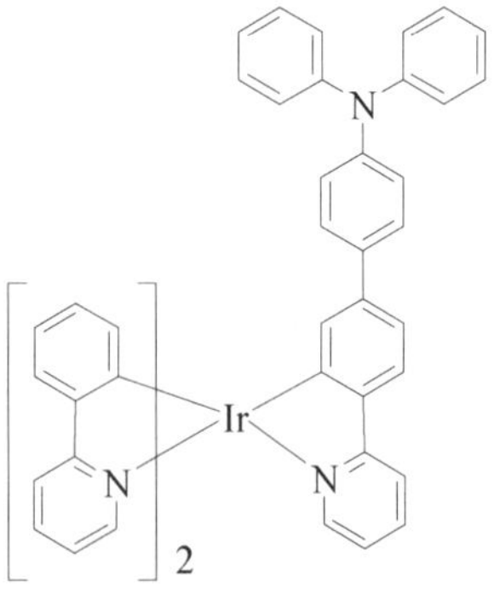

Ir-50

resulted in the unwanted phosphorescence quenching. A strong blue emission from the oligofluorene-substituted ligands besides the emission from the Ir chromophore was observed in **Ir-53** due to the inefficient energy transfer from the oligofluorene-substituted ligands to the Ir(III) complex core. On the contrary, negligible blue emission associated with the oligocarbazole was observed from **Ir-51** and **Ir-52**. Higher PL efficiencies of both of them resulted in better device performance than **Ir-53**. The best performance was obtained from **Ir-52**-based electrophosphorescent OLED with a L_{max} of 13 060 cd m^{-2} and a peak η_{L} of 4.3 cd A^{-1}, owing to its high PL efficiency and efficient energy transfer between the Ir(III) complex core and the ligands.

1.3 Platinum(II) Complexes

To date, numerous Pt(II) complexes have been used for yellow or orange EL devices. The main striking difference of Pt(II) complexes from the Ir(III) congeners lies in the fact that the emission of the resulting OLEDs may arise from a combination of their own emission as well as the excimeric component in the visible spectrum for the Pt(II) species. Modification of the aryl or pyridyl rings on the Pt(II) ion not only shifts the EL maximum but also makes the white-colour emission possible. This can minimise the complexity of device fabrication in achieving a good white colour balance using multiple emitters instead. Hence, well-balanced EL spectra can easily be obtained in WOLEDs with Pt(II) complexes by employing only one, or at most two dopants showing emission bands of the excimer or exciplex.

There are several types of Pt(II) complexes commonly applied for EL devices. The most common one involves cyclometallated Pt(II) complexes chelated with aromatic pyridyl ligands or similar derivatives and acac as the ancillary ligand. Complexes **Pt-1** and **Pt-2** bearing 2-phenylbenzothiazolate with different substituents were reported by Chen and coworkers[42] For complex **Pt-2** containing the electron-withdrawing fluoro group, it underwent ipsochromic shifts of the PL and EL emissions with respect to the unsubstituted **Pt-1**, inferring that the position occupied by the substituent in the ligand is dominated by the LUMO character. They emit tunable bright yellow-orange light at room temperature (548 and 540 nm for **Pt-1** and **Pt-2**, respectively), and show a bathochromatically shifted emission peak together with the excimer shoulder peak (625–630 nm) with increasing doping concentration. The device with **Pt-2** as the dopant shows relatively better EL performance (L_{max} of 11 320 cd m^{-2} and η_{L} of 11.3 cd A^{-1}) than that for **Pt-1**.

Efficient fluorescent-based OLEDs were recently reported by using highly fluorescent indolizino[3,4,5-*ab*]isoindoles as the emitting layer.[43] Because of their attractive PL quantum yields that are critical for OLED application, Wudl and coworkers complexed this ligand with Pt(II) ion in **Pt-3**.[44] **Pt-3** with 2-pyridin-2-ylindolizino[3,4,5-*ab*]isoindole (pin) was observed to have high quantum yield of 51% in DMSO at 546 nm, emitting mainly from a ligand-centred triplet π-π* state. Inappreciable degrees of charge-transfer character in

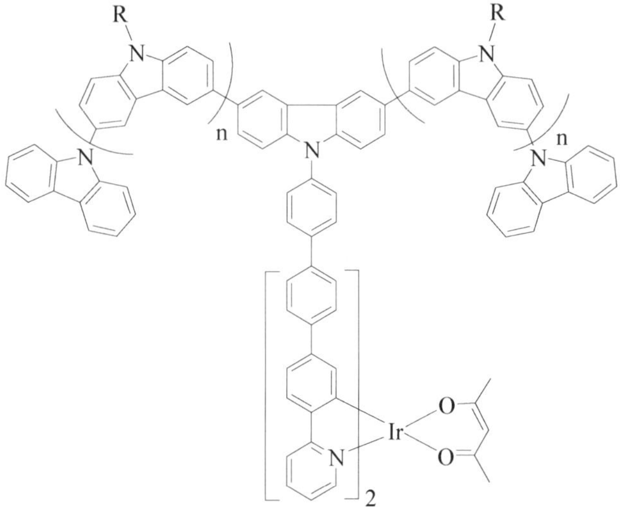

R = n-C$_8$H$_{17}$
n = 1   **Ir-51**
n = 3   **Ir-52**

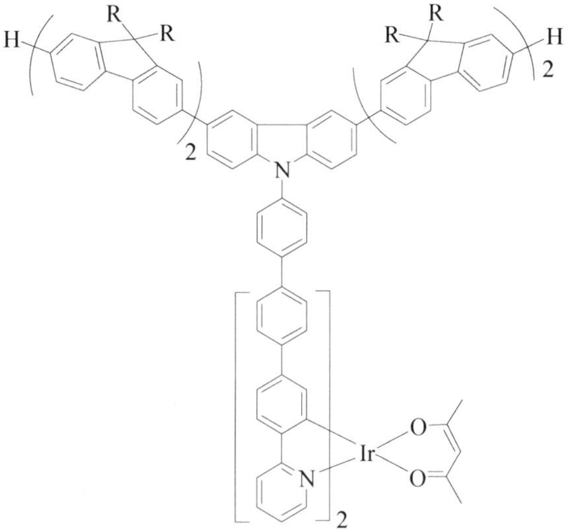

R = n-C$_8$H$_{17}$ **Ir-53**

X = H **Pt-1**
X = F **Pt-2**

Pt-3

this complex were confirmed by both the absorption and emission wavelength shifts with the polarity of the solvent used. It is a potential bifunctional emitter because of its relatively low oxidation potential (*ca.* 0.5 V) and this value is among some of the lowest values for other complexes reported.[45] It emits at 553 nm (CIE = 0.46, 0.49) with a FWHM of 63 nm without any aggregation in the EL spectrum.

A novel class of multicomponent neutral complexes **Pt-4** to **Pt-8** was nicely developed recently. **Pt-4**[46] and **Pt-5**[47] bearing carbazole and triphenylamine units exhibit improved thermal or glass-state durability. More importantly, their HI and HT abilities were also improved, catering for more efficient charge transport in their EL process. The introduction of electron-donating carbazole or triphenylamine moiety into the electron-deficient pyridine moiety is expected to increase the electronic conjugation and intramolecular donor–acceptor charge-transfer character of the ligand, causing a redshift in their absorption and emission features as compared to that in [Pt(Fl-py)(acac)] (**Fl-py** = 2-(9,9-diethyl-2-fluorenyl)pyridyl).[47] The greater delocalisation over the aryl ligands also leads to a lesser extent of metal d-orbital mixing and gives a smaller HOMO−LUMO gap, so that both complexes emit in the orange region, while [Pt(Fl-py)acac] emits in the greenish-yellow region. **Pt-4** showed the main emission at 548 nm with a broad shoulder at around 640 nm and revealed a clear aggregation phenomenon as the doping concentration increases. Therefore, the CIE coordinates would vary with a change of concentration and are located at $x = 0.49$, $y = 0.51$ at 6 wt.% doping concentration. A maximum η_{ext} of 0.40%, η_P of 0.19 lm W^{-1} and η_L of 1.13 cd A^{-1} were achieved at 3 wt.% doping level. In contrast to many other devices based on similar Pt(II) emitters, however, the problem of aggregation becomes neglectable in **Pt-5**. The devices based on **Pt-5** emit a strong pure orange light with stable CIE colour coordinates of (0.55, 0.45). The use of **Pt-5** suggests a new avenue towards obtaining good colour purity of Pt-based OLEDs by eliminating such drawbacks as strong intramolecular interactions and poor emission colour purity of the cyclometallated Pt(β-diketonato) complexes. The best device performance was achieved with the structure of ITO/NPB/5% Pt:mCP/TPBI/LiF/Al, leading to η_{ext} of 4.65%, η_L of 11.75 cd A^{-1} and η_p of

5.27 lm W^{-1}. Much higher efficiencies were obtained if mCP was replaced by CBP, and a L_{max} of 4195 cd m^{-2}, η_{ext} of 6.64%, η_L of 15.41 cd A^{-1} and η_p of 7.07 lm W^{-1} were achieved. Simple WOLEDs were fabricated by the same research group using **Pt-5** and 9,10-di(2-naphthyl)anthracene (ADN) by replacing 1,3-bis(9*H*-carbazol-9-yl)benzene (mCP) or CBP with ADN. Although the preliminary results are not yet attractive, these single-colour EL devices based on **Pt-5** are good candidate for WOLED applications. A blueshift of the EL peak to 540 nm was observed in the yellow region if two triphenylamine groups are attached at the 9-position of fluorene in **Pt-6**.[31] An encouraging performance with L_{max} of 16 070 cd m^{-2} at 17 V, maximum η_{ext} of 3.36%, η_L of 9.55 cd A^{-1} and η_P of 2.31 lm W^{-1} was detected. T–T annihilation only plays its role at a much higher current density (360 $\pm$ 85 mA cm^{-2}) at a high doping concentration and it is probably due to its long phosphorescence lifetime (8.2 µs at room temperature).

A novel trifunctional cyclometallated Pt(II) complex **Pt-7** integrating the HT triarylamine, ET oxadiazole and EL metallated group into a single molecule was reported by Wong and coworkers.[48,49] This metal chelate is thermally and morphologically stable with respect to sublimation during the device fabrication. It displays a bipolar character since the LUMO level is lower than the most widely used ET/HB material PBD and comparable to Alq$_3$, whereas the HOMO is also close to NPD, showing its good HT ability. This molecule was used to fabricate a bilayer vacuum-deposited device of ITO/CuPc/**Pt-7**/Ca/Al in which **Pt-7** acts as the neat emissive film without any ET, HT and EI layers. This device exhibited strong orange-yellow EL at 538 and 578 nm with the CIE coordinates at (0.52, 0.47). The multifunctional nature of **Pt-7** renders it a suitable candidate to act as an efficient dopant-free electrophosphorescent emitter. Further enhancement of the charge transfer within the trifunctional molecule was observed in **Pt-8**, a molecule that incorporates a bulky electron-deficient boron moiety and HT arylamine donor as the cyclometallating chelate.[50] The OLEDs based on this emitter greatly improve the efficiencies by the boron functionality that increases the energy

Pt-4   **Pt-5**

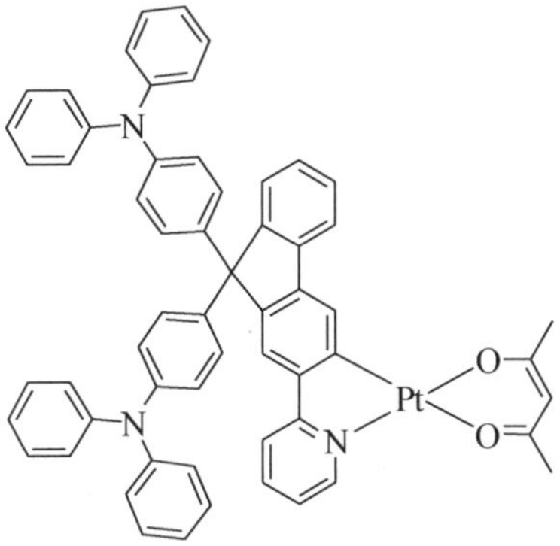

Pt-6 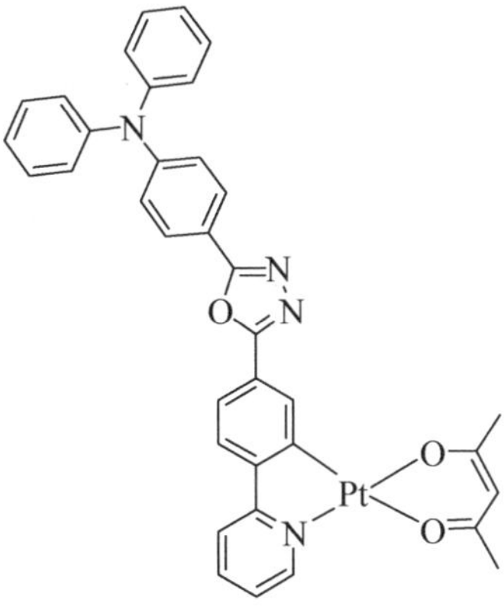   **Pt-7**

separation between the emissive state and radiatively quenched d-d states, thus providing a large increase in Φ_P and emission brightness despite the large redshift in the emission profile. A remarkably high Φ_P was noted in both the solid ($\Phi_P = 0.46$) and solution ($\Phi_P = 0.91$) states. The highest device efficiency was detected using CBP as the host in the absence of any HB layer. Orange electrophosphorescence peaking at 580 nm (CIE = 0.51, 0.48) was achieved with maximum η_L and η_P of 35.0 cd A^{-1} and 36.6 lm W^{-1} and peak η_{ext} of 10.6%. The result is among the highest reported for a device using a triarylboron-based phosphorescent emitting layer and one of the highest reported using Pt(II) complexes in orange OLEDs.

A good design aimed at colour tuning and suppressing the self-aggregation of Pt(II) complexes is needed to exploit the full potential of Pt(II) emitters. A series of readily sublimable and thermally robust Pt(II) pyridyl azolate complexes (**Pt-9** and **Pt-10**) were prepared by Chang *et al.* that exhibited high structural tuning capability and flexibility at the ligand sites.[51] The steric hindrance from *tert*-butyl and trifluoromethyl functionalities effectively

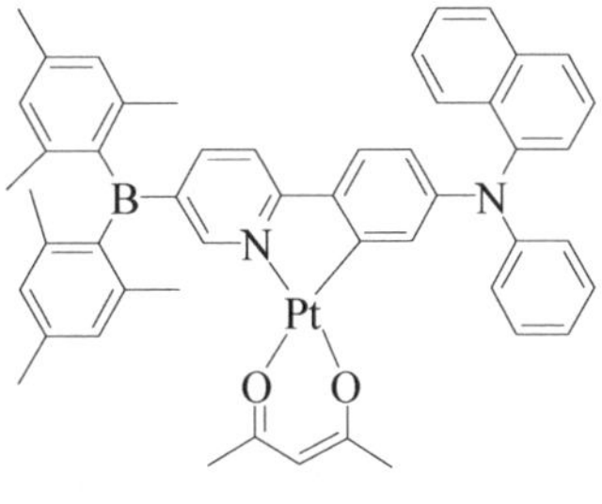

Pt-8

R = t-Bu **Pt-9**
R = CF$_3$ **Pt-10**

suppresses the self-quenching activities caused by the planar geometry of the Pt(II) complexes and maintains the amorphous property essential for device fabrication. Very bright emissions were observed at all doping concentrations for the multilayer devices with the configuration of ITO/NPB/CBP:**Pt-9**/BCP/Alq$_3$/LiF/Al. Upon increasing doping concentration, there is an increase in current density and a large redshift in emission maximum. The EL emission shifted from 502 nm (blue-green) to 556 nm (yellow) as the dopant concentration rose from 6% to 100%. A broadening of the emission band was also observed. A similar redshift from 552 nm (yellow) to 616 nm (red) was apparent in the OLEDs fabricated with **Pt-10** with increasing dopant concentration. For emitters such as **Pt-9** that possess ^{3}MLCT/$^3\pi-\pi$ states (MLCT = metal-to-ligand charge transfer), the best device performance was realised using a 6% doping level. The η_{ext} of 2.9%, η_L of 8.6 cd A^{-1} at 20 mA cm^{-2}, and a luminance of 27 114 cd m^{-2} at 15 V were realised. With the emitter **Pt-10** possessing ^{3}MMLCT state (MMLCT = metal-metal-to-ligand charge transfer), the best device performance was achieved at 20% dopant level, along with the η_{ext} of 6.0% and η_L of 19.7 cd A^{-1} both at 20 mA cm^{-2}. These promising results make them remarkable for colour-tuning applications in OLEDs by simply changing the doping concentration of the Pt(II) materials.

Research into the ligand design for EL materials has been dominated by Alq$_3$ due to its dual function as emissive and ET materials. Attempts to develop alternative quinolinato-substituted Pt(II) complexes **Pt-11** and **Pt-12** for EL applications were made by Che and coworkers.[52] The complexes ligated by tetradentate auxiliaries bis(2'-phenol)-bipyridine and -phenanthroline

R$_{1,2}$ = HC CH, X = Ph **Pt-11**
R$_1$ = R$_2$ = H, X = t-Bu **Pt-12**

showed good thermal stability. Both complexes are highly luminescent in solution and display a structureless emission in CH_2Cl_2 at λ_{PL} = 586 and 595 nm, respectively, and the emission properties are influenced by solvent polarity. OLEDs based on 10% of **Pt-11** gave yellow EL and was dominated by its PL at λ_{EL} = 588 nm. Maximum L_{max} and η_P of 850 cd m^{-2} and 0.26 lm W^{-1} were obtained, respectively. Yellow emission (CIE 0.42, 0.56) was also generated for **Pt-12** when its dopant content was increased to 2% with optimal luminance of 4480 cd m^{-2} and η_P of 0.51 lm W^{-1}.

Another family of Pt(II) complexes commonly used for OLED applications are those associated with C^N^N substituted tridentate chelates. They own moderate σ-donating and π-accepting abilities, thus satisfying the demand of the square planar Pt(II) coordination geometry to discourage the D_{2d} distortion that is likely to result in a nonradiative decay and makes them useful in materials design.[53] There are two pathways to fine tune the PL and EL properties of this kind of complexes, namely, modification of the substituent in the C^N^N ligand (**Pt-13**)[54] or attachment of acetylide as the σ-donor ligand with different substituents at the fourth coordination site (**Pt-14**).[55] The donor–acceptor interaction can be strengthened by combining the electron-rich diphenylamine group with the electron-deficient pyridine moiety in **Pt-13**, which shifts the absorption and triplet emission maxima bathochromically to λ_{PL} = 595 nm. The optimal OLED using this complex shows an efficient orange emission with peak maximum at 588 nm and CIE coordinates of (0.570, 0.427) without any voltage-dependence problem. The device turned on at 4.0 V and a luminance of 9000 cd m^{-2} was achieved at 13 V. The peak η_L was shown to be about 11.3 cd A^{-1} and the corresponding η_{ext} is 5.7%. **Pt-14** with the *para*-substituted phenylacetylene shows MLCT PL emission at 582 nm. This is in accordance with the strong σ-donating strength of the alkynyl ligand, which destabilises the dπ(Pt) HOMO to yield relatively low-energy MLCT 5d(Pt)→π*(C^N^N) transitions. A L_{max} of 7800 cd m^{-2} at

Pt-13   **Pt-14**

11 V and a maximum efficiency of 2.4 cd A^{-1} at 30 mA cm^{-2} were obtained for an orange-emitting OLED (λ_{EL} = 564 nm) using **Pt-14** at the 4% doping level.

Che also exploited a novel series of naphthyl-substituted cyclometallated C^N^N-type Pt(II) complexes **Pt-15** and **Pt-16** that showed yellow or orange electrophosphorescence.[56] They are strongly emissive, with λ_{PL} depending on the nature of the substituent group R. They possess high thermal stabilities and can be easily sublimed in *vacuo* that render them suitable for application in OLEDs. The devices were fabricated with the architecture of ITO/NPB/Pt dye:CBP/BCP/LiF/Al. For the device containing dopant **Pt-15**, it emitted bright orange light and the EL spectrum showed a peak at 553 nm with a shoulder at 594 nm, and maximum η_{ext} = 12.4%, η_L = 32.3 cd A^{-1} and η_P = 11.2 lm W^{-1} were achieved at a current density of 3.1 mA cm^{-2}, corresponding to a brightness of 1100 cd m^{-2}. The device exhibited a L_{max} of 21 700 cd m^{-2} with excellent colour stability. A better performance was achieved for the device containing dopant **Pt-16**, and it emitted yellow light with a major peak at 540 nm accompanied by a shoulder peak of 580 nm. Maximum η_{ext} = 16.1%, η_L = 51.8 cd A^{-1} and η_P = 23.2 lm W^{-1} were achieved at a current density of 0.44 mA cm^{-2}, corresponding to a brightness of 230 cd m^{-2}. The device exhibited a maximum brightness of 23 500 cd m^{-2} at 16 V and nearly constant CIE coordinates of x = 0.44, y = 0.54.

Another similar class of Pt(II) complexes with O^N^N chelates **Pt-17–Pt-21** were reported by the same group.[57] They are robust yellow emitters with good thermal stabilities. Excitation of **Pt-17** to **Pt-21** in DMF solutions resulted in a broad orange-red emission (593–618 nm) at room temperature but blueshifted EL emission bands were observed because of solvent relaxation and doping effects. Their EL maximum is independent of their doping concentration and the device performance follows a trend similar to those for the emission quantum yield and thermal stability (**Pt-21** > **Pt-19** > **Pt-18** > **Pt-20** ~ **Pt-17**). The complex **Pt-21** with bulky *tert*-butyl group and fluoro substituent gave better performance with L_{max} of 37 000 cd m^{-2} and η_L = 7.8 cd A^{-1}.

R = CF$_3$ **Pt-15**
R = t-Bu **Pt-16**

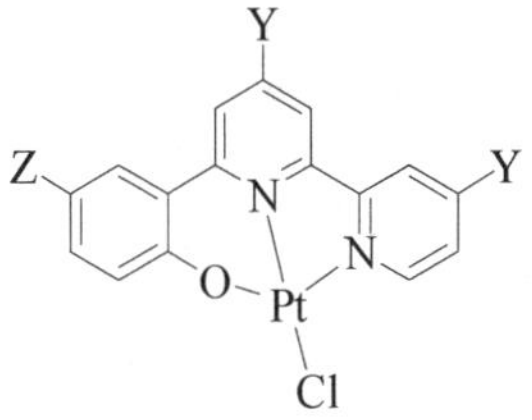

Z = H; Y = Me   **Pt-17**

Z = H; Y = H   **Pt-18**

Z = H; Y = t-Bu   **Pt-19**

Z = F; Y = H   **Pt-20**

Z = F; Y = t-Bu   **Pt-21**

1.4   Other Metals

Besides Ir(III) and Pt(II) complexes, some other heavy-metal complexes also demonstrate yellow or orange EL bands. The Au(III) alkynyl-based complex **Au-1** with strongly electron-releasing amino substituent was employed as a dopant in multilayer colour-tunable OLEDs.[58] The tunable colour of its EL (from 500 nm to 580 nm upon increasing the dopant concentration) is mainly attributed to the higher order and better packing of the molecules, leading to a stronger π-stacking of the C^N^N ligand and hence resulting in a dimeric or excimeric intraligand EL at the lower energy. Its optimal device gave a L_{max} of 10 000 cd m^{-2}, $\eta_L = 17.6$ cd A^{-1}, $\eta_P = 14.5$ lm W^{-1} and $\eta_{ext} = 5.5\%$.

Osmium(II) complexes are also suitable for OLED applications and three orange OLEDs were reported by Shu *et al.* Due to the lower positive charge at the central Os(II) metal cation, the Os$^{2+/3+}$ oxidation is shifted to a lower potential in comparison to the Ir$^{3+/4+}$ couple of its isoelectronic Ir(III) complexes. Likewise, the lowest-energy excited state of Os(II) complexes should contain an increased proportion of the MLCT character. The increase of the MLCT participation may render a significant reduction of its radiative lifetime compared with the respective Ir(III) system and the cathodically shifted oxidation potential pushes up the energy level of HOMO and makes the

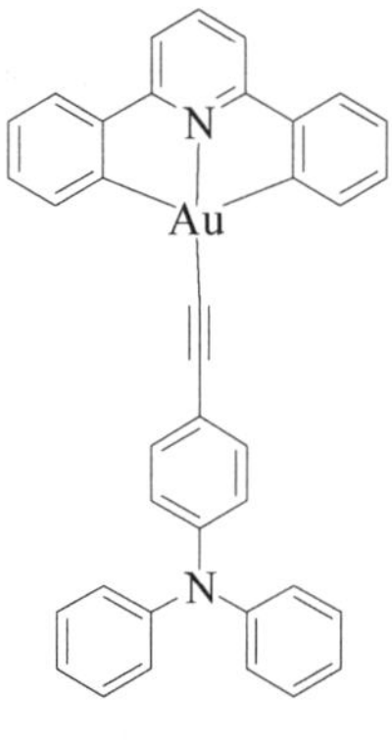

Au-1

neutral Os(II) complexes suitable to serve as the sites for direct charge trapping. In addition, the shortened radiative lifetime reduces the T–T annihilation in the devices, especially at high dopant concentration or higher driving voltage. The solution-processed PLEDs doped with **Os-1** has the EL peak at 602 nm with η_{ext} = 18.7% and η_L = 45 cd A^{-1} and CIE coordinates of (0.58, 0.41).[59] This metallophosphor was also used in combination with a blue fluorescent emitter 4,4′-bis[2-{4-(N,N-diphenylamino)phenyl}vinyl]biphenyl for the fabrication of WOLEDs. The CIE was recorded at x = 0.33, y = 0.34 at 9.0 V. The short triplet excited state lifetime of **Os-1** (*ca.* 1.0 μs) is beneficial to the remarkable colour stability. This doubly doped device exhibited a pure white-light emission having a high η_{ext} of 6.12% (13.2 cd A^{-1}), and a L_{max} of 11 306 cd m^{-2}. Charge-neutral complexes **Os-2** and **Os-3** can be used to make orange OLEDs using the solution-processing technique.[60] Their emissions appear in the longer-wavelength region and reveal significantly higher quantum yield (**Os-2**: λ_{PL} = 584 nm, Φ_P = 0.80; **Os-3**: λ_{PL} = 572 nm, Φ_P = 0.90). Typical device configurations of ITO/PEDOT/PVK:PBD:Os dye/ TPBI/Mg:Ag/Ag were employed. Characteristic emission band at *ca.* 548 nm from **Os-2** and blue emission from PVK/PBD host were totally suppressed, but this is not the case for the **Os-3** based devices, which still presents a substantial amount of the PVK host emission at the 0.4 mol% dopant level. This implies complete Förster energy transfer from the host to the dopant sites in the former case. Increasing the doping concentration of **Os-3** to 1.6 mol% resulted in complete energy transfer and improved the device performance. The maximum η_{ext} of the devices doped with 0.4 mol% of **Os-2** and **Os-3** are 11.7% (40.4 cd A^{-1} at 2.1 mA cm^{-2}) and 11.7% (38.5 cd A^{-1} at 2.2 mA cm^{-2}); whereas the CIE coordinates are located at (0.45, 0.53) and (0.42, 0.52), respectively, and do not show much variation upon changing the applied voltage. By increasing the doping concentration of **Os-3** to 1.6 mol%, the peak η_{ext} reached 13.3% (48.9 cd A^{-1} at 6.3 mA cm^{-2}) with a L_{max} of 28 816 cd m^{-2} at 16 V (96 mA cm^{-2}). The **Os-3**-based device exhibited prominent performance on η_P, which reached as high as 16.8 lm W^{-1} and is 1.7 times higher than that of the **Os-2**-based device due to the significantly lowered driving voltage The use of Os(II) complexes as the active emitting materials in OLEDs should be a

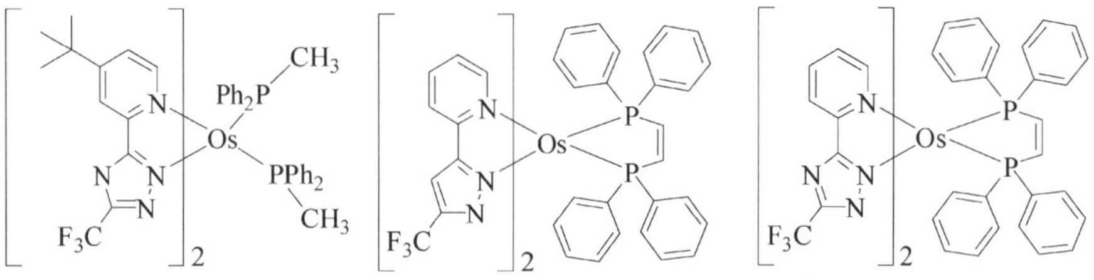

Os-1   **Os-2**   **Os-3**

valuable option for solid-state lighting applications because of the simple device architecture and the promise of low-cost manufacturability.

1.5   Concluding Remarks

With the aim of optimising the performance of OLEDs, research on the transition-metal complexes evidently still offers great prospects in terms of molecular design and materials properties. In this chapter, we present a library of heavy-metal complexes that can show yellow or orange EL features that are suitable for the development of highly efficient WOLEDs. It is clear that these results should spur new interest to the scientists in the future design and preparation of luminescent materials incorporating heavy transition elements. We are grappling with how to increase the efficiency of these devices (for both monochromatic and white OLEDs) and ensure that they can withstand wear and tear for practical use. We are looking forward these exciting and promising developments with substantial optimism.

Acknowledgements

The authors would like to thank all postgraduate students, postdoctoral associates and collaborators whose names appear in the references. This work was supported by a grant from the University Grants Committee of HKSAR, China (AoE/P-03/08), Hong Kong Research Grants Council (HKBU202709 and HKUST2/CRF/10) and Hong Kong Baptist University (FRG2/10-11/101).

References

1. E. Baranoff, J.-H. Yum, M. Grätzel and M. K. Nazeeruddin, *J. Organomet. Chem.*, 2009, **694**, 2661.
2. C. J. Humphreys, *MRS Bull.*, 2008, **33**, 459.
3. (a) G. Schwartz, M. Pfeiffer, S. Reineke, K. Walzer and K. Leo, *Adv. Mater.*, 2007, **19**, 3672; (b) S.-J. Su, E. Gonmori, H. Sasabe and J. Kido, *Adv. Mater.*, 2008, **20**, 4189.
4. (a) B. W. D'Andrade and S. R. Forrest, *Adv. Mater.*, 2004, **16**, 1585; (b) A. Misra, P. Kumar, M. N. Kamalasanan and S. Chandra, *Semicond. Sci. Technol.*, 2006, **21**, R35; (c) R. F Service, *Science*, 2005, **310**, 1762.
5. (a) S. C. Stinson, *Chem. Eng. News*, 2000, **78**, 22; (b) B. Johnstone, *Technol. Rev.*, 2001, **104**, 80.
6. S. Reineke, F. Lindner, G. Schwartz, N. Seidler, K. Walzer, B. Lussem and K. Leo, *Nature*, 2009, **459**, 234.
7. (a) L. Flamigni, A. Barbieri, C. Sabatini, B. Ventura and F. Barigelletti, *Top. Curr. Chem.*, 2007, **281**, 143; (b) J. A. G. Williams, *Top. Curr. Chem.*, 2007, **281**, 205; (c) Y. Chi and P.-T. Chou, *Chem. Soc. Rev.*, 2010, **39**, 638;

(d) J. A. G. Williams, A. J. Wilkinson and V. L. Whittle, *Dalton Trans.*, 2008, 2081; (e) S. Lamansky, P. Djurovich, D. Murphy, F. Abdel-Razzaq, H. E. Lee, C. Adachi, P. E. Burrows, S. R. Forrest and M. E. Thompson, *J. Am. Chem. Soc.*, 2001, **123**, 4304.

8. M. A. Baldo, M. E. Thompson and S. R. Forrest, *Nature*, 2000, **403**, 750.

9. G. Zhou, C.-L. Ho, W.-Y. Wong, Q. Wang, D. Ma, L. Wang, Z. Lin, T. B. Marder and A. Beeby, *Adv. Funct. Mater.*, 2008, **18**, 499.

10. S.-L. Lai, S.-L. Tao, M.-Y. Chan, M.-F. Lo, T.-W. Ng, S.-T. Lee, W.-M. Zhao and C.-S. Lee, *J. Mater. Chem.*, 2011, **21**, 4983.

11. C.-L. Li, Y.-J. Su, Y.-T. Tao, P.-T. Chou, C.-H. Chien, C.-C. Cheng and R.-S. Liu, *Adv. Funct. Mater.*, 2005, **15**, 387.

12. J. Qiao, L. Duan, L. Tang, L. Wang and Y. Qiu, *J. Mater. Chem.*, 2009, **19**, 6573.

13. W.-C. Chang, A. T. Hu, J.-P. Duan, D. K. Rayabarapu and C.-H. Cheng, *J. Organomet. Chem.*, 2004, **689**, 4882.

14. L. Chen, C. Yang, J. Qin, J. Gao and D. Ma, *Inorg. Chim. Acta*, 2006, **359**, 4207.

15. H.-W. Hong and T.-M. Chen, *Mater. Chem. Phys.*, 2007, **101**, 170.

16. N. Rehmann, C. Ulbricht, A. Köhnen, P. Zacharias, M. C. Gather, D. Hertel, E. Holder, K. Meerholz and U. S. Schubert, *Adv. Mater.*, 2008, **20**, 129.

17. D.-S. Leem, S. O. Jung, S.-O. Kim, J.-W. Park, J. W. Kim, Y.-S. Park, Y.-H. Kim, S.-K. Kwon and J.-J. Kim, *J. Mater. Chem.*, 2009, **19**, 8824.

18. G. Zhang, F. Wu, X. Jiang, P. Sun and C.-H. Cheng, *Synth. Met.*, 2010, **160**, 1906.

19. G. Ge, J. He, H. Guo, F. Wang and D. Zou, *J. Organomet. Chem.*, 2009, **694**, 3050.

20. G. Ge, G. Zhang, H. Guo, Y. Chuai and D. Zou, *Inorg. Chim. Acta*, 2009, **362**, 2231.

21. T. Peng, Y. Yang, H. Bi, Y. Liu, Z. Hou and Y. Wang, *J. Mater. Chem.*, 2011, **21**, 3551.

22. C.-H. Lin, Y. Chi, M.-W. Chung, Y.-J. Chen, K.-W. Wang, G.-H. Lee, P.-T. Chou, W.-Y. Hung and H.-C. Chiu, *Dalton Trans.*, 2011, **40**, 1132.

23. X. Ren, M. E. Kondakova, D. J. Giesen, M. Rajeswaran, M. Madaras and W. C. Lenhart, *Inorg. Chem.*, 2010, **49**, 1301.

24. H. J. Bolink, S. G. Santamaria, S. Sudhakar, C. Zhen and Z. Sellinger, *Chem. Commun.*, 2008, 618.

25. J. D. Slinker, A. A. Gorodetsky, M. S. Lowry, J. Wang, S. Parker, R. Rohl, S. Bernhard and G. G. Malliaras, *J. Am. Chem. Soc.*, 2004, **126**, 2763.

26. (a) D. Neher, *Macromol. Rapid Commun.*, 2001, **22**, 1365; (b) W.-Y. Wong, *Coord. Chem. Rev.*, 2005, **249**, 971.

27. X.-M. Yu, G.-J. Zhou, C.-S. Lam, W.-Y. Wong, X.-L. Zhu, J.-X. Sun, M. Wong and H.-S. Kwok, *J. Organomet. Chem.*, 2008, **693**, 1518.

28. C.-L. Ho, W.-Y. Wong, G.-J. Zhou, B. Yao, Z. Xie and L. Wang, *Adv. Funct. Mater.*, 2007, **17**, 2925.
29. (a) W.-Y. Wong, G.-J. Zhou, X.-M. Yu, H.-S. Kwok and B.-Z. Tang, *Adv. Funct. Mater.*, 2006, **16**, 838; (b) X.-M. Yu, H.-S. Kwok, W.-Y. Wong and G.-J. Zhou, *Chem. Mater.*, 2006, **18**, 5097.
30. W.-Y. Wong, G.-J. Zhou, X.-M. Yu, H.-S. Kwok and Z. Lin, *Adv. Funct. Mater.*, 2007, **17**, 315.
31. G.-J. Zhou, W.-Y. Wong, B. Yao, Z. Xie and L. Wang, *J. Mater. Chem.*, 2008, **18**, 1799.
32. J. H. Yao, C. Zhen, K. P. Loh and Z.-K. Chen, *Tetrahedron*, 2008, 10814.
33. C. Huang, C.-G. Zhen, S. P. Su, Z.-K Chen, X. Liu, D.-C. Zou, Y.-R. Shi and K. P. Loh, *J. Organomet. Chem.*, 2009, **694**, 1317.
34. (a) K. Brunner, A. van Dijken, H. Borner, J. J. A. M. Bastiaansen, N. M. M. Kiggen and B. M. W. Langeveld, *J. Am. Chem. Soc.*, 2004, **126**, 6035; (b) S.-J. Yeh, M.-F. Wu, C.-T. Chen, Y.-H. Song, Y. Chi, M.-H. Ho, S.-F. Hsu and C.-H. Chen, *Adv. Mater.*, 2005, **17**, 285; (c) M.-H. Tsai, H.-W. Lin, H.-C. Su, T.-H. Ke, C.-C. Wu, F.-C. Fang, Y.-L. Liao, K.-T. Wong and C.-I. Wu, *Adv. Mater.*, 2006, **18**, 1216; (d) K. R. Justin Thomas, J. T. Lin, Y.-T. Tao and C.-W. Ko, *J. Am. Chem. Soc.*, 2001, **123**, 9404; (e) P. Strohriegl and J. V. Grazulevicius, *Adv. Mater.*, 2003, **13**, 445.
35. (a) W.-Y. Wong, C.-L. Ho, Z.-Q. Gao, B.-X. Mi, C.-H. Chen, K.-W. Cheah and Z. Lin, *Angew. Chem., Int. Ed.*, 2006, **45**, 7800; (b) C.-L. Ho, W.-Y. Wong, Q. Wang, D. Ma, L. Wang and Z. Lin, *Adv. Funct. Mater.*, 2008, **18**, 928; (c) C.-L. Ho, M.-F. Lin, W.-Y. Wong, W.-K. Wong and C.-H. Chen, *Appl. Phys. Lett.*, 2008, **92**, 083301; (d) C.-L. Ho, W.-Y. Wong, Z.-Q. Gao, C.-H. Chen, K.-W. Cheah, B. Yao, Z. Xie, Q. Wang, D. Ma, L. Wang, X.-M. Yu, H.-S. Kwok and Z. Lin, *Adv. Funct. Mater.*, 2008, **18**, 319; (e) C.-L. Ho, Q. Wang, C.-S. Lam, W.-Y. Wong, D. Ma, L. Wang, Z.-Q. Gao, C.-H. Chen, K.-W. Cheah and Z. Lin, *Chem. Asian J.*, 2009, **4**, 89.
36. S. Bettington, M. Tavasli, M. R. Bryce, A. Beeby, H. Al-Attar and A. P. Monkman, *Chem. Eur. J.*, 2007, **13**, 1423.
37. Y. Tao, Q. Wang, C. Yang, K. Zhang, Q. Wang, T. Zou, J. Qin and D. Ma, *J. Mater. Chem.*, 2008, **18**, 4091.
38. H. Tang, Y. Li, B. Chen, H. Wu, W. Yang and Y. Cao, *Opt. Mater.*, 2011, **33**, 1291.
39. D. Wang, J. Wang, H.-L. Fan, H.-F. Huang, Z.-Z. Chu, X.-C. Gao and D.-C. Zou, *Inorg. Chim. Acta*, 2011, **370**, 340.
40. J. Lu, Q. Liu, J. Ding and Y. Tao, *Synth. Met.*, 2008, **158**, 95.
41. Q.-D. Liu, J. Lu, J. Ding and Y. Tao, *Macromol. Chem. Phys.*, 2008, **209**, 1931.
42. I. R. Lasker, S.-F. Hsu and T.-M Chen, *Polyhedron* 2005, **24**, 881.
43. (a) C. J. Tonzola, A. P. Kulkarni, A. P. Gifford, W. Kaminsky and S. A. Jenekhe, *Adv. Funct. Mater.*, 2007, **17**, 863; (b) R. C. Chiechi, R. J. Tseng, F. Marchioni, Y. Yang and F. Wudl, *Adv. Mater.*, 2006, **18**, 325.

44. T. Mitsumori, L. M. Campos, M. A. Garcia-Garibay, F. Wudl, H. Sato and Y. Sato, *J. Mater. Chem.*, 2009, **19**, 5826.
45. W. Lu, C. W. Chan, K. Cheung and C. M. Che, *Organometallics*, 2001, **20**, 2477.
46. C.-L. Ho, W.-Y. Wong, B. Yao, Z. Xie, L. Wang and Z. Lin, *J. Organomet. Chem.*, 2009, **694**, 2735.
47. G.-J. Zhou, X.-Z. Wang, W.-Y. Wong, X.-M. Yu, H.-S. Kwok and Z. Lin, *J. Organomet. Chem.*, 2007, **692**, 3461.
48. W.-Y. Wong, Z. He, S.-K. So, K.-L. Tong and Z. Lin, *Organometallics*, 2005, **24**, 4079.
49. Z. He, W.-Y. Wong, X. Yu, H.-S. Kwok and Z. Lin, *Inorg. Chem.*, 2006, **45**, 10922.
50. Z. M. Hudson, M. G. Helander, Z.-H. Lu and S. Wang, *Chem. Commun.*, 2011, **47**, 755.
51. S.-Y. Chang, J. Kavitha, S.-W. Li, C.-S. Hsu, Y. Chi, Y.-S. Yeh, P.-T. Chou, G.-H. Lee, A. J. Carty, Y.-T. Tao and C.-H. Chien, *Inorg. Chem.*, 2006, **45**, 137.
52. Y.-Y. Lin, S.-C. Chan, M. C. W. Chan, Y.-J. Hou, N. Zhu, C.-M. Che, Y. Liu and Y. Wang, *Chem. Eur. J.*, 2003, **9**, 1264.
53. (a) S.-W. Lai, M. C.-W. Chan, T.-C. Cheung, S.-M. Peng and C.-M. Che, *Inorg. Chem.*, 1999, **38**, 4046; (b) W. Lu, M. C. W. Chan, N. Zhu, C.-M. Che, C. Li and Z. Hui, *J. Am. Chem. Soc.*, 2004, **126**, 7639.
54. D. Qiu, J. Wu, Z. Xie, Y. Cheng and L. Wang, *J. Organomet. Chem.*, 2009, **694**, 737.
55. W. Lu, B.-X. Mi, M. C. W. Chan, Z. Hui, N. Zhu, S.-T. Lee and C.-M. Che, *Chem. Commun.*, 2002, 206.
56. B.-P. Yan, C. C. C. Cheung, S. C. F. Kui, V. A. L. Roy, C.-M. Che and S.-J. Xu, *Appl. Phys. Lett.*, 2007, **91**, 063508.
57. C.-C. Kwok, H. M. Ngai, S.-C. Chan, I. H. T. Sham, C.-M. Che and N. Zhu, *Inorg. Chem.*, 2005, **44**, 4442.
58. K. M.-C. Wong, X. Zhu, L.-L. Hung, N. Zhu, V. W.-W. Yam and H.-S. Kwok, *Chem. Commun.*, 2005, 2906.
59. P.-I. Shih, C.-F. Shu, Y.-L. Tung and Y. Chi, *Appl. Phys. Lett.*, 2006, **88**, 251110.
60. Y.-M. Cheng, G.-H. Lee, P.-T. Chou, L.-S. Chen, Y. Chi, C.-H. Yang, Y.-H. Song, S.-Y. Chang, P.-I. Shih and C.-F. Shu, *Adv. Funct. Mater.*, 2008, **18**, 183.

Photochemically Degradable Polymers; Synthesis of Polymers with Metal–Metal Bonds Along the Backbone Using Click Chemistry

SARAH E. BRADY AND DAVID R. TYLER*

Department of Chemistry, University of Oregon, Eugene, Oregon, USA 97403
*E-mail: dtyler@uoregon.edu

2.1 Introduction

Photochemically degradable polymers[1–6] fall into two general categories: polymers that are intended to degrade and those that are not intended to degrade. Among the former polymers are those used as photodegradable plastics[7] and as photoresists.[8] Virtually every other polymer is in the latter category because every polymer is eventually susceptible to some type of photochemical degradation. Mechanistic knowledge of photochemical degradation processes is helpful in designing new polymers that are intended to degrade and, conversely, in devising strategies to prevent unintended degradation. However, a major problem in designing polymers with specific degradation properties is that the photochemical degradation processes in polymers are mechanistically complicated.[4] This is not to say that the mechanisms are not understood; in fact, many are understood in detail.

RSC Polymer Chemistry Series No. 2
Molecular Design and Applications of Photofunctional Polymers and Materials
Edited by Wai-Yeung Wong and Alaa S Abd-El-Aziz
© The Royal Society of Chemistry 2012
Published by the Royal Society of Chemistry, www.rsc.org

Rather, the mechanisms are intricate, often involving multiple steps, cross-linking, and side reactions; this makes pinpointing the effects that various environmental and molecular parameters have on the degradation process difficult. For example, one formidable complication is that oxygen diffusion is the rate-limiting step in many photo-oxidative degradations.[9,10] This adds to the intricacy of the analysis because oxygen diffusion rates are frequently time-dependent.[10,11]

To circumvent many of the experimental and mechanistic complexities in the photodegradation process and therefore make it less difficult to interpret data and obtain fundamental insights, we study model polymers with metal–metal bonds along the backbone of the polymer.[12,13] Such polymers are photodegradable because the metal–metal bonds can be cleaved with visible light (eqn (2.1)) and the resulting metal radicals captured with an appropriate radical trap, typically a carbon–halogen bond or O_2 (Scheme 2.1).[12] By studying the photodegradation of these model polymers, we have been able to extract information without the mechanistic complications inherent in the degradation mechanisms of purely organic polymers. For example, metal radicals do not lead to crosslinking, so this complicating feature found in the degradation of other polymers can be avoided.[12] An additional advantage is that the metal–metal bond is a chromophore, and its distinctive absorption bands in the visible region can be used to monitor the photodegradation reactions by electronic absorption spectroscopy. The use of UV-vis methods to quantify and compare the various degradation rates is a time-saving technique because polymer degradation reactions are typically monitored by stress testing, molecular weight measurements, or attenuated total reflection (ATR) spectroscopy, all of which can be laborious and time consuming. The utility of the metal–metal bond-containing polymers in photodegradation studies has been demonstrated in previous investigations into the effects of mechanical stress,[14] polymer morphology,[15] and temperature[16] on polymer photodegradation rates.

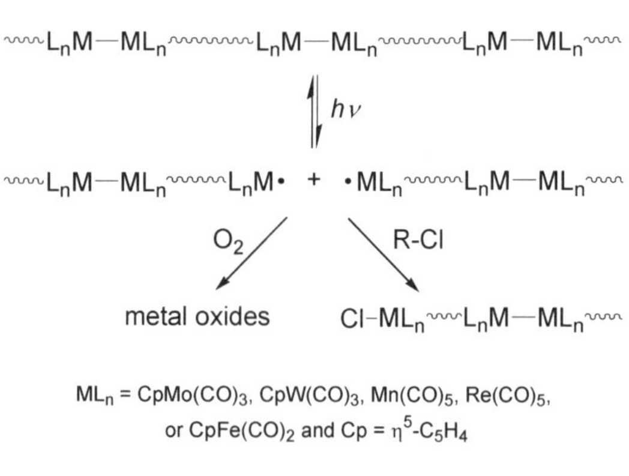

$L_nM—ML_n$ $L_nM—ML_n$ $L_nM—ML_n$
$h\nu$
$L_nM—ML_n$ $L_nM\cdot$ + $\cdot ML_n$ $L_nM—ML_n$
O_2
R-Cl
metal oxides
$Cl–ML_n$ $L_nM—ML_n$
$ML_n = CpMo(CO)_3, CpW(CO)_3, Mn(CO)_5, Re(CO)_5,$
or $CpFe(CO)_2$ and $Cp = \eta^5\text{-}C_5H_4$

Scheme 2.1 Photochemical reactivity of a polymer with metal–metal bonds along its backbone.

$$L_nM-ML_n \xrightarrow{\text{hv}} L_nM\cdot + \cdot ML_n$$

$$\cdot ML_n = CpMo(CO)_3 \ (Cp = \eta^5\text{-}C_5H_5), \ CpW(CO)_3, \ Mn(CO)_5, \ Re(CO)_5, \ CpFe(CO)_2$$

(2.1)

In prior papers and reviews, we described metal–metal bond-containing polymers synthesized by step polymerization, chain polymerization, ADMET, and ROMP reactions.[12,13] A recurring problem is that the polymers synthesized by these methods generally have low molecular weights. In order to increase the molecular weight of the polymers and to expand the repertoire of polymer types and morphologies available for study, we have been investigating the use of click chemistry to synthesize polymers. This chapter summarizes some recent work in our laboratory on photochemically degradable polymers synthesized by click reactions. It is important to point out that our research is not necessarily intended to arrive at new commercial photodegradable polymers. Rather, our objectives are to explore the photochemical decomposition mechanisms in various types of polymers (*e.g.*, linear, dendrimers, star, comb, *etc.*) and to explore the effect of environmental and polymer properties on these degradation processes. With this knowledge, we and other workers will be able to design more suitable photodegradable polymers or more suitable light-stable polymers. Before discussing click polymerization reactions, an overview of the methods used to synthesize metal–metal bond-containing polymers is provided. The photochemistry of these polymers is also reviewed.

2.2 Functionalized Metal–Metal Bonded Dimers

Polymers with metal–metal bonds along the backbone have been synthesized using step polymerization, chain polymerization, ROMP, ADMET, and click methods. All of these methods require a functionalized metal–metal bonded dimer as a synthetic starting point, and in our laboratory, we use organometallic dimers with substituted cyclopentadienyl (Cp) rings as the starting materials in these polymerization reactions. A selection of these metal–metal bond-containing synthons is shown in Figure 2.1.[13]

The synthesis of the metal–metal dimers with substituted Cp ligands is challenging due to the weak bond energy of the metal–metal bond ($BDE_{Mo-Mo} \approx 134 \text{ kJ mol}^{-1}$; $BDE_{W-W} \approx 235 \text{ kJ mol}^{-1}$).[13] Because the metal–metal bond is relatively weak, the $Cp_2Mo_2(CO)_6$ framework generally cannot withstand the harsh experimental conditions that are necessary to functionalize the Cp ligands. For that reason, it is necessary to synthesize the functionalized Cp ligands first and then coordinate these rings to the metals, followed by formation of the metal–metal bond. Two sample syntheses of metal–metal dimers with substituted Cp ligands are shown in Schemes 2.2 and 2.3.[17] (This synthetic method is emphasized here because it stands in contrast to the methods typically used in the preparation of monomers used to prepare polyferrocenes, perhaps the best-known category of metal-containing poly-

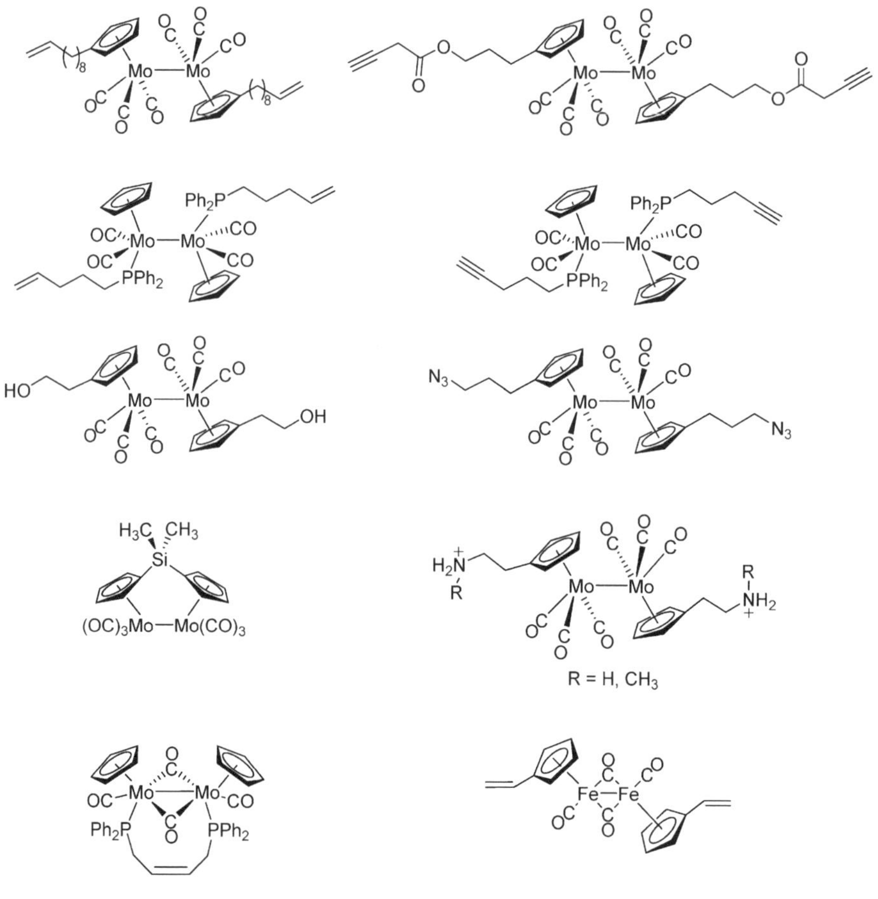

Figure 2.1 Functionalized metal–metal dimers for use in various types of polymerization reactions.

mers. The ferrocene molecule is so robust that the Cp ligands are typically substituted with functional groups after assembly of the Cp_2Fe unit.[18–20]).

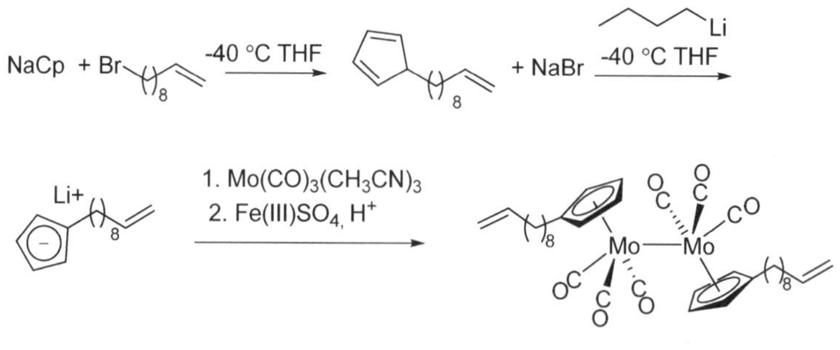

Scheme 2.2 Synthetic route for the preparation of $[(\eta^5\text{-}C_5H_4(CH_2)_8CH{=}CH_2)Mo(CO)_3]_2$.

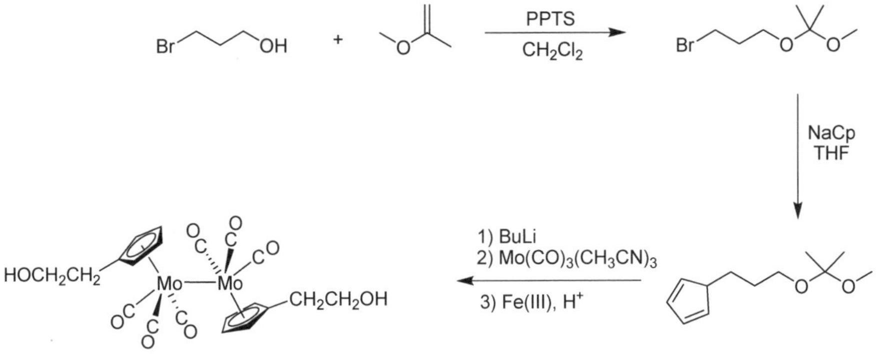

Scheme 2.3 Synthetic route for the preparation of $[(Cp(CH_2)_3OH)Mo(CO)_3]_2$.

2.3 Polymer Syntheses

Just as the comparatively weak metal–metal bonds pose problems for the synthesis of the functionalized-Cp dimers, they cause similar problems in the synthesis of the polymers. The relative weakness of the metal–metal bonds makes them more reactive than the bonds found in standard organic polymers; thus, under many standard polymerization reaction conditions, metal–metal bond cleavage would result. For example, reaction of $(Cp(CH_2)_2OH)_2Mo_2(CO)_6$ with a diacyl halide will typically not yield a high molecular weight polyester because acyl halides can react with metal–metal bonds to form a metal halide complex (eqn (2.2)).[12,21]

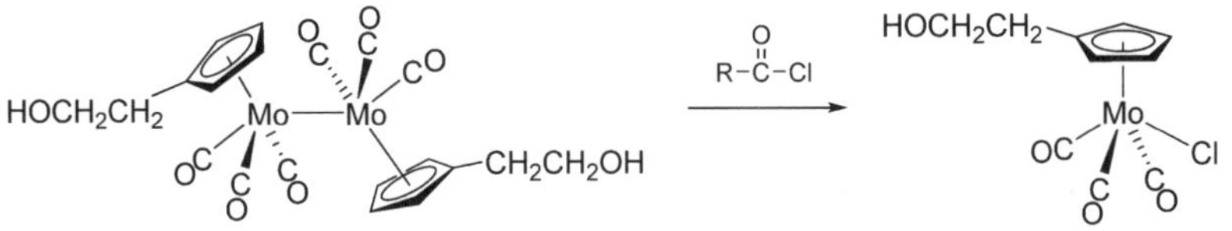

$$\text{(2.2)}$$

In another example, it is well known that Lewis bases react with metal–metal bonded dimers to form disproportionation products $(Mo^I\text{-}Mo^I \rightarrow Mo^0 + Mo^{II};$ eq 2.3).[22] Thus, many condensation polymerizations, in which bases are used to neutralize acid condensates, cannot be used in the polymerization reactions involving the metal–metal dimers. To summarize, polymerization reactions that involve metal–metal bonded precursors must generally utilize non-nucleophilic and noncoordinating reagents and solvents, and they must operate under mild reaction conditions. Ideally, they must also be functional group specific. As discussed in the next section, one of the advantages of click chemistry is that these rather stringent reaction conditions are met, and consequently high molecular weight polymers can be obtained.

$$Cp_2Mo_2(CO)_6 + 2\,L \xrightarrow{\ h\nu\ } CpMo(CO)_3^- + CpMo(CO)_2L_2^+ + CO \qquad \text{(2.3)}$$

Sample polymer syntheses that demonstrate the points above are shown in eqns (2.4)–(2.6).[23] Note, however, that in the homopolymerization reactions (*e.g.*, eq 2.4) molecular weights rarely exceed about 10 000 – 15 000 amu. High molecular weights are generally obtained only in the case of reactions with prepolymers or other copolymerization reactions.

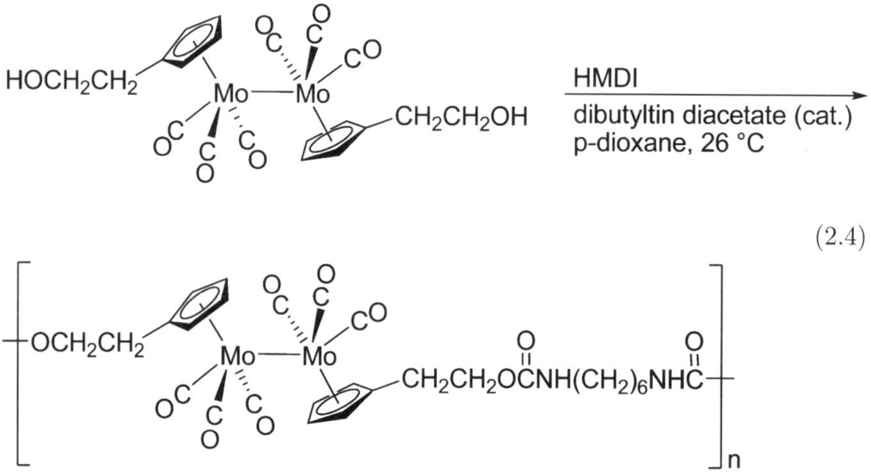

$$(2.4)$$

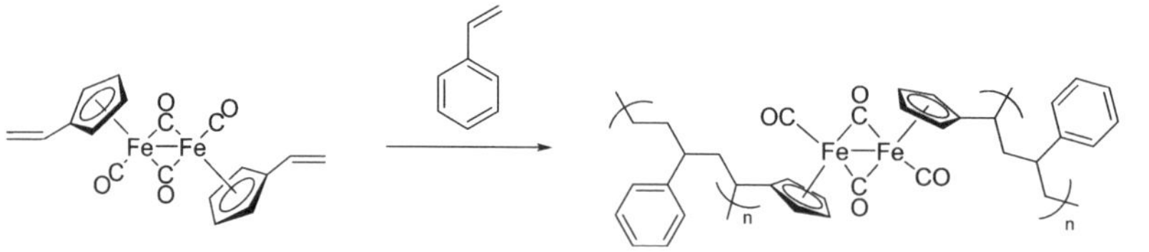

$$(2.5)$$

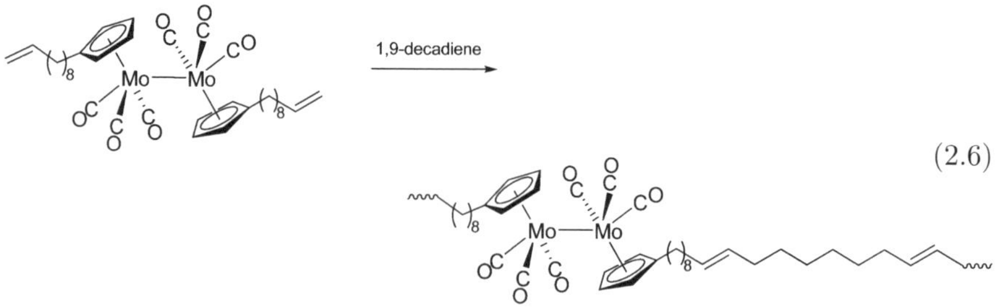

$$(2.6)$$

2.4   Polymer Photochemistry

As anticipated, the polymers and copolymers with metal–metal bonds in their backbones are photochemically reactive with visible light, and they degrade according to the general reactivity outlined in Scheme 2.1. In general, the photochemical reactivity of the metal–metal bond-containing polymers is qualitatively analogous to the reactions of the metal–metal bonded dimers in solution.[12,23–26] To understand the photochemical reactions of the polymers, the photochemistry of the metal–metal bonded dimers is described first.

2.4.1　Photochemical Reactions of the Polymers in Solution

Irradiation of metal–metal bonded complexes into their lowest-energy absorption band ($\approx$ 500 nm) leads to one of the three fundamental types of reactivity:[23,27,28] (1) The metal radicals produced by photolysis react with radical traps to form monomeric complexes (*e.g.*, eqn (2.7)). (2) The complexes react photochemically with ligands (nucleophiles) to form ionic disproportionation products (*e.g.*, eqn (2.8)). (3) The complexes react with oxygen to form metal oxides (eqn (2.9)). Higher-energy (UV) excitation can lead to M–CO bond dissociation, but this photoprocess does not necessarily lead to chain scission so it will not be discussed here.

$$Cp_2Mo_2(CO)_6 + 2\,CCl_4 \xrightarrow{h\nu} 2\,CpMo(CO)_3Cl + 2\,[\cdot CCl_3] \qquad (2.7)$$

$$Cp_2Mo_2(CO)_6 + 2\,PR_3 \xrightarrow{h\nu} CpMo(CO)_3^- + CpMo(CO)_2(PR_3)_2^+ + CO \qquad (2.8)$$

$$Cp_2Mo_2(CO)_6 \xrightarrow[O_2]{h\nu} Mo\ oxides \qquad (2.9)$$

　　As mentioned above, the photochemistry of the metal–metal bond-containing polymers is qualitatively analogous to the reactions of the metal–metal bonded dimers in solution.[12,23–26] Sample reactions of the polymers showing the three types of reactivity are shown in eqns (2.10)–(2.12). The quantum yields for these reactions (of either the discrete dimers or the polymers) are in the range $\approx$ 0.1 to 0.6, depending on the specific polymer and the identity of the metal in the M–M bond.[24]

$$(2.10)$$

$$\text{(2.11)}$$

$$\text{hv} \mid \text{P(OEt)}_3$$

$$[\text{P(OEt)}_3]_2(\text{CO})_2\overset{+}{\text{Mo}}-\text{CH}_2\text{CH}_2\text{OCNH(CH}_2)_6\text{NHCOCH}_2\text{CH}_2-\overset{-}{\text{Mo}}(\text{CO})_3$$

$$\text{(2.12)}$$

$$\text{hv} \mid \text{O}_2$$

metal oxides

Photochemical reactivity in the absence of exogenous radical traps is possible in the case of polymers that have carbon–halogen bonds along their backbones. For example, irradiation of polymers **1** – **3** in solution in the absence of CCl_4 or O_2 led to net metal–metal bond cleavage.[12] Spectroscopic monitoring of the reaction showed that metal–metal bond cleavage is accompanied by an increase in the concentration of $\text{CpMo(CO)}_3\text{Cl}$ units. Photochemical reactions analogous to the radical-trapping reaction in Scheme 2.4 were proposed.

1

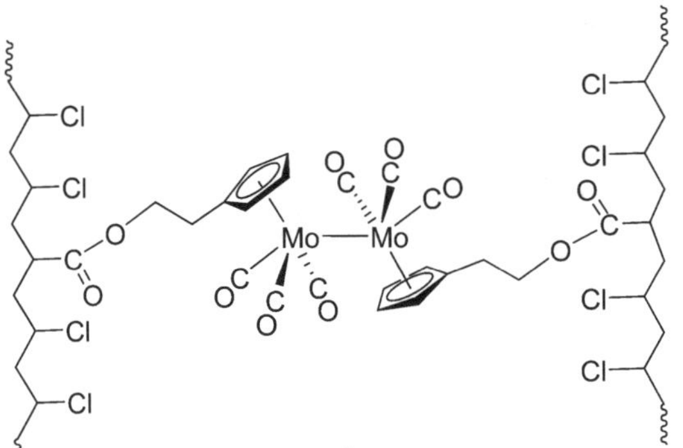

2

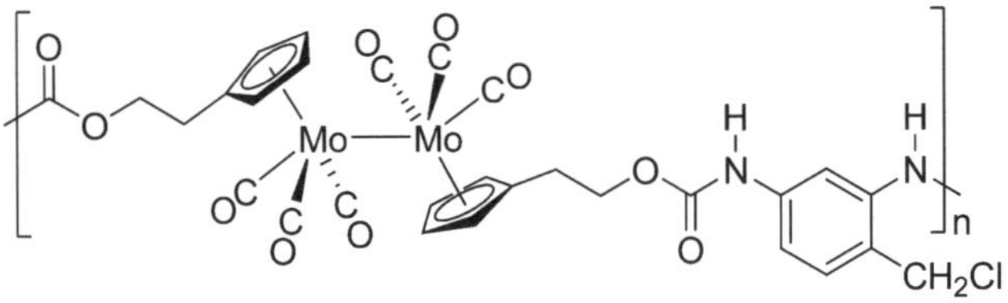

3

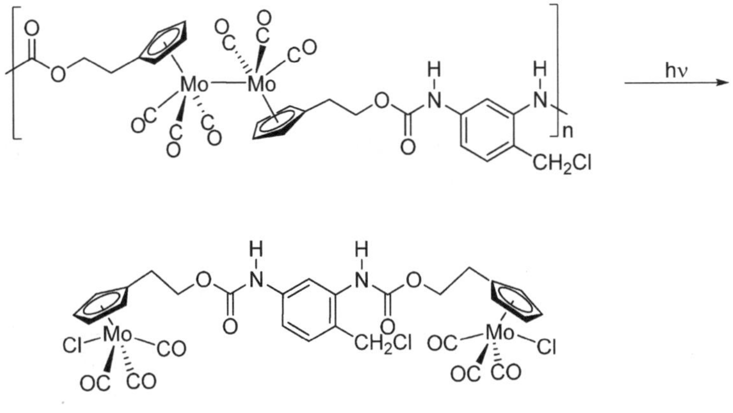

Scheme 2.4 Photochemical reaction of polymer **3** in the absence of an external trapping reagent.

2.4.2 Photochemical Reactions of the Polymers in the Solid State

In the presence of oxygen, thin films of the metal–metal bond-containing polymers react when they are exposed to visible light, whether from the overhead fluorescent lights in the laboratory, from sunlight, or from the filtered output of a high pressure Hg arc lamp.[12,23–26] Analysis of these solid-state reactions showed the products indicated in eqn (2.13).[29] Note that the Mo(I) in the polymer is oxidized to Mo(VI) (the exact species was not identified but it is presumably a Mo oxide species), CO is evolved (which, of course, implies it was dissociated from the Mo center), and the molecular weight of the polymer decreased. From these results, it was concluded that the polymer backbone is cleaved in the photochemical reaction, a consequence of the complete degradation of the $Cp_2Mo_2(CO)_6$ unit in the polymer backbone. Control experiments showed that the photochemical reaction of the discrete $Cp_2Mo_2(CO)_6$ molecule in the solid state likewise gave a Mo(VI) product and evolved CO. Thus, as in solution, the photochemistry of the polymers in the solid state is analogous to that of the $Cp_2Mo_2(CO)_6$ molecule. Control experiments also showed that oxygen was necessary for the photochemical degradation of the polymers (except, of course, for polymers **1 – 3**). This result is logical because in the absence of metal-radical traps there can be no net trapping reaction and the photochemically generated radicals recombine to re-form the metal–metal bond, eqn (2.14).[30–32]

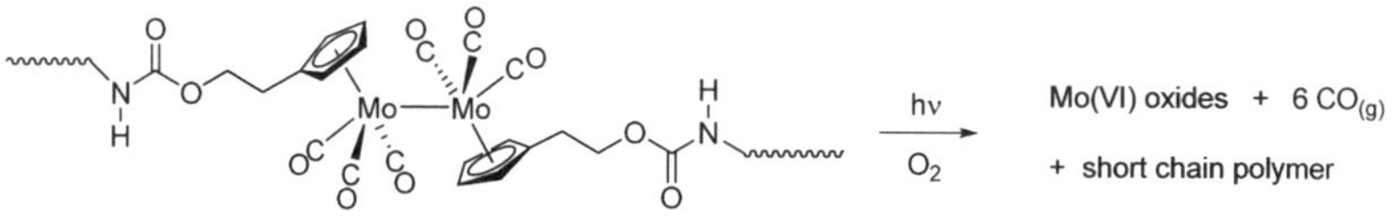

$$\text{(2.13)}$$

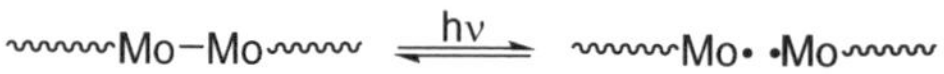

$$\text{(2.14)}$$

Because they have built-in C–Cl radical traps, polymers **1 – 3** reacted in the solid state in the absence of oxygen. The reactions were analogous to the reaction in Scheme 2.4.[15]

2.4.3 Dependence of Solid-State Photochemical Degradation Reactions on Temperature

The rates of polymer degradations in the solid state are temperature dependent.[16] To illustrate this effect, the plot in Figure 2.2 shows the temperature dependence for the photochemical degradation of polymer **2**.[16] (Recall this polymer degrades in the absence of oxygen by reacting with the internal C–Cl radical traps, analogous to the reaction for polymer **3** shown in Scheme 2.4.)

It was proposed that the large temperature dependence of the photodegradation efficiency arose from the temperature dependence of the reaction step that involves diffusive separation of the radical cage pair generated by photolysis of the Mo–Mo bond (see the k_d step in Scheme 2.5). Note there are two ways for the newly formed radicals to become separated: they can diffuse

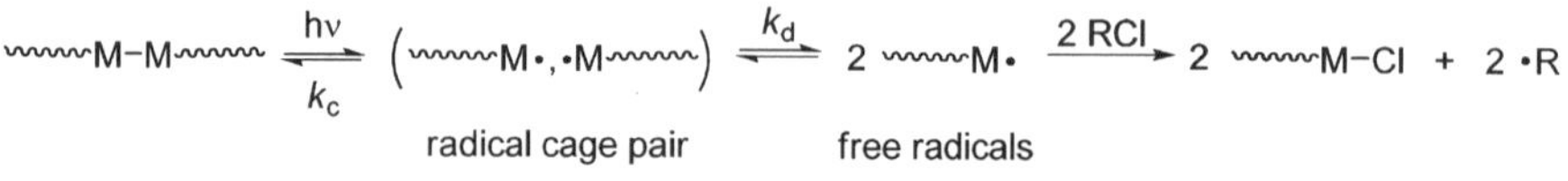

Scheme 2.5   Reaction of a metal–metal bond-containing polymer to form a caged radical pair followed by a radical trapping reaction.

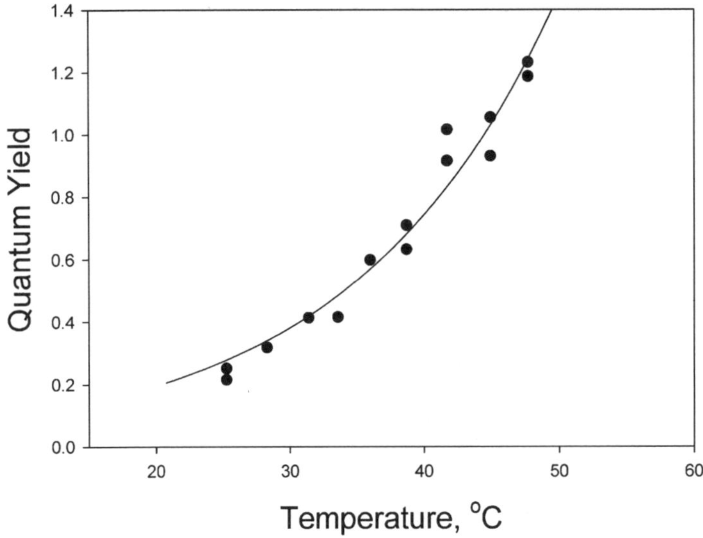

Figure 2.2   Plot of the quantum yield as a function of temperature for the disappearance of the $Cp_2Mo_2(CO)_6$ unit in polymer **2**.

apart translationally or rotationally. The increased thermal energy facilitates both the rotational and translational diffusion processes, which effectively increases the efficiency of free-radical formation. This leads to decreased radical–radical recombination and consequently an increase in photodegradation efficiency.

The quantum yield data for polymer **2** in Figure 2.2 have an exponential dependence on the inverse temperature, and it is tempting therefore to extract activation parameters from the natural log plots of quantum yield *vs.* inverse temperature. Balzani and Carassiti, however, has cautioned that the relationship between the temperature and activation parameters in a photochemical reaction is a complex one,[33] and the "apparent activation energies" thus obtained must be interpreted with care. With that disclaimer in mind, the activation energy obtained from the $\ln\Phi$ *vs.* T^{-1} plot is 14.1 ± 0.3 kcal mol^{-1}. This value is typical for secondary relaxation chain movements in polymers (which generally fall in the range $10 - 20$ kcal mol^{-1} [34–36]) and is consistent with the proposal that the temperature dependence of Φ results from chain movements involved in recoil and rotation processes.

Note that all of the temperature-dependent quantum yield data for polymer **2** were collected below the glass-transition temperatures of the polymer film ($T_g \approx 65$ °C). This point is noted because Dan and Guillet observed a sudden jump in the quantum yield to a value similar to that in solution ($\Phi = 0.24$) for the photochemical reactions of phenyl vinyl ketone polymer when the temperature increased above T_g.[37] (Below the glass-transition temperature, the effect of temperature on the quantum yield was minimal.) The sudden increase in Φ at the glass-transition temperature was attributed to facile chain movements above T_g and the ability of the polymer chains to form the proper conformation for the Norrish Type II reaction. An important conclusion from these results is that, in polymers that react by changing their chain conformations, the effect of temperature on the quantum yield will be relatively small below T_g but then discontinuous to a near-solution value at T_g. In contrast, in polymers that react by bond homolysis (such as the metal–metal bond-containing polymers) the effect of temperature below T_g is monotonic and rather large.

2.4.4 Dependence of Solid-State Photochemical Degradation Reactions on Tensile Stress

An interesting outcome of artificial weathering studies on polymers is the finding that tensile and shear stress can accelerate the rate of photodegradation.[38] For example, recent studies of this phenomenon have shown that tensile stress will accelerate the degradation of numerous polyolefins[39-43] as well as polycarbonates,[41] nylon,[44] and acrylic-melamine coatings.[45,46] These observations are of practical importance because most polymers are subjected to light and some form of temporary or permanent stress during their lifetime. In order to control the onset of degradation and the rate of degradation in these materials, it is important to understand the mechanistic origins of the synergism between light and stress in these systems so that lifetimes can be predicted.

Few mechanistic studies have probed the origin of stress-dependent photodegradation rates. What studies there are have generally been hampered by the mechanistic complexity of the degradation reactions.[4] As discussed above, photochemical degradation pathways generally involve multiple steps, crosslinking, and side reactions; these features make pinpointing the origin of stress-induced rate accelerations difficult. Another formidable complication is that oxygen diffusion is the rate-limiting step in many photo-oxidative degradations.[47] This adds to the intricacy of the analysis because oxygen diffusion rates are frequently time dependent.[10,11] To circumvent these experimental and mechanistic complexities, a study in our lab investigated the effects of tensile stress on polymer **2**.[14,48] Recall that polymer **2** will degrade in the absence of oxygen, thus eliminating the kinetically complicating effects of rate-limiting oxygen diffusion. The application of tensile stress changed the photodegradation efficiency of **2**, and a plot of relative quantum yield *vs.* stress is shown in Figure 2.3. Note that tensile stress initially caused the quantum yield to increase, but after a certain point additional stress caused a decrease in the quantum yield.

The results in Figure 2.3 are consistent with a hypothesis called the "decreased radical recombination" hypothesis.[49] According to this hypothesis, the application of tensile stress initially causes the straightening of polymer chains in the amorphous regions. These straightened chains contain taut tie molecules. (Tie molecules are those in the interlamellar or intercrystal fibrils.) When bonds in the taut tie molecules are cleaved by light, the probability of radical–radical recombination (Scheme 2.5) is decreased relative to nonstressed samples because entropic relaxation of the chain drives the radicals apart and prevents their efficient recombination because of their increased separation. As the tensile stress is increased, the chains are not only straightened but also "stretched," and mechanical recoil aids in the separation of the radicals (much like the midpoints of a stretched spring would fly apart if it were cut in the middle). According to this model, the role of stress is to increase the separation of the radical fragments produced by photolysis. An increased separation leads to slower radical–radical recombination, which increases the probability of radical trapping and thus of degradation.

Finally, as the stress increases even more, ordered regions develop in the polymer as segments of different chains align. Diffusion in ordered and crystalline regions is retarded relative to the amorphous material, and the efficiency of degradation is expected to decrease because of decreased diffusion apart of the radical pair and decreased radical-trap mobility. (Note that X-ray scattering and infrared spectroscopy experimentally confirmed the increase in chain order in polymer **2** at high stress.[14]) The conclusion, at least in this system, is that the role of stress is to increase the separation of the photochemically generated radical pair, which decreases their probability of

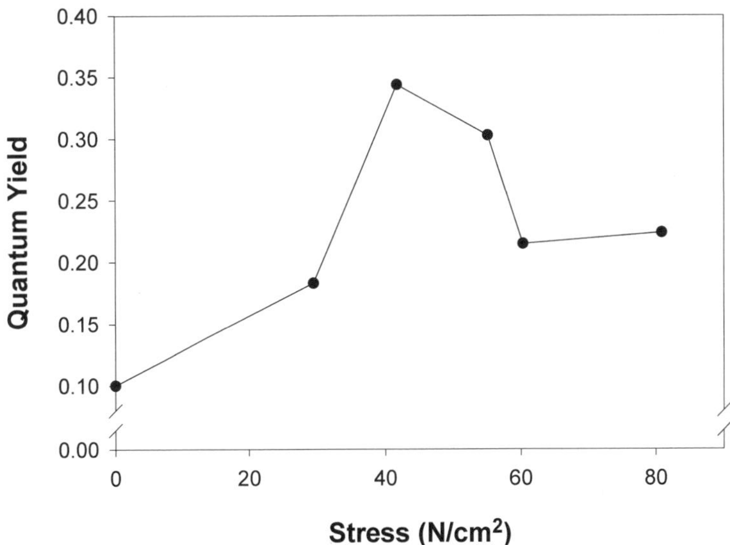

Figure 2.3 Quantum yields for degradation of polymer **3** *vs.* applied tensile stress.

recombination. Just as important, however, is the point that this study would not have been possible without the polymers that have metal–metal bonds along their backbone; acquiring quantum yield data as a function of stress would be nearly impossible in other polymers. The experimental utility of the metal–metal bond-containing polymers was thus demonstrated once again.

2.5 Overview of Click Chemistry

Click chemistry refers to a collection of reactions that meet specific requirements: the reactions must be exceptionally selective, involve mild reaction conditions, require little purification, and be high yielding.[50] The most often used and well-known example of a click reaction is the copper(I)-catalyzed 1,3-dipolar cycloaddition reaction between terminal acetylenes and azides to form 1,2,3-triazoles (commonly called the Huisgen reaction). In the absence of a catalyst, this reaction does not qualify as a click reaction because it yields a mixture of 1,4- and 1,5-adducts and must be run under high temperatures (Scheme 2.6).[51] With an appropriate Cu catalyst, however, the 1,4-adduct is formed selectively and the reaction can be carried out at room temperature. Two different copper catalyst systems are used: 1) a mixture of copper(II) sulfate and sodium ascorbate (to generate the copper(I) in situ), or 2) a copper(I) halide with a stabilizing ligand such as N,N,N′,N′,N′′-pentamethyldiethylenetriamine (PMDETA) or bipyridine. Alternatively, the ruthenium(II) catalysts Cp*Ru(PPh$_3$)$_2$Cl and Cp*Ru(COD)Cl selectively yield the 1,5-adduct.[52] Based on the proposed mechanisms, the selectivity of these reactions is dependent on how the acetylene coordinates to the metal catalyst.[53,54]

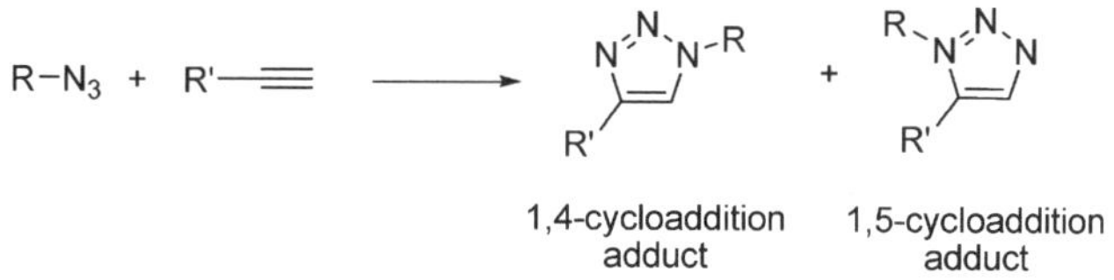

$$R-N_3 \ + \ R'-\!\!\equiv\!\!- \longrightarrow$$

1,4-cycloaddition adduct   1,5-cycloaddition adduct

Scheme 2.6　Huisgen 1,3-dipolar cycloaddition.

Click chemistry has been used for a myriad of applications from biochemistry to drug delivery to surface functionalization.[55–63] Polymer chemistry has also taken advantage of the simplicity and specificity of the Huisgen reaction, often by combining it with living/controlled polymerization reactions to synthesize a wide range of well-defined polymer architectures. In particular, well-defined block copolymers, cyclic polymers, side-chain functionalized polymers, dendrimers, star polymers, and crosslinked polymeric networks have all been achieved using the Huisgen reaction.[51,64–71] Step polymerization methods can also be employed using bifunctional alkyne and azide monomers to synthesize alternating copolymers under mild reaction conditions.[72,73] This is the strategy that has been used by our laboratory to prepare photoactive polymers containing metal–metal bonds.[74]

2.6 Synthesis of Metal–Metal Bond-Containing Click Synthons

Click synthons containing metal–metal bonds and Cp ligands with acetylene or azide functionalities can be synthesized from the previously reported [(η^5-$C_5H_4(CH_2)_3OH)Mo(CO)_3]_2$ molecule. (For the synthesis of this starting material, see Scheme 2.3.[75]) As an example, the acetylene functionalized dimer, [(Cp(CH$_2$)$_3$OCO(CH$_2$)$_2$C≡CH)Mo(CO)$_3$]$_2$ was prepared by a Steglich esterification of [(Cp(CH$_2$)$_3$OH)Mo(CO)$_3$]$_2$, as shown in Scheme 2.7.[69]

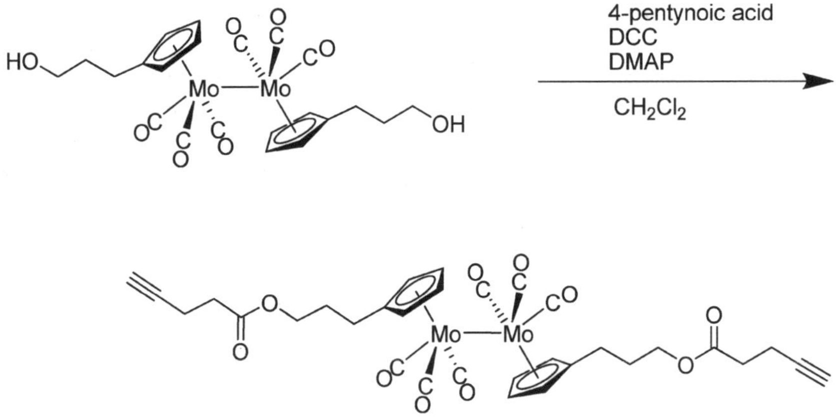

Scheme 2.7 Synthesis [(Cp(CH$_2$)$_3$OCO(CH$_2$)$_2$C≡CH)Mo(CO)$_3$]$_2$.

The azide functionalized molecule [(Cp(CH$_2$)$_3$N$_3$)Mo(CO)$_3$]$_2$ was synthesized by the reaction shown in Scheme 2.8.[69] In this reaction scheme, the alcohol functionalized dimer was tosylated, followed by azide substitution of

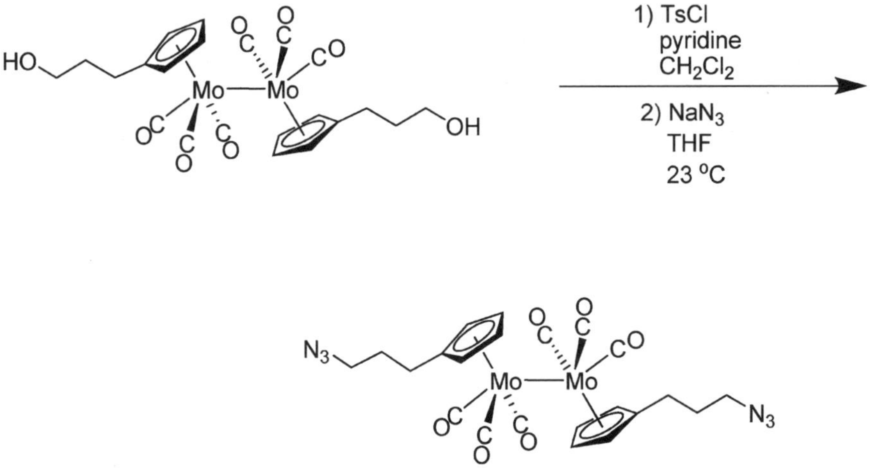

Scheme 2.8 Synthesis of [(Cp(CH$_2$)$_3$N$_3$)Mo(CO)$_3$]$_2$.

the tosylate group. Note that both the acetylene functionalized dimer and the azide functionalized dimers still exhibited the distinctive carbonyl stretching bands at 1954(vs) and 1913(s) cm^{-1}, indicative of an intact metal–metal bond in a $Cp_2Mo_2(CO)_6$ unit.[23]

2.7 Model Click Reactions

Model click reactions were carried out with the metal–metal bonded synthons in order to determine the best reaction conditions and how to characterize the products formed in the click polymerization reactions. The model reactions (Scheme 2.9) were performed with both the azide and acetylene synthons described in the preceding section with four different catalysts: 1) CuBr/ PMDETA; 2) CuBr/bipyridine; 3) Cp*Ru(PPh$_3$)$_2$Cl; and 4) Cp*Ru(COD)Cl (Cp* = η^5-C$_5$(CH$_3$)$_5$). The choice of catalyst turned out to be critical. For example, when [(Cp(CH$_2$)$_3$OCO(CH$_2$)$_2$C≡CH)Mo(CO)$_3$]$_2$ was reacted with benzyl azide using the CuBr/PMDETA or CuBr/bipyridine catalysts, disproportionation of the Mo–Mo bond occurred before the click coupling reaction had gone to completion (Scheme 2.10). That disproportionation had occurred was indicated by the appearance of new bands in the infrared spectrum at 1793(vs) and 1817(sh) cm^{-1}, assigned to the (Cp(CH$_2$)$_3$OCO(CH$_2$)$_2$C≡CH)Mo(CO)$_3^-$ product of the disproportionation reaction.[76] Control reactions demonstrated that no disproportionation occurred in the absence of free ligand (either PMDETA or bipyridine), which suggests that these amine ligands are acting as the ligands in the

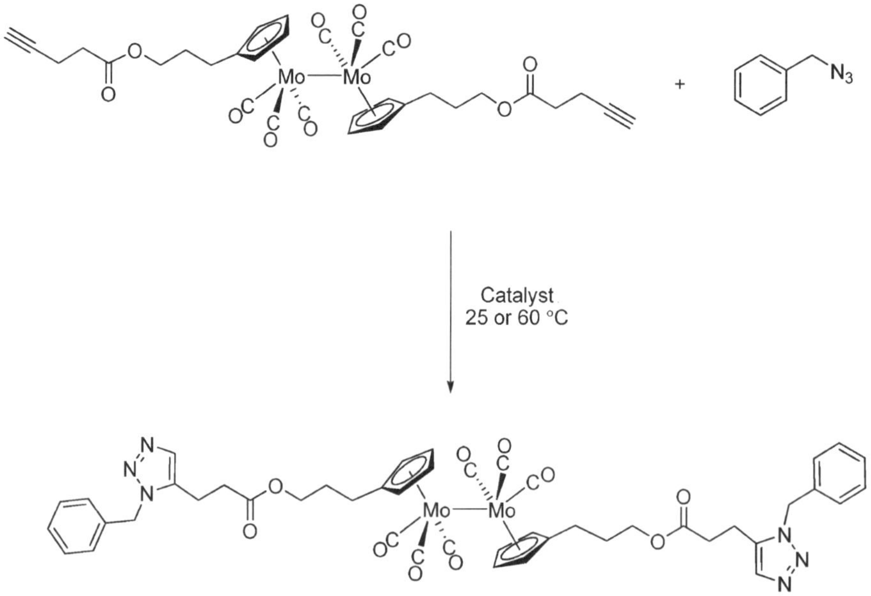

Scheme 2.9   Reaction scheme for model click reaction with the acetylene dimer.

$$Cp_2Mo_2(CO)_6 + L \longrightarrow CpMo(CO)_3^- + CpMo(CO)_2L^+ + CO$$

$$L = PMEDTA \text{ or bipyridine}$$

Scheme 2.10 Disproportionation reaction of a hexacarbonyl dimeric molybdenum complex.

disproportionation reaction (Scheme 2.10). Note that it is known that $Cp_2Mo_2(CO)_6$ will react with amines to give disproportionation products. Thus, this result was not too surprising. The disproportionation reaction is always a concern when working with $Cp_2Mo_2(CO)_6$ and its functionalized derivatives, and in fact disproportionation is one of the main obstacles to making high molecular weight polymers with metal–metal bonds in their backbones. In summary, although the CuBr/ligand system is the catalyst of choice in much of the current click-chemistry literature, it is not an appropriate choice when using reactants containing metal–metal bonds.[74]

The same model reaction of $[(Cp(CH_2)_3OCO(CH_2)_2C\equiv CH)Mo(CO)_3]_2$ with benzyl azide was studied using $Cp*Ru(PPh_3)_2Cl$ as the catalyst because this catalyst does not require the coaddition of a free ligand. Using this catalyst, the click coupling reaction (in Scheme 2.9) went to completion with no disproportionation at both 25 °C and 60 °C.[74] $Cp*Ru(PPh_3)_2Cl$ was also used to catalyze the click coupling of $[(Cp(CH_2)_3N_3)Mo(CO)_3]_2$ with phenylacetylene using the same reaction conditions (Scheme 2.11). Unexpectedly, no click coupling (or disproportionation) occurred in this case. ^{31}P NMR spectroscopic analysis of the reaction solution showed that the $[(Cp(CH_2)_3N_3)Mo(CO)_3]_2$ synthon underwent a Staudinger reaction (eqn (2.15)) with a triphenylphosphine molecule from the catalyst.[77] Note that, according to the proposed mechanism for the click reaction with $Cp*Ru(PPh_3)_2Cl$ catalyst, both triphenylphosphine ligands dissociate from the catalyst during the catalytic cycle.[53] Thus, it is reasonable that some free triphenylphosphine is present in solution and that it can participate in a Staudinger reaction.

$$R{-}N_3 + PPh_3 \longrightarrow RN{=}PPh_3 \qquad\qquad (2.15)$$

It is reasonable to ask why the Staudinger reaction was not observed in the reaction of $[(Cp(CH_2)_3OCO(CH_2)_2C\equiv CH)Mo(CO)_3]_2$ with benzyl azide and $Cp*Ru(PPh_3)_2Cl$ catalyst. There is no definitive answer to this question yet, but it was suggested that the azide functional group is more reactive in $[(Cp(CH_2)_3N_3)Mo(CO)_3]_2$ than in benzyl azide.

Because of the undesired Staudinger reactivity of $[(Cp(CH_2)_3N_3)Mo(CO)_3]_2$ with phenylacetylene, a third catalyst was tested, namely $Cp*Ru(COD)Cl$. Note that this complex can be used as a catalyst without the addition of any free ligand and that no free phosphines will be present. With this catalyst, click coupling was achieved and the Mo–Mo bond of the synthon remained intact.

All the model reactions described above demonstrated that temperature control is important. Past studies on $Cp_2Mo_2(CO)_6$ and its functionalized

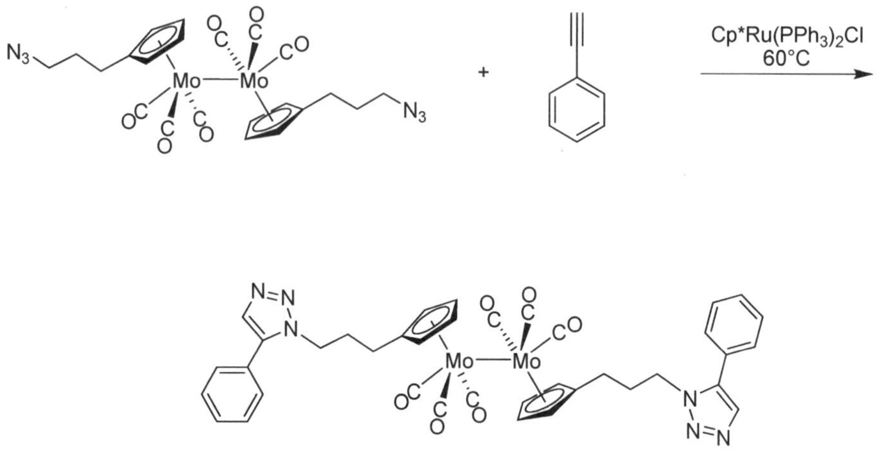

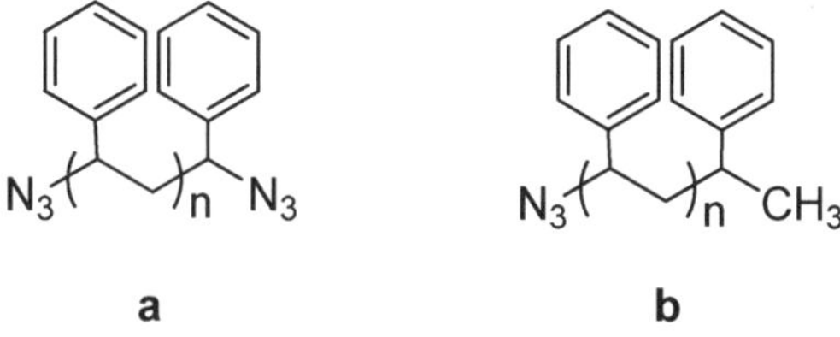

Scheme 2.11 Reaction scheme for the model click reaction of phenylacetylene with [(Cp(CH$_2$)$_3$N$_3$)Mo(CO)$_3$]$_2$.

derivatives established that the Mo–Mo bond can be thermally cleaved in the temperature range 70–135 °C, resulting in decomposition of the dimer.[78] Not unexpectedly, therefore, decomposition of the Mo–Mo synthon occurred when the model reactions were carried out at 80–90 °C. The decomposition was indicated by the disappearance of the ν(C≡O) bands of the dimer molecule at around 1954(vs) and 1913(s) cm^{-1} in the infrared spectrum.

2.8   Step Polymerizations Using the Huisgen Reaction

With the success of the Cp*Ru(PPh$_3$)$_2$Cl- and Cp*Ru(COD)Cl-catalyzed model click reactions, step polymerizations of the [(Cp(CH$_2$)$_3$N$_3$)Mo(CO)$_3$]$_2$ and [(Cp(CH$_2$)$_3$OCO(CH$_2$)$_2$C≡CH)Mo(CO)$_3$]$_2$ synthons were studied. For these polymerization studies, several azide-terminated prepolymers were synthesized for the purpose of reacting them with the [(Cp(CH$_2$)$_3$OCO(CH$_2$)$_2$C≡CH)Mo(CO)$_3$]$_2$ synthon. Specifically, homotelechelic polystyrene (M_n = 3700 g mol^{-1}; Figure 2.4, structure a) and heterotelechelic polystyrene (M_n = 3200 g mol^{-1}; Figure 2.4, structure b) were both synthesized by a literature method.[79] The [(Cp(CH$_2$)$_3$OCO(CH$_2$)$_2$C≡CH)Mo(CO)$_3$]$_2$ dimer reacted over seven days with

a   b

Figure 2.4 Homotelechelic polystyrene (**a**) and heterotelechelic polystyrene (**b**) used in the step polymerization reactions.

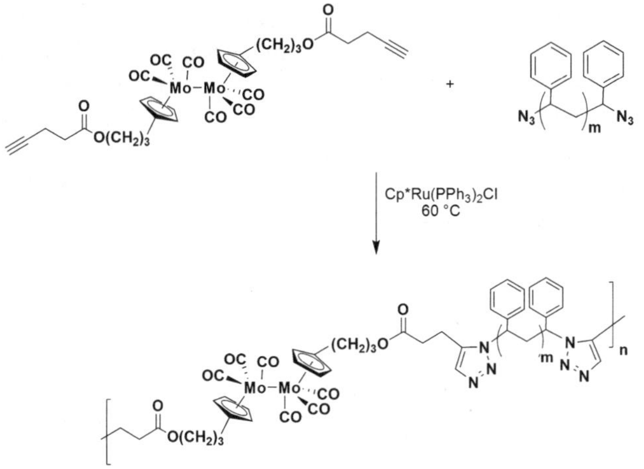

Scheme 2.12 Step polymerization of $[(Cp(CH_2)_3OCO(CH_2)_2C{\equiv}CH)Mo(CO)_3]_2$ with homotelechelic azide-terminated polystyrene.

homotelechelic polystyrene using the $Cp^*Ru(PPh_3)_3Cl$ catalyst (Scheme 2.12) to form a high molecular weight polymer, as evidenced by GPC. The GPC showed a bimodal distribution. The lower intensity peak ($M_n = 120\,000\,\mathrm{g\,mol^{-1}}$) corresponded to the click polymer, while the higher intensity peak ($M_n = 8900\,\mathrm{g\,mol^{-1}}$) likely corresponded to a simple coupled product or possibly to a side reaction that removed $Cp^*Ru(PPh_3)_2Cl$ from the catalytic cycle (see below for further discussion).[74] In contrast, the reaction of $[(Cp(CH_2)_3OCO(CH_2)_2C{\equiv}CH)Mo(CO)_3]_2$ with heterotelechelic polystyrene using the $Cp^*Ru(PPh_3)_3Cl$ catalyst showed no evidence of a click reaction, and GPC analysis indicated no increase in molecular weight.[74]

In the reaction with homotelechelic polystyrene, it was suggested that $Cp^*Ru(PPh_3)_2Cl$ was removed from the catalytic cycle because the reaction took seven days and only yielded a small amount of high molecular weight click polymer. It was further suggested that the side product could be a ruthenium tetraazadiene complex (Scheme 2.13). Such complexes are known to form readily with aryl azides. However, these complexes do not form with aliphatic azides because aliphatic azides are more stable than aryl azides.

Because the homotelechelic azide-terminated polystyrene appeared to be reacting with the catalyst, it was hypothesized that carrying out the polymerization with a polymeric aliphatic azide would decrease the reaction

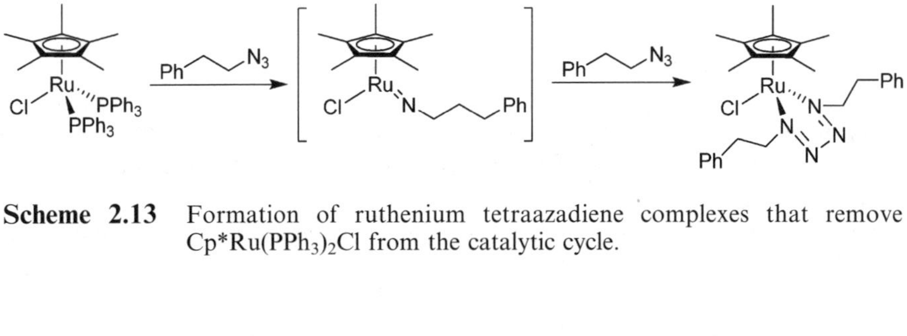

Scheme 2.13 Formation of ruthenium tetraazadiene complexes that remove $Cp^*Ru(PPh_3)_2Cl$ from the catalytic cycle.

time and increase the yield of high molecular weight polymer. To test this hypothesis, the azide-terminated poly(ethylene glycol) (PEG) in Scheme 2.14 was synthesized. In this scheme, the starting material PEG (M_n = 300 g mol^{-1}) was tosylated and the tosylate was substituted with an azide using sodium azide. Using the same reactions conditions determined by the step polymerization with homotelechelic polystyrene, [(Cp(CH$_2$)$_3$OCO(CH$_2$)$_2$C≡CH)Mo(CO)$_3$]$_2$ was reacted with the azide-terminated PEG using Cp*Ru(PPh$_3$)$_2$Cl as the catalyst (Scheme 2.15). After 15 h, GPC analysis showed two peaks (plus an additional peak corresponding to the unreacted azide-terminated PEG) with the first having M_n = 73 000 g mol^{-1} and the other having M_n = 9700 g mol^{-1}. The IR spectrum showed the Mo-Mo bond was still intact; however, it also showed unreacted [(Cp(CH$_2$)$_3$OCO(CH$_2$)$_2$C≡CH)Mo(CO)$_3$]$_2$. Further reaction time led to the decomposition of the unreacted [(Cp(CH$_2$)$_3$OCO(CH$_2$)$_2$C≡CH)Mo(CO)$_3$]$_2$.

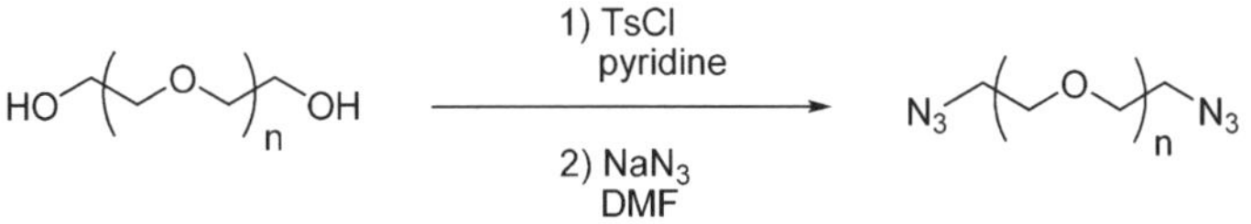

Scheme 2.14 Synthesis of an azide-terminated poly(ethylene glycol).

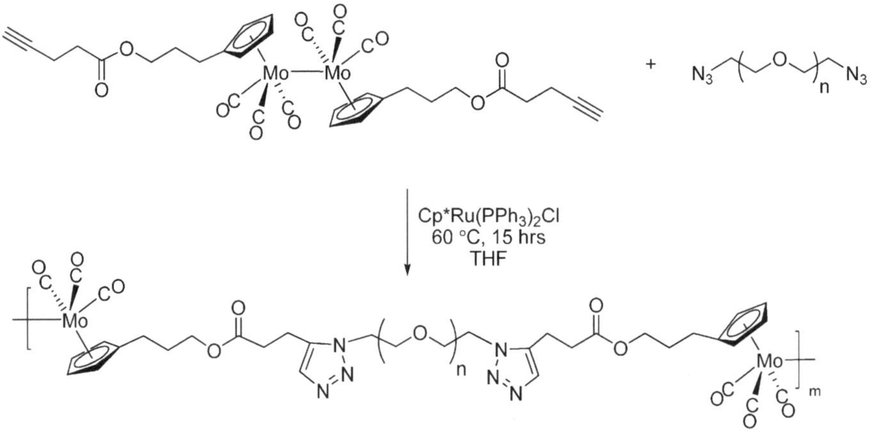

Scheme 2.15 Step polymerization of [(Cp(CH$_2$)$_3$OCO(CH$_2$)$_2$C≡CH)Mo(CO)$_3$]$_2$ with the acetylene-functionalized dimer and azide-terminated poly(ethylene glycol).

Current research in this area takes advantage of the specificity of the Huisgen reaction for the synthesis of nonlinear polymers. Consider, for example, the synthesis of a star polymer with metal–metal bonds along its radiating chains. Because of the required reaction conditions, such polymers would be difficult to synthesize using any of the other standard polymerization methods. However, such polymers should be straightforward to prepare with click chemistry. To this end, triethynylbenzene was reacted with the $[(Cp(CH_2)_3N_3)Mo(CO)_3]_2$ synthon using Cp*Ru(COD)Cl as the catalyst to synthesize a tripodal star core containing metal–metal bonds (Scheme 2.16).[80] Unexpectedly, no subsequent click reactions (or disproportionation reactions) occurred but a high molecular weight hydrocarbon polymer was obtained. It was hypothesized that 1,3,5-triethynylbenzene preferentially underwent cyclotrimerization instead of the Huisgen cycloaddition reaction. As mentioned earlier, phenylacetylene successfully undergoes click reactions with azides using the Cp*Ru(COD)Cl catalyst, as does 1,3-diethynylbenzene. Therefore, the formation of the polymer must be due to the 1,3,5-substitution on the benzene ring and not the catalyst itself.

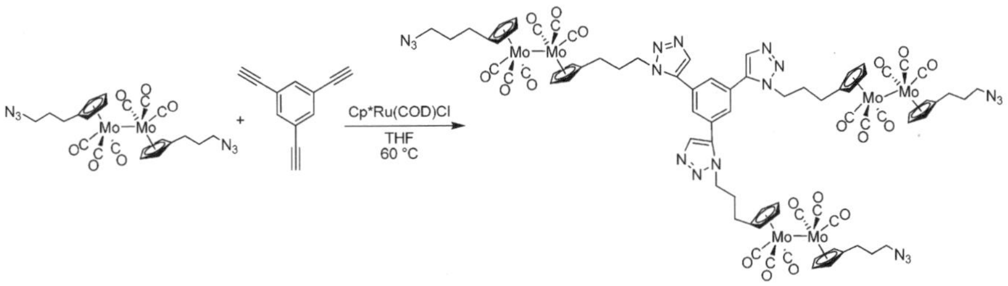

Scheme 2.16 Initial steps in the synthesis of a star polymer using click methods.

2.9 Summary

Polymers with metal–metal bonds along their main chain have been synthesized using step polymerization reactions, chain reactions, ADMET reactions, and ROMP reactions. To this list of polymerization methods can now be added click reactions, specifically the Huisgen cycloaddition reaction between terminal acetylenes and azides to form 1,2,3-triazoles. With the exception of the click reactions, a drawback to all of the other polymerization methods is that high molecular weight polymers are generally not obtained. Click chemistry, however, does give polymer products with high molecular weight. The photochemistry of the polymers synthesized using the Huisgen cycloaddition click reaction has not yet been investigated, but there is no reason to believe that the photochemical reactions of the metal–metal bonds in these polymers will qualitatively be any different from the photoreactions observed in the polymers synthesized by the other methods.

Current work in our laboratory is focused on examining the reactivity of a fifth catalyst, chloro[1,3-bis(2,4,6-trimethylphenyl)imidazol-2-ylidene]copper(I) (molecule **4**),[81,82] to synthesize a star core with 1,3,5-triethynylbenzene. Subsequent reaction of this star core with, for example, an acetylene-terminated poly(ethylene glycol) prepolymer should yield a high molecular weight star polymer containing Mo–Mo bonds along the arms of the star. This is just one of many interesting polymer architectures that should be possible, and it is difficult to imagine the syntheses of such polymers without click chemistry.

4

Acknowledgement

Acknowledgement is made to the National Science Foundation IGERT Fellowship Program under Grant No. DGE-0549503 for support of the recent research described herein.

References

1. N. Grassie and G. Scott, *Polymer Degradation and Stabilization*, Cambridge University Press, New York, 1985.
2. J. Guillet, *Polymer Photophysics and Photochemistry: An Introduction to the Study of Photoprocesses in Macromolecules*, Cambridge University Press, New York, 1985.
3. J. F. Rabek, *Mechanisms of Photophysical Processes and Photochemical Reactions in Polymers*, Wiley, New York, 1987.
4. G. Geuskens, *Compr. Chem. Kinet.*, 1975, **14**, 333.
5. S. H. Hamid and ed, *Handbook of Polymer Degradation*, Marcel Dekker, New York, 2000.
6. J. F. Rabek, *Polymer Photodegradation: Mechanisms and Experimental Methods*, Chapman & Hall, London, 1995.
7. J. E. Guillet, in *Degradable Materials*, ed. S. A. Barenberg, J. G. Brash, R. Narayan and A. E. Redpath, CRC Press, Boston, 1990, pp. 55.
8. R. West, *J. Organomet. Chem.*, 1986, **300**, 327.
9. M. Igarashi and K. L. DeVries, *Polymer*, 1983, **24**, 1035.
10. A. V. Cunliffe and A. Davis, *Polym. Degrad. Stab.*, 1982, **4**, 17.
11. J. Malik, A. Hrivik and D. Q. Tuan, *Advances in Chemistry Series*, 1996, **249**, 455. See page 459 in particular.
12. D. R. Tyler, *Coord. Chem. Rev.*, 2003, **246**, 291.
13. G. V. Shultz and D. R. Tyler, *J. Inorg. Organomet. Polym. Mater.*, 2009, **19**, 423.
14. R. Chen and D. R. Tyler, *Macromolecules*, 2004, **37**, 5430.
15. B. C. Daglen and D. R. Tyler, *J. Inorg. Organomet. Polym. Mater.*, 2009, **19**, 91.

16. B. C. Daglen, J. D. Harris and D. R. Tyler, *J. Inorg. Organomet. Polym. Mater.*, 2007, **17**, 267.
17. G. V. Shultz, J. M. Zemke and D. R. Tyler, *Macromolecules*, 2009, **42**, 7644.
18. P. Nguyen, P. Gomez-Elipe and I. Manners, *Chem. Rev.*, 1999, **99**, 1515.
19. C. U. Pittman, Jr. and M. D. Rausch, *Pure Appl. Chem.*, 1986, **58**, 617.
20. K. Gonsalves, L. Zhan-Ru and M. D. Rausch, *J. Am. Chem. Soc.*, 1984, **106**, 3862.
21. D. R. Tyler, in *Macromolecules Containing Metal and Metal-like Elements*, ed. A. S. Aziz, J. Carraher, C. E., J. Pittman, C. U., J. Sheats and M. Zeldin, John Wiley, New York, 2005, vol. 6.
22. A. E. Stiegman and D. R. Tyler, *Coord. Chem. Rev.*, 1985, **63**, 217.
23. S. C. Tenhaeff and D. R. Tyler, *Organometallics*, 1991, **10**, 473.
24. S. C. Tenhaeff and D. R. Tyler, *Organometallics*, 1991, **10**, 1116.
25. S. C. Tenhaeff and D. R. Tyler, *Organometallics*, 1992, **11**, 1466.
26. G. F. Nieckarz and D.R. Tyler, *Inorg. Chim. Acta.*, 1996, **242**, 303.
27. T. J. Meyer and J. V. Caspar, *Chem. Rev.*, 1985, **85**, 187.
28. G. L. Geoffroy and M. S. Wrighton, *Organometallic Photochemistry*, Academic Press, New York, 1979.
29. R. Chen, J. Meloy, B. C. Daglen and D. R. Tyler, *Organometallics*, 2005, **24**, 1495.
30. J. L. Male, B. E. Lindfors, K. J. Covert and D. R. Tyler, *Macromolecules*, 1997, **30**, 6404.
31. J. L. Male, M. Yoon, A. G. Glenn, T. J. R. Weakley and D. R. Tyler, *Macromolecules*, 1999, **32**, 3898.
32. E. Schutte, T. J. R. Weakley and D. R. Tyler, *J. Am. Chem. Soc.*, 2003, **125**, 10319.
33. V. Balzani and V. Carassiti, *Photochemistry of Coordination Compounds*, Academic Press, New York, 1970.
34. J. R. Fried, *Polymer Science and Technology*, Prentice Hall PTR, Upper Saddle River, New Jersey, 1995.
35. A. Elicegui, J. J. Del Val, J. L. Millan and C. Mijangos, *J. Non-Cryst. Solids*, 1998, **235–237**, 623.
36. L. David, C. Girard, R. Dolmazon, M. Albrand and S. Etienne, *Macromolecules*, 1996, **29**, 8343.
37. E. Dan and J. E. Guillet, *Macromolecules*, 1973, **6**, 230.
38. J. R. White and N. Y. Rapoport, *Trends Polym. Sci. (Cambridge, UK)*, 1994, **2**, 197.
39. B. O'Donnell and J. R. White, *J. Mater. Sci.*, 1994, **29**, 3955.
40. L. Tong and J. R. White, *Polym. Degrad. Stab.*, 1996, **53**, 381.
41. C. T. Kelly, L. Tong and J. R. White, *J. Mater. Sci.*, 1997, **32**, 851.
42. B. O'Donnell and J. R. White, *Polym. Degrad. Stab.*, 1994, **44**, 211.
43. W. K. Busfield and M. J. Monteiro, *Mater. Forum*, 1990, **14**, 218.
44. M. Igarashi and K. L. DeVries, *Polymer*, 1983, **24**, 769.

45. H. Nguyen Truc Lam and C. E. Rogers, *Polym. Mater. Sci. Eng.*, 1987, **56**, 589.

46. R.W. Appleby and W.K. Busfield, *J. Mater. Sci.*, 1994, **29**, 151.

47. L. Audouin, V. Langlois, J. Verdu and J. C. M. de Bruijn, *J. Mater. Sci.*, 1994, **29**, 569.

48. R. Chen, M. Yoon, A. Smalley, D. C. Johnson and D. R. Tyler, *J. Am. Chem. Soc.*, 2004, **126**, 305.

49. D. R. Tyler, *J. Macromol. Sci. Polym. Rev.*, 2004, **44**, 351.

50. H. C. Kolb, M. G. Finn and K. B. Sharpless, *Angew. Chem., Int. Ed. Engl.*, 2001, **40**, 2004.

51. D. Fournier, R. Hoogenboom and U. S. Schubert, *Chem. Soc. Rev.*, 2007, **36**, 1369.

52. L. Zhang, X. Chen, P. Xue, H. H. Y. Sun, I. D. Williams, K. B. Sharpless, V. V. Fokin and G. Jia, *J. Am. Chem. Soc.*, 2005, **127**, 15998.

53. B.C. Boren, S. Narayan, L.K. Rasmussen, L. Zhang, H. Zhao, Z. Lin, G. Jia and V.V. Fokin, *J. Am. Chem. Soc.*, 2008, **130**, 8923.

54. V. O. Rodionov, V. V. Fokin and M. G. Finn, *Angew. Chem., Int. Ed. Engl.*, 2005, **44**, 2210.

55. R. Alvarez, S. Velazquez, A. San-Felix, S. Aquaro, E. De Clercq, C.-F. Perno, A. Karlsson, J. Balzarini and M. J. Camarasa, *J. Med. Chem.*, 1994, **37**, 4185.

56. L. V. Lee, M. L. Mitchell, S.-J. Huang, V. V. Fokin, K. B. Sharpless and C.-H. Wong, *J. Am. Chem. Soc.*, 2003, **125**, 9588.

57. R. Manetsch, A. Krasinski, Z. Radic, J. Raushel, P. Taylor, K.B. Sharpless and H.C. Kolb, *J. Am. Chem. Soc.*, 2004, **126**, 12809.

58. C. W. Tornoe, C. Christensen and M. Meldal, *J. Org. Chem.*, 2002, **67**, 3057.

59. H. C. Kolb and K. B. Sharpless, *Drug Discovery Today*, 2003, **8**, 1128.

60. J. P. Collman, N. K. Devaraj and C. E. D. Chidsey, *Langmuir*, 2004, **20**, 1051.

61. S. Kinge, T. Gang, W. J. M. Naber, W. G. van der Wiel and D. N. Reinhoudt, *Langmuir*, **27**, 570.

62. T. Lummerstorfer and H. Hoffmann, *J. Phys. Chem. B.*, 2004, **108**, 3963.

63. M. Slater, M. Snauko, F. Svec and J. M. J. Frechet, *Anal. Chem.*, 2006, **78**, 4969.

64. H. Durmaz, A. Dag, O. Altintas, T. Erdogan, G. Hizal and U. Tunca, *Macromolecules*, 2007, **40**, 191.

65. H. Gao and K. Matyjaszewski, *Macromolecules*, 2006, **39**, 4960.

66. B. Helms, J. L. Mynar, C. J. Hawker and J. M. J. Frechet, *J. Am. Chem. Soc.*, 2004, **126**, 15020.

67. M. J. Joralemon, R. K. O'Reilly, J. B. Matson, A. K. Nugent, C. J. Hawker and K. L. Wooley, *Macromolecules*, 2005, **38**, 5436.

68. B. A. Laurent and S. M. Grayson, *J. Am. Chem. Soc.*, 2006, **128**, 4238.

69. J. A. Opsteen and J. C. M. van Hest, *Chem. Commun.*, 2005, 57.

70. D. A. Ossipov and J. Hilborn, *Macromolecules*, 2006, **39**, 1709.

71. P. Wu, A. K. Feldman, A. K. Nugent, C. J. Hawker, A. Scheel, B. Voit, J. Pyun, J. M. J. Frechet, K. B. Sharpless and V. V. Fokin, *Angew. Chem., Int. Ed. Engl.*, 2004, **43**, 3928.
72. N. V. Tsarevsky, B. S. Sumerlin and K. Matyjaszewski, *Macromolecules*, 2005, **38**, 3558.
73. D. J. V. C. van Steenis, O. R. P. David, G. P. F. van Strijdonck, J. H. van Maarseveen and J. N. H. Reek, *Chem. Commun.*, 2005, 4333.
74. S. E. Brady, G. V. Shultz and D. R. Tyler, *J. Inorg. Organomet. Polym. Mater.*, 2010, **20**, 511.
75. G. V. Shultz, 2009, Ph.D. Thesis, University of Oregon.
76. A. E. Stiegman, M. Stieglitz and D. R. Tyler, *J. Am. Chem. Soc.*, 1983, **105**, 6032.
77. S. Braese, C. Gil, K. Knepper and V. Zimmermann, *Angew. Chem., Int. Ed. Engl.*, 2005, **44**, 5188.
78. S. Amer, G. Kramer and A. Poe, *J. Organomet. Chem.*, 1981, **209**, C28.
79. S. G. Alvarez and M. T. Alvarez, *Synthesis*, 1997, 413.
80. W. Uhl, H. R. Bock, F. Breher, M. Claesener, S. Haddadpour, B. Jasper and A. Hepp, *Organometallics*, 2007, **26**, 2363.
81. S. Diez-Gonzalez, A. Correa, L. Cavallo and S. P. Nolan, *Chem. Eur. J.*, 2006, **12**, 7558.
82. S. Diez-Gonzalez and S. P. Nolan, *Angew. Chem., Int. Ed. Engl.*, 2008, **47**, 8881.

Metal Acetylide Complexes, Oligomers and Polymers in Photofunctional Materials Chemistry

MAREK JURA, PAUL R. RAITHBY* AND
PAUL J. WILSON

Department of Chemistry, University of Bath, Bath BA2 7BA, UK
*E-mail: Raithby@bath.ac.uk

3.1 Introduction

The issue of providing energy-efficient lighting is becoming ever more imperative in today's energy consuming world. Organic light-emitting diodes (OLEDs) have already entered the market place as replacement lighting to the traditional incandescent bulbs. The efficiency of the organic polymers used in OLEDs is effectively limited to a maximum of 25%, due to the laws governed by the spin-selection rules. In attempts to overcome these limitations there has been a rapid expansion, over recent years, in the incorporation of metals into these organic polymers. The inclusion of metal-containing units into these systems brings a range of beneficial properties. Heavy transition metals have significant spin-orbit coupling associated with them. The spin-orbit coupling allows for the partial breaking of the selection rules and facilitates intersystem crossing to triplet states that give rise to a greater luminescent efficiency. The

RSC Polymer Chemistry Series No. 2
Molecular Design and Applications of Photofunctional Polymers and Materials
Edited by Wai-Yeung Wong and Alaa S Abd-El-Aziz
© The Royal Society of Chemistry 2012
Published by the Royal Society of Chemistry, www.rsc.org

inclusion of metal-containing units in the polymers often also enhances their solubility in organic solvents and aids their processability.

The desirable properties provided by the presence of transition-metal-containing units has led to additional applications, including the use of transition-metal acetylide oligomers and polymers in light-emitting diodes (LEDs),[1,2] lasers,[3] photocells,[4] field-effect transistors (FETs),[5] liquid-crystal displays (LCDs),[6] sensors,[7] optical information storage,[8] optical switches and signal processing devices[9] as well as having nonlinear optical (NLO) applications,[10] and, perhaps most importantly, in optical power-limiting applications.[11] In this chapter the development of transition-metal acetylide chemistry will be traced and some of the applications of these fascinating materials in materials chemistry discussed.

3.1.1 OLEDs

Before beginning the discussion on luminescent complexes, oligomers and polymers that contain transition-metal acetylide units it is worth placing these systems in the wider context, which involves the development of organic light-emitting diodes (OLEDs) where the fundamental science revolves around the exploitation of the delocalisation present in the conjugated aromatic molecules and polymers present in OLED materials. OLEDs have been produced that utilise the conjugation of aromatic molecules for the production of light. OLEDs work by sandwiching several layers of conjugated polymeric materials between two electrodes, one of which is transparent and allows light to pass out of the device. The first observation of luminescence from an organic molecule was reported by Pope *et al.*,[12] on luminescence of anthracene. Research, however, remained largely focused on inorganic materials due to the low efficiency of organic light emitters. Organic dyes as light emitters regained popularity in the late 1980s, but these systems required expensive vapour deposition. This problem was overcome by Friend and coworkers, [13] who published a paper revealing a solution to this problem by using the conjugated polymer poly(*p*-phenylenevinylene) (PPV). Friend's group, based in the Cavendish Laboratory, at Cambridge, and a group in the Department of Chemistry, led by Holmes, recognised the fascinating properties of this material and were able to develop a simple device using a spin-coating process to produce the very first OLED.

Subsequent publications from many groups using PPV-based molecules that incorporated side chains and heteroatoms further illustrated the versatility of the system. By choice of substituent or molecular conformation it was possible to tune the bandgap of the materials and thereby the colour of the emitted light. Figure 3.1 shows some of the examples of the PPV-derived polymers that have been reported.

The red-orange emission of polymer A poly[2-methoxy-5-(2-ethylhexyloxy)-p-phenylenevinylene] (MEH-PPV), was reported by Braun and Heeger.[14] Polymer B was synthesised in 1996 by Phillips, and led to multilayer OLEDs

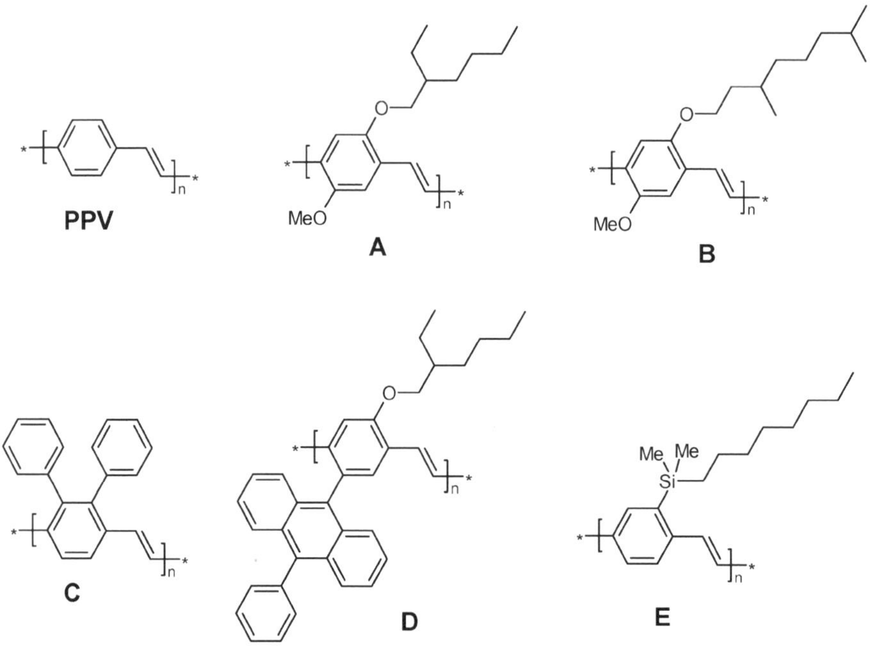

Figure 3.1 PPV and derivative polymers.

that were more efficient due to improved hole injection. The phenyl groups observed in polymer C poly(2,3-diphenylenevinylene) serve to increase the solubility of the polymer and therefore increase its processability, but the electroluminescent efficiency of this molecule as a single-layer device is relatively low, although it was found that the introduction of pendant groups such as that in polymer D behave as interchain spacers and delay the excited state decay, thereby increasing the electroluminescent efficiency. The silicon in polymer E served to increase the solubility of the polymer and increased the bandgap, allowing access to green photoemissions.

However, despite all these developments one of the key shortcomings of OLED devices arises because of their poor efficiency. The ratio of singlet to triplet states generated in conventional OLED devices has been determined as 1:3, which in turn limits the maximum efficiency to 25%.[15] The efficiency of OLEDs is limited by the statistics of singlet states recombining. Triplet states can emit photons but the probability of this is quite small and energy loss from the triplet state usually occurs by intersystem crossing, followed by internal conversion to the ground state. To overcome this problem the triplet state needs to be accessed. However, decay from the triplet excited state back to the singlet ground state is spin forbidden, as governed by the Pauli exclusion principle. This governing physical law can be relaxed by the inclusion of heavy atoms.

In very heavy atoms, relativistic shifting of the energies of the electron energy levels accentuates the spin-orbit coupling effect. In typical organic molecules with weak spin-orbit coupling and highly forbidden triplet–singlet transitions, the triplet-state population is transferred into heat. Only the singlet state can emit radiatively (fluorescence). On the other hand, in organo-transition-metal compounds, fast intersystem crossing induced by spin-orbit coupling effectively depopulates the excited singlet into the lowest triplet state. Again due to spin-orbit coupling, the triplet can decay radiatively as phosphorescence even with high emission quantum yield at ambient temperature. In the case of validity of spin statistics only 25% of the excitons can be exploited by organic emitters, while for triplet emitters additional 75% of the excitons are harvested. Thus, the efficiency of light emission in an electroluminescent device with triplet emitters can be up to a factor of four higher than with singlet emitters.[16]

To combine the properties of the highly conjugated aromatic compounds as exploited in OLEDs, with the spin-orbit coupling properties of transition metals, a method was needed for the synthesis of these new "organo-transition-metal compounds". One such method is the use of acetylenes as a bridging unit between the metal centre and the aromatic system. The *sp*-hybridised carbons of the acetylene allow for good orbital overlap with the *d*-orbital of transition metals.

3.1.2   Transition-Metal σ-Acetylide Complexes

Transition-metal σ-acetylide complexes (Figure 3.2) have been studied extensively because of their many interesting chemical and physical properties.[17,18]

These properties find many valuable potential applications in modern technology, from light-emitting diodes,[19] liquid-crystalline materials[20] to materials with nonlinear optical properties[21] and one-dimensional conductors.[22] Depending on the requirements of the application, either monomeric, oligomeric or polymeric complexes can have their properties tuned to optimise the functionality of the material.[23,24]

This tuning process can occur by the adjustment of both the acetylene ligand and the metal environment. Functional groups on the acetylene spacer ligands (R) can alter the electronic properties of complex as well as their solubility. In addition, the number and type of auxiliary ligands (L_n) (*i.e.* spectator ligands; often phosphines), the coordination geometry and oxidation state of the metal (M) have all be varied to investigate the effects on the properties of the acetylide complexes.[25]

It is clearly recognised that the metal to carbon bond, in metal σ-acetylide complexes, is generally thermally stable due to stabilisation through the all important dπ–pπ interactions along the metal acetylide axis. The designed omission of β-hydrogen atoms on the acetylene spacer ligand removes the

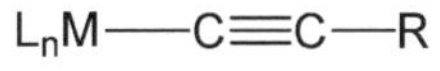

$$L_nM\!\!-\!\!C\!\equiv\!C\!\!-\!\!R$$

Figure 3.2   Transition-metal σ-acetylide complexes.

possibility of decomposition of these compounds by β-hydride elimination. Photoelectron spectroscopy and calculations indicate a significant overlap between the filled metal d-orbitals and the filled π-system of the acetylene moiety.[26] The energy difference between the HOMO metal d orbitals and the LUMO π^* acetylene orbitals appears to be too large (*ca.* 15 eV) for π-accepting behaviour. Thus, in these complexes, the acetylene behaves more like a chloride π-donor ligand rather than like a carbonyl π^*-acceptor ligand.

The metal geometry is also an important factor in the control of the properties of the acetylide complexes. Both Pt(II) and Pd(II) metal centres usually exhibit well-defined square planar geometries with *cis/trans* isomers possible, therefore, if linear rigid-rod complexes are desired, the *trans* geometry is employed and if angular materials are required then the *cis* geometry is used. Both Ru(II) and Rh(III) centres exhibit octahedral geometry and may also be used to make *trans* linear rod complexes, or *cis* angular complexes. Chelating ligands such as diphosphines and diimines can be used to control stereochemistry in these compounds. Ag(I), Au(I) and Hg(II) centres are generally limited to a linear, two-coordinate geometry and are therefore employed in the synthesis of linear rigid rod acetylide complexes. The choice of metal centre can have a significant effect on the bandgap, with four coordinate d^8 square planar metal centres favoured over six coordinate d^6 octahedral units, because of the greater optical gap energy of the latter (*ca* 0.4 eV).[27] Depending on the desired application, the tuning of the HOMO–LUMO bandgap in complexes and the valence-conductance band in polymers is perhaps the most important feature of these acetylide-containing transition-metal systems.

The chemistry of these various transition-metal alkynes are discussed in more detail in the following sections, with particular emphasis on their synthesis and photophysical properties.

3.2 Platinum and Palladium Acetylide Complexes, Oligomers and Polymers

The square-planar geometries, associated with the d^8 electron configuration of the metal centre, along with the kinetic inertness of Pt(II) and Pd(II) centres, because of the large ligand field splitting, gives rise to linear and stable metal acetylene complexes. For these reasons platinum(II) di-ynes and poly-ynes, in particular, have been investigated extensively, and the area has been reviewed comprehensively by Wong and Ho.[28] Only a brief survey of some of the basic results is included here.

Platinum(II) acetylide complexes were first synthesised by Hagihara *et al.*[29], using the condensation reaction between terminal or di-terminal acetylides and platinum chlorides. Subsequently this method was adapted,[30] for example as shown in Figure 3.3.

The copper(I) iodide in the reaction is used as a catalyst, and it is thought that a copper(I) acetylide intermediate is formed during the reaction. The

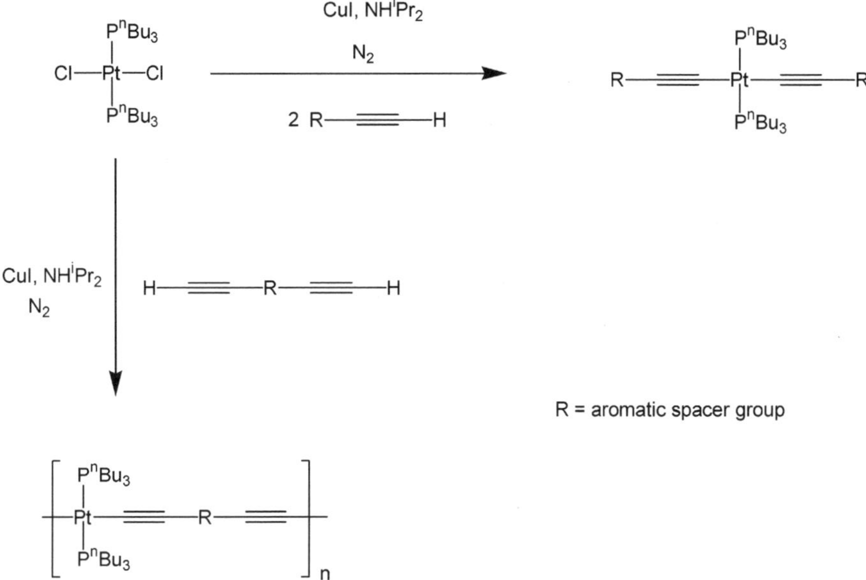

Figure 3.3   Synthesis of Pt(II) di-ynes and poly-ynes.

copper(I) acetylide is a highly reactive species, and the acetylene is readily able to transmetallate with the platinum-containing starting material to produce the desired product. The diisopropylamine is used as the solvent and also acts as a base to accept excess protons. This route has only been successfully used for the Group 10 metals palladium and platinum. It is unsuccessful with elements earlier in the transition series, including nickel since they tend to react with the basic amine solvent to form amine complexes.

Alternative synthetic routes have also been reported, involving the use of trimethylstannyl alkynyl derivatives,[31] for example as shown in Figure 3.4. The trimethylstannyl derivatives have been shown to be more stable than the free acetylenes,[32] although worries about the health implication of Sn(IV) have made this route less popular. An extra step is also required in the synthesis of the final product; this involves the treatment of the acetylene with ^{n}BuLi followed by Me_3SnCl.[33]

This method does have some advantages, namely the avoidance of amine solvents, allowing the synthesis of nickel analogues. Stoichiometric treatment of *bis*(trimethylstannyl)alkynes with *trans*-$[MCl_2(PBu_3)_2]$ (M = Ni, Pd, Pt) affords polymers in excellent yield. In the case of M = Pt average molecular weights up to M_w = 210 000 were observed,[34] which is on average 30 000 Daltons higher than the dehydrohalogenation route of Hagihara *et al.*

A variety of platinum(II) and palladium(II) species have been investigated with respect to their photoluminescence properties.[35] The majority of these are mononuclear species, with a terdentate pincer ligand,[36] such as terpyridine, blocking three of the coordination sites on the metal centre, while the fourth

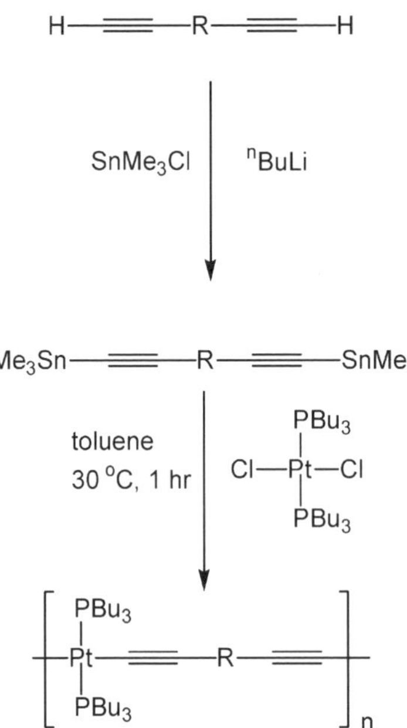

Figure 3.4 Synthesis of a platinum acetylide polymer *via* trimethylstannyl alkynyl derivatives.

site is occupied by a substituted acetylene ligand, as recently exemplified by the cation $[^{t}Bu_{3}tpyPt(C\equiv Cbpy)]^{+}$.[37] Of greater relevance to the current discussion are the monomeric complexes *trans*-$[Pt(C\equiv CH)_{2}(PEt_{3})_{2}]$ and *trans*-$[Pt(C\equiv CPh)_{2}(PEt_{3})_{2}]$ that give intense vibronically structured emission in rigid glasses at 77 K.[23] The emissive state has been assigned as $Pt(d)\rightarrow\pi^{*}(C\equiv C)$ MLCT transition in character upon excitation into the low-energy metal-to-ligand charge-transfer (MLCT) absorption bands. Polymeric metal acetylide complexes have also been studied such as the materials shown in Figure 3.5.

Studies on thin films at liquid helium temperatures have shown long-lived phosphorescence, assigned to a triplet–singlet transition. The platinum species have much longer phosphorescence lifetimes than isostructural palladium species and this is likely to be due to the greater effect of spin-orbit coupling for platinum.[27]

The polymeric metal acetylide complexes, with their extended conjugation lengths have shown low energies of absorption. The polymeric systems, however, are difficult to study by conventional chemical spectroscopic and diffraction methods because of their relatively large size and molecular weights. Di-yne complexes have been synthesised as dimeric models for the poly-yne systems, as they are easier to study and establish structure–property relationships for the systems. Examples of such di-ynes in the literature are given in Figure 3.6 along with the general synthetic route to the complexes. The same dehydrohalogenation reaction is employed in the synthesis of the

where M = Pt (II) or Pd (II); R and R' are aryl spacer groups, as shown below

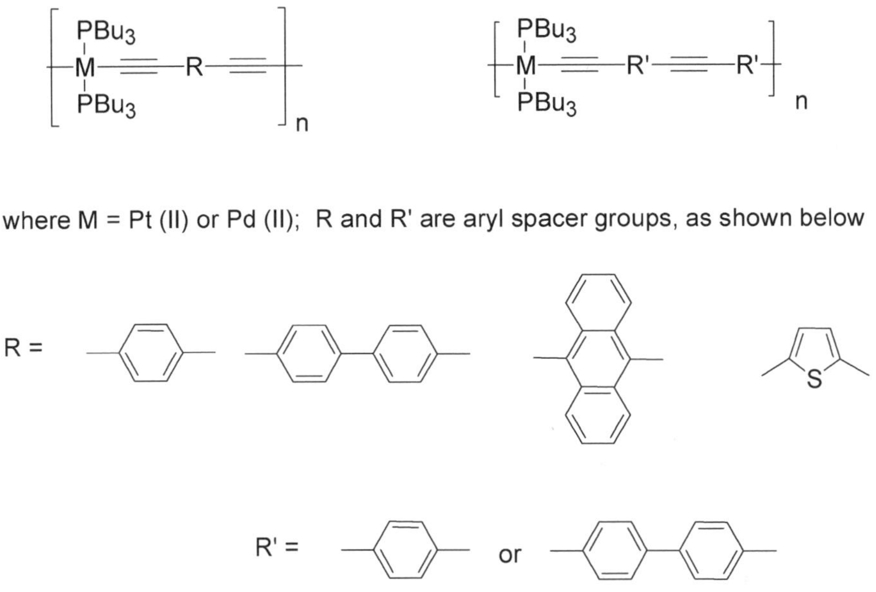

Figure 3.5   Pt(II) and Pd(II) poly-ynes investigated for photoluminescent behaviour.

dimeric system as for the poly-yne analogues, but using a different platinum starting material. *Trans*-Pt(PBu$_3$)$_2$Cl$_2$ is used for the synthesis of the polymeric systems, where the two chlorides offer two active sites leading to polymerisation. The butyl phosphines are employed to aid solubility leading to extensive degrees of polymerisation. The di-yne complexes are synthesised using *trans*-Pt(PEt$_3$)$_2$(Ph)Cl, this starting material only contains one chloride and therefore one reactive site towards the acetylene ligand. The phenyl group effectively "caps off" one site of the platinum centre forcing the synthesis of the discrete di-platinum di-yne complex when a di-acetylide linker H-C≡C-R-C≡C-H or a trimethylsilyl terminated di-acetylide, Me$_3$Si-C≡C-R-C≡C-SiMe$_3$, linker is used (R = an aromatic or heteroaromatic species). Once the di-ynes have been characterised fully, it is then possible to prepare the analogous polymers, and this methodology has been used successfully to produce di-ynes and polymers with a range of aromatic and heteroaromatic spacer groups.[38]

In order to optimise the electronic properties of the poly-yne polymers, work focused on fully understanding the electronic properties of the di-yne systems by monitoring the change in electronic properties as the nature of the spacer group was changed. For these initial studies the platinum unit, (Ph)Pt(PBu^n$_3$), was kept constant to eliminate changes in other electronic influences on the systems. Several series of di-acetylene spacer groups were investigated.

Raithby and coworkers investigated a series of systems containing thienyl-pyridine rings, with adjacent electron donating and accepting units as the aromatic spacer groups (Series A, Figure 3.6).[39] The aim of the investigation was to establish the affect of the donor-acceptor unit of the energies of the

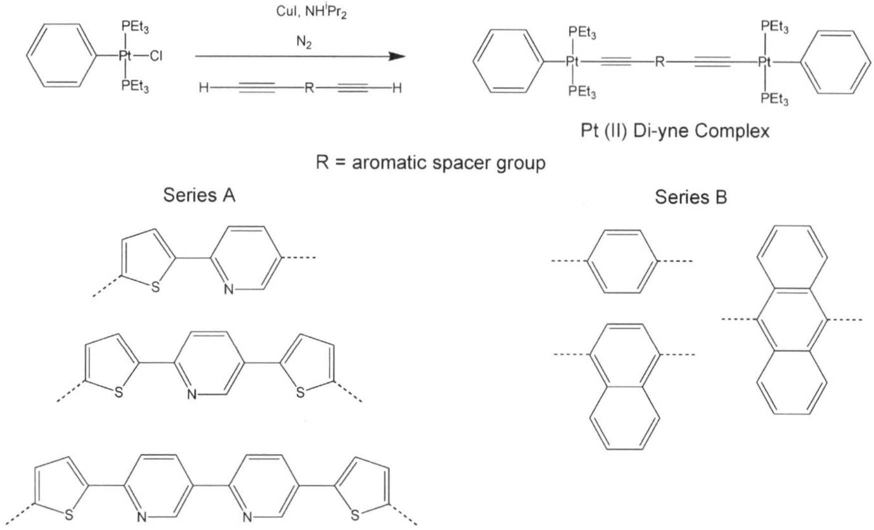

Figure 3.6 Synthesis of platinum di-yne complexes.

singlet and triplet excited states. Absorption spectra were shown to be dominated by π–π^* transitions, with the platinum fragments acting as net electron donors to the electron-withdrawing spacer groups. The pyridine rings in the systems were thought to withdraw electron density toward the lone pair, away from the system backbone. This reduction in electron density was rationalised as the reason for the higher optical gaps as compared to the purely thiophene-containing systems.

The use of fused ring spacers has been found to be a powerful approach to the production of some low-bandgap conjugated organic polymers.[40] The Raithby group took inspiration from this, and designed a series of fused ring spacers in platinum acetylide complexes (Figure 3.6, Series B).[41] Optical spectroscopic measurements showed that the electron-rich naphthalene and anthracene spacers create strong donor–acceptor interactions between the Pt(II) centres and conjugated ligands along the rigid backbone of the polymers. The bandgaps decrease as the size of the aromatic linker group increases, consistent with there being greater delocalisation within the anthracene linker group compared to the benzene linker group.

In a systematic study, optical bandgaps for the di-platinum poly-yne polymers shown in Figure 3.7 were measured. A clear trend was observed in that there was a close correlation between the bandgap and the electron-withdrawing ability of the spacer unit,[42] with the most electron-withdrawing group, **6**, having the lowest bandgap of only 1.7 eV[43] compared to that of 3.0 eV for spacer group **1**. Thus, while the energy of the S_1 singlet excited state decreased across the series, it was also found that the $S_1 - T_1$ energy gap remained constant at 0.7 $\pm$0.1 eV. Also, the lifetimes and the intensities of the

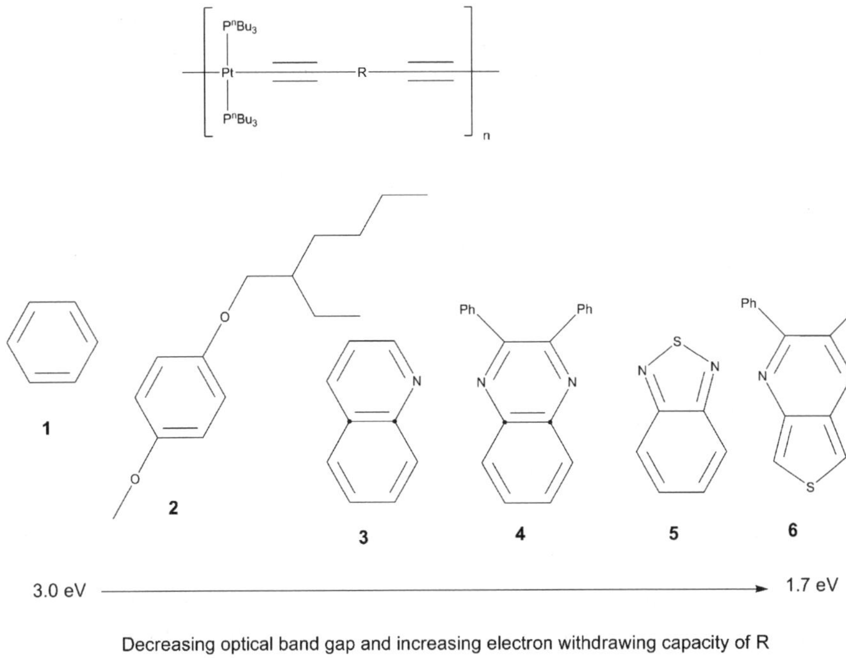

Figure 3.7 The relative optical gaps and the electron withdrawing abilities of a series of spacer units, R, in a series of rigid-rod platinum poly-ynes.

triplet emissions were found to reduce drastically with decreasing triplet energy. This is a result of nonradiative recombination.

The lowest bandgap observed for a Pt(II) poly-yne was subsequently reported by Wong and coworkers,[44] Figure 3.8. The group synthesised a Pt(II) polyyne with a 9-dicyanomethylenefluorene-2,7-diyl as a donor–acceptor spacer with a bandgap of 1.58 eV. The authors suggest that electron-withdrawing groups at the periphery exert a more pronounced influence on narrowing the bandgaps of these polymers than electron-donating substituents.

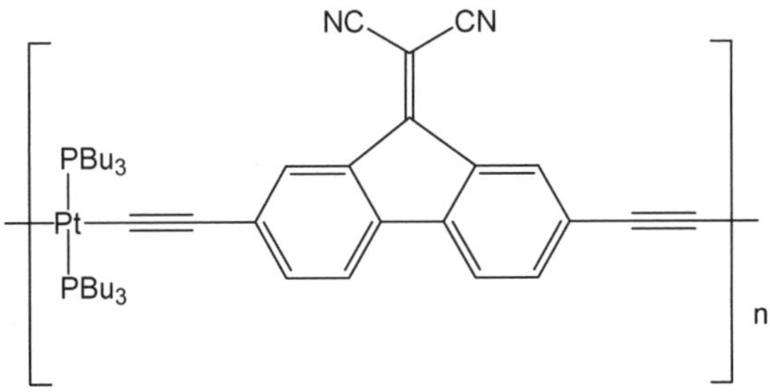

Figure 3.8 Low-bandgap Pt(II) poly-yne containing 9-dicyanomethylenefluorene-2,7-diyl spacer group.

The inclusion of polythiophene ligands as spacers in the platinum(II) rigid-rod di-ynes and poly-ynes has also proved a rich area of research for adding to our knowledge of the subtle structural and electronic variations in these organometallic systems.[45,46] In a recent study two series of polythiophene-containing platinum di-ynes and poly-ynes have been compared, one series in which the thiophenes are fused and in the other where they are not, as illustrated in Figure 3.9.[47] It was found that the singlet S_1 triple T_1 and T_n excited states are at higher energy in thin films made from the fused systems than from the nonfused systems. For ligands with the same number of rings the decreased number of double bonds in the fused system and the presence of an additional sulfur atom in spacers is thought to be responsible for this effect. These conclusions are supported by DFT calculations. The key feature is the number of double bonds in the linker group, the greater the number of double bonds the more extended is the conjugation and, as is usual in these materials, this is consistent with the lowering of the energy of both the singlet and triplet states.

Much of the work in the diplatinum poly-yne area has centred on systems with extensive delocalisation and, as a result, low-bandgap systems. However, a series of diplatinum poly-ynes that contain main-group elements as part of the linker group have been prepared, and in these the delocalisation through this main-group centre is reduced, giving larger bandgaps. An example of a silicon-containing and a germanium-containing system are illustrated in

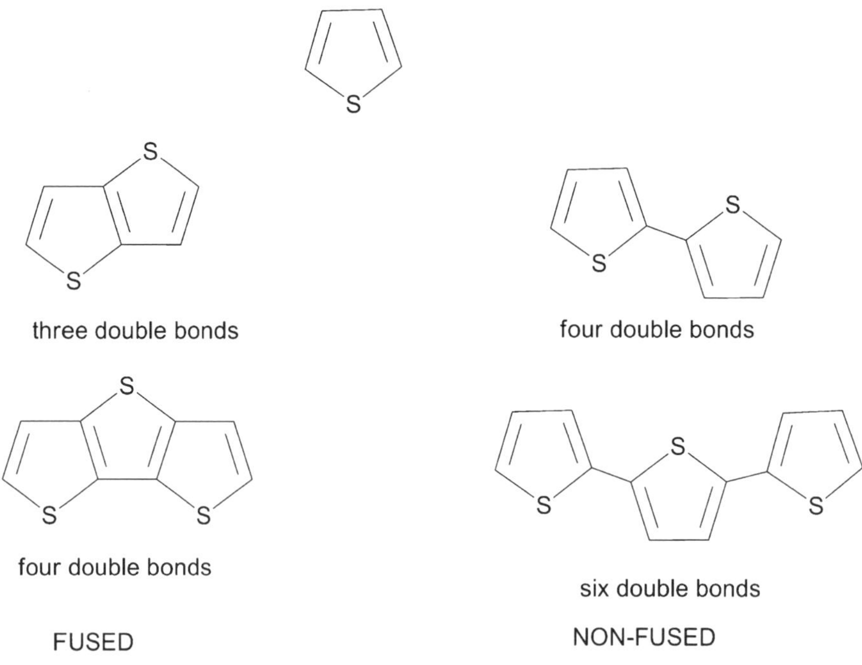

Figure 3.9 A comparison of fused and nonfused thiophene linker groups.

Figure 3.10. The silicon polymer has a bandgap of 3.70 eV[48] and the germanium polymer a gap of *ca.* 2.9 eV.[49] Thus, using this methodology, it is possible to tune the bandgap over an even wider range of energies.

The range of optical bandgaps that have been measured for the majority of known platinum(II) poly-yne polymers have been summarised by Wong in a recent review and are presented in Figure 3.11.[50]

In a recent development, with the intent of being able to switch "off" and "on" the electronic communication through the dimetallic di-yne or the poly-yne polymer, attempts have been made to incorporate dithienylethene linker groups into the molecular backbone since this group is known to undergo reversible ring closure and opening on exposure to light of different wavelengths.[51] The "open" form of the di-platinum di-yne (Figure 3.12) has been prepared and been shown to undergo ring closure on exposure to light of wavelength 365 nm in solution, in thin films and even in the single crystal. The process is reversible in solution when 530-nm light is used. Unusually, the ring closure occurs even in the presence of visible light ($<$ 400 nm). The results of a resonance Raman investigation confirm the involvement of the acetylide unit in the frontier orbitals of both closed and open forms in the photocyclisation process. This investigation represents a potential strategy for generating

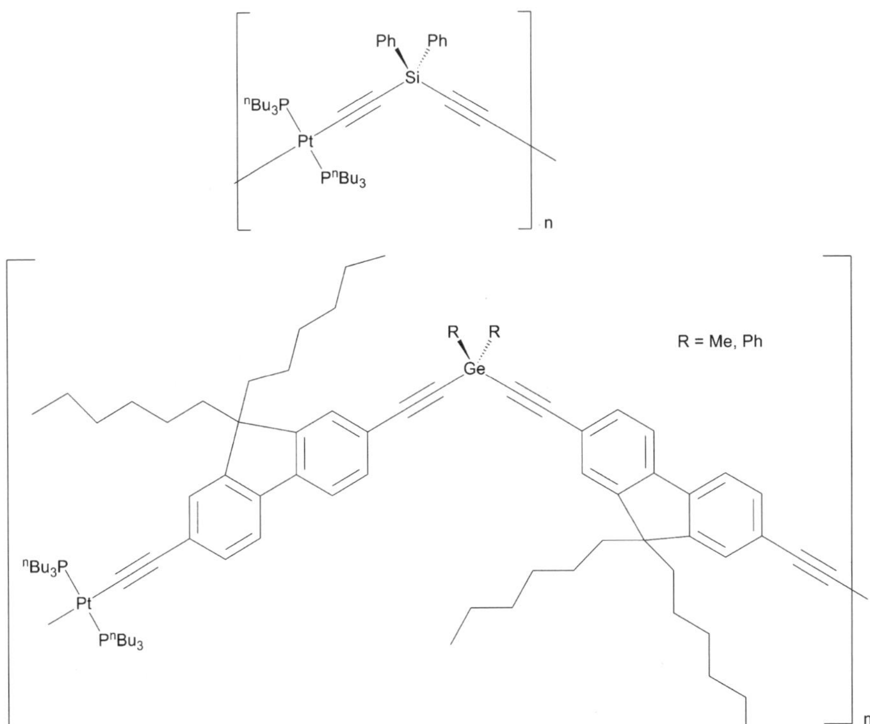

Figure 3.10 Silicon- and germanium-containing platinum(II) poly-yne polymers.

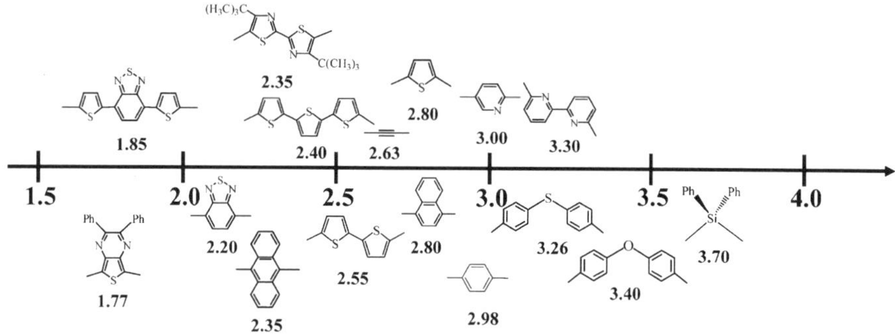

Figure 3.11 The range of optical bandgap values (eV) for platinum(II) poly-ynes with the spacer groups shown in the Figure. Taken from ref. 49 (W.-Y. Wong, *Dalton Trans.*, 2007, 4495) with permission from the author.

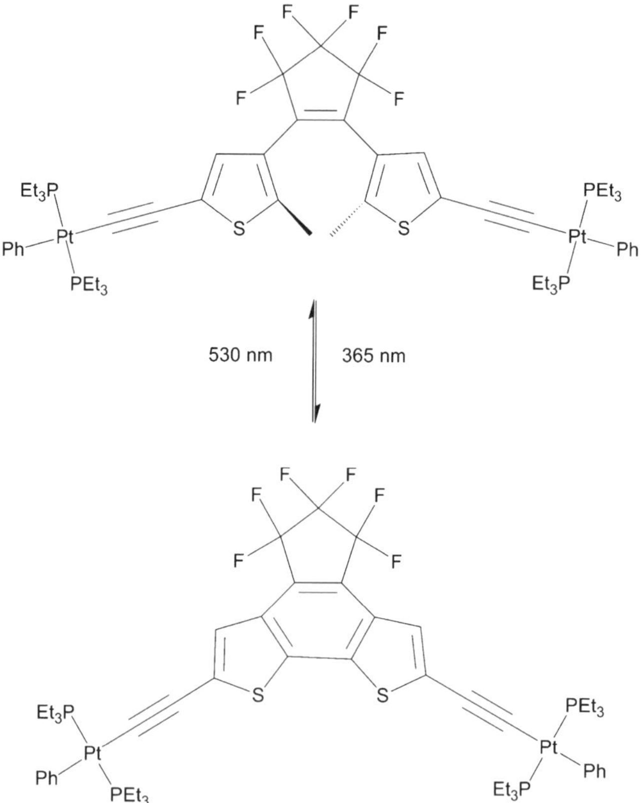

Figure 3.12 The reversible photochemical ring closure and ring opening of the dithienylethene unit in the diplatinum di-yne complex.

efficient solid-state photoswitches in which modification of the Pt(II) units can tune absorption properties and hence operational wavelength across the visible range.

Palladium(II) di-yne complexes and poly-yne complexes are much less common than those for platinum(II), partly because of their enhanced kinetic lability compared to the highly kinetically inert platinum systems. However, a series of dipalladium 3,6-carbazole acetylide complexes have been prepared, along with their platinum(II) analogues. A systematic comparison study has been carried out and the emission of the platinum(II) complexes was assigned to originate from an admixture of ^{3}IL/^{3}MLCT excited states that are predominantly intraligand (IL) in character.[52]

3.3 Gold, Silver and Mercury Acetylide Complexes, Oligomers and Polymers

Acetylide compounds are amongst the most stable gold(I) organometallic compounds known.[53] The preference of gold(I) for two-coordinate, linear coordination makes it ideal for the synthesis of rigid-rod materials with acetylide ligands. Several linear complexes [Au(C≡CR)L] have been synthesised (L = tertiary phosphine[54], arsine[55], isocyanide[56] or amine[57]) either from the polymeric [{Au(C≡CR)}$_n$] or by the reaction of [LAuCl] with Grignard reagents. A more convenient synthesis of gold(I) ethynyls makes use of the simple reaction between [Au(PR$_3$)Cl] and terminal alkynes in basic alcoholic solution,[58] as shown in Figure 3.13.

This method is generally applicable to a wide range of terminal acetylenes, easily precipitating the sparingly soluble gold(I) acetylide in high yield. Methanolic solutions also prove to be effective using either Na(OMe) or KOH as the base.[59] This approach has been used to synthesise a wide range of acetylide complexes.[60]

Many gold(I) complexes exhibit short Au···Au contacts between 2.7 – 3.5 Å. This tendency to aggregate was termed aurophilicity by Schmidbaur and coworkers,[61] and is a consequence of relativistic effects of the heavy-atom d^{10} system. The aurophilic effect can lead to long-chain arrangements of gold(I) complexes in the solid state – if these complexes are already rigid-rod linear structures, then this motif will be extended throughout the crystalline material to give linear polymers.

The photoluminescence of gold(I) complexes has been widely studied in the last decade. The first luminescent gold(I) acetylide reported was [Au$_2$(μ-dppe)$_2$(C≡C-Ph)$_2$].[62] Figure 3.14 shows intermolecular (but no intramolecu-

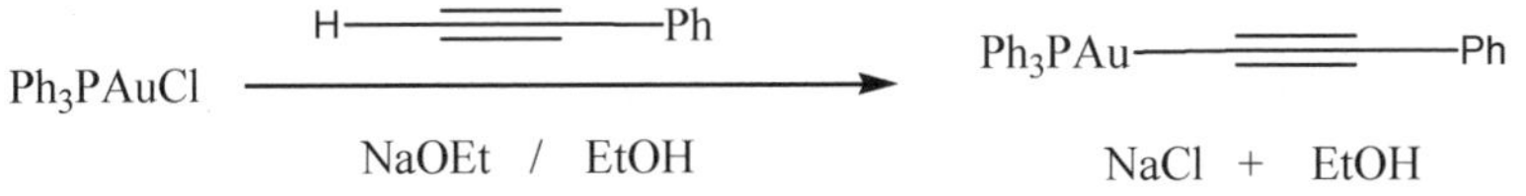

Ph$_3$PAuCl   H————≡————Ph   ⟶   Ph$_3$PAu————≡————Ph

NaOEt / EtOH   NaCl + EtOH

Figure 3.13 Synthesis of gold acetylide complex under basic ethanolic conditions.

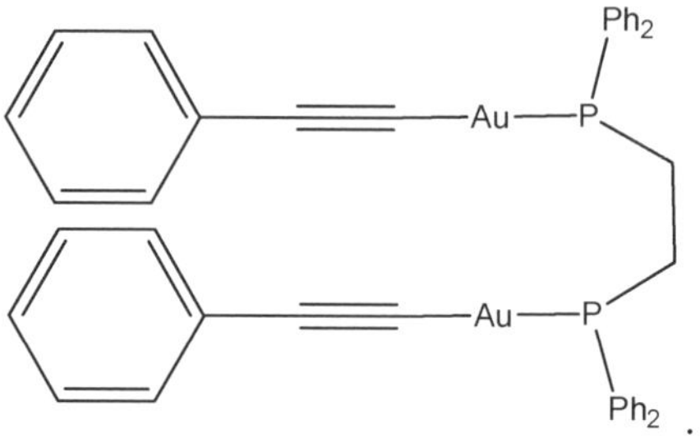

Figure 3.14 The first luminescent Au(I) acetylide complex reported, [Au$_2$(μ-dppe)$_2$(C$\equiv$C-Ph)$_2$].

lar) Au$\cdots$Au interaction between two units leading to an Au$\cdots$Au separation of 3.153(2) Å.

The complex displays a ligand-centred emission at 420 nm in dichloromethane solution at 298 K and the solid sample emits at 550 nm at 298 K. It was suggested that the Au$\cdots$Au interactions may influence the photoluminescence of these gold(I) complexes. Mononuclear, dinuclear and polymeric rigid-rod gold(I) acetylide containing complexes have also been synthesised and investigated for their luminescence properties, with examples shown in Figure 3.15.[56]

The compounds all show strong emission on excitation ($\lambda_{ex} > 350$ nm). The mono- and dinuclear complexes are soluble in dichloromethane and emit at 424–504 nm and 415 nm, respectively. These emissions have been assigned as σ–π^* or π–π^* transitions. The dinuclear complex emits at 540 nm in the solid state, strongly redshifted compared to the solution emission. This is suggested to be due to short Au$\cdots$Au contacts (3.163(1) Å) in the solid state. The polymeric complexes are insoluble and can only be investigated in the solid

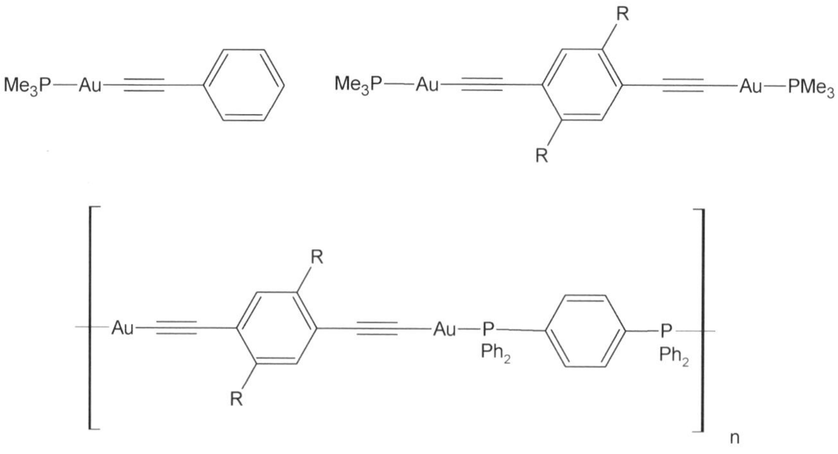

Figure 3.15 Gold(I) alkynyl compounds studied for their luminescent properties by Puddephatt and coworkers, where R = H or Me.

state; they show broad emission at 600 nm. This low-energy emission can be attributed to extensive delocalised π-conjugation along the polymer chain, as well as the short Au$\cdots$Au contacts in the solid state.

Yam *et al.* have extensively studied the photophysical properties of an extensive series of gold(I) acetylide complexes.[63] A series of mono-, di- and trinuclear gold(I) phosphine acetylide complexes have been synthesised and structurally characterised. In general, they assigned the intense absorption spectra as high-energy absorptions due to intraligand (IL) phosphine-centred and π–π*(C$\equiv$C) transitions, and the low-energy absorptions were assigned as metal-perturbed IL π–π*(C$\equiv$C) transitions mixed with metal-to-acetylide MLCT transitions. Both the electronic absorption and emission energies are found to depend on the nature of the acetylide ligands. The nuclearity of the metal complexes also has an influence on the electronic absorption as well as emission behaviour. The emission energies are generally found to be lowest in the trinuclear complexes, attributed to the presence of weak Au$\cdots$Au interactions.

Of interest, through comparison to the analogous platinum di-yne systems, are a series of gold(I) acetylides that contain polythiophene spacer groups, having the formula $[(R_3P)Au(C\equiv C(C_4H_2S)_nC\equiv C)Au(PR_3)]$ (n = 1, 2, 3; R = Ph, Cy).[64] In these systems the gold centres adopt the linear two-coordinate geometry, but only in the system, $[(PPh_3)Au(C\equiv C(C_4H_2S)C\equiv C)Au(PPh_3)]$, with the monothiophene spacer group, are the gold centres linked through Au$\cdots$Au interactions [Au$\cdots$Au 3.2915(10) and 3.2347(9) Å] to form a network structure. In the species with the longer spacer groups Au$\cdots$Au interactions are absent but the di-gold units are linked by hydrogen-bonding interactions. The solution absorption and emission spectra of these complexes display an increasing red shift as the length and planarity of the polythiophene spacer groups increases. This is consistent with the absorptions and emissions being dominated by ligand-centred π–π* transitions, and there is certainly much less interaction between the ligand systems and the gold centres than there is between the ligand and metal centres in the analogous platinum complexes. In a subsequent systematic study of a wide range of gold(I) di-yne and poly-yne complexes, including that of the two- and three-ring fused thiophene complexes, $[(Ph_3P)Au(C\equiv C)(C_6H_2S_2)(C\equiv C)Au(PPh_3)]$ and $[(Ph_3P)Au(C\equiv C)(C_8H_2S_3)(C\equiv C)Au(PPh_3)]$, it was found that Au$\cdots$Au interactions were only observed in systems where shorter spacer groups were present, and for longer spacers other intermolecular interactions were prominent.[65]

Silver acetylides are much less well known and studied although examples have been reported.[66] A series of trinuclear Ag complexes were synthesised by the Che[67] and Yam groups.[68] The groups reported Ag(I) alkynyl complexes $[Ag_3(\mu\text{-dppm})_3(\mu_3\text{-}\eta^1\text{-}C\equiv C\text{-}R)]^{2+}$ and studied their luminescence. Ag$\cdots$Ag interactions were also shown to influence the photoluminescent properties.

Although mercury acetylides have been known for almost a century, and although Hg(II) has a d^{10} electron configuration, the chemistry of Hg(II) acetylides has not been exploited to anything like the same extent as the analogous Au(I) d^{10} systems as discussed above.[69] However, the dimercury di-

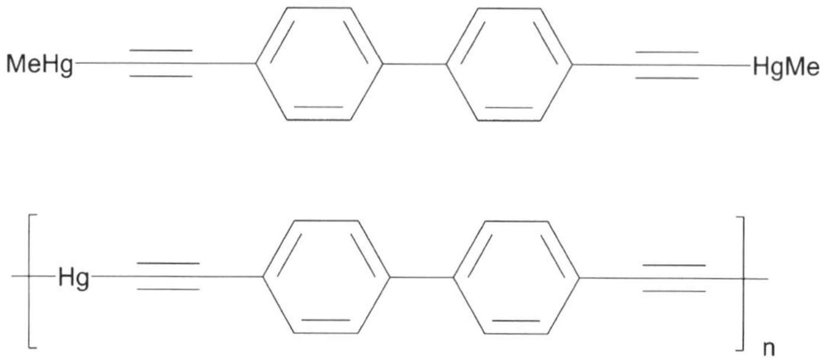

Figure 3.16   Structures of mercury(II) di-yne and poly-yne systems.

yne and poly-yne systems (Figure 3.16) have been prepared. The di-yne complex exhibits d^{10}–d^{10} Hg$\cdots$Hg contacts in the solid state to give an extended three-dimensional structure. The spatial extent of the lowest singlet (S_1) and triplet (T_1) excited states with variation in chain length was investigated and the S_1–T_1 energy gap was found to be 0.70–0.76 eV.[70] The results are in close agreement with those obtained for related triplet emitting platinum and gold complexes.

Wong *et al.* extended the chemistry of Hg(II) acetylides to include trimetallic systems,[71] such as (MeHg-C≡C)$_3$(C$_6$H$_3$), which is structurally analogous to the gold complex, (Ph$_3$PAu-C≡C)$_3$(C$_6$H$_3$), along with many analogues of the Au(I) systems described earlier in this section.[72] Of particular interest are the Hg(II) acetylides with fluorine groups as the linkers. This is because the resultant complexes have relatively high thermal and chemical stability, and this, coupled with their high emission quantum yields makes them good candidates as LED materials. The simplest member of the series is the direct analogue of the basic hydrocarbyl complexes, MeHg-C≡C-(C$_{13}$H$_8$)-C≡C-HgMe.[73]

3.4 Ruthenium and Osmium Acetylide Complexes, Oligomers and Polymers

Ruthenium and osmium acetylide complexes have been extensively investigated by Lewis and coworkers utilising trimethylstannnyl acetylides. Incorporation of Ru(II),[74] and Os(II)[75] metal centres into mono-, di- and polymeric alkynyl systems were studied. Synthetic methods were found to be particularly reliable for the synthesis of *trans*-oriented di- and polyalkynyl Ru(II) complexes (with stabilising dppe and dppm ligands), the synthesis of the monomeric and polymeric Ru alkynyl is shown in Figure 3.17. Similar schemes have also been reported for Os(II) (with dppm).

Dixneuf and coworkers developed a synthetic strategy to mono- and unsymmetrical diacetylide ruthenium complexes *via* a vinylidene intermediate.[76] Initial findings by the group showed that *cis*-[RuCl$_2$(Ph$_2$PCH$_2$PPh$_2$)$_2$]

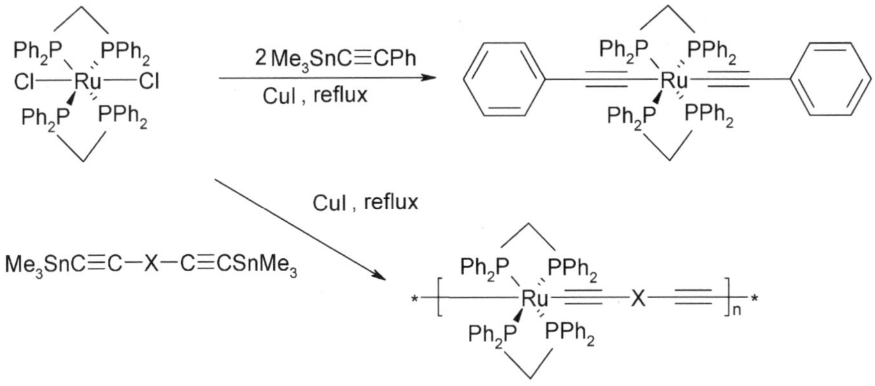

Figure 3.17   Synthesis of Ru acetylides using trimethylstannyl reagents.

readily activates terminal alkynes to produce stable and isolable vinylidene complexes in excellent yields. Sodium hexafluorophosphate is used for the production of a 16-electron Ru intermediate by the displacement of a chloride ligand. The vinylidene species is then deprotonated by the addition of one molar equivalent of DBU (1,8-diazabicyclo[5.4.0]undec-7-ene) in dichloromethane. The reaction is carried out at room temperature, and gives the *trans*-chloro acetylide metal derivative; the general reaction scheme is illustrated in Figure 3.18.

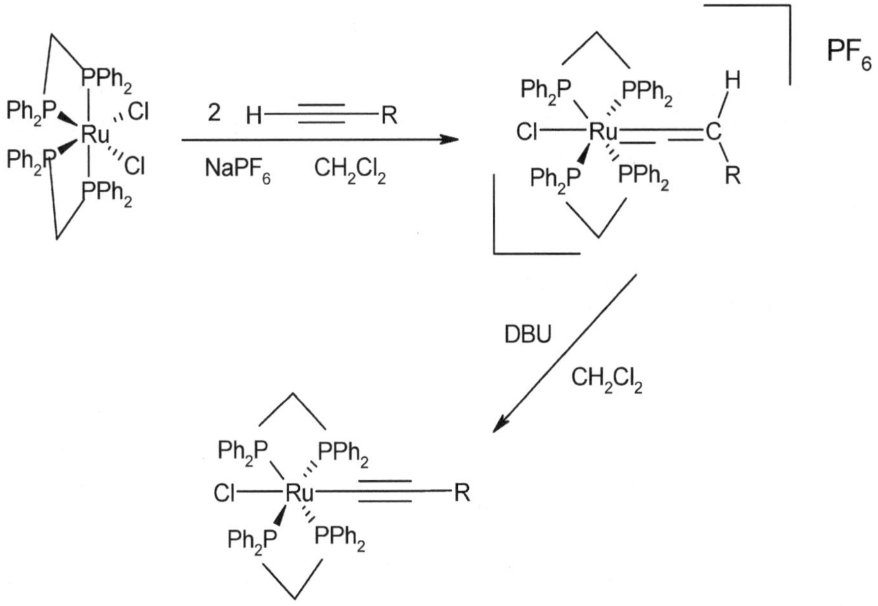

Figure 3.18   Formation of vinylidene intermediates in the synthesis of ruthenium acetylide complexes.

3.5 Properties of Transition-Metal σ-Acetylide Complexes

Transition-metal σ-acetylide complexes and polymers have many of the electronic and chemical properties to make them good materials for potential use in light-emitting diode (LED) technology. The conjugated aromatic systems linked to the metal centres *via* the acetylene groups have been used to tune the optical gaps of these materials. In addition, the heavy metal centres provide a route into efficient radiative use of triplet excited state, *via* spin-orbit coupling. These properties provide tuneability of absorption and emission right across the visible spectrum, thus making rigid-rod acetylide complexes and polymers containing the heavy metals, Pt, Au and Hg, ideal for use in optical-limiting devices. In optical-limiting devices the materials may be transparent to visible light in their ground states, but they become dark in colour, blocking transmitted light, when they are photoactivated. This area has been reviewed elegantly and comprehensively by Zhou and Wong recently, and the potential for these materials has been highlighted effectively.[11]

The rigid-rod metal acetylide complexes and polymers possess other interesting physical and electronic properties that may have applications in modern technology, such as nonlinear optics and liquid crystalline phases, and these will be discussed briefly in this final section of this chapter.

As light travels through a material a variety of nonlinear optical effects may occur. The interaction of light with such a material will cause its properties to change. As light travels through the material, its electric field interacts with charges in the material. The time-dependent electron density distribution in the material, resulting from this interaction, can affect the propagation of subsequent light waves travelling through the material. These interactions can cause the original optical beam to have its frequency, phase, polarisation, or path changed significantly. The study of these changes is termed nonlinear optics (NLO) since it describes the deviations of behaviour away from classical optics.

This ability to manipulate light has many important technological applications in optical signal processing, generation of variable frequency laser light, tuneable filters, and optical data storage. In order to control the properties of light, chemists must design and synthesise molecules within that the combination of photons can take place, and in which both the magnitude and response time of these optical processes can be controlled.[77]

Metal acetylide complexes provide new opportunities for engineering materials with nonlinear optical properties. It is possible to change the transition-metal element, its oxidation state and hence the number of *d*-electrons, examine the difference between diamagnetic and paramagnetic complexes and the effect of novel bonding geometries and coordination patterns with a view to altering the nonlinear properties of the material. The materials discussed in the previous sections are exceptionally useful for nonlinear optical studies because:

- Metal to ligand or ligand to metal charge-transfer bands are often observed in the UV to visible region of the spectrum. These optical absorption bands are often associated with large optical nonlinearities.
- Chromophores containing metals are among the most intensely coloured materials known. The intensity of the optical absorption band is related to its transition dipole moment. Large optical nonlinearities are associated with such strongly allowed transitions.[78]
- The presence of electron density away from the nuclei within a molecule increases its polarisability, and therefore its potential for nonlinear optical behaviour, hence systems with extensive π-conjugation are ideal.

Interest in nonlinear optics of square-planar metal aromatic complexes arose from the knowledge that the fragments, $MX(PEt_3)_2$, where M=Ni, Pd, and Pt; X=I, Br and Cl, are good electron donors. Experiments revealed that this class of compounds substituted with various aromatic acceptors could exhibit relatively large nonlinear optical behaviour.[79]

More recently, rigid-rod polymers have been used to study their nonlinear optical behaviour. The presence of low-energy MLCT transitions and the possibility of extended delocalisation through the polymer chain provide the rationale for studying these materials for NLO applications.[80] The susceptibilities of the polyacetylide organometallic polymers are greater than those of the equivalent organic polyacetylide. A strong dependence between NLO response and the metal employed (Ni, Pd, Pt) is seen. The nickel monomers and polymers tend to give larger hyperpolariseabilities than the platinum complexes. The full involvement of the metal in the polarisability of the polymer electron backbone is not yet known.[81]

Cifuentes and Humphrey have summarised the studies of the NLO properties of metal alkynyl complexes.[82] The review focuses on the synthesis and structure–property relationship of Ru, Ni and Au complexes. The paper also reviews the recent progress with switching the nonlinearity of acetylide complexes using protonation/deprotonation and, particularly, oxidation/reduction processes. This is an area in which metal acetylide complexes may well prove superior to organic molecules, and has been seen as a niche for organometallics in nonlinear optics. An example of a switchable NLO system is given in Figure 3.19.

Another potential application for rigid-rod metal acetylide complexes is that of liquid-crystalline materials. Liquid-crystalline compounds demonstrate molecular order between that of an ordered solid and a disordered liquid or solution. These intermediate phases are referred to as mesophases and can be split into two subdivisions (Figure 3.20).

- *Nematic* – the molecules align parallel in the direction *n*, but can move within the N phase and rotate freely around the long molecular axis. There is orientational, but no positional order.
- *Smectic* – this phase has a higher degree of order. Molecules are arranged in layers and are parallel within the layers. The S_A phase has molecules aligned

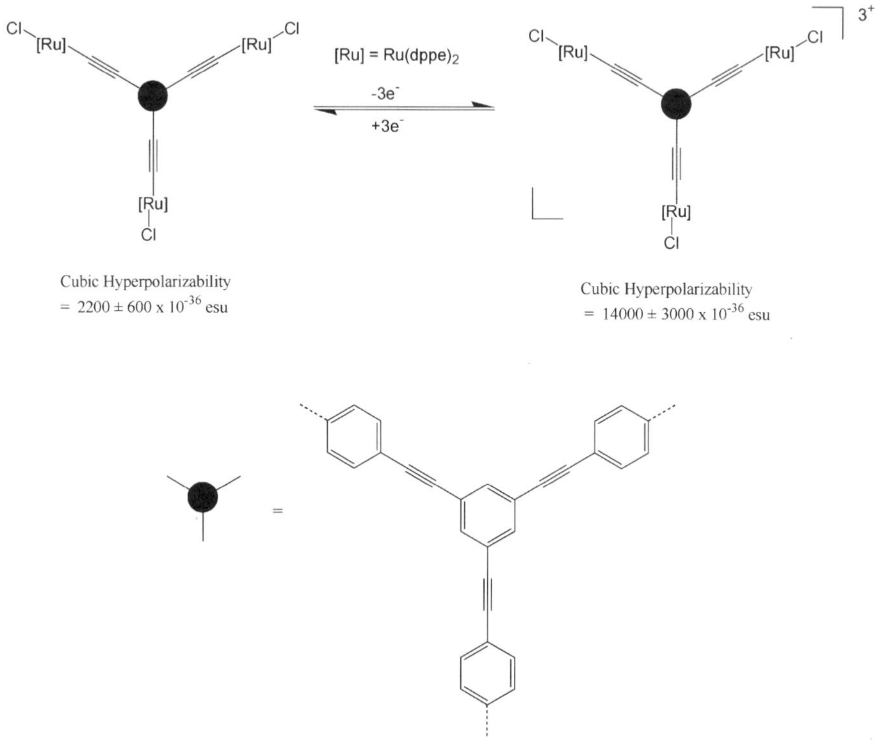

Cl
[Ru]
[Ru] = Ru(dppe)2
-3e-
+3e-
3+
Cubic Hyperpolarizability
= 2200 ± 600 x 10-36 esu
Cubic Hyperpolarizability
= 14000 ± 3000 x 10-36 esu
=

Figure 3.19 Switchable NLO system based on Ru acetylide complexes.

parallel to the layer normal, but with no positional order within the layer. Other Smectic phases include S_C, where the molecules are arranged as in the S_A phase but with the layers tilted relative to each other.

Since the pioneering work of Gray and coworkers in the early 1970s it has been possible to develop materials with NLO properties and sufficient stability

n
Nematic
Smectic (SA)
Smectic (SC)

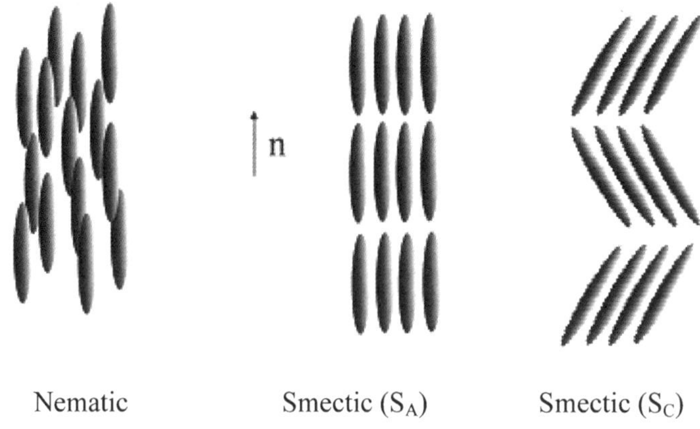

Figure 3.20 Liquid-crystalline phases.

to be used widely in commercial applications.[83] This has resulted in a multimillion-pound industry devoted to liquid-crystal displays (LCDs).

Studies on homometallic (Figure 3.21) and heterometallic rigid-rod acetylide polymers of the Group 10 metals (Figure 3.22) by ^{31}P NMR spectroscopic techniques have shown that all the systems investigated possessed a negative diamagnetic anisotropy. This means that the molecules or polymers align with their long molecular axes perpendicular to the applied magnetic field. This was interpreted in terms of the negative diamagnetic anisotropy of the carbon–carbon triple bond.[84]

This work with polymeric acetylides prompted studies of related low molar mass compounds, and the Pt(II) complex in Figure 3.22 was found to possess two types of liquid-crystalline phases.[85]

Particular attention has recently been directed towards a better understanding of the relationships between the shape and nature of a complex and the structure of the liquid crystal phase. Bruce and coworkers have studied a series of complexes of palladium(II) and platinum(II) based on trialkoxystilbazole esters and showed how varying the ligand bulk effects the various liquid crystal phases, Figure 3.23.[86]

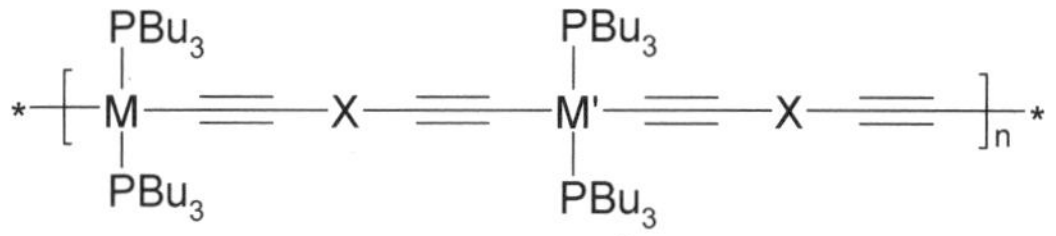

Figure 3.21   Rigid-rod polymers, where M' = Pd or Ni.

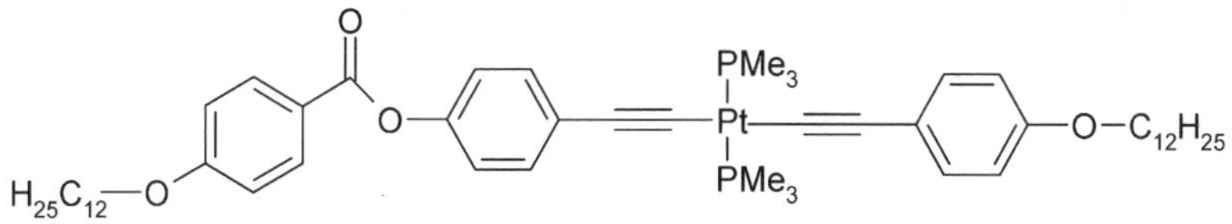

Figure 3.22   Pt(II) complex, which was found to posses liquid-crystalline phases.

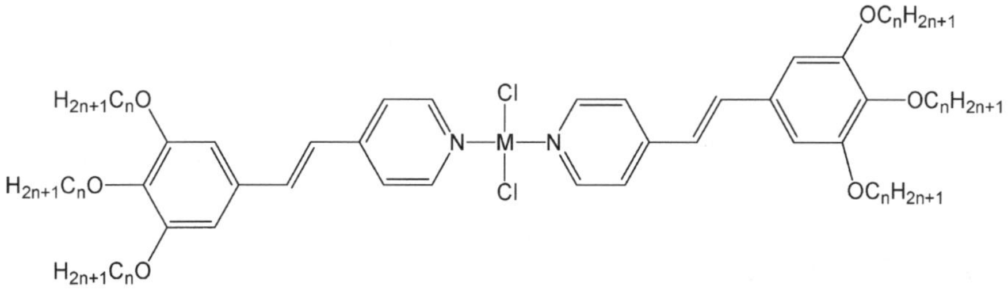

Figure 3.23   Complexes of Pd(II) and Pt(II) based on trialkoxystilbazole esters. M = Pd(II) or Pt(II), *n* = 6, 8, 10, 12, or 14.

The linear Au(I) acetylide complexes of the type illustrated in Figure 3.24 have been shown to form nematic and smectic A phases without decomposition.[87]

The isonitrile ligand is used in preference to much more sterically bulky phosphines for stabilisation of the gold acetylide. The molecules, therefore, have a more rod-like structure and are more suitable for the formation of liquid-crystal mesophases. A representative crystal packing diagram for the di-acetylide Au(I) complexes with the R groups below is illustrated in Figure 3.25.

It is clearly seen above how these linear rigid-rod gold(I) complexes order in the solid state as parallel rows. As the crystal melts some order is retained within the planes so as to give liquid-crystalline states.

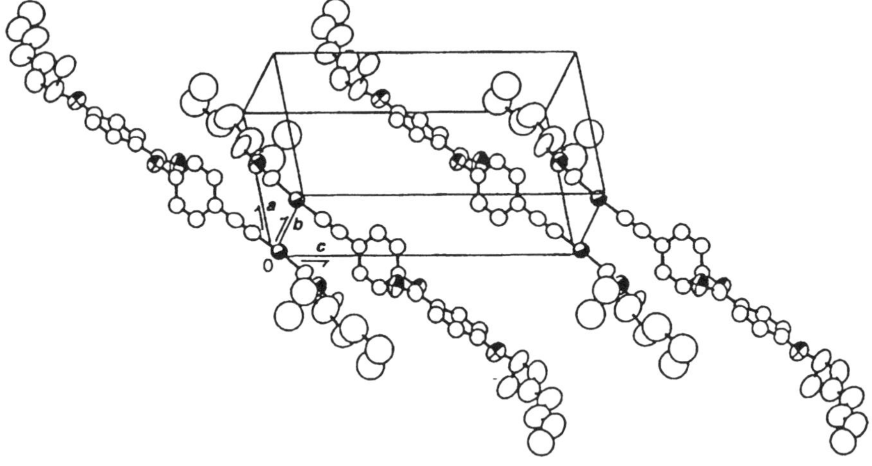

Figure 3.24   Gold(I) alkynyl complex studied for liquid-crystalline behaviour.

Figure 3.25   Crystal packing diagram of an Au(I) acetylide complex.

3.6   Conclusions

Over the last two decades it has been shown that a wide variety of metal acetylide complexes, oligomers and polymers can be prepared when the metal is one of the heavy group 8, 9 or 10 elements and that, in the majority of cases, these oligomers and polymers adopt a linear, rigid-rod structure. Delocalisation within this rigid rod structure that involves the overlap between π orbitals on the acetylene unit and π orbitals on the aromatic or heteroaromatic linker group, and to an extent d orbitals on the metals, induces a range of novel physical and optoelectronic properties on these materials, particularly in the solid state. The presence of the heavy-metal elements also enhances the evolution of accessible excited triplet states within these systems, which cannot be accessed in organic analogues. The heavy metals, with their associated spin-orbit coupling, allow the relaxation of quantum-mechanical selection rules and facilitate intersystem crossing between excited singlet and triplet states. Thus, phosphorescence is prevalent in species, which adds an extra dimension to the electronic properties of the materials. In addition, because of the variety of heavy transition metals and aromatic and heteroaromatic linker groups that can be incorporated into these systems it is possible, by chemical means, to tune the bandgaps in the polymeric materials over a significant energy range, from 1.7 eV to above 3 eV. Recent work has shown that it is also possible to introduce photoactivated switches into the systems that have the potential to allow the switching "on" or "off" of the communication between the metal centres in oligomeric systems.[51]

Another benefit of incorporating metal-containing fragments into the oligomers and polymers is that the auxiliary ligands, such as phosphines, that are coordinated to the metal allow the solubility and processability of the materials to be modified. In this way, the solubility of the materials and the ease with which they can be spun as thin films can be enhanced.

The enhancement of both the physical and electronic properties of oligomers and polymers that contain transition-metal acetylide units means that they have already shown the potential for applications in the areas of optoelectronics, including nonlinear optics and optical-limiting devices, and in the development of new LEDs and sensor devices. Their liquid-crystal properties may also result in novel applications in the future.

Acknowledgements

PRR is grateful to the EPSRC for the award of a Senior Research Fellowship and MJ thanks the University of Bath for the award of a Ph D studentship. PRR is also grateful to Professor the Lord Lewis of Newnham and Professor Brian Johnson who introduced him to the area of metal acetylides, and to the many group members in Cambridge and subsequently Bath who have worked in the area and whose work is incorporated into this chapter.

References

1. A. Kraft, A.C. Grimsdale and A.B. Holmes, *Angew. Chem. Int. Ed.*, 1998, **37**, 403.

2. A. Montali, P. Smith and C. Weder, *Syn. Met.*, 1998, **97**, 123

3. (a) N. Tessler, G. J. Denton and R. H. Friend, *Nature*, 1996, **382**, 695; (b) H. Meier, *Angew. Chem. Int. Ed. Engl.*, 1992, **31**, 1399.

4. (a) J. M. Halls, C. A. Walsh, N. C. Greenham, E. A. Marseglia, R. H. Friend, S. C. Moratti and A. B. Holmes, *Nature*, 1995, **376**, 498; (b) G. Yu, J. Gao, J. C. Hummelen, F. Wudl and A. J. Heeger, *Science*, 1995, **270**, 1789; (c) C. Weder, A. Sarwa, A. Montali, C. Bastiaansen and P. Smith, *Science*, 1998, **279**, 835; (d) A. Kohler, H. F. Wittmann, R. H. Friend, M. S. Khan and J. Lewis, *Synth. Met.*, 1994, **67**, 245; (e) A. Kohler, H. F. Wittmann, R. H. Friend, M. S. Khan and J. Lewis, *Synth. Met.*, 1996, **77**, 147.

5. (a) F. Garnier, R. Hajlaoui, A. Yasser and P. Svwastra, *Science*, 1994, **265**, 498; (b) N. Tessler and R. H. Friend, *Science*, 1998, **280**, 1741.

6. A. Montali, C. Bastiaansen, P. Smith and C. Weder, *Nature*, 1998, **392**, 261.

7. (a) C. Caliendo, I. Fratoddi, M. V. Russo and C. L. Sterzo, *J. App. Phys.*, 2003, **93**, 10071; (b) C. Caliendo, E. Verona, A. D'Amico, A. Furlani, G. Infante and M. V. Russo, *Sens. Acuators B*, 1995, **25**, 670; (c) T. M. Swager, *Acc. Chem. Res.*, 1998, **31**, 201; (d) D. T. McQuade, A. E. Pullen and T. M. Swager, *Chem. Rev.*, 2000, **100**, 2537.

8. (a) L. Feringa, W. F. Jager and B. de Lange, *Tetrahedron*, 1993, **49**, 8267; (b) G. H. W. Buning in *Organic Materials for Photonics,* ed. G. Zerbi, Elsevier, Amsterdam, 1993, pp. 367–387; (c) K. Lösch, *Macromol. Symp.*, 1995, **100**, 65; (d) J.-M. Lehn. *Supramolecular Chemistry: Concepts and Perspectives*, VCH, Weinheim, 1995, pp. 124–138; (e) P. J. Martin in *Introduction to Molecular Electronics*, ed. M. C. Petty, M. R. Bryce and D. Bloor, Edward Arnold, London, **1995**. pp. 112–141; (f) J.-J. Kim and E.-H. Lee, *Mol. Cryst. Liq. Cryst.*, 1993, **227**, 71; (g) *Photochromism, Molecules and Systems,* ed. H. Durr and H. Bouas-Laurent, Elsevier, Amsterdam, 1990; H. Durr, *Angew. Chem. Int. Ed. Engl.*, 1989, **28**, 413.

9. S. R. Marder, B. Kippelen, A. K.-Y. Jen and N. Peyghambarian, *Nature*, 1997, **388**, 845.

10. (a) A. J. Heeger and J. Long, Jr., *Opt. Photon. News*, 1996. **7**, 24; (b) H. S. Nalwa, in *Nonlinear Optics of Organic Molecules and Polymers*, ed. H. S. Nalwa and S. Miyata, CRC, New York, 1997, pp 611–787.

11. G.-J. Zhou and W.-Y. Wong, *Chem. Soc. Rev.*, 2011, **40**, 2541.

12. M. Pope, H. P. Kallman and P. Magnante, *J. Chem. Phys.*, 1963, **38**, 2042.

13. J. H. Burroughes, D. D. C. Bradley, A. R. Brown, R. N. Marks, K. Mackay, R. H. Friend, P. L. Burn and A. B. Holmes. *Nature*, 1990, **347**, 539.

14. D. Braun and A. J. Heeger, *Appl. Phys. Lett.*, 1991, **58**, 1982.

15. J. Ho, (2002) *Improving the External Extraction Efficiency of Organic LEDs [WWW]* http://www.mtl.mit.edu/researchgroups/MEngTP/ John_Ho_Proposal.pdf (January 17th 2006)

16. J. Brooks, Y. Babayan, S. Lamansky, P.I. Djurovich, I. Tsyba, R. Bau and M. E. Thompson, *Inorg. Chem.*, 2002, **41**, 3055.
17. N. J. Long, *Angew. Chem., Int. Ed. Engl.*, 1995, **34**, 21.
18. V. W.-W. Yam and K. M.-C. Wong, *Top. Curr. Chem.*, 2005, **257**, 1.
19. N. Chawdhury, M. Younus, P. R. Raithby, J. Lewis and R. H. Friend, *Opt. Mater.*, 1998, **9**, 498.
20. S. Takahashi, M. Kariya, T. Yatake, K. Sonogashira and N. Hagihara, *Macromolecules*, 1978, **11**, 1063.
21. W. J. Blau, H. J. Byrne, D. J. Cardin and A. P. Davey, *J. Mater. Chem.*, 1991, **1**, 245.
22. O. Lhost, J. M. Toussaint, J. L. Brédas, H. F. Wittmann, K. Fuhrmann, R. H. Friend, M. S. Khan and J. Lewis, *Synth. Met.*, 1993, **55–57**, 56.
23. N. J. Long and C. K. Williams, *Angew. Chem. Int. Ed. Engl.*, 2003, **42**, 2586.
24. P. Nguyen, P. Gómez-Elipe and I. Manners, *Chem. Rev.*, 1999, **99**, 1515.
25. V. W.-W. Yam, K. K.-W. Lo and K. M.-C. Wong, *J. Organomet. Chem.*, 1999, **578**, 3.
26. J. Manna, K. J. John and M. D. Hopkins, *Adv. Organomet. Chem.*, 1996, **38**, 79.
27. G. Frapper and M. Kertz, *Inorg. Chem.*, 1993, **32**, 732.
28. W.-Y. Wong and C.-L. Ho, *Coord. Chem. Rev.*, 2006, **250**, 2627.
29. (a) K. Sonogashira, Y. Toda and N. Hagihara, *Tetrahedron Lett.*, 1975, 4467; (b) N. Hagihara, K. Sonogashira and S. Takahashi, *Ado. Poivrn. Sci.*, 1981, **41**, 149.
30. M. S. Khan, A. K. Kakkar, N. J. Long, J. Lewis, P. R. Raithby, P. Nguyen, T. B. Marder, F. Wittmann and R. H. Friend, *J. Mater. Chem.*, 1994, **4**, 1227.
31. B. Cetinkaya, M. F. Lappert, J. McMeeking and D. E. Palmer, *J. Chem. Soc. Dalton Trans.*, 1973, 1202.
32. B. F. G. Johnson, A. K. Kakkar, M. S. Khan, J. Lewis, A. E. Dray, R. H. Friend and F. Wittmann, *J. Mater. Chem.*, 1991, **1**, 485.
33. M. E. Wright, *Macromolecules*, 1989, **22**, 3256.
34. S. J. Davies, B. F. G. Johnson, M. S. Khan and J. Lewis, *J. Chem. Soc. Chem. Commun.*, 1991, 187.
35. (a) D. Beljonne, H. F. Wittmann, A. Kohler, S. Graham, M. Younus, J. Lewis, P. R. Raithby, M. S. Khan, R. H. Friend and J. L. Bredas, *J. Chem. Phys.*, 1996, **105**, 3868; (b) A. Kohler, J. S. Wilson, R. H. Friend, M. K. Al-Suti, M. S. Khan, A. Gerhard and H. Bassler, *J. Chem. Phys.*, 2002, **116**, 9457; (c) E. E. Silverman, T. Cardolaccia, X. M. Zhao, K. Y. Kim, K. Haskins-Glusac and K. S. Schanze, *Coord. Chem. Rev.*, 2005, **249**, 1491; (d) K. Glusac, M. E. Kose, H. Jiang and K. S. Schanze, *J. Phys. Chem. B*, 2007, **111**, 929; (e) W.-Y. Wong, *Dalton Trans.*, 2007, 4495; (f) T. Cardolaccia, A. M. Funston, M. E. Kose, J. M. Keller, J. R. Miller and K. S. Shanze, *J. Phys. Chem. B*, 2007, **111**, 10871; (g) L. Yang, J.-K. Feng, W.-Y. Wong and S.-Y. Poon, *Polymer*, 2007, **48**, 6457; (h) G. R. Whittel and I. Manners, *Adv. Mater.*, 2007, **19**, 3439; (i) F. Guo, Y.-G. Kim, J. R.

Reynolds and K. S. Schanze, *Chem. Commun.*, 2006, 1887; (j) V. W.-W. Yam, *Acc. Chem. Res.*, 2002, **35**, 555; (k) M. P. Cifuentes and M. G. Humphrey, *J. Organomet. Chem.*, 2004, **689**, 3968; (l) A. Köhler and D. Beljonne, *Adv. Funct. Mater.* 2004, **14**, 11; (m) J. S. Wilson, N. Chawdhury, M. R. A. Al-Mandhary, M. Younus, M. S. Khan, P. R. Raithby, A. Kohler and R. H. Friend, *J. Am. Chem. Soc.*, 2001, **123**, 9412.

36. K. M.-C. Wong and V. W.-W. Yam, *Coord. Chem. Rev.*, 2007, **251**, 2477.

37. M. L. Muro, S. Diring, X. Wang, R. Ziessel and F. N. Castellano, *Inorg. Chem.*, 2009, **48**, 11533.

38. (a) M. S. Khan, M. R. A. Al-Mandhary, M. K. Al-Suti, A. K. Hisham, P. R. Raithby, B. Ahrens, M. F. Mahon, L. Male, E. A. Marseglia, E. Tedesco, R. H. Friend, A. Köhler, N. Feeder and S. J. Teat, *J. Chem. Soc., Dalton Trans.*, 2002, 1358; (b) M. S. Khan, M. K. Al-Suti, M. R. A. Al-Mandhary, B. Ahrens, J. K. Bjernemose, M. F. Mahon, L. Male, P. R. Raithby, R. H. Friend, A. Köhler and J. A. Wilson, *Dalton Trans.*, 2003, 65; (c) M. S. Khan, M. R. A. Al-Mandhary, M. K. Al-Suti, P. R. Raithby, B. Ahrens, M. Mahon, L. Male, C. E. Boothby and A. Köhler, *Dalton Trans.*, 2003, 74.

39. M. S. Khan, M. R. A. Al-Mandhary, M. K. Al-Suti, N. Feeder, S. Nahar, A. Köhler, R. H. Friend, P. J. Wilson and P. R. Raithby, *J. Chem. Soc. Dalton Trans.*, 2002, 2441.

40. K. Ogawa and S. Rasmussen, *J. Org. Chem.*, 2003, **68**, 2921.

41. M. S. Khan, M. R. A. Al-Mandhary, M. K. Al-Suti, F. R. Al-Battashi, S. Al-Saadi, B. Ahrens, J. K. Bjernemose, M. F. Mahon, P. R. Raithby, M. Younus, N. Chawdhury, A. Köhler, E. A. Marseglia, E. Tedesco, N. Feeder and S. J. Teat, *Dalton Trans.*, 2004, 2377.

42. J. S. Wilson, A. Köhler, R. H. Friend, M. K. Al-Suti, M. R. A. Al-Mandhary, M. S. Khan and P. R. Raithby, *J. Chem. Phys.*, 2000, **113**, 7627.

43. M. Younus, A. Köhler, S. Cron, N. Chawdhury, M. R. A. Al-Mandhary, M. S. Khan, J. Lewis, N. J. Long, R. H. Friend and P. R. Raithby, *Angew. Chem., Int. Ed. Engl.*, 1998, **37**, 3026.

44. W.-Y. Wong, K.-H. Choi, G.-L. Lu and J.-X. Shi, *Macromol. Rapid Commun.*, 2001, **22**, 461.

45. J. Lewis, N. J. Long, P. R. Raithby, G. P. Shield, W.-Y. Wong and M. Younus, *J. Chem. Soc. Dalton Trans.*, 1997, 4283.

46. N. Chawdhury, A. Kohler, R. H. Friend, W.-Y. Wong, M. Younus, P. R. Raithby, J. Lewis, T. C. Corcoran, M. R. A. Al-Mandhary and M. S. Khan, *J. Chem. Phys.*, 1999, **110**, 4963.

47. L. S. Devi, M. K. Al-Suti, N. Zhang, S. J. Teat, L. Male, H. A. Sparks, P. R. Raithby, M. S. Khan and A. Kohler, *Macromolecules*, 2009, **42**, 1131.

48. W.-Y. Wong, C.-K. Wong, G.-L. Lu, A. W.-M. Lee, K.-W. Cheah and J.-X. Shi, *Macromolecules*, 2003, **36**, 983.

49. S.-Y. Poon, W.-Y. Wong, K.-W. Cheah and J.-X. Shi, *Chem. Eur. J.*, 2006, **12**, 2550.

50. W.-Y. Wong, *Dalton Trans.*, 2007, 4495.

51. S. K. Brayshaw, S. Schiffers, A. J. Stevenson, S. J. Teat, M. R. Warren, R. D. Bennet, I. V. Sazanovich, A. R. Buckley, J. A. Weinstein and P. R. Raithby, *Chem. Eur. J.*, 2011, **17**, 4385.
52. C.-H. Tao, N. Zhu and V. W.-W. Yam, *J. Photochem. Photobiol. A*, 2009, **207**, 94.
53. G. K. Anderson, *Adv. Organomet. Chem.*, 1982, **20**, 39.
54. H. Lang, S. Koecher, S. Back, G. Rheinwald and G. Van Koten, *Organometallics*, 2001, **20**, 1968.
55. A. Abu-Salah and M. Omar, *J. Organomet. Chem.*, 1990, **387**, 123.
56. M. J. Irwin, J. J. Vittal and R. J. Puddephatt, *Organometallics*, 1997, **16**, 3541.
57. J. Vicente, M. T. Chicote, M. M. Alvarez-Falcon and P. G. Jones, *Organometallics*, 2005, **24**, 4666.
58. R. J. Cross and M. F. Davidson, *J. Chem. Soc., Dalton Trans.*, 1986, 411.
59. M. I. Bruce, E. Horn, J. G. Matisons and M. R. Snow, *Aust. J. Chem.*, 1984, **37**, 1163.
60. V. W.-W. Yam, K. K.-W. Lo and K. M.-C. Wong, *J. Organomet. Chem.*, 1999, **571**, 1.
61. F. Sherbaum, A. Grohmann, B. Huber, C. Krüger and H. Schmidbaur, *Angew. Chem., Int. Ed.*, 1988, **27**, 1544.
62. D. Li, X. Hong, C. M. Che, W. C. Lo and S. M. Peng, *J. Chem. Soc., Dalton Trans.*, 1993, 2929.
63. V. W.-W. Yam, C. K. Li and C. L. Chan, *Angew. Chem., Int. Ed.*, 1998, **37**, 2857.
64. P. Li, B. Ahrens, K.-H. Choi, M. S. Khan, P. R. Raithby and W.-Y. Wong, *CrystEngComm.*, 2002, **4**, 405.
65. P. Li, B. Ahrens, N. Feeder, P. R. Raithby, S. J. Teat and M. S. Khan, *Dalton Trans.*, 2005, 874.
66. R. J. Lancashire, *Comprehensive Coordination Chemistry*, 1987, **5**, Pergamon, Oxford, 861.
67. C. F. Wang, C. M. Chan and C. M. Che, *Polyhedron*, 1996, **15**, 1853.
68. V. W.-W Yam, W. K. M. Fung and K. K. Cheung, *Organometallics*, 1997, **16**, 2032.
69. W.-Y. Wong, *Coord. Chem. Rev.*, 2007, **251**, 2400.
70. L. Liu, S.-Y. Poon and W.-Y. Wong, *J. Organomet. Chem.*, 2005, **690**, 5036.
71. L. Liu, W.-Y. Wong and C.-L. Ho, *Aust. J. Chem.*, 2006, **59**, 1.
72. (a) L. Liu, M.-X. Li and W.-Y. Wong, *Aust. J. Chem.*, 2005, **58**, 799; (b) S.-Y. Poon, W.-Y. Wong, K.-W. Cheah and J.-X. Shi, *Chem. Eur. J.*, 2006, **12**, 2550; (c) W.-Y. Wong, G.-L. Lu, L. Liu, J.-X. Shi and Z. Lin, *Eur. J. Inorg. Chem.*, 2004, 2066.
73. (a) W.-Y. Wong, K.-H. Choi, G.-L. Lu, J.-X. Shi, P.-Y. Lai and S.-M. Chan, *Organometallic*s, 2001, **20**, 5446; (b) L. Liu, Z.-X. Chen, S.-Z. Liu and W.-Y. Wong, *Acta Chim. Sin.*, 2006, **64**, 884; (c) W.-Y. Wong, G.-L. Lu, K.-H. Choi and J.-X. Shi, *Macromolecules*, 2002, **25**, 3506.

74. A. J. Hodge, S. L. Ingham, A. K. Kakkar, M. S. Khan, J. Lewis, N. J. Long, D. G. Parker and P. R. Raithby, *J. Organomet. Chem.*, 1995, **488**, 205.

75. M. Younus, N. J. Long, P. R. Raithby and J. Lewis, *J. Organomet. Chem.*, 1998, **570**, 55.

76. O. Lavastre, J. Plass, P. Bachmann and P. H. Dixneuf, *Organometallics*, 1997, **16**, 184.

77. D. W. Bruce and D. O'Hare, *Inorganic Materials*, 2nd edn, Wiley, Chichester, 1997, 122.

78. G. L. Geoffroy and M. S. Wrighton, *Organometallic Photochemistry*, Academic Press, New York, 1979.

79. W. Tam and J. C. Calabrese, *Chem. Phys. Lett.*, 1988, **144**, 79.

80. L. Plasseraud, L. Gonzalez Cuervo, D. Guillon, G. Suss-Fink, R. Deschenaux, D. W. Bruce and B. J. Donnio, *J. Mater. Chem.*, 2002, **12**, 2653.

81. N. Chawdhury, M. Younus, P. R. Raithby, J. Lewis and R. H. Friend, *Opt. Mater.*, 1998, **9**, 498.

82. M. P. Cifuentes and M. G. Humphrey, *J. Organomet. Chem.*, 2004, **689**, 3968.

83. G. W. Gray, K. J. Harrison and J. A. Nash, *Electron Lett.*, 1973, **9**, 130.

84. S. Takahashi, Y. Takai, H. Morimoto, K. Sonogashira and N. Hagihara, *Mol. Cryst. Liq. Cryst.*, 1982, **82**, 139.

85. T. Kaharu, H. Matsubara and S. Takahashi, *J. Mater. Chem.*, 1992, **2**, 43.

86. L. Plasseraud, L. Gonzalez Cuervo, D. Guillon, G. Suss-Fink, R. Deschenaux, D. W. Bruce and B. J Donnio, *J. Mater. Chem.*, 2002, **12**, 2653.

87. T. Kaharu, R. Ishii, T. Adachi, T. Yoshida and S. Takahashi, *J. Mater. Chem.*, 1995, **5**, 687.

Metal σ-Alkynyl Complexes as Molecular Wires and Devices: A Comparative Study of Electron Density and Delocalisation

MICHAEL S. INKPEN AND NICHOLAS J. LONG*

Department of Chemistry, Imperial College London, South Kensington, London SW7 2AZ, UK
*E-mail: n.long@imperial.ac.uk

4.1 Introduction

The transfer of electrons through molecular species is a subject of fundamental study, important in numerous subject areas including, but not limited to, chemical biology, molecular catalysis and materials science. In the last 20 years, charge transport through molecules of well-defined lengths and structure has been specifically considered in terms of 'molecular electronics', the development of single, or groups of, molecules that may serve as nanoscale analogues of common electronic components.[1] The electron-transfer properties of such species may ultimately be studied through their self-assembly (*via* suitable terminal surface-binding moieties) into 'molecular junctions' – assemblies in which the molecular component bridges donor/accepter electrodes.[2] Current flow across the junction can subsequently be measured as a function of applied bias voltage, and through careful experiment the junction conductance ($G = 1/R$), the mechanism of electron transfer through

RSC Polymer Chemistry Series No. 2
Molecular Design and Applications of Photofunctional Polymers and Materials
Edited by Wai-Yeung Wong and Alaa S Abd-El-Aziz

the molecular bridge (tunnelling, hopping, *etc.*), and related parameters can be ascertained. Whilst interesting in their own right, the results of such investigations may also guide work in related areas such as the development of organic light-emitting materials, solar cells, and conducting polymers.

Efforts towards improving the efficiency of molecular 'wires' (electron- or hole-conducting species demonstrating a more efficient route to electron transfer than through space) have resulted in the synthesis and study of a vast array of different structure types from alkanes,[3] alkenes[4] and alkynes[5] through to carbon nanotubes,[6] porphyrins[7] and DNA.[8] As might be expected from their delocalised electronic structure, unsaturated species are generally observed to be 'more conducting' than their saturated equivalents,[9] yet this is an oversimplification. In reality, numerous factors including i) the nature of the molecule–electrode contact,[3,10] ii) the energies of the molecular bridge HOMO (highest occupied molecular orbital) and LUMO (lowest unoccupied molecular orbital) relative to the Fermi energy (E_F) of the electrodes,[11] iii) the degree of localisation or delocalisation of the HOMO/LUMO over the entire molecular component,[12] iv) temperature,[11b,13] and v) the local molecular environment (single adsorbed molecule/monolayer,[14] vacuum/condensed matter,[15] *etc.*), require consideration. Whilst not straightforward, the identification of structure–property relationships in single-molecule electronics is of particular interest and importance.

The last decade has seen a particular surge of activity in exploring the incorporation of metals into components – enabling redox or photochemical addressability,[16] facile modification of electronic properties (through changing metal or ancillary ligand framework) and device functionality due to accessible (redox) molecular orbital energy levels.[11b,17] As shown in previous studies, the inclusion of redox-active moieties (and/or inherent electronic asymmetry) allows exploration beyond wire-like (symmetrical current–voltage) behaviour, where a wide variety of intriguing electronic phenomena such as conductance/stochastic switching,[14c,18] hysteresis loops,[19] negative differential resistance,[13a] current rectification,[20] Coulomb blockades[11b] and Kondo resonances[21] may be observed.

Of particular interest in this regard are alkynyl σ complexes, readily prepared using a wide variety of transition metals.[22] Typical acetylide ligands are π-conjugated, linear and rigid, facilitating the synthesis of complexes to well-defined molecular lengths as well as guaranteeing through-space (or through-bond) electron transfer between donor and acceptor (rather than by diffusion), making their electronic properties easier to study. Though only a small number have yet been studied in molecular junctions, they have previously seen significant investigation as nonlinear optical,[23] luminescent[24] and liquid-crystalline[25] materials. Crucially, a great many have been also examined as mixed-valence (MV) compounds, in which the extent of electron delocalisation between metal centres linked by a bridging ligand may be evaluated from near-infrared (near-IR), electrochemical and other studies. Such works are vital for current understandings of how molecular structure influences electron transfer (supplementing sparse molecular conductance data) – particularly following the direct relation of electron transfer rate and molecular conductance by theoretical treatments.[26]

In discussing the use of alkynyl σ complexes as molecular electronic components this review initially considers their synthesis, highlighting approaches of particular interest. Relevant molecular conductance studies are subsequently surveyed, with the role of the metal centre critically examined. Exploring this topic further, investigations into relevant mixed-valence (MV) complexes are addressed. In what we believe is the first attempt to do so, the link between electron density (at the metal centre) and the extent of electronic delocalisation in such systems – often qualitatively proposed[27] – is here quantified by correlation of easily measurable experimental parameters. Our initial study is readily expandable and may prove particularly useful in predicting and guiding future work in this area.

4.2 Synthetic Considerations

4.2.1 Metal Alkynyl σ-Bond Formation

Reviewed by our group in 2003,[22] M–C≡C–R σ-bond forming routes include: i) dehydrohalogenation (often Cu^I catalysed);[28] ii) metathesis using trimethyl-

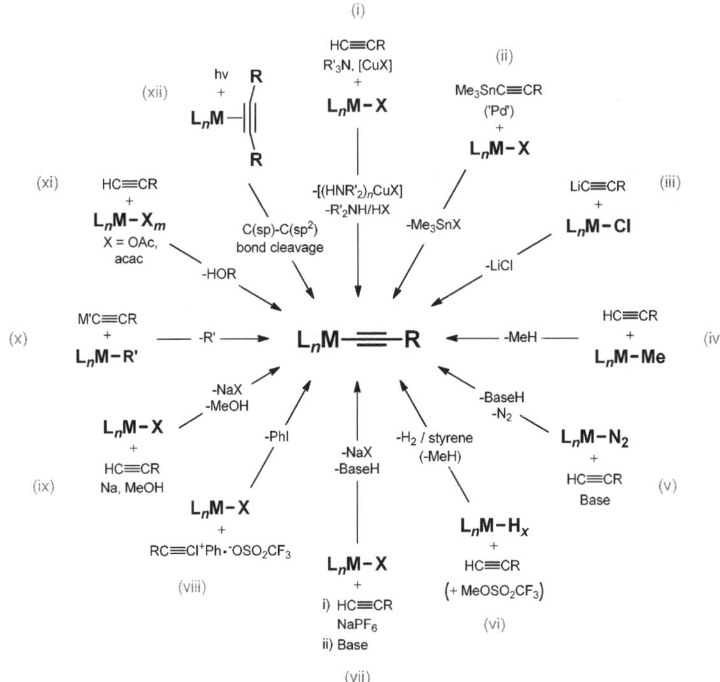

Scheme 4.1   A survey of M–C≡C–R σ-bond-forming routes. (Roman numerals refer to descriptions in the bulk text.)

stannyl reagents (sometimes Pd^0 catalysed);[29] iii) displacement using alkali-/ alkali-earth-metal alkynyl complexes;[30] iv) methane,[31] v) N_2,[32] or vi) H_2 elimination;[33] vii) vinylidene formation;[34] viii) use of $RC{\equiv}C^+$ synthons (from alkynyliodinium triflates);[35] ix) the sodium methoxide method;[36] x) alkynyl ligand exchange;[37] xi) the OAc^-/acac method,[38] and xii) C–C bond activation;[39] amongst others (Scheme 4.1). In preparing complexes, the method of synthesis is largely determined by the metal, or selected to achieve a particular degree of substitution. Not discussed in detail here, the synthesis of main-group acetylides has recently been reviewed by Gleiter and Werz.[40]

4.2.2 Complexes for Molecular Electronics

Initial studies into the molecular conductance of alkynyl σ complexes have focused on linear structures such as those shown in Figure 4.1, considering Ru and Pt centres only (see Section 4.3). In preparing analogous compounds with different metals, opportunities to expand this series are limited only by a few key considerations. A route to the *trans*-bis-alkynyl species must be known for the metal of interest (so that the complex may be extended linearly in two directions), with surface-binding moieties readily incorporated to bind the molecule to electrodes and facilitate electron transfer. Ideally, resulting complexes should be thermally, air and moisture stable for ease of subsequent study and application.

Terminal binding groups thus far utilised for metal-free wires include thiol,[3–4,10,41] selenium,[42] isocyanide,[10,43] amine,[3,44] pyridine,[5,45] carboxylic acid[3] and diazonium salts (forming direct Si–C bonds),[19] though this list is not exhaustive and more elaborate bi- and tri-coordinate anchors such as cyclopentadithiophene[46] and tripodal pyridyls[47] are also now being considered. Some of these have already been incorporated into metal-containing systems, with thiol-containing Ru^{II} and Pt^{II} complexes,[41c,48] and pyridyl-containing Ru^{II} and Hg^{II} materials known.[49] Their inclusion may not always be straightforward however, as indicated by recent syntheses of di- and tri-metallic {Ru(dppe)$_2$} compounds by Olivier *et al.*, where protected isocyanide

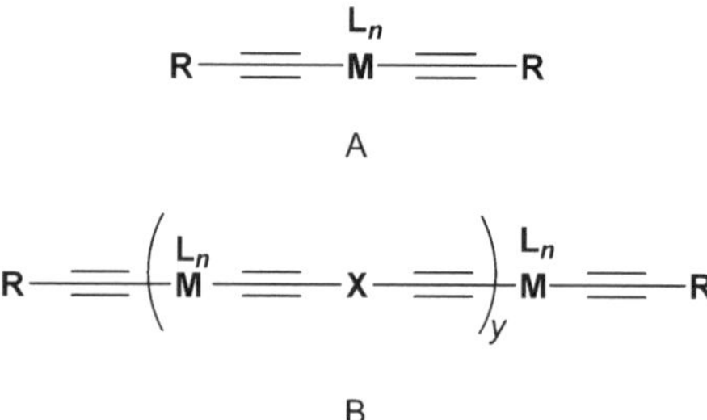

Figure 4.1 Typical structure types for molecular electronics applications (X = bridging ligand, R = suitable surface binding moiety).

termini were introduced as formamide groups to avoid complications during M–C σ-bond formation.[50]

For the majority of transition metals, their *trans*-bis-acetylides have precedent. Examples are provided in Table 4.1, with notable absences discussed in the text.

Group 4 analogues exist only in tetrahedral geometries such as the $Ti(Cp)_2(C\equiv CPh)_2$ complex[63] and analogous Zr/Hf derivatives. In the vast majority of cases the alkyne ligands are bridged by π-bonded transition metals forming so-called "tweezer" systems.[64] Octahedral geometries such as the recently reported *cis*-alkynyl $Zr(Bn_2Cyclam)(C\equiv CPh)_2$ ($Bn_2Cyclam$ = *trans*-N,N'{PhCH_2}_2Cyclam),[65] or the hexa-alkynyl $M\{Li(^tBu_3SiC\equiv C)_3\}_2$ (M = Zr, Hf),[52] are rare.

Group 11 acetylides are typically employed as alkynylating agents,[61] such as within Hay/Sonogashira coupling reactions, isolated examples being uncom-

Table 4.1   Selected examples of known *trans*-bis-alkynyl species of the d block transition metals

group	*trans*-metal complex	*oxid^n state*	d^n	*synthesis*	*stability^a*	*ref*
5	$V(C\equiv C^tBu)_2(TMEDA)_2$[b]	II	d^3	displacement	air/H_2O sensitive	51
	$Ta(C\equiv CSi^tBu_3)_3\text{-}\{Li(^tBu_3SiC\equiv C)_3\}$	V	d^0	displacement	light sensitive	52
6	$Cr(dmpe)2(C\equiv CPh)2$[c]	II	d^4	displacement	/	30b
	$Mo(dppe)_2(C\equiv CPh)_2$[d]	II	d^4	$-N_2$	/	53
	$W(dppe)_2(C\equiv CPh)_2$	II	d^4	$-N_2$ / H_2	/	53
7	$Mn(depe)_2(C\equiv CPh)_2$[e]	II	d^5	displacement	$[M]^+$ air stable	54
8	$Fe(dmpe)_2(C\equiv CPh)_2$	II	d^6	$-H_2$/styrene	/	55
	$Ru(dppe)_2(C\equiv CC_7H_4N)_2$	II	d^6	vinylidene	/	50
	$Os(dppm)_2(C\equiv CPh)_2$[f]	II	d^6	metathesis	/	56
9	$Co(N_4)(C\equiv CPh)_2$[g]	III	d^6	metathesis	air stable	57
	$Rh(PMe_3)_3(SnMe_3)\text{-}(C\equiv CPh)_2$	III	d^6	metathesis	/	29b
	$Ir(C\equiv CH)_2(CH=CH_2)\text{-}(CO)(P^iPr_3)$	III	d^6	$-H_2$	air stable	58
10	$Ni(PPh_3)_2(C\equiv CPh)_2$	II	d^8	displacement	/	59
	$Pd(PPh_3)_2(C\equiv CPh)_2$	II	d^8	dehydrohalogenation	/	60
	$Pt(PPh_3)_2(C\equiv CC_6H_4SAc)_2$	II	d^8	dehydrohalogenation	/	48c
11	$[Ag(C\equiv CPh)_2]^-$	I	d^{10}	ligand substitution	/	61
	$[Au(C\equiv CPh)_2]^-$	I	d^{10}	ethynylation by $[Ag(C\equiv CR)_2]^-$	/	61
12	$Hg(C\equiv CC_5H_4N)_2$	II	d^{10}	displacement	poss. explosive[62]	49d

^{a}Where explicitly stated by the authors. ^{b}TMEDA = N,N,N',N'-tetramethyl-ethane-1,2-diamine. ^{c}dmpe = 1,2-bis(dimethylphosphino)ethane. ^{d}dppe = 1,2-bis(diphenylphosphino)ethane. ^{e}depe = 1,2-bis(diethylphosphino)ethane. ^{f}dppm = 1,2-bis(diphenylphosphino)methane. ^{g}N$_4$ = 3,9-dimethyl-4,8-diazaundecane-2,10-dione dioxime.

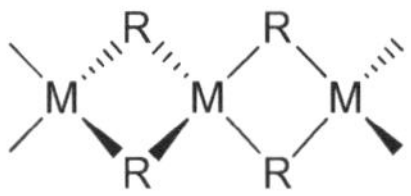

Figure 4.2 Associated polymeric structure suggested for M(C≡CR)$_2$ (M = Zn, Cd) complexes (R = C$_2$Ph).[68c]

mon. Their bonding with the C≡C unit is varied, with η^2-bridged species[66] and trinuclear *trans*-alkynyl triangles[67] often observed. Zn and Cd *trans*-acetylides (group 12) are known but their application suffers somewhat from their relative instability and insolubility.[68] For example, Zn(C≡CPh)$_2$ reportedly forms diphenylbutadiyne in the presence of CuBr,[68b] and it has been suggested that this complex and the Cd analogue have an associated polymeric structure in the solid state (Figure 4.2).[68c]

A handful of polynuclear complexes analogous to Figure 4.1-B have been prepared, including thiol-terminated {Ru$_2$(ap)$_4$} (ap = 2-anilinopyridinate),[17b,69] amine-terminated {Ru(dppe)$_2$}/{Fe(dppe)$_2$Cp*},[70] and the aforementioned isocyanide-terminated {Ru(dppe)$_2$} materials.[50] Di-nuclear species without surface-binding moieties are relatively abundant, extensively studied as mixed valence species (see section 4). Related polynuclear compounds without surface-binding groups include RuRuRu, RuPdRu, FePdFe and FeWWFe systems.[27d,71]

Whereas symmetrical bis-acetylides of the type Figure 4.1-A may be prepared through any aforementioned robust method of M–C bond formation (*e.g.* dehydrohalogenation, metathesis or displacement), complexes of the type Figure 4.1-B often require the synthesis of asymmetric σ-alkynyl precursors. Whilst this may be achieved *via* stoichiometric control of reagents, lengthy purification is often required in isolating desired materials from the mixtures that readily form. Relatively few alternative methods are yet known for the controlled, stepwise production of unsymmetrical bis–alkynyl species, and as such they will now be discussed. These become ever more valuable when considering asymmetrical variants of the structures in Figure 4.1 (*e.g.* complexes with different terminal binding moieties or substituted phenyl rings, or type-B heterometallic analogues), useful in the pursuit of molecular electronic devices.

4.2.3   Stepwise Bis-Acetylide Syntheses

4.2.3.1   *Vinylidene Formation*

Pioneered by Dixneuf and coworkers in the 1990s,[34b,c] unsymmetrical bis-acetylides of *cis*-RuCl$_2$(dppe)$_2$ are readily prepared *via* stepwise chloride substitution using a suitable halide abstracting agent (NaPF$_6$ or AgOTf) and base (1,8-diazabicycloundec-7-ene, Et$_3$N or KO^{t}Bu). This approach is well established and highly convenient, utilising readily synthesised chloride

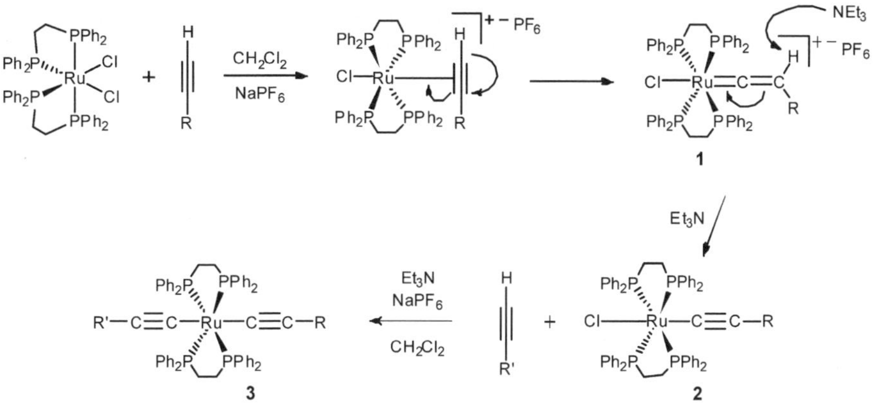

Scheme 4.2 Suggested mechanism of unsymmetrical bis-acetylide formation *via* the approach of Dixneuf *et al.*[34c,50,72–73]

complex precursors (rather than hydride, dinitrogen, or methylated variants) and terminal alkynes (rather than their lithiated or stannylated equivalents). In addition to their convenient preparation, *trans*-acetylides of {Ru(dppe)$_2$} typically have good air and solution stability, with electronic properties highly suited for applications within molecular electronics (see Section 4.3).

In the absence of a base the reaction likely proceeds *via* initial η^2-coordination of alkyne to a 5-coordinate Ru centre (generated by abstraction of chloride by NaPF$_6$ or similar salt, and loss of NaCl) (Scheme 4.2). The reactive species then undergoes a 1,2-hydrogen shift[72] to form a vinylidene intermediate (**1**) that can subsequently be isolated. This can be deprotonated, affording the mono-alkynyl species (**2**), and/or reacted on with additional alkyne in the presence of a base (typically NEt$_3$) to form the desired *trans*-bis-alkynyl complex (**3**).

Recently, a useful synthesis of reactive ruthenium starting materials *cis*-RuCl$_2$(dppe)$_2$ and [RuCl(dppe)$_2$]OTf was presented by Fox *et al.*[73] Their approach bypasses problems preparing *cis*-RuCl$_2$(dppe)$_2$ from RuCl$_2$(dmso)$_4$ *via* the often-cited procedure for preparation of *cis*-RuCl$_2$(dppm)$_2$.[74] Ligand exchange of dmso typically yields a difficult to separate mixture of *cis*- and *trans*-complexes when dppe is used, *trans*-RuCl$_2$(dppe)$_2$ being the thermodynamically stable isomer.

Mono-alkynyl complexes of the type *trans*-[MCl(dppm)$_2$(C≡CR)] (M = Ru, Os) may also be prepared *via* vinylidene formation,[75] though their bis-acetylides are not readily formed. Unsymmetrical bis-acetylides have, however, been made from subsequent reaction of these mono-alkynyl complexes with stannylated alkynes.[76]

4.2.3.2 Elimination of Methane

Routes to unsymmetrical Fe and Ru bis-acetylides have been reported by Field *et al,*[71f,77] using depe and dmpe ancillary ligands. In initial work, it was found that symmetrical and unsymmetrical Ru bis-acetylides could be formed as mixtures from reaction of two different terminal acetylenes with $RuCl_2(depe)_2$ in MeO^-/MeOH (separated by fractional recrystallisation/chromatography).[77b] Recently, a vastly improved approach exploits the discovery that at room temperature in benzene, *trans*-$Ru(CH_3)_2(dmpe)_2$ (**4**) reacts with phenylacetylene to give only the mono-acetylide, *trans*-$Ru(C\equiv CPh)(CH_3)(dmpe)_2$ (**5**) (Scheme 4.3).[77a] The remaining methyl group could be substituted by rapid reaction of **5** and $HC\equiv CR$ in methanol/benzene, yielding complexes of the type *trans*-$[Ru(C\equiv CPh)(C\equiv CR)(dmpe)_2]$ (**7**). Use of methanol in the solvent mixture appears crucial to displacement of the second methyl ligand – it was speculated by the authors that the relative acidity of the alcohol promoted dissociation of methane, leaving a methoxide ligand in its place (**6**) that is more readily displaced by a terminal acetylide. This approach has been subsequently applied to the synthesis of more elaborate bi- and tri-nuclear systems.[71e]

Photochemical metathesis of $[Fe(dmpe)_2(C\equiv CR)(CH_3)]$ (**8**) and $HC\equiv CR'$ provides entry to unsymmetrical bis-acetylides containing $\{Fe(dmpe)_2\}$ centres (Scheme 4.4).[71f] Here, mono-acetylides were formed by reaction of $[Fe(dmpe)_2(CH_3)(Cl)]$ and magnesium bis-acetylides, with unsymmetrical complexes of the type *trans*-$[Fe(dmpe)_2(CCR)(CCR')]$ (**11**), plus some di- and tri-nuclear analogues, produced from reaction of **8** with terminal acetylides in benzene/THF. Interestingly, no reaction was observed unless the solution was irradiated with UV light. It was suggested that this generates a vacant coordination site at the metal centre *via* dissociation of one arm of the diphosphine ligand (**9**), enabling η^2-coordination of the alkyne (**10**), subsequent loss of methane, and re-coordination of the previously displaced -PMe_2 group.

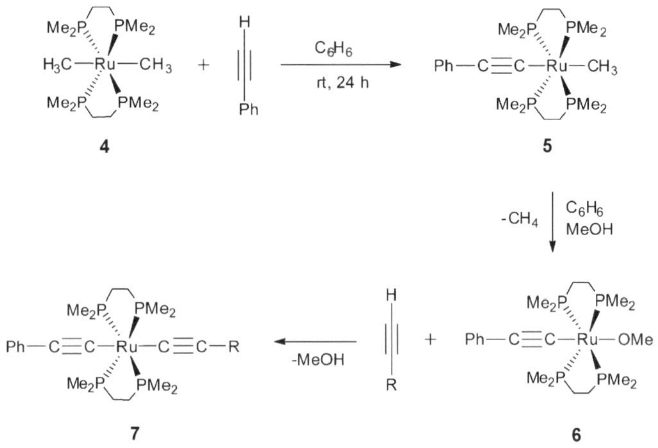

Scheme 4.3 Proposed mechanism of stepwise condensation reaction between terminal acetylenes and methyl ruthenium complexes.[77a]

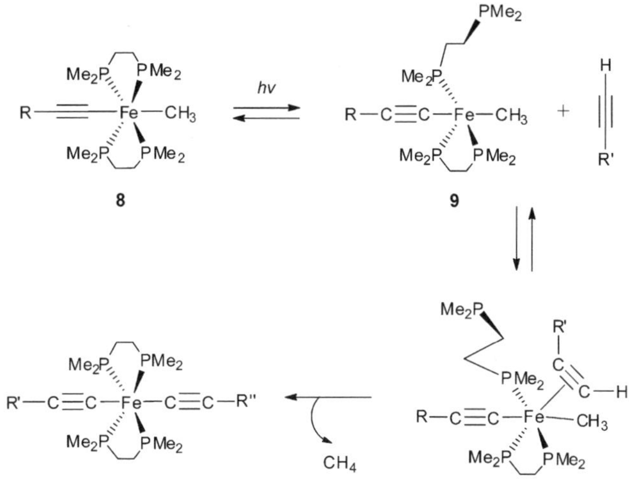

Scheme 4.4 Photochemical metathesis of [Fe(dmpe)$_2$(C≡CR)(CH$_3$)] centres.[71f]

4.2.3.3 Reaction of fac–W(CO)$_3$(dppe)(THF)

The stepwise substitution of W^{IV} centres with all carbon ligands has been demonstrated recently (Scheme 4.5).[71a,78] Starting with *fac*–W(CO)$_3$(dppe)(THF) (**12**), the dinuclear complex Li$_2$[(CO)$_3$(dppe)$_2$WC$_4$W(dppe)$_2$(CO)$_3$] (**13**) was prepared through reaction with approximately 0.5 equivalents of Li–C≡C–C≡C–Li. Whilst it is unclear whether the lithiated acetylide preferentially or exclusively displaced the coordinated THF in **12**, the authors reported the product easily isolated from the reaction solution. Subsequent oxidation of **13** with I$_2$ resulted in the air-stable [I(CO)$_2$(dppe)$_2$WC$_4$W(dppe)$_2$(CO)$_2$I] (**14**) in excellent yield (91%). A binuclear complex containing metal centres analogous to [W(CO)$_3$(dppe)$_2$(C≡CPh)(I)] was suggested as a likely intermediate in the conversion of **13** to **14**, this mononuclear complex being formed in a model reaction between I$_2$ and Li[W(CO)$_3$(dppe)$_2$(C≡CPh)].

Prompting further inquiry into substitution of the iodide to acetylides, the axial iodide ligands of **14** appeared relatively labile in initial synthetic studies, being readily exchanged for isothiocyanate (52% yield with AgNCS) or triflate (83% yield with AgOTf).[78] Suitable conditions were eventually found, where **15a** was reacted with a catalytic amount of CuI and a threefold excess of thallium triflate and LiC$_4$SiMe$_3$, to form [(Me$_3$SiC$_4$)(dppe)$_2$W(C$_4$)W(dppe)$_2$(C$_4$SiMe$_3$)] (**15b**) in 52% yield.[71a] Desilylation provided the terminal alkyne (**15c**) and subsequent stannylation resulted in **15d** (a stable, but reactive, intermediate). This was followed by reaction with FeCl$_2$(depe)$_2$ to provide the tetranuclear {[Cl(depe)$_2$Fe]C$_4$[(dppe)$_2$W]C$_4$[W(dppe)$_2$]C$_4$[Fe(depe)$_2$Cl]} (**15e**). Work towards incorporation of terminal surface-binding moieties (into **15c/e**) for molecular conductance studies are reportedly in progress.

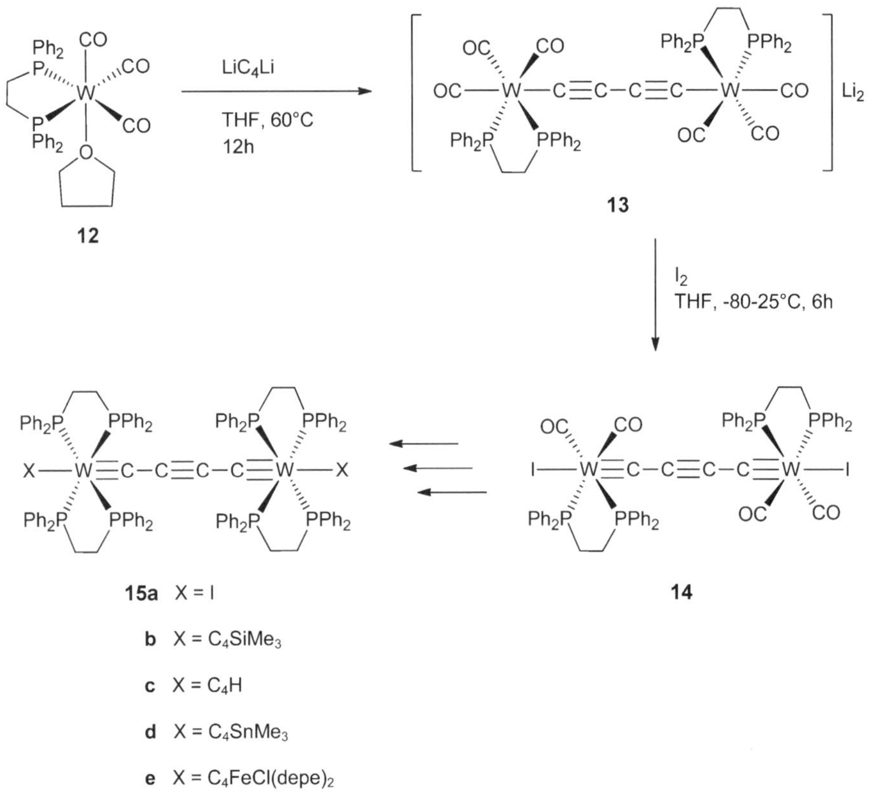

Scheme 4.5   Route to asymmetrical W^{IV} complexes.[71a,78]

4.2.3.4   *Selective Desilylation of {Mn(dmpe)$_2$} Bis-acetylides*

Unsymmetrical Mn^{II} and Mn^{III} σ-alkynyl complexes may be prepared from $[Mn(dmpe)_2(C\equiv CR)_2]$ (**16**) (R = $SiEt_3$, Si^iPr_3, $Si(^tBu)Me_2$) (Scheme 4.6).[79] Reduction of **16** with sodium in toluene gives the corresponding *cis*-bis-acetylide (likely through sodium chelation of the alkynyl ligands), which upon addition of 1 eq. MeOH or the more forcing conditions of excess MeOH/KOH (or THF/TBAF) affords the analogous *trans*-$[Mn(dmpe)_2(C\equiv CR)(C=CHR)]$ (**17**) or *trans*-$[Mn(dmpe)_2(C\equiv CR)(C=CH_2)]$ (**18**) complexes, respectively. Reaction of **18** with $[Cp_2Fe][PF_6]$ and quinuclidine yields the unsymmetrical $[Mn(dmpe)_2(C\equiv CR)(C\equiv CH)][PF_6]$ (**19**). This can subsequently be reacted on through the terminal alkyne and reduced with $[Co(Cp^*)_2]$ to generate neutral Mn^{II} species, a technique recently employed to prepare binuclear C_4 bridged Mn polyyne complexes.[79b]

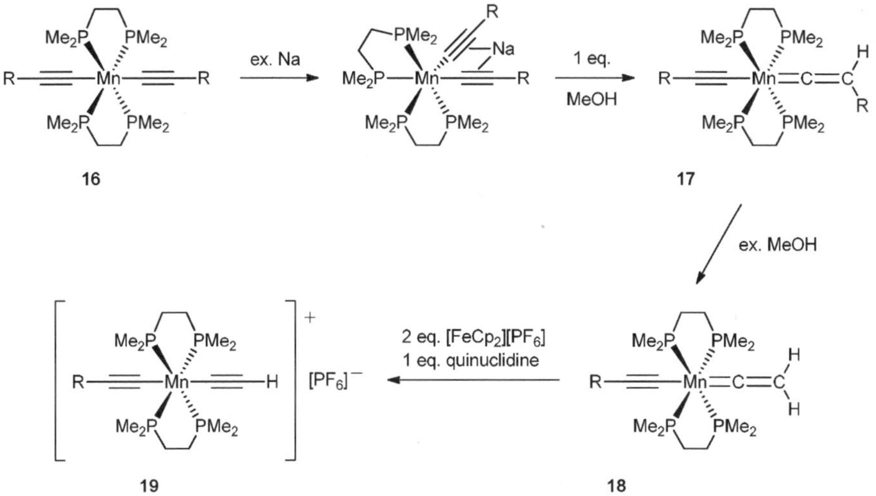

Scheme 4.6 Route to unsymmetrical Mn^{III} bis-acetylide complexes.[79]

4.2.3.5 Steric Effects at {Ru₂(L–L)₄} Centres

There has been a renewed interest in the synthesis and properties of 'paddlewheel' complexes, *i.e.* $M_2(L–L)_4$, in recent years.[80] Their mono- and (unsymmetrical[81]) bis-acetylides may be formed *via* displacement, with selectivity largely dependent on the nature of L–L.

The complexes $Ru_2(ap)_4(C{\equiv}CPh)_n$ (**21a**: $n = 1$, $Ru^{II}Ru^{III}$;[82] **21b**: $n = 2$, $Ru^{III}Ru^{III}$ [83]) were prepared from $Ru_2(ap)_4Cl$ (**20**) as shown in Scheme 4.7. In these compounds, the ap ligands are aligned with all the pyridines bound at one metal centre (Ru_A) and all the anilino groups coordinated to the other (Ru_B). This renders the two metal centres inequivalent and reaction of **20** with 5 equivalents of $LiC{\equiv}CR$ (R = Ph, C_2SiMe_3, *etc.*) generates only the mono-alkyne (**21a**) – chloride substitution and M–C σ-bond formation occurring exclusively at Ru_A. More forcing conditions (10–20 equivalents of $LiC{\equiv}CR$) are required to add the second acetylide (at Ru_B), producing **21b**. Steric bulk of the anilino phenyl groups at Ru_B was cited to explain the observed difference in reactivity at the Ru centres, and in support of this hypothesis, it was found that ligands with larger bulk than C_2Ph (*i.e.* $C_2SiMe_3/C_2Si^iPr_3$) could not be bound at Ru_B, whereas extended C_n analogues such as C_4SiMe_3 were readily incorporated. Furthermore, if the paddlewheel framework was changed from ap to the more symmetric DArF (diarylformamdinate) or DABA (N,N'-dialkylbenzamidinate) ligands, isolation of mono-acetylides and entry to unsymmetrical complexes became slightly[84] and significantly[85] more difficult, respectively.

In addition to their convenient preparation, complexes **21a–b** are reported to have very high air and moisture stabilities, making the {Ru₂(ap)₄} unit highly suitable as a building block for molecular electronic materials.

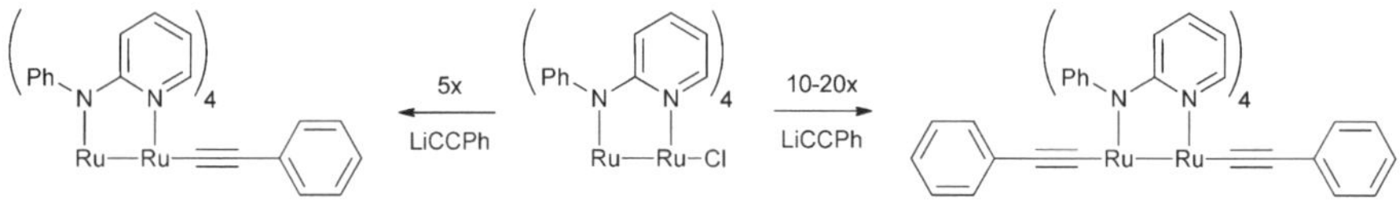

Scheme 4.7   Synthesis of mono- and bis-acetylides with a {Ru(ap)$_4$} centre.[82–83]

Analogous complexes known as extended metal atom chains (EMACs) can be constructed using a multitude of other metals, as exemplified by the series of trinuclear complexes M$_3$(dpa)$_4$Cl$_2$ (where dpa = 2,2′–dipyridylamine; M = Cr,[86] Co,[87] Ni,[88] Cu,[89] Ru,[90] Rh,[90a] and Pd;[91]). Limits to the extension of M$_n$ is unclear in light of recent reports of the undecanickel [Ni$_{11}$(tentra)$_4$X$_2$](PF$_6$)$_4$ [tentra = the deprotonated trianion of N^2-(2-(1,8-naphthyridin-7-ylamino)-1,8-naphthyridin-7-yl)-N^7-(1,8-naphthy-ridin-2-yl)-1,8-naphthyridine-2,7-diamine; X = Cl, NCS;],[92] and decanickel [Ni$_{10}$(μ_{10}-bdpdany)$_4$-(NCS)$_2$](PF$_6$)$_2$ [μ_{10}-bdpdany = the deprotonated tetraanion of 2,7-Bis(dipyridyldiamino)-1,8-naphthyridine] complexes.[93] Though methods have not yet been published to synthesise alkynyl σ complexes of each EMAC mentioned here, preparations of M$_3$(dpa)$_4$(C≡CPh)$_2$ (M = Ni, Co, Cr) and Co$_3$(dpa)$_4$(C≡CFc)$_2$ by Berry *et al.* in 2004 are extremely encouraging.[94]

4.3   Molecular Conductance Measurements

Despite their straightforward synthesis with many metals, conductance measurements of σ-alkynyl complexes have so far been limited to materials containing {M(PR$_3$)} (M = Pt, Pd; R = Ph, Cy, OEt;),[27d,48c,d] {Ru$_2$(ap)$_4$},[17b,69] {Ru(dppe)$_2$},[11b] and {Ru(dppm)$_2$}[12b] centres (Figure 4.3). These reports will now be examined, also using electrochemical and electron paramagnetic resonance (EPR) data where relevant. Studies into the charge-transport properties of related metal-containing systems including EMACs,[18b,95] terpyridyl complexes,[15a,96] and ferrocene-containing structures[97] are of additional interest but beyond the scope of this review.

For the purpose of the following discussions, it is assumed that the electrochemically determined value $\Delta E_{1/2}$ (proportional to the comproportionation constant, K_c) provides an experimental measure of electronic delocalisation between two redox centres. The larger this is, by analogy, the greater the expected molecular conductance through the system. It should be noted, however, that the use of $\Delta E_{1/2}$ in such a way is potentially erroneous, the validity of such assumptions (and the full definition of these terms) is discussed in Section 4.4.1.3.

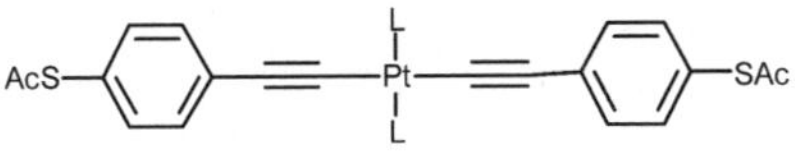

22a L = PCy$_3$

b L = PPh$_3$

c L = P(OEt)$_3$

d L = P(OPh)$_3$

e L = PBu$_3$

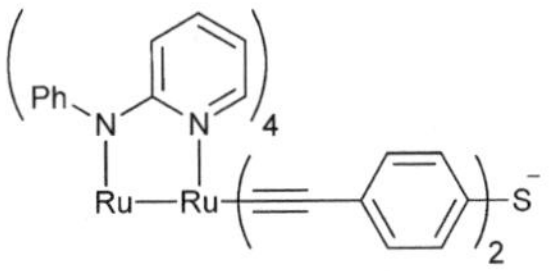

23

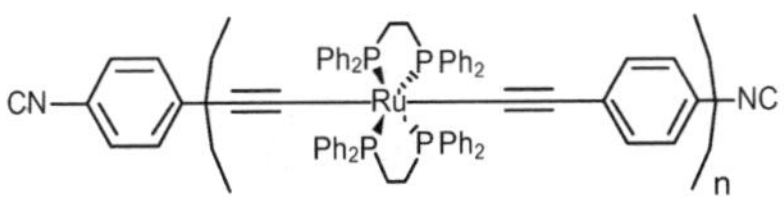

24a n=1

b n=2

c n=3

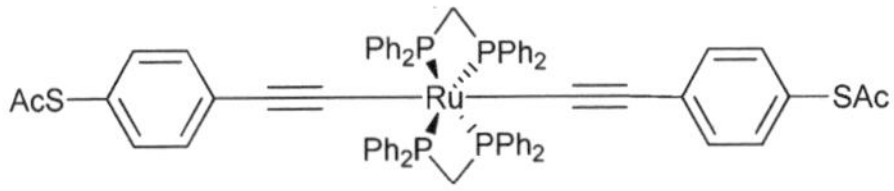

25

26

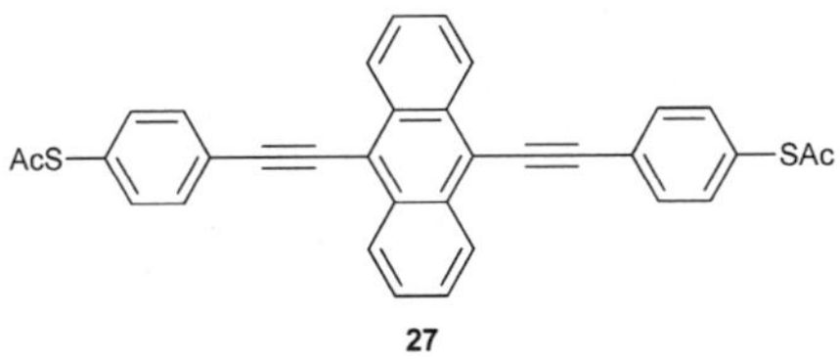

27

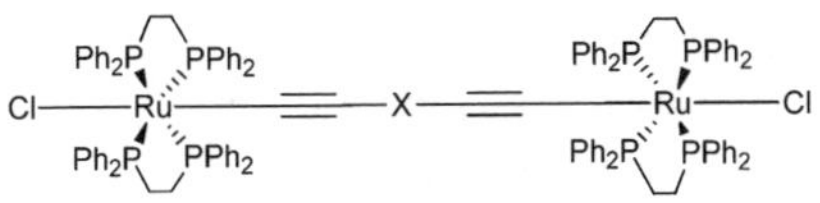

28a X = –C$_6$H$_4$–

b X = –C$_6$H$_4$–C≡C–C$_6$H$_4$–

c X = –C$_6$H$_4$–C≡C–C$_6$H$_4$–C≡C–C$_6$H$_4$–

d X = –C$_6$H$_4$–C≡C–[Pd(PBu$_3$)$_2$]–C≡C–C$_6$H$_4$–

Figure 4.3 Examples of metal σ-alkynyl complexes and organic analogues thus far investigated for their electronic properties.[11b,12b,17b,27d,48c,d,69]

4.3.1 Pt-Containing Systems

Several conflicting reports on the electronic properties of *trans*-platinum acetylides (Figure 4.3) have been published over the last decade. An early study of Pt-tetraethynylethene complexes revealed little π-electron delocalisation along the oligomeric backbone, the authors subsequently declaring Pt atoms 'true insulating centres'.[98] However, the sequential {Pt(PEt)$_3$} units were here separated by a significant distance, and no alternative metal centres were integrated into similar structures for comparison. Very little electronic communication ($K_c = 7$) is observed between ruthenium centres separated by two aryl diethynyl units at similar distances, for example.[27d] Furthermore, EPR spectroscopy of a diradical system bridged by a Pt(PPh$_3$)$_2$ unit found that electronic communication through {Pt(PPh$_3$)$_2$} was indeed possible.[99]

Molecular conductance measurements offered no absolute clarification on this issue, with somewhat contradictory results published by different groups in 2002. Mayor *et al.* studied **22b** in a mechanically controlled break junction showing it to be 2–3 times more insulating than a 9,10–anthracene bridged organic analogue (**27**).[48c] However, Schull *et al.* found that complexes **22a-c** were 2-3 times more conductive than the phenyl bridged dithiol **26** in crosswire junction measurements.[48d] Whereas these results appear incompatible in the first instance, it should be noted that 9,10-anthracene systems may be *expected* to exhibit a higher conductance than phenyl versions – $\Delta E_{1/2} = 0.26$ V for [Fe(dppe)Cp*]$_2$(μ–C≡C–C$_6$H$_4$–C≡C),[100] *vs.* 0.36 V for [Fe(dppe)Cp*]$_2$(μ–C≡C-9,10-ant-C≡C).[101] The conductance of **22a-c** might therefore simply be intermediate between **26** and **27**.

Fuelling the debate, it is suggested from molecular orbital arguments that Pt complexes should be more insulating than other metals due to a lack of π-character in Pt–C σ-alkynyl bonds.[48c,102] Interesting in this regard (and to the discussions in Section 4.4), from studies of complexes **22a–e** it was noted that changing ancillary ligand σ-donor strength showed no correlation with Pt–C bond length, HOMO–LUMO gap, or ultimately molecular conductance (within experimental error). However, a clear linear relationship was observed with molar absorptivity, carbon–carbon triple bond lengths and IR stretching frequencies.

Other investigations have considered how a platinum bridge may affect the degree of electronic communication between two redox centres as measured by cyclic voltammetry. A report in 2004 using redox active triarylamine groups found a small decrease in electron delocalisation when a platinum bridge was used, compared to a phenyl-containing analogue.[103] This result was echoed by Vives *et al.* in studies of acetylenic systems linking ferrocene moieties, their experiments showing that substitution of one acetylene for a {Pt(PR$_3$)$_2$} unit again limited the electronic communication (by a factor of $\sim$1.5).[102] Comparisons to Ru-containing bridges also suggest reduced communication (Table 4.2).

Conversely, but in line with the findings of Schull *et al.*, an electrochemical study (this time employing {Ru(dppe)$_2$} units as redox centres) indicated group

Table 4.2 Electrochemical data[a] for selected complexes of the type [{*trans*-M(L)$_n$}(C≡CFc)$_2$][b].

#	$M(L)_n$	CV data				
		$E_{1/2}^{I}$	$E_{1/2}^{II}$	$\Delta E_{1/2}$ (V)	*conditions*[c]	*ref*
1	Ru(dppm)$_2$	-0.39	-0.17	0.22	Bu$_4$N$^+$PF$_6^-$	104
		-0.40	-0.18	0.22	/	105
2	Ru(PBu$_3$)$_2$(CO)$_2$	-0.17^d	-0.08^d	0.09	Bu$_4$N$^+$PF$_6^-$	106
3	Ru(PBu$_3$)$_2$(CO)(C$_5$H$_5$N)	-0.24^d	-0.11^d	0.13	Bu$_4$N$^+$PF$_6^-$	106
4	Ru(PBu$_3$)$_2$(CO)(P(OMe)$_3$)	-0.26^d	-0.11^d	0.15	Bu$_4$N$^+$PF$_6^-$	106
5	Ru(dppe)$_2$	-0.36	-0.16	0.21	Bu$_4$N$^+$PF$_6^-$	107
6	Pt(L)$_2^e$	/	/	$\sim 0.8^f$	Bu$_4$N$^+$PF$_6^-$/THF	108
7	Pt(PPh$_2$C≡CFc)$_2$	-0.15	-0.06	0.9	Bu$_4$N$^+$PF$_6^-$	109
8	Pt(PEt$_3$)$_2$	-0.14	-0.06	0.8	/	102

[a]All redox potentials relative to [Cp$_2$Fe]$^+$/Cp$_2$Fe. [b]Fc = FeCp$_2$. [c]Bu$_4$N$^+$BF$_4^-$ in CH$_2$Cl$_2$ with Pt working electrode unless otherwise stated. [d]Assuming [Cp$_2$Fe]$^+$/Cp$_2$Fe = 0.43 V *vs.* SCE. [e]L = PPh$_3$, P(*p*-tol)$_3$, PMePh$_2$, PBu$_3$. [f]*Via* derivative neopolarography.[108]

10 centres might be superior to phenyl building blocks.[27d] Upon increasing the number of aryl units in a [–C≡C–C$_6$H$_4$–]$_n$ bridge, where $n=1$ (**28a**), 2 (**28b**) and 3 (**28c**), K_c values were calculated as 6 × 10^5, 7 and 0, respectively. When the central aryl unit in the $n=3$ bridged complex was replaced by {Pd(PBu$_3$)$_2$} (**28d**), K_c = 150 was obtained, suggesting a much greater electronic communication between the Ru centres than could be explained by changes in the molecular length alone.

4.3.2 Ru-Containing Systems

An initial study into the performance of Ru-containing molecular wires came in 2005 where Blum *et al.* measured the conductance of individual or small molecular bundles of Ru$_2$(ap)$_4$[σ(C≡CC$_6$H$_4$)$_2$S$^-$] (**23**) inserted into a self-assembled monolayer of undecanethiol *via* scanning tunnelling microscopy (STM).[69] Whilst only a moderately low value of β = 0.59−0.88 ($\pm$0.08−0.12) Å^{-1} was obtained (where β is the tunnelling decay constant, indicative of the efficiency of electron transport along a molecule), in their setup this was 15–45% smaller than that obtained for a purely organic analogue (**26**), suggesting **23** was a superior mediator of electron transport. A subsequent study in 2008 considered *trans*-Ru$_2$(ap)$_4$[σ(C≡CC$_6$H$_4$)$_n$S$^-$]$_2$ (n = 1, 2) complexes assembled in nanogap molecular junctions.[17b] Interestingly, conductance peaks were observed at voltages comparable to the energies of redox events as measured by solution voltammetry. The increased conductance was attributed to alignment of molecular HOMO and LUMO energy levels with the metal electrode E_F at those voltages (resonant tunnelling).

Kim and coworkers studied a series of conjugated mono-, bi- and trinuclear ruthenium complexes bound to Au electrodes *via* terminal isocyanide (–NC) groups (**24a–c**).[50] Studies of the electron-transfer properties of these structures were made (as self-assembled monolayers) at room temperature and 5 K using conducting probe atomic force microscopy (CP–AFM) and crossed wire junction techniques.[11b] These exhibited a value of $\beta = 0.09$ Å^{-1}, encouragingly low compared to that of typical conjugated organic wires ($\beta = 0.06$–0.63 Å^{-1}, Table 4.3). Whilst this is significantly smaller than the β value obtained for **23**, the former was obtained from studies of monolayers and the latter from single/small groups of molecules, hindering their comparison.

Regardless, this result suggests there is a small energy gap (ΔE_{DB}) between the metal electrode E_F and the energy of the molecular frontier orbitals (E_{HOMO} or E_{LUMO}) of **24a–c**, important for rapid electron transfer over long distances. Interestingly though, an exceptionally large contact resistance ($R_C = 4.2 \times 10^8$ Ω) was calculated (the term 'contact resistance' referring to effects localised at the molecule/electrode interface that affect the conductance of the entire junction). Such a value is not readily explainable (as the proposed small ΔE_{DB} would imply a small R_C), and contributes significantly to the total resistance of the molecular junction. Furthermore, a Coulomb blockade-like behaviour was observed in low-temperature measurements of **24c**. This was attributed to the large R_C measured, which would favour charge localisation on the molecule. *I–V* traces exhibited by **24a** and **b** are practically temperature independent, indicating that in these systems tunnelling plays the dominant role in electron transport. The effect of surface binding moieties (in this case – NC) on contact resistance has begun to be studied systematically, and an inverse correlation to metal–molecule binding strength noted.[3,10]

Liu *et al.* applied a crossplatform approach (combining STM apparent height, break-junction and CP-AFM measurements) to investigate the electronic properties of **25** and **26** (Figure 4.3).[12b] Comparable to the results of Blum and coworkers, values of $\beta = 1.01 \pm 0.25$ Å^{-1} and 1.11 ± 0.18 Å^{-1}

Table 4.3   Experimentally determined β-values[a] for selected structure types.

structure	formula	terminal group	β (Å^{-1})	ref
alkane	[–CH$_2$–]$_n$	SH	1.02–1.08 ($\pm$0.14)	3
		NH$_2$	0.81–0.88 ($\pm$0.01)	3
		COOH	0.77–0.81 ($\pm$0.01)	3
acene	(benzene, naphthalene–X, etc)	SH	0.50 ($\pm$0.09)	10
		NC	0.49 ($\pm$0.08)	10
phenylene-vinylene	[–CH=CH–C$_6$H$_4$–]$_n$	SAc	0.53–0.63 ($\pm$0.13)	41a
			0.4 ($\pm$0.12)	41b
alkene	[–CH=CH–CH=CMe–]$_n$	C$_6$H$_4$SH	0.22 ($\pm$0.04)	4
phenylene-ethynylene	[–C≡C–C$_6$H$_4$–]$_n$	SAc	0.21 ($\pm$0.01)	41c
alkyne	[–C≡C–]$_n$	C$_5$H$_4$N	0.06 ($\pm$0.03)	5

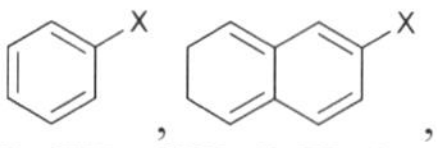

[a] β = the tunnelling decay constant (indicative of the efficiency of electron transport along a molecule)

(STM apparent height measurements) were extracted for the Ru-containing system and organic dithiol, respectively. Furthermore, analysis of the CP–AFM data yielded a tunnelling barrier height of 0.25 $\pm$ 0.03 eV, in good agreement with the E_F–E_{HOMO} (ΔE_{DB}) offset ($\leq$0.25 eV) proposed by Kim *et al.* for **24c**. From density functional theory (DFT) calculations of **25** a hole-mediated tunnelling mechanism was proposed on the basis that the HOMO spans the whole length of the molecule, in contrast to the LUMO that is localised on the {Ru(dppm)$_2$} centre. Further rationalisation of this data was provided by UV-Vis studies that showed **25** had a smaller HOMO–LUMO gap than **26** (3.47 eV compared to 3.69 eV) – in turn lowering ΔE_{DB} and β.

4.4   Role of the Metal Centre in Electron Delocalisation

From molecular conductance studies and related work undertaken thus far, structure–property relationships pertinent to {M(L)$_n$} are difficult, if not impossible, to establish. It may be concluded that Ru-containing σ alkynyl complexes are better mediators of electron transport than their Pt analogues, but *how much* better, and how they would compare to other metal-containing structures, or to Ru-containing materials with a different ancillary ligand set, is unclear. As mentioned in the introduction a large volume of work relating to electron delocalisation in MV complexes has been compiled, which might provide further insights. The data set is large and varied in comparison, such studies being relatively quick and easy compared to methods of measuring molecular conductance directly.

We have thus undertaken a survey of relevant complexes, with the intention of exploring further the role of {M(L)$_n$} on electron delocalisation, and through analogy, molecular conductance. Prior to the details of this investigation, crucial aspects relating to the study of MV complexes will be discussed.

4.4.1   Theoretical Aspects

4.4.1.1   *Mixed-Valence Complexes – The Creutz–Taube Ion*

The first characterised example of a MV complex was the Creutz-Taube ion, generated from [{(NH$_3$)$_5$Ru}(μ-py){Ru(NH$_3$)$_5$}]$^{4+}$ [110] (the oxidation states of the metal centres represented by the notation [Ru$_1^{II}$Ru$_2^{II}$]). Electronic coupling between the metals facilitates the one-electron oxidation to [{(NH$_3$)$_5$Ru}(μ-py){Ru(NH$_3$)$_5$}]$^{5+}$ (denoted as [Ru$_1^{II}$Ru$_2^{III}$], or [Ru$_1^{III1/2}$Ru$_2^{III1/2}$] depending on whether the charge is considered localised or delocalised, respectively). Near-IR spectroscopy of this MV species revealed an absorption band not present in the fully reduced [Ru$_1^{II}$Ru$_2^{II}$] or oxidised [Ru$_1^{III}$Ru$_2^{III}$] states. This feature was thus attributed to the transition of an electron from the Ru^{II} centre to the Ru^{III} centre through the bridging bipyridyl ligand, *i.e.* [Ru$_1^{II}$Ru$_2^{III}$] $\rightarrow$ [Ru$_1^{III}$Ru$_2^{II}$]*.

Analogous systems of the type $[\{M_1(L)_n\}(\mu\text{-}BL)\{M_2(L)_n\}]$ (where M_1 and M_2 represent metal centres and L and BL represent the ancillary ligand framework and bridging ligand, respectively) have since been extensively studied, with 'intervalence charge-transfer (IVCT) bands' present in the spectra of all (class II and III) MV complexes. Over the last 40 years, a firm theoretical basis for these and other observations has been established, allowing useful information to be extracted from a variety of physical measurements.

4.4.1.2 Near-IR Spectroscopy

The extent of electronic coupling between redox centres may be deduced from near-IR studies using either the Hush[111] or CNS[112] models to analyse spectral features of the MV state. Whereas the former describes metal–metal electronic coupling directly (two-state model), the latter considers the metal–metal coupling mediated by the bridging ligand (*i.e.* a superexchange-type mechanism[113]) through metal-to-ligand (electron transfer) and ligand-to-metal (hole transfer) transitions (three-state model). Key aspects of these approaches are summarised here for context, comprehensive discussions may be found elsewhere.[27e,113–114]

In Hush theory, the extent of electron delocalisation is provided by the electronic coupling parameter, H_{ab}. Depending on specific indicators, complexes may be assigned according to the Robin and Day classification system – indicating whether electrons are considered fully localised (or valence trapped – Class I), delocalised (Class III), or somewhere in between these extremes (Class II).[115] The calculation of H_{ab}, and features specific to each class are summarised – from discussions by Aguirre-Etcheverry and O'Hare[27e] – in Table 4.4.

Class I and classes II/III complexes are easily distinguished by the absence or presence of an IVCT band, respectively, yet assignment of class II or III is often difficult. An initial indicator is the width of the IVCT band, with class II complexes exhibiting a broader (and class III a narrower) band than predicted by eqns (4.1) or (4.2) (at room temperature) – the Hush limits.

$$\Delta v_{1/2} = (16RT \ln 2\lambda)^{1/2} \tag{4.1}$$

$$\Delta v_{1/2} = (2310 v_{max})^{1/2} \tag{4.2}$$

For borderline II/III complexes the delocalisation parameter (Γ),[116] calculated as per eqn (4.3), may provide further insight (as described in Table 4.4). Ultimate conclusions can be drawn depending on whether redox centres are distinguishable from each another by any form of spectroscopy. If they are not then the complex is almost certainly class III, as complete electron delocalisation results in equivalence of the redox sites.

Table 4.4 [7e] Details resulting from classical Hush theory that distinguishes the three classes of MV complex[27e].

class	IVCT band	additional comments	calculation of H_{ab}, λ^a from IVCT band characteristics	H_{ab}	$\Delta G^{\#b}$	Γ^c
I	not observed	/	/	$=0$	$=\dfrac{\lambda}{4}$	0–0.1
II	observed, $\Delta v_{1/2} >$ Hush limit[d]	λ often solvent dependant	$\lambda = v_{max}$ $H_{ab} = \dfrac{(2.06 \times 10^{-2})}{r_{ab}}(v_{max}\Delta v_{1/2}\varepsilon_{max})^{1/2e}$ $H_{ab} = \dfrac{\mu_{ge}}{er_{ab}}v_{max}^{\ f}$	$<\dfrac{\lambda}{2}$	$=\dfrac{\lambda-2H_{ab}^2}{4\lambda}$	0.1–0.5
III	observed, $\Delta v_{1/2} <$ Hush limit[d]	higher peak intensities (ε), weak solvent dependence	$hv_{max} = 2H_{ab}$	$\geq\dfrac{\lambda}{2}$	$=0$	0.5 (borderline class II/II) $\gg 0.5$ (class III)

[a] λ = the reorganisation energy. [b] $\Delta G^{\#}$ = free energy of activation. [c] Γ = delocalisation parameter. [d] See eqns (4.1) and (4.2), $\Delta v_{1/2}$ = bandwidth at half-height for a Gaussian-shaped IVCT band (cm^{-1}). [e] r_{ab} = effective separation between donor and acceptor in the case of nonadiabatic states (Å), ε_{max} = extinction coefficient at the band maximum (M^{-1}cm^{-1}). [f] In the case of symmetric Gaussian-shaped IVCT bands, μ_{ge} = transition dipole moment of the IVCT band, e = electronic charge.

$$\Gamma = \left[1 - \frac{\Delta v_{1/2}}{(2310 v_{\text{max}})^{1/2}} \right] \tag{4.3}$$

Despite offering predictions for strongly coupled (class III) systems, the Hush approach is only accurate in the weakly coupled limit. It has been suggested that the CNS model (by Creutz, Newton and Sutin) is more appropriate for comparing donor–acceptor systems with a wide range of coupling.[113] Here, the effective (*i.e.* bridge-mediated) coupling of the metal centres ($H_{\text{MM}'}$) is given by eqn (4.4), reducing to eqn (4.5) when the hole-transfer pathway dominates.

$$H_{\text{MM}'} = \left(\frac{H_{\text{ML}} H_{\text{M}'\text{L}}}{2\Delta E_{\text{ML}}^{\text{eff}}} \right) + \left(\frac{H_{\text{LM}} H_{\text{LM}'}}{\Delta E_{\text{LM}}^{\text{eff}}} \right) \tag{4.4}$$

$$H_{\text{MM}'} = \left(\frac{H_{\text{ML}}^2}{\Delta E_{\text{LM}}^{\text{eff}}} \right) \tag{4.5}$$

Here, $H_{\text{ML}}/H_{\text{LM}}$ are coupling elements associated with metal–ligand interactions of the electron-transfer/hole-transfer pathways, and $\Delta E_{\text{ML}}^{\text{eff}}/\Delta E_{\text{LM}}^{\text{eff}}$ denote reduced energy gaps between metal and ligand orbitals. H_{LM}^2 and $\Delta E_{\text{LM}}^{\text{eff}}$ are evaluated using spectroscopic parameters of both the ligand-to-metal and IVCT bands.

Whilst the use of near-IR spectroscopic information as an indication of electron delocalisation is widely accepted and implemented, it carries several associated problems aptly discussed recently by D'Alessandro and Keene.[114a] Solvent, electrolyte and temperature must be standardised for meaningful comparisons to be made, and even then accurate determination of H (H_{ab} or $H_{\text{MM}'}$) can be difficult. In various applications the value of r_{ab} is generally taken as the distance between metal centres (r_{geom}), however, this often provides only lower limits for electronic coupling – effective charge-transfer distances are usually smaller due to delocalisation of charge onto the bridging ligand. Furthermore, multiple overlapping IVCT bands (from nondegenerate d orbitals leading to additional degrees of freedom) and additional non-IVCT bands in the near-IR (for example from d–d transitions when spin-orbit coupling is significant) may be observed, complicating analysis.[117]

4.4.1.3 *Electrochemical Methods*

In addition to IR investigations, the properties of MV complexes may be studied electrochemically. Compounds containing two identical redox centres that are *not* electronically coupled show reduction and oxidation of both redox centres simultaneously ($2e^-$ waves) in cyclic voltammetry studies. In materials where electronic coupling is significant, sequential redox events at $E_{1/2}^{\text{I}}$ and $E_{1/2}^{\text{II}}$

($1e^-$ waves) are observed. For $[M_1^{II}M_2^{II}]$ complexes these are consistent with the generation of first $[M_1^{II}M_2^{III}]$ then $[M_1^{III}M_2^{III}]$ states. From a practical point of view, large values of $\Delta E_{1/2}$ (where $\Delta E_{1/2} = E_{1/2}^{I} - E_{1/2}^{II}$) enable chemical oxidation and isolation of MV complexes for near-IR spectral studies. This measurable quantity may also be utilised as an indicator of electron delocalisation, as discussed below.

$\Delta E_{1/2}$ is related to the comproportionation constant, K_c, by eqn (4.6), where K_c is a measure of the thermodynamic stability of the mixed valence state relative to the fully oxidised and reduced states (eqn (4.7)).

$$\Delta E_{1/2} = \frac{-RT}{F}\ln K_c \tag{4.6}$$

$$[M_1^{II}M_2^{II}] + [M_1^{III}M_2^{III}] \overset{K_c}{\rightleftharpoons} 2[M_1^{II}M_2^{III}]$$

where,

$$K_c = \frac{[M_1^{II}M_2^{III}]^2}{[M_1^{II}M_2^{II}][M_1^{III}M_2^{III}]} \tag{4.7}$$

When $\Delta E_{1/2}$ is expressed in terms of the free energy of comproportionation (ΔG_c) (eqn (4.8)), its contributing terms are given by equation 9.[113,114,117,118]

$$-\Delta E_{1/2}F = -RT \ln K_c = \Delta G_c \tag{4.8}$$

$$\Delta G_c = \Delta G_s + \Delta G_e + \Delta G_i + \Delta G_r + \Delta G_{ex} + \Delta G_{ip} \tag{4.9}$$

Here, ΔG_s represents the statistical distribution of the comproportionation equilibrium, ΔG_e the decreased electrostatic repulsion of $M^{II}M^{III}$ relative to the $M^{II}M^{II}$ and $M^{III}M^{III}$ states, ΔG_i is an inductive factor dealing with competitive coordination of the bridging ligand by the metal ions and ΔG_r (the only component representing metal–metal coupling) is the free energy of resonance exchange. If an antiferromagnetic exchange significantly stabilises one of the reactants of eqn (4.7), it is accounted for by ΔG_{ex}.[113,118a] Finally, ΔG_{ip} denotes an ion-pairing contribution that may vary with changes to the reaction medium conditions.[118b]

From eqn (4.9), it is clear that the magnitude of $\Delta E_{1/2}$ cannot be used as a direct measure of ΔG_r – as is rightly cautioned against by many authors. However, if the nonexchange contributions to ΔG_c, *i.e.* $\Delta G_{ne} = \Delta G_s + \Delta G_e + \Delta G_i + \Delta G_{ex} + \Delta G_{ip}$, are accounted for (or arguably constant), adjusted $\Delta E_{1/2}$ values may then be used to invoke electron delocalisation trends within a series of MV complexes.[113]

To this end, ΔG_s (for a symmetrical system) and ΔG_e can be calculated using eqns (4.10) and (4.11), respectively, and ΔG_{ex} can be estimated using the Van

Vleck expression for magnetic susceptibility.[113,118a]

$$\Delta G_{\mathrm{s}} = \frac{1}{2} RT \ln \frac{1}{4} \qquad (4.10)$$

$$\Delta G_{\mathrm{e}} = \frac{1}{4} \pi \varepsilon \varepsilon_0 r_{\mathrm{geom}} \qquad (4.11)$$

Here, r_{geom} is again the distance between metal centres (an approximation for effective charge separation), ε is the solvent's dielectric constant and ε_0 is the vacuum permittivity.

As $\Delta G_{\mathrm{r}}^{\circ}$ is linked to H (H_{ab} or $H_{\mathrm{MM'}}$) (through eqns (4.12) and (4.13) for localised and delocalised systems, respectively), the extent of electron delocalisation as measured by near-IR analyses or electrochemical methods may be compared.[119]

$$-\Delta G_{\mathrm{r}}' = \frac{2H^2}{\lambda} = \frac{2H_{\mathrm{ab}}^2}{v_{\mathrm{max}}} \qquad (4.12)$$

$$-\Delta G_{\mathrm{r}}' = 2\left(H_{\mathrm{ab}} - \frac{\lambda}{4} \right) = v_{\mathrm{max}} - \frac{\lambda}{2} \qquad (4.13)$$

Of the various approaches towards analysis of near-IR spectra (including the PKS[120] and Ondrechen[121] models, not discussed here), as yet none suitably provide definitive conclusions regarding the electronic delocalisation of complexes over a wide range of coupling. Unfortunately too, the utility of the electrochemical method suffers greatly from a lack of attention to ΔG_{ne} – measured values of $\Delta E_{1/2}$ for numerous complexes are often incomparable due to the wide array of medium conditions used. For these reasons perhaps, the relative merits of alternative methods to probe electron delocalisation – *e.g.* EPR/Mössbauer spectroscopy and DFT calculations – have been discussed recently.[117]

4.4.2 Investigations into the Correlation of Electron Density with $\Delta E_{1/2}$

4.4.2.1 *Background and Inspiration*

From MV complex and molecular conductance studies, trends have emerged relating electronic interactions between redox centres (or electrodes) to the nature and length of the bridging component (increasing donor–acceptor separation). For example, wire-like properties increase with increasing conjugation, resulting in the series $[-CH_2-]_n$ < $[-CH=CH-]_n$ < $[-C\equiv C-C_6H_4-]_n$ < $[-C\equiv C-]_n$ for hydrocarbons. Such trends have been quantified through molecular conductance measurements, with β values for relevant structure types given in Table 4.3. As discussed, the specific effect of an incorporated metal centre on molecular conductance is less well defined. Inferences are often made from the few direct measurements conducted, or from analogous MV systems (Table 4.2). Certain metals (*e.g.* Ru) are often claimed superior to others (*e.g.* Pt), yet this is an oversimplification and is misleading. Clearly ancillary ligands (Table 4.2) and the metal-bridging ligand interaction contribute significantly to the electronic properties of the metal centre.

During preparatory reading for this work, it was noted that changes in $\Delta E_{1/2}$ for complexes of the type $[\{M(L)_n\}(\mu\text{-BL})\{M(L)_n\}]$, were occasionally qualitatively attributed to differences in the electron density of the metal centre.[27] Furthermore, a correlation between electron density, from relevant $v(C\equiv C)/v(C\equiv O)$ values, and $\Delta E_{1/2}$ was observed for a small number of bisferrocenylacetylide complexes (1–4, Table 4.2)[106] – and a linear dependence found between solvent donor number and $\Delta E_{1/2}$ in Ru pentamine complexes (the greater Lewis basicity of the solvent, the greater the electron density at the metal centre).[122] To the best of our knowledge no attempts to broadly quantify this relationship have been made (particularly with reference to metal σ alkynyl systems), and the potential association was intriguing.

To explore this further, a convenient measure of electron density at the metal fragment $\{M(L)_n\}$ was required, and for this it was considered that the IR stretching frequencies of diagnostic ligands such as carbonyl ($C\equiv O$) or nitrile ($C\equiv N^-$) bound to mononuclear analogues of $[\{M(L)_n\}(\mu\text{-BL})\{M(L)_n\}]$ species, *i.e.* $\{M(L)_n\}$-X (X = $C\equiv O$, $C\equiv N^-$), might prove good indicators. Following Tolman's work on the electron-donating abilities of phosphines the inverse relationship between $v(C\equiv O)$ and the electron density of a metal centre (resulting from metal-to-ligand backbonding) is well established.[123]

We therefore sought to compare $\Delta E_{1/2}$ for complexes of the type $[\{M(L)_n\}(\mu\text{-BL})\{M(L)_n\}]$ with the electron density of $\{M(L)_n\}$ – quantified using $v(C\equiv O)/v(C\equiv N)$ values for analogous $\{M(L)_n\}$-X species – in different series of complexes containing the same (acetylide-terminated) bridging ligand. $C\equiv O$ and $C\equiv N^-$ ligands are isoelectronic with acetylides ($^-C\equiv CR$), further supporting their diagnostic application here. As well as measuring electron-density differences between $\{M(L)_n\}$, where M = M and $(L)_n \neq (L)_n$, $v(C\equiv O)/v(C\equiv N)$ were thought likely indicative of orbital energetic/spatial changes where M $\neq$ M and $(L)_n = (L)_n$.

4.4.2.2 Scope and Datasets

Four sets of complexes were considered, where μ-BL = $-C\equiv C-$ or $\equiv C-C\equiv$ (C_2), $-C\equiv C-C\equiv C-$ or $\equiv C-C\equiv C-C\equiv$ (C_4), $-C\equiv C-C\equiv C-C\equiv C-C\equiv C-$ (C_8) and $-C\equiv C-C_6H_4-C\equiv C-$ (phenylene). Only $\Delta E_{1/2}$ vs $v(C\equiv O)/v(C\equiv N)$ correlations are considered in this work – whilst the link between $v(C\equiv O)/v(C\equiv N)$ and other measures of electronic delocalisation is also of interest, available near-IR and EPR data for these series is relatively limited. The $\Delta E_{1/2}$

vs. $\nu(C\equiv O)/\nu(C\equiv N)$ dataset was itself restricted with many known complexes not studied electrochemically – and of those that have, some showing only quasireversible or irreversible features ($i_a/i_c \neq 1$). Furthermore, details of appropriate $\{M(L)_n\}$-X (X = C$\equiv$O, C$\equiv$N$^-$) species have not been published for a number of homobimetallic complexes exhibiting reversible features.

Known complexes omitted from this inquiry are detailed in Table 4.5. Electrochemical data for eligible complexes is tabulated in Tables 4.6–4.8 with IR data for their analogous mononuclear complexes given in Table 4.9. As only one data point could be established for complexes with μ-BL = C$_2$ – where $\{M(L)_n\}$ = Ru(dppe)Cp ($\Delta E_{1/2}$ = 0.85 V)[124] – inquiry into this series was suspended.

The majority of cyclic voltammetry studies were undertaken in CH_2Cl_2 with a $Bu_4N^+BF_4^-$ supporting electrolyte and platinum working electrode (referred to hereafter as the standard conditions). Deviations from the standard conditions were noted, as changes in either solvent, supporting electrolyte and (to a lesser extent) supporting electrolyte concentration are known to significantly alter $\Delta E_{1/2}$ through their impact on ΔG_e and ΔG_{ip} (eqn (4.9)) (Table 4.10).

Occasionally, more than one value of $\Delta E_{1/2}$ for the same homobimetallic complex was reported. When plotting the datasets, multiple values in the phenylene series were averaged as they were obtained under comparable CV conditions with values separated by $\leq$0.02 V. For $\{M(L)_n\}$ = Re(NO)(PPh$_3$)Cp* in the C$_4$ series, both results were plotted to exemplify solvent effects.

Larger data sets were available for $\Delta E_{1/2}$ *vs.* $\nu(C\equiv O)$ series as analogous $\{M(L)_n\}$-X complexes were more numerous where X = C$\equiv$O than X = C$\equiv$N$^-$. $\nu(C\equiv O)$ frequencies were typically obtained from PF$_6^-$ salts in CH_2Cl_2 (standard conditions), with deviations again noted for completeness. Whereas the variation of $\nu(C\equiv O)/\nu(C\equiv N)$ with sample preparation method is unclear, where $\nu(C\equiv O)$ frequencies were obtained for different salts of the same complex – $\{M(L)_n\}$ = *trans*-RuCl(dppm)$_2$ (Table 4.9) – a strong counterion dependence is observed.

To avoid useful data being omitted from the study, $\nu(C\equiv N)$ values for $\{M(L)_n\}$-C$\equiv$N complexes where $\{M(L)_n\}$ = Fe(dppe)Cp* and Ru(dppm)Cp* were estimated from values obtained for complexes where $\{M(L)_n\}$ = Fe(dppe)Cp and Ru(dppm)Cp (Table 4.9). $\nu(C\equiv N)$ values of the latter were adjusted by -10 cm^{-1} to provide $\nu(C\equiv N)_{est}$ values for the former, with the quantity -10 cm^{-1} chosen through analogy to the difference between the $\nu(C\equiv N)$ values for Ru(CN)(dppe)Cp* (2065 cm^{-1}) and Ru(CN)(dppe)Cp (2075 cm^{-1}).

Plots of $\Delta E_{1/2}$ *vs.* $\nu(C\equiv O)$ and $\nu(C\equiv N)$ for each set of complexes are presented in Figure 4.4 and Figure 4.5, respectively.

Table 4.5 Known complexes of the type $[\{M(L)_n\}(\mu\text{-BL})\{M(L)_n\}]$ not included in this study

μ-BL	$M(L)_n$		
	Electrochemical experiments not conducted	Nonreversible electrochemistry ($i_a/i_c \neq 1$)	$\{M(L)_n\}$-X^a analogues unknown
C_2	$MX(PR_3)$ (M = Pt, Pd; X = Cl, I; R = Me, Et, ^nBu, Ph;),[125] $Fe(CO)_2Cp^*$,[126] HgR (R = Me, alkyl),[127] $Au(PR_3)$,[128] $[AuR]PPh_4$ (R = CN, PhC≡C, MeC≡C, HC≡C),[129] $Mn(CO)_5$,[130] $[V(mes)_3]^-$,[131] $Cr(CO)_3Cp$,[132] $Ti(PMe_3)Cp_2$,[133] $Sm(thf)Cp^*_2$,[134] $ScCp^*_2$,[135] $[Pt(^tBu_3\text{-tpy})]OTf$[136]	$Ru(CO)_2Cp$,[137] $Re(CO)_5$,[138] $W(^tBuO_3)_3$[139]	$Mn(dmpe)(MeC_5H_4)$,[140] $WCl(dmpe)_2$[141]
C_4	$Fe(CO)_2Cp^*$,[142] $Mo/W(CO)_2Tp'$,[b,143] $Rh/Ir(P^iPr_3)_2$-type centres,[144] $PtCl(PR_3)_2$,[145] $[Pt(^tBu_3\text{-tpy})]OTf$,[136] $Au(P(p\text{-tol})_3)_3$[38b]	$Re(CO)_3(^tBu_2bpy)$,[146] $W(CO)_3Cp^*/W(O)_2Cp^*$,[147] $Au(PCy_3)$[128d]	$Os(dppe)Cp^*$,[148] $Mn(dppe)_2R$ (R = I,[149] $SiEt_3$, Si^iPr_3, Si^tBuMe_2[79b]), $WI(dppe)$[78]
C_8	$Fe(CO)_2Cp^*$,[126c] $M(CO)_3Cp$ (M = Mo, W),[150] $[Pt(^tBu_3\text{-tpy})]PF_6$,[136] $Au(P(p\text{-tol})_3)_3$[38b]	$Ru(dpf)_4(C_4SiMe_3)$,[c,151] Pt $(PY_3)_2(R)$ (Y = p-tol, PPh_2X^d;[152] R = p-tol,[153] C_6F_5,[154] Cl^{155};), $Re(NO)(PR_3)Cp^*$ (R = Ph, p-C_6H_4-^tBu, p-C_6H_4-C_6H_5, PPh_2X^d)[27c,156]	$Ru(PPh_3)_2Cp$,[157] $Ru(PPh_3)_2(R)^{e,158}$
phenylene	$MX(PR_3)_2$ (M = Pd,[159] Pt;[159c,160] X = Cl, Br, I, NCS, H, OTf, C_6H_5; R = Et, Bu, p-tol, Ph;), $[M(PR_3)_2L]X$ (M = Pd,[159a] Pt;[160a] R = Et; L = PEt_3, Py, CO; X = ClO_4;), $Pt(dppe)(C≡C\text{-}C_6H_4\text{-}C≡CH)$,[161] MR_3 (M = Sn, Pb; R = CH_3, C_6H_5;),[162] $V(EtMe_4C_5H_5)_2$,[163] $M(PR_3)$ (M = Cu, Ag; R = Et, ^nBu, Ph;),[164] $Rh(P^nBu_3)_4$,[165] $[Rh(PMe_3)(H)]Cl$,[165] $[MCl(CO)(PPh_3)_2(CH_3CN)]OTf$ (M = Ir, Rh),[166] $Au(PR_3)$ (R = Me,[167] p-tol[168]), $Cu(PR_3)_2$ (R = Et, Ph),[169] $RuCl(CO)(dppf)$,[f,170] $Fe(CO)_2Cp$,[171] $[M(P(OEt)_3)_5]X$ (M = Ru,[172] Fe;[173] X = PF_6, BPh_4;), $Ru(PPh_3)_2Cp$,[174] $Re(CO)_2P_3$ (P = $PPh(OEt)_2$, $PPh_2(OEt)$),[175] $Ir(\eta^3\text{-}CH_2CHCHPh)Cp^*$,[176] $Ir(CO)_2(PPh_3)_2(CHCH_2)_2$[177]	$Pt(NCN)$,[g,178] $Au(PCy_3)$,[179] $Pt(P(^nBu)_3)_2(C≡C\text{-}C_6H_5)$,[180] $U(NN'_3)^{h,181}$	$Ru(PPh_3)_2(R)$,[158] $TiCp_2(CH_2SiMe_3)$[182]

^{a}X = C≡O, C≡N. ^{b}Tp′ = hydridotris(3,5-dimethylpyrazolyl)borate). ^{c}dpf = N,N′-diphenylformamidine. ^{d}X = alkyl link to PPh_2 ligand bound on other metal centre. ^{e}R = N-(benzoyl)-N′-(picolinylidene)-hydrazine or 4′-phenyl-2,2′:6′,2″-terpyridine. ^{f}dppf = 1,1′-bis(diphenyl phosphino)ferrocene. ^{g}NCN = $[C_6H_3(Me_2NCH_2)_2\text{-}2,6]^-$. ^{h}NN′$_3$ = $N(CH_2CH_2NSi^tBuMe_2)_3$.

Table 4.6 Electrochemical data[a] for complexes of the type [{M(L)$_n$}(μ-C$_4$){M(L)$_n$}]

$M(L)_n$	CV data				
	$E_{1/2}^{I}$ (V)	$E_{1/2}^{II}$ (V)	$\Delta E_{1/2}$ (V)	conditions[b]	ref
Fe(dppe)Cp*	−1.14	−0.42	0.72	Bu$_4$N$^+$PF$_6^-$	183
Fe(dippe)Cp* [c]	−1.43	−0.64	0.79	Bu$_4$N$^+$PF$_6^-$	27a
Ru(dppe)Cp*	−0.89	−0.24	0.65	/	184
Ru(dppm)Cp*	−0.94	−0.31	0.63	/	184
Ru(dppe)Cp	−0.70	−0.11	0.59	Bu$_4$N$^+$PF$_6^-$	27b
Ru(PPh$_3$)$_2$Cp	−0.69	−0.05	0.64	[glassy carbon]	185
Ru(PPh$_3$)(PMe$_3$)Cp	−0.72	−0.13	0.59	[glassy carbon]	185
Os(PPh$_3$)$_2$Cp	−0.79	−0.30	0.49	THF	186
Re(NO)(PPh$_3$)Cp*	−0.50	−0.06	0.44	Et$_4$N$^+$ClO$_4^-$/ CH$_3$CN	187
	−0.45	0.08	0.53	/	27c
Re(NO)(P(*p*-tol)$_3$)Cp*	−0.68	−0.15	0.53	/	27c

[a]All redox potentials relative to [Cp$_2$Fe]$^+$/Cp$_2$Fe. [b]Bu$_4$N$^+$BF$_4^-$ in CH$_2$Cl$_2$ with Pt working electrode unless otherwise stated. [c]dippe = 1,2-bis(diisopropylphosphino)ethane.

4.4.2.3 *Identification of Outliers*

$\Delta E_{1/2}$ for outlier 1 in the C$_4$ (C≡O and C≡N) data series {M(L)$_n$} = Re(NO)(PPh$_3$)Cp* was obtained from CV experiments using Et$_4$N$^+$ClO$_4^-$/ CH$_3$CN. The change of solvent from CH$_2$Cl$_2$ to CH$_3$CN (*ca.* −21 mV through analogy to values in Table 4.10), and electrolyte cation from Bu$_4$N$^+$ to Et$_4$N$^+$ (*ca.* −31 mV), would result in a lower value, only partially compensated for by changing the electrolyte anion from the smaller BF$_4^-$ to the larger ClO$_4^-$ (*ca.* +12 mV). Outlier 1 and 2 from the C$_8$ (C≡N) and C$_4$ (C≡N) series, respectively, correspond to $\Delta E_{1/2}$ values for {M(L)$_n$} = Os(PPh$_3$)$_2$Cp obtained from CV experiments in THF. These would be significantly lower (*ca.* −67 mV) than the rest of the dataset as a result of solvent effects.

Table 4.7 Electrochemical data[a] for complexes of the type [{M(L)$_n$}(μ-C$_8$){M(L)$_n$}]

$M(L)_n$	CV data				
	$E_{1/2}^{I}$ (V)	$E_{1/2}^{II}$ (V)	$\Delta E_{1/2}$ (V)	conditions[b]	ref
Fe(dppe)Cp*	−0.65	−0.22	0.43	Bu$_4$N$^+$PF$_6^-$	188
Ru(dppe)Cp*	−0.38	−0.03	0.35	/	186
Ru(PPh$_3$)$_2$Cp	−0.22	0.12	0.34	/	186
Os(PPh$_3$)$_2$Cp	−0.32	−0.11	0.21	THF	186
Re(NO)(P(*p*-tol)$_3$)Cp*	−0.30	−0.01	0.29	/	27c
Re(NO)(PCy$_3$)Cp*	−0.35	−0.03	0.32	/	27c

[a]All redox potentials relative to [Cp$_2$Fe]$^+$/Cp$_2$Fe. [b]Bu$_4$N$^+$BF$_4^-$ in CH$_2$Cl$_2$ with Pt working electrode unless otherwise stated.

Table 4.8　Electrochemical data[a] for complexes of the type $[\{M(L)_n\}(\mu\text{-phenylene})\{M(L)_n\}]$

$M(L)_\text{n}$	CV data				
	$E_{1/2}{}^{I}$ (V)	$E_{1/2}{}^{II}$ (V)	$\Delta E_{1/2}$ (V)	conditions[b]	ref
Fe(dppe)Cp*	−0.74	−0.48	0.26	$Bu_4N^+PF_6{}^-$	100
	−0.76	−0.50	0.26	$Bu_4N^+PF_6{}^-$	189
Fe(dppe)Cp	−0.37	−0.15	0.22	[Pt disk]	190
	0.00	0.21	0.21	$Bu_4N^+PF_6{}^-$	191
Ru(dppe)Cp*	−0.50	−0.22	0.28	[Pt microdisk]	192
Ru(dppe)Cp	−0.32	−0.09	0.23	[Pt microdisk]	192
Ru(PPh$_3$)$_2$Cp	−0.30	−0.01	0.29	[Pt microdisk]	192
Mo(dppe)(η-C$_7$H$_7$)	−0.84	−0.67	0.17	Unknown	193
trans-FeCl(depe)$_2$	−0.63	−0.47	0.16	/	194
trans-FeCl(dmpe)$_2$	−0.70	−0.50	0.20	$Bu_4N^+ClO_4{}^-$	36a
trans-RuCl(dppe)$_2$	−0.33	0.01	0.34	$Bu_4N^+PF_6{}^-$	27d
	−0.34	0.02	0.36	$Bu_4N^+PF_6{}^-$	195
trans-RuCl(dppm)$_2$	−0.30	0.00	0.30	/	194
	−0.34	−0.02	0.32	[Pt disk]	196
trans-OsCl(dppm)$_2$	−0.51	−0.21	0.30	/	194

[a]All redox potentials relative to $[Cp_2Fe]^+/Cp_2Fe$. [b]$Bu_4N^+BF_4{}^-$ in CH_2Cl_2 with Pt working electrode unless otherwise stated.

With the exception of $\{M(L)_n\}$ = *trans*-FeCl(dmpe)$_2$ in the phenylene series – where the electrolyte anion is $ClO_4{}^-$ (outlier 2) – no significant deviation resulting from changing CV conditions is expected. Certainly, whilst experiments using $Bu_4N^+PF_6{}^-$ might be expected to yield $\Delta E_{1/2}$ values higher than their $Bu_4N^+BF_4{}^-$ counterparts (*ca.* +70 mV), this is not observed when measurements in both electrolytes are made with $\{M(L)_n\}$ = Fe(dppe)Cp. Notably, all of the Fe-containing complexes demonstrate a significantly lower $\Delta E_{1/2}$ than expected from the values obtained for their analogues in the C$_4$ and C$_8$ series. They are thus identified as asterisked outliers (1*, 2*, 3* and 5*). Finally, outlier 4 corresponds to $\{M(L)_n\}$ = Mo(dppe)(η-C$_7$H$_7$) where the medium conditions are as yet unreported (and tentatively predicted to deviate from $CH_2Cl_2/Bu_4N^+BF_4{}^-$, on the basis of this analysis).

There is, however, a wide variation in conditions used to obtain $v(C\equiv O)$ frequencies in the phenylene series. Certainly, the remarkably low $v(C\equiv O)$ = 1906 cm^{-1} for $\{M(L)_n\}$ = *trans*-FeCl(depe)$_2$ (outlier 1*) may be attributed to counterion effects, being obtained from measurement of the chloride salt. These are further exemplified by the data points marked as I and II (Figure 4.4) – representing $\{M(L)_n\}$ = *trans*-RuCl(dppm)$_2$ where $v(C\equiv O)$ was measured for $[\{M(L)_n\}\text{-}C\equiv O]^+$ as the $BF_4{}^-$ and $PF_6{}^-$ salts, respectively.

4.4.2.4　Discussion

Each of the series follows the accepted trend for electronic communication through a bridge based upon its length and chemical structure, that is, C$_4$ > C$_8$

Table 4.9   Selected IR data for $\{M(L)_n\}$-X complexes (X = C≡O,[a] C≡N$^-$ [b])

$M(L)_n$	IR data[c]					
	$v(C{\equiv}O)$	complex (conditions)	ref	$v(C{\equiv}N)$	complex (conditions)	ref
Fe(dppe)Cp*	1940	(unknown)	197	<2060	Fe(CN)(dppe)Cp (unknown)	198
Fe(dippe)Cp*	1928	[M′]BPh$_4$ (Nujol mull)	199	/	/	/
Fe(dppe)Cp	1981	/	200	2060	(unknown)	198
Ru(dppe)Cp*	1972	/	201	2065	/	202
Ru(dppm)Cp*	/	/	/	<2076	Ru(CN)(dppm)Cp	202
Ru(dppe)Cp	1990	/	203	2075	/	202
Ru(PPh$_3$)$_2$Cp	1986	(CHCl$_3$)	204	2072	/	202
Ru(PPh$_3$)(PMe$_3$)Cp	1995	(Nujol mull)	205	/	/	/
Os(PPh$_3$)$_2$Cp	/	/	/	2065	(KBr pellets)	206
Re(NO)(PPh$_3$)Cp*	2002	[M′]BF$_4$ (CD$_2$Cl$_2$)	207	2090	(thin film)	208
Re(NO)(P(p-tol)$_3$)Cp*	2002	[M′]BF$_4$	209	/	/	/
Re(NO)(PCy$_3$)Cp*	1993	[M′]BF$_4$	209	/	/	/
Mo(dppe)(η-C$_7$H$_7$)	1958	/	200	2073	/	210
trans-FeCl(depe)$_2$	1906	[M′]Cl (Nujol mull)	211	/	/	/
trans-FeCl(dmpe)$_2$	1938	[M′]BPh$_4$ (Nujol mull)	211	/	/	/
trans-RuCl(dppe)$_2$	1946	(KBr)	212	/	/	/
trans-RuCl(dppm)$_2$	1965	[M′]BF$_4$ (Nujol mull)	213	/	/	/
	1980	(Nujol mull)	213			
trans-OsCl(dppm)$_2$	1962	[M′]SbF$_6$	214	/	/	/

[a]As the [M′]$^+$[PF$_6$]$^-$ salt in CH$_2$Cl$_2$ unless otherwise stated, where [M′]$^+$ = [$\{M(L)_n\}$−C≡O]$^+$. [b]In CH$_2$Cl$_2$ unless otherwise stated. [c]In cm^{-1}.

Table 4.10   Selected examples of the variation of $\Delta E_{1/2}$ with reaction medium for bis(fulvalene)dinickel, taken from a systematic study of effects by Barrière and Geiger[118b]

solvent	electrolyte	$\Delta E_{1/2}$ (mV)
THF	$Bu_4N^+BF_4^-$	343
CH_2Cl_2	$Bu_4N^+BF_4^-$	410
CH_2Cl_2	$Bu_4N^+ClO_4^-$	422
CH_2Cl_2	$Bu_4N^+PF_6^-$	480
CH_3CN	$Bu_4N^+PF_6^-$	459
acetone	$Bu_4N^+PF_6^-$	441
acetone	$Et_4N^+PF_6^-$	410

> phenylene. With omission of the aforementioned outliers, clear correlations between $\Delta E_{1/2}$ measurements for complexes of the type $[\{M(L)_n\}(\mu\text{-}BL)\{M(L)_n\}]$ and $v(C\equiv O)/v(C\equiv N)$ frequencies for analogous complexes $\{M(L)_n\}\text{-}X$ ($X = C\equiv O$, $C\equiv N^-$) are observed for both the C_4 and C_8 series – with the phenylene dataset showing greater ambiguity. Such trends quantify the proposed link between $\Delta E_{1/2}$ of MV complexes and electron density at $\{M(L)_n\}$.

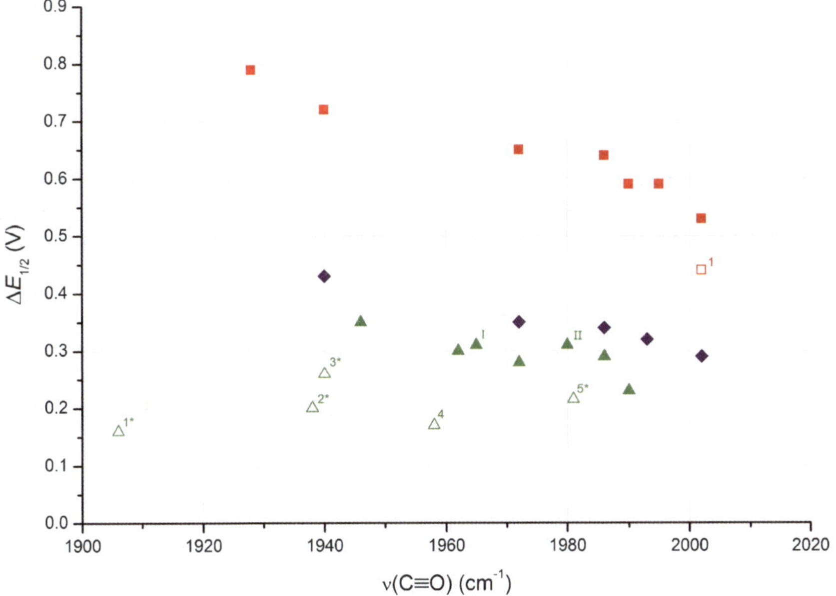

Figure 4.4 $\Delta E_{1/2}$ (V) *vs.* $v(C\equiv O)$ (cm^{-1}) for the C_4 (red squares), C_8 (blue diamonds) and phenylene (green triangles) series. Selected outliers are denoted by hollow shapes, numbered, and discussed in the text. Green triangles marked as I and II represent $\{M(L)_n\}$ = *trans*-RuCl(dppm)$_2$ where $v(C\equiv O)$ was measured for $[\{M(L)_n\}\text{-}C\equiv O]^+$ as the BF_4^- and PF_6^- salt, respectively.

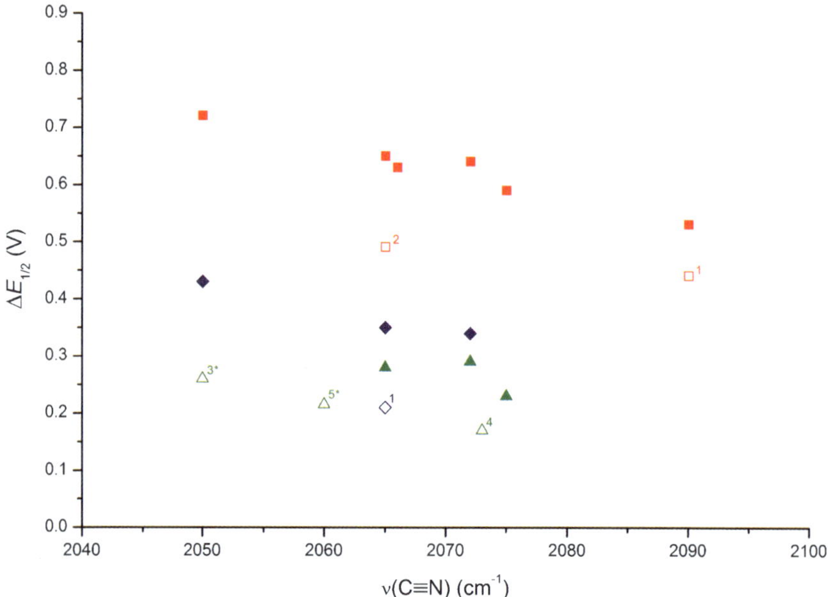

Figure 4.5 $\Delta E_{1/2}$ (V) *vs.* $v(C\equiv N)$ (cm^{-1}) for the C_4 (red squares), C_8 (blue diamonds) and phenylene (green triangles) series. Selected outliers are denoted by hollow shapes, numbered, and discussed in the text.

These observations may be reasonably explained by the notion that the more electron-rich a redox centre, the easier it is to oxidise. Mono-cations of such compounds will additionally feature a higher electron density, and should therefore exhibit a higher stability. In line with eqns (4.6) and (4.7), increased stability of a mono-cation of the type $[M_1^{II}M_2^{III}]$ (with respect to the fully reduced $[M_1^{II}M_2^{II}]$ and oxidised $[M_1^{III}M_2^{III}]$ forms) will ultimately increase $\Delta E_{1/2}$ and K_c by shifting the thermodynamic equilibrium in its favour. This reasoning has recently been used to explain an increase in $\Delta E_{1/2}/K_c$ with the electron donating ability of μ-BL in the series of complexes $[\{Fe(L)_n\}(-C\equiv C$-p-$C_6H_2X_2$-$C\equiv C$-$)\{Fe(L)_n\}]$, where $\{Fe(L)_n\}$ = Fe(dppe)Cp* and X = NMe_2, OMe, CH_3, H, F and CF_3,[189] – a similar trend was also noted for their corresponding H_{ab} values.[27e]

Such an explanation is useful, though one based on how electron density affects the metal orbitals relative to those of the bridging ligand is more desirable, certainly from a molecular electronics/superexchange perspective. As discussed by Evans *et al.*,[113] increased electron density at $\{M(L)_n\}$ would increase $E_{D/A}$ (the energy of the metal d orbitals) relative to E_{LUMO} (the energy of the bridge LUMO) and E_{HOMO}, increasing coupling between donor and acceptor in a electron transfer (LUMO-mediated) mechanism (and *vice versa*), for example.

Ultimately though, such arguments do not specifically rationalise the magnitude of $\Delta E_{1/2}$ with ΔG_r, nor do they explain how the quantities of the other contributions to $\Delta E_{1/2}$ may be affected by increasing electron density at the metal centre (eqns (4.8) and (4.9)). Whilst outliers may indeed be identified in terms of deviation from the standard reaction medium conditions as set by the series (and the subsequent effects on ΔG_{ip}), more consideration is needed to explain the unusually small values of $\Delta E_{1/2}$ for the Fe-containing complexes of the phenylene series, and the reduced correlation of that dataset in general. A partial explanation of the latter may be that correlations would be expected to decrease in proportion to the extent of electronic communication between redox centres, if the correlation is due to ΔG_r specifically. ΔG_r is expected to dominate the terms in eqn (4.9) when electronic communication is high (*i.e.* when $\Delta E_{1/2} > 180$ mV).[113,119] Furthermore, it is possible that with identical medium conditions in the C_4/C_8 series ΔG_{ne} is constant, and so the changes in $\Delta E_{1/2}$ represent a real change in ΔG_r, implying in turn a direct correlation between ΔG_r and the corresponding $v(C\equiv O)/v(C\equiv N)$ values/electron density.

4.4.2.5 Predictions

Regardless of the meaning of the observed correlation, we believe that it is useful as a diagnostic tool where outliers (due to medium effects or other phenomena) may be identified and analysed. Furthermore, the relation may be used to predict unknown $\Delta E_{1/2}$ values for given $[\{M(L)_n\}(\mu\text{-}BL)\{M(L)_n\}]$ complexes from the $v(C\equiv O)/v(C\equiv N)$ frequencies of their $\{M(L)_n\}$-X analogues (and *vice versa*). Values suggested from linear fits of nonoutlier data points (parameters given in Table 4.11) for C_4- and C_8-bridged complexes are shown in Table 4.12. Whilst only $\{M(L)_n\}$ centres used in this study have been included here, the approach is readily extendable to other systems using any known $\{M(L)_n\}$-X species.

Homobimetallic compounds containing acetylide bridges not directly bound to the metal centre $[\{M^*\}(\mu\text{-}BL)\{M^*\}]$ (*e.g.* where $M^* = $ Fc) do not have $\{M(L)_n\}$-X analogues yet may also be considered. $\Delta E_{1/2}$ values for these complexes can be incorporated into existing linear fits (of the same μ-BL) whereby the M^* moiety is assigned an *effective* $v(X)$ frequency. This can be used to predict $\Delta E_{1/2}$ values with other μ-BL. Not extensively explored here, this concept is supported by the following example. When the measured $\Delta E_{1/2}$

Table 4.11 Parameters from linear fits of nonoutlier datapoints for $\Delta E_{1/2}$ *vs.* $v(X)$ (Figure 4.4 and Figure 4.5)[a].

	$v(C\equiv O)$ plots		$v(C\equiv N)$ plots	
	C_4	C_8	C_4	C_8
m	-3.1×10^{-3}	-2.1×10^{-3}	-4.7×10^{-3}	-4.3×10^{-3}
c	6.7588	4.5663	10.356	9.2158
R^2	0.9409	0.9762	0.9542	0.9544

[a]$\Delta E_{1/2} = mv(X) + c$, where X $=$ C$\equiv$O, C$\equiv$N.

Table 4.12 Predicted $\Delta E_{1/2}$ values[a] for unknown complexes of the type $[\{M(L)_n\}(\mu\text{-BL})\{M(L)_n\}]$ based on linear fits of selected data plots for known complexes.

	$\Delta E_{1/2}$ (V)			
	C_4		C_8	
ML_n	$v(C{\equiv}O)$ *plots*	$v(C{\equiv}N)$ *plots*	$v(C{\equiv}O)$ *plots*	$v(C{\equiv}N)$ *plots*
Fe(dppe)Cp*	–	–	–	–
Fe(dippe)Cp*	–	/	–	/
Fe(dppe)Cp	0.62	0.67	0.41	0.36
Ru(dppe)Cp*	–	–	–	–
Ru(dppm)Cp*	/	–	/	0.33
Ru(dppe)Cp	–	–	0.39	0.29
Ru(PPh$_3$)$_2$Cp	–	–	–	–
Ru(PPh$_3$)(PMe$_3$)Cp	–	/	0.38	/
Os(PPh$_3$)$_2$Cp	/	0.65	/	0.34
Re(NO)(PPh$_3$)Cp*	–	–	0.36	0.23
Re(NO)(P(p-tol)$_3$)Cp*	–	/	–	/
Re(NO)(PCy$_3$)Cp*	0.58	/	–	/
Mo(dppe)(η-C$_7$H$_7$)	0.69	0.61	0.45	0.30
trans-FeCl(depe)$_2$	0.85	/	0.56	/
trans-FeCl(dmpe)$_2$	0.75	/	0.50	/
trans-RuCl(dppe)$_2$	0.73	/	0.48	/
trans-RuCl(dppm)$_2$	0.67	/	0.44	/
	0.62	/	0.41	/
trans-OsCl(dppm)$_2$	0.68	/	0.45	/

[a]For values obtained in Bu$_4$N$^+$BF$_4^-$/CH$_2$Cl$_2$ with Pt working electrode. Calculated by inputting $v(X)$ values for $\{ML_n\}$-X into the equation $\Delta E_{1/2} = mv(X) + c$ using the relevant parameters from Table 4.11. "/" = no $v(X)$ value known for calculation, "–" = value already measured in a comparable electrochemical system.

for $[\{Fc\}(\mu\text{-}C_4)\{Fc\}]$ (0.09 V[215]) is incorporated into the $v(C{\equiv}O)$ C_4 linear fit (Figure 4.4, Table 4.11), a $v(C{\equiv}O)_{eff}$ of 2151 cm^{-1} is obtained. If this $v(C{\equiv}O)_{eff}$ value is then used to predict $\Delta E_{1/2}$, where μ-BL = C_8, a value of 0.05 V is suggested, and a lower value still would be expected where μ-BL = phenylene. This is consistent with the experimentally observed lack of electronic communication in the $[\{Fc\}(\mu\text{-phenylene})\{Fc\}]$ complex.[27d,195]

Furthermore, the fact that numerous Pt complexes exhibit nonreversible electrochemistry for the C_8 series (as noted in Table 4.5), and are generally thought to exhibit wire-like properties inferior to analogous Ru systems (Section 4.3), could also be addressed by this electron density–$\Delta E_{1/2}$ relation. When using $v(C{\equiv}O)$ frequencies (Nujol mulls) to predict $\Delta E_{1/2}$ for the μ-BL = C_8 complexes $[\{Pt(PPh_3)_2(Ph)\}\text{-}C{\equiv}O]PF_6$ (2118 cm^{-1}),[216] $[\{Pt(PPh_3)_2(p\text{-tol})\}\text{-}C{\equiv}O]PF_6$ (2105 cm^{-1}),[216] $[\{Pt(PPh_3)_2(C_6F_5)\}\text{-}C{\equiv}O]BF_4$ (2115 cm^{-1}),[217] values between 0.12 and 0.15 V are obtained, small in comparison to other metal centres, including $\{Ru(L)_n\}$ (for μ-BL = C_4, $\Delta E_{1/2}$ predictions are between 0.19 and 0.23 V).

4.4.2.6 Extensions to the Current Study

Whilst further analyses of observed correlations are beyond the remit of this work, preliminary results are extremely encouraging and directing.

Ideally, existing datasets should be extended in both directions through the studies of more electron-rich/deficient metal centres. Currently, with only small variations in electron density, errors due to solvent effects and electrolyte changes/concentration are significant (and will extend to predictions). Of particular interest is the identification of extremely electron-rich metal centres, $v(C\equiv O) \ll 1928$ cm^{-1}, to extend $\Delta E_{1/2}$ beyond what is obtainable for $\{M(L)_n\}$ = Fe(dippe)Cp*. The gap between $\{M(L)_n\}$ = Re(NO)(PPh$_3$)Cp* and Pt(PPh)$_2$(L) (>100 cm^{-1}) is sizable, and should be closed to further explore the lower limits of electron density required to observe electronic communication through a given bridge (*i.e.* where linear extrapolation arrives at $\Delta E_{1/2}$ = 0). These currently stand at $v(C\equiv O)$ = 2174, 2180 cm^{-1} and $v(C\equiv N)$ = 2143, 2203 cm^{-1} for C$_8$ and C$_4$ bridged complexes, respectively.

Addition of further datasets for different μ-BL would help ratify this study's preliminary conclusions, as would the correlation of $v(C\equiv O)/v(C\equiv N)$ with other measures of electron delocalisation (*e.g.* H_{ab} or $H_{MM'}$). Future studies should focus on this (in addition to the adjustment of $\Delta E_{1/2}$ closer to ΔG_r by consideration of ΔG_{ne}) to explore further the possible link between electron density and electron delocalisation in these systems.

4.5 Conclusion

The incorporation of transition-metal centres into materials for molecular electronics is of particular interest, lending addressability to the nanoscale components and offering intriguing device functionality. Typical synthetic routes to metal σ alkynyl complexes have been summarised, with key methods for preparing unsymmetrical bis-alkynyl species highlighted. To assess the role of the metal centre in electron delocalisation and molecular conductance, studies of appropriate Ru- and Pt-containing systems were critically examined. The inclusion of Ru centres generally appears to improve the wire-like properties of materials relative to their Pt and organic analogues, which are roughly equivalent.

In an attempt to expand these discussions to include all metals and ancillary ligand frameworks, we have quantified the relationship between electron density at $\{M(L)_n\}$ and $\Delta E_{1/2}$ (a value linked to the extent of electronic delocalisation) in complexes of the type $[\{M(L)_n\}(\mu\text{-BL})\{M(L)_n\}]$ (μ-BL = C$_4$, C$_8$) from analysis of relevant data taken from studies of their MV complexes and mono-nuclear complexes $\{M(L)_n\}$-X (X = C$\equiv$O, C$\equiv$N$^-$). These initial correlations suggest that the magnitude of $\Delta E_{1/2}$ may be predicted for analogous unknown homobimetallic systems using $v(X)$ for the mononuclear $\{M(L)_n\}$–X complexes. Compounds where μ-BL = phenylene exhibit a less well defined $\Delta E_{1/2}$ *vs.* $v(X)$ relationship, suggesting $\Delta G_{ne} \neq$ constant.

Acknowledgements

The authors would like to thank Dr. Tim Albrecht for reading the chapter and providing helpful comments, and to thank the EPSRC for funding a PhD studentship (to M.I.).

References

1. (a) N. Robertson and C. A. McGowan, *Chem. Soc. Rev.*, 2003, **32**, 96; (b) R. L. Carroll and C. B. Gorman, *Angew. Chem., Int. Ed.*, 2002, **41**, 4379; (c) N. Weibel, S. Grunder and M. Mayor, *Org. Biomol. Chem.*, 2007, **5**, 2343; (d) N. J. Tao, *Nature Nanotechnol.*, 2006, **1**, 173.

2. (a) F. Chen, J. Hihath, Z. Huang, X. Li and N. J. Tao, *Annu. Rev. Phys. Chem.*, 2007, **58**, 535; (b) H. Haick and D. Cahen, *Prog. Surf. Sci.*, 2008, **83**, 217; (c) R. J. Nichols, W. Haiss, S. J. Higgins, E. Leary, S. Martin and D. Bethell, *Phys. Chem. Chem. Phys.*, 2010, **12**, 2801.

3. F. Chen, X. Li, J. Hihath, Z. Huang and N. J. Tao, *J. Am. Chem. Soc.*, 2006, **128**, 15874.

4. J. He, F. Chen, J. Li, O. F. Sankey, Y. Terazono, C. Herrero, D. Gust, T. A. Moore, A. L. Moore and S. M. Lindsay, *J. Am. Chem. Soc.*, 2005, **127**, 1384.

5. C. Wang, A. S. Batsanov, M. R. Bryce, S. Martín, R. J. Nichols, S. J. Higgins, V. M. García-Suárez and C. J. Lambert, *J. Am. Chem. Soc.*, 2009, **131**, 15647.

6. T. Rueckes, K. Kim, E. Joselevich, G. Y. Tseng, C.-L. Cheung and C. M. Lieber, *Science*, 2000, **289**, 94.

7. M. U. Winters, E. Dahlstedt, H. E. Blades, C. J. Wilson, M. J. Frampton, H. L. Anderson and B. Albinsson, *J. Am. Chem. Soc.*, 2007, **129**, 4291.

8. (a) A. K. Mahapatro, G. U. Lee, K. J. Jeong and D. B. Janes, *Appl. Phys. Lett.*, 2009, **95**, 083106; (b) M. Taniguchi and T. Kawai, *Physica E*, 2006, **33**, 1.

9. (a) L. T. Cai, H. Skulason, J. G. Kushmerick, S. K. Pollack, J. Naciri, R. Shashidhar, D. L. Allara, T. E. Mallouk and T. S. Mayer, *J. Phys. Chem. B*, 2004, **108**, 2827; (b) A. Salomon, D. Cahen, S. Lindsay, J. Tomfohr, V. B. Engelkes and C. D. Frisbie, *Adv. Mater.*, 2003, **15**, 1881.

10. B. Kim, J. M. Beebe, Y. Jun, X.-Y. Zhu and C. D. Frisbie, *J. Am. Chem. Soc.*, 2006, **128**, 4970.

11. (a) A. Nitzan and M. A. Ratner, *Science*, 2003, **300**, 1384; (b) B. Kim, J. M. Beebe, C. Olivier, S. Rigaut, D. Touchard, J. G. Kushmerick, X.-Y. Zhu and C. D. Frisbie, *J. Phys. Chem. C*, 2007, **111**, 7521.

12. (a) J. M. Seminario, A. G. Zacarias and J. M. Tour, *J. Am. Chem. Soc.*, 2000, **122**, 3015; (b) K. Liu, X. Wang and F. Wang, *ACS Nano*, 2008, **2**, 2315.

13. (a) J. Chen, W. Wang, M. A. Reed, A. M. Rawlett, D. W. Price and J. M. Tour, *Appl. Phys. Lett.*, 2000, **77**, 1224; (b) Y. Selzer, M. A. Cabassi, T. S. Mayer and D. L. Allara, *Nanotechnology*, 2004, **15**, S483.

14. (a) N. Gergel-Hackett, N. Majumdar, Z. Martin, N. Swami, L. R. Harriott, J. C. Bean, G. Pattanaik, G. Zangari, Y. Zhu, I. Pu, Y. Yao and J. M. Tour, *J. Vac. Sci. Technol. A*, 2006, **24**, 1243; (b) Y. Selzer, L. Cai, M. A. Cabassi, Y. Yao, J. M. Tour, T. S. Mayer and D. L. Allara, *Nano Lett.*, 2005, **5**, 61; (c) Z. J. Donhauser, B. A. Mantooth, K. F. Kelly, L. A. Bumm, J. D. Monnell, J. J. Stapleton, D. W. Price Jr., A. M. Rawlett, D. L. Allara, J. M. Tour and P. S. Weiss, *Science*, 2001, **292**, 2303; (d) Z.-F. Shi, L.-J. Wang, H. Wang, X.-P. Cao and H.-L. Zhang, *Org. Lett.*, 2007, **9**, 595; (e) Y. Ie, M. Endou, S. K. Lee, R. Yamada, H. Tada and Y. Aso, *Angew. Chem., Int. Ed.*, 2011, 50.

15. (a) T. Albrecht, K. Moth Poulsen, J. B. Christensen, J. Hjelm, T. Bjørnholm and J. Ulstrup, *J. Am. Chem. Soc.*, 2006, **128**, 6574; (b) S. Kubatkin, A. Danilov, M. Hjort, J. Cornil, J.-L. Brédas, N. Stuhr-Hansen, P. Hedegård and T. Bjørnholm, *Nature*, 2003, **425**, 698; (c) A. M. Kuznetsov, P. Sommer-Larsen and J. Ulstrup, *Surf. Sci.*, 1992, **275**, 52.

16. P. J. Low, *Dalton Trans.*, 2005, 2821.

17. (a) T. Albrecht, A. Guckian, J. Ulstrup and J. G. Vos, *IEEE Trans. Nanotechnol.*, 2005, **4**, 430; (b) A. K. Mahapatro, J. Ying, T. Ren and D. B. Janes, *Nano Lett.*, 2008, **8**, 2131.

18. (a) M. A. Reed, J. Chen, A. M. Rawlett, D. W. Price and J. M. Tour, *Appl. Phys. Lett.*, 2001, **78**, 3735; (b) I.-W. P. Chen, M.-D. Fu, W.-H. Tseng, J.-Y. Yu, S.-H. Wu, C.-J. Ku, C.-h. Chen and S.-M. Peng, *Angew. Chem., Int. Ed. Engl.*, 2006, **45**, 5814.

19. J. He, B. Chen, A. K. Flatt, J. J. Stephenson, C. D. Doyle and J. M. Tour, *Nature Mater.*, 2006, **5**, 63.

20. (a) M. Elbing, R. Ochs, M. Koentopp, M. Fischer, C. v. Hänisch, F. Weigend, F. Evers, H. B. Weber and M. Mayor, *Proc. Natl. Acad. Sci. U. S. A.*, 2005, **102**, 8815; (b) I. Díez-Pérez, J. Hihath, Y. Lee, L. Yu, L. Adamska, M. A. Kozhushner, I. I. Oleynik and N. J. Tao, *Nature Chem.*, 2009, **1**, 635.

21. W. Liang, M. P. Shores, M. Bockrath, J. R. Long and H. Park, *Nature*, 2002, **417**, 725.

22. N. J. Long and C. K. Williams, *Angew. Chem., Int. Ed. Engl.*, 2003, **42**, 2586.

23. (a) N. J. Long, *Angew. Chem., Int. Ed. Engl.*, 1995, **34**, 21; (b) C. E. Powell and M. G. Humphrey, *Coord. Chem. Rev.*, 2004, **248**, 725.

24. V. W.-W. Yam and K. M.-C. Wong, *Top. Curr. Chem.*, 2005, **257**, 1.

25. D. W. Bruce in *Inorganic Materials*, ed. D. W. Bruce and D. O'Hare, Wiley, Chichester, 2nd edn, 1996 p. 429.

26. (a) A. Nitzan, *J. Phys. Chem. A*, 2001, **105**, 2677; (b) A. Nitzan, *Isr. J. Chem.*, 2002, **42**, 163; (c) M. C. Traub, B. S. Brunschwig and N. S. Lewis, *J. Phys. Chem. B*, 2007, **111**, 6676.

27. (a) M. Guillemot, L. Toupet and C. Lapinte, *Organometallics*, 1998, **17**, 1928; (b) M. I. Bruce, B. G. Ellis, M. Gaudio, C. Lapinte, G. Melino, F.

Paul, B. W. Skelton, M. E. Smith, L. Toupet and A. H. White, *Dalton Trans.*, 2004, 1601; (c) W. E. Meyer, A. J. Amoroso, C. R. Horn, M. Jaeger and J. A. Gladysz, *Organometallics*, 2001, **20**, 1115; (d) A. Klein, O. Lavastre and J. Fiedler, *Organometallics*, 2006, **25**, 635; (e) P. Aguirre-Etcheverry and D. O'Hare, *Chem. Rev.*, 2010, **110**, 4839.

28. (a) K. Sonogashira, S. Takahashi and N. Hagihara, *Macromolecules*, 1977, **10**, 879; (b) P. J. Kim, H. Masai, K. Sonogashira and N. Hagihara, *Inorg. Nucl. Chem. Lett.*, 1970, **6**, 181; (c) K. Sonogashira, T. Yatake, Y. Tohda, S. Takahashi and N. Hagihara, *Chem. Commun.*, 1977, 291.

29. (a) B. Cetinkaya, M. F. Lappert, J. McMeeking and D. E. Palmer, *J. Chem. Soc., Dalton Trans.*, 1973, 1202; (b) S. J. Davies, B. F. G. Johnson, M. S. Khan and J. Lewis, *J. Chem. Soc., Chem. Commun.*, 1991, 187; (c) Z. Atherton, C. W. Faulkner, S. L. Ingham, A. K. Kakkar, M. S. Khan, J. Lewis, N. J. Long and P. R. Raithby, *J. Organomet. Chem.*, 1993, **462**, 265.

30. (a) M. L. H. Green in *Organometallic Compounds*, ed. G. E. Coates, M. L. H. Green and K. Wade, Methuen, London, 3rd. edn, 1968; (b) A. López-Hernández, K. Venkatesan, H. W. Schmalle and H. Berke, *Monatsh. Chem.*, 2009, **140**, 845.

31. (a) D. Zargarian, P. Chow, N. J. Taylor and T. B. Marder, *Chem. Commun.*, 1989, 540; (b) H. B. Fyfe, M. Mlekuz, D. Zargarian, N. J. Taylor and T. B. Marder, *Chem. Commun.*, 1991, 188; (c) H.-F. Klein and H. H. Karsch, *Chem. Ber.*, 1975, **108**, 944.

32. C. Bianchini, M. Peruzzini, A. Vacca and F. Zanobini, *Organometallics*, 1991, **10**, 3697.

33. (a) C. Bianchini, F. Laschi, F. Ottaviani, M. Peruzzini and P. Zanello, *Organometallics*, 1988, **7**, 1660; (b) L. D. Field and A. V. George, *J. Organomet. Chem.*, 1993, **454**, 217; (c) C. Bianchini, P. Frediani, D. Masi, M. Peruzzini and F. Zanobini, *Organometallics*, 1994, **13**, 4616.

34. (a) P. Haquette, N. Pirio, D. Touchard, L. Toupet and P. H. Dixneuf, *Chem. Commun.*, 1993, 163; (b) D. Touchard, C. Morice, V. Cadierno, P. Haquette, L. Toupet and P. H. Dixneuf, *Chem. Commun.*, 1994, 859; (c) D. Touchard, P. Haquette, S. Guesmi, L. L. Pichon, A. Daridor, L. Toupet and P. H. Dixneuf, *Organometallics*, 1997, **16**, 3640.

35. P. J. Stang and C. M. Crittell, *Organometallics*, 1990, **9**, 3191.

36. (a) L. D. Field, A. V. George, F. Laschi, E. Y. Malouf and P. Zanello, *J. Organomet. Chem.*, 1992, **435**, 347; (b) M.-Y. Choi, M. C.-W. Chan, S. Zhang, K.-K. Cheung, C.-M. Che and K.-Y. Wong, *Organometallics*, 1999, **18**, 2074.

37. (a) R. J. Cross and M. F. Davidson, *J. Chem. Soc., Dalton Trans.*, 1986, 411; (b) W. M. Khairul, M. A. Fox, N. N. Zaitseva, M. Gaudio, D. S. Yufit, B. W. Skelton, A. H. White, J. A. K. Howard, M. I. Bruce and P. J. Low, *J. Chem. Soc., Dalton Trans.*, 2009, 610.

38. (a) J. Vicente and M. T. Chicote, *Coord. Chem. Rev.*, 1999, **193-195**, 1143; (b) M. I. Bruce, M. Jevric, B. W. Skelton, M. E. Smith, A. H. White and N. N. Zaitseva, *J. Organomet. Chem.*, 2006, **691**, 361.

39. C. Müller, C. N. Iverson, R. J. Lachicotte and W. D. Jones, *J. Am. Chem. Soc.*, 2001, **123**, 9718.

40. R. Gleiter and D. B. Werz, *Chem. Rev.*, 2010, **110**, 4447.

41. (a) K. Moth-Poulsen, L. Patrone, N. Stuhr-Hansen, J. B. Christensen, J.-P. Bourgoin and T. Bjørnholm, *Nano Lett.*, 2005, **5**, 783; (b) D. S. Seferos, A. S. Blum, J. G. Kushmerick and G. C. Bazan, *J. Am. Chem. Soc.*, 2006, **128**, 11260; (c) K. Liu, G. Li, X. Wang and F. Wang, *J. Phys. Chem. C*, 2008, **112**, 4342.

42. (a) L. Patrone, S. Palacin and J. P. Bourgoin, *Appl. Surf. Sci.*, 2003, **212–213**, 446; (b) S. Yasuda, S. Yoshida, J. Sasaki, Y. Okutsu, T. Nakamura, A. Taninaka, O. Takeuchi and H. Shigekawa, *J. Am. Chem. Soc.*, 2006, **128**, 7746.

43. J. M. Beebe, V. B. Engelkes, L. L. Miller and C. D. Frisbie, *J. Am. Chem. Soc.*, 2002, **124**, 11268.

44. L. Venkataraman, J. E. Klare, I. W. Tam, C. Nuckolls, M. S. Hybertsen and M. L. Steigerwald, *Nano Lett.*, 2006, **6**, 458.

45. B. Xu and N. J. Tao, *Science*, 2003, **301**, 1221.

46. G. A. Koutsantonis, P. A. Schauer and B. W. Skelton, *Organometallics*, 2011, **30**, 2680.

47. Y. Ie, T. Hirose, H. Nakamura, M. Kiguchi, N. Takagi, M. Kawai and Y. Aso, *J. Am. Chem. Soc.*, 2011, **133**, 3014.

48. (a) H.-W. Lin, X.-H. Wang, X.-J. Zhao, J. Li and F.-S. Wang, *Synth. Met.*, 2003, **135–136**, 239; (b) H. W. Lin, X. H. Wang, X. J. Zhao, J. Li and F. S. Wang, *Chin. Chem. Lett.*, 2003, **14**, 35; (c) M. Mayor, C. von Hänisch, H. B. Weber, J. Reichert and D. Beckmann, *Angew. Chem., Int. Ed.*, 2002, **41**, 1183; (d) T. L. Schull, J. G. Kushmerick, C. H. Patterson, C. George, M. H. Moore, S. K. Pollack and R. Shashidhar, *J. Am. Chem. Soc.*, 2003, **125**, 3202.

49. (a) J.-L. Zuo, E. Herdtweck, F. F. d. Biani, A. M. Santos and F. E. Kühn, *New J. Chem.*, 2002, **26**, 889; (b) F. E. Kühn, J.-L. Zuo, F. F. d. Biani, A. M. Santos, Y. Zhang, J. Zhao, A. Sandulache and E. Herdtweck, *New J. Chem.*, 2004, **28**, 43; (c) Q. Ge and T. S. A. Hor, *J. Chem. Soc., Dalton Trans.*, 2008, 2929; (d) M. Ferrer, L. Rodríguez, O. Rossell, F. Pina, J. C. Lima, M. F. Bardia and X. Solans, *J. Organomet. Chem.*, 2003, **678**, 82.

50. C. Olivier, B. Kim, D. Touchard and S. Rigaut, *Organometallics*, 2008, **27**, 509.

51. H. Kawaguchi and K. Tatsumi, *Organometallics*, 1995, **14**, 4294.

52. T. P. Vaid, A. S. Veige, E. B. Lobkovsky, W. V. Glassey, P. T. Wolczanski, L. M. Liable-Sands, A. L. Rheingold and T. R. Cundari, *J. Am. Chem. Soc.*, 1998, **120**, 10067.

53. A. Hills, D. L. Hughes, N. Kashef, M. A. N. D. A. Lemos, A. J. L. Pombeiro and R. L. Richards, *J. Chem. Soc., Dalton Trans.*, 1992, 1775.

54. V. V. Krivykh, I. L. Eremenko, D. Veghini, I. A. Petrunenko, D. L. Pountney, D. Unseld and H. Berke, *J. Organomet. Chem.*, 1996, **511**, 111.

55. L. D. Field, A. V. George, T. W. Hambley, E. Y. Malouf and D. J. Young, *Chem. Commun.*, 1990, 931.

56. Z. Atherton, C. W. Faulkner, S. L. Ingham, A. K. Kakkar, M. S. Khan, J. Lewis, N. J. Long and P. R. Raithby, *J. Organomet. Chem.*, 1993, **462**, 265.

57. M. S. Khan, N. A. Pasha, A. K. Kakkar, P. R. Raithby, J. Lewis, K. Fuhrmann and R. H. Friend, *J. Mater. Chem.*, 1992, **2**, 759.

58. H. Werner, A. Höhn and M. Schulz, *J. Chem. Soc., Dalton Trans.*, 1991, 777.

59. J. Chatt and B. L. Shaw, *J. Chem. Soc.*, 1960, 1718.

60. R. D'Amato, A. Furlani, M. Colapietro, G. Portalone, M. Casalboni, M. Falconieri and M. V. Russo, *J. Organomet. Chem.*, 2001, **627**, 13.

61. O. M. Abu-Salah, A. R. Al-Ohaly and H. A. Al-Qahtani, *Inorg. Chim. Acta*, 1986, **117**, L29.

62. R. D. Dewhurst, A. F. Hill and M. K. Smith, *Organometallics*, 2006, **25**, 2388.

63. F. G. Kirchbauer, P.-M. Pellny, H. Sun, V. V. Burlakov, P. Arndt, W. Baumann, A. Spannenberg and U. Rosenthal, *Organometallics*, 2001, **20**, 5289.

64. H. Lang, D. S. A. George and G. Rheinwald, *Coord. Chem. Rev.*, 2000, **206–207**, 101.

65. R. F. Munhá, M. A. Antunes, L. G. Alves, L. F. Veiros, M. D. Fryzuk and A. M. Martins, *Organometallics*, 2010, **29**, 3753.

66. (a) D. L. Reger and M. F. Huff, *Organometallics*, 1990, **9**, 2807; (b) D. Michael, P. Mingos, J. Yau, S. Menzer and D. J. Williams, *Angew. Chem., Int. Ed. Engl.*, 1995, **34**, 1894.

67. (a) V. W.-W. Yam, *J. Organomet. Chem.*, 2004, **689**, 1393; (b) M. P. Gamasa, J. Gimeno, E. Lastra, A. Aguirre and S. Garcia-Granda, *J. Organomet. Chem.*, 1989, **378**, C11; (c) V. W.-W. Yam, W. K.-M. Fung and K.-K. Cheung, *Organometallics*, 1997, **16**, 2032; (d) C. F. Wang, S. M. Peng, C. K. Chan and C. M. Che, *Polyhedron*, 1996, **15**, 1853.

68. (a) R. Nast, O. Künzel and R. Müller, *Chem. Ber.*, 1962, **95**, 2155; (b) O. Y. Okhlobystin and L. I. Zakharkin, *J. Organomet. Chem.*, 1965, **3**, 257; (c) E. A. Jeffery and T. Mole, *J. Organomet. Chem.*, 1968, **11**, 393.

69. A. Blum, T. Ren, D. A. Parish, S. A. Trammell, M. H. Moore, J. G. Kushmerick, G.-L. Xu, J. R. Deschamps, S. K. Pollack and R. Shashidhar, *J. Am. Chem. Soc.*, 2005, **127**, 10010.

70. H. Qi, S. Sharma, Z. Li, G. L. Snider, A. O. Orlov, C. S. Lent and T. P. Fehlner, *J. Am. Chem. Soc.*, 2003, **125**, 15250.

71. (a) S. N. Semenov, S. F. Taghipourian, O. Blacque, T. Fox, K. Venkatesan and H. Berke, *J. Am. Chem. Soc.*, 2010, **132**, 7584; (b) J.-W. Ying, I. P.-C. Liu, B. Xi, Y. Song, C. Campana, J.-L. Zuo and T. Ren, *Angew. Chem., Int. Ed. Engl.*, 2010, **49**, 954; (c) C. Olivier, S. Choua, P. Turek, D. Touchard and S. Rigaut, *Chem. Commun.*, 2007, 3100; (d) C. Olivier, K. Costuas, S. Choua, V. Maurel, P. Turek, J.-Y.

Saillard, D. Touchard and S. Rigaut, *J. Am. Chem. Soc.*, 2010, **132**, 5638; (e) L. D. Field, A. M. Magill, T. K. Shearer, S. B. Colbran, S. T. Lee, S. J. Dalgarno and M. M. Bhadbhade, *Organometallics*, 2010, **29**, 957; (f) L. D. Field, A. J. Turnbull and P. Turner, *J. Am. Chem. Soc.*, 2002, **124**, 3692.

72. M. Bassetti, V. Cadierno, J. Gimeno and C. Pasquini, *Organometallics*, 2008, **27**, 5009.

73. M. A. Fox, J. E. Harris, S. Heider, V. Pérez-Gregorio, M. E. Zakrzewska, J. D. Farmer, D. S. Yufit, J. A. K. Howard and P. J. Low, *J. Organomet. Chem.*, 2009, **694**, 2350.

74. B. Chaudret, G. Commenges and R. Poilblanc, *J. Chem. Soc., Dalton Trans.*, 1984, 1635.

75. A. J. Hodge, S. L. Ingham, A. K. Kakkar, M. S. Khan, J. Lewis, N. J. Long, D. G. Parker and P. R. Raithby, *J. Organomet. Chem.*, 1995, **488**, 205.

76. M. Younus, N. J. Long, P. R. Raithby, J. Lewis, N. A. Page, A. J. P. White, D. J. Williams, M. C. B. Colbert, A. J. Hodge, M. S. Khan and D. G. Parker, *J. Organomet. Chem.*, 1999, **578**, 198.

77. (a) L. D. Field, A. M. Magill, T. K. Shearer, S. J. Dalgarno and P. Turner, *Organometallics*, 2007, **26**, 4776; (b) L. D. Field, A. V. George, D. C. R. Hockless, G. R. Purches and A. H. White, *J. Chem. Soc., Dalton Trans.*, 1996, 2011.

78. S. N. Semenov, O. Blacque, T. Fox, K. Venkatesan and H. Berke, *J. Am. Chem. Soc.*, 2010, **132**, 3115.

79. (a) F. J. Fernández, M. Alfonso, H. W. Schmalle and H. Berke, *Organometallics*, 2001, **20**, 3122; (b) K. Venkatesan, T. Fox, H. W. Schmalle and H. Berke, *Organometallics*, 2005, **24**, 2834.

80. T. Ren, *Organometallics*, 2005, **24**, 4854.

81. T. Ren, *Organometallics*, 2002, **21**, 732.

82. A. R. Chakravarty and F. A. Cotton, *Inorg. Chim. Acta*, 1986, **113**, 19.

83. G. Xu and T. Ren, *J. Organomet. Chem.*, 2002, **655**, 239.

84. (a) J. L. Bear, B. Han, S. R. Huang and K. M. Kadish, *Inorg. Chem.*, 1996, **35**, 3012; (b) K. P. Baldwin, R. S. Simons, D. A. Scheiman, R. Lattimer, C. A. Tessier and W. J. Youngs, *J. Chem. Crystallogr.*, 1998, **28**, 353.

85. G. Xu, C. Campana and T. Ren, *Inorg. Chem.*, 2002, **41**, 3521.

86. F. A. Cotton, L. M. Daniels, C. A. Murillo and I. Pascual, *J. Am. Chem. Soc.*, 1997, **119**, 10223.

87. (a) E.-C. Yang, M.-C. Cheng, M.-S. Tsai and S.-M. Peng, *J. Chem. Soc., Chem. Commun.*, 1994, 2377; (b) R. Clérac, F. A. Cotton, L. M. Daniels, K. R. Dunbar, K. Kirschbaum, C. A. Murillo, A. A. Pinkerton, A. J. Schultz and X. Wang, *J. Am. Chem. Soc.*, 2000, **122**, 6226.

88. (a) S. Aduldecha and B. Hathaway, *J. Chem. Soc., Dalton Trans.*, 1991, 993; (b) R. Clérac, F. A. Cotton, K. R. Dunbar, C. A. Murillo, I. Pascual and X. Wang, *Inorg. Chem.*, 1999, **38**, 2655.

89. (a) L.-P. Wu, P. Field, T. Morrissey, C. Murphy, P. Nagle, B. Hathaway, C. Simmons and P. Thornton, *J. Chem. Soc., Dalton Trans.*, 1990, 3835; (b) G. J. Pyrka, M. El-Mekki and A. A. Pinkerton, *Chem. Commun.*,

1991, 84; (c) J. Berry, F. A. Cotton, P. Lei and C. A. Murillo, *Inorg. Chem.*, 2003, **42**, 377.

90. (a) J.-T. Sheu, C.-C. Lin, I. Chao, C.-C. Wang and S.-M. Peng, *Chem. Commun.*, 1996, 315; (b) C.-K. Kuo, J.-C. Chang, C.-Y. Yeh, G.-H. Lee, C.-C. Wang and S.-M. Peng, *J. Chem. Soc., Dalton Trans.*, 2005, 3696.

91. R. E. Podobedov, T. A. Stromnova, A. V. Churakov, L. G. Kuzmina and I. A. Efimenko, *J. Organomet. Chem.*, 2010, **695**, 2083.

92. R. Ismayilov, W.-Z. Wang, G.-H. Lee, C.-Y. Yeh, S.-A. Hua, Y. Song, M.-M. Rohmer, M. Bénard and S.-M. Peng, *Angew. Chem., Int. Ed. Engl.*, 2011, **50**, 2045.

93. J. H. Kuo, T. B. Tsao, G. H. Lee, H. W. Lee, C. Y. Yeh and S. M. Peng, *Eur. J. Inorg. Chem.*, 2011, 2025.

94. (a) J. F. Berry, F. A. Cotton and C. A. Murillo, *Organometallics*, 2004, **23**, 2503; (b) J. F. Berry, F. A. Cotton, C. A. Murillo and B. K. Roberts, *Inorg. Chem.*, 2004, **43**, 2277.

95. D.-H. Chae, J. F. Berry, S. Jung, F. A. Cotton, C. A. Murillo and Z. Yao, *Nano Lett.*, 2006, **6**, 165.

96. (a) J. Park, A. N. Pasupathy, J. I. Goldsmith, C. Chang, Y. Yaish, J. R. Petta, M. Rinkoski, J. P. Sethna, H. D. Abruña, P. L. McEuen and D. C. Ralph, *Nature*, 2002, **417**, 722; (b) L. H. Yu, Z. K. Keane, J. W. Ciszek, L. Cheng, M. P. Stewart, J. M. Tour and D. Natelson, *Phys. Rev. Lett.*, 2004, **93**, 266802.

97. S. A. Getty, C. Engtrakul, L. Wang, R. Liu, S.-H. Ke, H. U. Baranger, W. Yang, M. S. Fuhrer and L. R. Sita, *Phys. Rev. B*, 2005, **71**, 241401.

98. P. Siemsen, U. Gubler, C. Bosshard, P. Günter and F. Diederich, *Chem. Eur. J.*, 2001, **7**, 1333.

99. C. Stroh, M. Mayor, C. v. Hänisch and P. Turek, *Chem. Commun.*, 2004, 2050.

100. N. Lenarvor and C. Lapinte, *Organometallics*, 1995, **14**, 634.

101. F. d. Montigny, G. Argouarch, K. Costuas, J.-F. Halet, T. Roisnel, L. Toupet and C. Lapinte, *Organometallics*, 2005, **24**, 4558.

102. G. Vives, A. Carella, S. Sistach, J.-P. Launay and G. Rapenne, *New J. Chem.*, 2006, **30**, 1429.

103. S. C. Jones, V. Coropceanu, S. Barlow, T. Kinnibrugh, T. Timofeeva, J. L. Brédas and S. R. Marder, *J. Am. Chem. Soc.*, 2004, **126**, 11782.

104. N. D. Jones, M. O. Wolf and D. M. Giaquinta, *Organometallics*, 1997, **16**, 1352.

105. M. C. B. Colbert, J. Lewis, N. J. Long, P. R. Raithby, A. J. P. White and D. J. Williams, *J. Chem. Soc., Dalton Trans.*, 1997, 99.

106. Y. Zhu, O. Clot, M. O. Wolf and G. P. A. Yap, *J. Am. Chem. Soc.*, 1998, **120**, 1812.

107. C. Lebreton, D. Touchard, L. L. Pichon, A. Daridor, L. Toupet and P. H. Dixneuf, *Inorg. Chim. Acta*, 1998, **272**, 188.

108. D. Osella, R. Gobetto, C. Nervi, M. Ravera, R. D'Amato and M. V. Russo, *Inorg. Chem. Commun.*, 1998, **1**, 239.

109. Á. Díez, E. Lalinde, M. T. Moreno and S. Sánchez, *Dalton Trans.*, 2009, 3434.

110. C. Creutz and H. Taube, *J. Am. Chem. Soc.*, 1967, **91**, 3988.

111. N. S. Hush, *Prog. Inorg. Chem.*, 1967, **8**, 391.

112. C. Creutz, M. D. Newton and N. Sutin, *J. Photochem. Photobiol., A*, 1994, **82**, 47.

113. C. E. B. Evans, M. L. Naklicki, A. R. Rezvani, C. A. White, V. V. Kondratiev and R. J. Crutchley, *J. Am. Chem. Soc.*, 1998, **120**, 13096.

114. (a) D. M. D'Alessandro and F. R. Keene, *Chem. Soc. Rev.*, 2006, **35**, 424; (b) A. Troisi and M. A. Ratner, *Small*, 2006, **2**, 172.

115. M. B. Robin and P. Day, *Adv. Inorg. Chem.*, 1968, **10**, 247.

116. B. S. Brunschwig, C. Creutz and N. Sutin, *Chem. Soc. Rev.*, 2002, **31**, 168.

117. K. Costuas and S. Rigaut, *J. Chem. Soc., Dalton Trans.*, 2011, **40**, 5643.

118. (a) C. Lapinte, *J. Organomet. Chem.*, 2008, **693**, 793; (b) F. Barrière and W. E. Geiger, *J. Am. Chem. Soc.*, 2006, **128**, 3980; (c) F. Salaymeh, S. Berhane, R. Yusof, R. d. l. Rosa, E. Y. Fung, R. Matamoros, K. W. Lau, Q. Zheng, E. M. Kober and J. C. Curtis, *Inorg. Chem.*, 1993, **32**, 3895.

119. B. S. Brunschwig and N. Sutin, *Coord. Chem. Rev.*, 1999, **187**, 233.

120. S. B. Piepho, E. R. Krausz and P. N. Schatz, *J. Am. Chem. Soc.*, 1978, **100**, 2996.

121. (a) M. J. Ondrechen, J. Ko and L. J. Root, *J. Phys. Chem.*, 1984, **88**, 5919; (b) J. Ko and M. J. Ondrechen, *J. Am. Chem. Soc.*, 1985, **107**, 6161; (c) M. J. Ondrechen, J. Ko and L.-T. Zhang, *J. Am. Chem. Soc.*, 1987, **109**, 1672.

122. J. P. Chang, E. Y. Fung and J. C. Curtis, *Inorg. Chem.*, 1986, **25**, 4233.

123. C. A. Tolman, *Chem. Rev.*, 1977, **77**, 313.

124. M. I. Bruce, K. Costuas, B. G. Ellis, J.-F. Halet, P. J. Low, B. Moubaraki, K. S. Murray, N. Ouddaï, G. J. Perkins, B. W. Skelton and A. H. White, *Organometallics*, 2007, **26**, 3735.

125. (a) H. Ogawa, K. Onitsuka, T. Joh, S. Takahashi, Y. Yamamoto and H. Yamazaki, *Organometallics*, 1988, **7**, 2257; (b) K. Suenkel, U. Birk and C. Robl, *Organometallics*, 1994, **13**, 1679.

126. (a) M. Akita, M. Terada, S. Oyama and Y. Moro-oka, *Organometallics*, 1990, **9**, 816; (b) M. Akita, M. Terada, S. Oyama, S. Sugimoto and Y. Moro-oka, *Organometallics*, 1991, **10**, 1561; (c) M. Akita, M.-C. Chung, A. Sakurai, S. Sugimoto, M. Terada, M. Tanaka and Y. Moro-oka, *Organometallics*, 1997, **16**, 4882.

127. R. J. Spahr, R. R. Vogt and J. A. Nieuwland, *J. Am. Chem. Soc.*, 1933, **55**, 2465.

128. (a) R. J. Cross, M. F. Davidson and A. J. McLennan, *J. Organomet. Chem.*, 1984, **265**, C37; (b) R.-Y. Liau, A. Schier and H. Schmidbaur, *Organometallics*, 2003, **22**, 3199; (c) M. I. Bruce, K. R. Grundy, M. J. Liddell, M. R. Snow and E. R. T. Tiekink, *J. Organomet. Chem.*, 1988, **344**, C49; (d) C. M. Che, H. Y. Chao, V. M. Miskowski, Y. Li and K. K. Cheung, *J. Am. Chem. Soc.*, 2001, **123**, 4985.

129. R. Nast, P. Schneller and A. Hengefeld, *J. Organomet. Chem.*, 1981, **214**, 273.

130. J. A. Davies, M. El-ghanam, A. A. Pinkerton and D. A. Smith, *J. Organomet. Chem.*, 1991, **409**, 367.

131. G. Kreisel, P. Scholz and W. Seidel, *Z. Anorg. Allg. Chem.*, 1980, **460**, 51.

132. N. A. Ustynyuk, V. N. Vinogradova and D. N. Kravtsov, *Metalloorg. Khim.*, 1988, **1**, 85.

133. P. Binger, P. Müller, P. Philipps, B. Gabor, R. Mynott, A. T. Herrmann, F. Langhauser and C. Krüger, *Chem. Ber.*, 1992, **125**, 2209.

134. W. J. Evans, G. W. Rabe and J. W. Ziller, *J. Organomet. Chem.*, 1994, **483**, 21.

135. M. S. Clair, W. P. Schaefer and J. E. Bercaw, *Organometallics*, 1991, **10**, 525.

136. V. W.-W. Yam, K. M.-C. Wong and N. Zhu, *Angew. Chem., Int. Ed. Engl.*, 2003, **42**, 1400.

137. G. A. Koutsantonis and J. P. Selegue, *J. Am. Chem. Soc.*, 1991, **113**, 2316.

138. J. Heidrich, M. Steimann, M. Appel, W. Beck, J. R. Phillips and W. C. Trogler, *Organometallics*, 1990, **9**, 1296.

139. K. G. Caulton, R. H. Cayton, M. H. Chisholm, J. C. Huffman, E. B. Lobkovsky and Z. Xue, *Organometallics*, 1992, **11**, 321.

140. S. Kheradmandan, K. Venkatesan, O. Blacque, H. W. Schmalle and H. Berke, *Chem. Eur. J.*, 2004, **10**, 4872.

141. J. Sun, S. E. Shaner, M. K. Jones, D. C. O'Hanlon, J. S. Mugridge and M. D. Hopkins, *Inorg. Chem.*, 2010, **49**, 1687.

142. M. Akita, M.-C. Chung, A. Sakurai and Y. Moro-oka, *Chem. Commun.*, 2000, 1285.

143. B. E. Woodworth, P. S. White and J. L. Templeton, *J. Am. Chem. Soc.*, 1997, **119**, 828.

144. (a) J. Gil-Rubio, M. Laubender and H. Werner, *Organometallics*, 1998, **17**, 1202; (b) J. Gil-Rubio, M. Laubender and H. Werner, *Organometallics*, 2000, **19**, 1365; (c) H. Werner, R. W. Lass, O. Gevert and J. Wolf, *Organometallics*, 1997, **16**, 4077; (d) B. Callejas-Gaspar, M. Laubender and H. Werner, *J. Organomet. Chem.*, 2003, **684**, 144.

145. K. Onitsuka, N. Ose, F. Ozawa and S. Takahashi, *J. Organomet. Chem.*, 1999, **578**, 169.

146. V. W. W. Yam, V. C. Y. Lau and K. K. Cheung, *Organometallics*, 1996, **15**, 1740.

147. R. L. Roberts, H. Puschmann, J. A. K. Howard, J. H. Yamamoto, A. J. Carty and P. J. Low, *Dalton Trans.*, 2003, 1099.

148. M. I. Bruce, K. Costuas, T. Davin, J.-F. Halet, K. A. Kramarczuk, P. J. Low, B. K. Nicholson, G. J. Perkins, R. L. Roberts, B. W. Skelton, M. E. Smith and A. H. White, *Dalton Trans.*, 2007, 5387.

149. S. Kheradmandan, K. Heinze, H. W. Schmalle and H. Berke, *Angew. Chem., Int. Ed. Engl.*, 1999, **38**, 2270.

150. M. I. Bruce, M. Ke, P. J. Low, B. W. Skelton and A. H. White, *Organometallics*, 1998, **17**, 3539.

151. K.-T. Wong, J.-M. Lehn, S.-M. Peng and G.-H. Lee, *Chem. Commun.*, 2000, 2259.

152. J. Stahl, J. C. Bohling, E. B. Bauer, T. B. Peters, W. Mohr, J. M. Martín-Alvarez, F. Hampel and J. A. Gladysz, *Angew. Chem., Int. Ed. Engl.*, 2002, **41**, 1872.

153. T. B. Peters, J. C. Bohling, A. M. Arif and J. A. Gladysz, *Organometallics*, 1999, **18**, 3261.

154. W. Mohr, J. Stahl, F. Hampel and J. A. Gladysz, *Chem. Eur. J.*, 2003, **9**, 3324.

155. Q. Zheng, F. Hampel and J. A. Gladysz, *Organometallics*, 2004, **23**, 5896.

156. (a) R. Dembinski, T. Bartik, B. Bartik, M. Jaeger and J. A. Gladysz, *J. Am. Chem. Soc.*, 2000, **122**, 810; (b) C. R. Horn and J. A. Gladysz, *Eur. J. Inorg. Chem.*, 2003, 2211.

157. M. I. Bruce, B. D. Kelly, B. W. Skelton and A. H. White, *J. Organomet. Chem.*, 2000, **604**, 150.

158. L.-B. Gao, J. Kan, Y. Fan, L.-Y. Zhang, S.-H. Liu and Z.-N. Chen, *Inorg. Chem.*, 2007, **46**, 5651.

159. (a) R. Nast and A. Beyer, *J. Organomet. Chem.*, 1980, **194**, 125; (b) K. Onitsuka, H. Ogawa, T. Joh, S. Takahashi, Y. Yamamoto and H. Yamazaki, *J. Chem. Soc., Dalton Trans.*, 1991, 1531; (c) A. L. Groia, A. Ricci, M. Bassetti, D. Masi, C. Bianchini and C. L. Sterzo, *J. Organomet. Chem.*, 2003, **683**, 406.

160. (a) R. Nast and J. Moritz, *J. Organomet. Chem.*, 1976, **117**, 81; (b) M. I. Bruce, J. Davy, B. C. Hall, Y. J. v. Galen, B. W. Skelon and A. H. White, *Appl. Organomet. Chem.*, 2002, **16**, 559; (c) B. Olenyuk, M. D. Levin, J. A. Whiteford, J. E. Shield and P. J. Stang, *J. Am. Chem. Soc.*, 1999, **121**, 10434; (d) I. Ara, J. R. Berenguer, E. Eguizábal, J. Forniés, J. Gómez, E. Lalinde and J. M. Sáez-Rocher, *Organometallics*, 2000, **19**, 4385; (e) A. Köhler, J. S. Wilson, R. H. Friend, M. K. Al-Suti, M. S. Khan, A. Gerhard and H. Bässler, *J. Chem. Phys.*, 2002, **116**, 9457.

161. N. J. Long, C. K. Wong and A. J. P. White, *Organometallics*, 2006, **25**, 2525.

162. R. Nast and H. Grouhi, *J. Organomet. Chem.*, 1979, **182**, 197.

163. F. H. Koehler, W. Proessdorf and U. Schubert, *Inorg. Chem.*, 1981, **20**, 4096.

164. E. C. Royer and M. C. Barral, *Inorg. Chim. Acta*, 1984, **90**, L47.

165. R. R. Tykwinski and P. J. Stang, *Organometallics*, 1994, **13**, 3203.

166. P. Stang and R. Tykwinski, *J. Am. Chem. Soc.*, 1992, **114**, 4411.

167. G. Jia, R. J. Puddephatt, J. D. Scott and J. J. Vittal, *Organometallics*, 1993, **12**, 3565.

168. V. W.-W. Yam and S. W.-K. Choi, *J. Chem. Soc., Dalton Trans.*, 1996, 4227.

169. K. Osakada, T. Takizawa, M. Tanaka and T. Yamamoto, *J. Organomet. Chem.*, 1994, **473**, 359.

170. A. Santos, J. Lopez, J. Montoya, P. Noheda, A. Romero and A. M. Echavarren, *Organometallics*, 1994, **13**, 3605.

171. L.-K. Liu, K.-Y. Chang and Y.-S. Wen, *J. Chem. Soc., Dalton Trans.*, 1998, 741.

172. G. Albertin, S. Antoniutti, E. Bordignon and M. Granzotto, *J. Organomet. Chem.*, 1999, **585**, 83.

173. G. Albertin, P. Agnoletto and S. Antoniutti, *Polyhedron*, 2002, **21**, 1755.

174. M. I. Bruce, B. C. Hall, B. D. Kelly, P. J. Low, B. W. Skelton and A. H. White, *J. Chem. Soc., Dalton Trans.*, 1999, 3719.

175. G. Albertin, S. Antoniutti, E. Bordignon and D. Bresolin, *J. Organomet. Chem.*, 2000, **609**, 10.

176. C. S. Chin, D. Chong, M. Kim and H. Lee, *Bull. Korean Chem. Soc.*, 2001, **22**, 739.

177. C. S. Chin, H. Lee, H. Park and M. Kim, *Organometallics*, 2002, **21**, 3889.

178. G. v. Koten, S. Back, M. H. Lutz, A. L. Spek and H. Lang, *J. Organomet. Chem.*, 2001, **620**, 227.

179. H.-Y. Chao, W. Lu, Y. Li, M. C. W. Chan, C.-M. Che, K.-K. Cheung and N. Zhu, *J. Am. Chem. Soc.*, 2002, **124**, 14696.

180. T. Cardolaccia, A. M. Funston, M. E. Kose, J. M. Keller, M. J. R. and K. S. Schanze, *J. Phys. Chem. B*, 2007, **111**, 10871.

181. B. S. Newell, A. K. Rappé and M. P. Shores, *Inorg. Chem.*, 2010, **49**, 1595.

182. S. Back, R. A. Gossage, G. Rheinwald, I. del Rı?o, H. Lang and G. van Koten, *J. Organomet. Chem.*, 1999, **582**, 126.

183. N. Lenarvor, L. Toupet and C. Lapinte, *J. Am. Chem. Soc.*, 1995, **117**, 7129.

184. M. I. Bruce, B. G. Ellis, P. J. Low, B. W. Skelton and A. H. White, *Organometallics*, 2003, **22**, 3184.

185. M. I. Bruce, L. I. Denisovich, P. J. Low, S. M. Peregudova and N. A. Ustynyuk, *Mendeleev Commun.*, 1996, **6**, 200.

186. M. Bruce, K. Kramarczuk, B. Skelton and A. White, *J. Organomet. Chem.*, 2010, **695**, 469.

187. J. W. Seyler, W. Weng, Y. Zhou and J. A. Gladysz, *Organometallics*, 1993, **12**, 3802.

188. F. Coat and C. Lapinte, *Organometallics*, 1996, **15**, 477.

189. Y. Matsuura, Y. Tanaka and M. Akita, *J. Organomet. Chem.*, 2009, **694**, 1840.

190. L. Medei, L. Orian, O. V. Semeikin, M. G. Peterleitner, N. A. Ustynyuk, S. Santi, C. Durante, A. Ricci and C. L. Sterzo, *Eur. J. Inorg. Chem.*, 2006, 2582.

191. C. Ornelas, J. Ruiz and D. Astruc, *J. Organomet. Chem.*, 2009, **694**, 1219.

192. D. J. Armitt, M. I. Bruce, M. Gaudio, N. N. Zaitseva, B. W. Skelton, A. H. White, B. Le Guennic, J. F. Halet, M. A. Fox, R. L. Roberts, F. Hartl and P. J. Low, *J. Chem. Soc., Dalton Trans.*, 2008, 6763.

193. E. C. Fitzgerald, A. Ladjarafi, N. J. Brown, D. Collison, K. Costuas, R. Edge, J.-F. o. Halet, F. Justaud, P. J. Low, H. Meghezzi, T. Roisnel, M. W. Whiteley and C. Lapinte, *Organometallics*, 2011, **30**, 4180.

194. M. C. B. Colbert, J. Lewis, N. J. Long, P. R. Raithby, M. Younus, A. J. P. White, D. J. Williams, N. N. Payne, L. Yellowlees, D. Beljonne, N. Chawdhury and R. H. Friend, *Organometallics*, 1998, **17**, 3034.

195. O. Lavastre, J. Plass, P. Bachmann, S. Guesmi, C. Moinet and P. H. Dixneuf, *Organometallics*, 1997, **16**, 184.

196. S. K. Hurst, M. P. Cifuentes, A. M. McDonagh, M. G. Humphrey, M. Samoc, B. Luther-Davies, I. Asselberghs and A. Persoons, *J. Organomet. Chem.*, 2002, **642**, 259.

197. D. Catheline and D. Astruc, *Organometallics*, 1984, **3**, 1094.

198. G. J. Baird and S. G. Davies, *J. Organomet. Chem.*, 1984, **262**, 215.

199. A. d. l. J. Leal, M. J. Tenorio, M. C. Puerta and P. Valerga, *Organometallics*, 1995, **14**, 3839.

200. E. Isaacs and W. A. G. Graham, *J. Organomet. Chem.*, 1976, **120**, 407.

201. M. I. Bruce, B. G. Ellis, B. W. Skelton and A. H. White, *J. Organomet. Chem.*, 2005, **690**, 792.

202. C. Nataro, J. Chen and R. J. Angelici, *Inorg. Chem.*, 1998, **37**, 1868.

203. M. I. Bruce, M. G. Humphrey, G. A. Koutsantonis and M. J. Liddell, *J. Organomet. Chem.*, 1987, **326**, 247.

204. M. I. Bruce, M. G. Humphrey and M. J. Liddell, *J. Organomet. Chem.*, 1987, **321**, 91.

205. S. G. Davies and S. J. Simpson, *J. Chem. Soc., Dalton Trans.*, 1984, 993.

206. W. M. Laidlaw and R. G. Denning, *J. Organomet. Chem.*, 1993, **463**, 199.

207. A. T. Patton, C. E. Strouse, C. B. Knobler and J. A. Gladysz, *J. Am. Chem. Soc.*, 1983, **105**, 5804.

208. T.-S. Peng, C. H. Winter and J. A. Gladysz, *Inorg. Chem.*, 1994, **33**, 2534.

209. W. E. Meyer, A. J. Amoroso, M. Jaeger, J. L. Bras, W.-T. Wong and J. A. Gladysz, *J. Organomet. Chem.*, 2000, **616**, 44.

210. G. M. Aston, S. Badriya, R. D. Farley, R. W. Grime, S. J. Ledger, F. E. Mabbs, E. J. L. McInnes, H. W. Morris, A. Ricalton, C. C. Rowlands, K. Wagner and M. W. Whiteley, *J. Chem. Soc., Dalton Trans.*, 1999, 4379.

211. J. M. Bellerby, M. J. Mays and P. L. Sears, *J. Chem. Soc., Dalton Trans.*, 1976, 1232.

212. B. Gómez-Lor, A. Santos, M. Ruiz and A. M. Echavarren, *Eur. J. Inorg. Chem.*, 2001, 2001, 2305.

213. J. T. Mague and J. P. Mitchener, *Inorg. Chem.*, 1972, **11**, 2714.

214. G. Smith, D. J. Cole-Hamilton, M. Thornton-Pett and M. B. Hursthouse, *J. Chem. Soc., Dalton Trans.*, 1985, 387.

215. A. Donoli, A. Bisello, R. Cardena, A. Ceccon, M. Bassetti, A. D'Annibale, C. Pasquini, A. Raneri and S. Santi, *Inorg. Chim. Acta*, 2011, **374**, 442.

216. M. Kubota, R. K. Rothrock and J. Geibel, *J. Chem. Soc., Dalton Trans.*, 1973, 1267.

217. W. J. Cherwinski and H. C. Clark, *Inorg. Chem.*, 1971, **10**, 2263.

CHAPTER 5

Luminescent Transition-Metal Complexes as Biomolecular and Cellular Probes

KENNETH KAM-WING LO* AND STEVE PO-YAM LI

Department of Biology and Chemistry, City University of Hong Kong, Tat
Chee Avenue, Kowloon, Hong Kong, P. R. China
*E-mail: bhkenlo@cityu.edu.hk

5.1 Introduction

The photophysical and photochemical properties of luminescent transition-
metal complexes have been attracting much interest for several decades.[1] Many
of these properties would allow the complexes to be exploited as useful
biological labels and probes; for example, the emission of many transition-
metal complexes is intense and occurs in the visible region. The emission is
usually phosphorescence in nature, giving rise to emission lifetimes in the
micro- or submicrosecond timescale, which is sufficiently long for time-
resolved applications. Additionally, due to their large Stokes' shifts,
luminescent transition-metal complexes do not suffer from self-quenching,
which is commonly observed in biological molecules multiply labelled with
organic dyes.[2] Furthermore, the emission of many transition-metal complexes
is very sensitive to their local surroundings, which allows them to act as
luminescent molecular reporters for biomolecular recognition.[3]

Many transition-metal complexes exhibit interesting photoredox activity
towards biomolecules; for example, photocleavage of DNA *via* oxidation of

RSC Polymer Chemistry Series No. 2
Molecular Design and Applications of Photofunctional Polymers and Materials
Edited by Wai-Yeung Wong and Alaa S Abd-El-Aziz
© The Royal Society of Chemistry 2012
Published by the Royal Society of Chemistry, www.rsc.org

guanine[4] and photoinduced electron-transfer (ET) in metalloproteins[5] have been reported. The possible use of various redox-active quenchers in the design of bioassays can also be explored.[6] Importantly, the ground-state and excited-state redox potentials of transition-metal complexes can be fine tuned using various metal centres and ligands. Thus, the reactivity and specificity of the photoredox reactions can be controlled and the development of new photoreactive labels can be facilitated.

In this chapter, we summarise recent reports on biological labels and probes derived from luminescent transition-metal complexes. Two major types of biomolecules, DNA (oligonucleotides) and proteins, are focused on here. Since there is an emerging interest in luminescent transition-metal complexes as cellular probes and imaging reagents, the cellular uptake properties, cytotoxicity and intracellular localisation of selected luminescent transition-metal complexes are also described.

5.2 DNA Labels and Probes

5.2.1 Covalent DNA Labels

The design of transition-metal complexes that can be covalently attached to nucleic acids has been the focus of extensive research since they are anticipated to increase the understanding of genetic information processing and lead to the development of methods to interfere with these mechanisms. In addition to the continuous design of artificial metallonucleases and metal-based chemotherapeutics such as cisplatin, photoactive transition-metal complexes have been successfully incorporated into nucleic acids *via* solid-phase synthesis of oligonucleotides using metal-complex-labelled building blocks (premodification) and the reaction of an amine- or sulfhydryl-modified oligonucleotide with a metal complex in the solution phase (postmodification).

5.2.1.1   Premodification

With the advances of nuclear acid synthetic strategies, photoactive transition-metal complexes can be introduced in a site-specific manner to the genetic materials by the premodification approach. This method is the attachment of metal complexes to different nucleobase building blocks, which are then used to construct the desired labelled oligonucleotide in a site-specific manner *via* solid-phase synthetic chemistry.[7] A number of metal complexes have been incorporated into nuclear acids by this method; for example, Lewis and coworkers have modified 4,4′-dicarboxy-2,2′-bipyridine and equipped it with a $[Ru(bpy)_2]^{2+}$ moiety to produce a luminescent and reactive building block for the synthesis of oligonucleotides including dT_4-$[Ru(bpy)_2(dabp)]^{2+}$-dA_4, dG_3-$[Ru(bpy)_2(dabp)]^{2+}$-dC_3 and $dGCAATTGC$-$[Ru(bpy)_2(dabp)]^{2+}$-$DGCAATTGC$ (**1**).[8] The sequences of both sides of the DNA are complementary to each other, allowing the formation of hairpin structures with the metal complex as a bridging unit. Upon irradiation, these conjugates display a ^{3}MLCT emission band at 665 nm (τ_o = 608 – 815 ns, Φ_{em} = 0.16 – 0.19). The occurrence of lower emission energy of these conjugates compared to the model complex [Ru(bpy)$_3$]$^{2+}$ is due to the electron-withdrawing dicarboxamide diimine ligand with lower-lying π^* orbitals.

Zubieta and coworkers have functionalised the C2′ and C5′ ribose positions of the thymidine and uridine nucleobases with a tridentate chelating ligand.[9] Upon excitation, the Re(CO)$_3$-bis(quinoline)-nucleoside derivatives (**2**) display ^{3}IL emission at 418 – 429 nm and ^{3}MLCT emission 554 – 559 nm (τ_o = 14.4 – 18.6 µs, Φ_{em} = 0.0004 – 0.0234). The spacers between the bis(quinolinyl) group and the thymidine/uridine moiety do not significantly perturb the absorption and emission properties of the complexes, rendering them potential luminescent probes for *in vitro* imaging studies. Owing to their high specificity and stability, a uridine phosphoramidite unit and its derivatives have been utilised in the synthesis of labelled oligonucleotides.

The design of metallooligonucleotides using ruthenium(II)-derivatised uridine phosphoramidites has been reported independently by Grinstaff's[10] and Tor's[11] groups. Grinstaff and coworkers have synthesised a series of site-specifically [Ru(bpy)$_2$(4-m-4′-pa-bpy)]$^{2+}$-labelled oligonucleotides,[10] in which the alkynyl-functionalised ruthenium(II) polypyridine complex (**3**) is located at various positions in the sequence. After the complex is incorporated into the oligonucleotide, its emission maximum is slightly blueshifted to 660 nm. Interestingly, the excited-state lifetime of the complex in the duplex (594 ns) is only slightly longer than that in the single-stranded (ss) DNA molecule (544 ns), indicating that the hybridisation does not substantially alter the electronic structure of the complex.

Hurley and Tor have reported nucleotides containing a [M(bpy)$_2$(3-ethynyl-1,10-phenanthroline)]$^{2+}$ (M = Ru, Os) unit covalently attached to the 5-position of 2′-deoxyuridine (**4**).[11] The rigid acetylene linker produces substantial electron delocalisation, allowing electronic communication between the base-pairing heterocycle and the metal chelator. The labelled nucleosides have been incorporated into oligonucleotides. Upon irradiation, the ruthenium-modified oligonucleotides emit at 627 – 632 nm (τ_o = 1.54 – 1.63 and 0.72 – 0.85 µs)

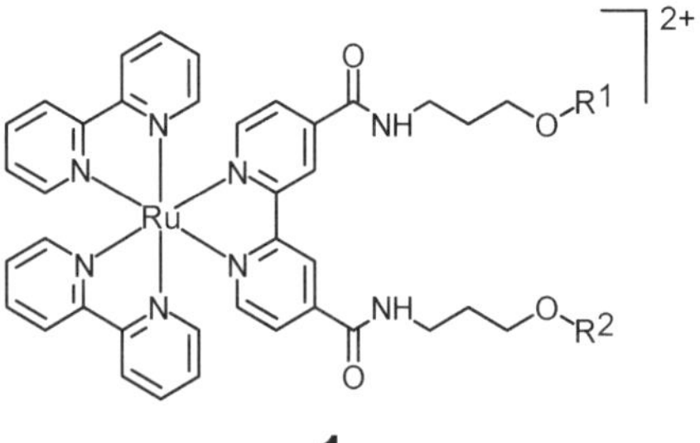

1

R1 = dT$_4$, R2 = dA$_4$
R1 = dG$_3$, R2 = dC$_3$
R1 = dGCAATTGC, R2 = dGCAATTGC

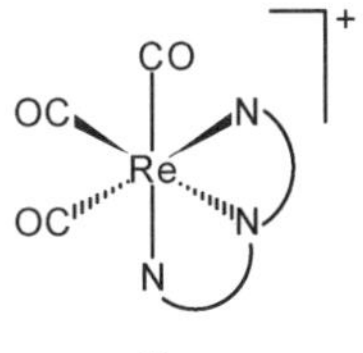

2

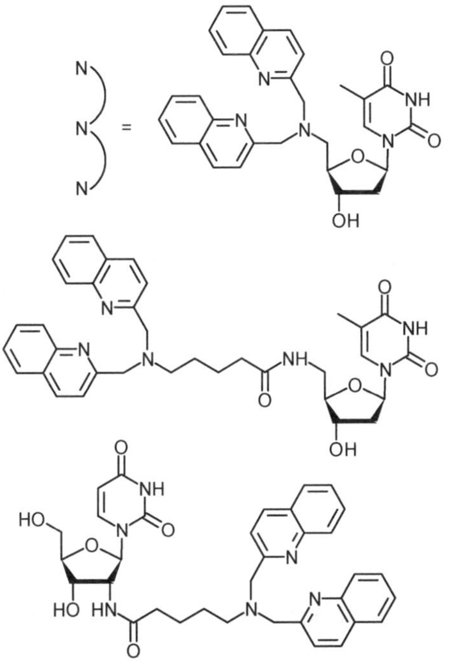

whereas the osmium-modified one emits at 740 nm ($\tau_o = 0.08$ and 0.024 μs). The diastereomeric ruthenium duplexes show a 10-nm difference in their emission wavelengths ($\lambda_{max} = 633$ nm for the Λ-duplex; $\lambda_{max} = 622$ nm for the Δ-duplex), while the emission of the epimeric mixture occurs at 629 nm. The larger contribution of the long-lived component in the excited Δ-ruthenium duplex, together with its shorter emission wavelength, suggest better accommodation of this stereoisomer within the major groove of DNA molecules compared to the Λ-ruthenium complex. The osmium duplex emits at 740 nm ($\tau_o = 0.08$ and 0.021 μs), which is very similar to the modified ss oligonucleotide.

In addition to labelling of building blocks for DNA, fundamental studies on the coordination of nucleobases to luminescent transition-metal complexes and their photoreactions with the complexes have been reported;[12] for example, Thorp and coworkers have synthesised the luminescent complex [Re(bpy)(CO)$_3$(EtG)]$^+$ (**5**),[13] which shows a distinctive emission band at 600

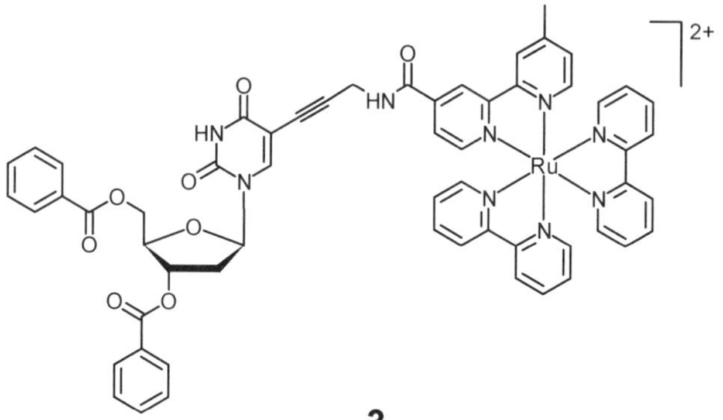

3

nm upon irradiation. The ^{3}MLCT ($d\pi$(Re) $\rightarrow \pi^*$(bpy)) emission energy of the complex is lower compared to that of its acetonitrile counterpart [Re(bpy)(CO)$_3$(CH$_3$CN)]$^+$ (λ_{max} = 536 nm), which is attributed to the stronger π-donating property of the nucleobase ligand. The high emission quantum yield of complex **5** (Φ_{em} = *ca.* 0.054) means that the complex serves as an excellent design of luminescent metal complexes that target guanine bases.

A transition-metal complex may be covalently attached to guanine residues by photocrosslinking reactions resulting from photoinduced ET; for example, Kirsch-De Mesmaeker and coworkers have studied the interactions and photoreactions of a series of ruthenium(II) tetraazaphenanthrene (tap) and hexaazatriphenylene (hat) complexes,[14] such as [Ru(hat)$_3$]$^{2+}$ and [Ru(tap)$_3$]$^{2+}$ (**6**) (and their mixed-ligand counterparts) with nucleotides and DNA. These complexes possess π-deficient hat and tap ligands with very low-lying lowest unoccupied molecular orbitals (LUMOs), and the excited complexes are sufficiently strong oxidising agents to remove an electron from guanosine-5'-monophosphate (GMP) and DNA molecules, yielding covalent adducts. The rate of emission quenching due to the photoreaction is dependent on the number of hat and tap ligands in the complexes and hence the excited-state reduction potentials of the metal complexes.

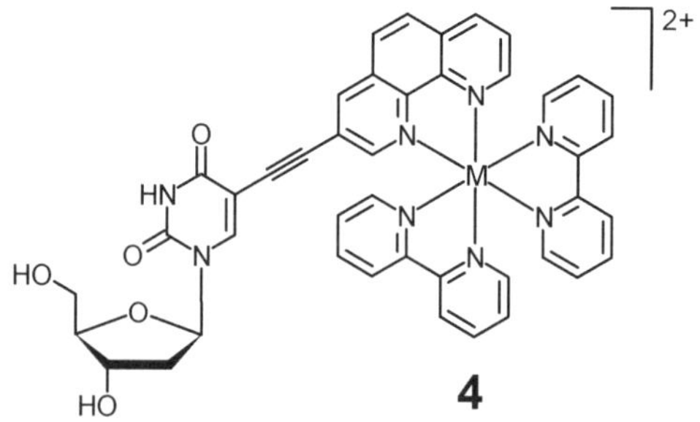

4

M = Ru, Os

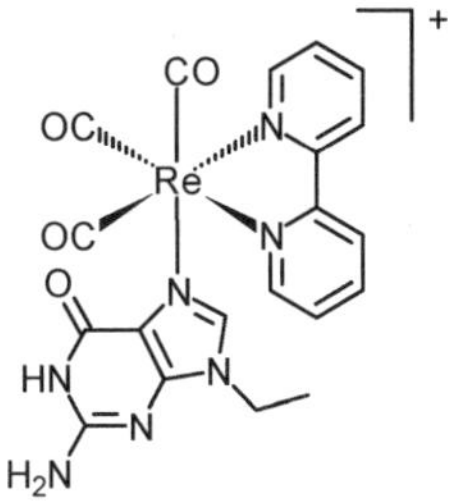

5

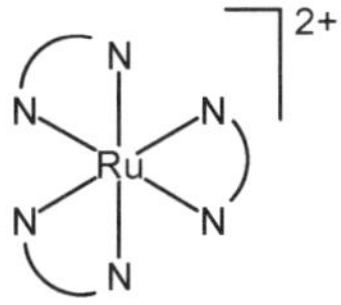

6

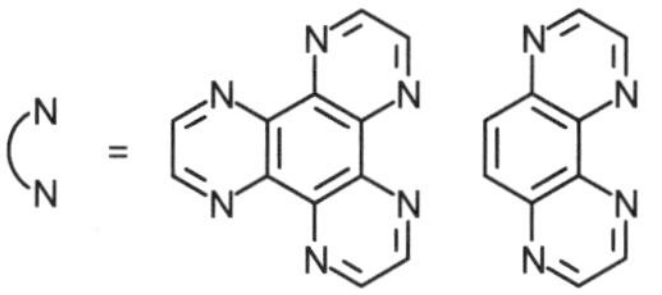

5.2.1.2 Postmodification

While modification of nucleobases has attracted much interest due to the high specificity and flexibility in the synthesis of labelled oligonucleotides, postmodification of an oligonucleotide has also been applied in a number of studies.[15] This synthetic strategy typically involves the reaction of an amine- or sulfhydryl-modified oligonucleotide with a transition-metal complex in the solution phase; for example, Barton and coworkers have linked a metallointercalator $[Rh(phi)_2(bpy)]^{3+}$ (**7**) to an oligonucleotide using an *N*-succinimidyl ester (NHS) coupling reaction.[16] The modified oligonucleotide forms DNA duplex with its complementary strand containing 5′-GG-3′ doublet oxidation sites. Upon irradiation, it shows excitation wavelength-dependent pathways for photoreactions. The sites of direct photocleavage (at the site of intercalation) and oxidative DNA damage (37 Å away from the site of intercalation) are marked at 313 and 365 nm, respectively.

Related studies on photoinduced ET through a covalently modified DNA helix have been reported;[17] for example, emission quenching of $[Ru(phen)_2(dppz)]^{2+}$ (**8**) by $[Rh(phi)_2(phen)]^{3+}$ (**9**) *via* an oligomeric assembly has been observed and examined.[18] Although the ruthenium(II)-modified oligonucleotide hybridised to an unmodified complementary strand shows intense ^{3}MLCT emission at *ca.* 610 nm, the ruthenium(II)-modified oligomer hybridised to a rhodium(III)-modified

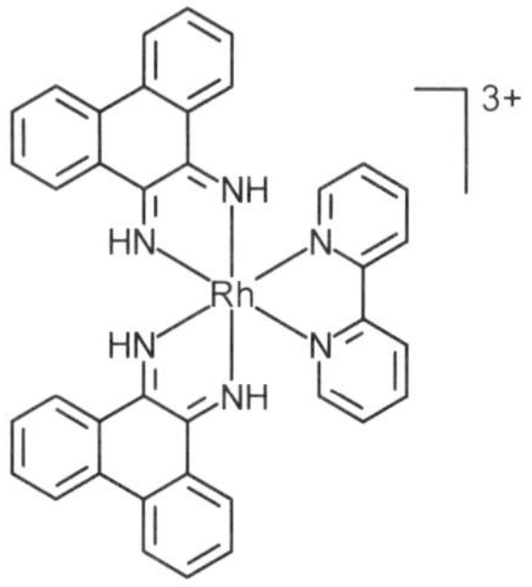

7

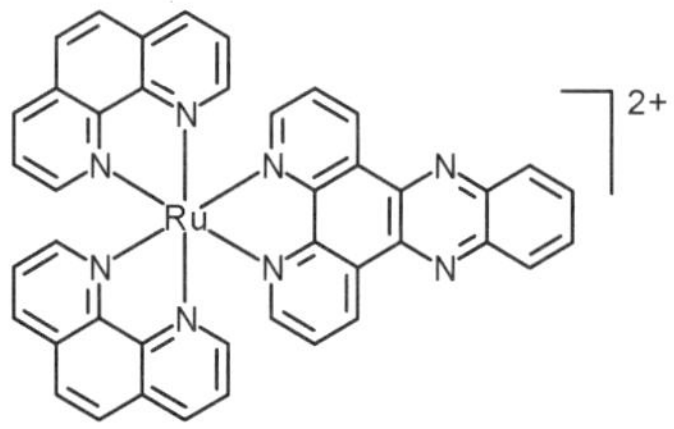

8

complementary strand reveals no detectable signal, which has been attributed to photoinduced ET from the ruthenium(II) donor to the rhodium(III) acceptor.

Luminescent transition-metal complexes containing an isothiocyanate moiety are labelling reagents that react readily with the primary amines of a biomolecule, such as DNA; for example, Lo and coworkers have labelled a universal M13 reverse sequencing primer (16 mer) modified with an aminohexyl group at the 5′-end, M13-NH$_2$, with the rhenium(I)

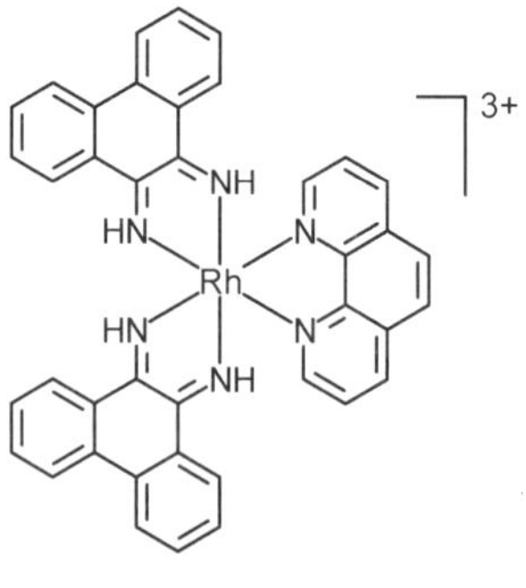

9

[Re(phen)(CO)$_3$(py-3-NCS)]$^+$ (**10**)[19] and iridium(III) isothiocyanate complexes [Ir(ppy)$_2$(phen-NCS)]$^+$ (**11**).[20] Upon irradiation, the rhenium(I)-labelled primer exhibits intense emission at *ca.* 548 nm (τ_o = 0.52 μs) whereas its iridium(III)-labelled counterpart emits at 580 nm (τ_o = 0.51 and 0.16 μs) in degassed Tris-Cl buffer at 298 K. The hybridisation properties of the oligonucleotide are retained after the modification. The intense emission of the modified M13 primers allows them to detect unmodified oligonucleotides containing a complementary sequence in polyacrylamide gel.

Sulfhydryl-specific labels involving the maleimide and iodoacetamide as the reactive functional group have been reported. Unlike the disulfide bond, which is susceptible to oxidative cleavage and exchange, the thioether moiety formed from the reaction of a sulfhydryl with a maleimide and iodoacetamide, respectively, is very stable; for example, Lo and coworkers have reacted the luminescent labels [Re(N^N)(CO)$_3$(py-3-mal)]$^+$ (**12**)[21] and [Ir(ppy)$_2$(phen-IAA)]$^+$ (**13**)[20] with the sulfhydryl-modified M13 sequencing primer M13-SH. The resultant DNA molecules display intense ^{3}MLCT (dπ(Re/Ir) $\rightarrow$ π*(diimine)) emission with a lifetime in the microsecond timescale. These modified oligonucleotides have been hybridised with their unmodified complementary strands to form luminescent duplexes, and the photophysical properties of the complexes are similar in both the ss and double-stranded (ds) DNA molecules.

Wiederholt and McLaughlin have incorporated a bipyridine linker into the backbone of oligonucleotides *via* the phosphoramidite chemistry.[22] The

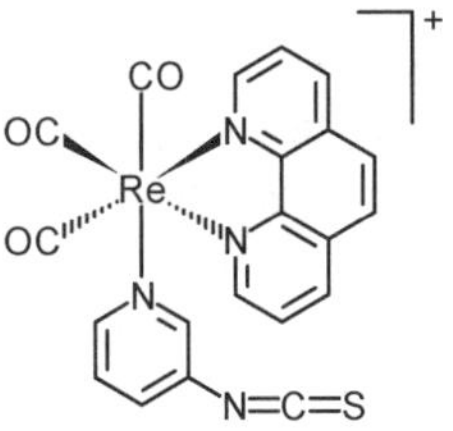

10

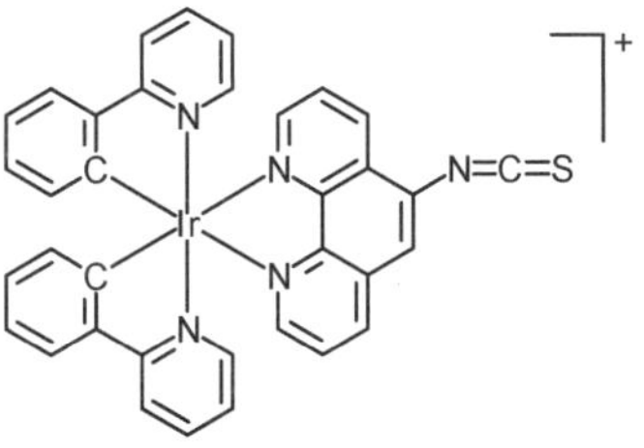

11

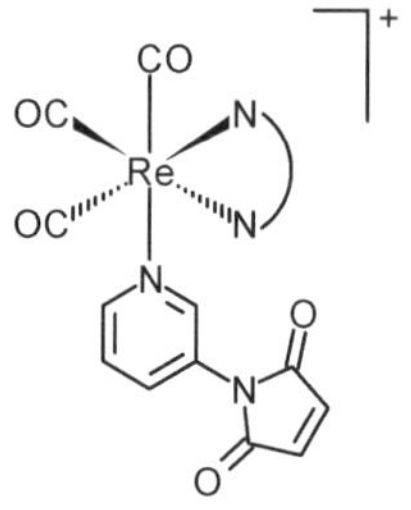

12

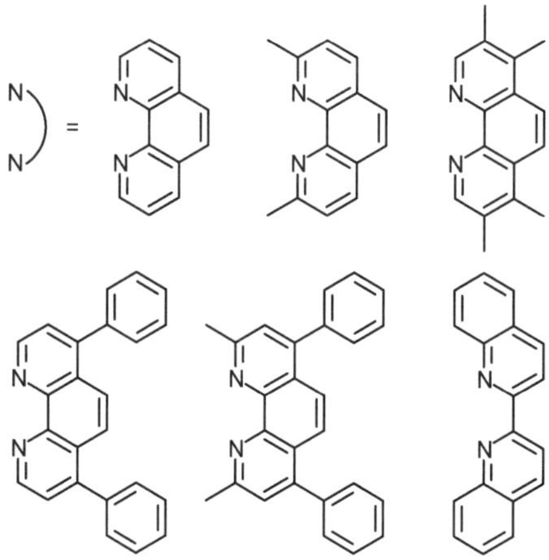

13

resultant oligonucleotides are then coordinated with a $[Ru(bpy)_2]^{2+}$ unit to yield luminescent DNA molecules (λ_{em} = 590 nm). The ruthenium(II)-containing sequence 5′-dTGTGTG-$[Ru(bpy)_3]^{2+}$-dTGTGTG-3′ (**14**) produced using this strategy shows no significant change in its emission quantum yield when it binds to either a complementary ss or ds sequence (hence generating the corresponding duplex and triplex, respectively). These results indicate that the luminescent ruthenium(II) metal complex does not interact with these ss, ds or triple-stranded (ts) DNA, but rather functions as an effective reporter group for the specific nucleic acid binding event.

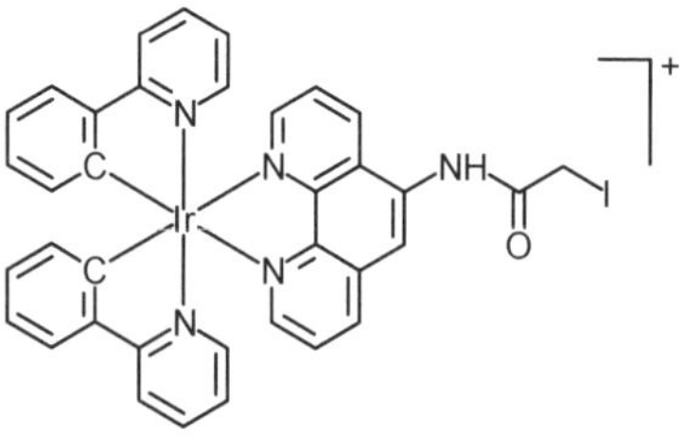

The iridium(III) intercalator [Ir(ppy)$_2$(dppz$'$)]$^+$ has been covalently tethered to different oligonucleotides (**15**) using phosphoramidite chemistry.[23] Shao and Barton have reported that the complex serves as both a strong photo-oxidant (+1.7 V *vs.* NHE) and photoreductant (–0.9 V *vs.* NHE) for studying DNA-mediated hole-transport (HT) and ET processes. Upon photolysis at 380 nm, the reduction of cyclopropylamine-substituted cytosine (^{CP}C) and oxidation of cyclopropylamine-substituted (^{CP}G) are promoted by DNA-mediated ET and HT, respectively, which can be probed on a picosecond timescale. Thus, this iridium–DNA assembly is a unique model system that allows comparison on the characteristics of DNA-mediated HT and ET directly in the same donor-bridge acceptor system.

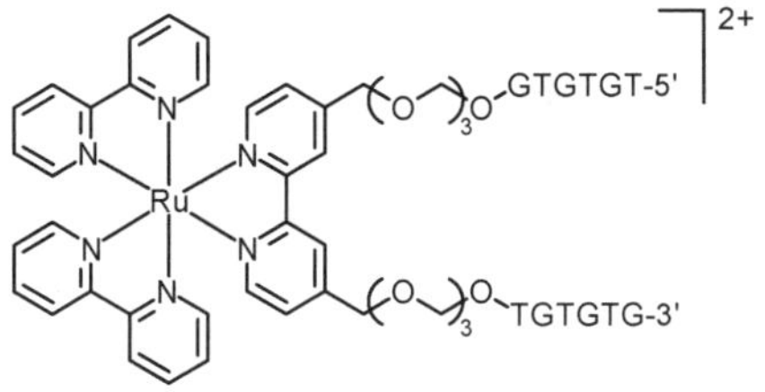

14

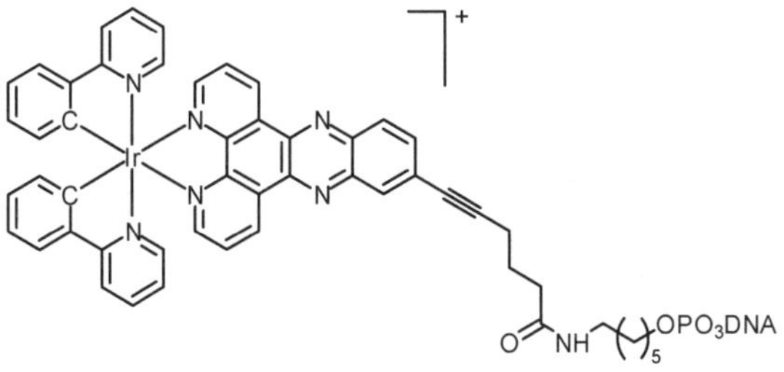

15

5.2.2   Noncovalent DNA Probes

In parallel to the evolving design of covalent adducts with DNA, luminescent transition-metal complexes have been exploited in the development of noncovalent DNA luminescent probes and light-triggered reactive agents. The DNA intercalative binding interactions of the model complexes [Pt(terpy)(SCH$_2$CH$_2$OH)]$^+$ [24] and [Ru(phen)$_3$]$^{2+}$ [25] have stimulated much interest in the design of DNA-binding agents through the use of a variety of polypyridine ligands. There are generally two strategies in the development of luminescent transition-metal complexes as DNA intercalators: (1) the coordination of an extended planar ligand to the transition-metal centre and

(2) the attachment of an organic intercalator to the luminescent metal complex *via* a space-arm. The changes in the emission properties of these two types of complexes resulting from the intercalation process may be different; details are discussed below.

5.2.2.1 *Complexes Coordinated with an Extended Planar Ligand*

Square-planar platinum(II) complexes containing planar aromatic fragments can bind to DNA by intercalation without direct interactions between the metal centre and DNA bases. Che and coworkers have reported two coordinatively unsaturated platinum(II) metallointercalators [Pt(dpp-C,N,N')(MeCN)]$^+$ (**16**) and [Pt(pby-C,N,N')(μ-dppm)]$^{2+}$ (**17**).[26] The ^{3}MLCT (dπ(Pt) $\rightarrow$ π*(dpp)) emission of complex **16** occurs at 549, 593, 632 (sh) nm and the ^{3}MMLCT (dσ*(Pt$_2$) $\rightarrow$ σ(π*(pby))) emission of complex **17** at 647 nm. Upon intercalating into the base pairs of calf thymus (ct) DNA in aqueous buffer, both complexes exhibit drastic emission enhancement (I/I_o = 271 for complex **16** and 1172 for complex **17**) (Figure 5.1). These observations are attributed to the suppression of solvent-induced quenching after the complexes have intercalated into the DNA base pairs.

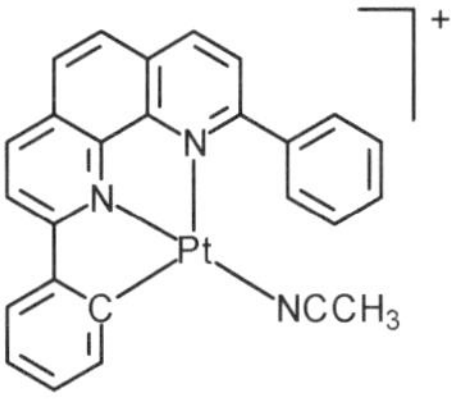

16

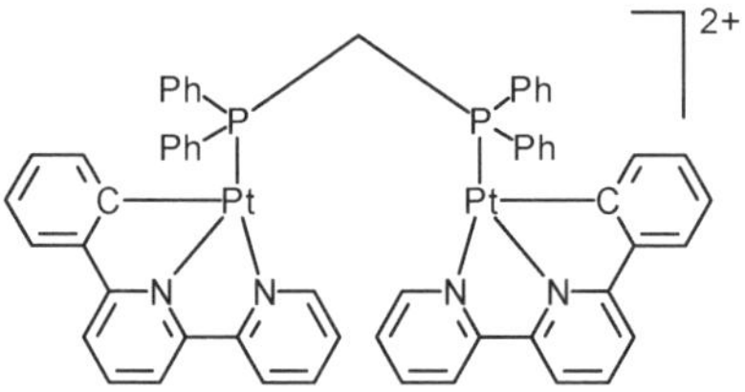

17

The platinum(II) complex [Pt(dppz)(^{t}Bu-ppy)]$^+$ (**18**) is a ^{3}IL emitter (λ_{em} = 477 nm, τ_o = 2.3 μs, Φ_{em} = 0.0012) with a moderate quantum yield in CH$_3$CN but becomes nonemissive in aqueous solution.[27] Interestingly, it shows an increase of emission intensity (I/I_o = 140) at λ_{max} = 650 nm upon addition of ct DNA. This phenomenon is attributed to a decrease in complex–solvent

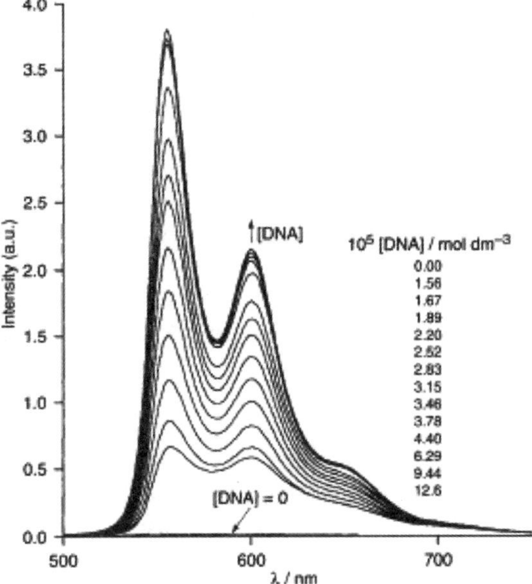

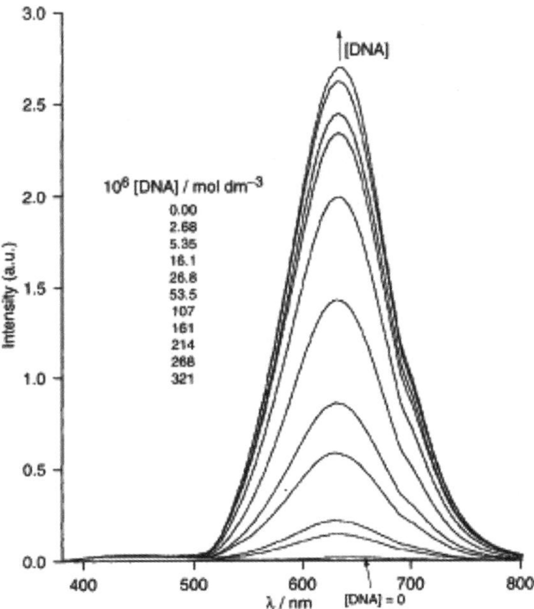

Figure 5.1 Emission spectral traces of [Pt(dpp-*C,N,N'*)(MeCN)]⁺ (top) and [Pt(pby-*C,N,N'*)(μ-dppm)]²⁺ (bottom) upon addition of ct DNA in 3% MeOH-Tris buffer. Reproduced from ref. 26 with permission of the Royal Society of Chemistry.

interactions and reduced mobility of the complex due to intercalation of the complex into DNA base pairs.

Cusumano and coworkers have reported that the weak ^{3}IL/^{3}MLCT emitting platinum(II) dppn complex [Pt(dppn)(py)$_2$]$^{2+}$ (**19**) (λ_{em} = 640 nm, τ_o = 8 ns) displays emission changes upon addition of ct DNA depending on different binding modes.[28] The emission of complex **19** is enhanced at a low [DNA]/[complex] ratio (where external binding is the preferred interaction mode) as a consequence of the protection effect exerted by the DNA against solvent or oxygen quenching. However, at a higher [DNA]/[complex] ratio (when intercalation dominates), the effect of photoinduced ET from guanine becomes predominant, resulting in emission quenching. For example, addition of poly(dG)-poly(dC) quenches the emission to 20%, whereas poly(dA)-poly(dT) gives a 35-fold emission enhancement.

In addition to planar complexes, octahedral complexes with extended planar polypyridine ligands such as phen, dpq, dppz, phi, phzi and chrysi, are known to display intercalative interactions with ds DNA.[29] A number of ruthenium(II), rhenium(I), osmium(II), rhodium(III) and iridium(III) complexes equipped with these polypyridine ligands have been identified to be DNA metallointercalators.[29] The most extensively studied system is the ruthenium(II) dppz complexes that show interesting DNA-binding properties. Although the complexes [Ru(N$^\wedge$N)$_2$(dppz)]$^{2+}$ (N$^\wedge$N = bpy, phen) (**20**) are emissive in organic solvents such as ethanol, propanol and CH$_3$CN (λ_{em} = 610 – 622 nm, τ_o = 2.3 μs, Φ_{em} = 0.0012), their emission is effectively quenched in aqueous solution.[30] Electrochemical and photophysical measurements of their ground and excited states show that the charge transfer is directed from the metal centre to the phenazine ring. It is likely that the major nonradiative deactivation pathway in aqueous solution involves protonation of the phenazine nitrogen atoms in the excited state. However, upon binding to ct DNA by intercalation, the emission is significantly enhanced (in some cases *I*/I_o > 10^4 fold) as a consequence of the protection of the ligand from quenching by water.

The emission properties and excited-state character of the complexes [Re(N$^\wedge$N)(CO)$_3$(py-R)]$^+$ (N$^\wedge$N = dppz, dppn; R = H, CH$_3$) (**21**) with an extended planar diimine ligand have been studied by Schanze *et al.*[31] and Yam *et al.*[32] The emissive state of the dppz complexes has been identified to be ^{3}IL ($\pi \rightarrow \pi^*$)(dppz) in nature by luminescence and transient absorption spectro-

18

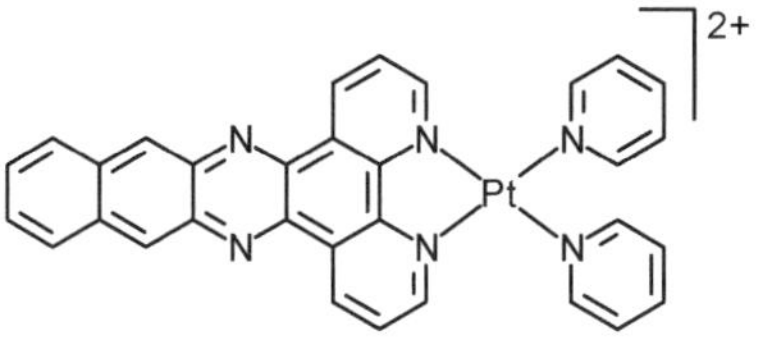

19

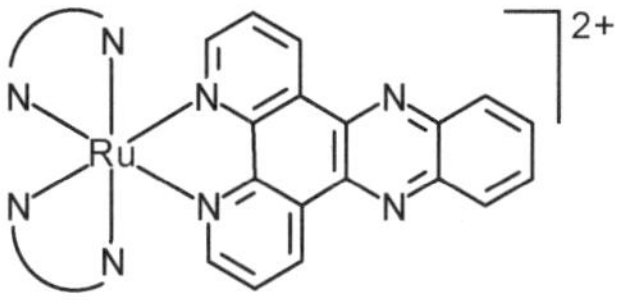

20

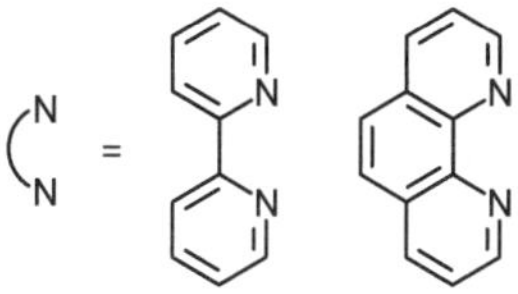

scopy. However, the emission of the dppn complex is derived from a state of substantial ^{3}MLCT (dπ(Re) $\rightarrow$ π*(dppn)) character. Absorption and emission titrations show that these complexes intercalate into ct DNA and synthetic oligonucleotides. Upon irradiation, these complexes cleave the DNA plasmid pBR322 efficiently through two mechanisms. The excited dppz complex [Re(dppz)(CO)$_3$(py)]$^{+*}$ oxidises the DNA molecule directly. However, in the case of the dppn complex [Re(dppn)(CO)$_3$(py)]$^+$, superoxide and hydroxyl radicals generated from the reaction of the excited complex and oxygen in the air are the cleavage reagents.

Barton and coworkers have reported the photophysical and DNA-binding properties of the osmium(II) dppz complexes [Os(bpy/phen)$_2$(dppz)]$^{2+}$ and their methylated analogues such as [Os(phen)$_2$(Me$_2$-dppz)]$^{2+}$ and [Os(5,6-dmp/4,7-dmp)$_2$(dppz)]$^{2+}$ (**22**).[33] Upon binding to ds DNA, the complexes display low-energy ^{3}MLCT emission bands at *ca.* 731 – 761 nm with multiexponential lifetimes (*ca.* 50 – 85 ns) in Tris-Cl buffer at 298 K. Methylation of the ancillary ligands decreases the DNA-induced emission enhancement factors, whereas complexes with methyl substituents on the dppz ligand show the highest emission quantum yields in the presence of DNA. Importantly, studies with poly(dA)-poly(dT), poly(dG)-poly(dC) and mixed-sequence DNA show that the emission quantum yields are dependent on the DNA sequence. The multiexponential decay of these complexes upon binding to ct DNA suggests the existence of a head-on and a side-on stacking modes, leading to different exposure of the phenazine nitrogens to water and hence difference degrees of quenching.

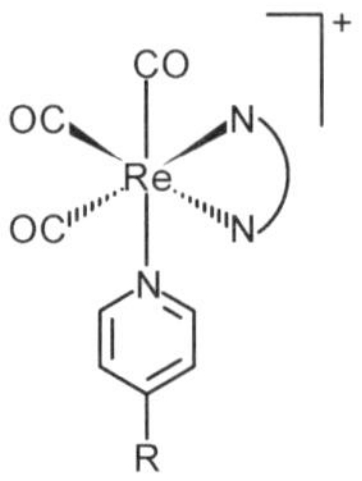

21

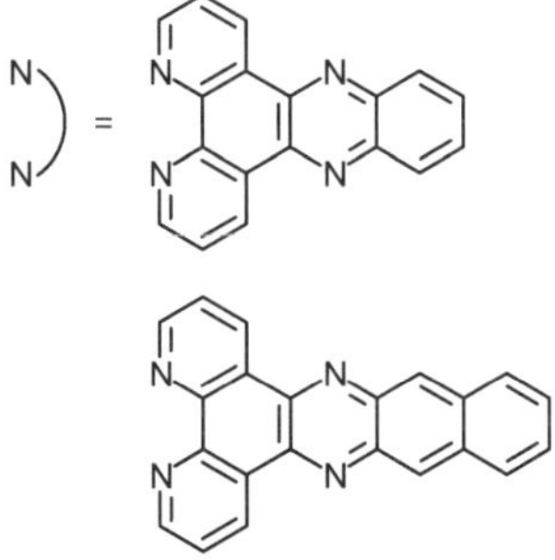

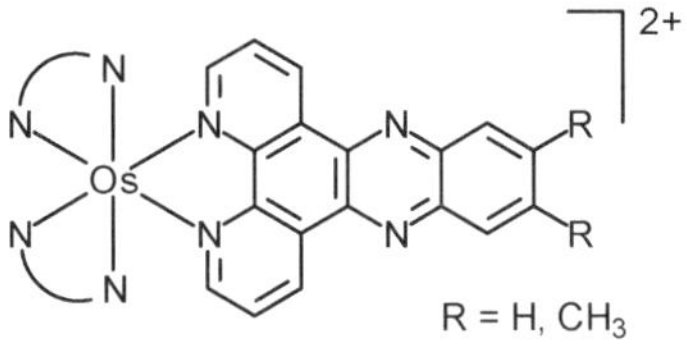

R = H, CH$_3$

R = H

22

R = H, CH$_3$

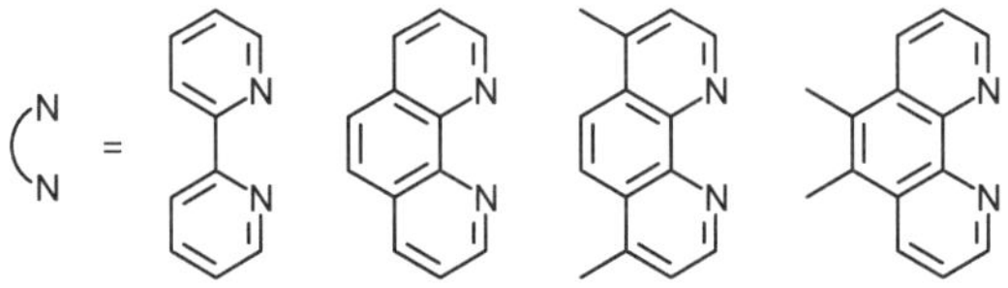

With the aims to enhance the DNA-recognition properties by the avidin–biotin system and to design luminescent probes for both DNA molecules and proteins, Lo and coworkers have designed bifunctional luminescent metallointercalators [Re(N^N)(CO)$_3$(L)]$^{2+}$ (N^N = dppz, dppn; L = py-spacer-biotin) (**23**)[34] and [Ir(ppy)$_2$(N^N)]$^+$ (N^N = dpq-spacer-biotin, dppz-spacer-biotin) (**24**)[35] containing a biotin moiety. The rhenium(I) complexes **23** show emission enhancement (I/I_o = 2.2 to 3.5) upon intercalation into ds DNA, whereas the iridium(III) complexes **24** do not show similar responses, probably due to the steric hindrance of the biotin unit attached to the end of the dpq and dppz ligands. All the biotin complexes exhibit binding interactions with the protein avidin, as revealed by absorption assays and emission titrations.

Interestingly, the attachment of an amide linkage to the ligand dpq significantly quenches the emission of the complexes in aqueous solution due to hydrogen bonding interactions with water molecules.[35–37] This is an advantage because suppression of the emission of the complexes in the free state will result in higher biomolecule-induced emission enhancement.

Electrostatic interactions between a cationic transition-metal complex and DNA molecules are the fundamental attractive force bringing the two entities

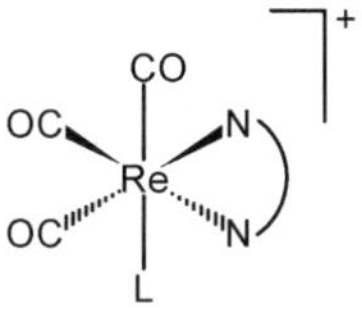

23

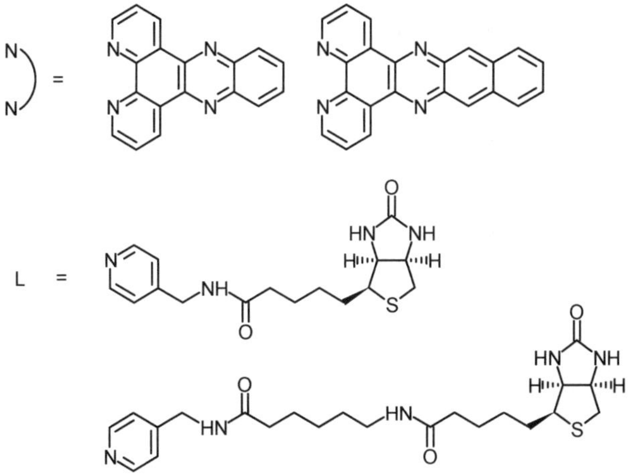

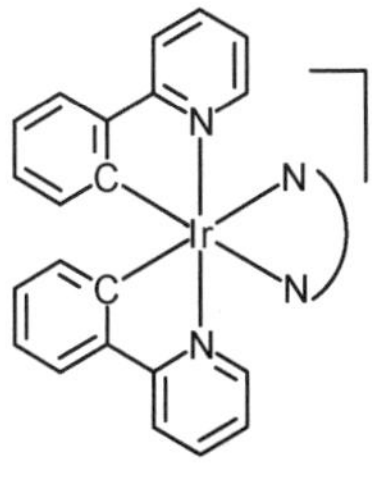

24

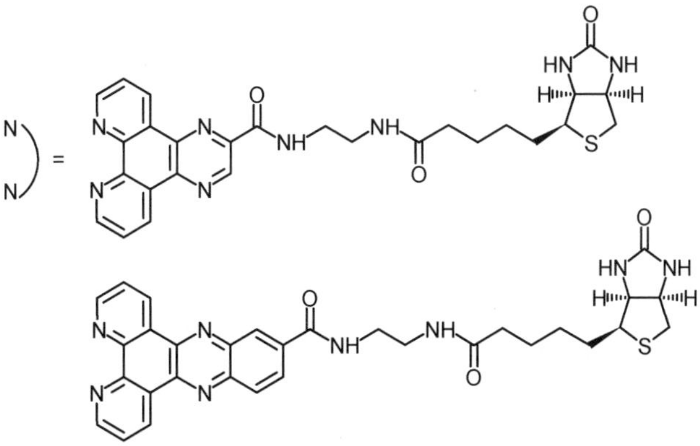

together before other modes of binding interactions may occur; for example, Yam and coworkers have developed a probe for both ss and ds DNA molecules by exploiting the metal–metal and π–π stacking interactions of the platinum(II) complex $[Pt(tpy)C{\equiv}C\text{-}C{\equiv}CCH_2OH]^+$ (**25**).[38] The initial electrostatic interactions between the cationic complex and negatively charge oligonucleotides induce the self-assembly of the metal complex molecules, as reflected by the appearance of a ^{3}MMLCT emission band with vibronic structures at *ca.* 800 nm. Remarkably, this emission band gradually disappears and is replaced by a strong one at *ca.* 557 nm at high DNA concentration, which has been ascribed to the intercalation of the complex into the duplex.

The interactions between bimetallic complexes and DNA molecules have been investigated; for example, O'Reilly and Kelly have reported that by tethering relatively weak binding systems such as $[Ru(bpy)_3]^{2+}$ and $[Ru(phen)_3]^{2+}$ into the bimetallic systems (**26**) using alkyl chains of different lengths,[39] the DNA-binding affinities are enhanced by two orders of magnitude. The emission enhancement factors are similar (I/I_o = 2.41 for monomeric $[Ru(phen)_2(Me_2\text{-}bpy)]^{2+}$ and 1.68 – 2.08 for the dimer). Only the C7 linker allows full interactions of both metal centres and this complex also shows the greatest resistance to increased ionic strength of the solution.

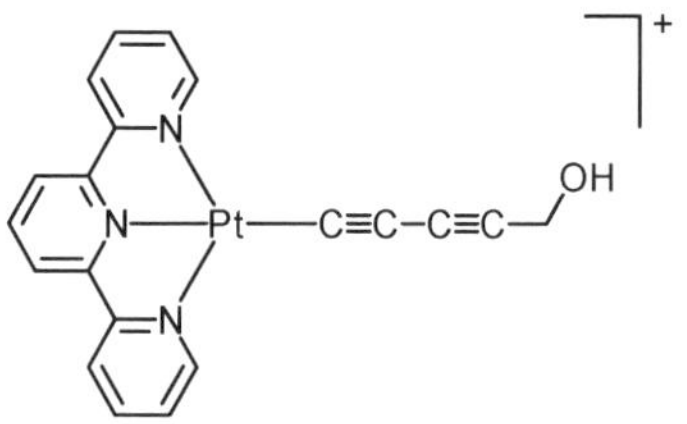

25

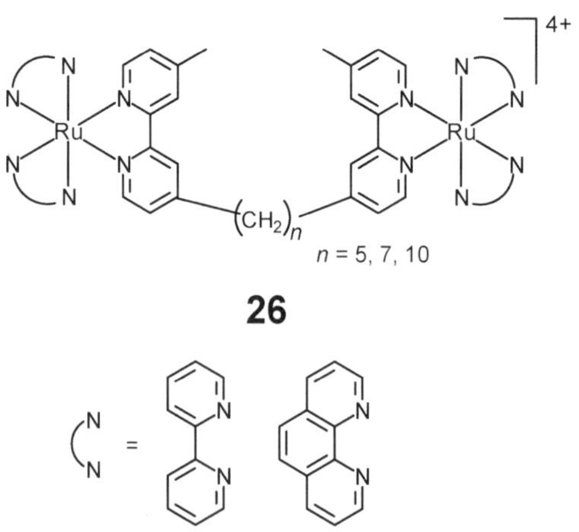

26

$n = 5, 7, 10$

Nordén and coworkers have demonstrated the unique slow dissociation of $[Ru_2(phen)_4(\mu\text{-bidppz})]^{4+}$ (**27**) from ct DNA and poly(dA-dT)$_2$, resulting from the extremely slow rearrangement of the initial groove-binding mode to an intercalative geometry.[40] This is achieved by threading one of the $[Ru(phen)_2]$ moieties through the DNA duplex and intercalating one of the bridging dppz ligands between the DNA base pairs and thus placing one metal centre in each groove. The emission quantum yield of the complex in its final binding mode is about 1500 times higher than that of the free probe in water, and triexponential decay is observed (τ_o = 406, 156 and 36 ns).

Thomas and coworkers have reported the DNA-photocleavage properties of the heterobimetallic complex $[\{Ru(dpm)(dppz)\}(\mu\text{-dpp5})\{Re(CO)_3(dppz)\}]^{3+}$ (**28**).[41] Excitation of the complex in CH_3CN at 430 nm leads to ^{3}MLCT $(d\pi(Ru) \rightarrow \pi^*(dppz))$ emission at *ca.* 680 nm. However, excitation at 370 nm populates the ^{3}MLCT and ^{3}IL states on both the rhenium(I) and ruthenium(II) centres, resulting in increased emission intensity (τ_o = 77 ns). The complex is nonemissive in aqueous solution but shows an emission band at *ca.* 650 nm upon addition of ct DNA. Irradiation of the complex and plasmid DNA mixture leads to the generation of relaxed singly nicked plasmid, while higher loading of the complex rapidly results in complete degradation of the DNA. Thus, this dinuclear complex functions as both a light switch and direct cleavage "scissors" of DNA molecules.

27

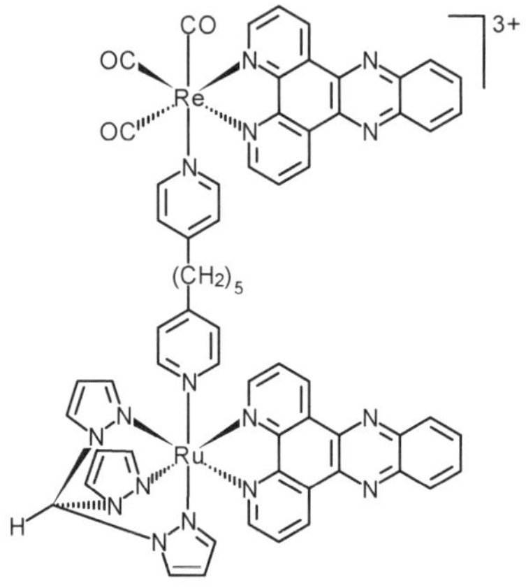

28

5.2.2.2 *Complexes Appended with an Organic Intercalator*

Organic DNA intercalators such as acridine, pyrene and anthracene have been incorporated into transition-metal complexes *via* a spacer-arm, to yield new DNA-binding agents. Lippard and coworkers have reported the DNA-binding properties of the platinum(II) acridine orange complexes with trimethylene and hexamethylene spacer-arms (**29**).[42] The platinum(II) unit forms a covalent adduct with DNA, whereas the appended acridine orange moiety intercalates into the ds DNA one or two base pairs away.

Santos and coworkers have reported that complex $[\text{Re}(N^\wedge N^\wedge N\text{-acridine-}N)(\text{CO})_3]^{2+}$ (**30**) binds to DNA by intercalation, as confirmed by the bathochromic shift of its absorption band from 496 to 506 nm upon addition of the biopolymer.[43] The emission intensity of this complex at *ca.* 525 nm is enhanced upon addition of ds DNA. The complex $[\text{Re}(N^\wedge N^\wedge N\text{-pyrene})(\text{CO})_3]^+$ (**31**) displays characteristic pyrene emission bands ($\tau_o < 1$ ns).[44] The absorption spectrum of this complex shows a drastic difference upon addition of fish sperm DNA, resulting from the intercalation of the pyrene pendant into the DNA ($K_b = 1.5 \pm 0.2 \times 10^5\,\text{M}^{-1}$, $n = 3.2 \pm 0.2$). The emission band of the complex $[\text{Re}(N^\wedge O\text{-COOH})(\text{CO})_3(\text{NC-acridine})]^+$ (**32**) at *ca.* 525 nm exhibits 3-fold emission intensity enhancement and a small redshift upon addition of ct DNA, indicative of intercalative binding ($K_b = 1.13 \times 10^5\,\text{M}^{-1}$).[45]

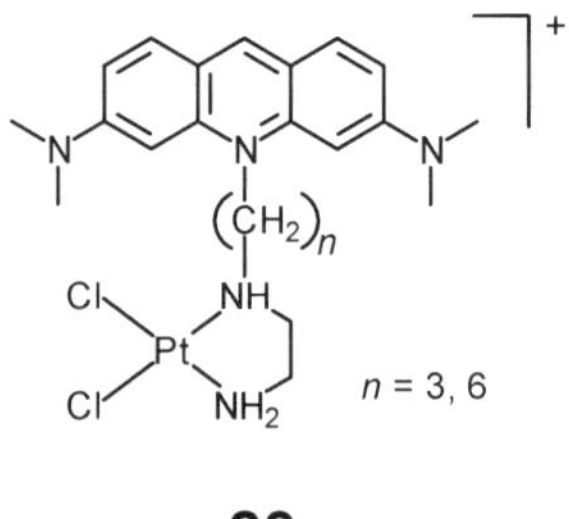

29

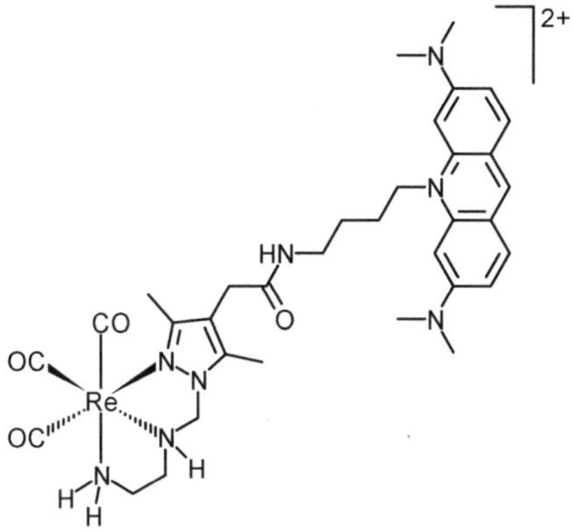

30

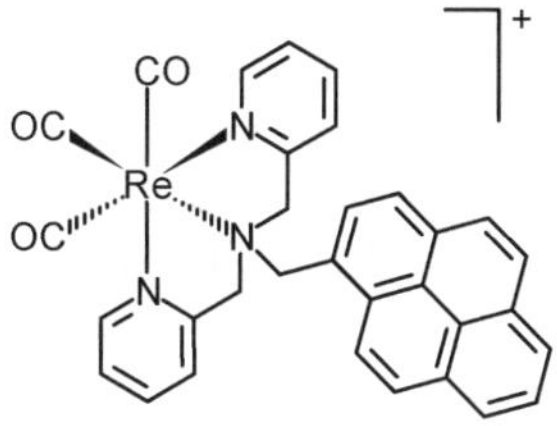

31

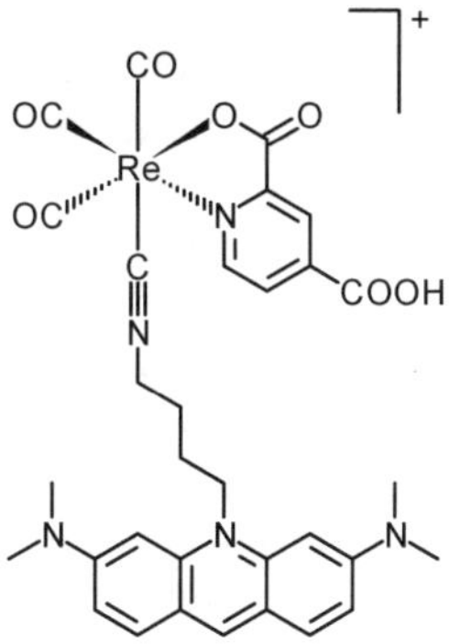

32

Thornton and Schanze have shown that the anthracene moiety of the rhenium(I) metal complex [Re(bpy)(CO)$_3$(py-4-An)]$^+$ (**33**) quenches the ^{3}MLCT (dπ(Re) $\rightarrow$ π*(bpy)) emission *via* triplet–triplet intramolecular energy transfer (E$_N$T).[46] However, addition of DNA leads to an increase in the intensity of the ^{3}MLCT emission at *ca.* 580 nm by 5 fold and a decrease of the intensity and broadening of the vibronic bands of the anthracene fluorescence. These observations are ascribed to intercalation of the anthracene unit into the DNA and the binding constant is determined to be 4.6×10^5 M^{-1}.

33

5.2.2.3 Groove Binders

A number of transition-metal complexes show DNA-binding interactions in the groove-binding mode. Most of these complexes contain hydrophobic ligands or pendants that are structurally related to organic DNA groove-binders; for example, Thomas and coworkers have introduced the catechol moiety to a [Ru(bpy)$_2$] core *via* a vinyl bridge to yield the complex [Ru(bpy)$_2$(bpy-CH=CH-catechol)]$^{2+}$ (**34**).[47] The diol moiety of catechol forms hydrogen bonds with the nucleobase-acceptor residues within the minor groove of the DNA duplex. Interestingly, the resulting hydrogen bonds quench the ^{3}MLCT emission at 620 nm upon initial addition of DNA. However, after these sites are occupied, the emission intensity increases as the luminophore migrates to more nonpolar and solvent-inaccessible binding sites. The binding constant of the first binding event is at least five times larger than that of the second binding event.

Hannon and coworkers have reported the synthesis and characterisation of the ts ruthenium(II) [Ru$_2$L$_3$]$^{4+}$ [48] and iron(II) [Fe$_2$L$_3$]$^{4+}$ [49] helicates using the ligand bis(pyridyl)imine (**35**). These helicates bind strongly and noncovalently in the major groove of DNA. The ruthenium(II) helicate exhibits an MLCT absorption band at 485 nm ($\varepsilon = 16\,900$ M^{-1} cm^{-1}) and an emission band at 705 nm. The ^{3}MLCT emission of the ruthenium(II) helicate is useful in probing the DNA binding as the emission intensity is enhanced by 2 fold and the emission maximum is blueshifted by 8 nm upon addition of ct DNA. The emission enhancement is very rapid and more striking than that for [Ru(bpy)$_3$]$^{2+}$, which shows little or no enhancement upon binding to DNA. Additionally, this complex is a groove binder for DNA because its binding towards DNA induces dramatic and unprecedented intramolecular DNA coiling in natural polymeric DNAs as revealed by flow-linear dichroism experiments.

Mukherjee and coworkers have reported the synthesis of the dinuclear copper(II) complex [Cu$_2$(L)$_2$(μ-SO$_4$)$_2$] (L = pyridine-2-carboxaldehyde-2-

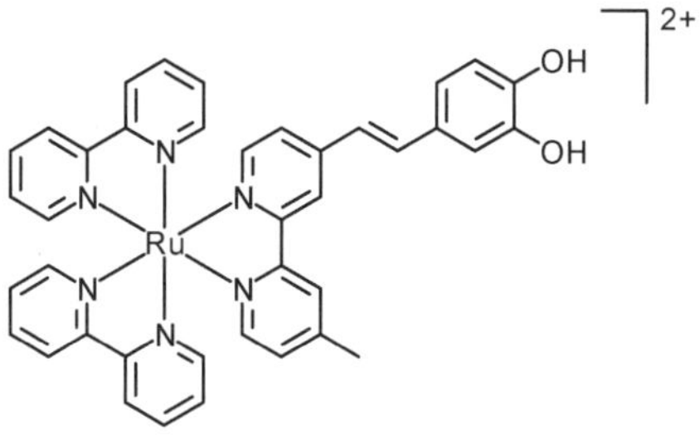

34

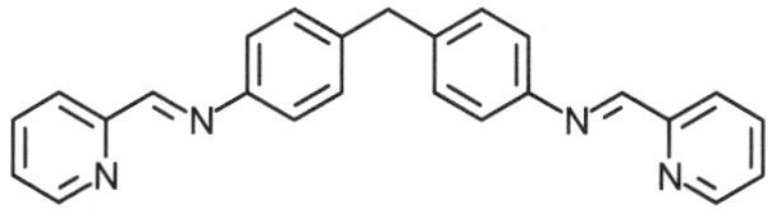

35

pyridylhydrazone) (**36**).[50] Upon excitation, the complex emits at 354 nm, which is slightly higher in energy compared to the free ligand L. Upon addition of ct DNA, clear hypochromism and hyperchromism are observed in the absorption spectra, suggesting that there are strong interactions between the complex and DNA. DNA gel retardation assays and viscosity measurements on the DNA/complex mixtures reveal nonintercalative binding, and the mode of interactions is assigned to groove binding.

Thomas and coworkers have reported that the dinuclear complex [Ru$_2$(bpy)$_4$(μ-bpy-azo-bpy)]$^{4+}$ (**37**) is a DNA groove-binding complex that shows distinctive colorimetric responses on DNA binding.[51] Although both bpy-based and 4-azo-based MLCT absorption bands display considerable hypochromicity ($<$ 10% and 17%, respectively), the most striking change upon addition of ct DNA is the large bathochromic shift of over 40 nm for the 4-azo based MLCT, leading to a distinctive change of colour from purple to green. This is probably due to the change of the environment for the azo group as it moves from the bulk aqueous solvent into the DNA groove. The nature of the interactions of this complex with DNA duplex is also investigated by viscosity measurements and UV-visible spectroscopic titrations on poly(dG)-poly(dC) and poly(dA)-poly(dT).

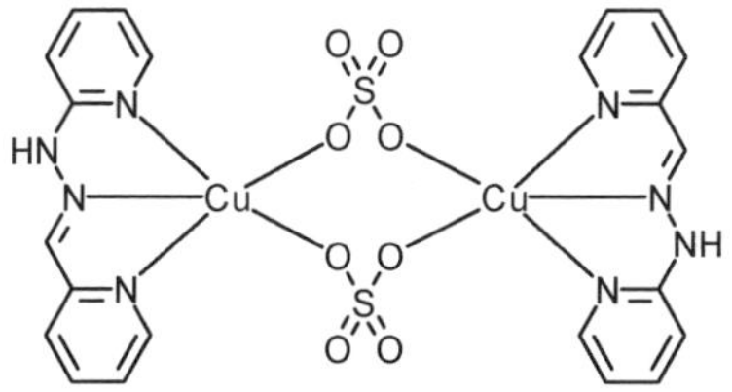

36

37

5.3 Protein Labels and Probes

5.3.1 Covalent Protein Labels

The common reactive sites for bioconjugation of proteins include the amine groups of the lysine residue and the N-terminal, the sulfhydryl group of the cysteine residue, the imidazole group of the histidine residue and the carboxy groups of the aspartate and glutamate residues and the C-terminal.[52] The majority of the labels reported to date are equipped with functional groups that target the amine, sulfhydryl and imidazole moieties.

5.3.1.1 Amine-Reactive Complexes

Covalent labelling of proteins generally involves the modification of amino acid side chains. Among the amino acids, the lysine residue is the most common site for modification owing to its high abundance and reactivity. Thus, a number of protein labels have been functionalised with amine-reactive groups such as isothiocyanate, aldehyde and NHS-ester; for example, Vos and coworkers have reported ruthenium(II) isothiocyanate complexes [Ru(bpy/phen)$_2$(phen-NCS)]$^{2+}$ (**38**).[53] The emission of these complexes occurs at 608 – 612 nm in 0.1 M carbonate buffer at pH 9.6. These complexes are used as excellent labels for amine-containing biomolecules including albumins, polylysine (PLL) and goat anti-mouse Immunoglobulin G (IgG). The labelling efficiencies of the complexes are generally similar to that of fluorescein isothiocyanate (FITC). The immunological activity of goat anti-mouse IgG is retained after the conjugation. All the conjugates are strongly emissive and the emission wavelengths are similar to those of the free labels. Interestingly, the ruthenium(II)-PLL conjugate shows significantly shorter lifetimes than the ruthenium(II)-bovine serum albumin (BSA) conjugate, suggesting the tertiary structure of PLL provides fewer hydrophobic pockets for the complex and a more open structure that allows easier access of water molecules for emission quenching.

Lo and coworkers have labelled human serum albumin (HSA) using the isothiocyanate complexes [Re(phen)(CO)$_3$(py-3-NCS)]$^+$ (**10**)[19] and [Ir(ppy)$_2$(phen-NCS)]$^+$ (**11**).[20] The resultant conjugates show intense and long-lived emission originating from a ^{3}MLCT (dπ(Re/Ir) $\rightarrow$ π*(diimine)) excited state. Biexponential emission decay is observed in all the conjugates, indicative of the heterogeneous microenvironments of the complexes. Additionally, BSA molecules labelled with a series of related iridium(III) isothiocyanate complexes [Ir(N$^\wedge$C)$_2$(phen-NCS)]$^+$ (**39**) with other cyclometal-lating ligands also show interesting emission properties.[54] It is worth mentioning that all these rhenium(I)- and iridium(III)-bioconjugates show different emission energy compared to the free labels, which is attributed to the decrease in electron-withdrawing ability of the resultant thiourea group compared to the isothiocyanate unit.

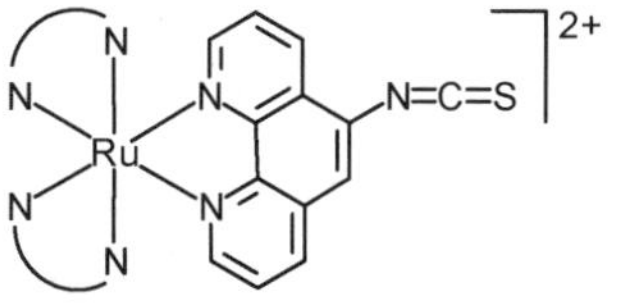

38

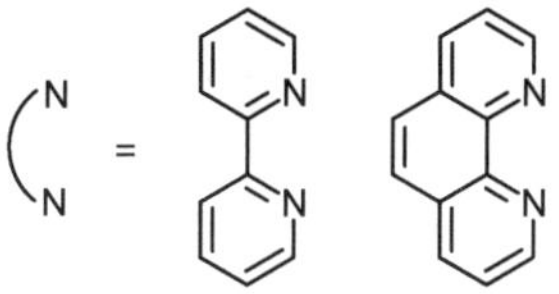

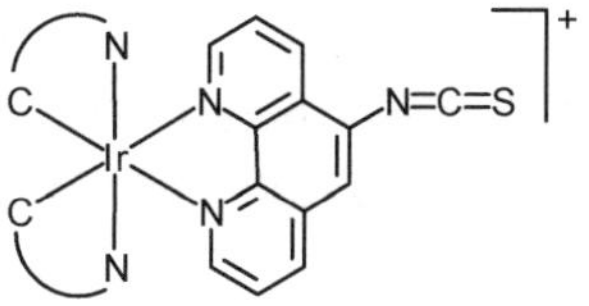

39

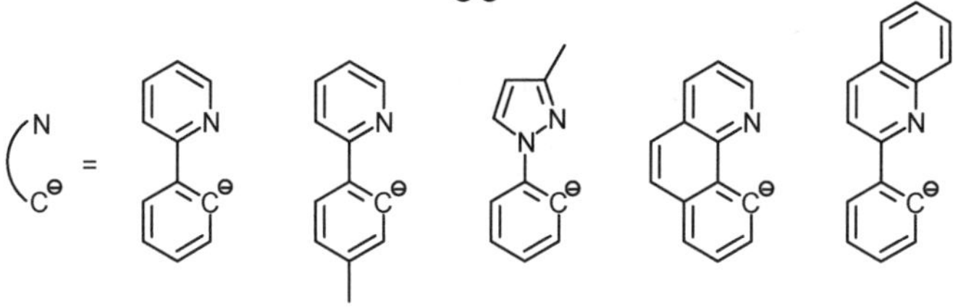

Yam and coworkers have reported the luminescent platinum(II) isothiocya-nate complex [Pt(^{t}Bu$_3$tpy)(C$\equiv$C-Ph-NCS)]$^+$ (**40**) which shows ^{3}MLCT emis-sion (λ_{em} = 586 nm, τ_o = 0.56 μs, Φ_{em} = 0.019) in CH$_3$CN at 298 K upon excitation.[55] The complex is conjugated to HSA and the bioconjugate shows emission at 650 nm in Tris-Cl buffer at 298 K. The emission is assigned to an excited state of mixed ^{3}MLCT/LLCT nature.

Since the aldehyde group undergoes facile reductive amination with a primary amine group to yield a stable secondary amine, the pyridylbenzaldehyde (pba) complexes $[M(pba)_2(N^\wedge N)]^+$ (M = Ir, Rh) (**41**) have been designed as labelling reagents and crosslinkers for biomolecules.[56,57] The ^{3}IL ($\pi \rightarrow \pi^*$)(pba) emission of the iridium(III) and rhodium(III) complexes occurs at 529 – 568 sh nm (τ_o = 4.32 – 7.19 µs, Φ_{em} = 0.21 – 0.27) and 506 – 586 sh nm (τ_o = 4.24 – 8.68 µs, Φ_{em} = 0.015 – 0.032), respectively. Upon conjugation to proteins, the emissive-state character changes to ^{3}MLCT ($d\pi(Ir) \rightarrow \pi^*(N^\wedge N)$) for the iridium conjugates and ^{3}IL ($\pi \rightarrow \pi^*$)($N^\wedge N$) for the rhodium conjugates, resulting in photophysical properties that are very different from those of the free labels. Recently, the more π-conjugated quinolylbenzaldehyde (qba) complexes $[Ir(qba)_2(N^\wedge N)]^+$ (**42**) featuring lower absorption and emission energies, and similar biological labelling behaviour have been reported.[58] The ^{3}IL ($\pi \rightarrow \pi^*$)(qba) emission of these complexes occurs at 573 – 624 (sh) nm (τ_o = 2.43 – 2.67 µs, Φ_{em} = 0.29 – 0.43). The biological crosslinking properties of these complexes are also demonstrated by SDS-PAGE analysis using BSA as a model protein (Figure 5.2). The crosslinked proteins emit at 586 – 593 nm (τ_o = 0.29 – 0.44 and 1.06 – 1.28 µs, Φ_{em} = 0.039 – 0.10) originating from a ^{3}MLCT ($d\pi(Ir) \rightarrow \pi^*(N^\wedge C)$)/^{3}IL ($\pi \rightarrow \pi^*$)($N^\wedge C$). Unlike the pba complexes, whose conjugates show different emission colours due to the ^{3}MLCT ($d\pi(Ir) \rightarrow \pi^*(N^\wedge N)$) state, the conjugates labelled with these qba complexes emit with a very similar colour.

Luminescent iridium(III) and rhodium(III) pba complexes appended with a biotin moiety $[M(pba)_2(bpy\text{-}spacer\text{-}biotin)]^+$ (M = Ir, Rh) (**43**) have been designed as novel biotinylation reagents.[59] The biotinylated Ir-BSA conjugates emit at *ca.* 570 nm (τ_o = 0.17 – 0.48 and 0.60 – 1.65 µs, Φ_{em} = 0.038 – 0.01), whereas their rhodium(III) counterpart at 529, 551 nm (τ_o = 0.95 µs, Φ_{em} = 0.002). One of the Ir-BSA conjugates shows avidin-binding properties as

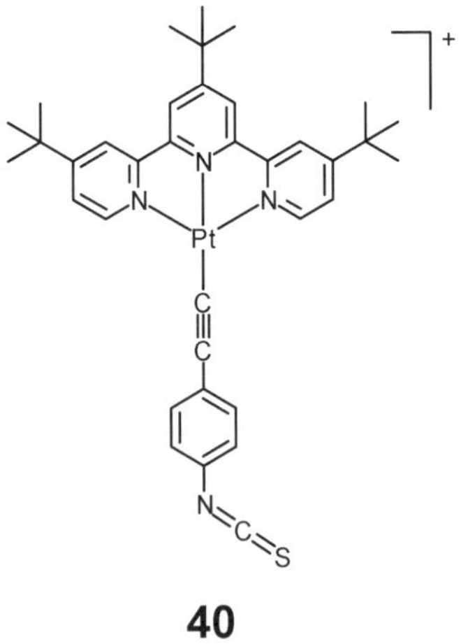

40

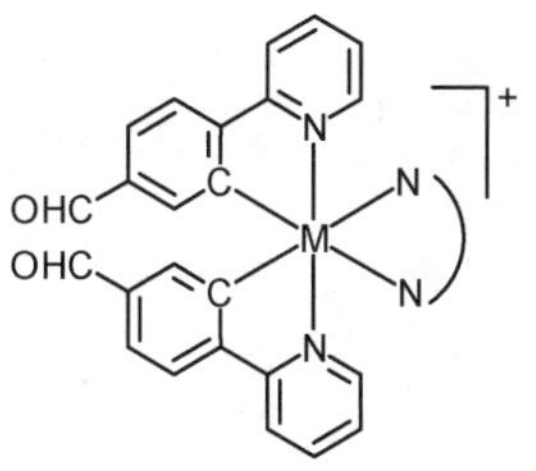

41

M = Ir, Rh

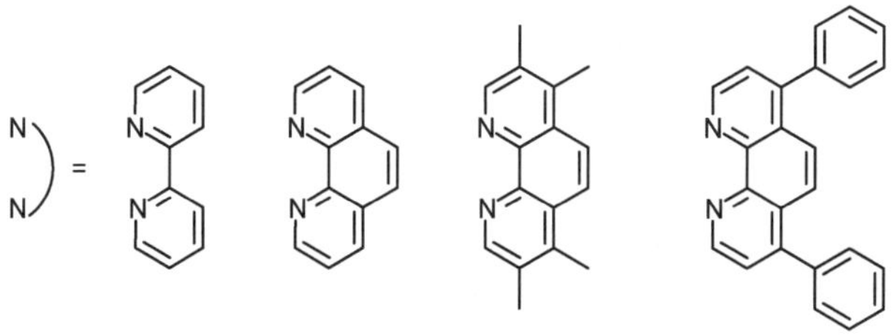

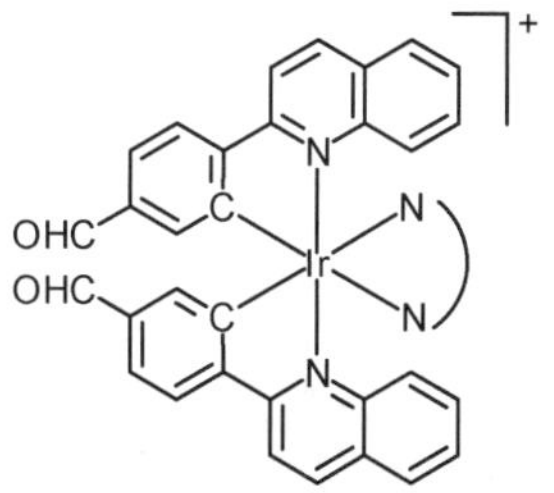

42

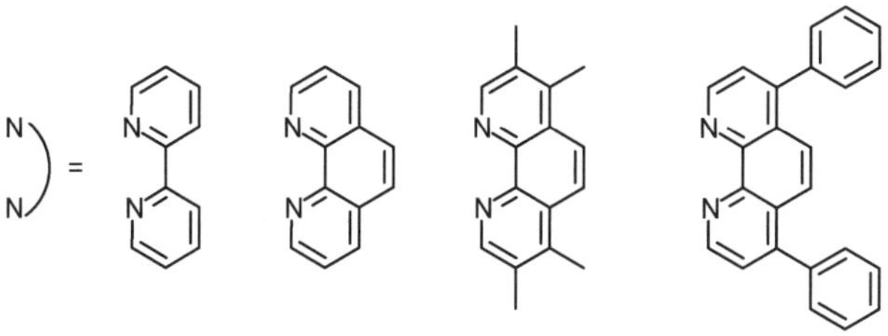

revealed by the HABA assay. These results render the complex a potential candidate for the design of new bioassays with enhanced reporting signals.

The NHS-ester is an important amine-reactive functional group that has been commonly used in bioconjugation. The resulting amide linkage in the bioconjugate is very stable in aqueous solution and does not significantly affect the photophysical properties. Thus, luminescent transition-metal complexes equipped with the NHS-ester are excellent labelling reagents; for example,

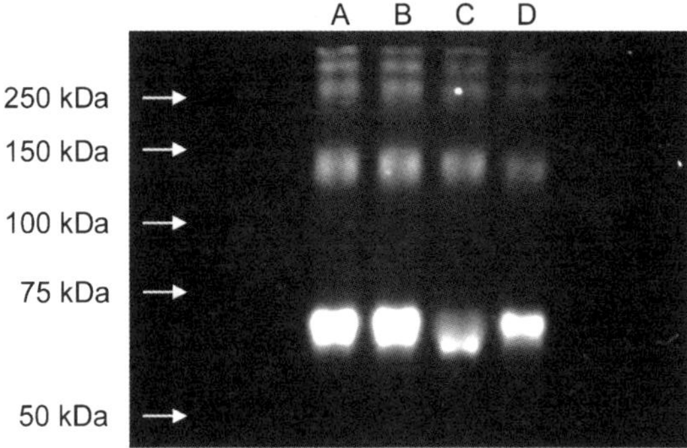

Figure 5.2 SDS-PAGE analysis of BSA conjugates crosslinked with [Ir(qba)$_2$(N$^\wedge$N)]$^+$ (N$^\wedge$N = bpy (Lane A), phen (Lane B), Me$_4$-phen (Lane C) and Ph$_2$-phen (Lane D)). Reproduced from ref. 58 with permission of the Royal Society of Chemistry.

Lakowicz and coworkers have described the protein-labelling properties of two ^{3}MLCT emitters including a ruthenium(II) complex [Ru(bpy)$_2$(dcbpy)]$^{2+}$ (**44**)[60] and a rhenium(I) complex [Re(bcp)(CO)$_3$(py-4-COOH)]$^+$ (**45**).[61] The photophysical data of the resultant bioconjugates are generally similar to those of the NHS labels except that oxygen quenching is much less efficient after the labels are bound to macromolecules.

43

M = Ir, Rh

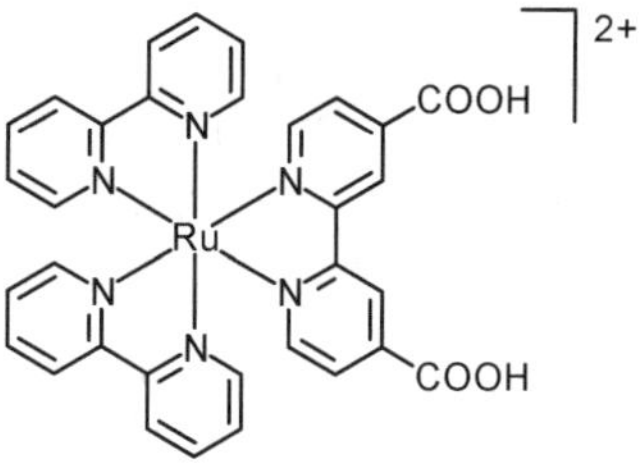

44

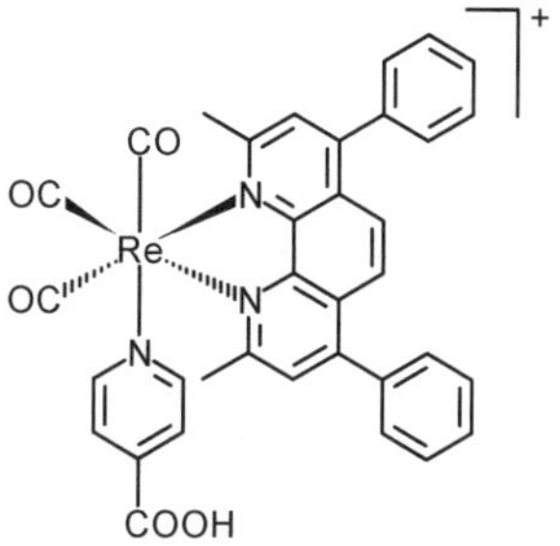

45

5.3.1.2 Sulfhydryl-Reactive Complexes

As the natural abundance of cysteine residues is low, the development of sulfhydryl-specific probes allows biological labelling with higher specificity. Lakowicz and coworkers have demonstrated that the rhenium(I) iodoacetamide complex [Re(phen-IAA)(CO)$_3$Cl] (**46**) (λ_{em} = 600 nm, τ_o = 436 ns, Φ_{em} = 0.039) can modify the cysteine residues of HSA, yielding a bioconjugate that displays a high anisotropy ratio and an elongated emission lifetime (1.1 μs).[62]

Lo and coworkers have reported the modification of HSA, glutathione and cysteine with a series of luminescent iridium(III) iodoacetamide complexes [Ir(N^C)$_2$(phen-IAA)]$^+$ (**47**) and the photophysical properties of the resultant bioconjugates.[54] The HSA conjugate [Ir(mppz)$_2$(phen-IAA)]-HSA emits at higher energy, with a higher emission quantum yield and longer lifetime (λ_{em} = 563 nm, τ_o = 0.68 and 0.12 μs, Φ_{em} = 0.05) than its glutathione and cysteine counterparts (λ_{em} = 588 – 594 nm, τ_o = 0.065 – 0.067 μs, Φ_{em} = 0.0005 – 0.014), which can be accounted for by the hydrophobicity of the serum albumin molecules.

The rhenium(I) maleimide complexes [Re(phen)(CO)$_3$(py-3-mal)]$^+$ (**12**)[21] and [Re(CO)$_3$(DQA-C6-mal)]$^+$ (**48**)[63] reported by Lo and coworkers and Zubieta and coworkers, respectively, react with the sulfhydryl groups of proteins in phosphate-buffered saline at pH 7.4 to yield stable conjugates. The photophysical properties of the labelled proteins are generally similar to those of the free complexes, except that the emission lifetimes are longer due to the relatively more hydrophobic protein matrix.

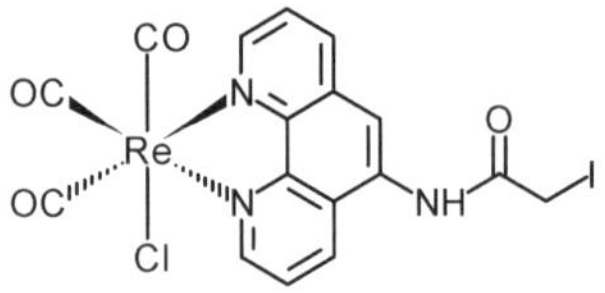

46

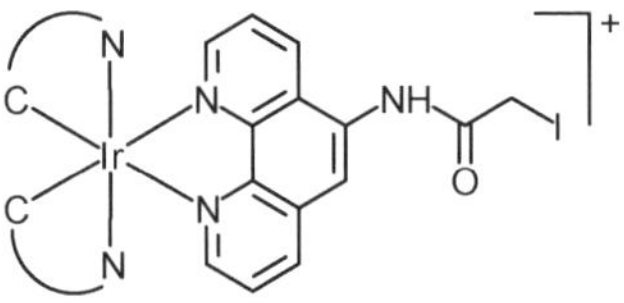

47

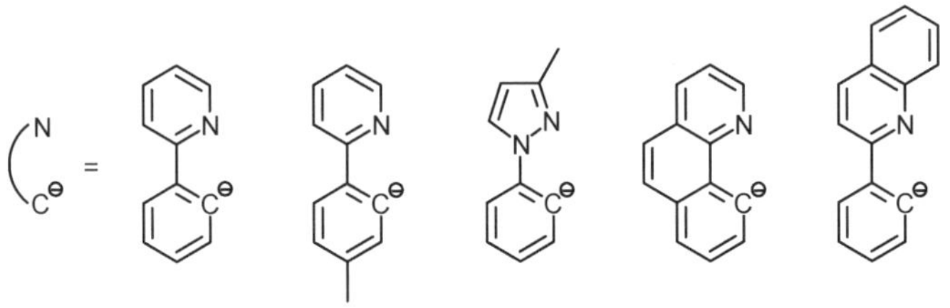

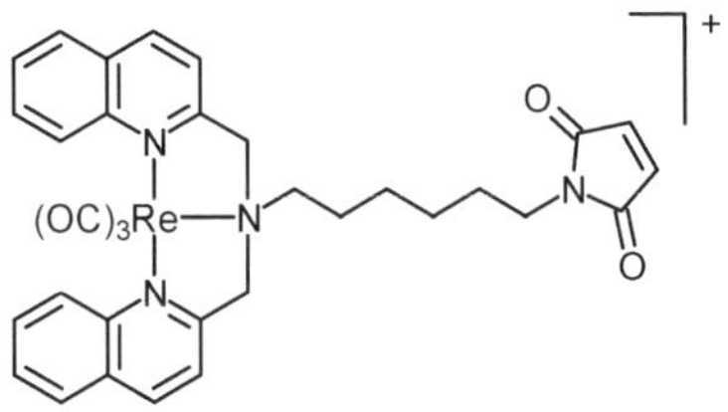

48

5.3.1.3   *Imidazole Complexes*

Weak field ligands such as H_2O and chloride are easily substituted by strong field π-accepting imidazole. Thus, complexes with such labile ligands may act as a label for proteins containing histidine (His) residues. Wong and coworkers have designed the iridium(III) complex $[Ir(ppy)_2(H_2O)_2]^+$ (**49**) as a biological probe.[64] Although this complex is weakly emissive in aqueous solution, it exhibits intense emission at 505 nm in the presence of His and BSA. The emission originates from the substitution of the coordinated water molecules by the imidazole unit of His. As this complex shows a luminescent switch-on effect towards His, it is a useful tool to detect His-rich biomolecules in SDS-PAGE gel.

Gray and coworkers have labelled azurin with the rhenium(I) aqua complex $[Re(phen)(CO)_3(H_2O)]^+$ **(50)** to yield $[Re(phen)(CO)_3(His-83)]^+$-AzCu$^+$ and studied the ET in this mutant.[65] Excitation of the rhenium(I) complex leads to direct oxidation of the copper(I) centre of the protein to copper(II). Using the flash-quench technique with $[Co(NH_3)_5Cl]^{2+}$ as a quencher, photogeneration of Re^{2+} followed by production of tryptophan and tyrosine radicals in these rhenium(I)-conjugated azurin proteins are observed. These phototriggered reactions originate from the strong photo-oxidising properties of the rhenium(I) polypyridine complex $[Re(phen)(CO)_3(imidazole)]^+$ ($E°(Re^{+*/0})$ = +1.2 V *vs.* SCE).

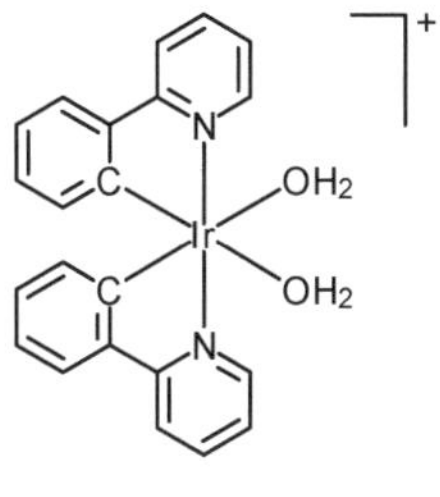

49

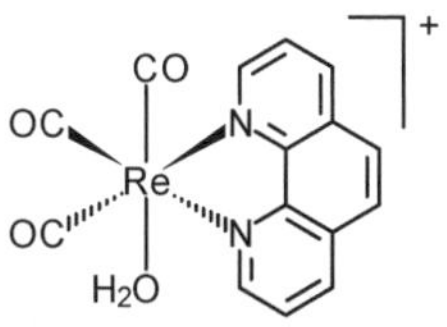

50

5.3.2   Noncovalent Protein Probes

5.3.2.1   *Nonspecific Probes*

Hydrophobic interactions are a key intermolecular force that brings transition-metal complexes to an inner pocket of a protein and thus gives subsequent alternations of photophysical and photochemical properties of the complexes. Gray and coworkers have designed rhenium(I) **(51)** and ruthenium(II) **(52)** diimine complexes containing a perfluorobiphenyl moiety.[66] These complexes bind tightly to the enzyme cytochrome P450cam and nitric oxide synthase due to hydrophobic interactions between the complexes and the substrate access channel of the protein. The emission of these complexes is partially quenched upon the protein-binding event, as a consequence of ET from the excited metal complex to the haem group of proteins.

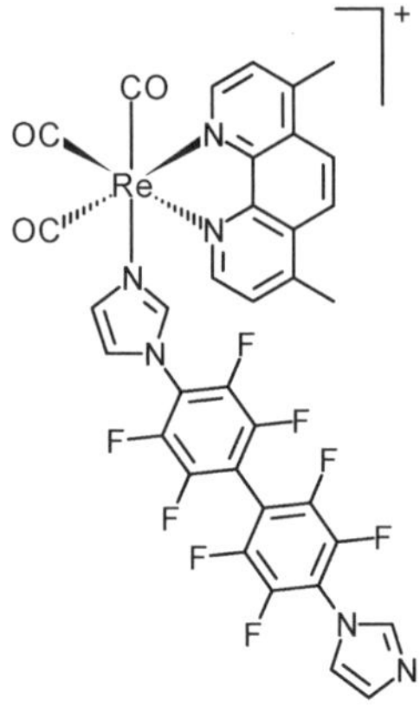

51

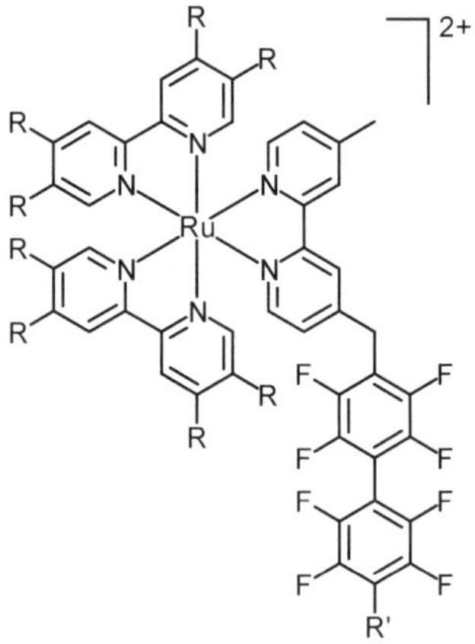

52

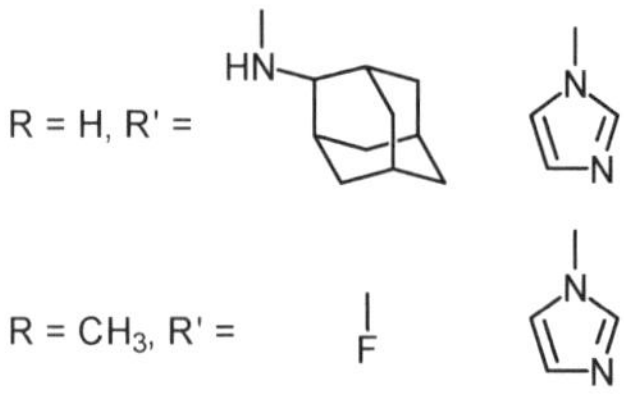

R = H, R' =

R = CH₃, R' =

Vaidyanathan and Nair have reported that the platinum(II) complex [Pt(bzimpy)Cl]$^+$ (**53**) functions as a biomolecular light switch with drastic emission enhancement ($I/I_o = 70$) and a blueshift in emission maxima from 625 to 553 nm upon binding to BSA.[67] In contrast, the complex displays emission quenching in the presence of haem-related proteins such as haemoglobin, which is attributed to fluorescence resonance energy-transfer (FRET) process.

The cyclometallated platinum(II) carboxyl complex [Pt(pby-Ph-COOH)Cl] (**54**), which is a typical ^{3}MLCT emitter, shows emission enhancement upon binding to protein molecules.[68] Density functional theory (DFT) calculation on its lowest triplet excited state shows structural distortion from the coplanar geometry, which exhibits a pronounced impact on the photophysical properties of the complex upon protein-binding.

The rhenium(I) alkyl complex [Re(bpy)(CO)$_3$(py-C14)]$^+$ (**55**) shows emission quenching when it interacts with a lipid-binding protein HSA.[69] This unusual property is attributed to the unfolding of the metal complex in aqueous solution once the appended alkyl chain binds to the hydrophobic pocket of the protein and no longer interacts with the bipyridine unit.

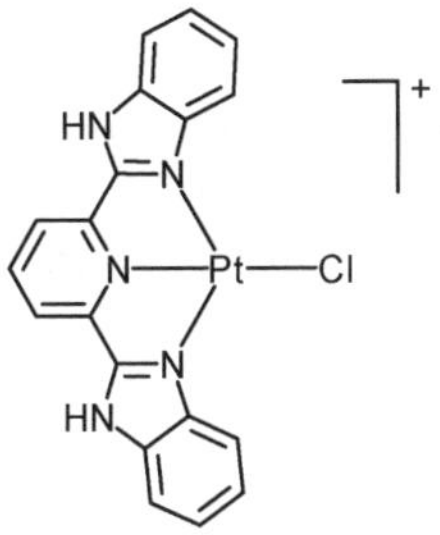

53

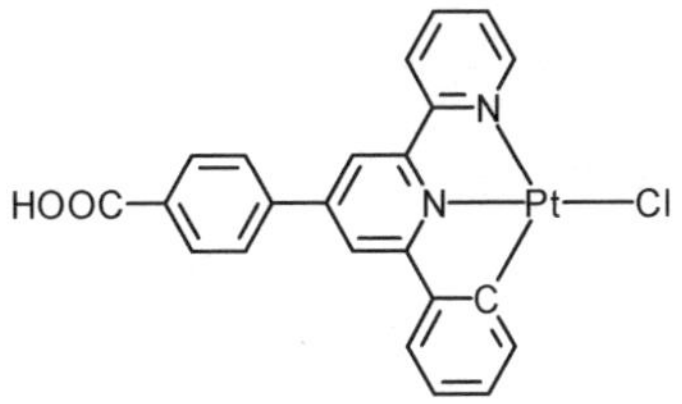

54

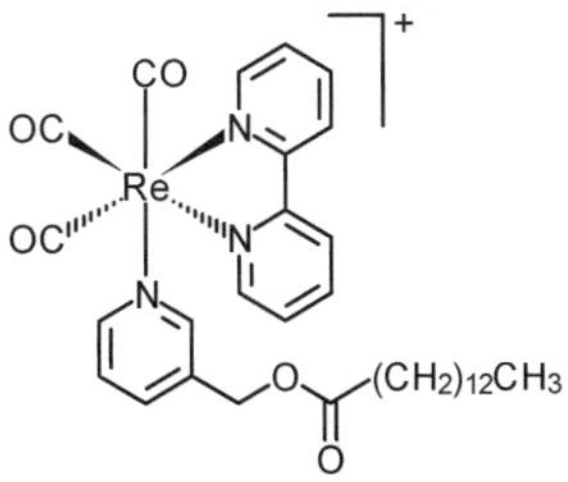

55

5.3.2.2 *Estradiol, Indole and Biotin Complexes*

Lo and coworkers have incorporated an estradiol unit into the luminescent rhenium(I) [Re(N^N)(CO)$_3$(py-spacer-est)]$^+$ (**56**),[70] ruthenium(II) [Ru(N^N)$_2$(bpy-spacer-est)]$^{2+}$ (**57**)[71] and iridium(III) [Ir(N^C)$_2$(bpy-spacer-est)]$^+$ (**58**)[72] polypyridine complexes. Upon visible-light irradiation, these complexes exhibit intense and long-lived ^{3}MLCT (dπ(Re/Ru/Ir) $\rightarrow$ π*(diimine)) emission in fluid solutions at 298 K and in low-temperature glass. Interestingly, these complexes display enhanced emission intensities and lifetime elongation upon binding to estrogen receptor α (ERα). The emission spectral traces of [Ir(N^C)$_2$(bpy-spacer-est)]$^+$ upon addition of ERα are shown in Figure 5.3. This is accounted for by the increased hydrophobicity and rigidity of the local environment of the complexes upon the binding event. A longer spacer-arm between the metal complex core and estradiol gives rise to stronger binding but lower emission enhancement as the environmental-sensitive luminescent metal complex core remains more exposed to the polar buffer after the binding. The

56

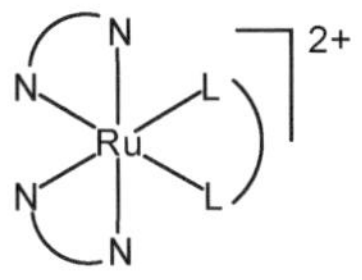

57

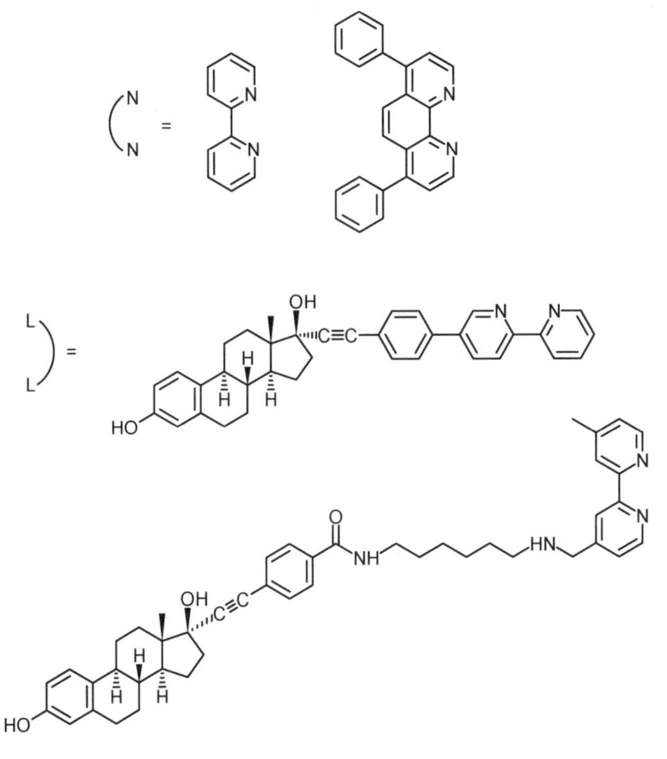

binding is specific to ERα as the estradiol complexes do not show noticeable changes in the presence of other proteins such as BSA.

Lo and coworkers have reported the synthesis, characterisation and photophysical and electrochemical properties of luminescent rhenium(I) diimine indole complexes [Re(N^N)(CO)$_3$(py-spacer-indole)]$^+$ (**59**).[73] Due to self-quenching *via* intermolecular ET, these rhenium-indole complexes exhibit much lower luminescence quantum yields and shorter emission lifetimes compared to those of their indole-free counterparts. Upon binding to an indole-binding protein BSA, the emission intensities of the indole complexes are enhanced by up to 17 fold. The ruthenium(II) analogues [Ru(bpy)$_2$(bpy-spacer-indole)]$^{2+}$ (**60**) do not show emission self-quenching since the excited state of the complexes is not sufficiently oxidising.[74] Nevertheless, the emission intensities of the complexes are increased by 1.01 to 1.38 fold upon binding to BSA. In the presence of a quencher [Fe(CN)$_6$]$^{4-}$, which preferentially suppresses the emission of the free probes, the complexes display more significant emission enhancement upon

58

binding to BSA (*ca.* 2.5 to 19.7 fold). Related luminescent iridium(III) indole complexes (**61**) have also been studied.[75]

The avidin–biotin system is a classic protein–ligand model and is widely employed in bioanalytical applications due to the strong affinity of biotin towards avidin.[76] Lo and coworkers have incorporated biotin into various luminescent rhenium(I) $[Re(N^{\wedge}N)(CO)_3(py\text{-}spacer\text{-}biotin)]^+$ (**62**),[77] ruthenium(II) $[Ru(bpy)_2(N^{\wedge}N\text{-}spacer\text{-}biotin)]^{2+}$ (**63**)[78] and iridium(III) $[Ir(N^{\wedge}C)_2(bpy\text{-}spacer\text{-}biotin)]^+$ (**64**)[79] polypyridine complexes to yield sensors for avidin. These biotin complexes exhibit increased emission intensities and lifetimes upon binding to avidin; the emission titration curves for $[Re(Me_2\text{-}Ph_2\text{-}phen)(CO)_3(py\text{-}4\text{-}CH_2\text{-}NH\text{-}biotin)]^+$ is shown in Figure 5.4 as an example. These changes result from the increased hydrophobicity and rigidity of the local surroundings of the probes upon binding. The detection sensitivity of these probes can be increased by various strategies. First, an external quencher

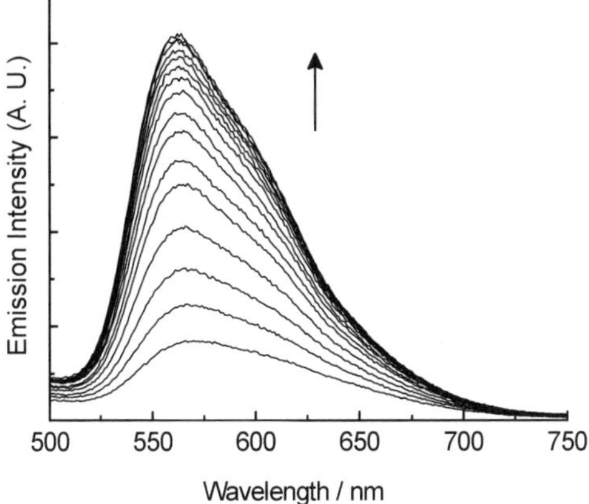

Figure 5.3 Emission spectral traces of [Ir(pq)$_2$(bpy-spacer-est)]$^+$ in potassium phosphate buffer (50 mM, pH 7.4)/methanol (9:1) at 298 K in the presence of 0 – 375 nM ERα. Reproduced from ref. 72 with permission of Wiley-VCH.

59

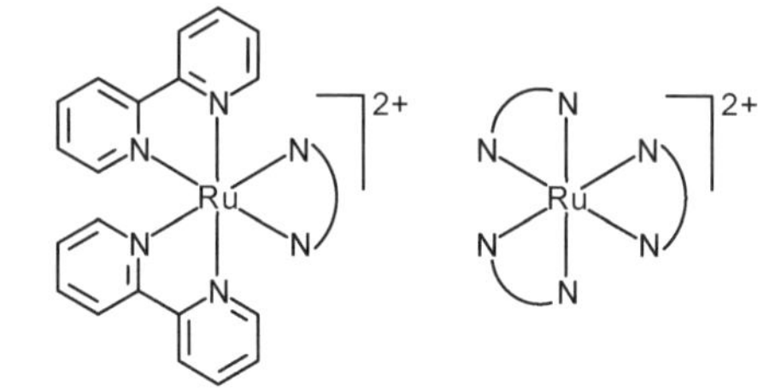

60

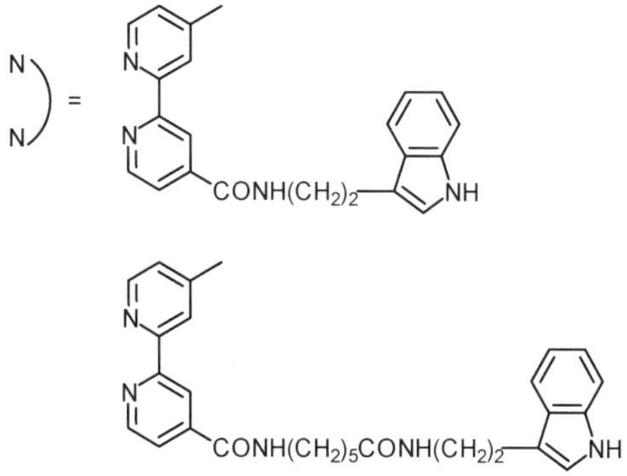

61

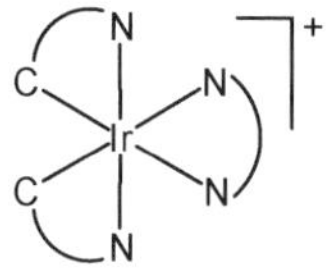

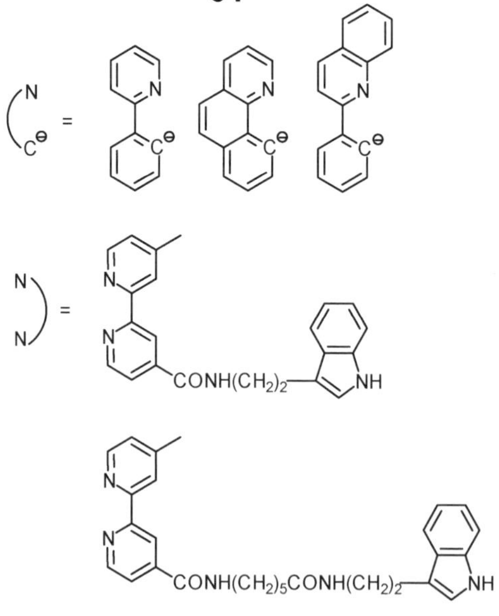

such as QSY7 and MV^{2+} that selectively suppresses the emission of free biotin complexes can be used. Second, avidin-induced emission enhancement can also be significantly increased by attaching the biotin moiety to dpq- and dppz-metal complexes (**23, 24**) that show extremely weak or no emission in the free form, but intense emission upon binding to avidin.[34,35] The third method is the introduction of an energy acceptor to the luminescent biotin complex; for example, Hong and coworkers have designed a tripod complex (**65**) consisting of an iridium(III) complex [Ir(ppy-F_2)$_2$(py-COO)] (energy acceptor), mCP (energy donor) and a biotin unit.[80] The emission spectrum of the donor and the absorption spectrum of the iridium complex show significant overlap, rendering efficient FRET to occur. Upon binding to avidin, the acceptor and donor are brought closer to each other, resulting in an increased FRET efficiency. Thus, the complex reveals a 14-fold increase in emission intensity when excited at the donor absorption peak at 310 nm.

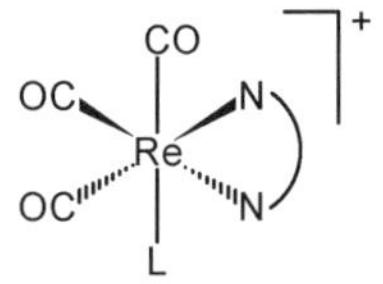

62

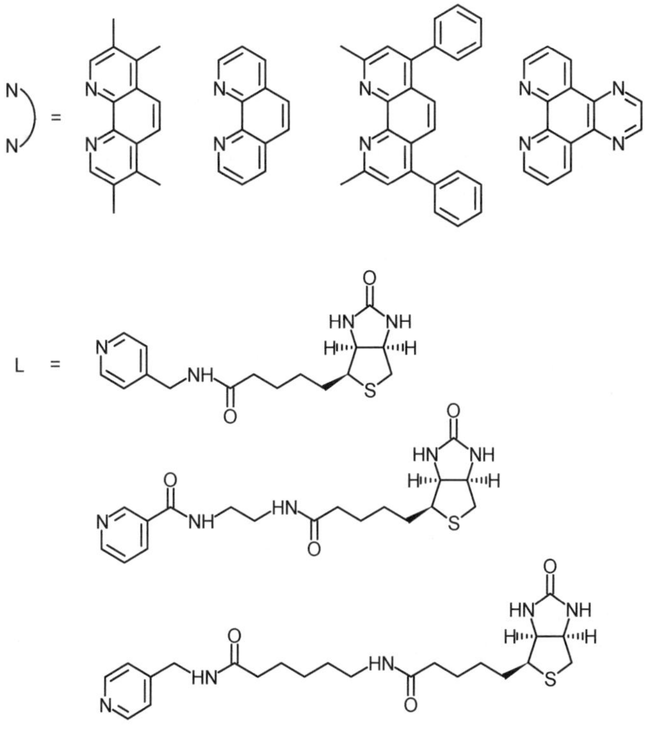

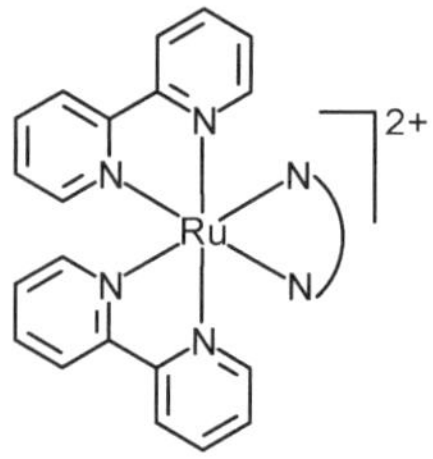

63

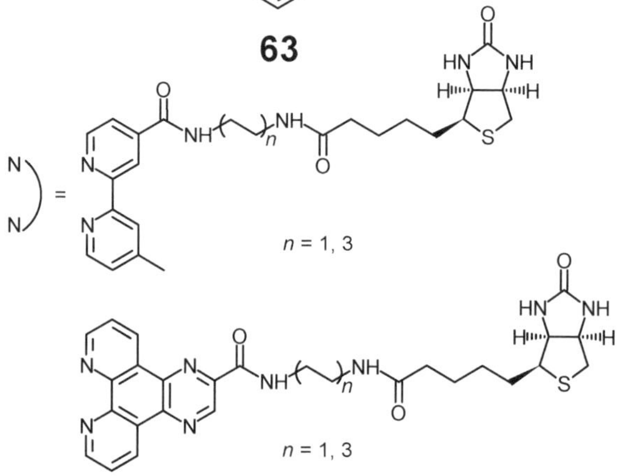

$n = 1, 3$

$n = 1, 3$

Luminescent crosslinkers for avidin are attractive candidates for the design of new avidin–biotin assays with enhanced sensitivity, as the number of labels can be significantly increased. Lo and coworkers have reported the luminescent rhenium(I) [Re(bpy{spacer-biotin}$_2$)(CO)$_3$(py)]$^+$ (**66**)[81] and iridium(III) [Ir(ppy-spacer-biotin)$_2$(N$^\wedge$N)]$^+$ (**67**)[82] bis-biotin complexes that show intense and long-lived ^{3}MLCT (dπ(Re/Ir) $\rightarrow$ π*(diimine)) emission. In spite of the small avidin-induced emission enhancement and lifetime elongation, these complexes effectively crosslink avidin to yield luminescent protein dimers, trimers and oligomers, which have been analysed by FRET emission-quenching experiments, microscopy studies on avidin-modified microspheres and size-exclusion HPLC.

In all the examples discussed thus far, the probes display changes in emission intensities and emission lifetimes upon binding to their protein hosts. In spite of small shifts of emission maxima, the emission profile and spectral characteristics of the luminescent probes basically remain the same. Lo and coworkers have discovered that the cyclometallated iridium(III) bipyridine complex [Ir(ppy-C4)$_2$(bpy-Et)]$^+$ (**68**) shows interesting dual emission in fluid solutions at room temperature (Figure 5.5).[83] The high-energy (λ_{em} = *ca.* 500 nm, τ_o = 1 – 3 µs) and low-energy (λ_{em} = *ca.* 600 nm, τ_o = 0.1 – 0.3 µs) features become predominant in solvents of high and low polarity, respectively. On the basis of these environmental-sensitive emission properties, dual-emissive complexes containing a biotin, an estradiol and a hydrophobic C18 chain (**69**) have been designed.[83] These complexes show apparent changes in the

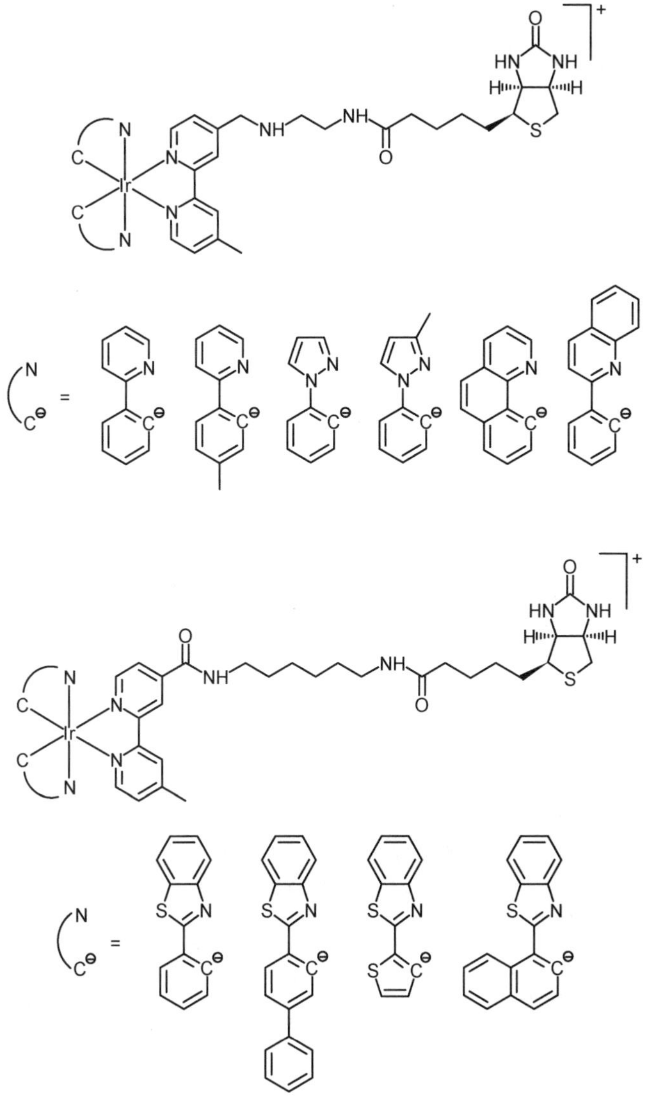

64

emission profile upon binding to their protein hosts, *i.e.* avidin, ERα and HSA, respectively, due to a change of the polarity of their local environment.

5.3.2.3   *Other Specific Probes*

Luminescent transition-metal complexes modified with a moiety that is specific to copper amine oxidase from *Arthrobacter globiformis* (AGAO), HSA and concanavalin A (Con A) have been designed. Gray and coworkers have

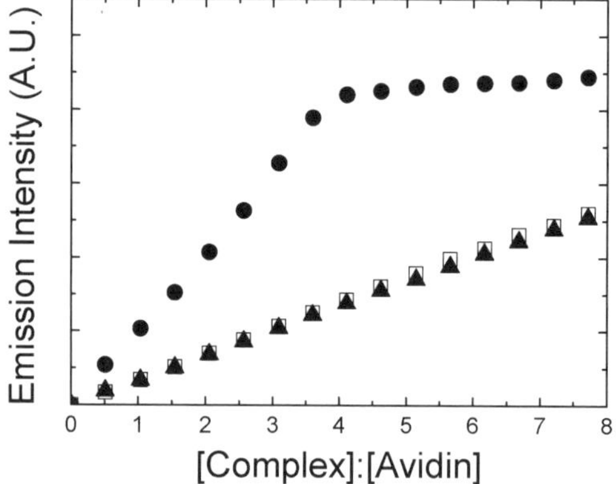

65

Figure 5.4 Emission titration curves for the titrations of (i) 3.8 μM avidin (solid circles), (ii) 3.8 μM avidin and 380.0 μM unmodified biotin (solid triangles) and (iii) a blank phosphate buffer solution (open squares) with [Re(Me$_2$-Ph$_2$-phen)(CO)$_3$(py-4-CH$_2$-NH-biotin)]$^+$. Reproduced from ref. 77 with permission of the American Chemical Society.

designed the molecular wires [Re(phen-spacer-DMA)(CO)$_3$(Me-im)]$^+$ (**70**) and [Ru(bpy)$_2$(phen-spacer-DMA)]$^{2+}$ (**71**) comprising a luminescent rhenium(I)- or ruthenium(II)-complex head group, an aromatic tail group and an alkyl spacer-arm;[84] all these wires bind reversibly to AGAO. The complexes bind to the protein and inhibit its enzymatic properties. The ability of these complexes to act as luminescent probes for AGAO is evaluated by both steady-state and time-resolved emission spectroscopy.

Che and coworkers have linked phenylalanine, tryptophan and glycine to cyclometallated platinum(II) metal complexes [Pt(N^C)(amino acid)] (**72**) to yield protein probes.[85] The probes are weakly emissive in aqueous buffer but show emission enhancement up to 10 fold at 562 nm upon binding to HSA.

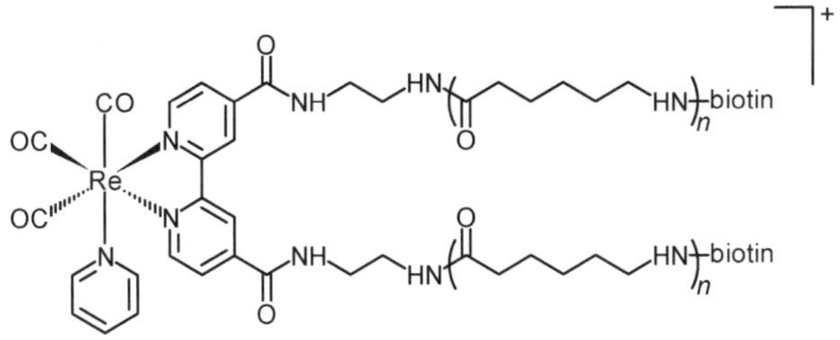

66

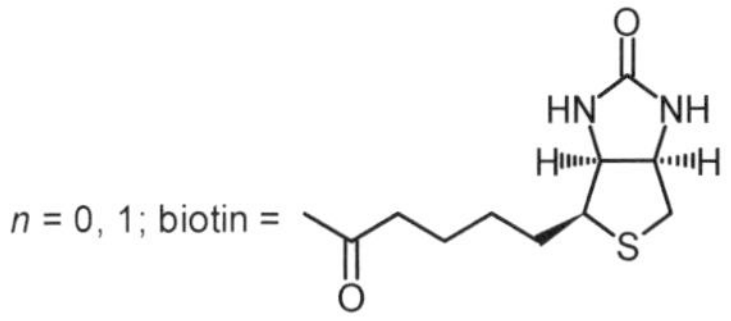

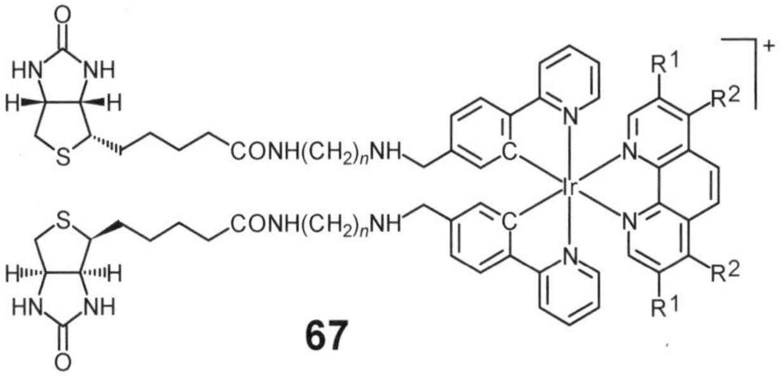

$n = 0, 1$; biotin =

67

$R^1 = R^2 = CH_3$; $n = 2, 6$
$R^1 = H$; $R^2 = C_6H_5$; $n = 2, 6$

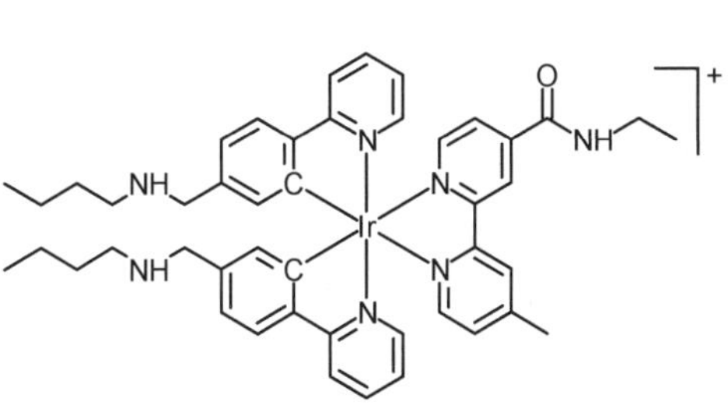

68

The high affinity (*ca.* 10^6 dm^3 mol^{-1}) towards HSA is ascribed to complementary hydrogen bonding interactions. In the control experiments, only small changes are detected when nonspecific proteins such as human globulins (HGs) are used.

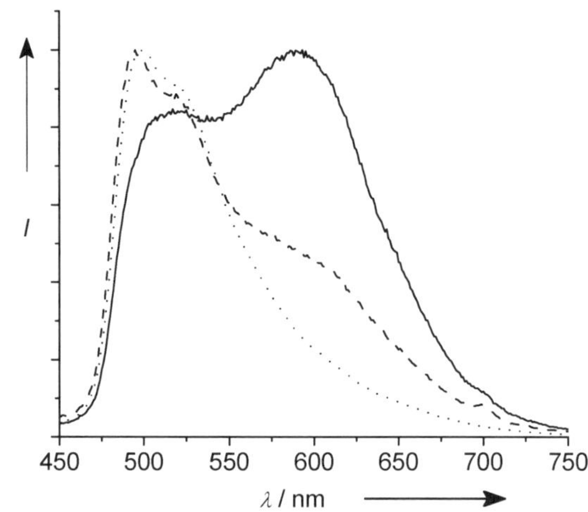

69

R = ...

Figure 5.5 Emission spectra of $[Ir(ppy-C4)_2(bpy-Et)]^+$ in degassed CH_2Cl_2 (solid line), CH_3OH (dashed line) and phosphate buffer (dotted line) at 298 K. Reproduced from ref. 83 with permission of Wiley-VCH.

The ruthenium(II) complexes $[Ru(bpy\{spacer-sugar\}_2)_3]^{2+}$ (**73**) appended with various saccharide pendants have been designed.[86] The ^{3}MLCT emission of these complexes occurs at 601 – 607 nm ($\Phi_{em} = 0.086 - 0.150$). The binding of these complexes to Con A has been studied with different techniques. It is found that the complexes appended with α-manno clusters display excellent

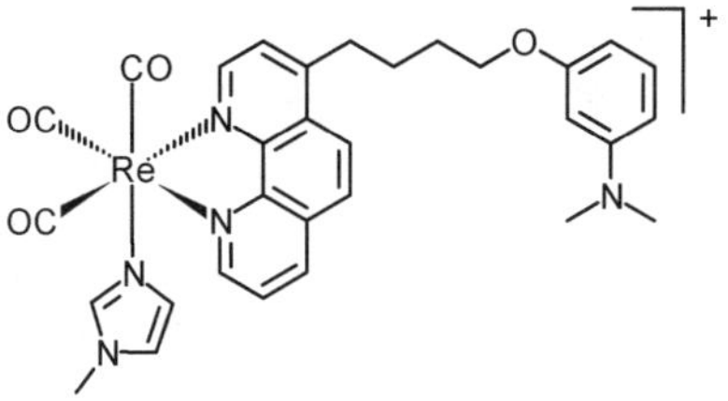

70

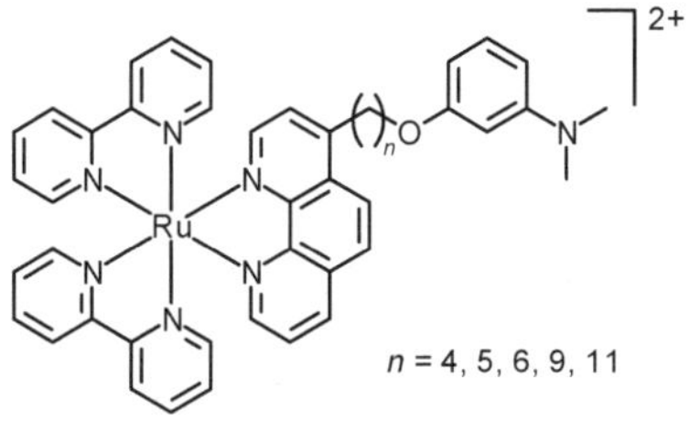

71

$n = 4, 5, 6, 9, 11$

binding affinities towards the lectin and emission quenching is observed upon the binding event.

Lo and coworkers have synthesised a series of rhenium(I) glucose complexes [Re(N^N)(CO)$_3$(py-3-glucose)]$^+$ (**74**).[87] The Ph$_2$-phen complex shows greenish-yellow ^{3}MLCT (dπ(Re) $\rightarrow$ π*(N^N)) emission at 543 – 557 nm (τ_o = 0.63 – 1.84 μs, Φ_{em} = 0.0023 – 0.034) upon excitation. The complex binds to Con A resulting in emission enhancement of *ca.* 3.2 fold and lifetime extension from 0.40 μs to (2.04 (33%), 0.40 (67%) μs). The binding constant (K_a) (4.4 $\times$ 10^5 M^{-1}) is comparable to that of the ruthenium(II) α-glucose complex **73** (K_a = 9.5 $\times$ 10^5 M^{-1}) but one order of magnitude larger than that of *p*-nitrophenyl-α-D-glucopyranoside (K_a = 1.0 $\times$ 10^4 M^{-1}). The higher binding affinity is due to the relatively hydrophobic rhenium(I) and ruthenium(II) polypyridine units.

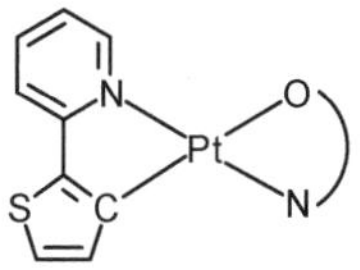

72

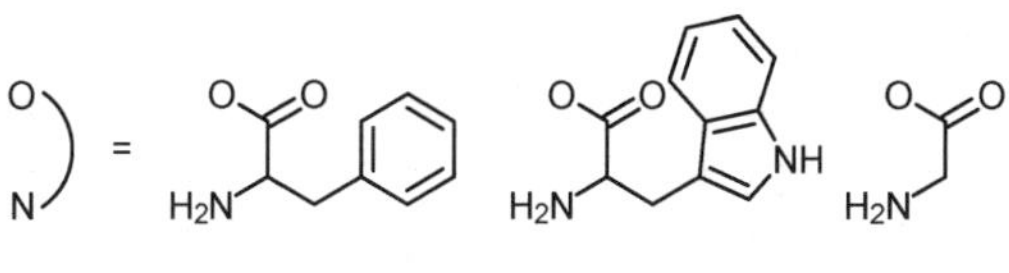

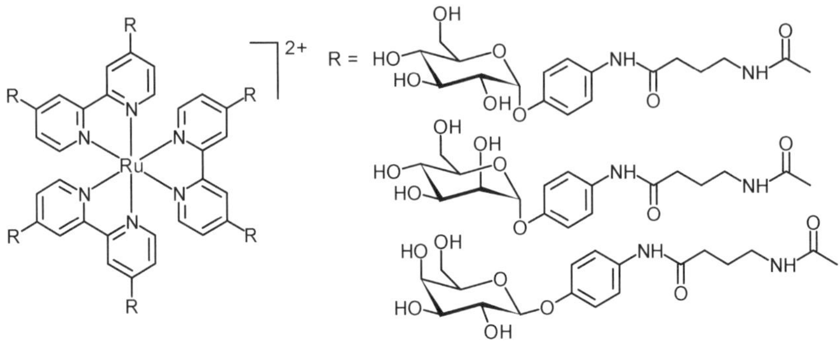

73

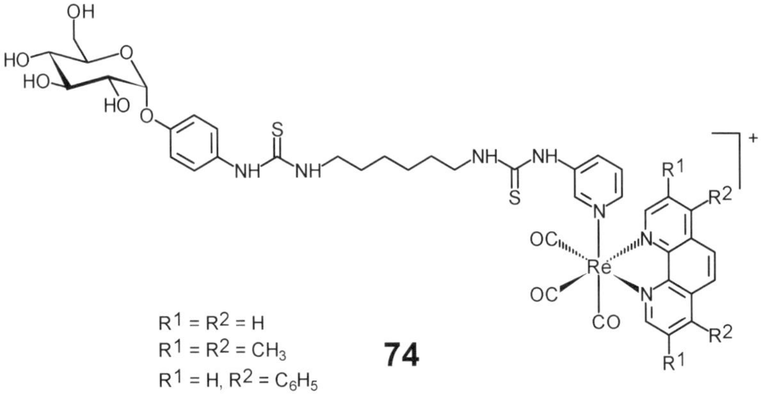

R¹ = R² = H
R¹ = R² = CH₃
R¹ = H, R² = C₆H₅

74

5.4 Cellular Probes

5.4.1 Cellular Uptake and Cytotoxicity

In the past few years, there has been a fast-growing interest in the cellular uptake properties of luminescent transition-metal complexes. Studies of the cellular uptake properties of these complexes have been focused on the effects of charges and lipophilicity of these complexes on their cellular uptake and cytotoxic activity and the possibility of using these complexes as bioimaging reagents.[88]

5.4.1.1 *Effects of Charge and Lipophilicity*

Due to the negative surface charge of a lipid bilayer of the cytomembrane, cationic transition-metal complexes can readily enter living cells, whereas anionic complexes show very limited or no cellular uptake at all; for example, Li and coworkers have reported that incubation of HeLa cells with the cationic iridium(III) complexes [Ir(dfpy)$_2$(N$^\wedge$N)]$^+$ (**75**) in DMSO/PBS (pH 7, 1:49, v/v)

for 10 min at 25 °C leads to efficient uptake.[89] Laser scanning confocal microscopy reveals intense intracellular emission that originates from the complexes. One of the advantages of these probes is that their emission colours can be tuned using various N^N ligands.

Coogan and coworkers have designed a trimeric rhenium(I)-based molecular vessel (**76**) that binds silver(I) ions strongly, yielding a loaded cationic cup.[90] Human adenocarcinoma (MCF-7) cells incubated with the empty neutral vessel in 4 °C show no evidence of uptake. However, when the cells are treated with the silver-loaded cationic complex, *ca.* 80% of the cells display emission, which is indicative of efficient uptake. This work illustrates the importance of the cationic charge of the complex on cellular internalisation.

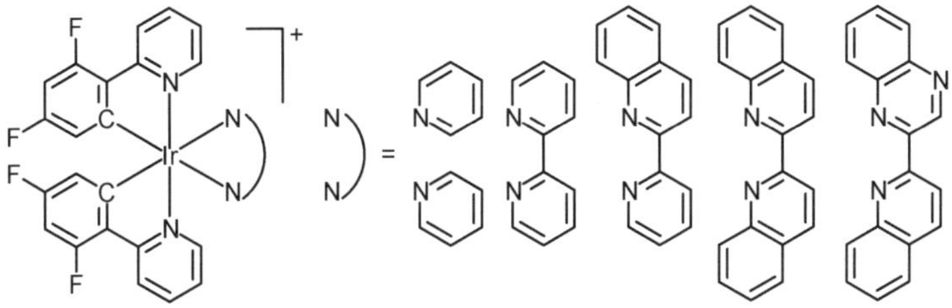

75

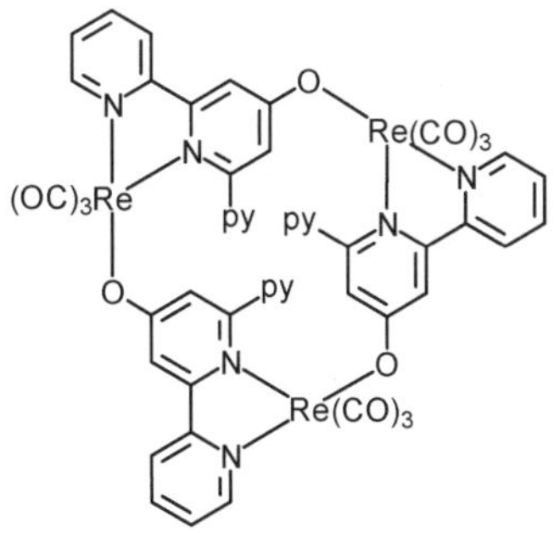

76

Pascu and coworkers have reported the different cellular uptake efficiencies of the copper(II) and zinc(II) diacetylbis(*N*-methylthiosemicarbazone) (ATSM) complexes [M(ATSM)] (M = Cu, Zn) (**77**).[91] The zinc(II) complexes are protonated more readily than the copper(II) analogues and thus show a higher uptake rate into the cytosol. In other studies, the anionic rhenium(I) [Re(phen{C$_6$H$_4$-SO$_3$}$_2$)(CO)$_3$(py-3-R)]$^-$ (**78**)[92] and neutral iridium(III) [Ir(N^C)$_2$(Hdcbpy)] (**79**)[93] complexes show insignificant cellular uptake. Thus, the introduction of a negative charge to transition-metal complexes is a means to reduce the cellular uptake efficiency of small molecules.

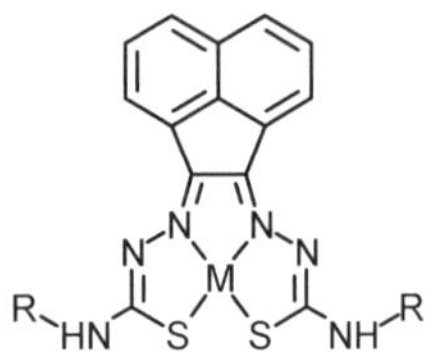

77

M = Cu, R = CH$_3$, CH$_2$CH$_3$
Zn, R = CH$_3$, CH$_2$CH$_3$

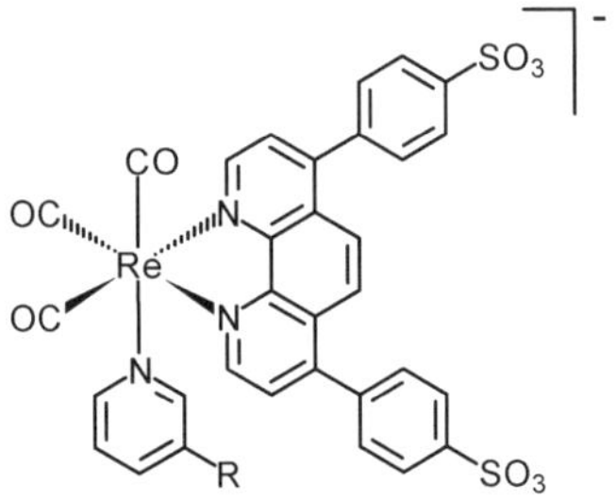

78

R = H, CH$_2$OH

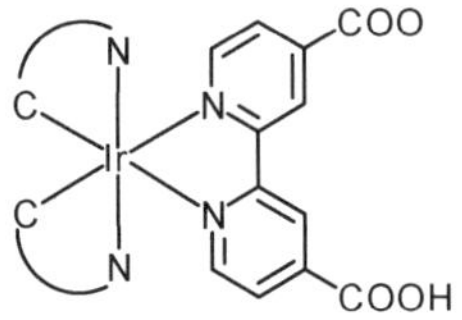

79

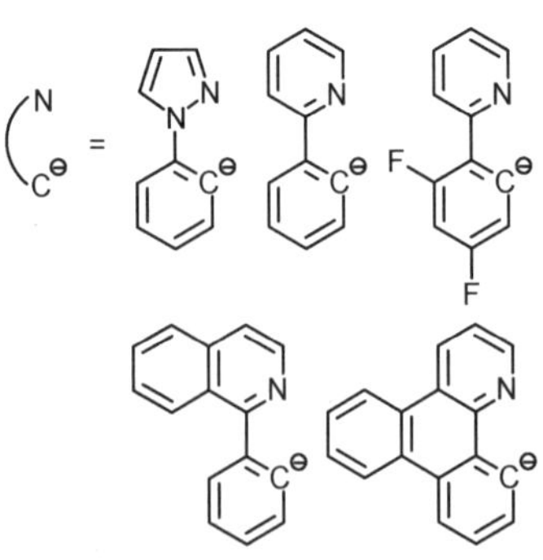

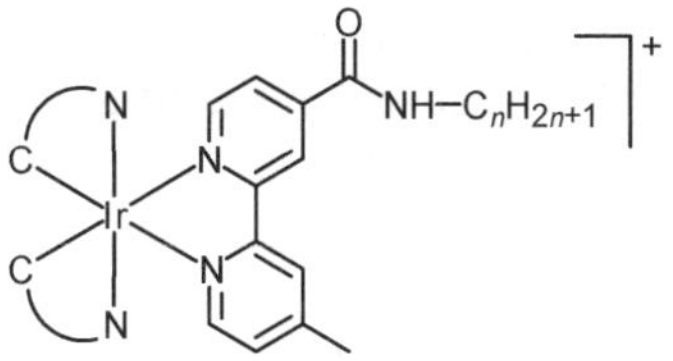

80

n = 18, 10, 2

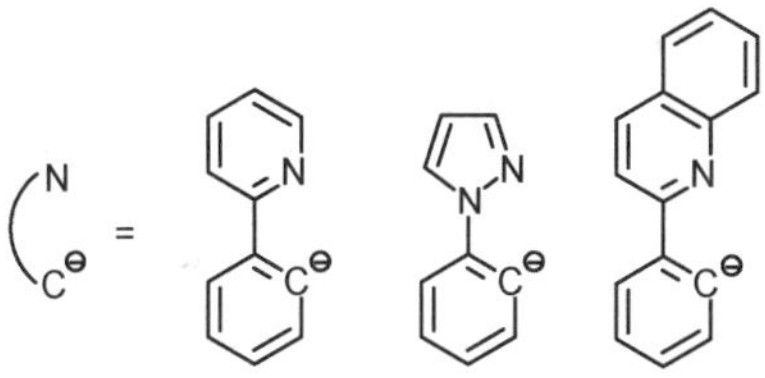

The lipophilicity of transition-metal complexes plays a key role in their cellular uptake efficiency; for example, Lo and coworkers have compared the lipophilicity and cellular uptake of three iridium(III) complexes $[Ir(N^\wedge C)_2(bpy\text{-}C_nH_{2n+1})]^+$ (n = 18, 10, 2) (**80**) appended with alkyl chains of different lengths.[94] These complexes are typical ^{3}MLCT emitters (λ_{em} = 568 – 626 nm, τ_o = 0.040 – 0.71 µs, Φ_{em} = 0.0069 – 0.25). The partition coefficients (log $P_{o/w}$ as a measure of lipophilicity) increase from 2.01 to 9.89 with increasing alkyl chain length from n = 2, 10 to 18. Interestingly, the cellular uptake efficiencies of these complexes follow the order: n = 10 > n = 2 > n = 18. The lowest uptake rate of the octadecyl complex is probably due to the diminished effect of the positive charge. Another possibility is the higher aggregation tendency of the complex in aqueous solution.

5.4.1.2 Specific Cellular Uptake

Modification of a transition-metal complex with a molecular transporter is an effective strategy to facilitate entry of the metal complexes into living cells in a controlled manner. An example of these molecular transporters is cell-penetrating peptides such as the HIV Tat peptide and oligoarginine, which are expected to enhance the cellular uptake of their conjugates; for example, Puckett and Barton have appended a fluorescein to a ruthenium(II) dppz complex $[Ru(phen)(dppz)(bpy\text{-}D\text{-}R8\text{-}fluro)]^{2+}$ (**81**) *via* an octaarginine linker.[95] This conjugate can enter HeLa cells readily revealing strong emission in cytoplasm and in the nucleus at relatively low concentration (5 µM). A control experiment involving a ruthenium(II)-fluorescein conjugate without the octaarginine linker does not show sufficient cellular uptake under the same conditions.

Keyes and coworkers have reported the cellular uptake properties of two ruthenium(II) polypyridine complexes $[Ru(bpy)_2(pic\text{-}R5/R8)]^{2+}$ (**82**) equipped with a polyarginine pendant.[96] The conjugates emit at 607 nm (τ_o = 775 ns,

Φ_{em} = 0.067) upon photoexcitation. The pentaarginine conjugate cannot penetrate the cell membrane and enter the human blood platelets. However, the octaarginine conjugate transports passively through the cell membrane and preconcentrates inside the cell. This reflects the importance of the chain length of these polyarginine peptides on the cell-penetrating ability of these probes.

The rhenium(I) complexes $[Re(CO)_3(SAACQ\text{-}fMLF)]^+$ (**83**)[97] and $[Re(CO)_3(BQAV\text{-}FA)]^+$ (**84**),[98] equipped with tripodal N^N^N ligands modified with fMLF (a targeting sequence to formyl peptide receptor, FPR) and a folic acid (FA) moiety, respectively, have been reported by Zubieta and coworkers. The most important findings are the high specificity of the fMLF- and FA-conjugates to target leukocytes and FR-expressing (FR = folate receptor) cells (A2780/AD), respectively. The subsequent cellular uptake of these complexes towards their specific cell lines is efficient, rendering them excellent imaging reagents and potential chemotherapeutics.

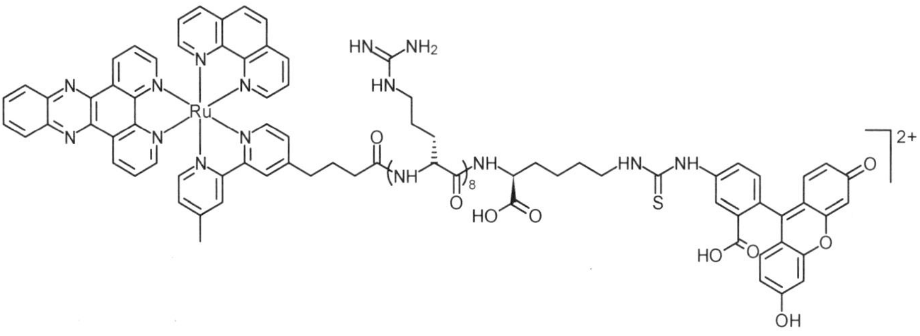

81

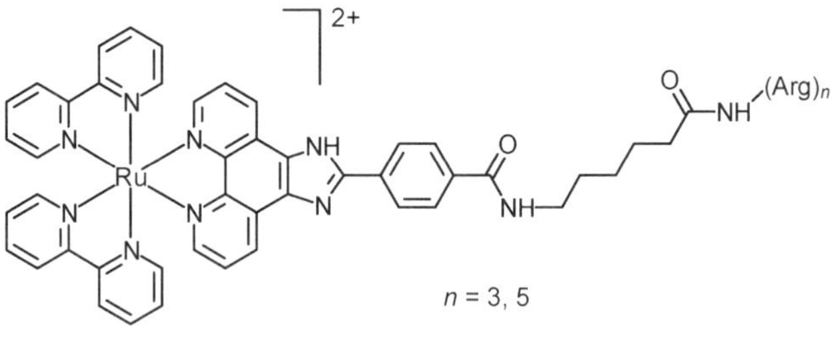

82

5.4.1.3 Uptake Mechanisms

Cellular uptake pathways include passive diffusion, transport proteins and endocytosis. Passive diffusion involves the movement of molecules directly through the lipid bilayer down their concentration gradient and is energy

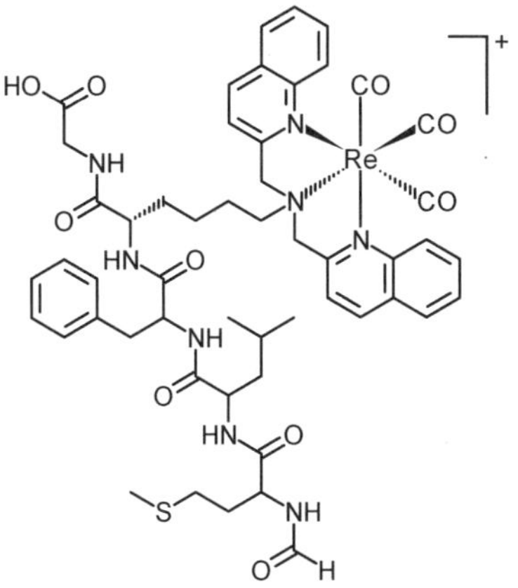

83

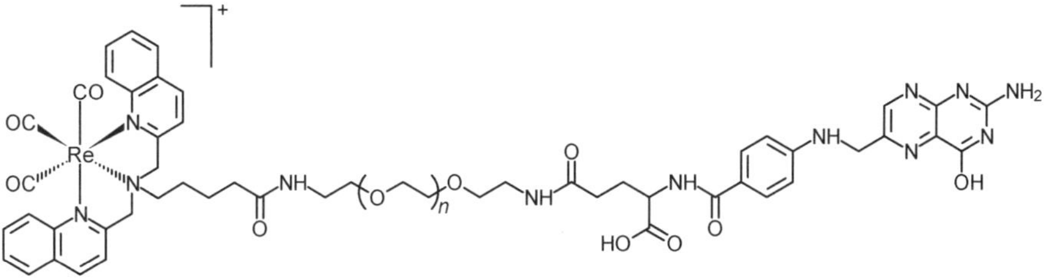

84

independent. Uptake *via* transport proteins may be energy independent (for channels and passive carriers) or energy dependent (ATP-powered pumps). These proteins carry a particular class of cargo, and their expression varies by tissue type. Endocytosis is energy-requiring uptake of macromolecules and solutes by vesicles derived from the plasma membrane.

Cationic lipophilic transition-metal complexes may passively diffuse across the cytomembrane in response to the membrane potential; for example, a ruthenium(II) complex $[Ru(Ph_2\text{-}phen)_2(dppz)]^{2+}$ (**85**) containing the lipophilic Ph_2-phen ligand enters cells by passive diffusion in a membrane-potential-dependent manner.[99] This finding is supported by the negative feedback from the inhibition experiments on cell metabolism and cation transporters as revealed by flow cytometry.

Cellular studies on cyclometallated iridium(III) polypyridine complexes by Zhang and Lo indicate that the cellular uptake of these complexes may occur *via* more than one mechanism; for example, the uptake of $[Ir(ppy)_2(bpy\{C4\}_2)]^+$ (**86**) into cytoplasmic foci is diminished at 4 °C, while the diffuse cytoplasmic staining is maintained, indicating that both passive diffusion and endocytosis occurs simultaneously.[100]

Endocytosis can be mediated by triggering the receptors located on the cell surface. One of these receptors is cubilin, which functions together with the intrinsic factor (IF), transporting vitamin B_{12} across the intestinal enterocyte. Doyle and coworkers have appended a B_{12} unit to a rhenium(I) *N,N*-bis(quinolinoyl) luminescent unit to form complex **87**.[101] This complex enters cells *via* receptor-mediated endocytosis, which is confirmed by experiments involving IF labelled by a pH-sensitive cyanine dye CypHer5E. Transition-metal complexes functionalised with other biologically relevant substrates such as biotin, FA and BSA are also potential markers for endocytic pathways during cellular uptake.

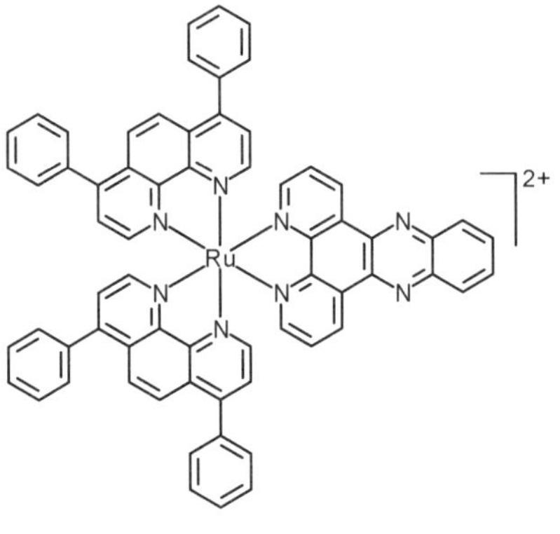

85

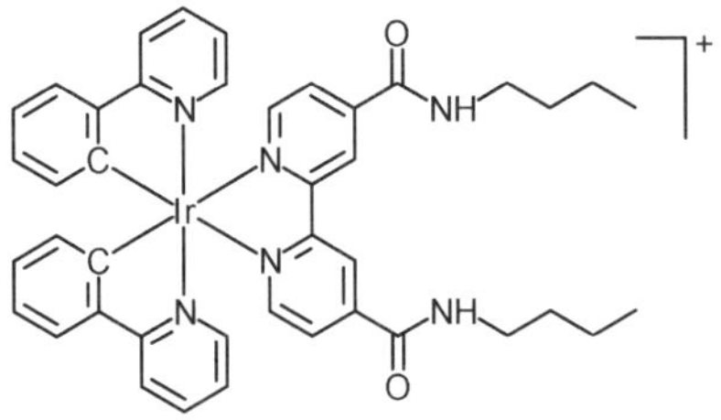

86

5.4.1.4 *Cytotoxicity*

The cytotoxicity (usually expressed as the IC_{50} values) of transition-metal complexes is usually related to their cellular uptake efficiencies as well as their lipophilicity; for example, the relationship of lipophilicity and cytotoxicity of a series of iridium(III) indole $[Ir(N^\wedge C)_2(bpy\text{-}spacer\text{-}indole)]^+$ **(61)**[75] and biotin $[Ir(N^\wedge C)_2(N^\wedge N)]^+$ (HN$^\wedge$C = Hppy-spacer-biotin, N$^\wedge$N = bpy; HN$^\wedge$C = Hppy, N$^\wedge$N = bpy-(spacer-biotin)$_2$; HN$^\wedge$C = Hppy-spacer-biotin, N$^\wedge$N = bpy-

87

spacer-biotin) (**88**)[100] complexes has been investigated. The IC_{50} values of the lipophilic indole complexes range from 1.1 to 6.3 μM, whereas the hydrophilic biotin complexes are basically noncytotoxic ($IC_{50} > 400$ μM). It is likely that the high cytotoxicity of the indole complexes originates from their higher lipophilicity and more efficient cellular uptake by the HeLa cells.

For luminescent transition-metal complexes to be used as imaging reagents, they should ideally be noncytotoxic. Incorporation of biocompatible poly(ethylene glycol) (PEG) into transition-metal complexes can significantly lower their cytotoxicity; for example, Lo and coworkers have reported a series of cyclometallated iridium(III) PEG complexes $[Ir(N^{\wedge}C)_2(bpy\text{-}PEG1/PEG3)]^+$ (**89**).[102] Results of the MTT assay reveal that the IC_{50} values of these complexes (*ca.* 286.5 to 1180.0 μM) are much higher than those of their PEG-free counterparts and cisplatin (*ca.* 4 to 15 μM) (Figure 5.6), indicative of their noncytotoxic nature. The high biocompatibility of these complexes can also be reflected by the fact that HeLa cells remain viable even after treatment with a high concentration of the complex (200 μM) for a long incubation time (2 h).

5.4.2   Intracellular Localisation

5.4.2.1   Cytomembrane and Cytoplasm

A highly lipophilic transition-metal complex may be used to probe membrane structures owing to the hydrophobic interactions between the complex and the lipid bilayer; for example, Lam and coworkers have reported an amphiphilic and water-soluble platinum(II) complex $[Pt(C^{\wedge}N^{\wedge}N\text{-}C18)(P\{C_6H_4\text{-}SO_3\}_3)]^{2-}$ (**90**) that emits at 528 nm with a lifetime of 1.95 μs.[103] With the hydrophobic

88

octadecyl chain, the complex shows rapid and efficient localisation in the plasma membrane of live HeLa cells. Time-lapse experiments indicate that the complex can remain in the plasma membrane for at least 3 h with no subcellular migration to the cytoplasm.

Since the cytoplasm contains numerous membrane structures, a lipophilic probe may show significant subcellular staining; for example, Coogan and coworkers have reported that the rhenium(I) complex [Re(bpy)(CO)$_3$(py-C14)]$^+$ (**55**) bearing a lipophilic carbon chain at the edge of the bpy ligand can interact with internal membranes partitioning cell compartments and with the constituents within organelles.[105] The intracellular localisation properties are dependent on the lengths of the lipophilic chains. Svensson and coworkers

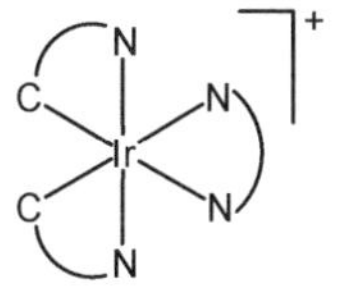

89

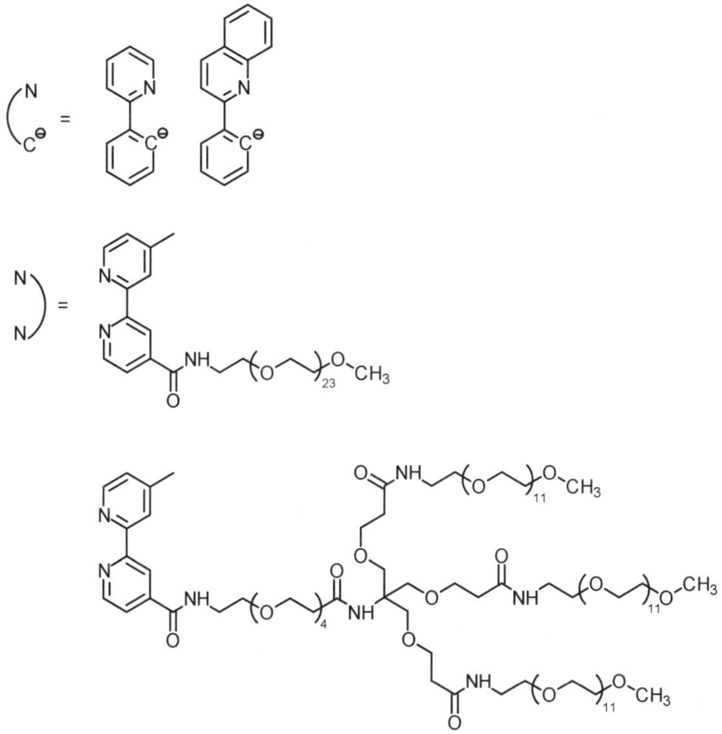

have reported a lipophilic ruthenium(II) complex $[Ru(phen)_2(dppz\{C6\}_2)]^{2+}$ (**91**) that preferentially stains membrane-rich parts of the cell.[104]

Many transition-metal complexes with simple and hydrophobic ligands are able to stain the cytoplasm; for example, the rhenium(I) $[Re(N^\wedge N)(CO)_3(py\text{-}3\text{-}TU\text{-}C2)]^+$ (**92**)[87] and platinum(II) $[Pt(C^\wedge N^\wedge N)Cl]$ (**93**)[106] complexes are mostly distributed inside the cytoplasm, with a lower level of nuclear uptake. Thus, the staining usually appears in the form of a perinuclear ring. The higher degree of localisation of the complex in the perinuclear region suggests that these complexes interact with hydrophobic organelles such as the endoplasmic reticulum and Golgi apparatus inside the cell.

5.4.2.2 Mitochondrion

Coogan and coworkers have functionalised a rhenium(I) complex $[Re(bpy)(CO)_3(py\text{-}3\text{-}CH_2Cl)]^+$ with a chloromethyl unit (**94**) that is reactive

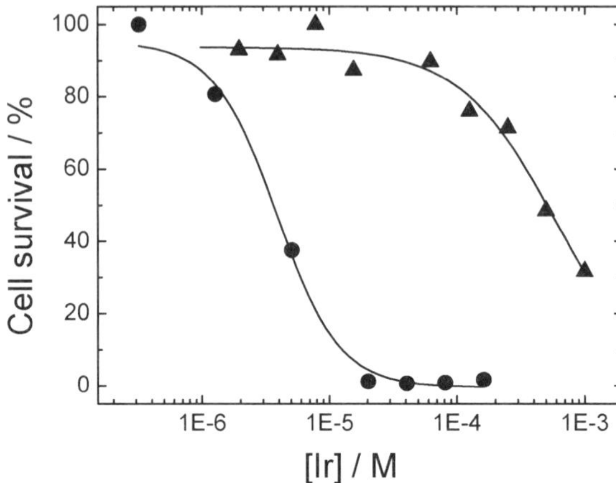

Figure 5.6   Dose dependence of surviving HeLa cells after exposure to [Ir(pq)₂(bpy-PEG1)]⁺ (solid circles) and its PEG-free counterpart [Ir(pq)₂(bpy-Et)]⁺ (solid triangles) for 48 h.

90

91

92

93

towards the thiol group of biomolecules including the cysteine of glutathione.[107] The complex emits at 551 nm with a lifetime of 131 ns. On the basis of its cationic, lipophilic and thiol-reactive nature, electrophoretic uptake of the complex should lead to localisation in the mitochondria, and the reactivity towards thiol should result in immobilisation of the complex at the organelle. Indeed, the staining pattern shows the 'grainy' nature of the organelles and is essentially identical to the mitotracker TMRE, rendering the complex a new mitochondrial dye.

Glucose is an essential substrate that is oxidised in mitochondria for the production of energy for the growth of cells. The luminescent rhenium(I) tricarbonyl glucose complexes $[Re(N^\wedge N)(CO)_3(py\text{-}3\text{-glucose})]^+$ (**74**) can act as glucose-uptake indicators for mammalian cells and new imaging reagents targeting mitochondria.[87] Due to the overexpression of glucose transporters (GLUTs) in transformed cell lines (HeLa and MCF-7), these complexes are

94

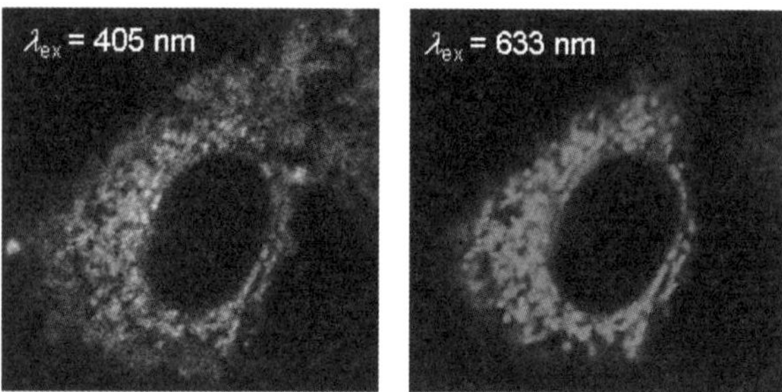

Figure 5.7   Laser-scanning confocal microscopy images of a HeLa cell treated successively with MitoTracker® Deep Red FM (100 nM, 20 min, λ_{ex} = 633 nm) and [Re(Ph$_2$-phen)(CO)$_3$(py-3-glucose)]$^+$ (100 μM, 5 min, λ_{ex} = 405 nm) in glucose-free medium at 37 °C. The excitation wavelengths are 405 nm (left) and 633 nm (right). Reproduced from ref. 87 with permission of Wiley-VCH.

more effectively internalised than in nontransformed cell lines (human embryonic kidney cells, HEK293T, and mouse embryonic fibroblast, NIH/3T3). Unlike their glucose-free counterparts, these glucose complexes show granular appearance inside the cells and colocalise with MitoTracker Deep Red FM (Figure 5.7). Thus, the glucose pendant of the complexes plays a role in the accumulation of the complexes in mitochondria.

5.4.2.3   Golgi Apparatus and Endoplasmic Reticulum

The Golgi apparatus and endoplasmic reticulum (ER) are a stack of membrane-bound vesicles and an interconnecting network of tubules, respectively. Both of them are responsible for processing and packaging macromolecules, such as proteins and lipids. Recently, Lo and coworkers have reported two luminescent iridium(III) complexes [{Ir(N$^\wedge$C)$_2$}$_n$(bpy-n)]$^{n+}$ (n = 8, 4) with the dendritic ligands (**95**) that can bind to the Golgi apparatus.[108] These complexes exhibit a lower cellular uptake efficiency compared to their monomeric counterparts, probably because of their higher cationic charge and much larger molecular size. The highly branched structure and slow uptake rate render them a phosphorescent marker for Golgi apparatus, which is confirmed by costaining experiments involving anti-golgin-97 mouse IgG$_1$ and Alexa 635 goat anti-mouse IgG(H+L).

The zinc(II) salen complex [Zn(Salen)-R^1,R^2] (**96**) with a cyclic or linear alkyl conjugates at salicyladehyde have been reported by Zhang and coworkers.[109] These two complexes emit at *ca.* 618 – 633 nm with a quantum yield of 0.53 – 0.60 in DMSO. Cellular uptake experiments reveal that these complexes are specifically colocalised with an ER marker in HeLa cells. As the complexes show two-photon absorption properties, they are promising optical probes for ER in living cell imaging.

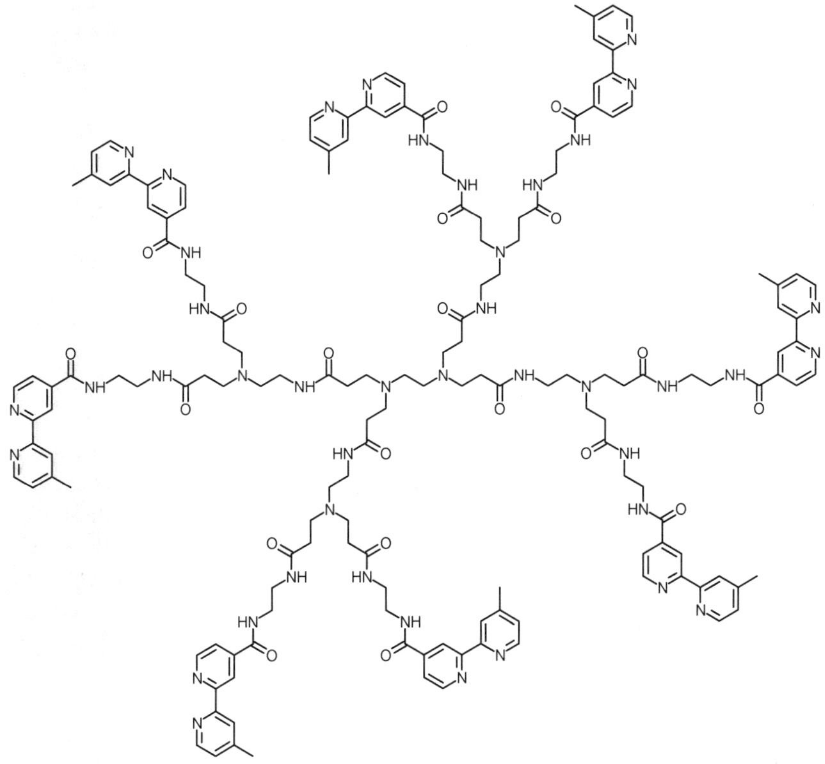

bpy-8

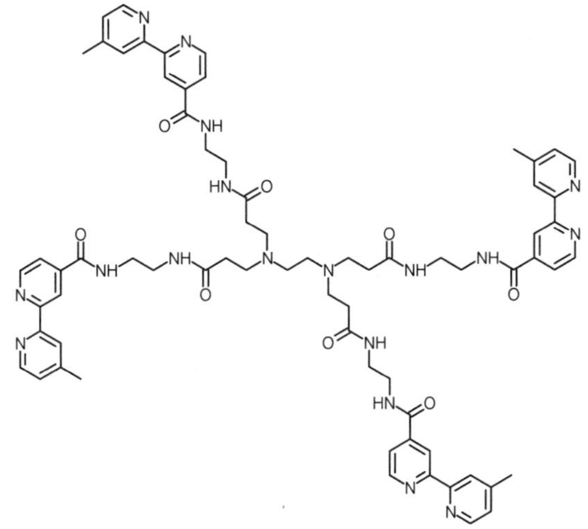

bpy-4

95

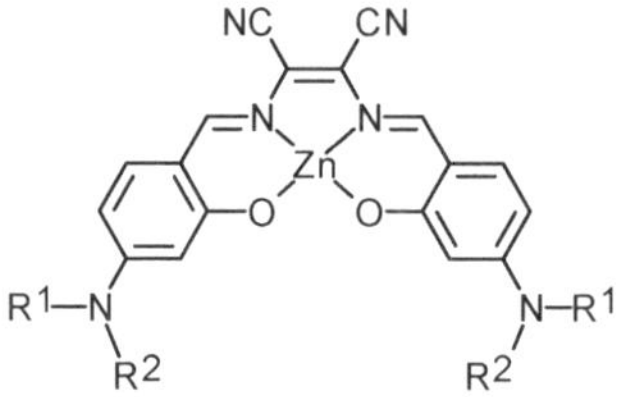

96

R^1-(CH$_2$)$_4$-R^2
R^1 = CH$_3$, R^2 = CH$_2$CH$_2$CH$_2$Cl

5.4.2.4 *Nucleus and Nucleolus*

Luminescent metal complexes that probe DNA *in vivo* are potential candidates for targeting nuclei in imaging studies; for example, Önfelt and coworkers have reported that the incubation of V79 Chinese hamster cells with the dinuclear ruthenium(II) dppz complex [Ru$_2$(phen)$_4$(μ-C4(cpdppz)$_2$)]$^{4+}$ (**97**) does not show emission inside the nucleus, indicating low nuclear membrane penetration.[110] However, when the cells are electroporated in the presence of the complex, emission is detected from the nucleus several hours later, suggesting that the complex is bound to DNA inside the nucleus.

Thomas and coworkers have reported a dinuclear ruthenium(II) complex [Ru$_2$(phen)$_4$(μ-tpphz)]$^{4+}$ (**98**) as a molecular "light switch" for DNA molecules and novel live-cell imaging tool.[111] High-resolution imaging and costaining with the membrane-permeable DAPI reveal that the complex is clearly targeting nuclear DNA but not the nucleolar regions as the staining pattern is in contrast to an organic dye SYTO 9 that preferentially binds to RNA over

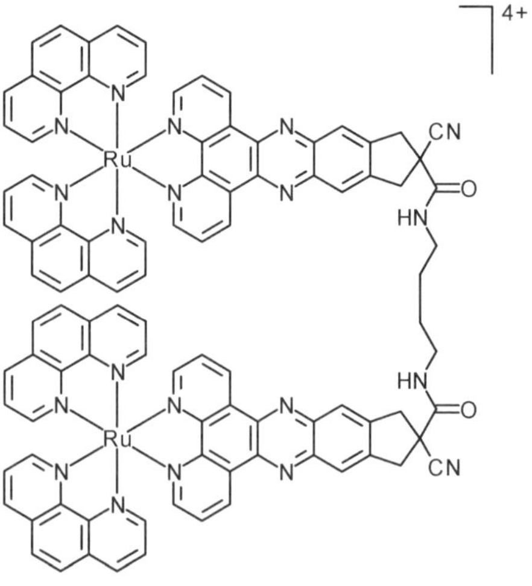

97

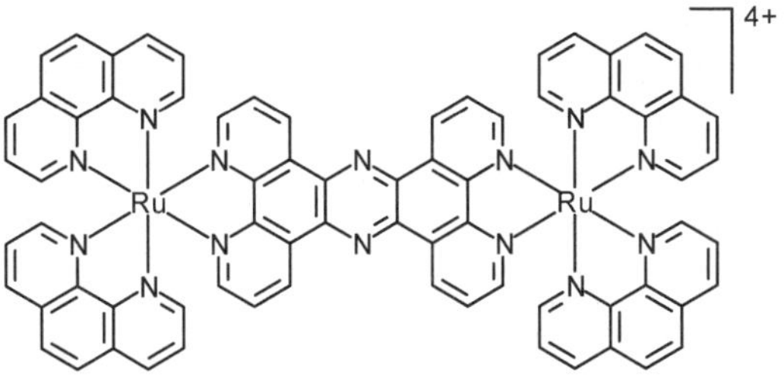

98

DNA. The complex is demonstrated to specifically target mitochondria in live cells by employing a biocompatible and pH-sensitive polymersome vector.

The luminescent dinuclear rhenium(I) complex–peptide nucleic acid (PNA) conjugate [Re$_2$(CO)$_6$(μ-pyridazine-PNA)(μ-Cl)$_2$] (**99**) has been synthesised by D'Alfonso and coworkers.[112] Although the PNA-conjugate stains both the cytoplasm and the nucleus, the emission of the conjugate inside the nucleus occurs at higher energy compared to that in the cytoplasm, resulting from the reduced mobility and the more hydrophobic character of the nuclear environment with respect to the cytoplasm.

Gold(I) phosphine complexes are known to enter the cell nuclei. Ott and coworkers have investigated the bio-distribution profile of gold(I) naphthalimide phosphine complexes [Au(Naphth)(PR$_3$)] (**100**) and found that the uptake of these complexes is significantly higher than that of their chloro counterpart;[113] for example, the nuclear uptake of the triphenylphosphine gold(I) naphthalimide complex by MCF-7 cells is about 25-fold higher than that of [Et$_3$PAuCl]. As a result of the enhanced nuclear uptake and environmental rigidity in the nuclei, nuclei of MCF-7 cells are effectively stained. However, further exposure to a concentration of this complex above its IC$_{50}$ value (10 μM) leads to a breakdown of the cell membrane integrity, rendering it a nucleus stain with a narrow concentration range for applications.

A nucleolus is a nonmembrane bound structure composed of proteins and nucleic acids (especially RNA) that are essential for transcription and

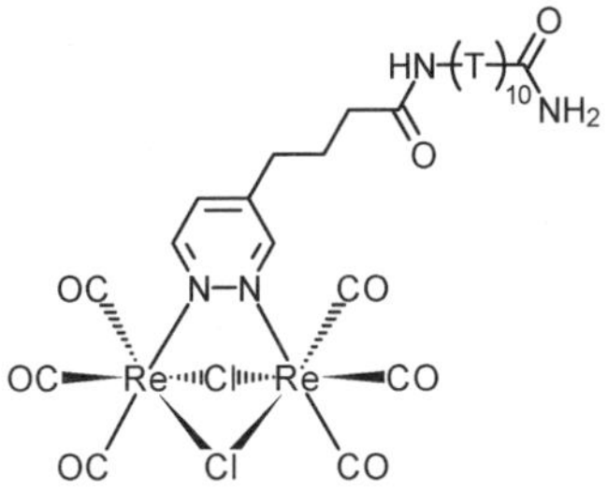

99

T = thymine PNA monomer

translation processes during the gene expression process. Luminescent metal complexes that bind to these proteins and RNA molecules are potential imaging reagents for nucleoli; for example, the iridium(III) complexes [Ir(N^C)$_2$(dpq)]$^+$ (**101**) are excellent reporters for the protein BSA and ds DNA.[114] Even though they do not appear to interact with RNA molecules, confocal microscopy images show clear nucleolar localisation in addition to some perinuclear staining (Figure 5.8). Since these complexes are nonemissive in aqueous medium, the emission observed in the stained cells is due to binding of these complexes to hydrophobic pockets of proteins and ds-DNA inside the nucleoli.

Cyclometallated platinum(II) complexes that stain cell nucleolus have been reported; for example, the characteristic emission from the triphenylphosphonium-functionalised cyclometallated platinum(II) complex [Pt(C^N^N-PPh$_3$)Cl]$^+$ (**102**) is detected in both nuclei and cytoplasm with some nuclear domains strongly enriched in both fixed and live cells.[115] Colocalisation experiments with fibrillarin confirm the nucleoli staining. This organelle specificity is probably related to its special affinity for proteins with cell nucleoli.

Williams and coworkers have designed the cyclometallated platinum(II) complex [Pt(N^C^N-COOMe)Cl] (**103**) with a labile chloro ligand.[116] The choro ligand can be replaced by the *N*-heterocyclic donors of nucleobases and thiol-related molecules. Binding of this complex towards DNA, RNA and

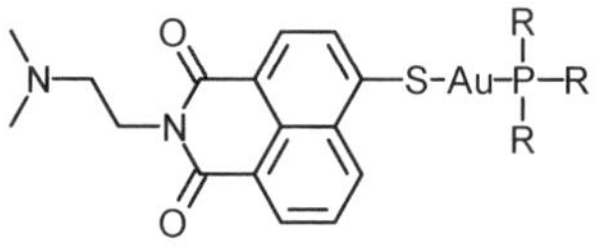

100

R = CH$_3$, CH$_2$CH$_3$, C(CH$_3$)$_3$, C$_6$H$_5$

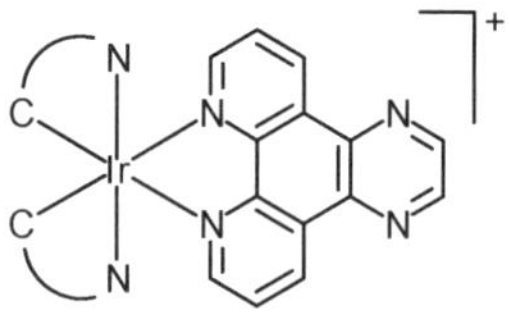

101

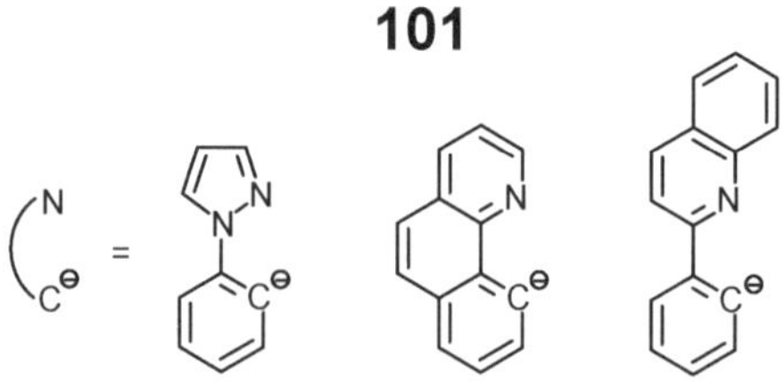

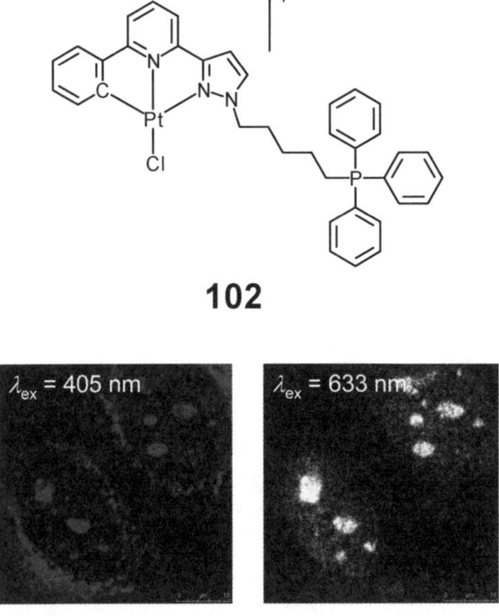

102

Figure 5.8　Laser-scanning confocal microscopy images of fixed MDCK cells treated successively with fibrillarin antibody (20 μL mL^{-1}, 1 h), Alexa 633 anti-rabbit IgG antibody (20 mg μL^{-1}, 30 min, λ_{ex} = 633 nm) and [Ir(ppz)$_2$(dpq)]$^+$ (5 μM, 30 min, λ_{ex} = 405 nm). The excitation wavelengths are 405 nm (left) and 633 nm (right). Reproduced from ref. 114 with permission of the American Chemical Society.

glutathione is a possible reason for the preferential accumulation of the complex in the nuclei and in particular, in the nucleoli. Another important feature for these platinum(II) complexes is the application of tissue-friendly NIR two-photon excitation.

Turro and coworkers have reported a ruthenium(II) complex [Ru(bpy)$_2$(phen-Eth)]$^{3+}$ (**104**) appended with a phenanthridine moiety.[117] In the absence of RNA molecules, the phenanthridine acts as an emission quencher for the ruthenium(II) complex and is nonemissive itself since the excited state of phenanthridine is efficiently quenched in an aqueous environment. However, upon intercalation of the phenanthridine moiety to the RNA molecules, the quantum yield of the complex **104** increases at least 9

103

104

fold as the quenching becomes inefficient. Confocal microscopy images of cells incubated with the complex show that the intensity of the complex is at a maximum in the nucleolus, rendering it an *in vivo* probe for RNA.

5.5 Conclusion

In this chapter, we review the exploitation of luminescent transition-metal complexes as biomolecular labels and probes. The cellular uptake properties, cytotoxicity and intracellular localisation of selected metal complexes have also been described. These examples illustrated that complexes of rhenium(I), ruthenium(II), osmium(II), iridium(III) and platinum(II) display interesting photophysical properties such as intense and long-lived emission in a wide range of emission wavelengths, as a result of a variety of emissive states. This behaviour is very useful in the design of luminescent labels and probes. Also, the rich photochemical properties of some of these complexes are anticipated to be applied in photoinduced ET and photodynamic therapeutic studies. Recently, there has been an emerging interest in the cellular uptake properties and intracellular localisation of luminescent transition-metal complexes. Different studies reveal that the cellular uptake is dependent on the formal charges and lipophilicity of the complexes. Thus, the possible use of a wide range of ligands can not only control the structural and photophysical properties of the complexes, but also their cellular uptake efficiency and specificity. Furthermore, the incorporation of various recognition groups and biologically relevant substrates into these complexes is also anticipated to result in the development of new cellular probes and imaging reagents for small molecules, organelles and tissues. With the use of various transition-metal centres, ligands, reactive functional groups and specific molecular substrates, luminescent transition-metal complexes will continue to contribute to the design of new biomolecular and cellular labels and probes.

Acknowledgements

We thank the Hong Kong Research Grants Councils (CityU 102410) and City University of Hong Kong (Project No. 7002679) for financial support. K.K.-

W.L. gratefully thanks members of his research group and the collaborators; all of whose names appear in the reference list. S.P.-Y.L. acknowledges the receipt of a Research Tuition Scholarship and an Outstanding Academic Performance Award administered by City University of Hong Kong.

References

1. (a) M. S. Wrighton and D. L. Morse, *J. Am. Chem. Soc.*, 1974, **96**, 998; (b) A. J. Lees, *Chem. Rev.*, 1987, **87**, 711; (c) V. Balzani and F. Scandola, *Supramolecular Photochemistry*, Ellis Horwood, New York, 1990; (d) K. Kalyanasundaram, *Photochemistry of Polypyridine and Porphyrin Complexes*, Academic Press, San Diego, 1992; (e) D. M. Roundhill, *Photochemistry and Photophysics of Metal Complexes*, Plenum Press, New York, 1994; (f) D. J. Stufkens and A. Vlček Jr., *Coord. Chem. Rev.*, 1998, **177**, 127.

2. I. A. Hemmila, *Applications of Fluorescence in Immunoassays*, John Wiley & Sons, New York, 1991, p 133.

3. (a) E. Terpetschnig, H. Szmacinski and J. R. Lakowicz, *Methods Enzymol.*, 1997, **278**, 295; (b) R. Blasius, C. Moucheron and A. Kirsch-De Mesmaeker, *Eur. J. Inorg. Chem.*, 2004, 3971; (c) K.K.-W. Lo, K. H.-K. Tsang, K.-S. Sze, C.-K. Chung, T. K.-M. Lee, K. Y. Zhan, W.-K. Hui, C.-K. Li, J. S.-Y. Lau, D. C.-M. Ng and N. Zhu, *Coord. Chem. Rev.*, 2007, **251**, 2292.

4. (a) S. Rickling, L. Ghisdavu, F. Pierard, P. Gerbaux, M. Surin, P. Murat, E. Defrancq, C. Moucheron and A. Kirsch-De Mesmaeker, *Chem. Eur. J.*, 2010, **16**, 3951; (b) S. Le Gac, S. Rickling, P. Gerbaux, E. Defrancq, C. Moucheron and A. Kirsch-De Mesmaeker, *Angew. Chem. Int. Ed.*, 2009, **48**, 1122; (c) B. Elias and A. Kirsch-De Mesmaeker, *Coord. Chem. Rev.*, 2006, **250**, 1627.

5. (a) H. B. Gray and J. R. Winkler, *Annu. Rev. Biochem.*, 1996, **65**, 537; (b) H. B. Gray and J. R. Winkler, *Biochim. Biophys. Acta*, 2010, **1797**, 1563; (c) J. L. Dempsey, J. R. Winkler and H. B. Gray, *Chem. Rev.*, 2010, **110**, 7024.

6. L. Chen, D. W. McBranch, H. L. Wang, R. Helgeson, F. Wudl and D. G. Whitten, *Proc. Natl. Acad. Sci. USA*, 1999, **96**, 12287.

7. H. Yang, K. L. Metera and H. F. Sleiman, *Coord. Chem. Rev.*, 2010, **254**, 2403.

8. F. D. Lewis, S. A. Helvoigt and R. L. Letsinger, *Chem. Commun.*, 1999, 327.

9. L. Wei, J. Babich, W. C. Eckelman and J. Zubieta, *Inorg. Chem.*, 2005, **44**, 2198.

10. S. I. Khan, A. E. Beilstein, M. T. Tierney, M. Sykora and M. W. Grinstaff, *Inorg. Chem.*, 1999, **38**, 5999.

11. D. J. Hurley and Y. Tor, *J. Am. Chem. Soc.*, 2002, **124**, 3749.

12. I. Ortmans, C. Moucheron and A. Kirsch-De Mesmaeker, *Coord. Chem. Rev.*, 1998, **168**, 233.

13. T. A. Oriskovich, P. S. White and H. H. Thorp, *Inorg. Chem.*, 1995, **34**, 1629.

14. (a) J.-P. Lecomte, M. M. Feeney, J. M. Kelly and A. Kirsch-De Mesmaeker, *Inorg. Chem.*, 1995, **34**, 6481; (b) L. Jacquet, R. J. H. Davies, A. Kirsch-De Mesmaeker and J. M. Kelly, *J. Am. Chem. Soc.*, 1997, **119**, 11763.

15. K. E. Erkkila, D. T. Odom and J. K. Barton, *Chem. Rev.*, 1999, **99**, 2777.

16. D. B. Hall, R. E. Holmlin and J. K. Barton, *Nature*, 1996, **382**, 731.

17. J. K. Barton, E. D. Olmon and P. A. Sontz, *Coord. Chem. Rev.*, 2011, **255**, 619.

18. C. J. Murphy, M. R. Arkin, Y. Jenkins, N. D. Ghatlia, S. H. Bossmann, N. J. Turro and J. K. Barton, *Science*, 1993, **262**, 1025.

19. K. K.-W. Lo, D. C.-M. Ng, W.-K. Hui and K.-K. Cheung, *J. Chem. Soc., Dalton Trans.*, 2001, 2634.

20. K. K.-W. Lo, D. C.-M. Ng and C.-K. Chung, *Organometallics*, 2001, **20**, 4999.

21. K. K.-W. Lo, W.-K. Hui, D. C.-M. Ng and K.-K. Cheung, *Inorg. Chem.*, 2002, **41**, 40.

22. K. Wiederholt and L. W. McLaughlin, *Nucl. Acids Res.*, 1999, **27**, 2487.

23. F. Shao and J. K. Barton, *J. Am. Chem. Soc.*, 2007, **129**, 14733.

24. K. W. Jennette, S. J. Lippard, G. A. Vassiliades and W. R. Bauer, *Proc. Natl. Acad. Sci. USA*, 1974, **71**, 3839.

25. J. K. Barton, A. Danishefsky and J. Goldberg, *J. Am. Chem. Soc.*, 1984, **106**, 2172.

26. H.-Q. Liu, T.-C. Cheung and C.-M. Che, *Chem. Commun.*, 1996, 1039.

27. C.-M. Che, M. Yang, K.-H. Wong, H.-L. Chan and W. Lam, *Chem. Eur. J.*, 1999, **5**, 3350.

28. F. Puntoriero, S. Campagna, M. L. Di Pietro, A. Giannetto and M. Cusumano, *Photochem. Photobiol. Sci.*, 2007, **6**, 357.

29. (a) B. M. Zeglis, V. C. Pierre and J. K. Barton, *Chem. Commun.*, 2007, 4565; (b) H.-K. Liu and P. J. Sadler, *Acc. Chem. Res.*, 2011, **44**, 349.

30. (a) A. E. Friedman, J.-C. Chambron, J.-P. Sauvage, N. J. Turro and J. K. Barton, *J. Am. Chem. Soc.*, 1990, **112**, 4960; (b) R. M. Hartshorn and J. K. Barton, *J. Am. Chem. Soc.*, 1992, **114**, 5919.

31. H. D. Stoeffler, N. B. Thornton, S. L. Temkin and K. S. Schanze, *J. Am. Chem. Soc.*, 1995, **117**, 7119.

32. (a) V. W.-W. Yam, K. K.-W. Lo, K.-K. Cheung and R. Y.-C. Kong, *J. Chem. Soc., Chem. Commun.*, **1995**, 1191; (b) V. W.-W. Yam, K. K.-W. Lo, K.-K. Cheung and R. Y.-C. Kong, *J. Chem. Soc., Dalton Trans.*, **1997**, 2067.

33. R. E. Holmlin, J. A. Yao and J. K. Barton, *Inorg. Chem.*, 1999, **38**, 174.

34. K. K.-W. Lo and K. H.-K. Tsang, *Organometallics*, 2004, **23**, 3062.

35. K. K.-W. Lo, C.-K. Chung and N. Zhu, *Chem. Eur. J.*, 2006, **12**, 1500.

36. K. K.-W. Lo, K. H.-K. Tsang and K.-S. Sze, *Inorg. Chem.*, 2006 **45**, 1714.
37. K. A. O'Donoghue, J. M. Kelly and P. E. Kruger, *Dalton Trans.*, 2004, 13.
38. C. Yu, K. H.-Y. Chan, K. M.-C. Wong and V. W.-W. Yam, *Proc. Natl. Acad. Sci. USA*, 2006, **103**, 19652.
39. F. M. O'Reilly and J. M. Kelly, *New J. Chem.*, 1998, 215.
40. L. M. Wilhelmsson, F. Westerlund, P. Lincoln and B. Nordén, *J. Am. Chem. Soc.*, 2002, **124**, 12092.
41. S. P. Foxon, T. Phillips, M. R. Gill, M. Towrie, A. W. Parker, M. Webb and J. A. Thomas, *Angew. Chem. Int. Ed.*, 2007, **46**, 3686.
42. B. E. Bowler, K. J. Ahmed, W. I. Sundquist, L. S. Hollis, E. E. Whang and S. J. Lippard, *J. Am. Chem. Soc.*, 1989, **111**, 1299.
43. T. Esteves, C. Xavier, S. Gama, F. Mendes, P. D. Raposinho, F. Marques, A. Paulo, J. C. Pessoa, J. Rino, G. Viola and I. Santos, *Org. Biomol. Chem.*, 2010, **8**, 4104.
44. L. A. Mullice, R. H. Laye, L. P. Harding, N. J. Buurma and S. J. A. Pope, *New J. Chem.*, 2008, **32**, 2140.
45. N. Agorastos, L. Borsig, A. Renard, P. Antoni, G. Viola, B. Spingler, P. Kurz and R. Alberto, *Chem. Eur. J.*, 2007, **13**, 3842.
46. N. B. Thornton and K. S. Schanze, *Inorg. Chem.*, 1993, **32**, 4994.
47. A. Ghosh, P. Das, M. R. Gill, P. Kar, M. G. Walker, J. A. Thomas and A. Das, *Chem. Eur. J.*, 2011, **17**, 2089.
48. G. I. Pascu, A. C. G. Hotze, C. Sanchez-Cano, B. M. Kariuki and M. J. Hannon, *Angew. Chem. Intl. Ed.*, 2007, **46**, 4374.
49. J. M. C. A. Kerckhoffs, J. C. Peberdy, I. Meistermann, L. J. Childs, C. J. Isaac, C. R. Pearmund, V. Reudegger, S. Khalid, N. W. Alcock, M. J. Hannon and A. Rodger, *Dalton Trans.*, 2007, 734.
50. S. Chowdhury, P. Mal, C. Basu, H. Stoeckli-Evans and S. Mukherjee, *Polyhedron*, 2009, **28**, 3863.
51. V. Gonzalez, T. Wilson, I. Kurihara, A. Imai, J. A. Thomas and J. Otsuki, *Chem. Commun.*, 2008, 1868.
52. G. T. Hermanson, *Bioconjugate Techniques*, Academic Press, San Diego, CA, 1996.
53. E. M. Ryan, R. O'Kennedy, M. M. Feeney, J. M. Kelly and J. G. Vos, *Bioconjugate Chem.*, 1992, **3**, 285.
54. K. K.-W. Lo, C.-K. Chung, T. K.-M. Lee, L.-H. Lui, K. H.-K Tsang and N. Zhu, *Inorg. Chem.*, 2003, **42**, 6886.
55. K. M.-C. Wong, W.-S. Tang, B. W.-K. Chu, N. Zhu and V. W.-W. Yam, *Organometallics*, 2004, **23**, 3459.
56. K. K.-W. Lo, C.-K. Chung and N. Zhu, *Chem. Eur. J.*, 2003, **9**, 475.
57. K. K.-W. Lo, C.-K. Li, K.-W. Lau and N. Zhu, *Dalton Trans.*, 2003, 4682.
58. P.-K. Lee, H.-W. Liu, S.-M. Yiu, M.-W. Louie and K. K.-W. Lo, *Dalton Trans.*, 2011, **40**, 2180.
59. S.-K. Leung, K. Y. Kwok, K. Y. Zhang and K. K.-W. Lo, *Inorg. Chem.*, 2010, **49**, 4984.

60. E. Terpetschnig, H. Szmacinski, H. Malak and J. R. Lakowicz, *Biophys. J.*, 1995, **68**, 342.

61. X.-Q. Guo, F. N. Castellano, L. Li, H. Szmacinski, J. R. Lakowicz and J. Sipior, *Anal. Biochem.*, 1997, **254**, 179.

62. J. D. Dattelbaum, O. O. Abugo and J. R. Lakowicz, *Bioconjugate Chem.*, 2000, **11**, 533.

63. S. R. Banerjee, P. Schaffer, J. W. Babich, J. F. Valliant and J. Zubieta, *Dalton Trans.*, 2005, 3886.

64. D.-L. Ma, W.-L. Wong, W.-H. Chung, F.-Y. Chan, P.-K. So, T.-S. Lai, Z.-Y. Zhou, Y.-C. Leung and K.-Y. Wong, *Angew. Chem. Int. Ed.*, 2008, **47**, 3735.

65. (a) W. B. Connick, A. J. Di Bilio, M. G. Hill, J. R. Winkler and H. B. Gray, *Inorg. Chim. Acta*, 1995, **240**, 169; (b) A. J. Di Bilio, B. R. Crane, W. A. Wehbi, C. N. Kiser, M. M. Abu-Omar, R. M. Carlos, J. H. Richards, J. R. Winkler and H. B. Gray, *J. Am. Chem. Soc.*, 2001, **123**, 3181.

66. A. R. Dunn, W. Belliston-Bittner, J. R. Winkler, E. D. Getzoff, D. J. Stuehr and H. B. Gray, *J. Am. Chem. Soc.*, 2005, **127**, 5169.

67. V. G. Vaidyanathan and B. U. Nair, *Eur. J. Inorg. Chem.*, 2005, 3756.

68. P. Wu, E. L.-M. Wong, D.-L. Ma, G. S.-M. Tong, K.-M. Ng and C.-M. Che, *Chem. Eur. J.*, 2009, **15**, 3652.

69. M. P. Coogan, V. Fernández-Moreira, J. B. Hess, S. J. A. Pope and C. Williams, *New J. Chem.*, 2009, **33**, 1094.

70. K. K.-W. Lo, K. H.-K. Tsang and N. Zhu, *Organometallics*, 2006, **25**, 3220.

71. K. K.-W. Lo, T. K.-M. Lee, J. S.-Y. Lau, W.-L. Poon and S.-H. Cheng, *Inorg. Chem.*, 2008, **47**, 200.

72. K. K.-W Lo, K. Y. Zhang, C.-K. Chung and K. Y. Kwok, *Chem. Eur. J.*, 2007, **13**, 7110.

73. K. K.-W. Lo, K. H.-K. Tsang, W.-K. Hui and N. Zhu, *Inorg. Chem.*, 2005, **44**, 6100.

74. K. K.-W. Lo, T. K.-M. Lee and K. Y. Zhang, *Inorg. Chim. Acta*, 2006, **359**, 1845.

75. J. S.-Y. Lau, P.-K. Lee, K. H.-K. Tsang, C. H.-C. Ng, Y.-W. Lam, S.-H. Cheng and K. K.-W. Lo, *Inorg. Chem.*, 2009, **48**, 708.

76. (a) M. Wilchek and E. A. Bayer, *Anal. Biochem.*, 1988, **171**, 1; (b) M. Wilchek and E. A. Bayer, *Methods Enzymol.*, 1990, **184**, 5.

77. (a) K. K.-W. Lo, W.-K. Hui and D. C.-M. Ng, *J. Am. Chem. Soc.*, 2002, **124**, 9344; (b) K. K.-W. Lo and W.-K. Hui, *Inorg. Chem.*, 2005, **44**, 1992.

78. K. K.-W. Lo and T. K.-M. Lee, *Inorg. Chem.*, 2004, **43**, 5275.

79. (a) K. K.-W. Lo, J. S.-W. Chan, L.-H. Lui and C.-K. Chung, *Organometallics*, 2004, **23**, 3108; (b) K. K.-W. Lo, C.-K. Li and J. S.-Y. Lau, *Organometallics*, 2005, **24**, 4594.

80. T.-H. Kwon, J. Kwon and J.-I. Hong, *J. Am. Chem. Soc.*, 2008, **130**, 3726.

81. M.-W. Louie, M. H.-C. Lam and K. K.-W. Lo, *Eur. J. Inorg. Chem.*, 2009, 4265.

82. K. K.-W. Lo and J. S.-Y. Lau, *Inorg. Chem.*, 2007, **46**, 700.

83. K. K.-W. Lo, K. Y. Zhang, S.-K. Leung and M.-C. Tang, *Angew. Chem., Int. Ed.*, 2008, **47**, 2213.

84. S. M. Contakes, G. A. Juda, D. B. Langley, N. W. Halpern-Manners, A. P. Duff, A. R. Dunn, H. B. Gray, D. M. Dooley, J. M. Guss and H. C. Freeman, *Proc. Natl. Acad. Sci. USA*, 2005, **102**, 13451.

85. P. K.-M. Siu, D.-L. Ma and C.-M. Che, *Chem. Commun.*, 2005, 1025.

86. T. Hasegawa, T. Yonemura, K. Matsuura and K. Kobayashi, *Bioconjugate Chem.*, 2003, **14**, 737.

87. M.-W. Louie, H.-W. Liu, M. H.-C. Lam, Y.-W. Lam and K. K.-W. Lo, *Chem. Eur. J.*, 2011, **17**, 8304.

88. (a) Q. Zhao, F. Li and C. Huang, *Chem. Soc. Rev.*, 2010, **39**, 3007; (b) Zhao, C. Huang and F. Li, *Chem. Soc. Rev.*, 2011, **40**, 2508; (c) K. K.-W. Lo, S. P.-Y. Li and K. Y. Zhang, *New J. Chem.*, 2011, **35**, 265; (d) K. Y. Zhang and S. P.-Y. Li, *Eur. J. Inorg. Chem.*, 2011, 3551.

89. Q. Zhao, M. Yu, L. Shi, S. Liu, C. Li, M. Shi, Z. Zhou, C. Huang and F. Li, *Organometallics*, 2010, **29**, 1085.

90. F. L. Thorp-Greenwood, V. Fernández-Moreira, C. O. Millet, C. F. Williams, J. Cable, J. B. Court, A. J. Hayes, D. Lloyd and M. P. Coogan, *Chem. Commun.*, 2011, **47**, 3096.

91. (a) S. I. Pascu, P. A. Waghorn, T. D. Conry, B. Lin, H. M. Betts, J. R. Dilworth, R. B. Sim, G. C. Churchill, F. I. Aigbirhio and J. E. Warren, *Dalton Trans.*, 2008, 2107; (b) S. I. Pascu, P. A. Waghorn, B. W. C. Kennedy, R. L. Arrowsmith, S. R. Bayly, J. R. Dilworth, M. Christlieb, R. M. Tyrrell, J. Zhong, R. M. Kowalczyk, D. Collison, P. K. Aley, G. C. Churchill and F. I. Aigbirhio, *Chem. Asian. J.*, 2010, **5**, 506.

92. A. J. Amoroso, M. P. Coogan, J. E. Dunne, V. Fernández-Moreira, J. B. Hess, A. J. Hayes, D. Lloyd, C. Millet, S. J. A. Pope and C. Williams, *Chem. Commun.*, 2007, 3066.

93. W. Jiang, Y. Gao, Y. Sun, F. Ding, Y. Xu, Z. Bian, F. Li, J. Bian and C. Huang, *Inorg. Chem.*, 2010, **49**, 3252.

94. K. K.-W. Lo, P.-K. Lee and J. S.-Y. Lau, *Organometallics*, 2008, **27**, 2998.

95. C. A. Puckett and J. K. Barton, *J. Am. Chem. Soc.*, 2009, **131**, 8738.

96. U. Neugebauer, Y. Pellegrin, M. Devocelle, R. J. Forster, W. Signac, N. Moran and T. E. Keyes, *Chem. Commun.*, 2008, 5307.

97. K. A. Stephenson, S. R. Banerjee, T. Besanger, O. O. Sogbein, M. K. Levadala, N. McFarlane, J. A. Lemon, D. R. Boreham, K. P. Maresca, J. D. Brennan, J. W. Babich, J. Zubieta and J. F. Valliant, *J. Am. Chem. Soc.*, 2004, **126**, 8598.

98. N. Viola-Villegas, A. E. Rabideau, J. Cesnavicious, J. Zubieta and R. P. Doyle, *ChemMedChem*, 2008, **3**, 1387.

99. C. A. Puckett and J. K. Barton, *Biochemistry*, 2008, **47**, 11711.

100. K. Y. Zhang and K. K. W. Lo, *Inorg. Chem.*, 2009, **48**, 6011.

101. N. Viola-Villegas, A. E. Rabideau, M. Bartholoma, J. Zubieta and R. P. Doyle, *J. Med. Chem.*, 2009, **52**, 5253.

102. S. P.-Y. Li, H.-W. Liu, K. Y. Zhang and K. K.-W. Lo, *Chem. Eur. J.*, 2010, **16**, 8329.

103. C.-K. Koo, K.-L. Wong, C. W.-Y. Man, H.-L. Tam, S.-W. Tsao, K.-W. Cheah and M. H.-W. Lam, *Inorg. Chem.*, 2009, **48**, 7501.

104. M. Matson, F. R. Svensson, B. Nordén and P. Lincoln, *J. Phys. Chem. B*, 2011, **115**, 1706.

105. V. Fernández-Moreira, F. L. Thorp-Greenwood, A. J. Amoroso, J. Cable, J. B. Court, V. Gray, A. J. Hayes, R. L. Jenkins, B. M. Kariuki, D. Lloyd, C. O. Millet, C. F. Williams and M. P. Coogan, *Org. Biomol. Chem.*, 2010, **8**, 3888.

106. C.-K. Koo, K.-L. Wong, C. W.-Y. Man, Y.-W. Lam, L. K.-Y. So, H.-L. Tam, S.-W. Tsao, K.-W. Cheah, K.-C. Lau, Y.-Y. Yang, J.-C. Chen and M. H.-W. Lam, *Inorg. Chem.*, 2009, **48**, 872.

107. A. J. Amoroso, R. J. Arthur, M. P. Coogan, J. B. Court, V. Fernández-Moreira, A. J. Hayes, D. Lloyd, C. Millet and S. J. A. Pope, *New J. Chem.*, 2008, **32**, 1097.

108. K. Y. Zhang, H.-W. Liu, T. T.-H. Fong, X.-G. Chen and K. K.-W. Lo, *Inorg. Chem.*, 2010, **49**, 5432.

109. Y. Hai, J.-J. Chen, P. Zhao, H. Lv, Y. Yu, P. Xu and J.-L. Zhang, *Chem. Commun.*, 2011, **47**, 2435.

110. B. Önfelt, L. Göstring, P. Lincoln, B. Nordén and A. Önfelt, *Mutagenesis*, 2002, **17**, 317.

111. M. R. Gill, J. Garcia-Lara, S. J. Foster, C. Smythe, G. Battaglia and J. A. Thomas, *Nature Chem.*, 2009, **1**, 662.

112. E. Ferri, D. Donghi, M. Panigati, G. Prencipe, L. D'Alfonso, I. Zanoni, C. Baldoli, S. Maiorana, G. D'Alfonso and E. Licandro, *Chem. Commun.*, 2010, **46**, 6255.

113. C. P. Bagowski, Y. You, H. Scheffler, D. H. Vlecken, D. J. Schmitz and I. Ott, *Dalton Trans.*, 2009, 10799.

114. K. Y. Zhang, S. P.-Y. Li, N. Zhu, I. W.-S. Or, M. S.-H. Cheung, Y.-W. Lam and K. K.-W. Lo, *Inorg. Chem.*, 2010, **49**, 2530.

115. C.-K. Koo, L. K.-Y. So, K.-L. Wong, Y.-M. Ho, Y.-W. Lam, M. H.-W. Lam, K.-W. Cheah, C C.-W. Cheng and W.-M. Kwok, *Chem. Eur. J.*, 2010, **16**, 3942.

116. S. W. Botchway, M. Charnley, J. W. Haycock, A. W. Parker, D. L. Rochester, J. A. Weinstein and J. A. G. Williams, *Proc. Natl. Acad. Sci. USA*, 2008, **105**, 16071.

117. N. A. O'Connor, N. Stevens, D. Samaroo, M. R. Solomon, A. A. Martí, J. Dyer, H. Vishwasrao, D. L. Akins, E. R. Kandel and N. J. Turro, *Chem. Commun.*, 2009, 2640.

Photoactive Multinuclear Metal-Containing Polymeric Systems

R. SAKAMOTO AND H. NISHIHARA*

Department of Chemistry, Graduate School of Science, The University of Tokyo, 7-3-1, Hongo, Bunkyo-ku, Tokyo 113-0033, Japan
*E-mail: nisihara@chem.s.u-tokyo.ac.jp

6.1 Introduction

This chapter discusses photoactive metallopolymers. This term is usually taken to describe coordination polymers and macromolecular metal complexes; however, in a broader sense it also covers metal-organic frameworks (MOFs) and porous coordination polymers (PCPs),[1–18] and Prussian Blue (PB)[19–22]-type inorganic crystalline materials. This chapter considers all of these compounds, as well as organic polymers doped with photoresponsive metal complexes.

Some of the most important manifestations of molecular photoactivity are seen in photoinduced structural changes. Organic photochromic compounds such as diarylethene[23] and azobenzene[24,25] are good examples of substances that show such phenomena. Structural changes in this series of compounds are always accompanied by significant alterations in their electronic structures, and by resultant drastic changes in colour. Recently, various kinds of conjugates between organic photochromics and coordination compounds have been synthesised, allowing control over the photo-, electro-, and magnetic properties of the latter.[26,27] Inversely, regulation of photochromism *via* the coordination compounds has also been achieved in these conjugates.[26,27] Photochromism in polymers can afford more profound benefits than in single

RSC Polymer Chemistry Series No. 2
Molecular Design and Applications of Photofunctional Polymers and Materials
Edited by Wai-Yeung Wong and Alaa S Abd-El-Aziz

molecules; cooperative and synergetic effects can be expected, such as modulation of interchain interactions, macroscopic structural changes, and ordering of the components. In fact, Ikeda and coworkers reported an azobenzene-containing polymer film that could roll up and spread out in response to photoirradiation and subsequent reversible photoisomerisation.[28] A couple of metallopolymers that contain organic photochromics are introduced in Section 6.2.

Significant colour changes can also be induced simply by oxidising or reducing the metal centres of coordination compounds. For instance, the colour of ferrocene is yellow in the Fe^{II} form, but this changes to blue after oxidation into the Fe^{III} form. This change is accounted for by the modulation of d–d (ligand-field, LF) transitions stemming from the alternation of the d-orbital configuration (d^5/d^6).[29] $Ru^{II}(bpy)_3^{2+}$ (bpy = 2,2′-bipyridine) shows an intense absorption in the visible region ($\lambda_{max} = 450$ nm, $\varepsilon_{max} = 14\ 500$ M^{-1} cm^{-1}), which is assignable to a metal-to-ligand charge transfer (MLCT) transition from $Ru(t_{2g})$ to $bpy(\pi^*)$; $Ru^{III}(bpy)_3^{3+}$ exhibits a weaker and redshifted absorption peak ($\lambda_{max} = 675$ nm, $\varepsilon_{max} = 750$ M^{-1} cm^{-1}) ascribable to a ligand-to-metal charge transfer (LMCT) transition from $bpy(\pi)$ to $Ru(t_{2g})$.[30] These molecules experience only slight changes in bond lengths and angles. Because electrochemical polarisation can be applied very conveniently (even to polymer materials), and because the redoxes of transition-metal complexes and organometallics tend to be more robust than those of organic compounds, a wide range of electrochromic materials based on metal-containing polymers have been investigated; these are treated in Section 6.3.

Section 6.4 covers important works on photomagnetisation in the crystalline solid state. PB analogues show reversible photochromism that is accompanied by modulations in the solid-state magnetic properties at low temperatures.[31–34] This series of phenomena was first reported by Hashimoto, Fujishima and coworkers in FeCo PB, the composition of which was $K_{0.2}Co_{1.4}[Fe(CN)_6]\cdot6.9H_2O$.[35] In the photomagnetisation of PB analogues, the key photochemical process consists of metal-to-metal charge-transfer (MMCT) transitions, also called intervalence charge-transfer (IVCT) transitions.[36,37] Another phenomenon capable of reversibly photoregulating magnetic properties is light-induced excited spin state trapping (LIESST).[38–41] This photomagnetic effect was first documented by Decurtins and coworkers in 1984 with $[Fe^{II}(ptz)_6](BF_4)_2$ (ptz = 1-propyltetra-zole).[42] At low temperatures, the ground low-spin $^1A_1$ state and metastable high-spin $^5T_2$ state could be reversibly switched upon excitation with the d–d transitions.

Recently, molecule-based optoelectronics, such as organic photovoltaic cells (OPVCs)[43,44] and organic light-emitting diodes (OLEDs)[45,46], have gained increasing attention from researchers. Light-emitting or light-absorbing semiconducting polymer materials afford various benefits. For example, they allow the use of spin-coating and printing methods suitable for the preparation of large devices at low cost. In combination with the peculiar photoproperties of coordination compounds (such as the generation of long-lived triplet excited states and subsequent phosphorescence),[47,48] the properties of these metal-containing polymer materials have facilitated the development of dozens of optoelectronic devices exhibiting excellent performances (Section 6.5).

Tremendous progress has been made in the MOF and PCP research field in the last decade.[1–18] These materials possess two- or three-dimensional order, resulting in highly crystalline structures. MOFs and PCPs feature highly porous networks. Motivated by this property, their host–guest chemistry has been rigorously investigated for applications such as fine gas-storage materials. Gas separation and molecular catalysis have also been extensively studied in MOFs and PCPs. The greatest advantage of MOFs and PCPs over other porous materials such as zeolites is their "soft" nature; several kinds of MOFs and PCPs transform reversibly with guest uptake and discharge, as well as outer chemical and physical stimuli, leading to more sophisticated, profound, and multiple functionalities. Another virtue of MOFs and PCPs is their ease of fabrication; solvothermal syntheses using metal salts and organic ligands typically produce materials with large crystals. From the viewpoint of photochemistry, MOFs and PCPs have not been sufficiently investigated. However, several impressive works have been published recently, some of which are described in Section 6.6.

6.2 Polymers Containing Organic Photochromics

6.2.1 Photomodulation of Magnetic Properties

Single-molecule magnets (SMMs) feature slow relaxation of the magnetisation, and thereby behave as nanometre-sized magnets below a blocking temperature T_B.[49–56] Their relaxation dynamics are attributed to their intrinsic molecular properties, which include a large easy-axis-type magnetoanisotropy (negative zero-field splitting parameter $D < 0$), and a large spin ground state (S_T). These parameters result in a large energy barrier (Δ) of $|D|S_T^2$ or $|D|(S_T^2 - 1/4)$ between spin-up and spin-down configurations ($m_s = \pm S_T$), for integer and half-integer spins, respectively. Much effort has been devoted to this research field by chemists and physicists, with the aim of realising potential SMM applications such as memory devices that exploit SMMs as bits,[57] and quantum computers based on quantum tunneling magnetisation (QTM).[58–62]

To achieve photoregulation of SMM behaviour, Miyasaka and coworkers designed one-dimensional coordination-polymer-type SMMs based on photochromic diarylethene ligand dae^{2-} (Figure 6.1a).[63,64] An example of such SMMs is depicted in Figure 6.1c. Mixed-valency tetranuclear $[Mn^{II}_2Mn^{III}_2]$ units (Figure 6.1b), which are responsible for the expression of magnetism, are bridged by the dae^{2-} ligands to form 1D-coordination chains. The authors differentially synthesised 1D SMM chains with open- and closed-isomers (**1o** and **1c**); these two complexes had different unit cells, and different ligands (ClO_4^- or H_2O) in the first coordination sphere. In both **1o** and **1c**, the $[Mn^{II}_2Mn^{III}_2]$ unit exhibited SMM behaviour with a ground state of $S_T = 9$,

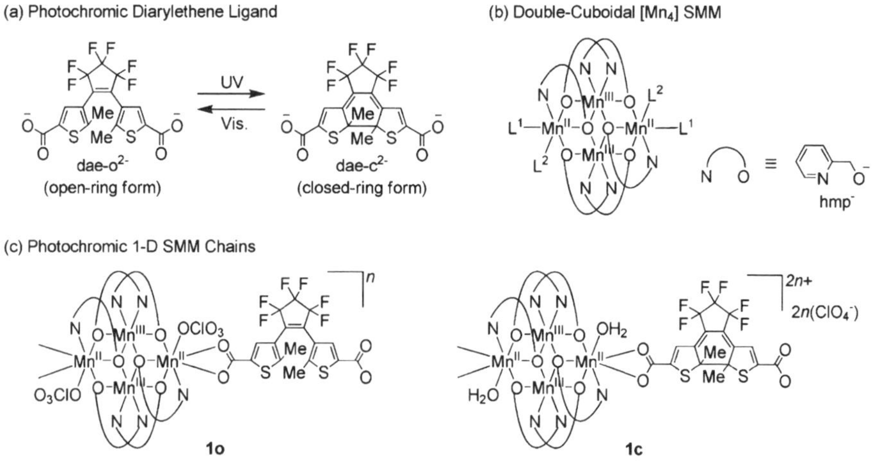

Figure 6.1 (a) Photochromic diarylethene ligand dae^{2-}; (b) Mn tetranuclear SMM unit; (c) Photochromic 1D SMM chain. Reproduced with permission from ref. 63. Copyright 2009, American Chemical Society.

which was ascribed to two kinds of intracluster ferromagnetic exchange interactions. Δ/k_B was estimated to be 25.1 and 26.7 K for **1o** and **1c**, respectively. Both **1o** and **1c** also showed reversible photochromism in the solid state, which was characteristic of the dae^{2-} ligand (Figure 6.2). However, the variation in magnetic properties with photochromism was observed only in **1c**; it showed quite a weak antiferromagnetic interaction of $zJ/k_B \approx -1 \times 10^{-2}$ K between the $[Mn^{II}_2Mn^{III}_2]$ units. In contrast, a remarkable enhancement in antiferromagnetism ($zJ/k_B \approx -0.19$ K) was induced in **1c-vis**, which was created by visible-light irradiation of **1c** to convert the dae^{2-} ligands into the open form. Two plausible explanations are proposed for this switching behaviour: direct switching of through-bond superexchange interactions among the $[Mn^{II}_2Mn^{III}_2]$ clusters in the same chain *via* the dae^{2-} ligand π-system, and indirect switching of through-space interactions among the SMM units in different chains, due to transformation. The former was discounted because **1o** did not show any significant changes in magnetism; the authors therefore concluded that the magnetic switching behaviour in **1c** was derived from intrachain interactions triggered by structural changes upon photo-isomerisation (Figure 6.3). In fact, slight variations in unit cell parameters were also observed between **1c** and **1c-vis**.

6.2.2 Changeover of Photochromic Wavelengths

Nishihara and coworkers reported that 3-ferrocenylazobenzene (3-FcAB) exhibited a redox-conjugated reversible photoisomerisation in solutions and on electrodes (Figure 6.4).[65–67] Excitation of the MLCT band with 546 nm green light caused the *trans*-to-*cis* isomerisation of the azobenzene moiety in the

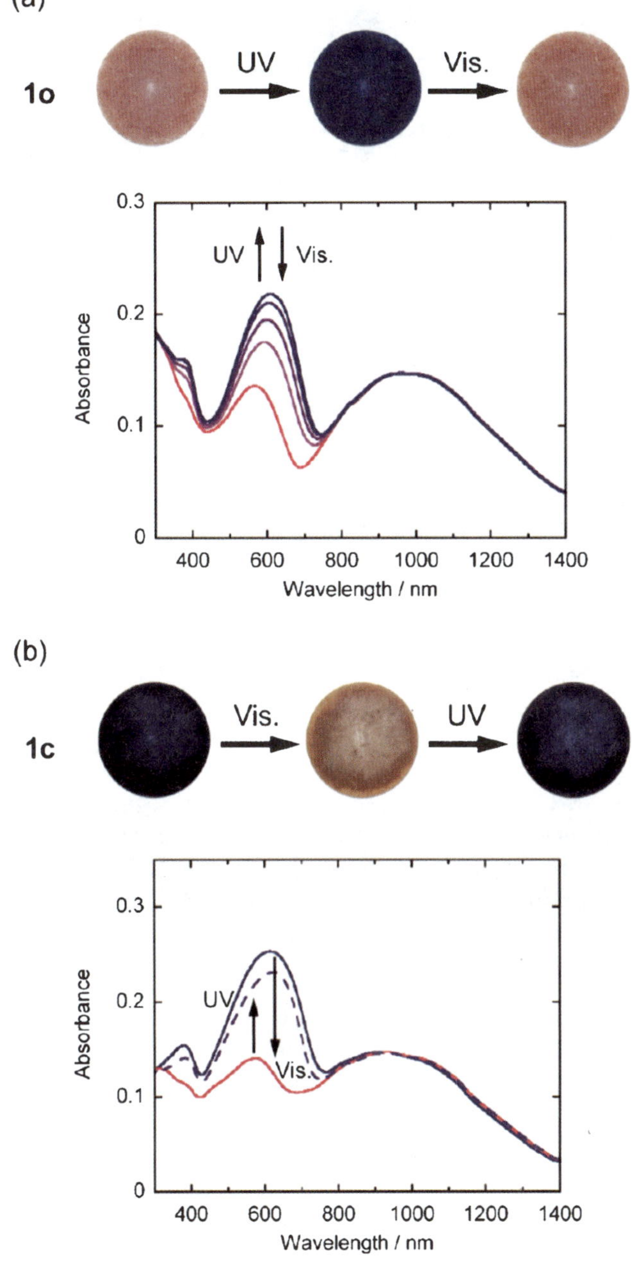

Figure 6.2 Photochromism of 1D SMM chain in KBr pellets: (a) Initially in the open state; (b) Initially in the closed state. Reproduced with permission from ref. 63. Copyright 2009, American Chemical Society.

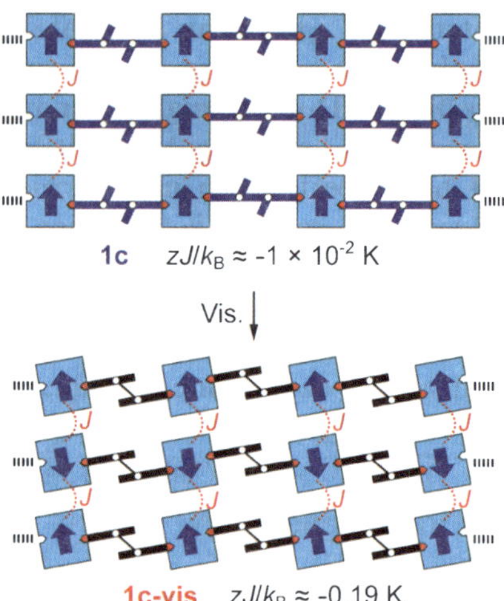

Figure 6.3   Proposed mechanism for the occurrence of the antiferromagnetic ordering in **1c-vis** after irradiation with visible light to **1c**, where blue arrows indicate the magnetisation of SMM aligned on an easy axis. Reproduced with permission from ref. 63. Copyright 2009, American Chemical Society.

reduced Fe^{II} state, in addition to the normal UV and blue-light response (*trans*-to-*cis* isomerisation by excitation of the azo π–π^* band with 365 nm UV light, and the reverse *cis*-to-*trans* isomerisation by excitation of the n–π^* band with 436 nm blue light). Once 3-FcAB was oxidised to 3-FcAB$^+$ with the Fe^{III} nucleus, the MLCT band disappeared, and the *cis*-to-*trans* reverse isomerisation occurred *via* excitation of the n–π^* band with the same 546 nm light. Using this feature, 3-FcAB could be controlled reversibly by a single green-light source coupled with the redox reaction between Fe^{II} and Fe^{III}. Working from these findings, the authors expected that assemblies of this compound had the

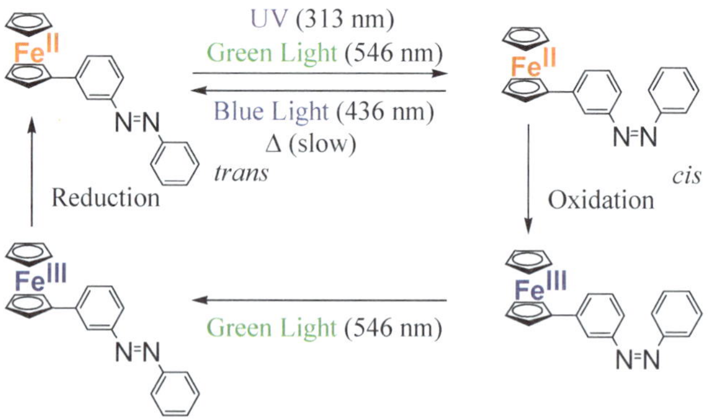

Figure 6.4   Redox-conjugated photoisomerisation of 3-FcAB.[65]

potential to be applied as small-size devices achieving photoswitching without dual light sources; this formed the next step in this study.

Solution polymerisation and emulsion polymerisation were used in this study to create size-controlled discrete polymer particles.[68] Figure 6.5 shows the fabrication procedure for 3-FcAB-containing polymers. The solution polymerisation reaction of 2-bromoethyl acrylate was carried out in both toluene and in cyclohexane. Polymerisation proceeded in both solvents almost quantitatively. In the emulsion polymerisation process, sodium dodecyl sulfate (SDS) was used as an emulsifier; the monomer was well dispersed in water as a stable monomer micelle, so that the mixture of the monomer and SDS in water was transparent. Modification of poly(2-bromoethyl acrylate) formed by solution polymerisation was performed *via* the addition of sodium *tert*-butoxide and 3-ferrocenyl-4′-hydroxymethylazobenzene, giving **Polymer A**. Regarding the modification of poly(2-bromoethylacrylate) formed by emulsion polymerisation with 3-FcAB, the reaction was carried out in the suspension to give **Polymer B**. The modification reactions occurred only at the outer sphere, since the solubility decreased rapidly as the reactions proceeded. The exchange ratio in the 3-FcAB-modification reaction was estimated (from the signal integration of a ^{1}H NMR spectrum) to be *ca.* 20% for **Polymer A**.

According to scanning electron microscope (SEM) analysis, the average sizes of the resultant polymer particles were *ca.* 300 nm and 20 nm for **Polymer A** and **Polymer B**, respectively. The diameter of the **Polymer A** particles was polydisperse; the larger particles may have been generated by the aggregation of polymer particles at the solvent evaporation stage. In contrast, the **Polymer B** particles were monodisperse. These results indicated that polymer particles prepared by emulsion polymerisation were rigid enough that their shapes and sizes were unchanged by the sample preparation process.

Green-light (546 nm) irradiation caused *trans*-to-*cis* isomerisation in **Polymer A** (*cis* ratio: 20%) and in **Polymer B** (11%) in dichloromethane. An addition of 10 eq of iodine oxidised the ferrocene moiety to ferrocenium. In this state, irradiation with green light converted the *cis*-isomer to the *trans*-isomer almost quantitatively. These facts demonstrated that the *cis*-isomers generated by green-light irradiation in their reduced states were reversibly converted to the *trans*-isomers by irradiation with the same green light.

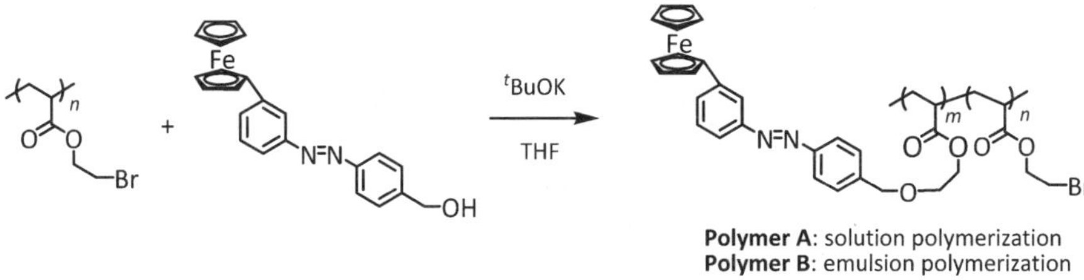

Figure 6.5 Fabrication procedure for 3-FcAB-containing polymers.[68]

6.2.3   ON/OFF Switching of Photochromism

Transition-metal complexes and organometallics can quench photoexcited molecules *via* electron transfer and energy-transfer quenching. For example, ferrocene has donor abilities (the HOMO with a high energy level), fast electron transfer abilities, and a low-lying ^{3}d–d excited state.[69–72] These characteristics trigger reductive electron-transfer quenching[71] and triplet–triplet energy transfer quenching *via* the Dexter mechanism.[71,72] Because all photochromic reactions proceed *via* photoexcited states, metal complexes and organometallics with quenching abilities have the potential to stop the reactions. This series of quenching abilities is quite sensitive to the oxidation state of the metal centre. ON/OFF switching of the photochromism can therefore be realised in organic-photochromic–transition-metal-complex or organic-photochromic–organometallic conjugates by taking advantage of quenching and reversible redox switching.

Nishihara and coworkers found that ferrocenylspiropyran featured redox-regulated photochromism (Figure 6.6).[73] The neutral species, with an iron oxidation number of +2, showed general photoisomerisation behaviour; upon irradiation with UV light it was converted to the merocyanine form, showing a blue colour in dichloromethane (the proportion of the merocyanine form was 56%). On the other hand, visible-light irradiation or standing in the dark facilitated a backreaction, yielding the colourless spiropyran form (the proportion of the merocyanine form was 0%). The ferrocene moiety could be easily and reversibly oxidised ($E^{0'} = 0.0$ V *vs.* Fc$^+$/Fc) with mild oxidants such as 1,1′-dichloroferrocenium hexafluorophosphate, or electrochemical oxidation. The oxidised spiropyran species underwent almost full conversion to the merocyanine form ($\sim$100%) upon irradiation with UV light; in contrast,

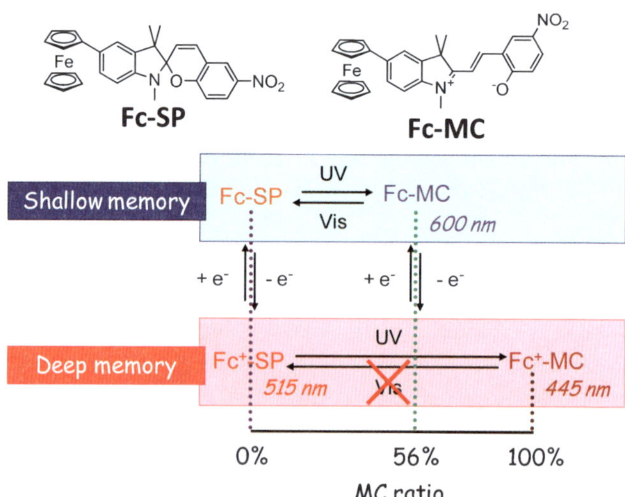

Figure 6.6   Redox-conjugated photoisomerisation of ferrocenylspiropyran.[73]

Figure 6.7 Photochromism of ferrocenylspiropyran in a PVC film.[73]

the backreaction to the spiropyran form took place neither under irradiation with visible light nor in the dark. This result showed that both the photo- and thermal isomerisations were locked in the oxidised form. The authors regarded this redox-regulated photochromism series as "shallow memory" and "deep memory" (Figure 6.6). Photochromism and thermochromism in the reduced state correspond to "shallow memory", where the merocyanine form (memorised state) can be converted to the spiropyran form (unmemorised state) with either photo- or thermal stimulus. In the oxidised state, however, the "memory" cannot be erased by these outer stimuli unless the ferrocenium moiety is reduced.

This ferrocenylspiropyran redox-conjugated photochromism series was also performed in a polyvinyl chloride (PVC) matrix. Ferrocenylspiropyran-

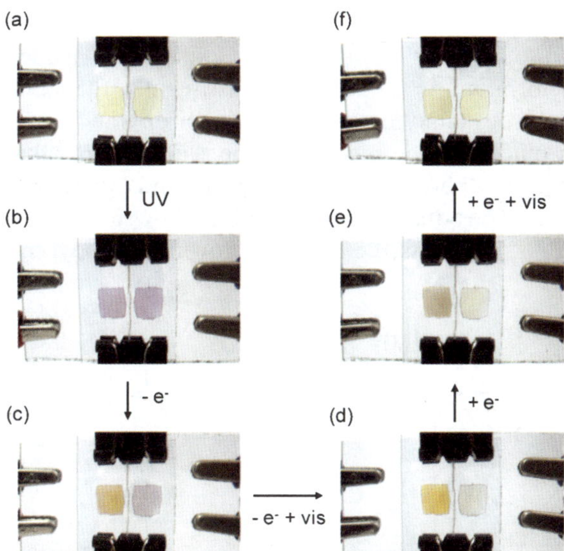

Figure 6.8 Photographs of the redox-conjugated photoisomerisation of ferrocenyl-spiropyran in PVC films doped with tetra-*n*-buthylammonium tetra-fluoroborate as a supporting electrolyte. The left-hand polymer film in each picture was applied a voltage, while the right-hand one was not: (a) as-prepared films; (b) Irradiation with UV light; (c) application of a voltage of 2 V; (d) Irradiation with visible light; (e) application of a voltage of 0 V; (f) irradiation with visible light.[73]

containing PVC films were fabricated as follows: A ferrocenylspiropyran solution was swollen in PVC films, and the solvent was then evaporated in vacuo. When the film covered with a T-shaped photomask was irradiated with UV light, a letter "T" was printed on the film (Figure 6.7); the photoisomerisation was therefore available even in the PVC matrix. In addition, by doping a supporting electrolyte (tetra-*n*-buthylammonium tetrafluoroborate) into the PVC matrix, redox-coupled photochromism was observed, similar to that seen in solution (Figure 6.8).

6.2.4 Photocontrol of Coordination Affinity and Environment

Photoisomerisation can be used to control the binding affinity of organic photochromics with metal ions and complexes,[74–79] and their coordination environment.[80] These phenomena can be exploited in a wide range of applications, such as ion extraction,[74,75] molecular sensors,[76,77] secret ink,[78] photoelectric conversion,[79] and ligand-driven light-induced spin change (LD-LISC).[80]

Kimura and coworkers fabricated a series of vinyl polymers that contained monoazacrowned spiropyran (Figure 6.9a).[81] The authors previously reported that monoazacrowned spiropyran captured alkali-metal ions in a fashion different from that of usual crown ether: In the coloured merocyanine form, the phenolate moiety coordinated to the uptaken cation to form a more rigid complex (Figure 6.10).[82] This peculiar complexation made the metastable merocyanine form more stable, and consequently the binding ability was higher in the merocyanine form than in the spiropyran form. In addition, the characteristic binding mode modulated the selectivity of alkali-metal ions. A vinyl polymer, where photochromic spiropyran and cation-capturing azacrown were separately introduced (Figure 6.9b), showed a normal affinity of $Na^+>K^+>Li^+$,[81,83] whereas the monoazacrowned spiropyran polymer encap-

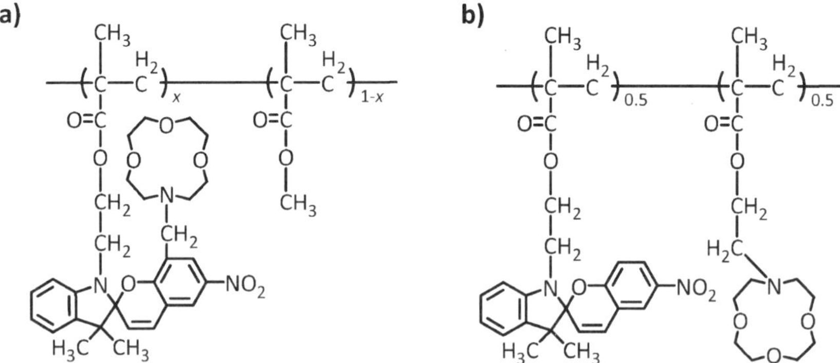

Figure 6.9 Vinyl polymers containing spiropyran and azacrown in the side chains: (a) Both components are conjugated with one another, (b) or are isolated. $x = 0.59, 0.16, 0.04$.[81]

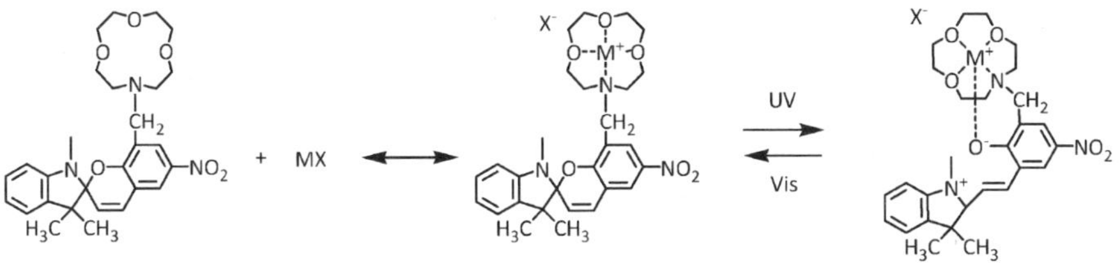

Figure 6.10 Photochromism and peculiar cation-binding ability of a spiropyran derivative modified with monoazacrown ether.[82]

sulating alkali metal cations showed an order of $Li^+ > Na^+ > K^+$. This strong Li^+ affinity was ascribed to the fact that the ion radius of Li^+ is the smallest, and it therefore produced the highest positive charge density. The authors demonstrated reversible photoisomerisation; the spiropyran→merocyanine conversion occurred under UV light, and the merocyanine→spiropyran transformation occurred with visible light.

Taking advantage of the monoazacrowned spirobenzopyran polymer's higher affinity with Li^+, the authors fabricated PVC-supported or self-supporting composite films containing $LiClO_4$, and demonstrated their photoswitchable ion conductivity. Lower ion conductivity in the merocyanine form was accounted for by its higher affinity with Li^+, which resulted in a lower concentration of free Li^+, reducing the number of free ions able to contribute to the ion conductivity.

6.2.5 Photonic Switching of Nonlinear Optical Properties

Quadric and cubic nonlinear optics (NLOs) are important among NLO phenomena from the viewpoint of practical applications. The former is dominated by the first hyperpolarisabilities. Its representative application is second-harmonic generation (SHG).[84] Cubic NLO is based on the second hyperpolarisabilities, and is, for example, exploited in two-photon absorption (TPA).[86] Molecular design criteria have been established; donor–acceptor dipolar, quadrupolar, and octupolar molecules are known to show excellent quadric and cubic NLO properties.[85–88]

For the construction of practical molecule-based SHG materials, it is essential to assemble them in bulk in a noncentrosymmetric fashion. Then, even if the first hyperpolarisability of a certain molecule is gigantic, centrosymmetric single crystals based on the molecule will exhibit no SHG. In the case of flexible polymers, the electrical poling method[89–93] is available, taking advantage of the electric dipole moments of dipolar NLO-phores. Quadrupolar and octupolar NLO-phores tend to possess NLO properties superior to corresponding dipolar ones. However, the electrical poling method cannot be applied to polymer materials containing these molecules, due to the absence of electric dipole moments.

Le Bozec and coworkers demonstrated another method – the "all-optical poling" technique – in polymers including an octupolar NLO-phore.[94] The octupolar NLO-phores used featured a Zn(bpy)$_3^{2+}$ core; this served as the origin of the D_3-symmetry 3D framework, as well as the acceptor site. On the other hand, azobenzene-inserted dialkylaminostyryl units functioned as both donor sites and photoresponsive moieties (Figure 6.11). This complex exhibited an exceptionally fine NLO activity, with β and β_0 values of 863 $\times$ 10^{-30} esu and 590 $\times$ 10^{-30} esu, respectively. The authors embedded this octupolar NLO-phore into poly(methyl methacrylate) (PMMA) by means of atom-transfer radical polymerisation (ATRP). The resultant star-shaped polymer grafted by the octupolar complex exhibited a number average molecular weight (M_n) of 63 000, a weight-average molecular weight (M_w) of 68 000, a polydispersity (PDI) of 1.07, and a glass-transition temperature (T_g) of 125 °C. The authors also prepared a polycarbonate sample doped with the complex as a reference.

Upon excitation of the ligand moieties' π–π^* transition with 488 nm light, the complexes and polymer films underwent *trans*-to-*cis* photoisomerisation,

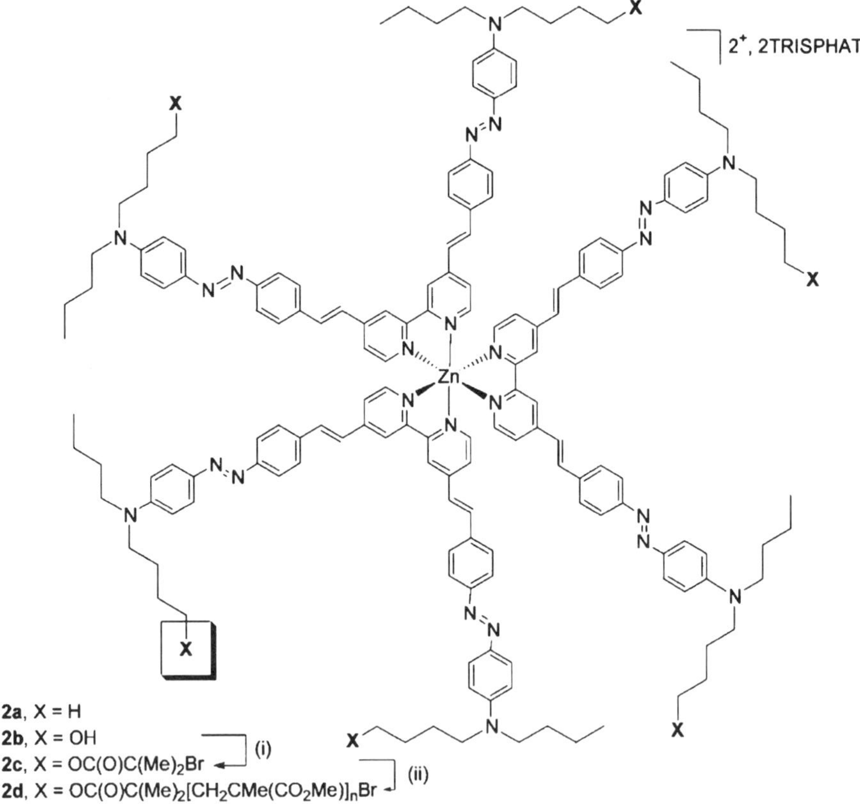

Figure 6.11 Azobenzene-conjugated octupolar Zn(bpy)$_3^{2+}$ complexes and corresponding polymer. Reproduced with permission from ref. 94. Copyright 2004, American Chemical Society.

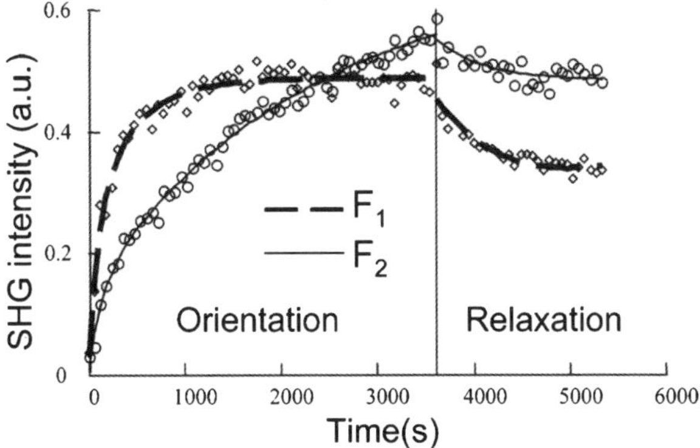

Figure 6.12 Optical orientation and relaxation of the polymer films doped (F$_1$) or grafted (F$_2$) with the octupolar NLO-phore under resonant one- and two-photon excitations (diamonds and circles). The lines indicate biexponential fittings. Reproduced with permission from ref. 94. Copyright 2004, American Chemical Society.

followed by a decrease in the π–π^* band intensity and an increase in the absorptivities around 370 nm. The generated *cis* form experienced a thermal *cis*-to-*trans* reaction.

The authors conducted optical poling of thin films of the polymers (Figure 6.12). After reaching the photostationary state, the grafted polymer possessed an SHG intensity that was 40% higher than the doped one. In addition, a 10% loss in the SHG intensity associated with the thermal relaxation was observed in the grafted polymer sample; this loss was far smaller than that measured in the doped polymer (25%). This experiment demonstrated that grafting the octupolar chromophores allowed molecular reorientation processes to more efficiently dominate the molecular Brownian motion, and thereby induce a higher and more stable photoinduced noncentrosymmetry.

6.3   Electrochromic Material

6.3.1   Redox-Active Metal Complexes and Organometallics

Higuchi and coworkers exploited metallosupramolecular coordination polymers comprising bis(tpy)-M^{II} complexes (tpy = 2,2′:6′,2″-terpyridine, M = Fe, Co, Ru).[95–98] These polyelectrolytes were fabricated using a facile method; di-tpy bridging ligands and metal salts were mixed in solution, giving rise to the corresponding coordination polymer *via* a self-assembly process (Figure 6.13). When the metal centres were divalent, this series of Fe and Ru polyelectrolytes exhibited strong absorption in the visible region (Figure 6.14), which was assigned to MLCT transitions from metal(d) to tpy(π^*). On the other hand, absorptions in the visible region with Co^{II} were tinted, due to Laporte-

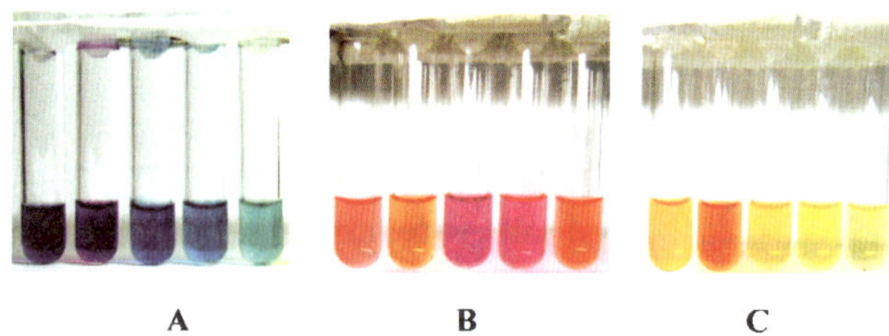

L1: R = H, n = 1

L2: R = H, n = 2

L3: R = OMe, n = 1

L4: R = OMe, n = 2

L5: R = Br, n = 1

ML1-MEPE: R = H, n = 1

ML2-MEPE: R = H, n = 2

ML3-MEPE: R = OMe, n = 1

ML4-MEPE: R = OMe, n = 2

ML5-MEPE: R = Br, n = 1

Figure 6.13 Schematic illustration of coordination polymers comprising bis(tpy)-M^{II} complexes (M = Fe, Ru, Co). Reproduced with permission from ref. 96. Copyright 2008, American Chemical Society.

forbidden d–d transitions. The colour tone was finely tuned using not only variations in the metal centre, but also modification of the bridging ligand. Specifically, these modifications involved substitution of electron-withdrawing and -donating substituents onto the tpy, and replacement of the central arene with longer ones. Intermediate colours could be realised by mixing different metal-containing polyelectrolytes in appropriate ratios. These MLCT and d–d bands diminished *via* the oxidation of the metal centre to a trivalent state. As a result, the polyelectrolytes became transparent. Thin films of these materials could be easily formed on electrode surfaces such as ITO using the solution-casting method, and the resultant thin films retained redox activity. In combination with the fine reversibility and rapidity of the redox switch between M^{II} and M^{III} in bis(tpy)-M complexes, this system underwent robust and quick electrochromism in thin films (Figure 6.15). This series of features is highly appreciated in rewritable electronic paper.

Figure 6.14 Colours of polyelectrolytes based on bis(tpy)-M^{II} complexes in methanol; A: M = Fe; B: M = Ru; C: M = Co. Reproduced with permission from ref. 96. Copyright 2008, American Chemical Society.

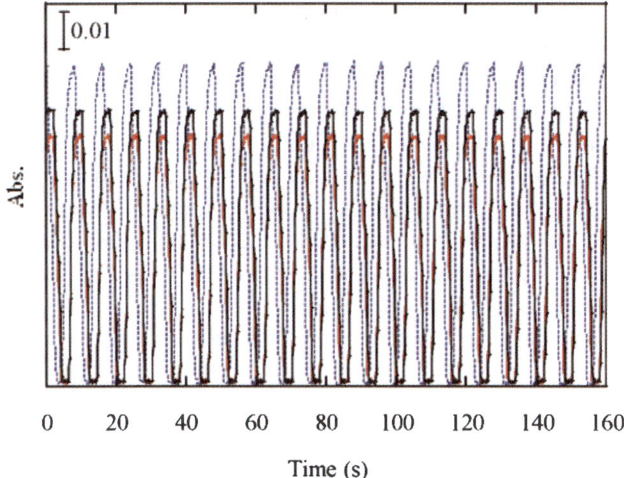

Figure 6.15 MLCT absorbance change of a thin film of Ru-based polyelectrolytes at potentials of +1.5 and 0.0 V vs Ag/AgCl with a 4-second delay between switching. First 40 (blue), middle 40 (black), and last 40 (red) of 4000 redox cycles are shown. Reproduced with permission from ref. 96. Copyright 2008, American Chemical Society.

Optoelectronic materials absorbing light efficiently in the near-infrared region (800–2000 nm) are potentially useful in telecommunications (1300–1550 nm), thermal imaging (>1500 nm), and solar cells (800–2000 nm). For example, a variable optical attenuator (VOA) is an essential component for telecommunications; it affords control over optical signals in wavelength-division multiplexing networks, and is applied for dynamic channel power regulation and equalisation in crossconnected nodes. One of the chief challenges in telecommunications is to develop inexpensive planar integrated VOA devices. Molecule-based materials that are electrically active in the near-infrared (NIR) region (*i.e.* at telecommunication wavelengths) can serve this purpose well, owing to their unique electrical and optical properties, low cost of fabrication, and feasibility for use in monolithically integrated optical devices.

Mixed-valent (MV) compounds, the representative example of which is the Creutz–Taube complex $[(NH_3)_5Ru\text{-pyrazine-}Ru(NH_3)_5]^{5+}$,[99,100] possess characteristic absorption bands, or IVCT bands.[101] In weakly interacting MV compounds (Class II in Robin and Day's classification[37]), an IVCT transition is associated with an electron transfer from an electron-rich redox site to an electron-deficient one. In those with strong interactions (Class III), however, it is assignable to a charge resonance among the redox sites. One of the characteristics of IVCT bands over other electronic transitions such as MLCT, LMCT, and π–π^*, is that they tend to emerge in the NIR region. MV states are generated with ease by the partial oxidation or reduction of redox sites;

incorporation of MV compounds in electrochromic materials can therefore expand these materials' field of performance, as detailed above.

A family of dinuclear ruthenium complexes with a dicarbonylhydrazine (DCH) bridging ligand is particularly interesting as a new class of NIR-absorbing materials. DCH-Ru complexes can occupy three forms depending on the oxidation state of the Ru centres: Ru^{II}/Ru^{II}, MV Ru^{III}/Ru^{II}, and Ru^{III}/Ru^{III}. The absorption bands of these species are typically centred at 550, 1600, and 800 nm; these are ascribable to the MLCT, IVCT, and LMCT transitions, respectively.[102,103] In addition, this class of complex is thermally stable up to 200 °C in the presence of oxygen and moisture.

Qi and Wang designed trimers of this dinuclear DCH-Ru complex bearing hydroxyl groups (Figure 6.16), and the corresponding crosslinked polymers were fabricated with triisocyanate *via* the urethane bonds.[104] The authors expected that the dendritic configuration of the trimers would afford a large free volume in the polymer network, producing rapid ion transportation, resulting in a rapid response time; in fact, thin films of the crosslinked polymer on ITO electrodes retained redox activity, undergoing fast and reversible oxidation and reduction. Figure 6.17 shows the absorption spectrum of the thin film in each oxidation state. As described above, only the MV Ru^{III}/Ru^{II} species featured a broad and intense NIR absorption. The authors also set up a prototype VOA device based on the crosslinked polymer, demonstrating a rapid response to the applied potentials in the NIR region; an attenuation of 5.4 dB at 1550 nm with a switching time of 3 s was achieved (Figure 6.18). The authors also synthesised polymers that were appended on the dinuclear DCH-Ru complexes as side chains,[105] and designed a similar type of dinuclear ruthenium complex as an NIR absorber.[106]

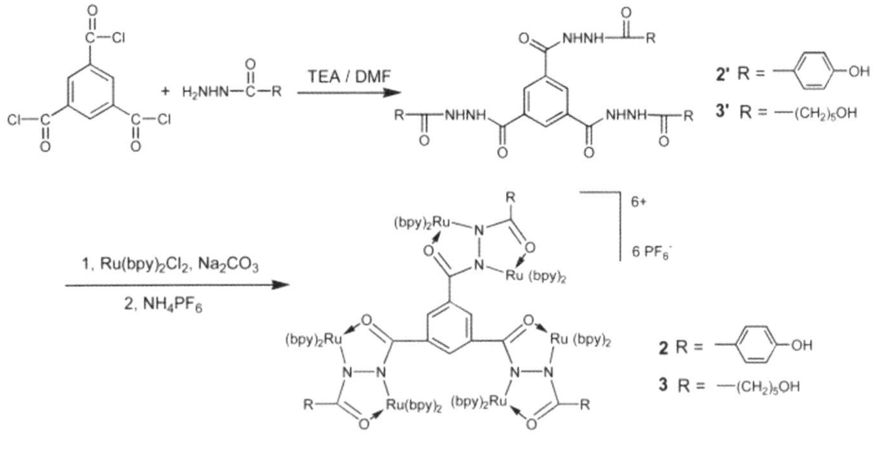

Figure 6.16 Dendritic trimers based on a dinuclear Ru-DCH complex. Reproduced with permission from ref. 104. Copyright 2003, American Chemical Society.

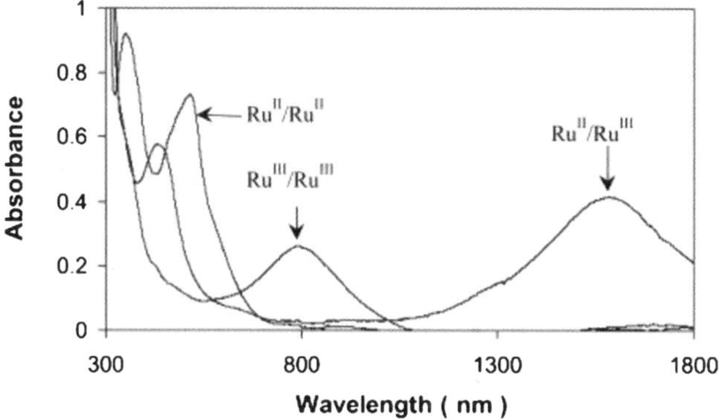

Figure 6.17 Absorption spectrum of thin film of the crosslinked polymer of the trimers in each oxidation state. Reproduced with permission from ref. 104. Copyright 2003, American Chemical Society.

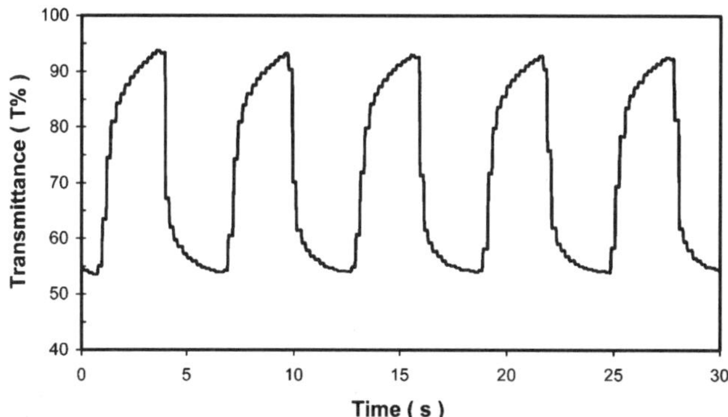

Figure 6.18 Variation of transmittance at 1550 nm for a prototype ECVOA device. Reproduced with permission from ref. 104. Copyright 2003, American Chemical Society.

6.3.2 Prussian Blue

PB[19–22] has been used as a colourant for more than three centuries, though it took a long time to determine its structure and composition. The structure of PB was first discussed by Keggin and Miles,[107] and was then reconsidered by Ludi and Güdel.[108] PB features a face-centred cubic (fcc) unit cell with a periodicity of 10.1–10.2 Å (Figure 6.19).[109] The basic component of PB is $[Fe^{III}Fe^{II}(CN)_6]^-$, where the Fe^{III} and Fe^{II} ions are bridged by cyanides. The Fe^{III} ions are coordinated by the nitrogen atoms, whereas the carbon atoms surround the Fe^{II} ions. Both metal centres adopt the octahedral coordination mode. For the neutralisation of the total charge, counter cations such as K^+ and Na^+ are included in a tetrahedral formation at the interstitial sites of the lattice. Fe^{III} can also be the countercation. In this case, 25% of the

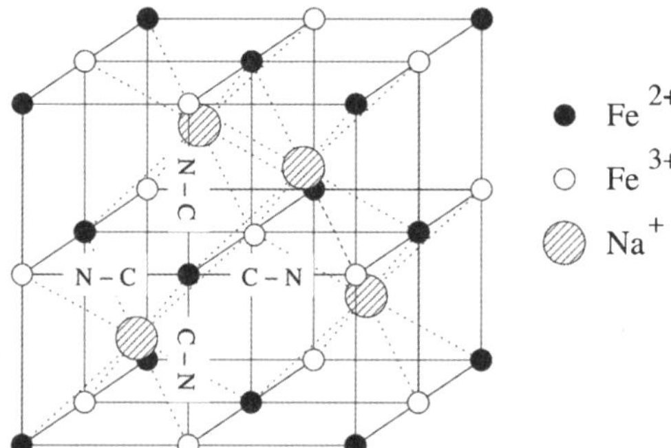

Figure 6.19   Ideal crystal structure of Prussian Blue with sodium ions for charge compensation. Reproduced with permission from ref. 109. Copyright 2004, Elsevier.

$[Fe^{II}(CN)_6]^{4-}$ sites (randomly arranged) are vacant, and the interstitial sites (Na^+ sites in Figure 6.19) are replaced by water molecules. The empty CN sites are also filled with water molecules, which compensate the vacant Fe^{III} coordination sphere with the oxygen atoms. PB features a deep blue colour, which is ascribed by Robin[36] to an IVCT transition, or a photoinduced electron transfer from Fe^{II} to Fe^{III}. The delocalisation of the valence is trivial (about 98% is localised on the Fe^{II} site), and PB is therefore classified by Robin and Day as a Class II system.[37] Instead of IVCT, the term "metal-to-metal charge transfer (MMCT)" is also frequently utilised for these kinds of materials.

Neff was the first to report the electrochromism of PB.[110] Following this work, enormous efforts have been devoted to electrochromism in thin films. PB possesses several oxidation states, which can be electrochemically transformed into PB in a reversible fashion. These oxidation states are Prussian White (PW) with a basic unit of $[Fe^{II}Fe^{II}(CN)_6]^{2-}$, PB, Prussian Green (PG) with $[Fe^{III}_3\{Fe^{III}(CN)_6\}_2\{Fe^{II}(CN)_6\}]^-$, and Prussian Brown (PX) with $[Fe^{III}Fe^{III}(CN)_6]$. Pure PG is obtained very infrequently. Instead, a mixture of PG and PX with a continuous composition ratio is generated in electrochemical treatments. It should be noted that this series of redox reactions is not accompanied by a significant transformation of the unit cell, but needs incursion and effusion of countercations (such as K^+ in a supporting electrolyte solution).

Figure 6.20 shows absorption spectra for PB thin films at various electrode potentials.[111] The PB form (i) exhibited intense absorptions in the visible and NIR region, assignable to the MMCT band. Upon reduction to the PW state (ii), the intense band almost disappeared. In contrast to the sharp change in the absorption spectra upon reduction, oxidation of the PB thin film yielded gradual spectral changes (Figures 6.20 (iii)–(vi)). This behaviour corresponded to the generation of a mixture of PG and PX. It was reported that this series of electrochromic behaviours occurred rapidly (in less than 100 ms) and robustly

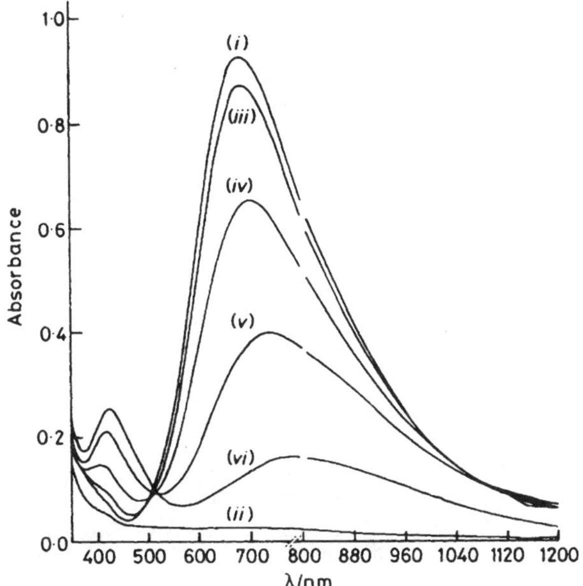

Figure 6.20 Absorption spectra of K^+ ion-containing PB thin film at various potentials [(i) +0.50, (ii) −0.20, (iii) +0.80, (iv) +0.85, (v) +0.90, and (vi) +1.20 V (*vs.* SCE) with a mixture of 0.2 M KCl and 0.01 M HCl as a supporting electrolyte. Reproduced with permission from ref. 111. Copyright 1984, Royal Society of Chemistry.

(repeated 5×10^6 times).[112,113] Several types of electrochromic devices based on PB thin films have been proposed and fabricated.[113–115] Another virtue of PB electrochromic materials lies in the fact that the colour tone can be tuned by the substitution of other metal ions for Fe^{II} and Fe^{III}.[116]

In view of applications such as electronic paper, one of the significant defects of PB is the lack of fabrication processes; PB is practically insoluble in all solvents. Kurihara and coworkers developed an innovative new methodology to process PB; this technique allows the solubilisation of nanocrystallised PB in organic solvents such as toluene, chloroform, diethyl ether, and hexane, *via* decoration of the surface of the PB nanocrystal with organic surfactants such as oleylamine.[117–119] According to DLS and XRD, the average diameter of the PB nanocrystals was 10 nm, and 90% of the particles had diameters of less than 20 nm. The nanoparticle dispersion in toluene underwent no precipitation, even after 6 weeks. The authors also succeeded in synthesising water-soluble PB nanoparticles by surrounding the nanoparticles with ferrocyanide anions.[120] This solubilisation trick drastically improved the processability of PB. For example, thin films of the PB nanocrystal could be easily formed using the conventional spin-coating technique.[118] This spin-coated film showed reversible electrochromism between the PB and PW forms, the same as that shown by PB films made *via* electroreductive deposition. The authors also demonstrated the robustness of the electrochromism, showing as many as 10

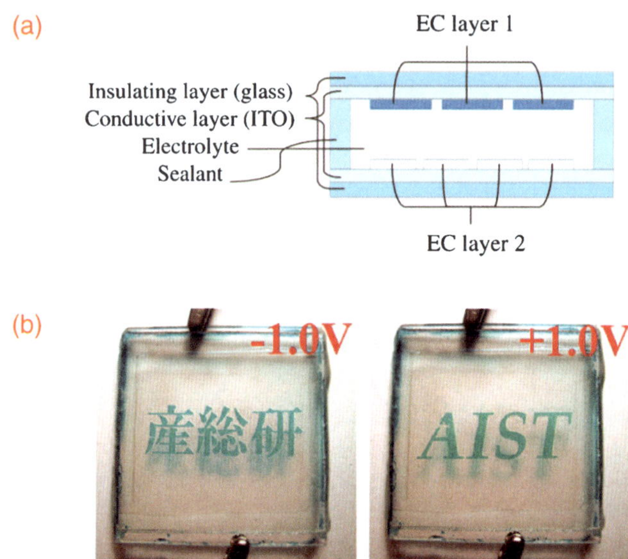

Figure 6.21 (a) Schematic illustration of electrochromic device; (b) Prototype of a display device. Chinese characters and alphabets were drawn on either electrode with PB nanoparticles, which alternatively emerged at voltages of −1.0 and 1.0 V, respectively. Reproduced with permission from ref. 119. Copyright 2008, The Japan Society of Applied Physics.

000 cycles. The authors also succeeded in patterning the PB nanocrystals, using photolithography.[118] Figure 6.21 shows a prototype of a display device made using the PB nanocrystals.[119]

Other pieces of distinctive research have studied the electrochromism of PB. DeLongchamp and Hammond fabricated polyaniline (PANI)-PB nanocomposites using layer-by-layer (LBL) assembly, exploiting the electrostatic attraction between positively charged PANI and negatively ionised PB.[121] This series of nanocomposites was electrochemically active even in thick, high-contrast films. In combination with the electrochromic properties of PANI, the nanocomposite film showed multiple-colour electrochromism. In another study, Lezec, and coworkers demonstrated an electrochemically driven optical switch based on the absorption modulation of surface plasmon polaritons (SPPs) propagating in a metallic nanoslit waveguide, which contained nanocrystals of PB.[122] The authors achieved a high optical switching contrast of 96% *via* the reversible electrochemical switching of PB.[122]

6.4   Photomagnetism

6.4.1   Prussian Blue Analogues

PB analogues with the general formula of $A_{2x}M'^{II}_{(1.5-x)}[M^{III}(CN)_6]\cdot zH_2O$ ($x = 0$–1; A = alkali cation; M'^{II}, M^{III} = transition-metal ions) show both thermal and photochemical charge-transfer-induced spin transitions (CTISTs).[31–34] Figure 6.22 shows their typical crystal structures.[31,107,108,123–128] Case (b) consists of an fcc lattice without lattice defects, where alkali cations are placed at the interstitial sites of the lattice. In case (a), where no alkali-metal ions are present, one-third of the $[Fe^{III}(CN)_6]$ units are removed for charge compensation; instead, water molecules are stored at the defects, some of which coordinate onto the neighbouring M'^{II} sites. This series of coordinated water molecules plays an important role in the expression of photomagnetisation.[32] Additional water molecules can be included at the interstitial sites of the lattice in a noncoordinated fashion.

At 340 K, $Na_{0.4}Co_{1.3}Fe(CN)_6 \cdot 5H_2O$, which is one of the FeCo PB family,[35,129–137] had a $\chi_m T$ value of 3.3 cm^3 K mol^{-1} ($\chi_m T$: magnetic susceptibility). This was attributed to low-spin Fe^{III} ($Fe^{III(LS)}$) and high-spin Co^{II} ($Co^{II(HS)}$) centres. When this material was cooled the $\chi_m T$ value abruptly decreased, reaching 0.8 cm^3 K mol^{-1} at 20 K. This behaviour was explained to result from a thermal charge transfer from $Co^{II(HS)}$ to $Fe^{III(LS)}$, giving rise to $Co^{III(LS)}$ and $Fe^{II(LS)}$.[129] Thus, $Na_{0.4}Co_{1.3}Fe(CN)_6 \cdot 5H_2O$ underwent a thermal phase transition, as follows:

$$Na_{0.4}Co_{0.3}^{II(HS)}Co^{III(LS)}\left[Fe^{II(LS)}(CN)_6\right] \leftrightarrow Na_{0.4}Co_{1.3}^{II(HS)}\left[Fe^{III(LS)}(CN)_6\right] \quad (6.1)$$

The left and right sides of eqn (6.1) represent the low-temperature (LT) and high-temperature (HT) phases, respectively. It should be noted that the phase transition is accompanied by a hysteresis loop,[130–132] where the largest hysteresis width observed was 40 K.[131] Such hysteresis is believed to derive from the cooperative effects afforded by the 3D network of PB analogues. These cooperative effects are regarded as one of the most important elements necessary for the expression of photomagnetic behaviour.

The LT phase of $K_{0.4}Co_{1.3}Fe(CN)_6 \cdot 5H_2O$ displayed a visible band around 550 nm, assignable to an MMCT band from $Fe^{II(LS)}$ to $Co^{III(LS)}$.[133] Upon

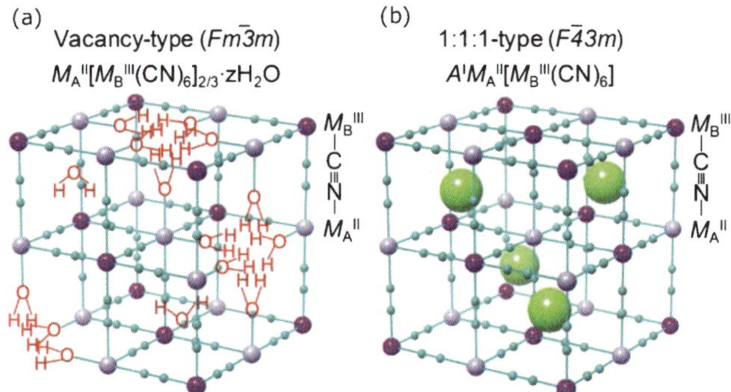

Figure 6.22 Schematic illustrations of the crystal structures of Prussian Blue analogues $A_{2x}M^{II}_{(1.5-x)}[M^{III}(CN)_6] \cdot zH_2O$: (a) $M'^{II}_{1.5}[M^{III}(CN)_6] \cdot zH_2O$ ($x = 0$); (b) $AM'^{II}[M^{III}(CN)_6]$ ($x = 0.5$, $z = 0$). Reproduced with permission from ref. 31. Copyright 2011, Royal Society of Chemistry.

excitation of this band with 500–750 nm visible light at 5 K, significant photomagnetism was induced (Figure 6.23); before illumination, the magnetisation value was negligible, whereas after illumination an increase in magnetisation was observed. This magnetisation was released at 26 K, while the lifetime of the metastable state was as long as several days (provided that the sample was stored at 5 K). In addition, the postillumination magnetisation versus external magnetic field plot featured hysteresis loops (Figure 6.24). Mössbauer,[134] IR,[131,135] X-ray-absorption fine-structure (XAFS),[136,137] and UV-vis-NIR[130,135] spectroscopies identified the photoinduced (PI) phase transition in the FeCo PB species to be responsible for the generation of an HT-like phase, $Fe^{III(LS)}$-CN-$Co^{II(HS)}$. In this PI-HT phase, $Fe^{III(LS)}(S = 1/2)$ and $Co^{II(HS)}(S = 3/2)$ were antiferromagnetically coupled, due to low temperatures, which resulted in ferrimagnetism throughout the 3D crystalline network. The PI-HT state had absorption peaks for $Fe^{III(LS)}$ and $Co^{II(HS)}$ at around 400 and 1300 nm, respectively.[133] Upon irradiation with IR light (with a maximum intensity at 1319 nm), the stimulated magnetisation was cancelled out (Figure 6.23).[133] These results showed that the photomagnetism in the FeCo PB family was reversible.

The photomagnetism can be drastically tuned by changing the ratios between A, M'^{II}, M^{III}, and water molecules, and by substituting components with other elements. The most profound effects are observed by changing between M'^{II} or M^{III}. Ohkoshi, Hashimoto, and coworkers investigated the $Rb_{2x}Mn^{II}_{(1.5-x)}[Fe^{III}(CN)_6]\cdot zH_2O$ series.[31,138–150] Similar to the FeCo PB

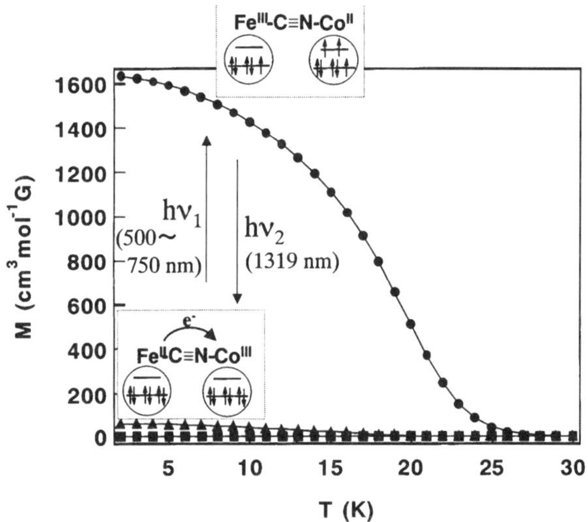

Figure 6.23 Field-cooled magnetisation *vs* temperature curves of $A_{0.4}Co_{1.3}Fe(CN)_6\cdot 5H_2O$ at $H = 5$ G (■) before light illumination, (●) after visible light (hv_1) illumination, and (▲) after near-IR light (hv_2) illumination. Reproduced with permission from ref. 33. Copyright 2003, American Chemical Society.

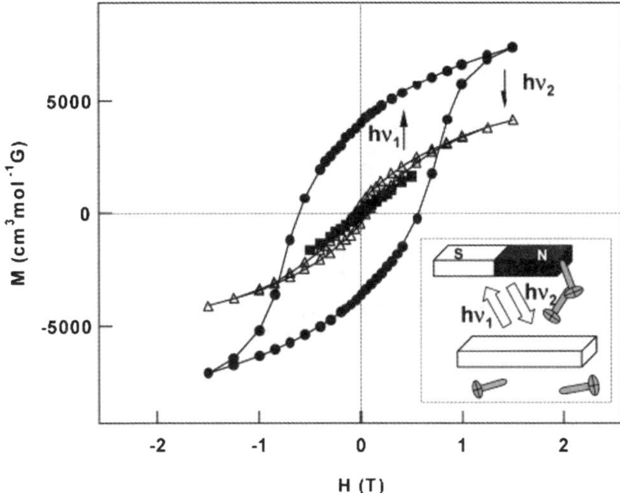

Figure 6.24 Hysteresis loops of $A_{0.4}Co_{1.3}Fe(CN)_6 \cdot 5H_2O$ at 2 K (■) before illumination, (●) after visible-light (hv_1) illumination, and (△) after near-IR-light (hv_2) illumination. Reproduced with permission from ref. 33. Copyright 2003, American Chemical Society.

family, $RbMn[Fe(CN)_6]$ underwent a thermal phase transition; the $\chi_M T$ value at 320 K was 4.65 cm^3 K mol^{-1}, whereas at 200 K it decreased to 3.16 cm^3 K mol^{-1}. The transition temperatures for HT→LT ($T_{1/2\downarrow}$) and LT→HT ($T_{1/2\uparrow}$) were 231 and 304 K, respectively, and the width of the thermal hysteresis loop ($\Delta T = T_{1/2\uparrow} - T_{1/2\downarrow}$) was 73 K.[138] This phase transition was accompanied by a colour change, from light brown (HT phase) to dark brown (LT phase).[138] A crystal lattice transformation also followed this transition, from cubic ($F\bar{4}3m$, $a = 10.533$ Å) to tetragonal ($I\bar{4}m2$, $a = b = 7.090$ Å, $c = 10.520$ Å).[139] From XPS and IR spectra, this HT→LT phase transition was determined to be associated with a charge transfer from Mn^{II} to Fe^{III}.[139] In addition, The $\chi_M T$ values, X-ray photoelectron spectroscopy (XPS), synchrotron radiation X-ray powder structural analysis (SR-XRD), and X-ray absorption near-edge structure (XANES) assigned the spin state of the HT and LT phases as $Fe^{III}(S = 1/2)$-CN-$Mn^{II}(S = 5/2)$ and $Fe^{II}(S = 0)$-CN-$Mn^{III}(S = 2)$, respectively.[140–142] It is noteworthy that Jahn–Teller distortion was responsible for the transformation of the crystal lattice in the LT phase.[139]

$Rb_{0.88}Mn[Fe(CN)_6]_{0.96} \cdot 0.5H_2O$,[143] which also exhibited a phase transition similar to that of $RuMn[Fe(CN)_6]$, showed ferromagnetism in the LT phase (Figure 6.25a). Upon irradiation of the LT phase with 532 nm light (corresponding to the excitation associated with the MMCT transition from Fe^{II} to Mn^{III}), a PI-HT phase emerged. This state featured much weaker magnetisation than the LT phase (Figure 6.25a). However, upon illumination of the PI-HT phase with 410 nm light, the magnetisation recovered to the previous level (Figure 6.25a). This backward reaction was induced by the

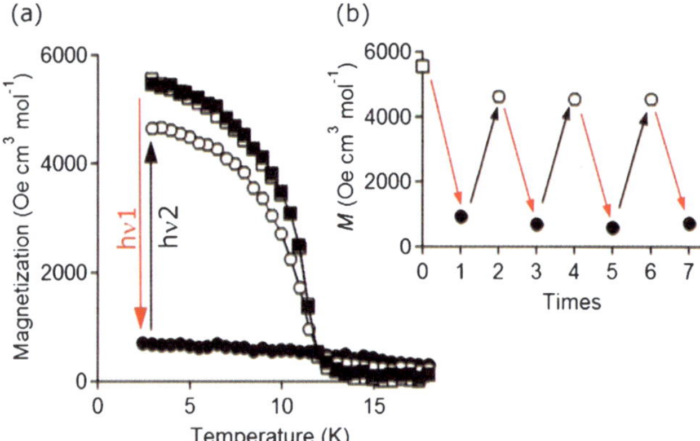

Figure 6.25   Reversible photomagnetism in $Rb_{0.88}Mn[Fe(CN)_6]_{0.96} \cdot 0.5H_2O$: (a) M–T plots under 200 Oe before (□) and after (●) $hv1$ (λ = 532 nm) irradiation, after $hv2$ (λ = 410 nm) irradiation (○), and after thermal annealing at 180 K (■); (b) M *vs.* irradiation-time plot at 3 K upon alternative irradiation with light at $hv1$ (red arrows) and $hv2$ (black arrows). Reproduced with permission from ref. 31. Copyright 2011, Royal Society of Chemistry.

excitation associated with an LMCT transition from CN to Fe^{III}. This series of photomagnetism behaviours proved to be reversible (Figure 6.25b). Neutron powder diffraction measurements[143] showed that the LT phase was a ferromagnet, due to ferromagnetic coupling between the Mn^{III} (S = 2) sites, whereas the PI-HT phase was an antiferromagnet, as shown in Figure 6.26. In summary, $Rb_{0.88}Mn[Fe(CN)_6]_{0.96} \cdot 0.5H_2O$ featured photoinduced magnetism that was opposite to that shown by the FeCo series.

Other analogues have also been described. Dunbar and coworkers synthesised another PB family, $Co^{II}_3[Os^{III}(CN)_6]_2 \cdot 6H_2O$, and disclosed its photomagnetic properties.[151] Cobalt octacyanotungstate[152,153] and copper octacyanomolybdate[154–157] were also reported to show photomagnetic effects. Strictly speaking, these two species are not PB-type complexes, but similarities can be found in the fact that they are bimetallic and comprise polycyanometallates.

6.4.2   Light-Induced Excited Spin State Trapping

One of the most significant challenges in LIESST chemistry lies in the realisation of ferromagnetic phase transitions. In order to bring such systems to reality, proper intermolecular interactions between LIESST complexes are indispensable. Ohkoshi and coworkers synthesised air-stable $Fe^{II}_2[Nb^{IV}(CN)_8](4\text{-}pyridinealdoxime)_8 \cdot 2H_2O$ by mixing two aqueous solutions ($FeCl_2 \cdot 4H_2O$ and 4-pyridinealdoxime) with $K_4[Nb(CN)_8] \cdot 2H_2O$ and stirring for 1 h under an argon atmosphere at room temperature.[158] Figure 6.27 shows the crystal structure of $Fe^{II}_2[Nb^{IV}(CN)_8](4\text{-}pyridinealdoxime)_8 \cdot 2H_2O$. Each Nb centre was appended with eight cyanides, and four of the eight participated in the ligation

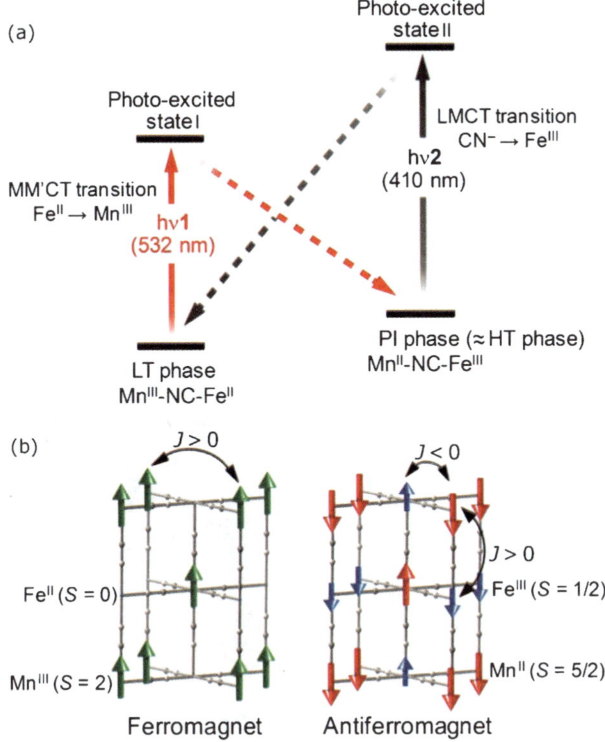

Figure 6.26 Schematic illustration of the reversible photomagnetism in $Rb_{0.88}Mn[Fe(CN)_6]_{0.96}\cdot 0.5H_2O$: (a) Mechanism of the reversible photoinduced charge transfer between the LT and photoinduced (PI) HT phases; (b) Spin ordering in the LT (left) and PI-HT (right) phases. Reproduced with permission from ref. 31. Copyright 2011, Royal Society of Chemistry.

with the Fe centres. The Fe centres featured pseudo-octahedral coordination; two of the six coordination sites were occupied in an axial fashion by the N atoms of the Nb-centre cyanide ligands, and 4-pyridinealdoxime ocuppied the rest. In addition to the Fe-CN-Nb linkage, three kinds of hydrogen bonds (involving the hydroxyl groups of 4-pyridinealdoxime) were responsible for the 3D bimetallic network: (1) with the N atoms of the nonbridged cyanide, (2) with noncoordinating water molecules, and (3) with the nitrogen atoms of other 4-pyridinealdoximes.

This material featured SCO behaviour. At 290 K, the $\chi_M T$ value was 7.15 K cm^3 mol^{-1}, whereas that at 50 K was 1.72 K cm^3 mol^{-1}. $T_{1/2}$ was 130 K, and no hysteresis was observed. UV-vis-NIR spectra (variation of d–d transitions) and XRD (decrease in the C–N bond lengths upon cooling) identified this change in the magnetic property as resulting from the phase transition between $Fe^{II(LS)}$ and $Fe^{II(HS)}$. ^{57}Fe Mössbauer spectroscopy determined the population of $Fe^{II(LS)}$ and $Fe^{II(HS)}$; $Fe^{II(HS)}_2$ and $Fe^{II(HS)}_{0.44}Fe^{II(LS)}_{1.56}$ for the HT and LT

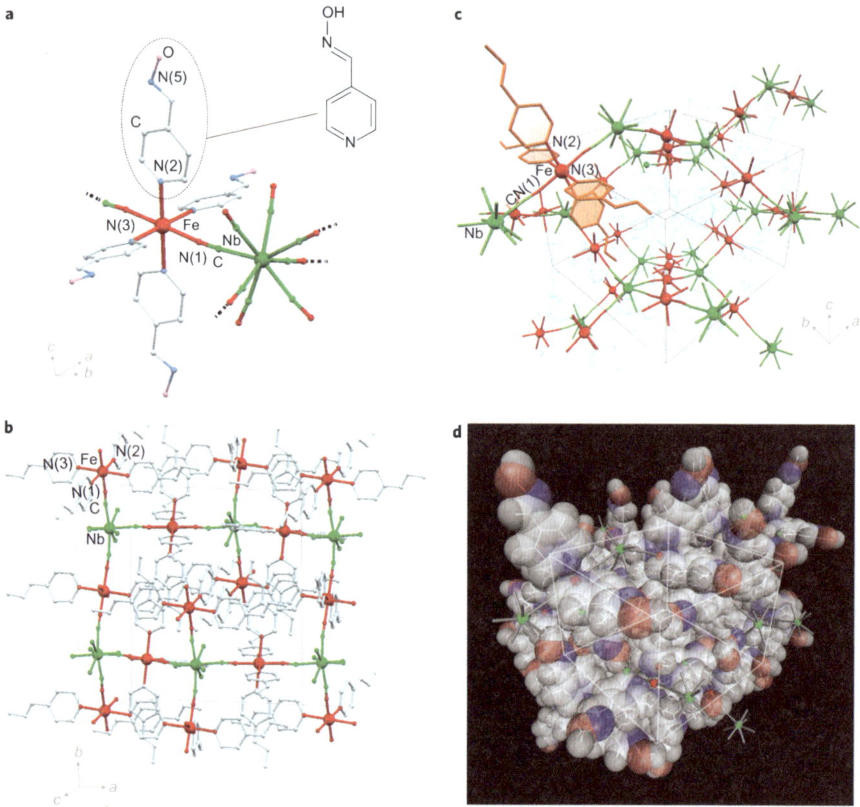

Figure 6.27 Crystal structure of $Fe^{II}_2[Nb^{IV}(CN)_8](4\text{-pyridinealdoxime})_8\cdot2H_2O$. Reproduced with permission from ref. 158. Copyright 2011, Nature Publishing Group.

phases, respectively. Upon irradiation with 473 nm diode laser light at 2 K (which excited the ^{1}d–d ($^1A_1\rightarrow{^1}T_2$) transition of the $Fe^{II(LS)}$ centre), a spontaneous magnetisation was induced with a Curie temperature (T_C) of 20 K (Figure 6.28a). In addition, a magnetic hysteresis loop with an H_c of 240 Oe was observed at 2 K (Figure 6.28b). The saturation magnetisation M_s was 7.4 μ_B at 7 T. This was in agreement with the theoretical value of M_s for ferromagnetic coupling between $Nb^{IV}(S = 1/2)$ and $Fe^{II(HS)}(S = 2)$, which is 7.4 μ_B. After photoirradiation the magnetisation decreased by 30%, but the remaining magnetisation persisted.

The proposed mechanism is shown in Figure 6.29. Before photoirradiation, there was no interaction between the diamagnetic $^1A_1$ $Fe^{II(LS)}$ and paramagnetic Nb^{IV} in the LT phase, resulting in paramagnetism. Upon visible-light irradiation, the excited $Fe^{II(LS)}$ relaxed into metastable $^5T_2$ $Fe^{II(HS)}$, which interacted with the Nb^{IV} in an antiferromagnetic fashion through the bridging cyanide ligands. As a result, ferrimagnetism was expressed.

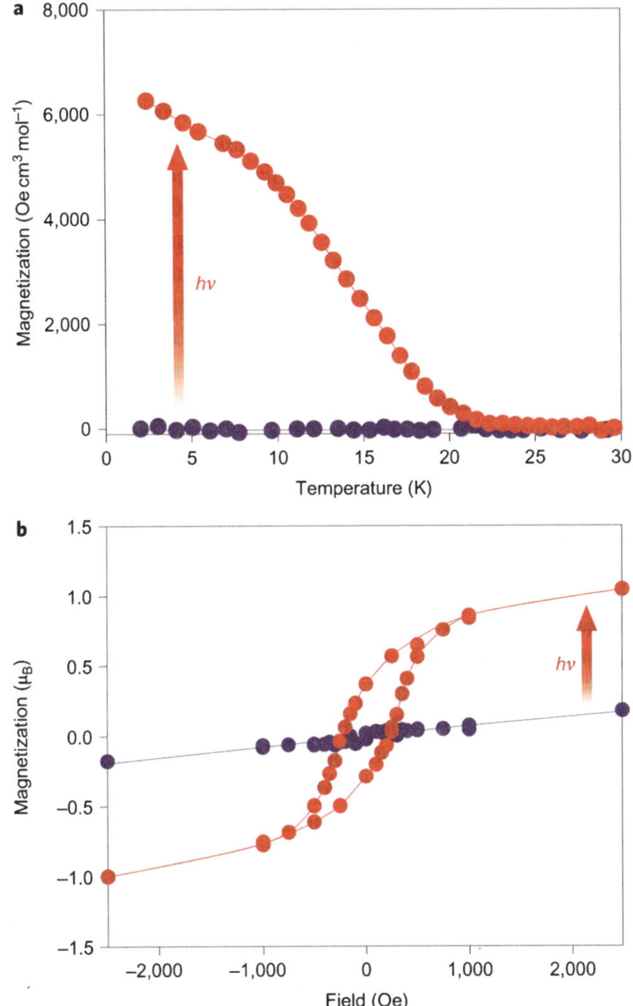

Figure 6.28 Photoinduced magnetisation caused by light-induced spin-crossover in $Fe^{II}_2[Nb^{IV}(CN)_8](4\text{-pyridinealdoxime})_8 \cdot 2H_2O$. (a) Magnetisation *vs.* temperature curves at 100 Oe. Light irradiation induces a spontaneous magnetisation with a Curie temperature of 20 K. (b) Magnetic hysteresis curves at 2 K. After irradiation, a magnetic hysteresis loop with a coercive field of 240 Oe appeared. Blue and red circles denote measurements before and after irradiation with 473-nm light. Reproduced with permission from ref. 158. Copyright 2011, Nature Publishing Group.

6.5 Applications in Optoelectronics

6.5.1 Organic Photovoltaics

Among the various kinds of organic photovoltaics (OPVs), bulk heterojunction solar cells in particular have gained increasing attention from

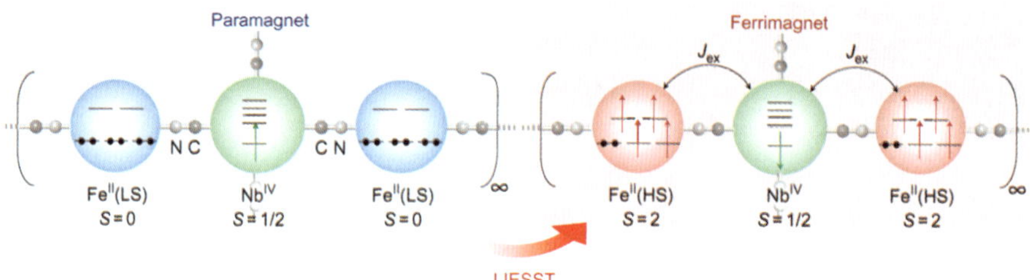

Figure 6.29   Light-induced ferromagnetism in $Fe^{II}_2[Nb^{IV}(CN)_8]$(4-pyridinealdoxime)$_8$·$2H_2O$. Reproduced with permission from ref. 158. Copyright 2011, Nature Publishing Group.

researchers.[159] They feature nanoscale phase separation between p-type and n-type molecule- or polymer-based semiconductors. This peculiar structure affords a broader interface between the two components, making it easier for excitons generated by photoirradiation to arrive at the interface and subsequently dissociate into electrons and holes. Bulk heterojunction solar cells are believed to realise large-area, low-cost, and flexible devices. One of the most well-known bulk heterojunction solar cells is based on p-type poly(3-hexylthiophene) (P3HT) and n-type [6,6]-phenyl-C_{61}-butyric acid methyl ester (PCBM). The power conversion efficiencies (PCEs) of this device reached 4–5%.[159] For the further improvement of PCEs, optimisation of p-type polymeric materials is desirable; this could, for example, provide a lower bandgap for more efficient acquisition of solar radiation.[160,161]

Wong and coworkers attempted to utilise platinum(II)-containing poly(aryleneethynylene)s as a p-type polymer.[162,163] Pioneering work on organometallic poly(aryleneethynylene)s was conducted by Hagihara and coworkers in the 1970s,[164–166] and many intensive studies were subsequently performed on this series of polymers. Pt-based systems in particular are known to be photochemically active,[167,168] and are therefore applied as phosphors in OLEDs[169–171] and in optical power limiting materials.[172] In this work, the authors took full advantage of the virtues of this class of materials; specifically, the photochemical properties of these materials, including the HOMO–LUMO gap, could be systematically tuned by the replacement of the arene spacers.[167,168] It should be noted that before this study, Reynolds and coworkers investigated bulk heterojunction solar cells based on a platinum(II)-containing poly(aryleneethynylene) with a thiophene linker, and disclosed that the photoelectric conversion proceeded *via* the triplet excited state.[173] However, the PCE was low ($\eta = ca.\ 0.27\%$).

To fit the absorptivity of platinum(II)-containing poly(aryleneethynylene)s with the solar spectrum, the authors installed 4,7-di-2′-thienyl-2,1,3-benzothia-diazole as the arene spacer (Figure 6.30). Corresponding monomers and polymers were synthesised by means of Sonogashira-type dehydrohalogena-tion (Figure 6.30); the latter had a degree of polymerisation of *ca.* 22. Figure 6.31a shows the absorption spectra for the free ligand, monomer, and polymer. The bandgap of the polymer was determined to be 1.85 eV from the onset of the absorption. This value was smaller than those of platinum(II)-containing poly(aryleneethynylene)s with electron-rich bithiophene (2.55 eV) or electron-deficient benzothiadiazole (2.20 eV) as the arene spacer. Therefore, the donor–acceptor interaction in the 4,7-di-2′-thienyl-2,1,3-benzothiadiazole unit, as well as the more extended π-electron delocalised system throughout the chain, was responsible for the narrower bandgap in this polymer. The emission lifetimes for the monomer and polymer were no more than nanoseconds, and were thus too short to be assigned to triplet phosphorescence. Stokes shifts (<0.40 eV) were also smaller than those of triplet phosphors (Figure 6.31b). In addition, the temperature dependence of the emission was also less prominent than those of Pt-polyyne compounds with triplet phosphorescence (Figure 6.31c). From these facts, the authors attributed this series of emissions to singlet fluorescence. This fluorescence showed positive solvatochromism, or a redshift in a polar solvent. The singlet excited state of the polymer therefore featured a donor–acceptor charge-transfer character.

Pt polymer-PCBM bulk heterojunction solar cells were fabricated by means of spin coating. The device configuration was Al electrode/Pt polymer-PCBM/PEDOT:PSS/ITO electrode, with an optimal ratio of Pt polymer to PCBM of 1:4. This device demonstrated excellent performance, with an open-circuit voltage of $V_{oc} = 0.82 \pm 0.03$ V, a short-circuit current density of $J_{sc} = 13.1 \pm 3.2$ mAcm^{-2}, a fill factor of FF $= 0.37 \pm 0.03$, and a PCE of $\eta = 4.1 \pm 0.9\%$.

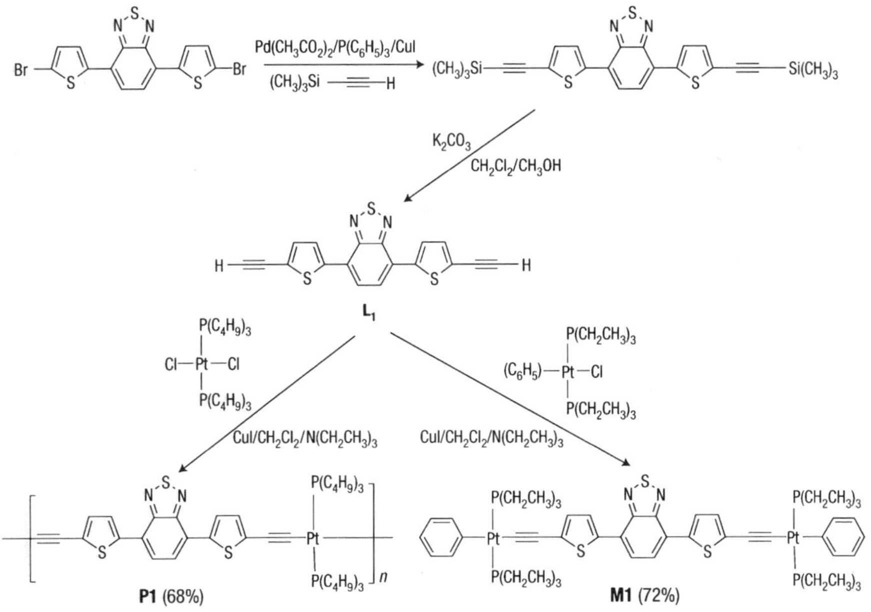

Figure 6.30 Synthetic scheme for the ligand (**L₁**), Pt monomer (**M1**), and Pt polymer (**P1**). Reproduced with permission from ref. 162. Copyright 2011, Nature Publishing Group.

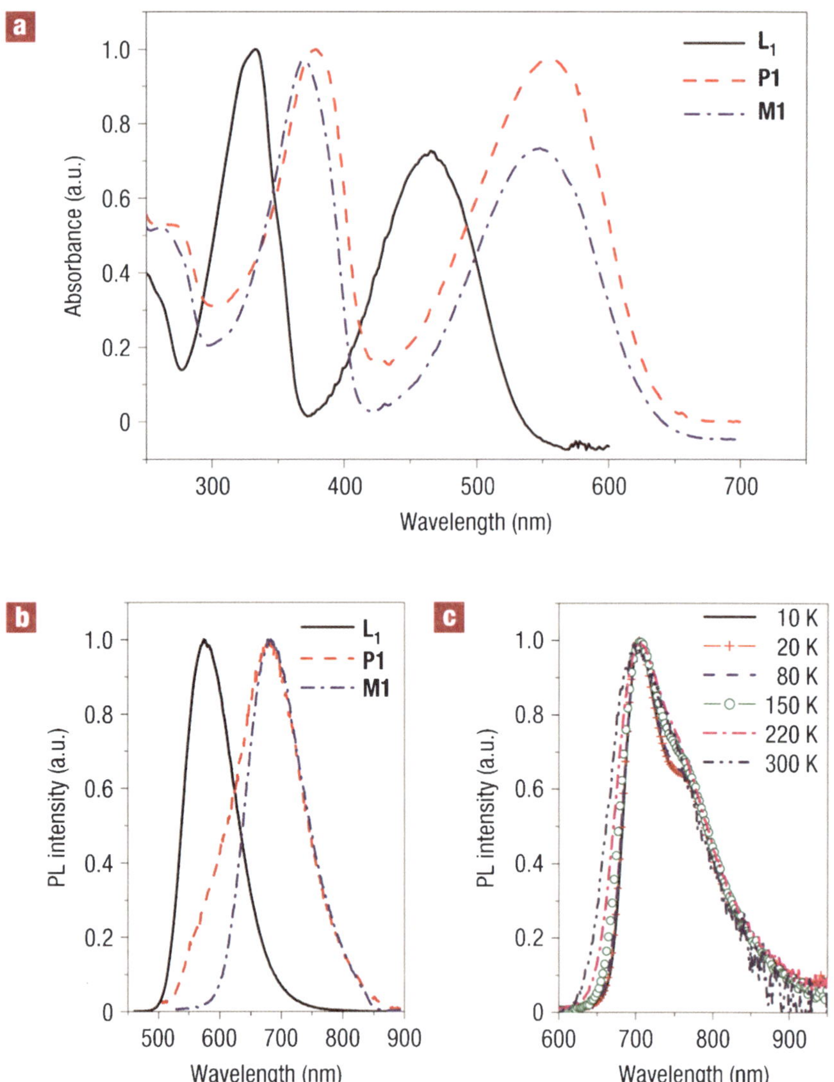

Figure 6.31 UV-vis (a) and emission (b) spectra of the ligand, Pt monomer, and Pt polymer in solution; (c) Emission spectra of the Pt polymer at different temperatures. Reproduced with permission from ref. 162. Copyright 2011, Nature Publishing Group.

The photoresponse was also extended to longer wavelengths compared with the P3HT-PCBM system (Figure 6.32), reflecting the narrower bandgap of the Pt polymer. The authors claimed that the performance of their device could be further enhanced with more elaborate device structures, a neater annealing process, and more careful fabrication under an inert atmosphere. They proposed that these factors were essential to achieve the desired standard of data with the P3HT-PCBM system.

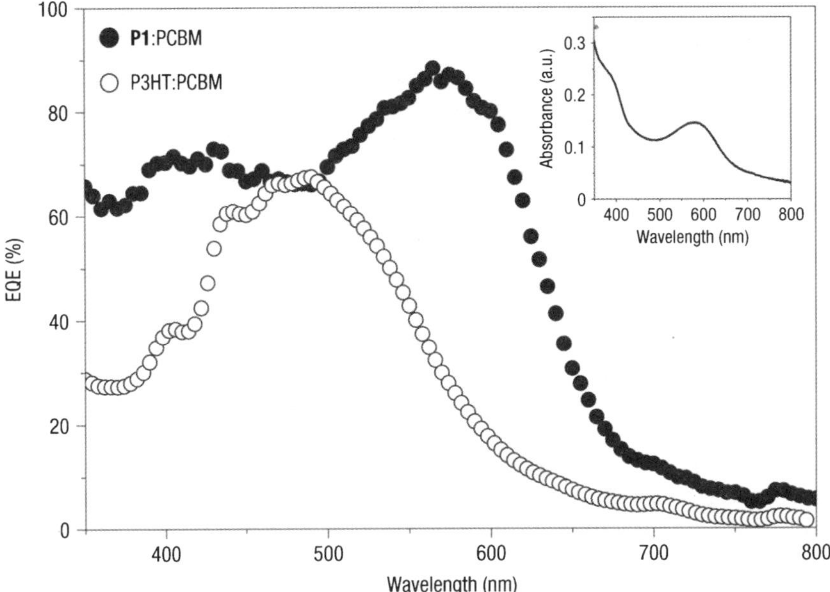

Figure 6.32 External quantum efficiency (EQE) wavelength dependencies of solar cells with P1/PCBM (1:4) and P3HT/PCBM (1:4) active layers. Inset shows the absorption of the P1/PCBM active layer. Reproduced with permission from ref. 162. Copyright 2011, Nature Publishing Group.

Schlz and Holdcroft designed a copolymer of $Ir(ppy)_3$ and 9,9-dioctylfluorene, and fabricated a bulk heterojunction solar cell (ppy = 2-phenylpyridine) with PCBM.[174] The authors found an increase in external quantum efficiency (EQE) compared with a device based on fully organic poly(9,9-dioctylfluorene), which was presumably due to the formation of a triplet excited state, and the longer diffusion length of the triplet exciton.

6.5.2 White Light-Emitting Diodes

White light-emitting diodes (WLEDs) have been intensively studied for applications such as illumination light sources, backlights for liquid crystal displays, and full-colour flat-panel displays with colour filters.[175] Among the different types of WLEDs, white polymer light-emitting diodes (WPLEDs) have been of particular interest, due to their low-cost, mass-production-compatible, large-area solution process, and facile fabrication, which can be achieved *via* techniques such as spin coating.[176–181] WPLEDs are often based on polymers doped with light-emitting small molecules or polymer blends. However, precise control over the concentration of the dopant molecules—which is essential for the realisation of true white colour—is arduous, and polymer blends suffer from the intrinsic drawback of phase separation, which results in lower stability and reproducibility. In contrast, single-component WPLEDs are free from these drawbacks.

Thus far, many kinds of organic single-component WPLEDs with organic fluorophores have been reported,[182–189] though the efficiencies of these systems have often been low. Recent interest has been aimed towards the incorporation of metal complexes into single-component WPLEDs, with the goal of improving the efficiency.[190] Yang and coworkers synthesised a single-component WPLED comprising fluorescent organic and phosphorescent inorganic species (Figure 6.33).[191] In this hybrid system, large-bandgap fluorene was used as the host species with blue light emission, and benzothiadiazole was selected as the green light emitter. These fragments were embedded in the polymer backbone. An Ir^{III} complex, which was appended on the side chains, was used as a triplet red-light emission species; phosphors are known to show much more efficient electroluminescence than fluorophors.[192]

White light-emitting copolymers were synthesised using the Suzuki polycondensation protocol (Figure 6.33). The resultant copolymers were soluble in organic solvents, and had number-average molecular weights of 34 300–65 500 g mol^{-1}. Figure 6.34 shows the EL spectra for the WPLED devices, the configuration of which was ITO/PEDOT/polyvinylcarbazole (PVK)/copolymer/CsF/Al. The spectra were made up of blue, green, and red emissions from each component of the copolymers. The best performance was obtained at a ratio of $x = 0.05$, $y = 0.2$, affording a maximum luminance efficiency of 6.1 cd A^{-1}, Commission Internationale de l'Éclairage (CIE) coordinates of (0.32, 0.44), and a maximum luminance of 10 110 cdm^{-2} at a current density of 345 mA cm^{-2}.

Neher and Shu embedded an Ir^{III} complex in the polymer backbone as a red phosphor.[193] In combination with fluorene and fluorenone applied, respec-

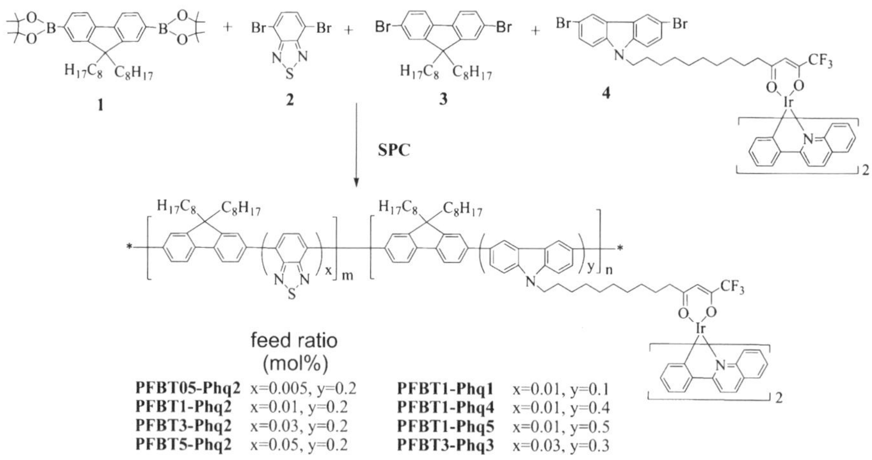

feed ratio
(mol%)

PFBT05-Phq2	x=0.005, y=0.2	PFBT1-Phq1	x=0.01, y=0.1
PFBT1-Phq2	x=0.01, y=0.2	PFBT1-Phq4	x=0.01, y=0.4
PFBT3-Phq2	x=0.03, y=0.2	PFBT1-Phq5	x=0.01, y=0.5
PFBT5-Phq2	x=0.05, y=0.2	PFBT3-Phq3	x=0.03, y=0.3

Figure 6.33 Synthetic route for the copolymers. SPC: Suzuki polycondensation; PFBT-Phq: the copolymer from fluorene (F), BT, and the 2-phenylquinoline iridium complex (Phq). Reproduced with permission from ref. 191. Copyright 2006, Wiley-VCH.

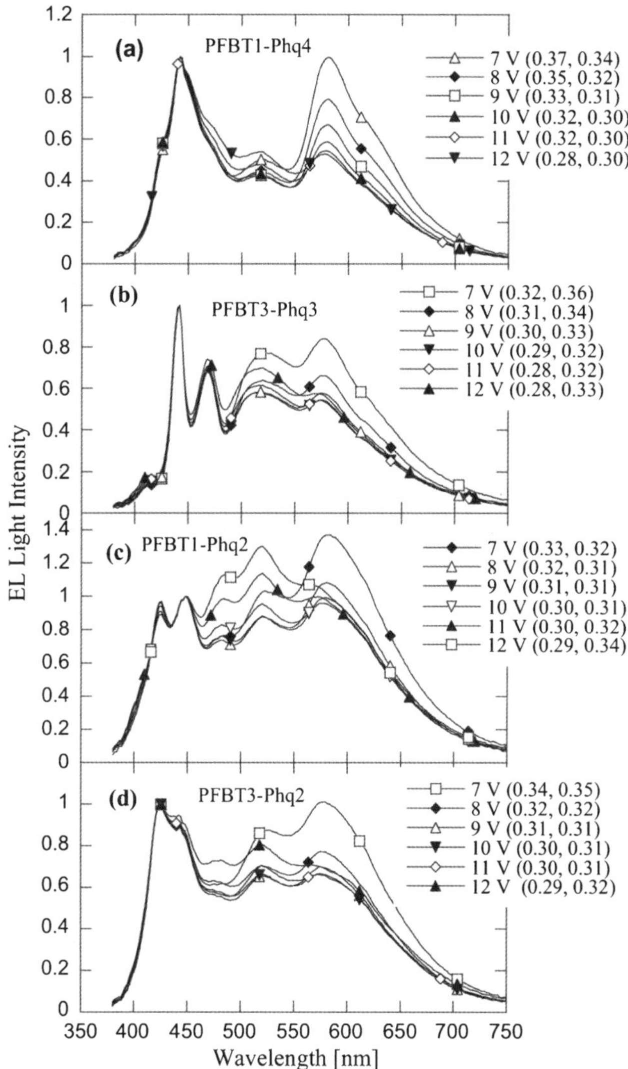

Figure 6.34 EL spectra of devices at various applied voltages with corresponding CIE coordinates: (a) copolymers PFBT1-Phq4; (b) PFBT3-Phq3; (c) PFBT1-Phq2; (d) PFBT3-Phq2. Reproduced with permission from ref. 191. Copyright 2006, Wiley-VCH.

tively, as blue and green fluorophores in the main chain, an WPLED device incorporating this copolymer emitted white light. In other work, Shu and coworkers covalently implanted a red phosphorescent Os^{II} complex in the main chain.[194] Together with the blue and green fluorescent fluorene and benzothiadiazole in the polymer backbone, this copolymer also served as a WPLED.

6.6 Photoactive Metal-Organic Frameworks and Porous Coordination Polymers

6.6.1 Photocatalysis

Lin and coworkers implanted catalytically active metal complexes into a Zr-based MOF, $Zr_6O_4(OH)_4(bpdc)_6$ (UiO-67, bpdc = parabiphenyldicarboxylate) (Figures 6.35 and 6.36).[195] $Ir^{III}(Cp^*)(dcppy)Cl$ (Cp* = pentamethylcyclopenta-dienyl, dcppy = 2-phenylpyridine-5,4'-dicarboxylic acid), $[Ir^{III}(Cp^*)(dcbpy)Cl]Cl$ (dcbpy = 2,2'-bipyridine-5,5'-dicarboxylic acid) and $[Ir^{III}(dcppy)_2(H_2O)]Cl$ acted as water oxidation catalysts (WOCs), $Re^{I}(CO)_3(dcbpy)Cl$ acted as a CO_2-reduction photocatalyst, and $[Ir^{III}(ppy)_2(dcbpy)]Cl$ and $[Ru^{II}(bpy)_2(dcbpy)]Cl$ (ppy = 2-phenylpyridine, bpy = 2,2'-bipyridine) acted as photocatalysts for organic reactions (aza-Henry reaction, aerobic amine coupling, and aerobic oxidation of thioanisole). The structural similarity between bpdc and the bidentate ligands with dicarboxylic acid functionalities in the catalysts (*i.e.* dcppy, dcbpy) meant that the catalyst-doped MOFs had structures similar to that of parental UiO-67, and therefore also featured large surface areas (1092–1497 m^2/g). The catalyst-embedded MOFs possessed catalytic activities similar to that of the

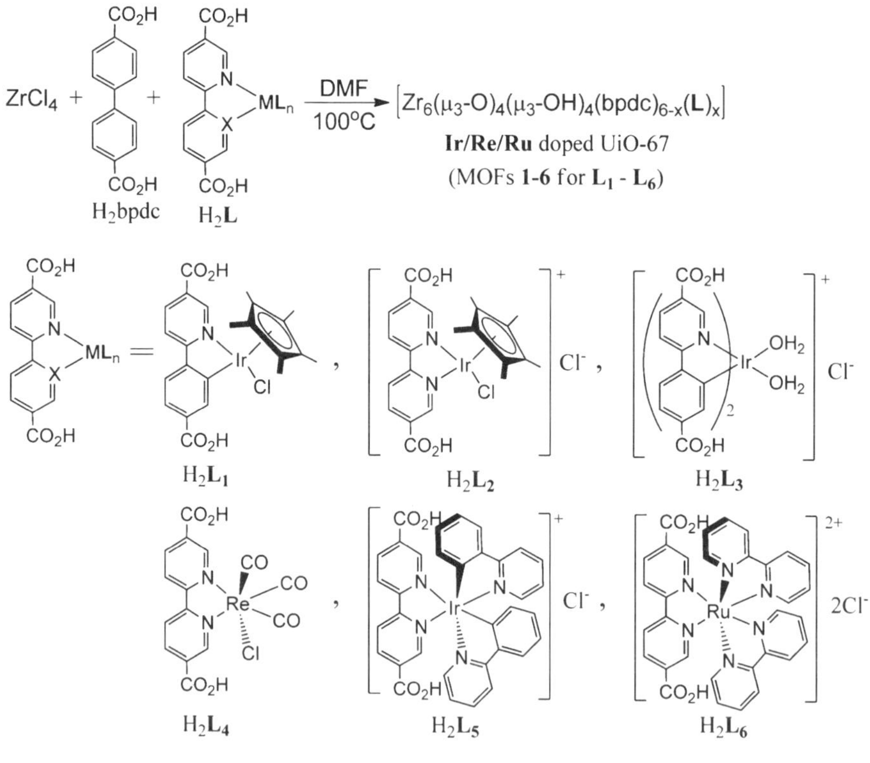

Figure 6.35 Fabrication method for catalyst-doped MOF UiO-67. Reproduced with permission from ref. 195. Copyright 2011, American Chemical Society.

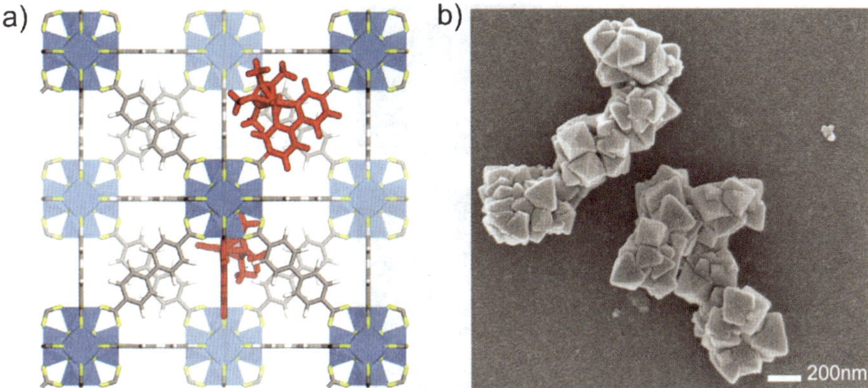

Figure 6.36   (a) Structure model of UiO-67 doped by $Ir^{III}(Cp^*)(dcppy)Cl$; (b) SEM image of UiO-67 doped by $Ir^{III}(Cp^*)(dcppy)Cl$. Reproduced with permission from ref. 195. Copyright 2011, American Chemical Society.

corresponding discrete transition-metal complexes, and facilitated their own reuse because of their heterogeneity.

6.6.2   Energy Transfer Among Hosts and Guests

Kim and coworkers synthesised a MOF that featured extremely high porosity (3.9 and 4.7 nm) and a guest-responsive luminescent property.[196] This mesoporous MOF was synthesised using a normal solvothermal procedure; 1,3,5-tribenzoic acid (H_3TATB) and $Tb(NO_3)_3 \cdot 5H_2O$ were heated at 105 °C for 2 days in a mixture of N,N-dimethylacetamide (DMA), methanol and water, giving truncated-octahedral crystals (Figure 6.37a). The Brunauer–Emmett–Teller (BET) and Langmuir surface areas were calculated to be 1419 and 2887 m^2/g, respectively. When the samples were activated at 160 °C, the surface areas expanded to 1783 m^2/g (BET) and 3855 m^2/g (Langmuir). Both the single crystal (Figure 6.37a) and bulk materials (Figure 6.37b) exhibited bright green photoluminescence, which was characteristic of Tb-centred emission. When single crystals were exposed to sublimed ferrocene at 100 °C, the Tb-based MOF stored the ferrocene molecules in its enormous voids; as many as 4420 ferrocene molecules were uptaken into each unit cell. This uptake was reflected in the appearance, and more importantly, the green luminescence disappeared (Figure 6.37c). The author ascribed this phenomenon to efficient energy transfer from the host framework to the guest ferrocene. In addition, the adsorbed ferrocene was reversibly expelled in vacuum at 50 °C; this was accompanied by the recovery of the green emission (Figure 6.37d).

6.6.3   Photonic Postsynthetic Modifications

Nanopore surface activity is of great importance for the porous properties of PCPs. The concept of postsynthetic modification (PSM)[197] has gathered

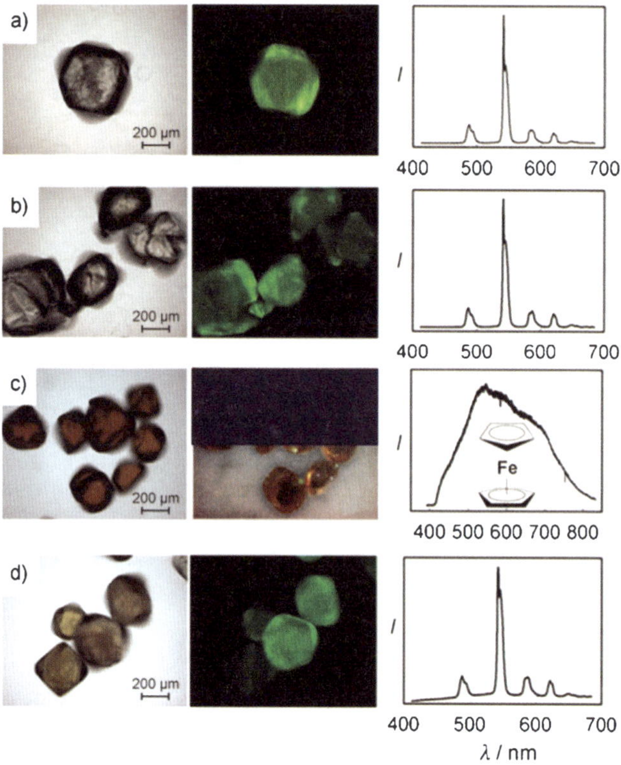

Figure 6.37   Microscope images of a crystal of the mesoporous Tb-containing MOF taken in the transmission mode (left) and in the fluorescence mode (middle), and its luminescence spectrum (right): (a) a single crystal; (b) a bulk sample; (c) ferrocene-introduced crystals. In the fluorescence image, half of the background color was subtracted to show the crystals; d) the crystals used in (c) after kickout of the ferrocene. Reproduced with permission from ref. 196. Copyright 2007, Wiley-VCH.

increasing interest as a useful synthetic approach. PSM has been found to be a powerful tool for introducing new functional groups on pore surfaces, and thereby acquiring new functionalities. PSM is especially invaluable for the implantation of unstable functional groups. Matsuda, Kitagawa, and coworkers integrated photoactive arylazide into a PCP.[198] Arylazide is known to show a denitrogen reaction at low temperatures upon UV irradiation, resulting in the formation of triplet aryl nitrene.[199] Slow diffusion of a methanol solution of 5-azidoisophthalic acid (H_2N_3-ipa), 4,4′-bipyridine (bpy) into a *N,N*-dimethylformamide (DMF) solution of $Zn(NO_3)_2 \cdot 6H_2O$ yielded block-shaped pale-yellow crystals of $[Zn_2(N_3\text{-ipa})_2(bpy)_2(DMF)_{1.5}]_n$ (CID-N_3) in a few days (Figure 6.38a).

Upon irradiation with UV light at 77 K, generation of nitrene was observed in both CID-N_3 and dried CID-N_3 (annealed at 120 °C in vacuum for 6 h); the nitrene was identified using IR (decrease of stretching bands of the azide group

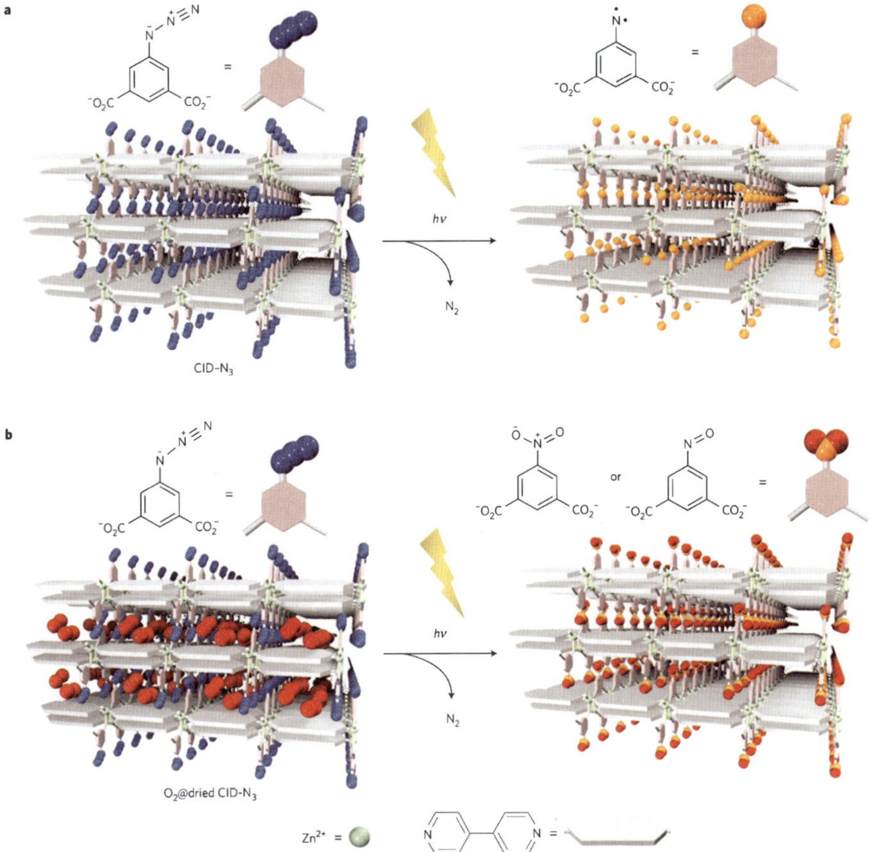

Figure 6.38 Schematic illustration of the photoactivation of a PCP with azide functionalities. The framework of CID-N$_3$ was formed by interdigitations of two-dimensional sheets composed of Zn^{2+}, 5-azidoisophthalate and 4,4′-bipyridine. (a) Photoactivation of the azide moieties led to the formation of triplet nitrenes on the pore surface. (b) Photochemical trapping of physisorbed oxygen molecules in the framework of dried CID-N$_3$ generated nitro and nitroso groups. Reproduced with permission from ref. 198. Copyright 2010, Nature Publishing Group.

at around 2110 cm^{-1}), and EPR (triplet and biradical forms in equilibrium at magnetic fields of 700 and 330 mT) spectroscopies. The conversion ratio in the photostationary state reached 70%. In addition, synchrotron X-ray diffraction revealed that this photoreaction was accompanied by a crystal transformation in the space group from *P2/n* to *C2/m*. Thus, the triplet nitrene anchored on the pore surface was well isolated and sufficiently kinetically stable.

The authors then investigated the photoexcitation of dried CID-N$_3$ in an O$_2$ atmosphere (80 kPa) at 120 K. The dried CID-N$_3$ was found to be nonporous for N$_2$, O$_2$ and CO at 77 K (shown by the slow diffusion of gaseous molecules into the micropore), whereas physisorption became active at 120 K for N$_2$, O$_2$

and CO, and at 195 K for CO_2 and C_2H_2. The photoproduct was different from that in the absence of oxygen, in that the azido groups were converted into nitro or nitroso groups, judging from ^{1}H-NMR spectral changes. It was also revealed that all of the photogenerated nitrene reacted with oxygen. On the other hand, photoirradiation of the free ligand H_2N_3-ipa under an O_2 atmosphere afforded a complicated mixture. The peculiar porous structure of CID-N_3 (which could encapsulate oxygen molecules) was therefore indispensable for the conversion of nitrenoarene into nitroarene and nitrosoarene (Figure 6.38b). It is noteworthy that the authors' new methodology successfully introduced unstable nitreno and nitroso groups in a PCP framework. The authors also confirmed that the adsorbed CO molecule could react with the photogenerated nitrene to give isocyanate.

References

1. S. Kitagawa and M. Kondo, *Bull. Chem. Soc. Jpn.*, 1998, **71**, 1739.
2. O. M. Yaghi, H. Li, C. Davis, D. Richardson and T. L. Groy, *Acc. Chem. Res.*, 1998, **31**, 474.
3. T. J. Barton, L. M. Bull, W. G. Klemperer, D. A. Loy, B. McEnaney, M. Misono, P. A. Monson, G. Pez, G. W. Scherer, J. C. Vartuli and O. M. Yaghi, *Chem. Mater.*, 1999, **11**, 2633.
4. M. Eddaoudi, D. B. Moler, H. Li, B. Chen, T. M. Reineke, M. O'Keeffe and O. M. Yaghi, *Acc. Chem. Res.*, 2001, **34**, 319.
5. N. L. Rosi, M. Eddaoudi, J. Kim, M. O'Keeffe and O. M. Yaghi, *CrystEngComm*, 2002, **4**, 401.
6. O. M. Yaghi, M. O'Keeffe, N. W. Ockwig, H. K. Chae, M. Eddaoudi and J. Kim, *Nature*, 2003, **423**, 705.
7. S. Kitagawa, R. Kitaura and S.-i. Noro, *Angew. Chem. Int. Ed.*, 2004, **43**, 2334.
8. S. Kitagawa and T. Uemura, *Chem. Lett.*, 2005, **34**, 132.
9. S. Kitagawa, S.-i. Noro and T. Nakamura, *Chem. Commun.*, 2006, 701.
10. J. L. C. Rowsell and O. M. Yaghi, *Angew. Chem. Int. Ed.*, 2005, **44**, 4670.
11. S. Kitagawa and K. Uemura, *Chem. Soc. Rev.*, 2005, **34**, 109.
12. S. Kitagawa and R. Matsuda, *Coord. Chem. Rev.*, 2007, **251**, 2490.
13. S. Kitagawa, *Nature*, 2006, **441**, 584.
14. O. M. Yaghi, *Nature Mater.*, 2007, **6**, 92.
15. D. J. Tranchemontagne, J. L. Mendoza-Cortes, M. O'Keeffe and O. M. Yaghi, *Chem. Soc. Rev.*, 2009, **38**, 1257.
16. S. Horike, S. Shimomura and S. Kitagawa, *Nature Chem.*, 2009, **1**, 695.
17. J. R. Long and O. M. Yaghi, *Chem. Soc. Rev.*, 2009, **38**, 1213.
18. J. Seo, H. Sakamoto, R. Matsuda and S. Kitagawa, *J. Nanosci. Nanotechnol.*, 2010, **10**, 3.
19. K. Itaya, I. Uchida and V. D. Neff, *Acc. Chem. Res.*, 1986, **19**, 162.
20. M. Verdaguer, A. Bleuzen, V. Marvaud, J. Vaissermann, M. Seuleiman, C. Desplanches, A. Scuiller, C. Train, R. Garde, G. Gelly, C. Lomenech,

I. Rosenman, P. Veillet, C. Cartier and F. Villain, *Coord. Chem. Rev.*, 1999, **190–192**, 1023.

21. R. J. Mortimer, *Chem. Soc. Rev.*, 1997, **26**, 147.

22. P. M. S. Monk, R. J. Mortimer and D. R. Rosseinsky, *Electrochromism: Fundamentals and Applications*, Wiley-VCH, Weinheim, 1995.

23. M. Irie, *Chem. Rev.*, 2000, **100**, 1685.

24. H. Rau in *Photochemistry and Photophysics, Vol. 2*, ed. J. F. Rabek, CRC Press, Boca Raton, 1990, p 119.

25. C. Dugave and L. Demange, *Chem. Rev.*, 2003, **103**, 2475.

26. S. Kume and H. Nishihara, *Dalton Trans.*, 2008, 3260.

27. C.-C. Ko and V. W.-W. Yam, *J. Mater. Chem.*, 2010, **20**, 2063.

28. Y. Yu, M. Nakano and T. Ikeda, *Nature*, 2003, **425**, 145.

29. Y. S. Sohn, D. N. Hendrickson and H. B. Gray, *J. Am. Chem. Soc.*, 1971, **93**, 3603.

30. S. M. Scott and K. C. Gordon, *Inorg. Chim. Acta*, 1997, **254**, 267.

31. H. Tokoro and S. Ohkoshi, *Dalton Trans.*, 2011, **40**, 6825.

32. A. Dei, *Angew. Chem. Int. Ed.*, 2005, **44**, 1160.

33. O. Sato, *Acc. Chem. Res.*, 2003, **36**, 692.

34. O. Sato, J. Photochem. Photobiol. C, 2004, **5**, 203.

35. O. Sato, T. Iyoda, A. Fujishima and K. Hashimoto, *Science*, 1996, **272**, 704.

36. M. B. Robin, *Inorg. Chem.*, 1962, **1**, 337.

37. M. B. Robin and P. Day, *Adv. Inorg. Chem. Radiochem.*, 1967, **10**, 247.

38. P. Gütlich, A. Hauser and H. Spiering, *Angew. Chem. Int. Ed.*, 1994, **33**, 2024.

39. In *Spin crossover in Transition Metal Compounds I-III*, ed. P. Gütlich and H. A. Goodwin, Springer-Verlag, Berlin, 2004.

40. J. A. Real, A. B. Gaspara and M. C. Muñoz, *Dalton Trans.*, 2005, 2062.

41. J.-F. Létard, *J. Mater. Chem.*, 2006, **16**, 2550.

42. S. Decurtines, P. Gütlich, C. P. Kohler, H. Spiering and A. Hauser, *Chem. Phys. Lett.*, 1984, **105**, 1.

43. *Organic Photovoltaics: Materials, Device Physics, and Manufacturing Technologies*, ed. C. Brabec, U. Scherf, V. Dyakonov, Wiley-VCH, Weinheim, 2008.

44. A. W. Hains, Z. Liang, M. A. Woodhouse, B. A. Gregg, *Chem. Rev.*, 2010, **110**, 6689.

45. *Organic Light-emitting Devices: A Survey*, ed. J. Shinar, Springer, New York, 2003.

46. H. Sasabe and J. Kido, *Chem. Mater.*, 2011, **23**, 621.

47. *Highly Efficient OLEDs with Phosphorescent Materials*, ed. H. Yersin, Wiley-VCH, Weinheim, 2007.

48. L. Xiao, Z. Chen, B. Qu, J. Luo, S. Kong, Q. Gong and J. Kido, *Adv. Mater.*, 2011, **23**, 926.

49. P. D. W. Boyd, Q. Li, J. B. Vincent, K. Folting, H.-R. Chang, W. E. Streib, J. C. Huffman, G. Christou and D. N. Hendrickson, *J. Am. Chem. Soc.*, 1988, **110**, 8537.

50. A. Caneschi, D. Gatteschi and R. Sessoli, *J. Am. Chem. Soc.*, 1991, **113**, 5873.

51. R. Sessoli, H.-L. Tsai, A. R. Schake, S. Wang, J. B. Vincent, K. Folting, D. Gatteschi, G. Christou and D. N. Hendrickson, *J. Am. Chem. Soc.*, 1993, **115**, 1804.

52. G. Christou, D. Gatteschi, D. N. Hendrickson and R. Sessoli, *MRS Bull.*, 2000, **25**, 66.

53. D. Gatteschi and R. Sessoli, *Angew. Chem., Int. Ed.*, 2003, **42**, 268.

54. S. K. Ritter, *Chem. Eng. News*, 2004, **82**, 29.

55. D. Gatteschi, R. Sessoli and J. Villain, *Molecular Nanomagnets*, Oxford University Press, New York, 2006

56. H. Miyasaka, M. Yamashita, *Dalton Trans.*, 2007, 399.

57. E. D. Dahlberg and J.-G. Zhu, *Phys. Today*, 1995, **48**, 34.

58. J. R. Friedman, M. P. Sarachik, J. Tejada, J. Maciejewski and R. Ziolo, *J. Appl. Phys.*, 1996, **79**, 6031.

59. L. Thomas, F. Lionti, R. Ballou, D. Gatteschi, R. Sessoli and B. Barbara, *Nature*, 1996, **383**, 145.

60. J. R. Friedman, M. P. Sarachik, J. Tejada and R. Ziolo, *Phys. Rev. Lett.*, 1996, **76**, 3830

61. J. R. Friedman, M. P. Sarachik, J. M. Hernández, X. X. Zhang, J. Tejada, E. Molins and R. Ziolo, *J. Appl. Phys.*, 1997, **81**, 3978

62. M. N. Leuenberger and D. Loss, *Nature*, 2001, **410**, 789.

63. M. Morimoto, H. Miyasaka, M. Yamashita and M. Irie, *J. Am. Chem. Soc.*, 2009, **131**, 9823.

64. T. Shiga, H. Miyasaka, M. Yamashita, M. Morimoto and M. Irie, *Dalton Trans.*, 2011, **40**, 2275.

65. M. Kurihara, A. Hirooka, S. Kume, M. Sugimoto and H. Nishihara, *J. Am. Chem. Soc.*, 2002, **124**, 8800.

66. A. Sakamoto, A. Hirooka, K. Namiki, M. Kurihara, M. Murata, M. Sugimoto and H. Nishihara, *Inorg. Chem.*, 2005, **44**, 7547.

67. K. Namiki, A. Sakamoto, M. Murata, S. Kume and H. Nishihara, *Chem. Commun.*, 2007, 4650.

68. K. Namiki, M. Murata, S. Kume and H. Nishihara, *New J. Chem.*, 2011, **35**, 2146.

69. *Ferrocenes*, ed. A. Togni and T. Hayashi, VCH Publishers, New York, 1995.

70. A. J. Fry, R. S. H. Liu and G. S. Hammond, *J. Am. Chem. Soc.*, 1966, **88**, 4781.

71. E. J. Lee and M. S. Wrighton, *J. Am. Chem. Soc.*, 1991, **113**, 8562.

72. M. Kikuchi, K. Kikuchi and H. Kokubun, *Bull. Chem. Soc. Jpn.*, 1974, **47**, 1331.

73. S. Nagashima, M. Murata and H. Nishihara, *Angew. Chem. Int. Ed.*, 2006, **45**, 4298.
74. S. Shinkai, T. Ogawa, Y. Kusano, O. Manabe, K. Kikukawa, T. Goto and T. Matsuda, *J. Am. Chem. Soc.*, 1982, **104**, 1960.
75. S. Shinkai, T. Minami, Y. Kusano and O. Manabe, *J. Am. Chem. Soc.*, 1983, **105**, 1851.
76. Z. Wang, A.-M. Nygård, M. J. Cook and D. A. Russell, *Langmuir*, 2004, **20**, 5850.
77. J. Zhang, M. Riskin, R. Tel-Vered, H. Tian and I. Willner, *Langmuir*, 2011, **27**, 1380.
78. M. Riskin, V. Gutkin, I. Felner and I. Willner, *Angew. Chem. Int. Ed.*, 2008, **47**, 4416.
79. S. Kume, M. Murata, T. Ozeki and H. Nishihara, *J. Am. Chem. Soc.*, 2005, **127**, 8800.
80. S. Venkataramani, U. Jana, M. Dommaschk, F. D. Sönnichsen, F. Tuczek and R. Herges, *Science*, 2011, **331**, 445.
81. K. Kimura, H. Sakamoto and R. M. Uda, *Macromolecules*, 2004, **37**, 1871.
82. K. Kimura, T. Yamashita and M. Yokoyama, *J. Chem. Soc. Perkin Trans. 2*, 1992, 613.
83. K. Kimura, M. Sumida and M. Yokoyama, *Chem. Commun.*, 1997, 1417.
84. *Polymers for Second-Order Nonlinear Optics*, G. A. Lindsay and K. D. Singer (eds), American Chemical Society, 1995.
85. D. R. Kanis, M. A. Ratner and T. J. Marks, *Chem. Rev.*, 1994, **94**, 195.
86. G. S. He, L.-S. Tan, Q. Zheng and P. N. Prasad, *Chem. Rev.*, 2008, **108**, 1245.
87. M. L. H. Green, S. R. Marder, M. E. Thompson, J. A. Bandy, D. Bloor, P. V. Kolinsky and R. J. Jones, *Nature*, 1987, **330**, 360.
88. M. Albota, D. Beljonne, J.-L. Brédas, J. E. Ehrlich, J.-Y. Fu, A. A. Heikal, S. E. Hess, T. Kogej, M. D. Levin, S. R. Marder, D. McCord-Maughon, J. W. Perry, H. Röckel, M. Rumi, G. Subramaniam, W. W. Webb, X.-L. Wu and C. Xu, *Science*, 1998, **281**, 1653.
89. K. D. Singer, J. E. Sohn and S. J. Lalama, *Appl. Phys. Lett.*, 1986, **49**, 248.
90. M. Eich, A. Sen, H. Looser, G. C. Bjorklund, J. D. Swalen, R. Tweig and D. Y. Yoon, *J. Appl. Phys.*, 1989, **66**, 2559.
91. K. D. Singer, M. G. Kuzyk, W. R. Holland, J. E. Sohn, S. J. Lalama, R. B. Comizzoli, H. E. Katz and M. L. Schillin, *Appl. Phys. Lett.*, 1988, **53**, 1800.
92. M. A. Mortazavi, A. Knoesen, S. T. Kowel, B. G. Higgins and A. Dienes, *J. Opt. Soc. Am. B.*, 1989, **6**, 733.
93. D. M. Burland, R. D. Miller and C. A. Walsh, *Chem. Rev.*, 1994, **94**, 31.
94. L. Viau, S. Bidault, O. Maury, S. Brasselet, I. Ledoux, J. Zyss, E. Ishow, K. Nakatani and H. Le Bozec, *J. Am. Chem. Soc.*, 2004, **126**, 8386.
95. F. S. Han, M. Higuchi and D. G. Kurth, *Adv. Mater.*, 2007, **19**, 3928.

96. F. S. Han, M. Higuchi and D. G. Kurth, *J. Am. Chem. Soc.*, 2008, **130**, 2073.
97. Higuchi, *J. Nanosci. Nanotech.*, 2009, **9**, 51.
98. R. R. Pal, M. Higuchi and D. G. Kurth, *Org. Lett.*, 2009, **11**, 3562.
99. C. Creutz and H. Taube, *J. Am. Chem. Soc.*, 1973, **95**, 1086.
100. H. Taube, *Angew. Chem., Int. Ed. Engl.*, 1984, **23**, 329.
101. N. S. Hush, *Coord. Chem. Rev.*, 1985, **64**, 135.
102. W. Kaim, V. Kasack, H. Binder, E. Roth and J. Jordanov, *Angew. Chem. Int. Ed. Engl.*, 1988, **27**, 1174.
103. V. Kasack, W. Kaim, H. Binder, J. Jordanov and E. Roth, *Inorg. Chem.*, 1995, **34**, 1924.
104. Y. Qi and Z. Y. Wang, *Macromolecules*, 2003, **36**, 3146.
105. S. Wang, X. Li, S. Xun, X. Wan and Z. Y. Wang, *Macromolecules*, 2006, **39**, 7502.
106. M. F. Rastegar, E. K. Todd, H. Tang and Z. Y. Wang, *Org. Lett.*, 2004, **6**, 4519.
107. J. F. Keggin and F. D. Miles, *Nature*, 1936, **137**, 577.
108. A. Ludi and H. U. Güdel, *Struct. Bonding*, 1973, **14**, 1.
109. J. C. Wojdeł and S. T. Bromley, *Chem. Phys. Lett.*, 2004, **397**, 154 and references therein.
110. V. D. Neff, *J. Electrochem. Soc.*, 1978, **125**, 886.
111. R. J. Mortimer and D. R. Rosseinsky, *J. Chem. Soc., Dalton Trans.*, 1984, 2059.
112. K. Itaya, T. Ataka and S. Toshima, *J. Am. Chem. Soc.*, 1982, **104**, 4767.
113. K. Itaya, K. Shibayama, H. Akahoshi and S. Toshima, *J. Appl. Phys.*, 1982, **53**, 804.
114. K. Honda, J. Ochiai and H. Hayashi, *J. Chem. Soc., Chem. Commun.*, 1986, 168.
115. M. K. Carpenter and R. S. Conell, *J. Electrochem. Soc.*, 1990, **137**, 2464.
116. N. R. de Tacconi, K. Rajeshwar and R. O. Lezna, *Chem. Mater.*, 2003, **15**, 3046.
117. A. Gotoh, H. Uchida, M. Ishizaki, T. Satoh, S. Kaga, S. Okamoto, M. Ohta, M. Sakamoto, T. Kawamoto, H. Tanaka, M. Tokumoto, S. Hara, H. Shiozaki, M. Yamada, M. Miyake and M. Kurihara, *Nanotechnology* 2007, **18**, 345609.
118. S. Hara, H. Tanaka, T. Kawamoto, M. Tokumoto, M. Yamada, A. Gotoh, H. Uchida, M. Kurihara and M. Sakamoto, *Jpn. J. Appl. Phys.*, 2007, **46**, L945.
119. S. Hara, H. Shiozaki1, A. Omura, H. Tanaka, T. Kawamoto, M. Tokumoto, M. Yamada, A. Gotoh, M. Kurihara and M. Sakamoto, *Appl. Phys. Exp.*, 2008, **1**, 104002-1.
120. H. Shiozaki, S. Hara, M. Tokumoto, A. Gotoh, T. Satoh, M. Ishizaki, M. Kurihara and M. Sakamoto, *Jpn. J. Appl. Phys.*, 2008, **47**, 1242.
121. D. M. DeLongchamp and P. T. Hammond, *Chem. Mater.*, 2004, **16**, 4799.

122. A. Agrawal, C. Susut, G. Stafford, U. Bertocci, B. McMorran, H. J. Lezec and A. A. Talin, *Nano Lett.*, 2011, **11**, 2774.

123. H. J. Buser, D. Schwarzenbach, W. Petter and A. Ludi, *Inorg. Chem.*, 1977, **16**, 2704.

124. D. Babel, *Comments Inorg. Chem.*, 1986, **5**, 285.

125. W. D. Griebler and D. Babel, *Z. Naturforsch. Teil B*, 1982, **37**, 832.

126. W.R. Entley and G. S. Girolami, *Inorg. Chem.*, 1994, **33**, 5165.

127. E. Ruiz, A. Rodríguez-Fortea, S. Alvarez and M. Verdaguer, *Chem. Eur. J.*, 2005, **11**, 2135.

128. H. Tokoro, M. Shiro, K. Hashimoto and S. Ohkoshi, *Z. Anorg. Allg. Chem.*, 2007, **633**, 1134.

129. O. Sato, Y. Einaga, T. Iyoda, A. Fujishima and K. Hashimoto, *J. Phys. Chem. B*, 1997, **101**, 3903.

130. N. Shimamoto, S. Ohkoshi, O. Sato, K. Hashimoto, *Mol. Cryst. Liq. Cryst.*, 2000, **344**, 95.

131. N. Shimamoto, S. Ohkoshi, O. Sato and K. Hashimoto, *Inorg. Chem.*, 2002, **41**, 678.

132. H.-W. Liu, K. Matsuda, Z.-Z. Gu, K. Takahashi, A.-L. Cui, R. Nakajima, A. Fujishima and O. Sato, *Phys. Rev. Lett.*, 2003, **90**, 167403-1.

133. O. Sato, Y. Einaga, T. Iyoda, A. Fujishima and K. Hashimoto, *J. Electrochem. Soc.*, 1997, **144**, L11.

134. Y. Einaga, O. Sato, T. Iyoda, K. Kobayashi, F. Ambe, K. Hashimoto and A. Fujishima, *Chem. Lett.*, 1997, **3**, 289.

135. O. Sato, Y. Einaga, T. Iyoda, A. Fujishima and K. Hashimoto, *Inorg. Chem.*, 1999, **38**, 4405.

136. T. Yokoyama, T. Ohta, T, O. Sato and K. Hashimoto, *Phys. Rev. B*, 1998, **58**, 8257.

137. T. Yokoyama, M. Kiguchi, T. Ohta, O. Sato, Y. Einaga and K. Hashimoto, *Phys. Rev. B*, 1999, **60**, 9340.

138. S. Ohkoshi, H. Tokoro, M. Utsunomiya, M. Mizuno, M. Abe and K. Hashimoto, *J. Phys. Chem. B*, 2002, **106**, 2423.

139. H. Tokoro, S. Ohkoshi, T. Matsuda and K. Hashimoto, *Inorg. Chem.*, 2004, **43**, 5231.

140. Y. Moritomo, M. Hanawa, Y. Ohishi, K. Kato, M. Takata, A. Kuriki, E. Nishibori, M. Sakata, S. Ohkoshi, H. Tokoro and K. Hashimoto, *Phys. Rev. B: Condens. Matter*, 2003, **68**, 144106.

141. H. Osawa, T. Iwazumi, H. Tokoro, S. Ohkoshi, K. Hashimoto, H. Shoji, E. Hirai, T. Nakamura, S. Nanao and Y. Isozumi, *Solid State Commun.*, 2003, **125**, 237.

142. H. Tokoro, S. Ohkoshi, T. Matsuda and K. Hashimoto, *Inorg. Chem.*, 2004, **43**, 5231.

143. H. Tokoro, T. Matsuda, T. Nuida, Y. Moritomo, K. Ohoyama, E. D. L. Dangui, K. Boukheddaden and S. Ohkoshi, *Chem. Mater.*, 2008, **20**, 423.

144. S. Ohkoshi, H. Tokoro and K. Hashimoto, *Coord. Chem. Rev.*, 2005, **249**, 1830.

145. K. Kato, Y. Moritomo, M. Sakata, M. Umekawa, N. Hamada, S. Ohkoshi, H. Tokoro and K. Hashimoto, *Phys. Rev. Lett.*, 2003, **91**, 255502.
146. H. Tokoro, S. Ohkoshi and K. Hashimoto, *Appl. Phys. Lett.*, 2003, **82**, 1245.
147. S. Ohkoshi, T. Nuida, T. Matsuda, H. Tokoro and K. Hashimoto, *J. Mater. Chem.*, 2005, **15**, 3291.
148. S. Ohkoshi, H. Tokoro, T. Matsuda, H. Takahashi, H. Irie and K. Hashimoto, *Angew. Chem. Int. Ed.*, 2007, **46**, 3238.
149. H. Tokoro and S. Ohkoshi, *Appl. Phys. Lett.*, 2008, **93**, 021906.
150. T. Mahfoud, G. Molńar, S. Bonhommeau, S. Cobo, L. Salmon, P. Demont, H. Tokoro, S. Ohkoshi, K. Boukheddaden and A. Bousseksou, *J. Am. Chem. Soc.*, 2009, **131**, 15049.
151. C. Avendano, M. G. Hilfiger, A. Prosvirin, C. Sanders, D. Stepien and K. R. Dunbar, *J. Am. Chem. Soc.*, 2010, **132**, 13123.
152. Y. Arimoto, S. Ohkoshi, Z. J. Zhong, H. Seino, Y. Mizohe and K. Hashimoto, *J. Am. Chem. Soc.*, 2003, **125**, 9240.
153. S. Ohkoshi, Y. Hamada, T. Matsuda, Y. Tunobuchi and H. Tokoro, *Chem. Mater.*, 2008, **20**, 3048.
154. G. Rombaut, M. Verelst, S. Golhen, L. Ouahab, C. Mathonière and O. Kahn, *Inorg. Chem.*, 2001, **40**, 1151.
155. S. Ohkoshi, N. Machida, Y. Abe, Z. J. Zhong and K. Hashimoto, *Chem. Lett.*, 2001, **30**, 312.
156. J. M. Herrera, V. Marvaud, M. Verdaguer, J. Marrot, M. Kalisz and C. Mathonière, *Angew. Chem. Int. Ed.*, 2004, **43**, 5468.
157. S. Ohkoshi, H. Tokoro, T. Hozumi, Y. Zhang, K. Hashimoto, C. Mathoniere, I. Bord, G. Rombaut, M. Verelst, C. C. D. Moulin and F. Villain, *J. Am. Chem. Soc.*, 2006, **128**, 270.
158. S. Ohkoshi, K. Imoto, Y. Tsunobuchi, S. Takano and H. Tokoro, *Nat. Chem.*, 2011, *3*, 564.
159. C. J. Brabec, S. Gowrisanker, J. J. M. Halls, D. Laird, S. Jia and S. P. Williams, *Adv. Mater.*, 2010, 22, 3839 and comprehensive references therein.
160. L. J. A. Koster, V. D. Mihailetchi and P. W. Bloom, *Appl. Phys. Lett.*, 2006, **88**, 093511.
161. M. C. Scharber, D. Mühlbacher, M. Koppe, P. Denk, C. Waldauf, A. J. Heeger and C. J. Brabec, *Adv. Mater.*, 2006, **18**, 789.
162. W.-Y. Wong, X.-Z. Wang, Z. He, A. B. Djurišić, C.-T. Yip, K.-Y. Cheung, H. Wang, C. S. K. Mak and W.-K. Chan, *Nature Mater.*, 2007, **6**, 521.
163. X.-Z. Wang, W.-Y. Wong, K.-Y. Cheung, M.-K. Fung, A. B. Djurišić and W.-K. Chan, *Dalton Trans.*, 2008, 5484.
164. Y. Fujikura, K. Sonogashira and N. Hagihara, *Chem. Lett.*, 1975, **4**, 1067.

165. K. Sonogashira, S. Takahashi and N. Hagihara, *Macromolecules*, 1977, **10**, 879.

166. S. Takahashi, M. Kariya, T. Yatake, K. Sonogashira and N. Hagihara, *Macromolecules*, 1978, **11**, 1063.

167. W.-Y. Wong and C.-L. Ho, *Coord. Chem. Rev.*, 2007, **251**, 2400.

168. W.-Y. Wong, *Dalton Trans.*, 2007, 4495.

169. J. S. Wilson, A. S. Dhoot, A. J. A. B. Seeley, M. S. Khan, A. Köhler and R. H. Friend, *Nature*, 2001, **413**, 828.

170. W.-Y. Wong, G.-J. Zhou, Z. He, K.-Y. Cheung, A. M.-C. Ng, A. B. Djurišić and W.-K. Chan, *Macromol. Chem. Phys.*, 2008, **209**, 1319.

171. C.-L. Ho, C.-H. Chui, W.-Y. Wong, S. M. Aly, D. Fortin, P. D. Harvey, B. Yao, Z. Xie and L. Wang, *Macromol. Chem. Phys.*, 2009, **210**, 1786.

172. G.-J. Zhou and W.-Y. Wong, *Chem. Soc. Rev.*, 2011, **40**, 2541.

173. F. Guo, Y.-G. Kim, J. R. Reynolds and K. S. Schanze, *Chem. Commun.*, 2006, 1887.

174. G. L. Schulz and S. Holdcroft, *Chem. Mater.*, 2008, **20**, 5351.

175. *WOLEDs and Organic Photovoltaics: Recent Advances and Applications*, ed. V. W.-W. Yam, Springer-Verlag, Berlin, 2010.

176. M. Strukelj, R. H. Jordan and A. Dodabalapur, *J. Am. Chem. Soc.*, 1996, **118**, 1213.

177. S. Tasch, E. J. W. List, O. Ekström, W. Graupner, G. Leising, P. Schlichting, U. Rohr, Y. Geerts, U. Scherf and K. Müllen, *Appl. Phys. Lett.*, 1997, **71**, 2883.

178. Z. Xie, J. S. Huang, C. N. Li, Y. Wang, Y. Q. Li and J. Shen, *Appl. Phys. Lett.*, 1999, **74**, 641.

179. C. W. Ko and Y. T. Tao, *Appl. Phys. Lett.*, 2001, **79**, 4234.

180. B. W. D'Andrade, M. E. Thompson and S. R. Forrest, *Adv. Mater.*, 2002, **14**, 147.

181. X. Gong, S. Wang, D. Moses, G. C. Bazan and A. J. Heeger, *Adv. Mater.*, 2005, **17**, 2053.

182. J. Y. Li, D. Liu, C. Ma, O. Lengyel, C.-S. Lee, C.-H. Tung and S. Lee, *Adv. Mater.*, 2004, **16**, 1538.

183. Y.-Z. Lee, X. Chen, M.-C. Chen and S.-A. Chen, *Appl. Phys. Lett.*, 2001, **79**, 308.

184. P. T. Furuta, L. Deng, S. Garon, M. E. Thompson and J. M. J. Fréchet, *J. Am. Chem. Soc.*, 2004, **126**, 15388.

185. M. Mazzeo, V. Vitale, F. D. Sala, M. Anni, G. Barbarella, L. Favaretto, G. Sotgiu, R. Cingolani and G. Gigli, *Adv. Mater.*, 2005, **17**, 34.

186. S. K. Lee, D. H. Hwang, B. J. Jung, N. S. Cho, J. Lee, J. D. Lee and H. K. Shim, *Adv. Funct. Mater.*, 2005, **15**, 1647.

187. J. Liu, Q. Zhou, Y. Cheng, Y. Geng, L. Wang, D. Ma, X. Jing and F. Wang, *Adv. Mater.*, 2005, **17**, 2974.

188. G. Tu, C. Mei, Q. Zhou, Y. Cheng, Y. Geng, L. Wang, D. Ma, X. Jing and F. Wang, *Adv. Funct. Mater.*, 2006, **16**, 101.

189. J. Liu, Y. Cheng, Z. Xie, Y. Geng, L. Wang, X. Jing and F. Wang, *Adv. Mater.*, 2008, **20**, 1357.

190. G. Zhoua, W.-Y. Wong and S. Suoc, *J. Photochem. Photobiol. C*, 2010, **11**, 133.

191. J. Jiang, Y. Xu, W. Yang, R. Guan, Z. Liu, H. Zhen and Y. Cao, *Adv. Mater.*, 2006, **18**, 1769.

192. M. A. Baldo, D. F. O'Brien, Y. You, A. Shoustikov, S. Sibley, M. E. Thompson and S. R. Forrest, *Nature*, 1998, **395**, 151.

193. F.-I. Wu, X.-H. Yang, D. Neher, R. Dodda, Y.-H. Tseng and C.-F. Shu, *Adv. Funct. Mater.*, 2007, **17**, 1085.

194. C.-H. Chien, S.-F. Liao, C.-H. Wu, C.-F. Shu, S.-Y. Chang, Y. Chi, P.-T. Chou and C.-H. Lai, *Adv. Funct. Mater.*, 2008, **18**, 1430.

195. C. Wang, Z. Xie, K. E. deKrafft and W. Lin, *J. Am. Chem. Soc.*, 2011, **133**, 2056.

196. Y. K. Park, S. B. Choi, H. Kim, K. Kim, B.-H. Won, K. Choi, J.-S. Choi, W.-S. Ahn, N. Won, S. Kim, D. H. Jung, S.-H. Choi, G.-H. Kim, S.-S. Cha, Y. H. Jhon, J. K. Yang and J. Kim, *Angew. Chem. Int. Ed.*, 2007, **46**, 8230.

197. Z. Wang and S. M. Cohen, *Chem. Soc. Rev.*, 2009, **38**, 1315.

198. H. Sato, R. Matsuda, K. Sugimoto, M. Takata and S. Kitagawa, *Nature Mater.*, 2010, **9**, 661.

199. N. P. Gritsan and M. S. Platz, *Chem. Rev.*, 2006, **106**, 3844.

Molecular Design and Synthesis of Photofunctional Materials

KEITH MAN-CHUNG WONG AND
VIVIAN WING-WAH YAM*

[†] Institute of Molecular Functional Materials and Department of Chemistry,
The University of Hong Kong, Pokfulam Road, Hong Kong
*E-mail: wwyam@hku.hk
[†] Areas of Excellence Scheme, University Grants Committee, Hong Kong

7.1 Introduction

The top-down fabrication approach, in particular of silicon-based technologies, is one of the most currently used techniques for the manufacture of products through the control of nanostructure.[1] Such a conventional approach for the generation of nanoscale devices usually requires microfabrication methods with larger, externally controlled tools in order to direct the materials into the assembly with desired shape and order. However, the inherent requirement of externally controlled microfabrication technique limits the ultimate development of this physical approach. On the other hand, chemical bottom-up processes represent the alternative approach for the fabrication of materials with optimum performances. In this approach, the chemical properties of single-molecular building block are first specified in great detail and such small components with predesigned functions at the molecular level might possess properties suitable for their applications such as in optoelectronic devices.[2]

RSC Polymer Chemistry Series No. 2
Molecular Design and Applications of Photofunctional Polymers and Materials
Edited by Wai-Yeung Wong and Alaa S Abd-El-Aziz

Published by the Royal Society of Chemistry, www.rsc.org

Optoelectronics involves the study and application of devices that possess the function of electrical-to-optical or optical-to-electrical conversions in their operations. Light-emitting diodes and solar cells/photovoltaics are representative devices of optoelectronic applications. However, the use of semiconducting materials such as silicon or gallium arsenide has been dominating this research field and application for many years. Since the first demonstration of efficient light-emitting diodes in multilayer thin films by Tang and van Slyke[3] in the 1980s, there have been tremendous growing interest in further development and exploitation of organic small molecules and organic polymer conductors as active materials in a new generation of optoelectronic devices such as organic light-emitting diodes (OLEDs)[4–10] and organic photovoltaics/solar cells.[11,12] The new efficient and potentially low-cost technique using thin-film processing for the fabrication of organic-based devices, which is not applicable for semiconductors, provides a huge potential for this research area to be developed for practical application. In view of the fact that the modification of chemical structure of the organic compounds with various molecular structures and functional groups is anticipated to readily tune their chemical and physical properties, such organic materials with great versatility have also been investigated for their application in nonlinear optics[13] and other optoelectronic devices.[14] The rational design and synthesis of molecular materials have attracted growing attention in recent years owing to the enormous and unpredictable potentials in their functional properties that could lead to the development of advanced technology.

Photofunctional materials and devices, which possess electroluminescence, solar energy conversion, nonlinear optical and photosensitising properties, are one of the most important research topics since the past decade. The development of key molecules by synthetic chemists is challenging and may represent the solution to the search for new functional materials. The incorporation of transition-metal centres into organic moieties to yield the corresponding transition-metal coordination and organometallic complexes, which represent an important class of building block, provides one of the promising strategies for the exploration of photofunctional materials. With the versatility and diversity of different transition-metal–ligand chromophores, extension of the absorption of the materials towards different spectral region would be readily achieved upon introduction of transition-metal centres. In addition, the polarisability could be readily tuned in the donor–acceptor system with various transition-metal centres of different oxidation states to give nonlinear optical properties.[15,16,17a–g] More readily accessible triplet excited states are anticipated due to relaxation of the spin selection rule through the spin-orbit coupling upon incorporation of heavy-metal centres into organic ligands. As a result, longer-lived phosphorescence derived from the triplet excited states would provide more important practical photofunctional applications in such metal-based materials.

Superior to the pure organic counterparts, the greater diversity and versatility of ligand design and synthesis as well as the variety of the different structural and bonding modes of metal complexation, offer the transition-metal–ligand systems a wide variety of attractive photofunctions with tunable electronic absorption, emission, excited-state redox and nonlinear optical properties. Through the systematic study on the variation of metal centres, molecular structures, bonding modes and ligands with different functionalities, the structure–property relationships of the various classes of transition-metal complexes could be established. By combining the advantages of high synthetic diversity and the understanding of the structure–property relationship established, molecular photofunctional materials based on the transition-metal complexes could be prepared by the approach of "function by design". The chemistry of metal alkynyls has attracted enormous attention, in particular, with the emerging interest in their potential applications in the field of materials science. The linear geometry and rigid structure of the alkynyl unit and its π-unsaturated nature have made metal alkynyls attractive building blocks, not only through σ-bonding but also *via* π-bonding interactions. The incorporation of alkynyl groups into transition-metal complexes may also lead to systems with unique optical, electronic and physical properties that may find potential applications as nonlinear optical materials, liquid crystals, molecular electronics and wires.[17] In this chapter, the design, synthesis and photophysical properties of several classes of alkynylplatinum(II) and alkynylgold(III) complexes prepared in our laboratory will be discussed. Their photofunctional properties, mainly including the two-photon absorption (TPA) and two-photon-induced luminescence (TPIL) nonlinear optical (NLO) properties, electroluminescence (EL) of organic light-emitting devices (OLEDs) and solar energy conversion of dye-sensitised solar cells (DSSC), of respective systems will also be highlighted.

7.2 Synthesis, Photophysical and Nonlinear Optical (NLO) Properties of Platinum(II) Bisphosphino Bis-Alkynyl Complexes

7.2.1 Background

Nonlinear optical (NLO) response refers to the change of electromagnetic field with different phase, frequency, amplitude, polarisation, and path from the incident light by the interaction of materials possessing NLO properties.[18] Nonlinear optics remained unexplored until the discovery of second-harmonic generation[19] shortly after the first advent of laser in 1960,[20] due to the requirement of such nonlinear optical observation in very high intensity of monochromatic light. In light of the potential applications including optical signal processing, optical data storage, optical communication and image processing, there have been extensive studies on the exploration and exploitation of NLO materials. Inorganic crystals, which are advantageous from the perspectives of their large transparency range, high stability as well as low optical loss, are the most widely used NLO materials for frequency mixing and third-order nonlinear processes.[18] On the other hand, many organic molecules and polymers have been found to possess NLO properties. Because of the synthetic versatility, ease of structural modification and thin-film formation, there have been numerous reports on the investigations of organic NLO materials.[21] In view of the greater synthetic diversity through the variations of ligand functionalities, metal centres, structural geometries, together with the presence of intramolecular charge-transfer transitions, related studies on NLO behaviours of the transition-metal coordination and organometallic complexes with such additional merits have attracted significant recent interest.[15,16,17a–g]

Two-photon absorption (TPA) is one of the NLO phenomena. In this process, a molecule is excited from the ground state to an excited state through the simultaneous absorption of two photons, while the energy difference between the ground state and the excited state is equal to the sum of the energies of these two photons. Such two-photon absorption is much weaker than the linear absorption and therefore the first direct experimental verification was only possible with the invention of the laser that was 30 years after the prediction of TPA by Maria Goeppert-Mayer in 1931.[22] Two-photon induced luminescence (TPIL) is the emission of light from the transition of excited state to the ground state, through the TPA process, and therefore the energy of emission is usually higher than that of absorption. There has been a growing interest in the development of TPA materials due to the potential applications, such as in optical data storage,[23] frequency upconverted lasing,[24] power limiting,[25] three-dimensional fluorescence imaging,[26] three-dimensional microfabrication,[27] nonlinear photonics,[28] pulse reshaping and stabilisation,[29] and photodynamic therapy.[30] Many research groups have focused their efforts on the design and synthesis of extended π-conjugated organic molecules with large two-photon absorption (TPA) cross section (σ_2) values, on the basis of the requirement of large differences in polarisation upon excitation. Organic materials that show two-photon activity usually achieve large polarisation changes by means of the structural modification of dipolar or quadrupolar molecules,[31,32] such as using the acceptor-π-donor, acceptor-π-acceptor; donor-π-donor; acceptor-π-donor-π-acceptor and donor-π-acceptor-π-donor approach. In addition, other extended π-conjugated star-shaped or octupolar multibranched molecules, such as 2,5-diphenyl-1,3,4-oxadiazole,[33] 1,3,5-tricyanobenzene,[34] 4,4′,4″-trisubstituted triphenylamine,[32a,35] 1,3,5-trisubstituted benzene,[36] 10,15-dihydro-5H-diindeno[1,2-a:1′,2′-c]fluorene (truxene),[37] have been reported to show very large two-photon absorption cross sections with remarkable improvements in the TPA properties. Preparation of novel materials with large σ_2 values could be achieved by such molecular design strategies.

In addition to the TPA studies of organic molecules, upconversion of light using transition metal coordination and organometallic complexes for potential NLO applications has attracted great attention in recent years.[38] The incorporation of organic electron-donor or electron-acceptor moieties through the formation of relatively strong metal–ligand bond has been found to be a promising strategy toward novel materials with NLO properties. Such hybrid systems consist of a metal centre coordinated to organic two-photon active ligands, in which the metal centre could provide a multidimensional template for increasing the number of NLO components. The metal centre could have an additional NLO effect owing to the possible involvement or modulation of the intramolecular charge-transfer process that governs the two-photon absorption process. A number of organotransition-metal complexes,[17d,39–41] including iron(II), nickel(II), ruthenium(II), manganese(I) and gold(I), have been reported by various research groups to possess interesting NLO responses.

Alkynylplatinum(II) phosphine complexes, containing two phosphine ligands and two alkynyl groups with square-planar geometry, exist in *cis-* or *trans*-configuration depending on the phosphine type and reaction conditions. A number of monomeric, oligomeric and polymeric platinum(II) compounds in *trans*-configuration, in which the platinum(II) metal centre and the two alkynyl ligands adopt a linear geometry, have been reported in view of their ideal conformations as potential candidates in the construction of molecular rods or wires.[42–44] Such platinum(II) alkynyl polymers and molecular materials represent one of the recent foci for the study of metal-containing NLO materials due to their strong excited-state absorptions in the visible range and relatively large spin-orbit coupling.[42g,42i,45] However, most of these complexes are mononuclear and are structurally related to the backbone of 4-(phenylethynyl)phenyl)ethyne. Incorporation of electron-donating or -accepting functionalities into this class of complexes for the enhancement of σ_2 values has also been demonstrated.[42i,45b–d] Although there have been extensive studies on dendritic or branched compounds, such as those based on oxadiazole, triphenylamine and truxene derivatives, in organic NLO materials, the use of multinuclear platinum(II) alkynyls with dendritic or branched backbones in the field of NLO studies has been relatively neglected and unexplored.[42i,46] As an extension of our research interests on the transition-metal alkynyl systems, a number of multinuclear dendritic or branched luminescent platinum(II) alkynyl complexes, with the use of various organic building blocks possessing NLO properties as the central core and/or peripheral ligands, have been designed and synthesised in our laboratory for exploration of their NLO activities. A series of branched luminescent multinuclear alkynylplatinum(II) complexes as multiphoton absorbing materials for two-photon absorption (TPA) and two-photon-induced luminescence (TPIL) will be discussed.

7.2.2 Synthesis and Characterisation

Reaction of organic alkynes and the complex precursor, *trans*-[Pt(PEt$_3$)$_2$Cl$_2$], using copper-catalyzed dehydrohalogenation reaction is the most common method to produce the platinum(II) alkynyl complexes, *trans*-[Pt(PEt$_3$)$_2$(C≡CR)$_2$].[42–44] Such synthetic methodology, involving the use of an amine as the base together with the trace amount of copper(I) halide, was developed by Hagihara and coworkers for the preparation of the first soluble transition metal alkynyl polymers of palladium(II) and platinum(II).[57] Extension of the corresponding alkynylplatinum(II) oligomers and polymers containing various bridging spacers, together with the study of their photophysical properties, have been extensively made by Lewis, Raithby, Wong and coworkers.[43,44] In our studies, the reaction of oligo-alkynes and complex precursor, *trans*-[Pt(PEt$_3$)$_2$Cl$_2$], for the construction of multinuclear platinum(II) alkynyls under similar reaction conditions afforded the corresponding insoluble polymeric materials. In order to prepare the multinuclear platinum(II) complexes with two nonidentical alkynyls, as well as to avoid the formation of such polymeric materials, a stepwise synthetic methodology involving two subsequent copper-catalyzed dehydrohalogenation reactions is employed *via* the formation of the chloroplatinum(II) monoalkynyl complex, *trans*-[Pt(PEt$_3$)$_2$(C≡CR)Cl],[45] which serves as a more flexible and versatile intermediate in both convergent and divergent approaches (Scheme 7.1). By utilisation of such a stepwise synthetic methodology, the dinuclear and trinuclear platinum(II) alkynyl complexes in branched backbones, **1–14**, have been successfully synthesised through the divergent approach with the respective dinuclear and trinuclear chloroplatinum(II) alkynyl intermediates

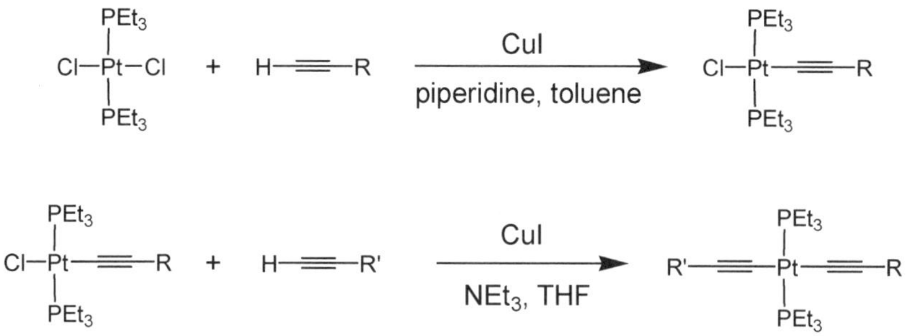

Scheme 7.1   Stepwise synthetic methodology for the preparation of unsymmetrical platinum(II) alkynyls of *trans*-[Pt(PEt$_3$)$_2$(C≡CR)(C≡CR′)].

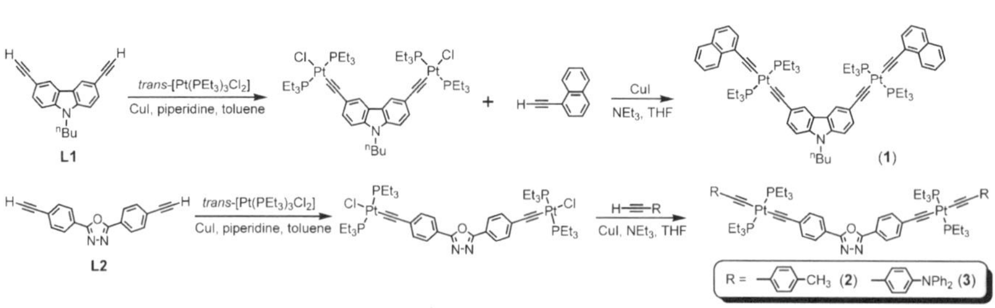

Scheme 7.2   Synthetic route of dinuclear platinum(II) alkynyls with V-shaped branched backbone.

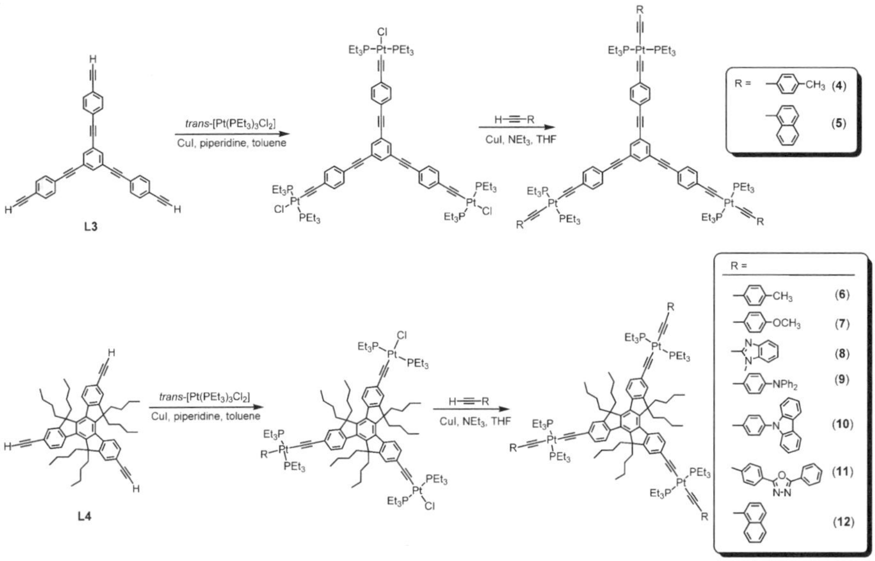

Scheme 7.3 Synthetic route of symmetrical trinuclear platinum(II) alkynyls with branched backbone.

(Scheme 7.2–7.4).[49–52] It is interesting to note that the unsymmetrical trinuclear complexes, **13** and **14**, which may provide larger dipole change with various substituents of electron donor and acceptor, were also prepared with the aid of the triisopropylsilyl protecting group (Scheme 7.4).[53] The tetranuclear platinum(II) alkynyl complexes **15** and **16** were synthesised through the convergent approach, while the preparation of the extended version of tetranuclear platinum(II) alkynyl complex **17** involved an oxidative homo-coupling reaction (Scheme 7.5).[53] In general, the target complexes, with the exception of complexes **15**, **18** and **19** reveal a singlet signal in their $^{31}P\{^{1}H\}$ NMR spectra, in the range of at δ 11.0–11.7 ppm for PEt$_3$ and 18.6 ppm for

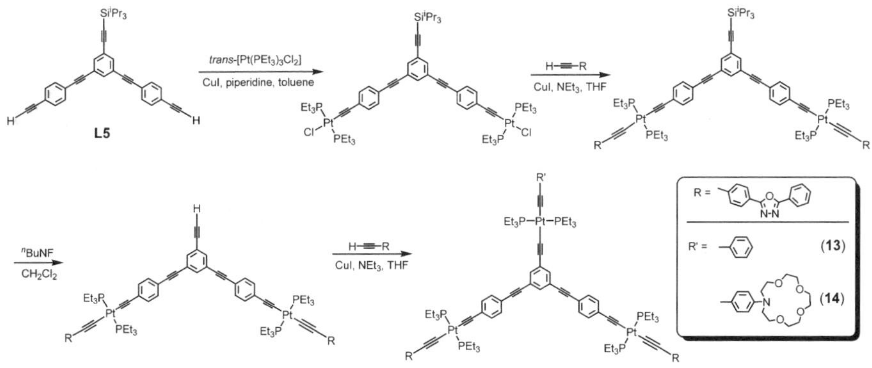

Scheme 7.4 Synthetic route of unsymmetrical trinuclear platinum(II) alkynyls with branched backbone.

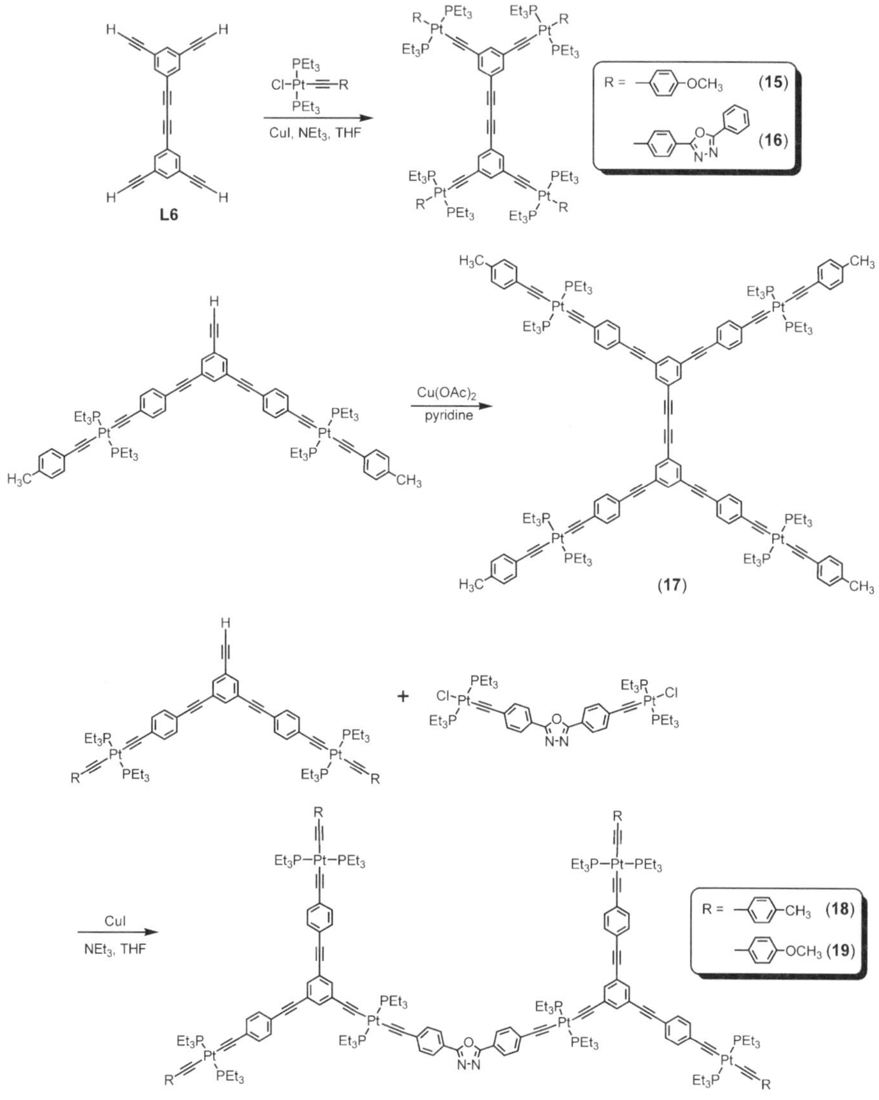

Scheme 7.5 Synthetic route of tetranuclear and hexanuclear platinum(II) alkynyl complexes.

PPh₃,[49–52] which indicates the highly symmetrical structure of the molecules, while the unsymmetrical complexes **15**, **18** and **19** exhibit two sets of signal at about δ 11.1 and 11.5 ppm,[53] suggestive of the nonequivalent environments of the phosphine ligands. The observation of platinum satellites with J_{Pt-P} of 2360–2640 Hz is characteristic of a *trans*-P–Pt–P configuration. The IR spectra of these multinuclear branched complexes showed a moderate band at about 2100 cm^{-1}, which is assigned as the $\nu(C\equiv C)$ stretch due to the alkynyl group directly attached to the platinum(II) centre.

7.2.3 Photophysical Studies

The electronic absorption properties of mononuclear platinum(II) alkynyl complexes with phosphine and stilbine ligands have been investigated by Masai *et al.*,[54] while the corresponding emission behaviour has been reported by Demas and coworkers.[55] In general, the lowest-energy excited states are assigned to be originated from the triplet $[d\pi(Pt) \rightarrow \pi^*(C \equiv CR)]$ metal-to-ligand charge-transfer (MLCT) state. The spectroscopic behaviour of the polymeric and oligomeric platinum(II) alkynyl complexes have also been extensively studied due to their possible potential applications for the construction of high-efficiency energy-harvesting systems and organic light-emitting diodes.[4] Since the first successful isolation of such classes of polymers in the late 1970s by Hagihara and coworkers,[47] their luminescence properties have been investigated by Lewis, Raithby, Friend, Wong and Schanze.[42h,i,43,44]

The electronic absorption spectra of these multinuclear branched platinum(II) alkynyl complexes have been recorded in benzene or dichloromethane solution at room temperature. Their photophysical data are summarised in Table 7.1. Figure 7.1 shows the representative electronic absorption spectra of **2**, **6** and **13** in benzene at room temperature. A

Table 7.1 Electronic absorption and emission data for **1–19**

Complex	λ_{abs}/nm (ε/dm^3 mol^{-1} cm^{-1})	λ_{em} /nm (τ_o/μs)	Φ_{em} [c]
	Absorption [a]	Emission [a]	
1	252 (94 920), 302 (46 990), 332 sh (74 260), 354 (91 540), 372 sh (66 640), 412 sh (2110) [b]	548 (< 0.1)	0.19
2	370 (106 100)	528 (80.0)	0.17
3	308 sh (53 250), 370 (107 990)	528 (32.0)	0.20
4	268 (79 610), 298 (71 610), 338 sh (134 590), 364 (213 650) [b]	532 (41.4)	0.30
5	308 (78 380), 370 (244 950) [b]	546 (68.0)	0.31
6	354 (178 300), 364 (171 800)	515 (22.0)	0.10
7	320 sh (68 200), 354 (188 900), 366 (187 400)	515 (21.0)	0.05
8	352 (261 100), 362 (251 200)	514 (56.0)	0.10
9	306 (91 700), 366 (212 900)	515 (45.0)	0.13
10	354 (197 200), 362 (183 800)	514 (38.0)	0.10
11	308 (103 400), 366 (305 700)	516 (42.0)	0.07
12	308 (60 500), 356 (203 700), 364 (208 500)	547 (60.0)	0.06
13	310 sh (75 800), 362 (196 490)	529 (54.0)	0.05
14	306 (82 830), 366 (204 310)	528 (49.0)	0.08
15	292 (121 200), 344 (173 360)	512 (70.0)	0.01
16	302 (82 390), 350 (201 090)	516 (70.0)	0.05
17	348 sh (211 840), 362 (222 640)	532 (98.0)	0.06
18	300 (119 200), 364 (365 900)	528 (100.0)	0.01
19	300 (148 500), 364 (410 200)	530 (78.0)	0.06

[a]Measured in C$_6$H$_6$ at 298 K. [b]Measured in dichloromethane at 298 K. [c]The luminescence quantum yield, measured at room temperature using quinine sulfate in 0.5 M H$_2$SO$_4$, as the reference (excitation wavelength = 365 nm, Φ_{lum} = 0.55).

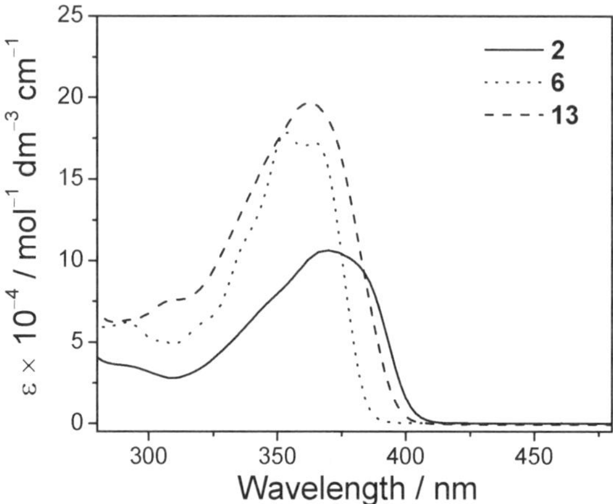

Figure 7.1 Electronic absorption spectra of **2**, **6** and **13** in benzene at room temperature.

vibronic-structured low-energy absorption band at about 348–370 nm, with the extinction coefficients on the order of 10^4–10^5 dm^3 mol^{-1} cm^{-1}, was observed in their electronic absorption spectra. In view of the much higher extinction coefficients than that of their complex precursors, *trans*-[Pt(PEt$_3$)$_2$Cl$_2$], and the close resemblance of their absorption patterns and similar extinction coefficients with the respective free alkynes of the central core ligands, such as diethynyl-3,6-butylcarbazole (**L1**), 2,5-bis-(4-ethynylphenyl)-l-1,3,4-oxadiazole (**L2**), triethynylhexabutyltruxene (**L4**), and 1,4-bis(3,5-diethynylphenyl)buta-1,3-diyne (**L6**), the assignment of transitions of predominantly intraligand (IL) [π→π*] character to such low-energy absorptions is suggested. In addition, very similar to identical electronic absorption spectra were also obtained in different solvents. The relatively insensitive nature of such absorption band toward the solvent polarity is in line with their largely intraligand character. On the other hand, previous spectroscopic studies on *trans*-[Pt(PEt$_3$)$_2$(C≡CR)$_2$] led to the assignment of the absorption band at 300–360 nm as platinum(II)-to-alkynyl [dπ(Pt)→π*(C≡CR)] metal-to-ligand charge-transfer (MLCT) transitions. Such an assignment has further been supported by the resonance Raman investigations of *trans*-[Pt(PEt$_3$)$_2$(C≡CH)$_2$] and *trans*-[Pt(PEt$_3$)$_2$(C≡CPh)$_2$].[42e] Taking into consideration this information, the intense low-energy absorptions of these multinuclear branched platinum(II) alkynyl complexes are best described as an admixture of IL [π→π*]/MLCT [dπ(Pt)→π*(C≡CR)] transition with predominantly IL character, or alternatively, as a metal-perturbed IL transition.[48–53] The low-energy absorption tails of such complexes were found to shift to the red, relative to those of the corresponding chloroplatinum(II) monoalkynyl intermediates. The incorporation of the peripheral alkynyl ligands would lower the transition energy, probably due to the increased electron richness of the platinum(II) metal centre as well as the increase in the extent of π-conjugation of the complexes. In general,

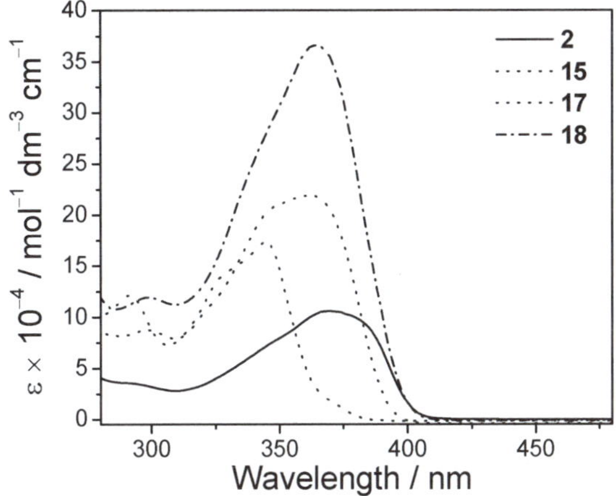

Figure 7.2 Electronic absorption spectra of **2**, **15**, **17** and **18** in benzene at room temperature.

the extinction coefficient of these low-energy absorption bands increases in a multiple fashion from the dinuclear to the tetranuclear and then to the hexanuclear species[49–53] (Figure 7.2), which is attributed to the increase in the number of chromophoric units within the molecules, and is in accordance with the assignment of an involvement of some MLCT character in these transitions.

Upon excitation at $\lambda \geq 365$ nm, all the multinuclear branched platinum(II) alkynyl complexes **1–19** were found to exhibit an intense vibronic-structured luminescence with peak maxima at 512–548 nm in degassed benzene at room temperature. The emission data are summarised in Table 7.1. The representative emission spectra of **2**, **6**, **12**, **13**, **15**, **17** and **18** are shown in Figure 7.3. In

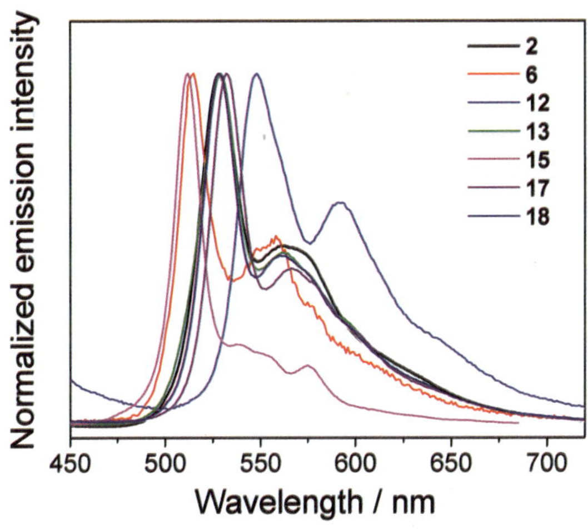

Figure 7.3 Normalised emission spectra of **2**, **6**, **12**, **13**, **15**, **17** and **18** in benzene at room temperature.

view of the large Stokes shifts and long-lived emission lifetimes in the microsecond range, such emission is attributed to a triplet parentage. By scrutinising the emission energies of the complexes with the same central core alkynyl ligands (**2** and **3** at 528 nm with a central core of **L2**; **6–9** at 514–516 nm with central core of **L4**; **13** and **14** at 529 nm with a central core of **L5**; **15** and **16** at 512–516 nm with a central core of **L6**; **18** and **19** at 528–530 nm),[48b,49–52] the emission energies in the same group were found to be insensitive to the peripheral aryl alkynyl ligands, which is supportive of a triplet IL emission mainly derived from the central core moieties. Accordingly, such intense emissions are likely to originate predominantly from the ^{3}IL [$\pi\rightarrow\pi$*] excited state of the central core with some mixing of ^{3}MLCT [dπ(Pt)$\rightarrow\pi$*(C$\equiv$CR)] character. The absence of the observation of the emission from the ^{3}IL state of the peripheral ligands is probably due to the facile energy transfer from the peripheral aryl alkynyl ligands to the central core moieties.[48b,c,49–52] On the other hand, the complexes with the peripheral naphthyl ligand, *i.e.* **1**, **5** and **12**, were found to exhibit similar yellowish green emission at 546–548 nm. The disappearance of the emission from the central core moieties in such complexes is attributed to the existence of a lower-lying ^{3}IL excited state associated with the peripheral naphthyl ligands.[48b,c,49,50,52] It could be concluded that the direction of the energy transfer, which is either from the peripheral to the central core or from the central core to the peripheral moiety, is determined by the relative energy levels between these two components. Through the incorporation of various alkynyl ligands at the periphery, such control of their relative energy levels could be achieved and hence would provide a convenient strategy to tune the emission properties of these multinuclear branched platinum(II) alkynyl complexes. Similar emission energies were observed in complexes **13**, **14**, **18** and **19** at 528–530 nm because of the presence of the same central core moiety of **L5** with the lowest-lying energy.[53]

7.2.4 NLO Properties of Two-Photon Absorption (TPA) and Two-Photon-Induced Luminescence (TPIL)

The mononuclear platinum(II) complexes of the structurally related bis((4-(phenylethynyl)phenyl)ethynyl)bis(tri-*n*-butylphosphine)platinum(II) have been reported to show NLO properties with the σ_2 values in the range of about 5–10 GM, and the TPIL was assigned as originated from the singlet or triplet excited state of the 4-(phenylethynyl)phenylethynyl backbone.[45b–d] Schanze and coworkers have revealed that some linear polymers and oligomers of multinuclear platinum(II) alkynyl complexes, having highly π-conjugated ligands substituted with π-donor or acceptor moieties, exhibit NLO behaviour with large σ_2 values in the visible and near-infrared region of the spectrum (600–800 nm).[42i] The incorporation of heavy-metal platinum(II) centres to the ligands possessing effective two-photon absorption cross sections would enhance the intersystem crossing to give rise to the long-lived triplet excited states.

Since the multinuclear branched platinum(II) alkynyl complexes were found to give readily tunable photophysical properties, their nonlinear optical properties and the corresponding structure–property relationships were also explored. In view of the better solubilities and processabilities, relative to their linear counterparts, such materials are anticipated to be advantageous for future applications. Upon excitation by a mode-locked femtosecond Ti:sapphire laser at 720 or 740 nm, all these multinuclear branched platinum(II) alkynyl complexes show yellowish green emission in degassed benzene solutions with concentrations of about 10^{-5} M.[49–53] Figure 7.4 shows the TPIL spectra of complexes **6** and **12** in degassed benzene at room temperature. The nonlinear photophysical data of all the complexes are tabulated in Table 7.2. Since there is no linear absorption in the wavelength range of 500–820 nm for these complexes, such emission generated from the 720 or 740 nm excitation is attributed to nonlinear processes, but not the linear processes. Such upconverted luminescence shows emission profiles and energies that are basically identical to that observed in their corresponding single-photon excited emission with higher energy excitation. The dependence of the upconverted emission intensity of all the complexes on the incident laser power was determined.[49–53] The plot of emission intensity of complex **6** against laser power is shown in Figure 7.5, which gives a typical power dependence curve as an example. The inset shows a linear relationship between log(emission intensity) and log(laser power) and a straight line with a slope of 2.11 was obtained. On the basis of the quadratic dependence of TPA on the incident laser intensity, the TPIL intensity would also be dependent on the excitation power. Theoretically, the plot of log(emission intensity) against log(laser power) should give a straight line with a gradient of 2.[45f] The observation of the quadratic dependence of the emission intensity on the

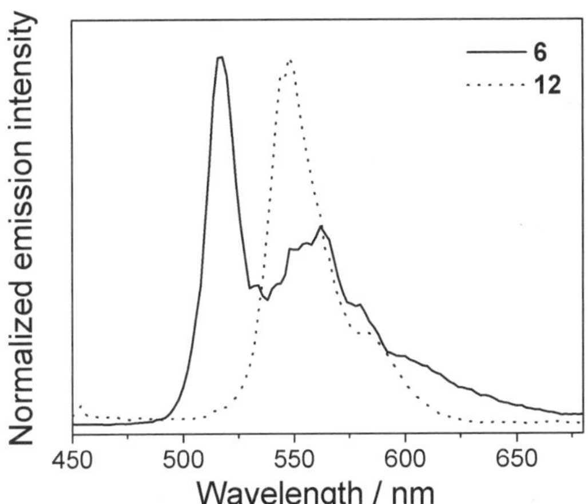

Figure 7.4 Two-photon-induced luminescence spectra of **6** and **12** in benzene at room temperature.

Table 7.2 Nonlinear optical data for **1–19**.

Complex	λ_{em} /nm [a]	σ_2 /GM	Power dependence
1	548 [b]	14	2.15
2	530	17	2.18
3	530	41	2.07
4	533 [b]	32	–[c]
5	547 [b]	13	–[c]
6	517	27	2.11
7	516	12	2.04
8	514	11	2.15
9	518	32	2.11
10	518	21	2.52
11	517	51	1.96
12	549	6	1.73
13	532	52	1.71
14	531	29	1.94
15	512	6.3	1.88
16	517	71	1.82
17	534	48	2.01
18	530	191	1.89
19	531	133	1.81

[a]Measured in C_6H_6 at 298 K, with excitation wavelength = 720 nm. [b]Measured in C_6H_6 at 298 K, with excitation wavelength = 740 nm. [c]Not determined.

incident laser power supports the emission upon excitation at 720 or 740 nm as a two-photon process.

The two-photon absorption cross section values of these multinuclear branched platinum(II) alkynyl complexes with excitation wavelengths at 720 or

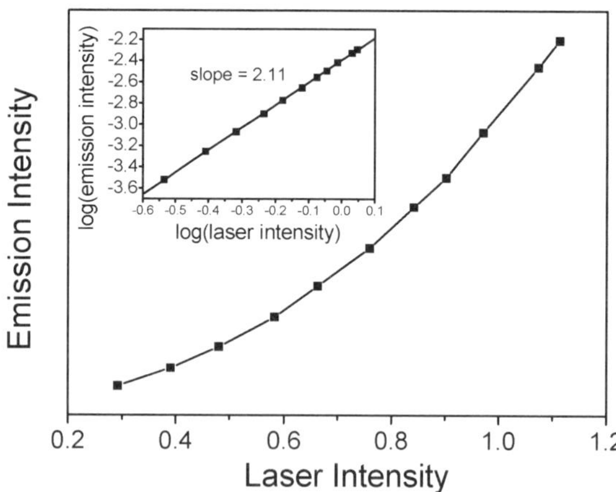

Figure 7.5 Power dependence of the upconverted luminescence intensity (■) of complex **6** in benzene at room temperature. The line shows the theoretical curve for the quadratic function. The inset shows the log(emission intensity) (▲) against log(laser power) and its linear regression.

740 nm have been determined using the TPIL method. By comparing their TPIL intensity with the two-photon fluorescence excitation cross section of a reference, the calculated σ_2 values were found to be in the range of 6–191 GM and are summarised in Table 7.2. The obtained σ_2 values could be up to 20 times greater than that for other mononuclear platinum(II) alkynyl complexes of bis((4-(phenylethynyl)phenylethynyl)bis(tri-*n*-butylphosphine)), which shows a σ_2 value of about 5–10 GM.[45b–d,f] These values are also comparable to those reported by Schanze and coworkers on some linear platinum(II) alkynyl complexes with extended π-conjugation in their donor-π or acceptor-π structural motifs within the alkynyl ligands.[42i] The structure–property relationship has been investigated through the systematic comparison study on the multinuclear branched platinum(II) alkynyl complexes with various central core and peripheral moieties as well as the number of platinum(II) chromophores. The σ_2 values were found to be insensitive to the central core moieties but rather dependent on the peripheral ligands. In general, a more electron-rich peripheral substituent would give rise to a relatively higher σ_2 value, which is similar to the observation in organic TPA chromphores.[49–53] On the other hand, the higher nuclearity of the complexes with more platinum(II) chromophores would enhance the σ_2 values. For the complexes with the same peripheral alkynyl ($C{\equiv}C\text{-}C_6H_4\text{-}CH_3$) ligands, two-photon absorption cross section values were found to increase dramatically from the dinuclear complex **2** (σ_2 = 17 GM), trinuclear complexes **4** and **6** (σ_2 = 27 and 32 GM, respectively) and tetranuclear complex **17** (σ_2 = 48 GM) to the hexanuclear complex **18** (σ_2 = 191 GM). From the understanding of such structure–property relationship, one can easily enhance the TPA cross section by a simple modification of the peripheral substituents as well as the number of chromophores. Moreover, such multinuclear branched platinum(II) alkynyl complexes also represent a new class of versatile TPIL materials with different emission energies, on the basis of the readily controllable energy-transfer direction by the tuning of the relative energies of the central core and the peripheral moieties.

7.3 Synthesis, Photophysical Properties of Bis-Cyclometalated Alkynylgold(III) Complexes and their Applications in Organic Light-Emitting Devices

7.3.1 Background

Organic light-emitting diodes (OLEDs) have attracted a great deal of attention due to their potential applications in both flat-panel displays and as lighting sources.[57–61] There have been significant improvements in OLED efficiencies by using phosphorescent materials to generate light emission from both singlet and triplet excitons. In view of the strong spin-orbit coupling of the heavy-metal atom that allows for efficient intersystem crossing (ISC) between singlet and triplet states as well as enhances the subsequent transition from the triplet to the ground state that can lead to a high quantum yield of emission from the triplet state, OLEDs with 100% internal quantum efficiencies could theoretically be obtained by harnessing both the triplet and singlet excitions.[4a–c] In this regard, considerable effort has been devoted to finding transition-metal complexes suitable for OLED applications. A recent advance has demonstrated the ability to make highly efficient electroluminescent devices using phosphorescent emitters based on heavy transition metals such as iridium(III),[4] platinum(II),[5] ruthenium(II)[6] and osmium(II)[7] complexes.

Among the complexes studied to date, cyclometalated tris-chelated Ir(III) complexes have emerged as particularly promising candidate materials due to the very high phosphorescence quantum efficiency and the relatively short phosphorescence lifetime (several microseconds) to provide high-efficiency electrophosphorescence. A variety of phosphorescent iridium(III) complexes have been synthesised, basically containing two cyclometalated ligands and a single monoanionic, bidentate ancillary ligand,[4c,f,h–k] or with three cyclome-talated ligands.[4a,b,d–k] Extensive synthetic studies have been carried out with a view to developing Ir(III) complexes with improved characteristics, in particular saturated phosphorescence colour, high phosphorescence efficiency, and physical compatibility.[4h] The nature of the organic ligand drastically influences the emission wavelength and EL efficiency of the iridium(III) complex.[4] In particular, well-defined molecular design methods for efficient phosphorescence colour tuning are currently under active investigation with the aim of achieving full-colour displays. In 2002, Thompson and coworkers reported the employment of the blue luminescent cyclometalated Pt(II) complex, which displays orange-red excimeric emission, as a dopant in white-light OLEDs.[5a] Che and coworkers reported different platinum(II) σ-alkynyl complexes as high-efficiency emitters that can be vacuum deposited into multilayer OLEDs to emit yellow to red light with high luminance and excellent efficiencies.[5c] Chi and coworkers reported the synthesis of a series of neutral Ru(II) complexes with two pyrazole and phosphine ligands[6c] and the utilisation of such complexes in the fabrication of highly efficient OLEDs by codeposition techniques. Recently, highly emissive Os(II) complexes have been applied in OLEDs[6d,7] and detailed characterisation of emissive processes of Os(II) complexes is strongly increasing. Electrophosphorescent devices were demonstrated by use of the Os(II) complexes with doped PVK:PBD or PVN:PBD as the emitting layer. When PVK:PBD was used as the host matrix, brightness of over 1400 cd/m^2 was achieved.[7a] The best external quantum efficiency of 2.2%, which corresponds to a photometric efficiency of 1.9 cd/A, was achieved when PVN: PBD was used as the host matrix.[7a]

On the other hand, the development of phosphorescent emitters with metal centres other than the most commonly employed iridium(III) and platinum(II) for OLED applications has been relatively less explored. Thus, the search for new classes of phosphorescent emitters with metal centres that are less expensive and of low toxicity and an environmentally benigned nature would be attractive. While there have been a number of gold(I) compounds that are strongly phosphorescent, gold(III) compounds, on the other hand, are rarely luminescent at room temperature, which may be ascribed to the presence of nonemissive low-energy d-d ligand field states and the electrophilicity of the metal centre.[62] In contrast to the isoelectronic platinum(II) compounds that are known to show rich luminescence properties, very few examples of luminescent gold(III) compounds have been reported.[63] Recently, as an extension of our efforts on luminescent metal alkynyls,[8,64–76] a novel class of luminescent neutral bis-cyclometalated alkynylgold(III) complexes, [Au(C^N^C)(C≡CR)], has been successfully prepared in our laboratory and the idea that incorporation of strong σ–donating alkynyl ligand into the [Au(C^N^C)] moiety would enrich the photoluminescence (PL) properties has also been demonstrated.[8,75] Their syntheses and photophysical behaviour will be highlighted. The use of a series of luminescent bis-cyclometalated alkynylgold(III) compounds, [Au(C^N^C)(C≡CR)] (HC^N^CH = 2,6-diphenylpyridine or substituted 2,6-diphenylpyridine, R = alkyl or aryl), as a new class of electrophosphorescent materials, both in the role of emitters and dopants of OLEDs to give electroluminescence (EL), will be discussed.

7.3.2 Synthesis and Characterisation

The complex precursor bis-cyclometalated chlorogold(III), [Au(C^N^C)Cl], was generally prepared in moderate yield, through a transmetalation reaction of [Hg(C^N^CH)] and KAuCl$_4$ or HAuCl$_4$ in acetonitrile solution for the coordination of Au(III) metal centre to the bis-cyclometalated C^N^C ligand. Two synthetic pathways could be employed for the incorporation of alkynyl ligand into the bis-cyclometalated gold(III) moiety, [Au(C^N^C)], *via* the replacement of chloro ligand with the organic terminal alkyne (Scheme 7.6).[75b] In both synthetic routes, pale yellow crude products were obtained after evaporation of the solvent to dryness. The desired products were isolated and purified by column chromatography on silica gel and subsequent recrystallisation by slow diffusion of diethyl ether vapor or by layering of *n*-hexane into the concentrated dichloromethane solution of the complexes. Because of the shorter reaction time, mild reaction conditions, higher product yield and formation of less side products, method (I) was the method of choice for the synthesis of this class of luminescent bis-cyclometalated alkynylgold(III) complexes (Chart 7.1).[75] Compared to the chloro complex precursors, the products were found to be more soluble in common organic solvents upon the incorporation of the alkynyl ligand. The IR spectra of the complexes show a

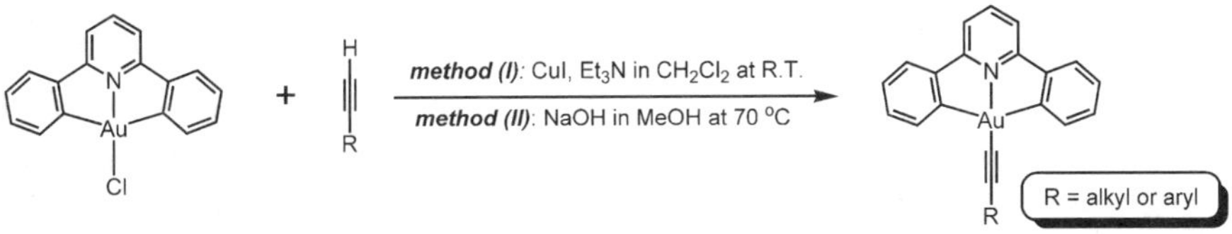

Scheme 7.6 General synthetic routes of bis-cyclometalated alkynylgold(III) complexes.

Chart 7.1 | R and R' = H; R" = | | R = ^{t}Bu; R' = H; R" = | |
| --- | --- | --- | --- |
| C_6H_5 | (20) | C_6H_5 | (28) |
| C_6H_4-Cl-4 | (21) | | |
| C_6H_4-NO_2-4 | (22) | R = H; R' = C_6H_4-CH_3-4; R" = | |
| C_6H_4-OCH_3-4 | (23) | | |
| C_6H_4-NH_2-4 | (24) | C_6H_4-C_6H_{13}-p | (29) |
| C_6H_4-N(C_6H_5)$_2$-4 | (25) | | |
| C_6H_4-C_6H_{13}-4 | (26) | R = H; R' = C_6H_4-F_2-2,5; R" = | |
| C_6H_{13} | (27) | C_6H_4-N(C_6H_5)$_2$-4 | (30) |

Chart 7.1   Luminescent alkynylgold(III) complexes.

weak band at 2143–2157 cm^{-1}, typical of the ν(C≡C) stretching frequency. Many of them have been structurally characterised by X-ray crystallography.[8,75] In each of their molecular structures, the gold(III) metal centre adopts a distorted square-planar coordination geometry, characteristic of d^8 metal complexes. The bis-cyclometalated (C^N^C) tridentate ligand and the alkynyl ligand occupy all the four coordination sites of the gold(III) metal centre. No considerable Au···Au interaction was observed in view of the fact that the closest Au···Au distances between adjacent molecules are found to be longer than the sum of van der Waals radii for two gold(III) centres.[8,75] Despite this, significant π–π stacking interactions, with the shortest interplanar distances between the [Au(C^N^C)] planes in the range from 3.4 to 3.6 Å, were revealed by their crystal packings for many of the structures.[8,75] It is interesting to note that their packing arrangements vary from one structure to another. Some of the complex molecules are stacked into an extended columnar array with identical Au···Au and π···π separations between them, while others are stacked together with alternating short and long Au···Au distances to give a "dimeric" configuration. Moreover, the [Au(C^N^C)] motifs are stacked with different extents of aromatic overlapping and in a variety of manners, such as head-to-tail, head-to-head or partial head-to-tail.

7.3.3 Photophysical Studies

The electronic absorption spectra of **20**−**30** in dichloromethane showed an intense absorption band at *ca.* 290−330 nm and a moderately intense vibronic-structured absorption band at *ca.* 360−420 nm with extinction coefficients of the order of 10^4 dm^3 mol^{-1} cm^{-1}. The electronic absorption spectral data are summarised in Table 7.3. The electronic absorption spectra of **22**, **24** and **28** in dichloromethane at room temperature are shown in Figure 7.6. For complexes **20**−**27**, the characteristic low-energy vibronic-structured absorption bands at *ca.* 360−400 nm were found to be rather insensitive to the nature of the monoynyl ligand. This eliminated the possibility of alkynyl-to-diarylpyridine ligand-to-ligand charge-transfer (LLCT) and alkynyl-to-gold ligand-to-metal charge-transfer (LMCT) transitions. The very low-lying energy of the dπ

Table 7.3 Electronic absorption and emission data for **20–30**

Complex	Absorption $\lambda_{max}/nm\ (\varepsilon_{max}/dm^3\ mol^{-1}\ cm^{-1})^a$	Emission Medium (T/K)	$\lambda_{max}/nm\ (\tau_o/\mu s)$	Φ_{em}
20	312 (19 890), 322 (19 980), 364 (5050), 381 (5870), 402 (4870)	CH$_2$Cl$_2$ (298)^d Solid (298) Thin film (298)^e	476, 506, 541, 582 (<0.05) 588 (<0.05) 568 (<0.05)	1.0×10^{-3} ^{b}
21	312 (19 400), 322 (19 640), 365(4640), 382 (5170), 402(4305)	CH$_2$Cl$_2$ (298)^d Solid (298)	476, 506, 539, 584 (<0.05) 550 (<0.05)	2×10^{-4} ^{b}
22	312 (26 800), 327 (36 100), 363 (18 545), 382 (10 260), 402 (5720)	CH$_2$Cl$_2$ (298)^d Solid (298)	478, 508, 540, 600 (0.3) 565 (<0.05)	1.9×10^{-3} ^{b}
23	312 (13 820), 322 (13 455), 362 (6400), 380 (6245), 400 (4190)	CH$_2$Cl$_2$ (298)^d Solid (298)	474, 505, 539, 584 (<0.05) 555 (<0.05)	5×10^{-4} ^{b}
24	310 (19 195), 322 sh (15 680), 365 (8855), 381 (10 100), 399 (8300), 415 sh (3410)	CH$_2$Cl$_2$ (298) Solid (298)	611 (0.3) 585 (0.2)	2.0×10^{-3} ^{b}
25	320 (38 270), 384 (9560), 400 (9565), 416 sh (5590)	CH$_2$Cl$_2$ (298) Solid (298)	620 (0.2) 566 (<0.1)	2.6×10^{-2} ^{b}
26	313 (18 010), 323 (18 240), 363 (5130), 381 (5250), 401 (3940)	CH$_2$Cl$_2$ (298)^d Solid (298)	474, 508, 546, 585 (<0.05) 562 (<0.05)	2.0×10^{-3} ^{b}
27	311 (14 860), 320 (14 040), 364 (3800), 380 (4840), 400 (4230)	CH$_2$Cl$_2$ (298)^d Solid (298)	471, 508, 535, 553 (<0.05) 561 (<0.05)	1.9×10^{-3} ^{b}
28	313 (26 190), 323 (24 460), 374 (6130), 392 (8035), 412 (7465)	CH$_2$Cl$_2$ (298)^d Solid (298)	484, 514, 548, 593 (0.1) 550 (<0.05)	6.0×10^{-3} ^{b}
29	290 (42 660), 304 (40 780), 330(25 220), 364 (6100), 386 (5120), 406 (3540)	CH$_2$Cl$_2$ (298)^d Solid (298)	475, 505, 547, 592 (<0.05) 563 (<0.05)	2.7×10^{-3} ^{b}
30	316 (45 915), 392 (7835), 410 (825)	CH$_2$Cl$_2$ (298) Solid (298) Thin film (298)^g 2% in PMMA 4% in PMMA 8% in PMMA 20% in PMMA 50% in PMMA	669 (<0.05) 590(<0.05, 0.2)^f 537 538 543 558 577	8.5×10^{-3} ^{c} 0.34

^{a}In dichloromethane at 298 K. ^{b}The luminescence quantum yield, measured at room temperature using [Ru(bpy)$_3$]$^{2+}$ as a standard. ^{c}The luminescence quantum yield, measured at room temperature using quinine sulfate as a standard. ^{d}Vibronic-structured emission band. ^{e}Prepared by vacuum deposition. ^{f}Double exponential decay. ^{g}Prepared by spin coating.

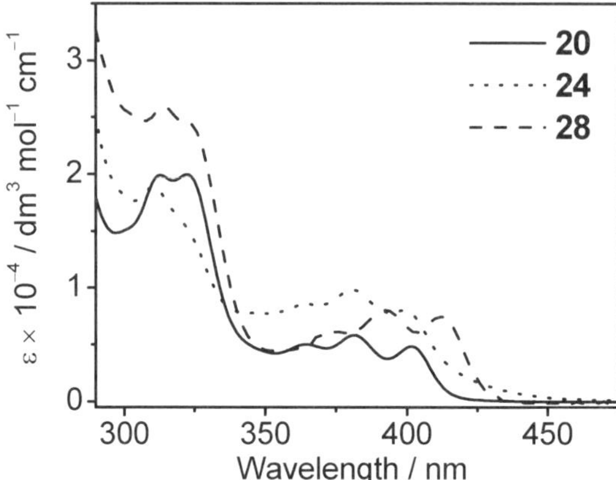

Figure 7.6 Electronic absorption spectra of **20**, **24** and **28** in dichloromethane at room temperature.

orbital of the gold(III) metal centre and its redox-inactive nature also precluded the assignment of a metal-to-ligand charge-transfer (MLCT) transition. The vibrational progressional spacings of *ca.* $1200-1300$ cm^{-1} are in close agreement with the skeletal vibrational frequency of the C^N^C ligand. With reference to the previous spectroscopic work on [Au(C^N^C)Cl],[77] the low-energy absorption was assigned as a metal-perturbed $\pi \rightarrow \pi^*$ intraligand (IL) transition of the C^N^C cyclometalated ligand, probably involving some charge-transfer character from the phenyl moiety to the pyridyl unit. It is also interesting to find that an additional shoulder was observed in the electronic absorption spectra of **25**, **26** and **30** with amino substituent on the ethynylbenzene at *ca.* $410-416$ nm.[8,75b] Due to the better electron-donating abilities of the electron-rich amino substituents on the aryl alkynyl ligands, low-lying alkynyl-to-C^N^C ligand-to-ligand charge-transfer (LLCT) transitions became feasible and could appear as shoulders at *ca.* $415-425$ nm. The low-energy absorptions in **25**, **26** and **30** were therefore assigned as an admixture of IL [$\pi \rightarrow \pi^*$(C^N^C)] / LLCT [π(C≡CC$_6$H$_4$NR$_2$) $\rightarrow \pi^*$(C^N^C)] transitions. The observation of lower absorption energy in **28**, relative to that of **20** with the same alkynyl ligand, is in line with the assignment of a metal-perturbed $\pi \rightarrow \pi^*$(C^N^C) intraligand (IL) transition. By the consideration of the fact that the energy level of the HOMO π orbital is raised to a larger extent than that of the LUMO π^* orbital, the electron-donating *tert*-butyl groups on the phenyl rings would therefore narrow the HOMO–LUMO energy gap. Better delocalisation over the C^N^C ligand upon introduction of the tolyl group on the pyridine ring in **29** would result in a narrowing of the $\pi-\pi^*$ energy gap, which accounts for the lower energy $\pi \rightarrow \pi^*$ IL transition in **29** relative to that of **26**.

Unlike the chloro precursor, [Au(C^N^C)Cl], which was reported to be only emissive in glass matrices,[77] photoexcitation at $\lambda > 360$ nm of these bis-cyclometalated alkynylgold(III) complexes resulted in strong luminescence at

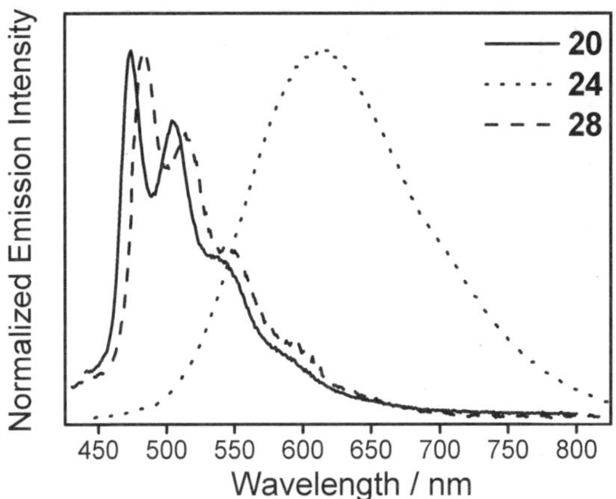

Figure 7.7 Normalised emission spectra of **20**, **24** and **28** in dichloromethane at room temperature.

470–655 nm in both the solid state and in fluid solutions. The photophysical data of **20–30** are summarised in Table 7.3. The representative emission spectra of **20**, **24** and **28** in dichloromethane at room temperature are depicted in Figure 7.7. The rich luminescence behaviour of **20–30** confirmed the idea that incorporation of strong σ-donating alkynyl ligands into the chlorogold(III) complex precursors could enlarge the d–d ligand field splitting, giving rise to an enhancement of their luminescence properties. The long-lived emission with lifetimes in the submicrosecond to microsecond range and the observed large Stokes' shift indicated that the emission was of a triplet parentage.

In dichloromethane solution at room temperature, the emission profiles and energies of complexes **20–23**, **26** and **27**, which contain the same tridentate C^N^C ligand, were rather similar and insensitive to the nature of the alkynyl ligands. The vibronically structured emission band with the vibrational progressional spacings of *ca.* 1300 cm^{-1} are characteristic of the C=C and C=N stretching frequencies of the tridentate ligand, indicative of the participation of the tridentate ligand in the excited state. The luminescence is assigned as originating from a metal-perturbed $^3[\pi \rightarrow \pi^*(RC^\wedge N(R')^\wedge CR)]$ IL state. It is interesting to note that complexes **24**, **25** and **30**, with an amino-phenyl alkynyl group, show a lower-energy structureless emission band centred at 610–669 nm. On the basis of the energetically higher-lying $\pi(C\equiv CC_6H_4-NR_2\text{-}p)$ orbital resulting from the strong electron-donating amino substituents, the lower energy emission bands in **24**, **25** and **30** were tentatively assigned as originating from an excited state of ^{3}LLCT $[\pi(C\equiv CC_6H_4NR_2) \rightarrow \pi^*(C^\wedge N^\wedge C)]$ origin. Such assignments were further corroborated by the computational studies[75b] as well as the assignment of the first oxidation wave in the electrochemical study as an alkynyl ligand-centred oxidation.[8b,75b] In parallel with the electronic absorption study, **28** exhibited

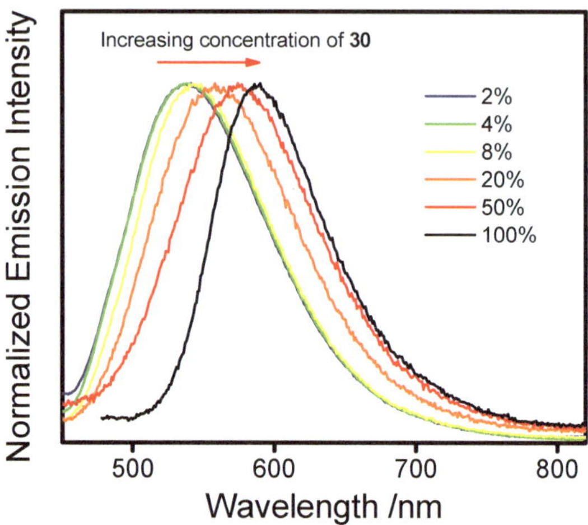

Figure 7.8 Normalised photoluminescence spectra of **30** at different concentrations (wt%) in PMMA thin films at 298 K. Reproduced with permission from ref. 8b.

an emission band at lower energy than **20**. Similarly, due to the attachment of the electron-rich *tert*-butyl groups on the phenyl rings, the phenyl-localised π orbital energy would be raised, leading to a lower $^3[\pi-\pi^*]$ IL emission energy (Figure 7.7). On the contrary, **20–30** exhibit a broad low-energy emission band at 550–590 nm in the solid state at ambient temperature. A concentration dependence of the emission energies was observed in the solid-state thin film of **30** doped in PMMA.[8b] The emission maxima were found to shift to the red from 538 to 575 nm with an increasing concentration from 2 to 50% in PMMA and further to 588 in the solid state (Figure 7.8). The redshift of the solid-state emission band compared to that in dichloromethane solution, as well as the observation of concentration-dependent emission energies, have been rationalised by the $\pi-\pi$ excimeric ^{3}IL emission due to the $\pi-\pi$ stacking interaction of the C^N^C ligand in the solid state, as shown by the ordered packing of the molecules in the X-ray crystal packings of the complexes. A better packing of the molecules would be anticipated at higher concentrations, and hence giving rise to a lower emission energy.

7.3.4 Electroluminescence Properties in Organic Light-Emitting Devices

The highest occupied molecular orbital (HOMO) and the lowest unoccupied molecular orbital (LUMO) as well as the triplet state energies of these bis-cyclometalated alkynylgold(III) complexes could be obtained from the results of photophysical and electrochemical studies, which provide useful knowledge for the judicious choices of hole- and electron-transporting materials for the fabrication of OLEDs.[8] Vacuum deposition is currently one of the common methods for the formation of thin films in advanced technology due to the controllable deposition rate and thickness, and homogeneous thin-film growth capability. Taking the advantage that these bis-cyclometalated alkynylgold(III) complexes are neutral and are thermally stable as revealed by the high decomposition temperature of > 460 °C in the thermogravimetric analysis, fabrication of OLEDs can be achieved by vacuum deposition of such luminescent compounds as phosphorescent emitters or dopants. In view of the fact that a relatively high luminescence quantum yield of 34% was found in a 2 wt% of **30** doped in PMMA,[8b] such class of complexes is anticipated to exhibit a promising potential as a candidate of phosphorescent emitting material for the fabrication of high-performance OLEDs in terms of high brightness and efficiency.

Apart from the well-studied systems of cyclometalated iridium(III)[4] and platinum(II),[5] other examples in the literature involving the use of d block transition-metal-containing materials as phosphorescent emitting material in OLEDs are relatively rare.[6,7] Multilayer OLEDs are fabricated by employing **20** and **25** as electrophosphorescent emitters and they exhibit intense electroluminescence upon applying a DC voltage. The EL spectra of the device [ITO/4,4′-bis[*N*-(1-naphthyl)-*N*-phenyl-amino]biphenyl (NPB) (60 nm)/ **20** (30 nm)/aluminum tris(8-hydroxyquinoline) (Alq$_3$) (10 nm)/LiF/Al] exhibit a blue vibronic-structured band at 452 nm and a red-orange band at about 585 nm, ascribed to the EL of NPB and **20**, respectively.[8a] With increasing DC voltage, the relative EL intensity ratio of **20**:NPB decreases gradually, indicating that some holes are blocked by **20** leading to light emission from the NPB layer. With the increasing applied bias, more electrons can pass through layer **20** and recombine with holes in the NPB layer, which results in more light emission from NPB. Pure electroluminescence could be obtained by using *N,N′*-diphenyl-*N,N′*-bis(3-methylphenyl)-[1,19-biphenyl]-4,4′-diamine (TPD) as the hole-transporting material from anther device [ITO/ TPD (70 nm)/ **20** (10 nm)/Alq$_3$ (40 nm)/LiF/Al].[8a] The lower HOMO level of TPD (~ 5.5 eV) is anticipated to reduce the energy barrier for hole transport from the TPD layer to layer **20**, leading to better recombination of the holes and electrons confined in the emitting layer of **20**, resulting in light emission only from **20**.

Apart from utilising these bis-cyclometalated alkynyl gold(III) complexes as an emissive layer of OLEDs, complexes **20**, **25** and **30** have been tested to function as dopant in the host material of 4,4′-*N,N′*-dicarbazole-biphenyl (CBP) in the emitting layer of other devices in order to improve the efficiency. The EL spectra exhibit the emission only from the bis-cyclometalated alkynylgold(III) complexes with the absence of any other residual emission from either the host or the adjacent hole- or electron-transporting layer even at high current density. This indicates a complete energy transfer from the CBP host exciton to the triplet state of such phosphorescent emitting material upon electrical excitation. By the insertion of the thin CBP layer (~ 5 nm thick) between the HTL and EML, the triplet excitons would be effectively confined within the EML, that is anticipated to suppress the unwanted energy transfer

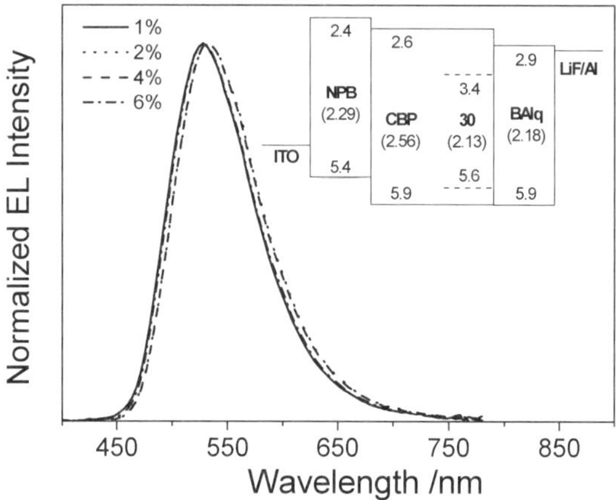

Figure 7.9 Normalised EL spectra of the devices with different concentrations of **30** doped into the CBP layer at a current density of 20 mA/cm^2. Inset: Proposed energy band diagram for the OLED. Note that numbers indicate the respective highest occupied and lowest unoccupied molecular orbital (HOMO and LUMO, respectively) energies relative to the vacuum as well as the triplet energy levels (in parentheses) in eV. Reproduced with permission from ref. 8b.

from the triplet excitons to the nonradiative NPB triplet states, owing to its higher triplet energy (2.56 eV) than NPB (2.29 eV).[78]

The devices of [ITO/NPB (70 nm)/CBP (5 nm)/CBP doped with x% (x = 1, 2, 4 or 6) **30** (30 nm)/aluminum(III) bis(2-methyl-8-quinolinato)-4-phenylphenolate (BAlq) (30 nm)/LiF/Al] was found to exhibit the highest brightness and the best efficiency. The EL spectra were very similar to the PL spectrum in the solid thin film, suggesting that they are originated from the same origin of dimeric or excimeric emission arising from the $\pi-\pi$ stacking of the C$^\wedge$N$^\wedge$C ligand. The normalised EL spectra of the devices with different concentrations of **30** doped into the CBP layer viewed at the normal direction at a current density of 20 mA/cm^2 are shown in Figure 7.9. In line with the observation of the concentration dependence of the PL spectra of **30** in the PMMA thin film, the EL band maxima are found to be slightly redshifted from 528 nm to 532 nm with increasing concentrations of **30**. This is rationalised by the stronger $\pi-\pi$ stacking of the C$^\wedge$N$^\wedge$C ligand and hence a lower energy dimeric/ oligomeric or excimeric emission in high dopant concentration of **30**, resulting from a higher molecular order and a better packing of the molecules. Figure 7.10 shows the current and power efficiencies and the EQE of the devices with different concentrations of **30** doped into the CBP layer as a function of current density. A maximum current efficiency of 37.4 cd/A, a maximum power efficiency of 26.2 lm/W and a peak EQE of 11.5% were achieved at an optimised dopant concentration of 4%. Such a high EQE is comparable to that of the optimised Ir(ppy)$_3$:CBP devices reported.[4j] This suggests that this class of bis-cyclometalated alkynylgold(III) complexes represents superior candidates as phosphorescent dopant for OLEDs.

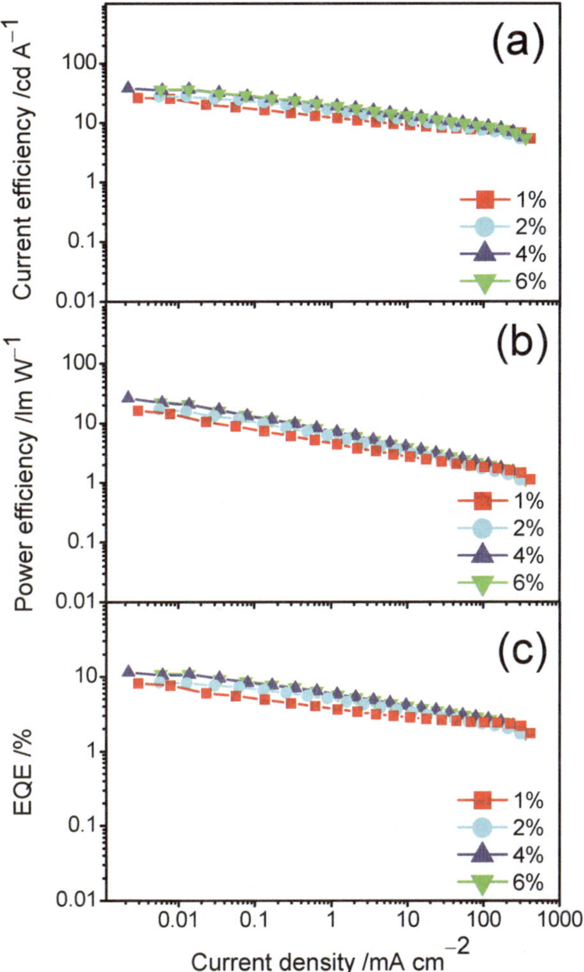

Figure 7.10 (a) Current and (b) power efficiencies and (c) external quantum efficiencies (EQE) of the devices with different concentrations of **30** doped into the CBP layer as a function of current density. Reproduced with permission from ref. 8b.

7.4 Synthesis, Photophysical Properties of Alkynylplatinum(II) Terpyridyl and Bipyridyl Complexes and their Use as Sensitisers in Dye-Sensitised Solar Cells

7.4.1 Background

In view of the environmental aspect and rising demand of energy, development of clean and renewable energy is one of the grand challenges in scientific research. There has been a fast growing interest in the search for organic and polymeric materials for photovoltaic applications as they show promise for the fabrication of low-cost, flexible devices for renewable sources of energy.[11,12] The diversity of metal–ligand chromophoric systems with tunable electronic absorption and excited-state redox properties has made transition-metal complexes attractive candidates for solar energy conversion. Since the first report on the Grätzel cell,[79] there has been a tremendous progress in the development of dye-sensitised solar cells (DSSCs) based on nanocrystalline TiO_2 semiconductor and well-designed panchromatic dyes such as the black dye of formula $[RuL(NCS)_3]^-$, where L is the anchoring 4,4′,4″-tricarboxy-2,2′:6′,2″-terpyridine ligand with reported efficiency of as high as 10.4%.[80] The well-studied photophysics and chemistry of ruthenium(II),[80] osmium(II),[81] rhenium(I)[82] and iridium(III)[83] polypyridine complexes extended their utilisation as the dye for photosensitising function with appropriate ligands for anchoring and charge separation functions. In the past few years, Grätzel and coworkers have made significant progress in improving the conversion efficiency of the DSSCs.[80e]

Apart from the platinum(II) phosphine systems, the photophysical properties of platinum(II) polypyridyl complexes with charge-transfer character have been widely studied due to their intriguing photophysical and spectroscopic properties resulting from their square planar coordination geometry.[84–90] Such a class of platinum(II) complexes usually displays an intense charge-transfer absorption band in the visible region, providing an advantage over the inorganic semiconductor in photovoltaic applications. Recent works by several research groups demonstrated the redshift of the electronic absorption band even to the near-infrared (NIR) region, which is one of the criteria to increase the power-conversion efficiency of DSSCs. This could be achieved by the incorporation of alkynyl ligands into the platinum(II) metal centre in order to introduce a lower-lying LLCT excited state[68b,c] or by the variation of the electronic properties of the polypyridyl ligands.[68c] Our group first reported the synthesis and photophysical behaviour of a series of luminescent alkynylplatinum(II) terpyridyl complexes, in which their electronic absorption and emission energies were found to be sensitive to the nature of the substituents on the alkynyl ligand.[68a] As an extension of these works, two classes of alkynylplatinum(II) terpyridyl and bipyridyl complexes have been explored for their functionalisation as photosensitisers in DSSCs. Their synthesis, photophysical behaviour and sensitising behaviour for TiO_2-based dye-sensitised solar cell applications will be discussed.

7.4.2   Synthesis and Characterisation

The donor−acceptor alkynylplatinum(II) terpyridyl and bipyridyl complexes have been successfully synthesised by the reaction of organic alkynes and the corresponding chloroplatinum(II) terpyridyl and bipyridyl complex precursors, respectively, in the presence of a catalytic amount of copper(I) iodide

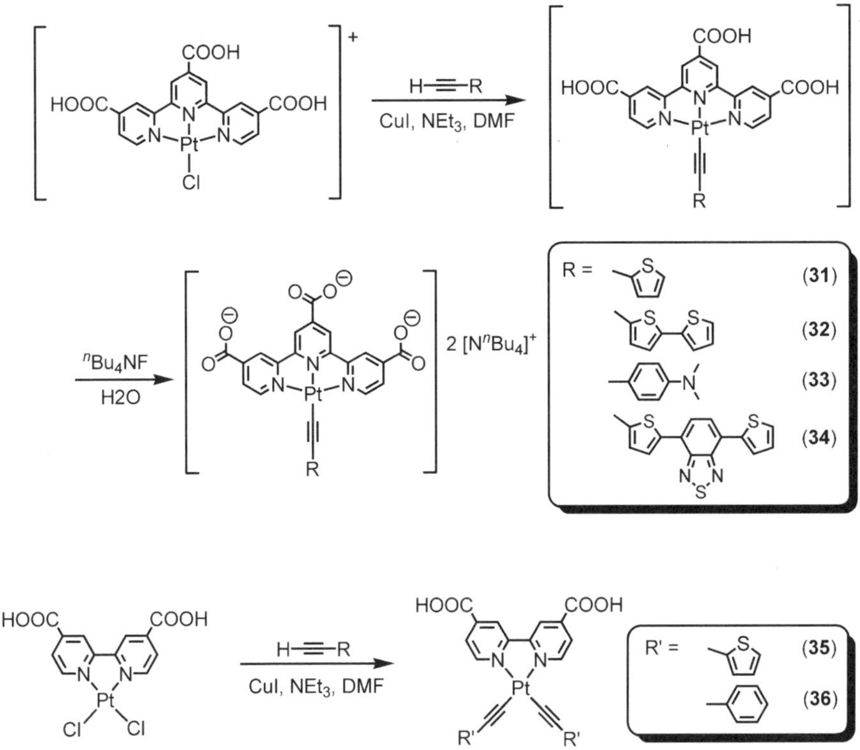

Scheme 7.7 Synthetic route of alkynylplatinum(II) terpyridiyl and bipyridyl sensitisers.

(Scheme 7.7). For the alkynylplatinum(II) terpyridyl complexes **31**−**34**, metathesis reactions with tetra-*n*-butylammonium hydroxide in aqueous solution were performed to afford the deprotonated form of the anionic complexes in tetra-*n*-butylammonium salt, which are soluble in common organic solvents.[68d] The alkynylplatinum(II) bipyridyl complexes **35** and **36** were isolated in the neutral form. The IR spectra of the alkynylplatinum(II) terpyridyl and bipyridyl complexes show a band at 2099 and 2110 cm^{-1}, respectively, typical of their $v(C\equiv C)$ stretching frequencies. All the complexes gave intense colour in the solid and solution states.

7.4.3 Photophysical Studies

The electronic absorption spectra of complexes **31**–**36** exhibit an intense absorption band at 302−388 nm and a less intense absorption band at 412−534 nm in methanol for the alkynylplatinum(II) terpyridyl complexes and in *N*,*N*-dimethylformamide for the alkynylplatinum(II) bipyridyl complexes at 298 K. Their electronic absorption spectra are depicted in Figure 7.11 and their photophysical data are summarised in Table 7.4. On the basis of the

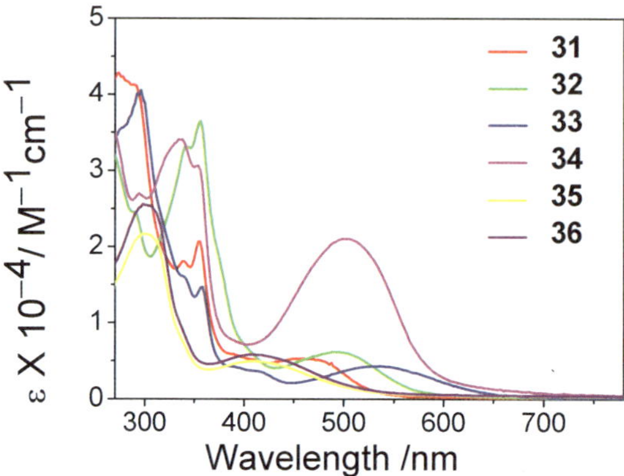

Figure 7.11 Electronic absorption spectra of complexes **31–34** in MeOH and complexes **35** and **36** in DM. Reproduced with permission from ref. 68d.

previous spectroscopic studies on the related systems of $[Pt(N^{\wedge}N^{\wedge}N)(C{\equiv}C{-}R)]^+$ and $[Pt(N^{\wedge}N)(C{\equiv}C{-}R)_2]$ complexes ($N^{\wedge}N^{\wedge}N$ = terpyridyl ligand; $N^{\wedge}N$ = bipyridyl ligand),[68a–c,88c,90b,91] the high-energy absorptions located at about 302–388 nm are assigned as the intraligand (IL) $\pi \rightarrow \pi^*(C{\equiv}C{-}R)$ and $\pi \rightarrow \pi^*$(tctpy/dcbpy) transitions of the respective alkynyl and substituted polypyridine ligands (tctpy = 4,4′,4″-tricarboxy-2,2′:6′,2″-terpyridine; dcbpy = 4,4′-dicarboxy-2,2′-bipyridine). In view of the fact that the complexes containing strong electron-donating groups on the alkynyl ligands, the low-energy band located in the visible region is assigned to an admixture of metal-to-ligand charge transfer (MLCT) [$d\pi(Pt) \rightarrow \pi^*$(tctpy/dcbpy)] and alkynyl-to-polypyridine ligand-to-ligand charge transfer (LLCT) [$\pi(C{\equiv}C{-}R) \rightarrow \pi^*$(tctpy/dcbpy)] transitions. Since the free ligand of Th-BTD-

Table 7.4 Electronic absorption and emission data for **31–36**.

Complex	Absorption [a] λ_{abs}/nm (ε/dm^3 mol^{-1} cm^{-1})	Emission [b] λ_{em}/nm (τ_o/μs)	Φ_{em} [d]
31	336 (18 260), 356 (20 680), 400 (5480), 460 (5690)	650 (0.64)	5.2×10^{-4}
32	356 (36 490), 406 (6265), 492 (6170)	–[c]	–
33	290 (40 550), 340 (16 000), 356 (14 600), 534 (4300)	–[c]	–
34	304 (27 000), 334 (34 000), 352 (31 000), 502 (21 000)	570	1.2×10^{-6}
35	302 (25 430), 412 (5800)	570 (0.85)	8.6×10^{-4}
36	330 (25 450), 414 (5790)	592 (0.83)	8.4×10^{-4}

[a]Electronic absorption data of complexes **31–34** were measured in MeOH; **35–36** were measured in DMF. [b]Measured in MeOH. [c]Nonemissive. [d]In butyronitrile glass. The luminescence quantum yield, measured at room temperature using [Ru(bpy)$_3$]Cl$_2$ as a standard.

Th absorbs strongly in the same region with molar extinction coefficients of the order of 10^4 $dm^3mol^{-1}cm^{-1}$, the observation of such higher molar extinction coefficients in this region suggests that IL [$\pi \rightarrow \pi^*(C{\equiv}C–R)$] transition of Th-BTD-Th would also contribute to this lowest energy absorption band in complex **34**.

Upon photoexcitation, complexes **31**, **35** and **36** displayed intense emission bands at *ca.* 570–650 nm in fluid solution at 298 K, an emission origin of a ^{3}MLCT [$d\pi(Pt) \rightarrow \pi^*(tctpy/dcbpy)$]/^{3}LLCT [$\pi(C{\equiv}C-R) \rightarrow \pi^*(tctpy/dcbpy)$] excited state is tentatively assigned. Complex **34** showed a weak emission band at *ca.* 570 nm, assignable to a metal-perturbed intraligand (IL) excited state.[68d] The photophysical data are tabulated in Table 7.1 and the emission spectra of **31**and **34** in MeOH, and **35** and **36** in DMF are shown in Figure 7.12. Complexes **32** and **33** were found to be nonemissive, probably attributed to the reductive quenching by photoinduced electron transfer (PET) from the strong electron-donating group, as well as the presence of lower-lying ^{3}LLCT state. Other related complexes bearing electron-donating substituents were also reported to be nonemissive or weakly emissive.[68a–c]

In order to obtain more information on the nature of the excited state, nanosecond transient absorption (TA) measurements were performed on complexes **31–36** in degassed dichloromethane solution at room temperature. Upon laser excitation, the platinum(II) terpyridyl and bipyridyl alkynyl complexes undergo electron-density redistribution toward the terpyridine or bipyridine moieties. Similar to the transient absorption results of other related platinum(II) polypyridyl systems,[92] complexes **31–33**, **35** and **36** all showed transient absorption signals typical of the ^{3}MLCT/^{3}LLCT excited-state absorption with the characteristics of the terpyridine and bipyridine radical anion absorptions and the decay profiles of such absorptions were found to match with those of the emissive state lifetimes. On the other hand, the transient absorption spectrum of complex **34** showed very long-lived strong

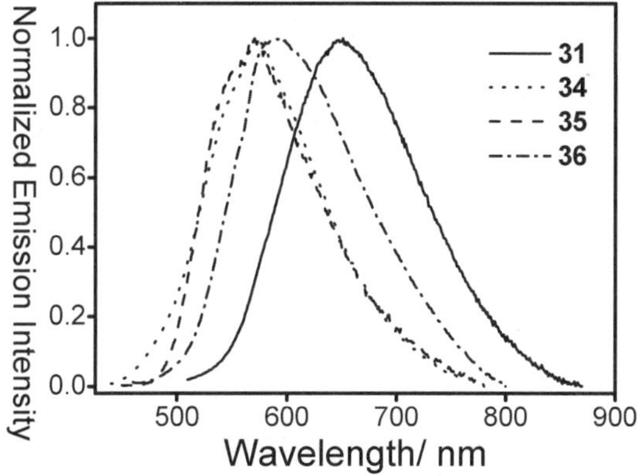

Figure 7.12 Normalised emission spectra of complexes **31** and **34** in MeOH, and **35** and **36** in DMF. Reproduced with permission from ref. 68d.

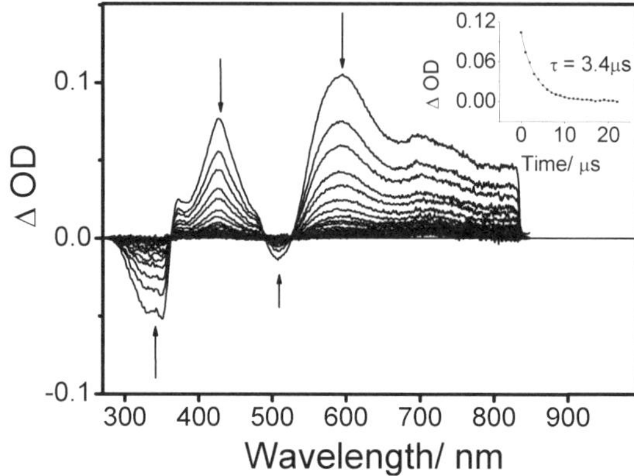

Figure 7.13   Transient absorption difference spectra of **34** in CH_2Cl_2 following a 355 nm laser pulse. Reproduced with permission from ref. 68d.

absorption bands at *ca.* 360–480 nm and 530–670 nm and a very broad band that extends into the NIR region (690–800 nm) (Figure 7.13). It is noteworthy that the transient signals were found to show a monoexponential decay of 3.4 μs, which is on a much longer timescale than that of the emission lifetime. On the basis of these findings, the generation of the bithienylbenzothiadiazole radical cation was suggested to be responsible for the absorption at *ca.* 580 nm, as a result of the formation of a charge-separated state that can be alternatively described as a $[Pt(tctpy)^{-\bullet}\text{–}C\equiv C\text{-Th-BTD-Th})^{+\bullet}]$ state.

7.4.4   Photocurrent–Voltage Characteristics of Pt-Dye-Coated TiO₂ Electrode

The functionalisation of carboxylate groups on the alkynylplatinum(II) terpyridyl and bipyridyl complexes **31–36** not only serves the purpose of providing the anchoring group but also acts as the electron-acceptor of donor-acceptor sensitisers for the dye-sensitised solar cells. On the other hand, the electron-rich substituents on the alkynyl moiety are anticipated to function as the electron donor, which gives an additional advantage of extension of the absorption energies to the red for better absorption and collection of solar sunlight in such sensitisers, as discussed in the previous section. In this study, the platinum(II) sensitisers **31–36** were coated on TiO_2 electrodes, and the current–voltage curves of the cells based these dye-sensitised TiO_2 electrode systems obtained both in the dark and under illumination of air mass (AM) 1.5G sunlight (100 mW/cm²) are shown in Figure 7.14a. The short-circuit current (J_{sc}) and open-circuit voltage (V_{oc}) values for each dye-sensitised TiO_2 electrode with an electrolyte containing 0.6 M 1,3-dimethylimidazolium iodide (DMII), 0.05 M I_2, 0.5 M *tert*-butylpyridine (TBP) and 0.1 M LiI in methoxyacetonitrile were recorded. On the basis of the spectroscopic E_{0-0} values and the electrochemical data $E°$, the estimated excited-state reduction

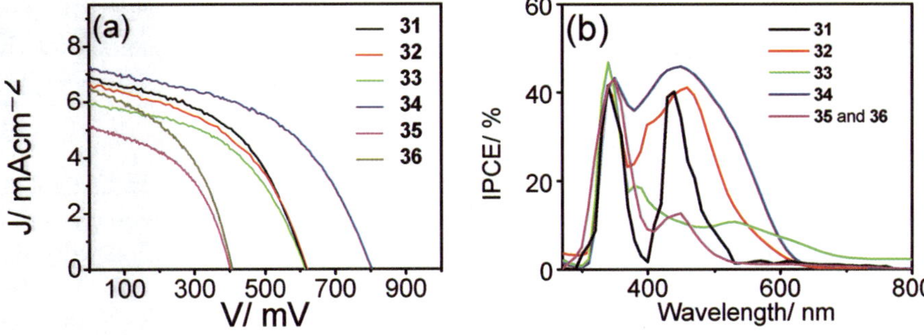

Figure 7.14 (a) *I–V* characteristics of the devices fabricated from **31–36** measured under AM 1.5G illumination. (b) Photocurrent action spectra of nanocrystalline TiO_2 films sensitised by complexes **31–36**. Reproduced with permission from ref. 68d.

potentials of **31–36** are sufficient to inject an electron into the conduction band of TiO_2 (–0.7 V *vs.* SCE).[93] The device performances of the platinum(II) terpyridyl system **31−34** are found to be higher than that of the platinum(II) bipyridyl system **35** and **36**. In general, the photocurrent density was found to increase with the extension of the π-conjugated system and the increase in electron-donating ability of the alkynyl donor functionality, from $-C{\equiv}C-C_6H_4NMe_2$ (6.04 mA/cm^2) to $-C{\equiv}C-C_8H_5S_2$ (6.85 mA/cm^2), and to the more π-conjugated and stronger electron-donating ability of $-C{\equiv}C-C_{14}H_7N_2S_3$ (7.12 mA/cm^2). Table 7.5 summarises the performance of the dye-sensitised solar cells for complexes **31–36**, with the most efficient sensitiser $[Pt(tctrpy)(C{\equiv}C-C_{14}H_7N_2S_3)][N^nBu_4]_2$ **34** exhibiting a photocurrent of 7.12 mA/cm^2 and an open-circuit potential of 780 mV with a fill factor (FF) of 0.65 under the illumination of AM 1.5G sunlight (100 mW/cm^2).

The photocurrent action spectra of each dye-coated TiO_2 electrode, with the incident photon-to-current conversion efficiency (IPCE) values plotted as a function of wavelength, collected under short-circuit conditions showed a strong correlation with the electronic absorption spectra of the metal complexes (Figure 7.14b). The introduction of the strongly electron-donating bithienylbenzothiadiazole alkynyl substituent in complex **34** would render the enhancement and redshifting of the MLCT/LLCT absorption band in the

Table 7.5 Performance of dye-sensitised solar cells for complexes **31–36**.

Complex	J_{sc} (mA cm^{-2})	V_{oc} (mV)	FF	η (%)	IPCE (%)
31	6.75	620	0.61	2.6	42 (340 nm); 40 (440 nm)
32	6.85	620	0.62	2.7	45 (340 nm); 40 (460 nm)
33	6.04	620	0.60	2.4	46 (340 nm); 10 (550 nm)
34	7.12	780	0.65	3.6	44 (340 nm); 46 (450 nm)
35	5.20	400	0.58	1.6	43 (340 nm); 13 (420 nm)
36	6.65	400	0.58	1.8	43 (340 nm); 13 (420 nm)

visible region. This, together with the long-lived charge-separated state observed in the TA study, the improved light-harvesting efficiency and the resulting photocurrent generation are responsible for the best efficiency of complex **34**. Compared to the power-conversion efficiency of the [Ru(dcbpy)$_2$(NCS)$_2$] dye (10.4%), complexes **31–36** (1.6–3.6%) showed relatively lower power conversion efficiencies, which is probably due to the high ratios of the rates of recombination and self-quenching phenomenon that counteract the light-harvesting behaviour of the dye. Similar findings were also observed and reported in other diimineplatinum(II) and dithiolatoplatinum(II) polypyridine complexes by Eisenberg and Sugihara and coworkers.[94] In order to increase the efficiency of such class of platinum(II) sensitiser-based solar cells, further optimisation of the molecular design and electrolyte combination with the modification of donor–acceptor ligands should be addressed.

7.5 Summary

This chapter summarises some of our works on the design, syntheses and photophysical studies of selected transition-metal alkynyl complexes, such as multinuclear branched platinum(II) bis-phosphino bis-alkynyl, bis-cyclometa-lated alkynylgold(III), and alkynylplatinum(II) terpyridyl and bipyridyl systems. The incorporation of alkynyl ligands into the transition-metal centres, in many cases, would enhance the photophysical properties of these systems through the variation or perturbation of the charge-transfer excited states. Their electronic absorption, emission and excited-state redox properties are anticipated to be readily tuned by making use of the diversity and versatility of ligand design and synthesis as well as the variety of the structural modifications. Through the systematic study on the variation of ligands with different functionalities, one can study the structure–property relationships of the various classes of luminescent transition-metal complexes. The approach of "function by design" could be employed to explore and exploit the potential applications of such luminescent transition-metal complexes. The approach of "function by design" has been demonstrated for the potential applications of such luminescent transition-metal complexes in photofunctional materials for nonlinear optics, organic light-emitting diodes and dye-sensitised solar cells. Numerous multinuclear branched platinum(II) complexes, which contain various central core and peripheral ligands possessing nonlinear optical properties, have been designed and synthesised. Their two-photon absorption and two-photon-induced luminescence have also been studied, with the correlation of the structure–property relationship. The direction of energy transfer within the branched molecules could be varied by controlling the relative energy levels between the central core and the peripheral ligands, leading to the tuning of emission energy using one-photon or two-photon excitation. A novel class of luminescent bis-cyclometalated alkynylgold(III) complexes has been demonstrated to possess EL properties and has been employed as electrophosphorescent emitter or dopant of OLEDs with high brightness and efficiency. This series of gold(III) complexes is an entirely new class of triplet emitters and is unique, and their electroluminescence (EL) properties have not been studied and exploited before. A series of alkynylplatinum(II) terpyridyl and bipyridyl complexes has been synthesised and found to show light-to-electricity conversion properties. These dyes were developed as sensitisers for the application in nanocrystalline TiO_2 dye-sensitised solar cells (DSSCs).

Acknowledgements

V.W.-W.Y. acknowledges support from The University of Hong Kong and the University Grants Committee Areas of Excellence Scheme (AoE/P-03/08). This work has also been supported by the General Research Fund (GRF) from the Research Grants Council of Hong Kong Special Administrative Region, China (HKU 7060/09P and HKU 7063/10P).

References

1. (a) *Part One: Top-Down Strategy in Nanotechnology, Volume 8: Nanostructured Surfaces*, ed. L.-F. Chi, Wiley-VCH, Weinheim, 2010, and the references therein; (b) X. Fang, Y. Bando, U. K. Gautam, C. Ye and D. Golberg, *J. Mater. Chem.*, 2008, **18**, 509; (c) S. Acharya, D. D. Sarma, Y. Golan, S. Sengupta and K. Ariga, *J. Am. Chem. Soc.*, 2009, **131**, 11282; (d) K. Ariga, X. Hu, S. Mandal and J. P. Hill, *Nanoscale*, 2010, **2**, 198.

2. (a) *Part Two: Bottom-Up Strategy in Nanotechnology, Volume 8: Nanostructured Surfaces*, ed. L.-F. Chi, Wiley-VCH, Weinheim, 2010, and the references therein; (b) F. Zeng and S. C. Zimmerman, *Chem. Rev.*, 1997, **97**, 1681; (c) T. Nakanishi, T. Michinobu, K. Yoshida, N. Shirahata, K. Ariga, H. Möhwald and D. G. Kurth, *Adv. Mater.*, 2008, **20**, 443; (d) M. Sathish, K. Miyazawa, J. P. Hill and K. Ariga, *J. Am. Chem. Soc.*, 2009, **131**, 6372.

3. C. W. Tang and S. A. van Slyke, *Appl. Phys. Lett.*, 1987, **51**, 913.

4. (a) M. A. Baldo, M. E. Thompson and S. R. Forrest, *Nature*, 2000, **403**, 750; (b) C. Adachi, M. A. Baldo, S. R. Forrest and M. E. Thompson, *Appl. Phys. Lett.*, 2000, **77**, 904; (c) C. Adachi, M. A. Baldo, M. E. Thompson and S. R. Forrest, *J. Appl. Phys.*, 2001, 90, 5048; (d) M. A. Baldo, S. Lamansky, P. E. Burrows, M. E. Thompson and S. R. Forrest, *Appl. Phys. Lett.*, 1999, **75**, 4; (e) B. W. D'Andrade, M. E. Thompson and S. R. Forrest, *Adv. Mater.*, 2002, **14**, 147; (f) X. Yang, D. C. Müller, D. Neher and K. Meerholz, *Adv. Mater.*, 2006, **18**, 948; (g) Y. Sun, N. C. Giebink, H. Kanno, B. Ma, M. E. Thomspon and S. R. Forrest, *Nature*, 2006, **440**, 908; (h) M. S. Lowry and S. Bernhard, *Chem. Eur. J.*, 2006, **12**, 7970; (i) N. R. Evans, L. S. Devi, C. S. K. Mak, S. E. Watkins, S. I. Pascu, A. Köhler, R. H. Friend, C. K. Williams and A. B. Holmes, *J. Am. Chem.*

Soc., 2006, **128**, 6647; (j) M. Ikai, Tokito, S. Y. Sakamoto, T. Suzuki and Y. Taga, *Appl. Phys. Lett.*, 2001, **79**, 156; (k) Y. Chi and P. T. Chou, *Chem. Soc. Rev.*, 2010, **39**, 638.

5. (a) B. W. D'Andrade, J. Brooks, V. Adamovich, M. E. Thompson and S. R. Forrest, *Adv. Mater.*, 2002, **14**, 1032; (b) V. Adamovich, J. Brooks, A. Tamayo, A. M. Alexander, P. I. Djurovich, B. W. D'Andrade, C. Adachi, S. R. Forrest and M. E. Thompson, *New J. Chem.*, 2002, **26**, 1171; (c) W. Lu, B. X. Mi, M. C. W. Chan, Z. Hui, C. M. Che, N. Zhu and S. T. Lee, *J. Am. Chem. Soc.*, 2004, **126**, 4958.

6. (a) S. Welter, K. Krunner, J. W. Hofstraat and D. De Cola, *Nature*, 2003, **421**, 54; (b) C. Y. Liu and A. J. Bard, *J. Am. Chem. Soc.*, 2002, **124**, 4190; (c) H. Rudmann, S. Shimada and M. F. Rubner, *J. Am. Chem. Soc.*, 2002, **124**, 4918; (d) Y. L. Tung, L. S. Chen, Y. Chi, P. T. Chou, Y. M. Cheng, E. Y. Li, G. H. Lee, C. F. Shu, F. I. Wu and A. J. Carty, *Adv. Funct. Mater.*, 2006, **16**, 1615; (e) Y. Chi and P. T. Chou, *Chem. Soc. Rev.*, 2007, **36**, 1421.

7. (a) B. Carlson, G. D. Phelan, W. Kaminsky, L. Dalton, X. Z. S. L. Jiang and A. K. Y. Jen, *J. Am. Chem. Soc.*, 2002, **124**, 14162; (b) S. Bernhard, X. Gao, G. G. Malliaras and H. D. Abruna, *Adv. Mater.*, 2002, **14**, 433; (c) Y. L. Tung, P. C. Wu, C. S. Liu, Y. Chi, J. K. Yu, Y. H. Hu, P. T. Chou, S. M. Peng, G. H. Lee, Y. Tao, A. J. Carty, C. F. Shu and F. I. Wu, *Organometallics*, 2004, **23**, 3745.

8. (a) K. M. C. Wong, X. Zhu, L. L. Hung, N. Zhu, V. W. W. Yam and H. S. Kwok, *Chem. Commun.*, 2005, 2906; (b) V. K. M. Au, K. M. C. Wong, D. P. K. Tsang, M. Y. Chan, N. Zhu and V. W. W. Yam, *J. Am. Chem. Soc.*, 2010, **132**, 14273.

9. (a) J. H. Burroughs, D. D. C. Bradley, A. R. Brown, N. Marks, R. H. Friend, P. L. Burn and A. B. Holmes, *Nature*, 1990, **347**, 539; (b) D. Braun, A. J. Heeger, *Appl. Phys. Lett.* 1991, **58**, 1982; (c) S. Doi, M. Kuawbara, T. Noguchi and T. Ohnishi, *Synth. Met.*, 1993, **55–57**, 4174.

10. (a) Q. G. Kong, D. Zhu, Y. W. Quan, Q. M. Chen, J. F. Ding, J. P. Lu and Y. Tao, *Chem. Mater.*, 2007, **19**, 3309; (b) B. J. Laughlin and R. C. Smith, *Macromolecules*, 2010, **43**, 3744; (c) A. Nagai, S. Kobayashi, Y. Nagata, K. Kokado, H. Taka, H. Kita, Y. Suzuri and Y. Chujo, *J. Mater. Chem.*, 2010, **20**, 5196; (d) K. T. Kamtekar, H. L. Vaughan, B. P. Lyons, A. P. Monkman, S. U. Pandya and M. R. Bryce, *Macromolecules*, 2010, **43**, 4481; (e) J. K. Lee, H. H. Fong, A. A. Zakhidov, G. E. McCluskey, P. G. Taylor, M. S. Berrios, H. D. Abruna, A. B. Holmes, G. G. Malliaras and C. K. Ober, *Macromolecules*, 2010, **43**, 1195; (f) H. Wu, L. Ying, W. Yang and Y. Cao, *WOLEDs and Organic Photovoltaics, Green Energy and Technology*, ed. V.W.W. Yam, Springer-Verlag, Berlin Heidelberg, 2010, pp 38–76.

11. (a) C. W. Tang, *Appl. Phys. Lett.*, 1986, **48**, 183; (b) P. Peumans and S. R. Forrest, *Appl. Phys. Lett.*, 2001, **79**, 126; (c) P. Peumans, S. Uchida and S. R. Forrest, *Nature*, 2003, **425**, 158; (d) H. Hoppe and N. S. Sariciftci, *J.*

Mater. Res., 2004, **19**, 1924; (e) F. Yang and S. R. Forrest, *ACS Nano*, 2008, **2**, 1022; (f) M. C. Scharber, D. Mühlbacher, M. Koppe, P. Denk, C. Waldauf, A. J. Heeger and C. J. Brabec, *Adv. Mater.*, 2006, **18**, 789; (g) B. P. Rand, D. P. Burk and S. R. Forrest, *Phys. Rev.*, 2007, **75**, 115327;

12. (a) G. Li, V. Shrotriya, J. Huang, Y. Yao, T. Moriarty, K. Emery and Y. Yang, *Nature Mater.*, 2005, **4**, 864; (b) J. Y. Kim, S. H. Kim, H. H. Lee, K. Lee, W. Ma, X. Gong and A. J. Heeger, *Adv. Mater.*, 2006, **18**, 572; (c) J. Peet, J. Y. Kim, N. E. Coates, W. L. Ma, D. Moses, A. J. Heeger and G. C. Bazan, *Nature Mater.*, 2007, **6**, 497; (d) S. Günes, H. Neugebauer and N. S. Sariciftci, *Chem. Rev.*, 2007, **107**, 1324; (e) Y. Liang, Y. Wu, D. Feng, S. T. Tsai, H. J. Son, G. Li and L. Yu, *J. Am. Chem. Soc.*, 2009, **131**, 56.

13. (a) D. S. Chemla and J. Zyss, *Nonlinear Optical Properties of Organic Molecules and Crystals*; Academic Press: Orlando, 1987; (b) S. Mukamel, *Principles of Nonlinear Optical Spectroscopy*; Oxford University Press, New York, 1995; (c) *Organic Materials for Nonlinear Optics II*; ed. R. A. Hann, D. Bloor, The Royal Society of Chemistry, Cambridge, 1991; (d) J. Seto, N. Asai, I. Fujiwara, T. Ishibashi, T. Kamei and S. Tamura, *Thin Solid Films*, 1996, **273**, 97; (e) Y. Maruyama, H. Hoshi, S. L. Fang and K. Kohama, *Synth. Met.*, 1995, **71**, 1653.

14. (a) J. Simon and J.-J. Andre, *Molecular Semiconductors: Photoelectrical Properties and Solar Cells*, Springer-Verlag, Berlin, 1985; (b) D. Wohrle and D. Meissner, *Adv. Mater.*, 1991, **3**, 129; (c) *Phthalocyanines: Properties and Applications*, ed. C. C. Leznoff, A. B. P. Lever, Wiley-VCH, New York, 1989; (d) R. B. Taylor, P. E. Burrows and S. R. Forrest, *IEEE Photonics Technol. Lett.*, 1997, **9**, 365.

15. G. S. He, L. S. Tan, Q. Zheng and P. N. Prasad, *Chem. Rev.*, 2008, **108**, 1245.

16. (a) I. R. Whittall, A. M. McDonagh, M. G. Humphrey and M. Samoc, *Adv. Organomet. Chem.*, 1998, **42**, 291; (b) R. Whittall, A. M. McDonagh, M. G. Humphrey and M. Samoc, *Adv. Organomet. Chem.*, 1999, **43**, 349.

17. (a) *Materials for Nonlinear Optics: Chemical Perspectives*, ed. S.R. Marder, J. E. Sohn, G. D. Stucky, ACS Symposium Series vol. **455**, American Chemical Society, Washington, DC, 1991; (b) *Inorganic Materials*, ed. D. W. Bruce, D. O'Hare, Wiley, Chichester, U.K, 1992; (c) N. J. Long, *Angew. Chem. Int. Ed. Engl.*, 1995, **34**, 21; (d) M. G. Humphrey and C. E. Powell, *Coord, Chem, Rev.*, 2004, **248**, 725; (e) J. Manna, K. D. John and M. D Hopkins, *Adv. Organomet. Chem.*, 1995, **38**, 79; (f) N. J. Long and C. K. Williams, *Angew. Chem. Int. Ed.*, 2003, **42**, 2586; (g) D. Beljonne, H. F. Wittman, A. Kohler, S. Graham, M. Younus, J. Lewis, P. R. Raithby, M. S. Khan, R. H. Friend and J. L. Bredas, *J. Chem. Phys.*, 1996, **105**, 3868; (h) F. Paul and C. Lapinte, *Coord. Chem. Rev.*, 1998, **178–180**, 431; (i) V. H. Houlding and V. M. Miskowski, *Coord. Chem. Rev.*, 1991, **111**, 145.

18. (a) Y.-R. Shen, *The Principles of Nonlinear Optics*, JohnWiley & Sons, Inc., New York, 1984; (b) R. W. Boyd, *Nonlinear Optics*, Academic Press, Boston, 2008.

19. P. Franken, A. Hill, C. Peters and G. Weinreich, *Phys. Rev. Lett.*, 1961, **7**, 118.

20. T. H. Maiman, *Nature* 1960, *187*, 493.

21. (a) *Nonlinear Optical Properties of Organic Molecules and Crystals I*, ed. D. S. Chemla, J. Zyss, Academic Press, Orlando, 1987; (b) *Nonlinear Optical Properties of Organic Molecules and Crystals II*, ed. D. S. Chemla, J. Zyss, Academic Press, Orlando, 1987; Nonlinear Optical Properties of Organic and Polymeric Materials, ed. D. J. Williams, American Chemical Society, Washington, DC, 1983; (c) B. Kirtman and B. Champagne, *Int. Rev. Phys. Chem.*, 1997, **16**, 389; (d) *Organic Materials for Non-Linear Optics*, ed. R. A. Hann, D. Bloor, Royal Society of Chemistry, London, 1989; (e) *Organic Materials for Non-Linear Optics II*, ed. R. A. Hann, D. Bloor, Royal Society of Chemistry, London, 1991; (f) T. Verbiest, S. Houbrechts, M. Kauranen, K. Clays and A. Persoons, *J. Mater. Chem.*, 1997, **7**, 2175.

22. M. Goppert-Mayer, *Ann. Phys.*, 1931, **9**, 273.

23. (a) J. H. Strickler and W. W. Webb, *Adv. Mater.*, 1993, **5**, 479; (b) D. A. Parthenopoulos and P. M. Rentzepis, *Science*, 1999, **245**, 843; (c) G. S. He, R. Gvishi, P. N. Prasad and B. Reinhardt, *Opt. Commun.*, 1995, **117**, 133.

24. (a) A. Mukherjee, *Appl. Phys. Lett.*, 1993, **62**, 3423; (b) Q. Zheng, G. S. He, T.-C. Lin and P. N. Prasad, *J. Mater. Chem.*, 2003, **13**, 2499.

25. (a) A. Mukherjee, *Appl. Phys. Lett.*, 1993, **62**, 3423; (b) Q. Zheng, G. S. He, T.-C. Lin and P. N. Prasad, *J. Mater. Chem.*, 2003, **13**, 2499.

26. (a) K. D. Belfield, K. J. Schafer, Y. Liu, J. Liu, X. Ren and E. W. Van Stryland, *J. Phys. Org. Chem.*, 2000, **13**, 837; (b) M. Gu, T. Tannous and C. J. R. Sheppard, *Opt. Commun.*, 1995, **117**, 406.

27. (a) X. L. Wang, J. Krebs, M. H. Al-Nuri, E. Pudavar, S. Ghosal, C. Liebow, A. A. Nagy, A. V. Schally and P. N. Prasad, *Proc. Natl. Acad. Sci. U.S.A.*, 1999, **96**, 11081; (b) S. Maruo, O. Nakamura and S. Kawata, *Opt. Lett.*, 1997, **22**, 132.

28. G. S. He, J. D. Bhawalkar, C. F. Zhao and P. N. Prasad, *Appl. Phys. Lett.*, 1995, **67**, 2433.

29. G. S. He, L. Yuan, J. D. Bhawalkar and P. N. Prasad, *Appl. Opt.*, 1997, **36**, 3387.

30. (a) J. D. Bhawalkar, N. D. Kumar, C. F. Zhao and P. N. Prasad, *J. Clin. Med. Surg.*, 1997, **37**, 510; (b) P. K. Frederiksen, M. Jorgensen and P. R. Ogilby, *J. Am. Chem. Soc.*, 2001, **123**, 1215.

31. (a) Q. Wang, *J. Mater. Chem.*, 2005, **15**, 4502; (b) P. Shao, Z. Li, J. Qin, H. Gong, S. Ding and Q. Wang, *Aust. J. Chem.*, 2006, **59**, 49; (c) Q. Zheng, G. S. He and P. N. Prasad, *Chem. Mater.*, 2005, **17**, 6004.

32. (a) S. R. Marder, *Chem. Commun.*, 2006, 131; (b) Z.-L. Huang, H. Lei, N. Li, Z.-R. Qiu, H.-Z. Wang, J.-D. Guo, Y. Luo, Z.-P. Zhong, X.-F. Liu and

Z.-H. Zhou, *J. Mater. Chem.*, 2003, **13**, 708; (c) M. Albota, D. Beljonne, J.-L. Bredas, J. E. Ehrlich, J.-Y. Fu, A. A. Heikal, S. E. Hess, T. Kogej, M. D. Levin, S. R. Marder, D. McCord-Maughon, J. W. Perry, H. Rockel, M. Rumi, G. Subramaniam, W. W. Webb, X.-L. Wu and C. Wu, *Science*, 1998, **281**, 1653; (d) B. A. Reinhardt, L. L. Brott, S. J. Clarson, A. G. Dillard, J. C. Bhatt, L. Yuan, G. S. He and P. N. Prasad, *Chem. Mater.*, 1998, **10**, 1863.

33. P. Shao, B. Huang, L. Chen, Z. Liu, J. Qin, H. Gong, S. Ding and Q. Wang, *J. Mater. Chem.*, 2005, **15**, 4502.

34. B. R. Cho, K. H. Son, S. H. Lee, Y.-S. Song, Y.-K. Lee, S.-J. Jeon, J. H. Choi, H. Lee and M. Cho, *J. Am. Chem. Soc.*, 2001, **123**, 10039.

35. H. J. Lee, J. Sohn, J. Hwang, S. Y. Park, H. Choi and M. Cha, *Chem. Mater.*, 2004, **16**, 456.

36. S.-I. Kato, T. Matsumoto, M. Shigeiwa, H. Gorohmaru, M. Shuuichi, T. Ishii and S. Mataka, *Chem. Eur. J.*, 2006, **12**, 2303.

37. Q. Zheng, G. S. He and P. N. Prasad, *Chem. Mater.*, 2005, **17**, 6004.

38. (a) J. W. Perry, K. Mansour, I.-Y. S. Lee, X.-L. Wu, P. V. Bedworth, C.-T. Chen, D. Ng, S. R. Marder, P. Miles, T. Wada, M. Tian and H. Sasabe, *Science* 1996, **273**, 1533; (b) Q. Zheng, G. S. He and P. N. Prasad, *J. Mater. Chem.*, 2005, **15**, 579; (c) R. P. Briñas, T. Troxler, R. M. Hochstrasser and S. A. Vinogradov, *J. Am. Chem. Soc.*, 2005, **127**, 11851; (d) S. Das, A. Nag, D. Goswami and P. K. Bharadwaj, *J. Am. Chem. Soc.*, 2006, **128**, 402.

39. (a) M. L. H. Green, S. R. Marder, M. E. Thompson, J. A. Bandy, D. Bloor, P. V. Kolinsky and R. J. Jones, *Nature*, 1987, **330**, 360; (b) S. Barlow and S. R. Marder, *Chem. Commun.*, 2000, 1555; (c) M. C. B. Colbert, A. J. Edwards, J. Lewis, N. J. Long, N. A. Page, D. G. Parker and P. R. Raithby, *J. Chem. Soc., Dalton Trans.*, 1994, **17**, 2589.

40. S. D. Bella, I. Fragala, I. Ledoux, M. A. Diaz-Garcia, P. G. Lacroix and T. J. Marks, *Chem. Mater.*, 1994, **6**, 881.

41. A. M. McDonagh, M. G. Humphrey, M. Samoc and B. Luther-Davies, *Organometallics*, 1999, **18**, 5195.

42. (a) H. B. Fyfe, M. Mlekuz, G. Stringer, N. J. Taylor and T. B. Marder, *Inorganic and Organometallic Polymers with Special Properties*, ed. R. M. Laine, NATO ASI Series, Kluwer Acad. Publ., Dordrecht, The Netherlands, p331, 1992; (b) A. Furlani, S. Licoccia, M. V. Russo, A. C. Villa and C. Guastini, *J. Chem. Soc., Dalton Trans*, 1982, 2449; (c) A. Furlani, S. Licoccia, M. V. Russo, A. C. Villa and C. Guastini, *J. Chem. Soc., Dalton Trans*, 1984, 2197; (d) L. Sacksteder, E. Baralt, B. A. DeGraff, C. M. Lukehart and J. N. Demas, *Inorg. Chem.*, 1991, **30**, 2468; (e) C. L. Choi, Y. F. Cheng, C. Yip, D. L. Phillips and V. W. W. Yam, *Organometallics*, 2000, **19**, 3192; (g) R. D'Amato, A. Furlani, M. Colapietro, G. Portalone, M. Casalboni, M. Falconieri and M. V. Russo, *J. Orgomet. Chem.*, 2001, **627**, 13; (h) Y. Liu, S. Jiang, K. Glusac, D. H. Powell, D. F. Anderson and K. S. Schanze, *J. Am. Chem. Soc.*, 2002,

124, 12412; (i) J. E. Rogers, J. E. Slagle, D. M. Krein, A. R. Burke, B. C. Hall, A. Fratini, D. G. McLean, P. A. Fleitz, T. M. Cooper, M. Drobizhev, N. S. Makarov, A. Rebane, K.-Y. Kim, R. Farley and K. S. Schanze, *Inorg. Chem.*, 2007, **46**, 6483.

43. (a) D. Beljonne, H. F. Wittman, A. Kohler, S. Graham, M. Younus, J. Lewis, P. R. Raithby, M. S. Khan, R. H. Friend and J. L. Bredas, *J. Chem. Phys.*, 1996, **105**, 3868; (b) J. Lewis, M. S. Khan, A. K. Kakkar, B. F. G. Johnson, T. B. Marder, H. B. Fyfe, F. Wittmann, R. H. Friend and A. E. Dray, *J. Organomet. Chem.*, 1992, **425**, 165; (c) H. F. Wittmann, R. H. Friend, M. S. Khan and J. Lewis, *J. Chem. Phys.*, 1994, **101**, 2694; (d) J. S. Wilson, A. Kohler, R .H. Friend, M. K. Al-Suti, M. R. A. Al-Mandhary, M. S. Khan and P. R. Raithby, *J. Chem. Phys.*, 2000, **113**, 7627; (e) A. Kohler, J. S. Wilson, R. H. Friend, M. K. Al-Suti, M. S. Khan, A. Gerhard and H. Bassler, *J. Chem. Phys.*, 2002, **116**, 9457.

44. (a) J. Lewis, N. J. Long, P. R. Raithby, G. P. Shields, W.-Y. Wong and M. Younus, *J. Chem. Soc., Dalton Trans.*, 1997, 4283; (b) J. Lewis, P. R. Raithby and W.-Y. Wong, *J. Organomet. Chem.*, 1998, **556**, 219; (c) W.-Y. Wong, W.-K. Wong and P. R. Raithby, *J. Chem. Soc. Dalton Trans.*, 1998, 2761; (d) N. Chawdhury, A. Köhler, R. H. Friend, W.-Y. Wong, J. Lewis, M. Younus, P. R. Raithby, T. C. Corcoran, M. R. A. Al-Mandhary and M. S. Khan, *J. Chem. Phys.*, 1999, **110**, 4963; (e) W.-Y. Wong and C.-L. Ho, *Coord. Chem. Rev.*, 2006, **250**, 2627.

45. (a) T. J. McKay, J. A. Bolger, J. Staromlynska and J. R. Davy, *J. Chem. Phys.*, 1998, **108**, 5537; (b) T. J. McKay, J. Staromlynska, P. Wilson and J. Davy, *J. Appl. Phys.*, 1999, **85**, 1337; (c) R. Vestberg, R. Westlund, A. Eriksson, C. Lopes, M. Carlsson, B. Eliasson, E. Glimsdal, M. Lindgren and E. Malmstrom, *Macromolecules*, 2006, **39**, 2238; (d) E. Gilmsdal, M. Carlsson, B. Eliasson, B. Minaev and M. Lindgren, *J. Phys. Chem. A*, 2007, **111**, 244; (e) P. Lind, D. Bostrom, M. Carlsson, A. Eriksson, E. Glimsdal, M. Lindgren and B. Eliasson, *J. Phys. Chem. A*, 2007, **111**, 1589; (f) R. Westlund, E. Glimsdal, M. Lindgren, R. Vestberg, C. Hawker, C. Lopes and E. Malmstrom, *J. Mater. Chem.*, 2008, **18**, 166.

46. C. Desroches, C. Lopes, V. Kessler and S. Parola, *Dalton Trans.*, 2003, 2085.

47. K. Sonogashira, S. Takahashi and N. Hagihara, *Macromolecules*, 1977, **10**, 879.

48. (a) V. W. W. Yam, C. H. Tao, L. Zhang, K. M. C. Wong and K. K. Cheung, *Organometallics*, 2001, **20**, 453; (b) C. H. Tao, N. Zhu and V. W. W. Yam, *Chem. Eur. J.*, 2005, **11**, 1647; (c) C. H. Tao, N. Zhu and V. W. W. Yam, *J. Photochem. Photobiol. A: Chem.*, 2009, **207**, 94.

49. (a) C. H. Tao, N. Zhu and V. W. W. Yam, *J. Photochem. Photobiol. C: Photochem. Rev.*, 2009, **10**, 13; (b) H. Yang, S. J. Xu, C. H. Tao, V. W. W. Yam and J. Zhang, *J. Appl. Phys.*, 2011, **110**, 043105; (c) H. Yang, S. J. Xu, C. H. Tao, V. W. W. Yam and J. Zhang, *J. Appl. Phys.*, 2011, **110**, 079901.

50. C. H. Tao, H. Yang, N. Zhu, V. W. W. Yam and S. J. Xu, *Organometallics*, 2008, **27**, 5453.

51. C. K. M. Chan, C. H. Tao, H. L. Tam, N. Zhu, V. W. W. Yam and K. W. Cheah, *Inorg. Chem.*, 2009, **48**, 2855.

52. C. K. M. Chan, C. H. Tao, K. F. Li, K. M. C. Wong, N. Zhu, K. W. Cheah and V. W. W. Yam, *J. Organomet. Chem.*, 2011, **696**, 1163.

53. C. K. M. Chan, C. H. Tao, K. F. Li, K. M. C. Wong, N. Zhu, K. W. Cheah and V. W. W. Yam, *Dalton Trans.*, 2011, **40**, 10670.

54. H. Masai, K. Sonogashira and N. Hagihara, *Bull. Chem. Soc. Jpn.*, 1971, **44**, 22.

55. L. Sacksteder, E. Baralt, B. A. DeGraff, C. M. Lukehart and J. N. Demas, *Inorg. Chem.*, 1991, **30**, 2468.

56. J. S. Wilson, A. S. Dhoot, A. J. A. B. Seeley, M. S. Khan, A. Kohler and R. H. Friend, *Nature*, 2001, **413**, 828.

57. S. M. Kelly, *Flat Panel Displays, Advanced Organic Materials*, ed. J. A. Connor, RSC Materials Monographs, The Royal Society of Chemistry, Cambridge, 2000.

58. *Organic Light-Emitting Devices, a Survey*, ed. J. Shinar, Springer, New York, 2004.

59. *Organic Electroluminescent Materials and Devices*, ed. S. Miyata, H. S. Nalwa, Gordon and Breach Publishers, Amsterdam, 1997.

60. *Organic Light-Emitting Devices: Synthesis, Properties, and Applications*, ed. K. Müllen, Wiley-VCH, Weinheim, 2006.

61. *Highly Efficient OLEDs with Phosphorescent Materials*, ed. H. Yersin, Wiley-VCH, Weinheim, 2007.

62. A. Vogler and H. Kunkely, *Coord. Chem. Rev.*, 2001, **219–221**, 489.

63. (a) V. W. W. Yam, S. W. K. Choi, T. F. Lai and W. K. Lee, *J. Chem. Soc., Dalton Trans.*, 1993, 1001; (b) C. W. Chan, W. T. Wong and C. M. Che, *Inorg. Chem.*, 1994, **33**, 1266; (c) K. H. Wong, K. K. Cheung, M. C. W. Chan and C. M. Che, *Organometallics*, 1998, **17**, 3505; (d) M. A. Mansour, R. J. Lachicotte, H. J. Gysling and R. Eisenberg, *Inorg. Chem.*, 1998, **37**, 4625.

64. V. W. W. Yam, K. K. W. Lo and K. M. C. Wong, *J. Organomet. Chem.*, 1998, **578**, 3.

65. V. W. W. Yam, *Chem. Commun.*, 2001, 789.

66. (a) V. W. W. Yam, *Acc. Chem. Res.*, 2002, **35**, 555; (b) K. M. C. Wong, C. K. Hui, K. L. Yu and V. W. W. Yam, *Coord. Chem. Rev.*, 2002, **229**, 123.

67. K. M. C. Wong and V. W. W. Yam, *Acc. Chem. Res.*, 2011, **44**, 424.

68. (a) V. W. W. Yam, R. P. L. Tang, K. M. C. Wong and K. K. Cheung, *Organometallics*, 2001, **20**, 4476; (b) K. M. C. Wong, W. S. Tang, X. X. Lu, N. Y. Zhu and V. W. W. Yam, *Inorg. Chem.*, 2005, **44**, 1492; (c) K. M. C. Wong and V. W. W. Yam, *Coord. Chem. Rev.*, 2007, **251**, 2477; (d) E. C. H. Kwok, M. Y. Chan, K. M. C. Wong, W. H. Lam and V. V. W. Yam, *Chem. Eur. J.*, 2010, **16**, 12244.

69. (a) V. W. W. Yam, K. M. C. Wong, N. Zhu, *J. Am. Chem. Soc.*, 2002, **12**, 6506; (b) V. W. W. Yam, K. H. Y. Chan, K. M. C. Wong, N. Zhu, *Chem. Eur. J.*, 2005, **11**, 4535.

70. (a) C. Yu, K. M. C. Wong, K. H. Y. Chan and V. W .W. Yam, *Angew. Chem., Int. Ed. Engl.*, 2005, **44**, 791; (b) C. Yu, K. H. Y. Chan, K. M. C. Wong and V. W. W. Yam, *Chem. Eur. J.*, 2008, **14**, 4577.

71. C. Yu, K. H. Y. Chan, K. M. C. Wong and V. W. W. Yam, *Proc. Natl. Acad. Sci. U.S.A.*, 2006, **103**, 19652.

72. (a) C. Yu, K. H. Y. Chan, K. M. C. Wong and V. W. W. Yam, *Chem. Commun.*, 2009, 3756; (b) M. C. L. Yeung, K. M. C. Wong, Y. K. T. Tsang and V. W. W. Yam, *Chem. Commun.*, 2010, **46**, 7709; (c) C. Y. S. Chung, K. H. Y. Chan and V. W. W. Yam, *Chem. Commun.*, 2011, **47**, 2000.

73. (a) V. W. W. Yam, K. H. Y. Chan, K. M. C. Wong and B. W. K. Chu, *Angew. Chem., Int. Ed. Engl.*, 2006, **45**, 6169; (b) K. H. Y. Chan, H. S. Chow, K. M. C. Wong, M. C. L. Yeung and V. W. W. Yam, *Chem. Sci.*, 2010, **1**, 477.

74. (a) A. Y. Y. Tam, K. M. C. Wong, G. Wang and V. W. W. Yam, *Chem. Commun.* 2007, 2028; (b) A. Y. Y. Tam, K. M. C. Wong and V. W. W. Yam, *J. Am. Chem. Soc.*, 2009, **131**, 6253; (c) A. Y. Y. Tam, K. M. C. Wong and V. W. W. Yam, *Chem. Eur. J.*, 2009, **15**, 4775; (d) A. Y. Y. Tam, K. M. C. Wong, N. Zhu, G. Wang and V. W. W. Yam, *Langmuir*, 2009, **25**, 8685.

75. (a) V. W. W. Yam, K. M. C. Wong, L. L. Hung and N. Zhu, *Angew. Chem., Int. Ed.*, 2005, **44**, 3107; (b) K. M. C. Wong, L. L. Hung, W. H. Lam, N. Zhu and V. W. W. Yam, *J. Am. Chem. Soc.*, 2007, **129**, 4350.

76. V. K. M. Au, K. M. C. Wong, N. Zhu and V. W. W. Yam, *Chem. Eur. J.*, 2010, **17**, 130.

77. K. H. Wong, K. K. Cheung, M. C. W. Chan and C. M. Che, *Organometallics*, 1998, **17**, 5305.

78. K. Goushi, R. Kwong, J. J. Brown, H. Sasabe and C. Adachi, *J. Appl. Phys.*, 2004, **95**, 7798.

79. B. O'Regan and M. Grätzel, *Nature*, 1991, **353**, 737.

80. (a) K. Kalyanasundaram and M. Grätzel, *Coord. Chem. Rev.*, 1998, **177**, 347; (b) M. Grätzel, *J. Photochem. Photobiol. C: Photochem. Rev.*, 2003, **4**, 145; (c) Md. K. Nazeeruddin, C. Klein, P. Liska and M. Grätzel, *Coord. Chem. Rev.*, 2005, **249**, 1460; (d) N. Robertson, *Angew. Chem. Int. Ed.*, 2006, **45**, 2338. (e) M. Grätzel, *Acc. Chem. Res.*, 2009, **42**, 1788.

81. G. Sauvé, M. E. Cass, G. Coia, S. J. Doig, I. Lauermann, K. E. Pomykal and N. S. Lewis, *J. Phys. Chem. A*, 2000, **104**, 6821.

82. G. M. Hasselmann and G. J. Z. Meyer, *Phys. Chem.*, 1999, **212**, 39.

83. E. I. Mayo, K. Kilså, T. Tirrell, P. I. Djurovich, A. Tamayo, M. E. Thompson, N. S. Lewis and H. B. Gray, *Photochem. Photobiol. Sci.*, 2006, **5**, 871.

84. (a) V. M. Miskowski and V. H. Houlding, *Inorg. Chem.*, 1989, **28**, 1529; (b) V. M. Miskowski and V. H. Houlding, *Inorg. Chem.*, 1991, **30**, 4446; (c) V. H. Houlding, V. M. Miskowski, *Coord. Chem. Rev.*, 1991, **111**, 145.

85. (a) J. A. Bailey, M. G. Hill, R. E. Marsh, V. M. Miskowski, W. P. Schaefer and H. B. Gray, *Inorg. Chem.*, 1995, **34**, 4591; (b) W. B. Connick, R. E. Marsh, W. P. Schaefer and H. B. Gray, *Inorg. Chem.*, 1996, **35**, 4585; (c) M. G. Hill, J. A. Baily, V. M. Miskowski and H. B. Gray, *Inorg. Chem.*, 1996, **35**, 4585; (d) W. B. Connick, R. E. Marsh, W. P. Schaefer and H. B. Gray, *Inorg. Chem.*, 1997, **36**, 913.

86. (a) K. W. Jennette, S. J. Lippard, G. A. Bassiliades and W. R. Bauer, *Proc. Natl. Acad. Sci. U.S.A.*, 1974, **71**, 3839; (b) K. W. Jennette, J. T. Gill, J. A. Sadownick and S. J. Lippard, *J. Am. Chem. Soc.*, 1976, **98**, 6159; (c) S. J. Lippard, *Acc. Chem. Res.*, 1978, **11**, 211.

87. H, Kunkely and A. Bogler, *J. Am. Chem. Soc.*, 1990, **112**, 5625.

88. (a) B. C. Tzeng, W. F. Fu, C. M. Che, H. Y. Chao, K. K. Cheung and S. M. Peng, *J. Chem. Soc., Dalton Trans.*, 1990, 1017; (b) H. K. Yip, L. K. Cheng, K. K. Cheung and C. M. Che, *J. Chem. Soc., Dalton Trans.*, 1993, 2933; (c) C. W. Chan, L. K. Cheng and C. M. Che, *Coord. Chem. Rev.*, 1994, **132**, 87; (d) S. W. Lai, M. C. W. Chan, K. K. Cheung and C. M. Che, *Inorg. Chem.*, 1999, **38**, 4262; (e) S. W. Lai and C. M. Che, *Top. Curr. Chem.*, 2004, **241**, 27; (f) S. C. F. Kui, Y. C. Law, G. S. M. Tong, W. Lu, M. Y. Yuen and C. M. Che, *Chem. Sci.*, 2011, **2**, 221.

89. (a) J. A. Zuleta, M. S. Burberry and R. Eisenberg, *Coord. Chem. Rev.*, 1990, **97**, 47; (b) W. Paw, S. D. Cummings, M. A. Mansour, W. B. Connick, D. K. Geiger and R. Eisenberg, *Coord. Chem. Rev.*, 1998, **171**, 125; (c) W. B. Connick, D. Geiger and R. Eisenberg, *Inorg. Chem.*, 1999, **38**, 3264; (d) M. Hissler, W. B. Connick, D. K. Geiger, J. E. McGarrah, D. Lipa, R. J. Lachicotte and R. Eisenberg, *Inorg. Chem.*, 2000, **39**, 447; (e) M. Hissler, J. E. McGarrah, W. B. Connick, D. K. Geiger, S. D. Cummings and R. Eisenberg, *Coord. Chem. Rev.*, 2000, **208**, 115; (f) S. Chakraborty, T.J. Wadas, H. Hester, C. Flaschenreim, R. Schmehl and R. Eisenberg, *Inorg. Chem.*, 2005, **44**, 6284; (g) T. J. Wadas, Q. M. Wang, Y. J. Kim, C. Flaschenreim, T. N. Blanton and R. Eisenberg, *J. Am. Chem. Soc.*, 2004, **126**, 16841.

90. (a) G. Arena, G. Calogero, S. Campagna, L. M. Scolaro, V. Ricevuto and R. Romeo, *Inorg. Chem.*, 1998, **37**, 2763; (b) C. Ed Whittle, J. A. Weinstein, M. W. George and K. S. Schanze, *Inorg. Chem.*, 2001, **40**, 4053; (c) C. J. Adams, S. L. James, X. Liu, P. R. Raithby and L. J. Yellowlees, *J. Chem. Soc., Dalton Trans.*, 2000, 63; (d) F. N. Castellano, I. E. Pomestchenko, E. Shikhova, F. Hua, M. L. Muro and N. Rajapakse, *Coord. Chem. Rev.*, 2006, **250**, 1819; (e) R. S. Osborn and D. Rogers, *J. Chem. Soc., Dalton Trans.*, 1974, 1002; (f) R. H. Herber, M. Croft, M. J. Coyer, B. Bilash and A. Sahiner, *Inorg. Chem.*, 1994, **33**, 2422; (g) J. A. G. Williams, *Top. Curr. Chem.*, 2007, **281**, 205; (h) E. M. A. Ratilla, H. M. Brothers II and N. M. Kostić, *J. Am. Chem. Soc.*, 1987, **109**, 4592; (i) I.

Eryazici, C. N. Moorefield and G. R. Newkome, *Chem. Rev.*, 2008, **108**, 1834; (j) C. S. Peyratout, T. K. Aldridge, D. K. Crites and D. R. McMillin, *Inorg. Chem.*, 1995, **34**, 4484.

91. (a) S. C. Chan, M. C.W. Chan, Y. Wang, C. M. Che, K. K. Cheung and N. Y. Zhu, *Chem. Eur. J.*, 2001, **7**, 4180; (b) F. Hua, S. Kinayyigit, J. R. Cable and F. N. Castellano, *Inorg. Chem.*, 2005, **44**, 471.

92. (a) D. Zhang, L.-Z. Wu, L. Zhou, X. Han, Q.-Z. Yang, L.-P. Zhang and C.-H. Tung, *J. Am. Chem. Soc.*, 2004, **126**, 3440; (b) P. Jarosz, J. Thall, J. Schneider, D. Kumaresan, R. Schmehl and R. Eisenberg, *Energy Environ. Sci.*, 2008, **1**, 573; (c) E. Shikhova, E. O. Danilov, S. Kinayyigit, I. E. Pomestchenko, A. D. Tregubov, F. Camerel, P. Retailleau, R. Ziessel and F. N. Castellano, *Inorg. Chem.*, 2007, **46**, 3038; (d) E. O. Danilov, I. E. Pomestchenko, S. Kinayyigit, P. L. Gentili, M. Hissler, R. Ziessel and F. N. Castellano, *J. Phys. Chem. A*, 2005, **109**, 2465; (e) F. N. Castellano, I. E. Pomestchenko, E. Shikhova, F. Hua, M. L. Muro and N. Rajapakse, *Coord. Chem. Rev.*, 2006, **250**, 1819; (f) P. Jarosz, P. Du, J. Schneider, S. H. Lee, D. McCamant and R. Eisenberg, *Inorg. Chem.*, 2009, **48**, 9653; (g) P. Jarosz, K. Lotito, J. Schneider, D. Kumaresan, R. Schmehl and R. Eisenberg, *Inorg. Chem.*, 2009, **48**, 2420.

93. (a) A. Islam, H. Sugihara, K. Hara, L. P. Singh, R. Katoh, M. Yanagida, Y. Takahashi, S. Murata and H. Arakawa, *Inorg. Chem.*, 2001, **40**, 5371; (b) A. Hagfeldt and M. Grätzel, *Chem. Rev.*, 1995, **95**, 49.

94. (a) M. Yanagida, T. Yamaguchi, M. Kurashige, K. Hara, R. Katoh, H. Sugihara and H. Arakawa, *Inorg. Chem.*, 2003, **42**, 7921; (b) W. B. Connick, D. Geiger and R. Eisenberg, *Inorg. Chem.*, 1999, **38**, 3264; (c) T. W. Hamann, R. A. Jensen, A. B. F. Martinson, H. V. Ryswyk and J. T. Hupp, *Energy Environ. Sci.*, 2008, **1**, 66.

Material Design and Applications of Rhenium(I)-Containing Macromolecules

WAI KEI CHEUNG AND CHRIS S. K. MAK*

Department of Chemistry, The University of Hong Kong, Pokfulam Road, Hong Kong, China
*E-mail: cskm@hku.hk

8.1 Introduction

Metal-containing polymers, or metallopolymers, are widely studied, stemming from their intriguing properties. This kind of materials combines the merits of the tunability of the redox and photophysical properties of metal complex and the processability of polymers, which are particularly important for low-cost large area device manufacturing in the field of optoelectronics. Selection of suitable metal complexes onto the polymer is crucial and the properties of the resulting polymers are greatly affected by the choice of the metal ion and its corresponding ligands. Complexes of heavy transition elements have been extensively studied for photo- or electroluminescent materials for OLEDs, photosensitisers for solar energy conversion systems, active materials for photoelectron-transfer systems, as well as sensors and luminescent probes for biological systems. Iridium(III), platinum(II) and ruthenium(II) complexes are the leading materials for these applications due to their versatility of coordination chemistry and rich electrochemistry, and the photophysics arise from their large variety of excited states. Rhenium(I) complexes, on the other

RSC Polymer Chemistry Series No. 2
Molecular Design and Applications of Photofunctional Polymers and Materials
Edited by Wai-Yeung Wong and Alaa S Abd-El-Aziz

Published by the Royal Society of Chemistry, www.rsc.org

hand, are more subtle and have drawn less attention. Halotricarbonyl rhenium(I) diimine complexes and their analogue species have been the focus of academic research due to their simple coordination chemistry and can be easily prepared by facile synthesis. Their photophysical and photochemical properties have been extensively studied and well documented.[1] The six-coordinated rhenium (I) complex with three carbonyl groups as coligands exhibited octahedral geometry with low-spin $5d^6$ electronic configurations. Many Re(N^N)(CO)$_3$L (where N^N is diimine ligand and L is monodentate ligand) complexes display phosphorescence at room temperature and the electrochemical and photophysical properties can be fine tuned by changing the diimine ligand or the monodentate ligand. Functionalised phenanthroline and bipyridine are the common coordination ligands in Re(I) complexes for fundamental photophysical investigations.[2,3] Most of the Re(I) tricarbonyl complexes exhibit broad and featureless emission band arising from the MLCT excited state.[4] The three carbonyl groups contribute π-back bonding to the metal centre significantly, and thus, the energy gap between the Re(I) d-orbitals and the π^* orbital of the ligand is reduced and low energy emission is observed. Generally, these types of complexes possess low quantum yield ($\Phi \sim 10^{-2} - 10^{-3}$) and emission lifetime is in the $10^2 - 10^{-1}$ μs regime. The emission energy, lifetime and quantum yield are sensitive to the change of the ancillary bidentate ligand, such as diimine, or the noncarbonyl spectator ligand of the emitter. Wrighton and Morse reported the first emissive metal carbonyls based on Re(I) and the lowest emissive state was assigned as triplet metal-to-ligand charge transfer (metal d orbital to N^N π^* orbital, MLCT) transition.[4] The energy of the MLCT transition is usually close to that of the $\pi - \pi^*$ ligand centred (LC) transition of the diimine ligand and thus the two energy levels are populated simultaneously upon photoexcitation. Emission from these nonequilibrated excited states was observed and the charge density of the metal centre can be systematically controlled by varying the N^N ligand.[5] Emission from $^3(\pi - \pi^*)$ was detected by changing the excitation wavelength in the vicinity of the onset of the absorption spectrum and dual emission from Re MLCT and $^3(\pi - \pi^*)$ excited states was observed at 77 K.[6] The energy levels of the triplet MLCT and LC states are so close and vary relative to each other. On the other hand, some of the Re(I) complexes have emission bands that lie in the visible spectrum, predominantly in the orange-yellow region with some vibronic features, which indicates that the lowest excited state possesses MLCT and LC admixture character or pure LC transition (Figure 8.1).

Unprecedented high quantum yield ($\Phi > 0.7$) and long excited-state lifetimes (>100 ms) were reported for a series of Re(I) diimine isonitrile complexes.[7] Density functional theory (DFT) and time-dependent density functional theory calculations on Re(I) complexes containing diimine ligands, like bipyridine or phenanthroline, revealed that the lowest transition of these complexes was metal-ligand-to-ligand charge transfer (MLLCT) or LC with the $\pi - \pi^*$ transition at higher energy.[8] Some studies have demonstrated that the metal complexes of the metallopolymer were capable of facilitating the

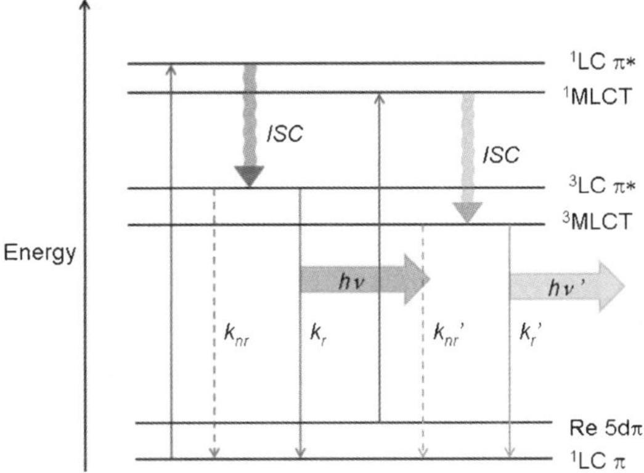

Figure 8.1   Jablonski diagram of [Re(N$^\wedge$N)(CO)$_3$L] complex.

charge carrier mobility of the polymers by functioning as extra charge carriers.[9,10] Nevertheless, most of the metal-containing polymers for optoelectronic applications in the literature are based on Ir(III), Ru(II) and Pt(II) complexes. Relatively little attention has been paid to the use of rhenium compounds as emissive materials, photosensitisers, *etc.*, and report on low-bandgap Re(I) complex containing polymer is sparse. In this contribution, the photophysical properties, nanostructures, optoelectronic and biological applications of Re(I)-containing polymers, macromolecules and bioconjugates will be reviewed. The material design, synthesis and applications of this class of materials will be presented and discussed in the following sections.

8.2   Photophysical Properties of Rhenium(I)-Containing Polymers

Considerable work has been carried out on the photochemical and photophysical properties of Re(I)-containing polymers spawned from their potential applications in optoelectronics, nanofabrications and catalysis. The optical and electronic properties of the materials will be strongly influenced by the introduction of a metal complex into the polymer chain, which leads to the generation of new redox states and new accessible electronic excited states. The study of the photophysics of rhenium-based luminecsent π-conjugated polymers was initiated by Schanze and coworkers. A series of poly(arylenethynylene) incorporated with different ratios of *fac*-(bpy)Re(CO)$_3$Cl chromophore onto the main chain (**2.1a – d**) were synthesised *via* Sonogashira coupling of diiodoaryl with the functionalised bipyridine Re complex using a palladium catalyst (Scheme 8.1).[11] Demetallation during the polymerisation reaction was not observed and the quantity of rhenium content was determined by the carbonyl absorption band in the infrared spectrum.

Two prominent absorption bands were assigned as π–π* and (Re)dπ to (polymer)π* MLCT transitions. The delocalised nature of the singlet exciton of

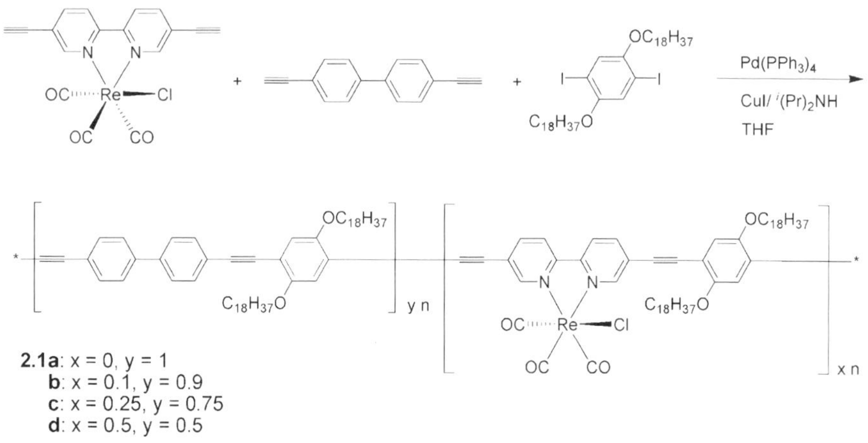

2.1a: x = 0, y = 1
 b: x = 0.1, y = 0.9
 c: x = 0.25, y = 0.75
 d: x = 0.5, y = 0.5

Scheme 8.1

the polymer backbone is evident by the significant redshift of the MLCT energy. Both fluorescence and phosphorescence were obtained at 80 K and the phosphorescence comprised $^3\pi-\pi^*$ and ^{3}MLCT transitions. The increase in metal content in the polymer decreased the yield of the $^3\pi-\pi^*$ state because of the quenching of $^1\pi-\pi^*$ by the metal complex. This dual chromophoric system showed an unusual feature that the two chromophores do not couple significantly. Energy transfer from the singlet IL state to MLCT state occurs, however, sensitisation of the $^3\pi-\pi^*$ from the MLCT is not observed. Slow internal conversion between IL $^3\pi-\pi^*$ and MLCT states has been reported in d^6 polypyridine-based metal complexes when the IL and MLCT states are nearly degenerate[12] and as a consequence, MLCT to $^3\pi-\pi^*$ energy transfer does not occur. Detailed photophysical studies were also performed on a series of monodispersed π-conjugated Re(I) oligomers of aryleneethynylenes **2.2 – 2.5** that contain the Re(2,2′-bipyridine)(CO)$_3$Cl (Scheme 8.2).

The photophysical properties of these compounds were dominated by the intense $\pi-\pi^*$ transition in the longer oligomeric aryleneethynylenes and the shorter ones were overwhelmed by the MLCT transitions, although these two states are in close energetic proximity.[13] Solvent and temperature effects on the photophysical properties of poly(vinylpyridine) attached with Re(2,2′-bipyridine)(CO)$_3$py$^+$ **2.6** were investigated.[14] Resonance energy transfer was investigated in the poly(vinylpyridine)-Re(I) phenanthroline complex system **2.7a – d** by Wolcan and coworkers (Scheme 8.3).[15]

The number of quenching sites within the Förster energy transfer radius of this Re(I) poly(vinylpyridine) polymer was calculated from the luminescent decay and the molecular modeling calculation. Two phenanthroline complexes with donor properties –Re(I)(tmphen)(CO)$_3$ and acceptor properties – Re(I)(NO$_2$-phen)(CO)$_3$, where tmphen and NO$_2$-phen stand for 3,4,7,8-tetramethyl-1,10-phenanthroline and 5-nitro-1,10-phenanthroline, respectively.

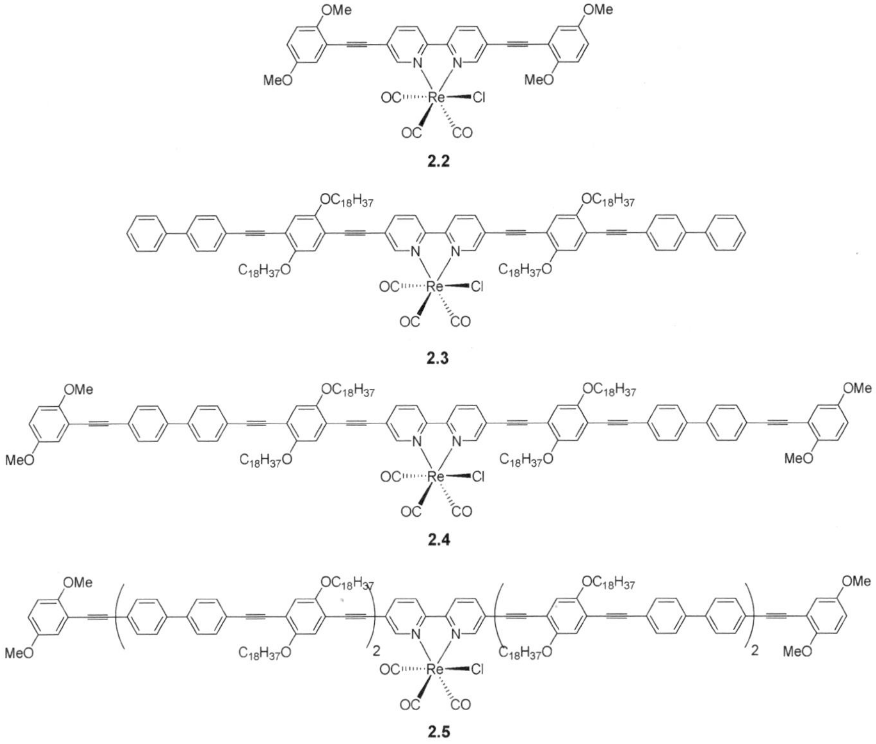

Scheme 8.2

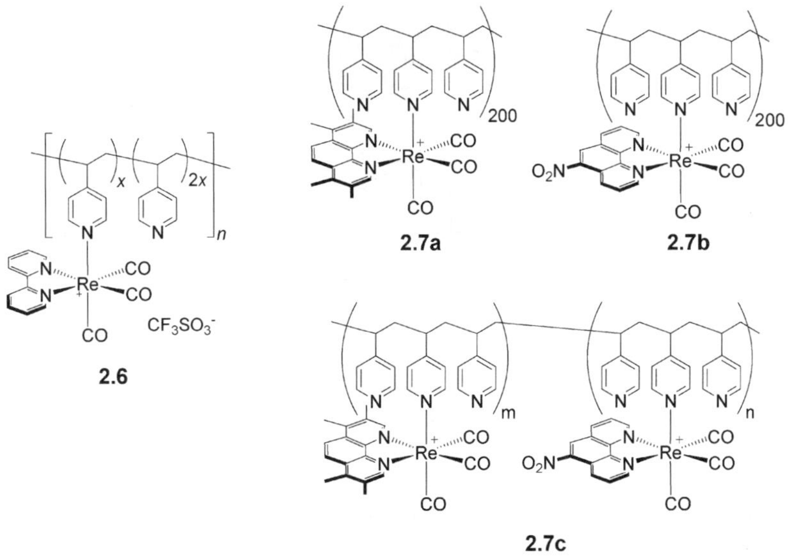

Scheme 8.3

Remarkable differences were observed between the metal complex Re(substituted phen)(CO)$_3$(py)$^+$ and the Re(I) bound polymer, because the extent of the photogenerated MLCT excited states is larger when Re(I) complex is bound. The energy-transfer process was reported from the two excited states MLCT$_{Re\rightarrow tmphen}$ and MLCT$_{Re\rightarrow NO2\text{-}phen}$ in the poly(vinylpyridine) containing both complexes. The coexistence of different Re(I) chromophores within a strand of polymer generated a more complex set of charge-transfer excited states originating from different chromophores in different coordination sites.

8.3 Nanostructures and Nanopatterning Based on Re(I)-Containing Block Copolymers

8.3.1 Polystyrene-Block-Poly(vinylpyridine) Tethered with Emissive Re(I) Complexes

Hou and Chan have demonstrated the material designs and synthesis of a series of Re(I) complex containing block copolymers. The emissive Re(I)

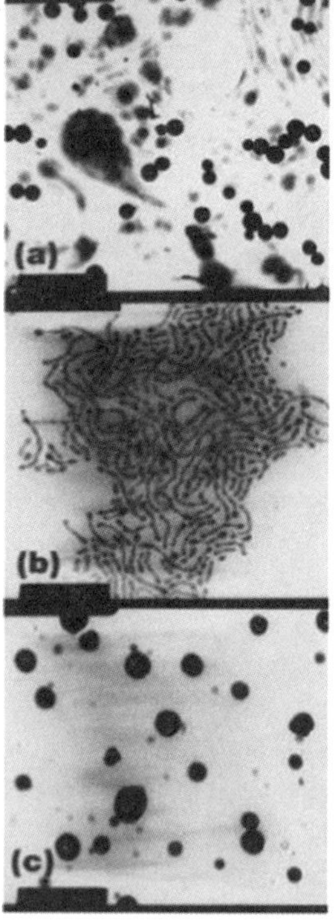

Figure 8.2 Transmission electron micrographs of (a) polymer **3.2** prepared from MeOH/CH$_2$Cl$_2$; (b) polymer **3.2** prepared from toluene/CH$_2$Cl$_2$; and (c) polymer **3.1** prepared from toluene/CH$_2$Cl$_2$. Reproduced from ref. 16 with permission from Wiley.

complexes tagged on the polymer can act as luminescent probes, which would indicate the interaction of the material with different environments. These types of materials display well-defined structures and morphological behaviour. Different nanostructures are formed by varying the chemical properties, solvent polarity, preparation procedures and size of each block of the block copolymers. Polystyrene-block-poly(4-vinylpyridine) (PS-*b*-P4VP) was synthesised by anionic polymerisation.[16] The pyridine moiety functioned as the monodentate ligand for the *fac*-(bpy)Re(CO)$_3$(L)$^+$ **3.1a, b**. Three micelle structures, disc, rod and sphere with particle size ranging from 20 – 400 nm, were formed in different block size and solvent systems (Figure 8.2).

Detailed morphological study of Re(I) complex containing PS-*b*-P2VP **3.2** based on polymer–solvent interaction parameters was performed by Xu *et al.* (Scheme 8.4).[17] The morphology and dimension of the Re(I) complexes within the nanosized micelles were controlled by changing the solvent system. Core-embedded Re(I) micelles were formed in THF/MeOH and corona-embedded Re(I) micelles appeared by dropping the THF solution of the block copolymer into water at low pH value, as shown in Figure 8.3.

In the study by Hou *et al.*, core–shell structure of **3.1a** was observed under high magnification of TEM images (Figure 8.4). The outer part of the nanostructures comprised the Re-P4VP blocks evidenced by the high contrast of the TEM images, and the hydrophobic PS block concentrated at the central part by the polar solvent. Since the Re(I) complex on the block copolymer is ionic in nature, the Re-P4VP block could self-assemble itself into different

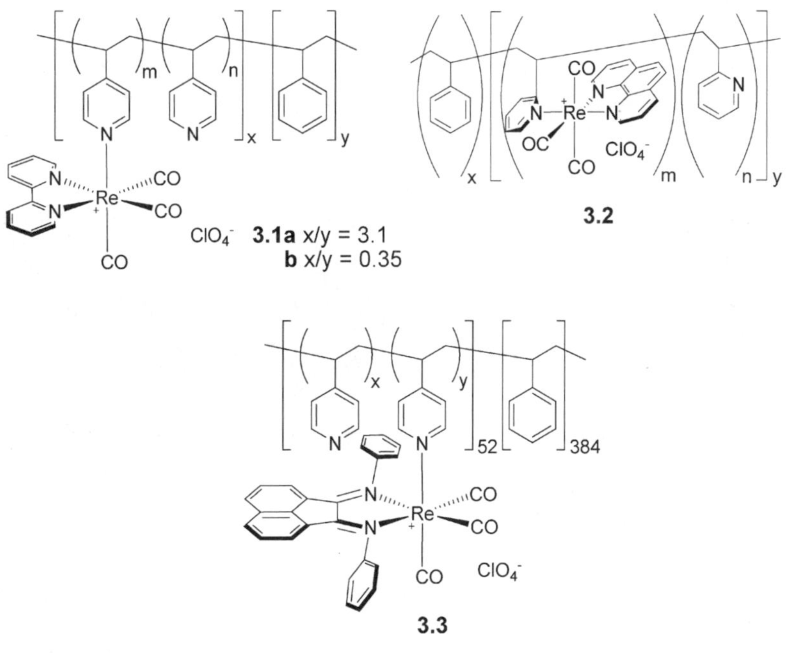

Scheme 8.4

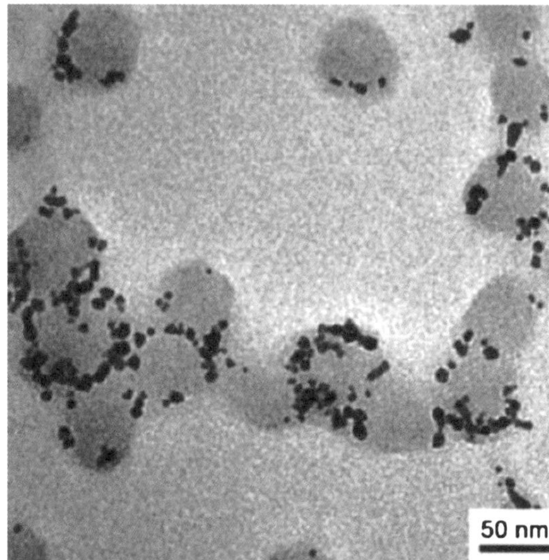

Figure 8.3 TEM image of corona-embedded structure of **3.3**. Reproduced from ref. 17 with permission from Wiley.

forms of micelles. More complex microstructures are formed instead of simple amphiphilic diblock copolymers as the Re(I) complex is randomly distributed within the P4VP block. Micelle formation was indicated by the significant change in luminescent properties and is shown in Figure 8.5. The intensity of the emission peak at 540 nm,[18] originating from the Re(I) complex, decreased dramatically while a new emission band at higher energy was developed. The changes in emission properties may be due to lack of dipole reorganisation aroused from the change in surroundings. These kinds of micelles can

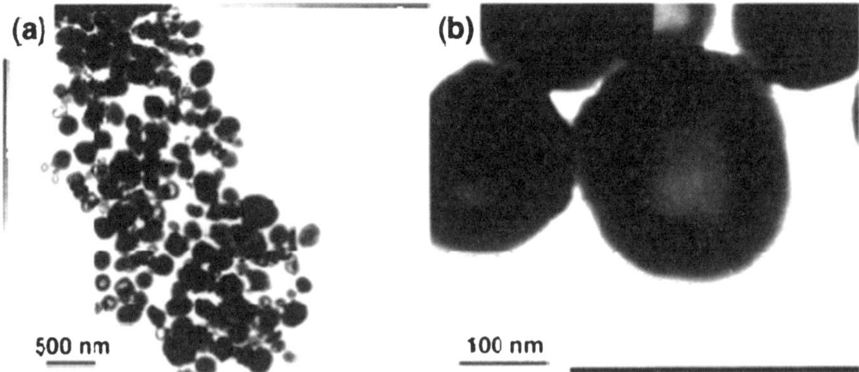

Figure 8.4 TEM images of the micelles formed by 3.1 in the hexane/dichloromethane system under different magnifications. Reproduced from ref. 18 with permission from the American Chemical Society.

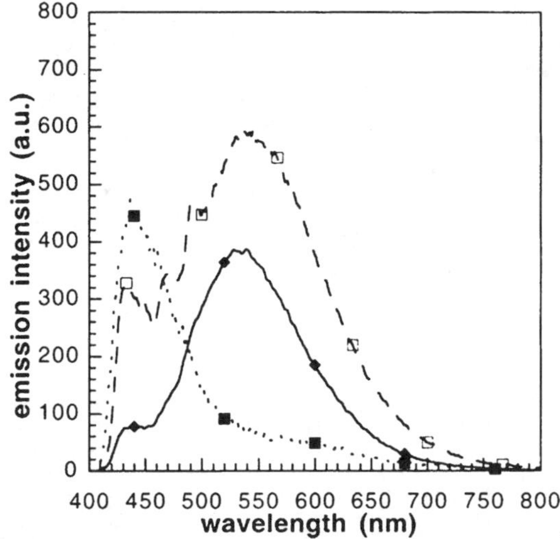

Figure 8.5   Solid-state emission spectra of **3.1** (◆) and **3.2** (□) and the emission spectrum of micelles formed by copolymer **3.2** in the MeOH-DCM system (■). Reproduced from ref. 18 with permission from the American Chemical Society.

potentially be used as luminescent probes for light and chemical sensing applications.

8.3.2   Polystyrene-Block-Poly(vinylpyridine) Tethered with Photosensitising Re(I) Complexes

Polymer morphology is also sensitive to the nature of the Re complex attached. The use of PS-*b*-P4VP block copolymer as a template for mesoporous materials has been reported.[19] Nanosized well-defined cylindrical or spherical domains (15 – 20 nm) were formed based on a series of PS-*b*-P4VP copolymers attached with ionic rhenium bis(phenylimino)acenaphthene (Re-DIAN) complex, which would largely increase the volume fraction of P4VP block **3.3** (Figure 8.6).

The presence of a charged complex in the block copolymer **3.3** has significantly changed the interaction parameter between the two polymer blocks and thus the copolymer bulk morphology. This ionic polymer could function as a template and therefore selectively deposited semiconducting nanoparticles for nanofabrication. Cadmium sulfide nanoparticles with sizes approximately 10 nm were deposited on the cylindrical domains by electrostatic attraction and the attachment was confirmed by XPS spectroscopy. Figure 8.7 shows the AFM image of the polymer film surface after the deposition of CdS nanoparticles. This layer-by-layer deposition method offers a potentially alternative way to the soft lithographic technique for nanofabrication.

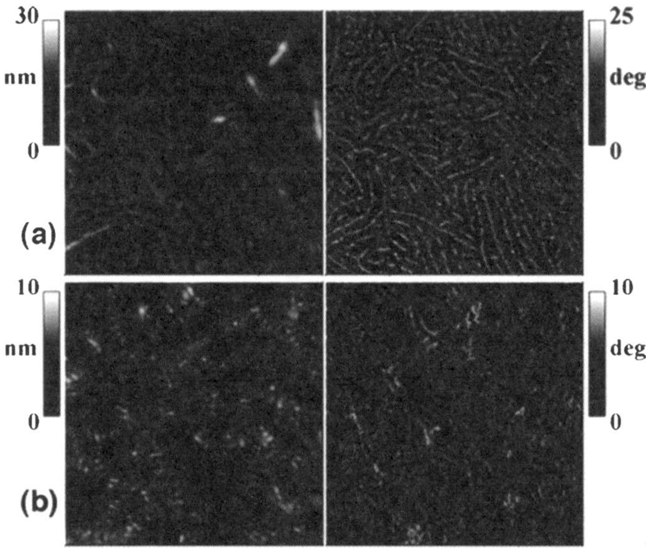

Figure 8.6   AFM images of **3.3** prepared by spin coating from (a) toluene solution and (b) chloroform solution on silicon wafer. The images on the left are tapping-mode height images and phase-contrast images are on the right with the scan size 1 × 1 × m. Reproduced from ref. 19 with permission from the American Chemical Society.

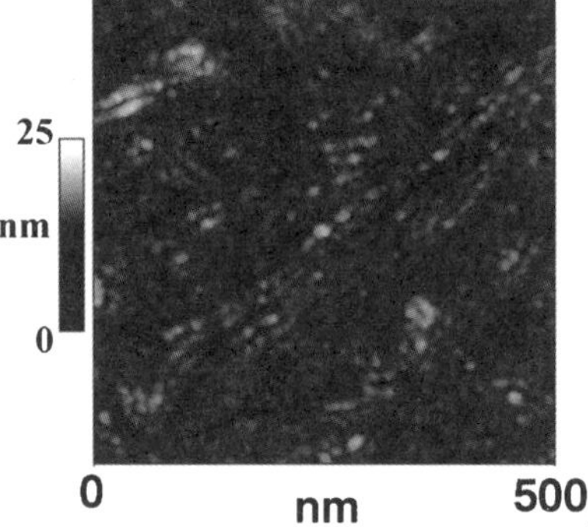

Figure 8.7   AFM image of **3.4** deposited with CdS nanoparticles. Reproduced from ref. 19 with permission from the American Chemical Society.

8.4   Development of Polymeric Luminescent Materials for OLEDs

8.4.1   Polymer-Based OLED Development

Research on conducting π-conjugated polymers has grown rapidly as a result of the feasibility of tailoring their chemical and physical properties to meet the demands of new applications. The discovery of electroluminescence from conjugated polymer by Friend and Holmes in the early 1990s provided a new impetus towards the intensive development of the polymer-based OLEDs.[20] Phosphorescent small-molecule OLEDs have demonstrated high efficiencies, however, the device fabrication of these materials involves complex and expensive techniques, while the manufacturing processes require high vacuum and control temperature. By contrast, light-emitting conjugated polymers can be readily deposited from solution as thin films over large areas by spin coating, the doctor-blade technique, or ink-jet printing. The size of the device is virtually unlimited and the large-area two-dimensional multilayer device can be manufactured by roll-to-roll coating technologies. Therefore, conjugated polymers are promising materials for optoelectronic devices as they can provide an easy and cheaper way for the device fabrication. Different emission colours from organic polymers are now accessible throughout the visible spectrum with readily tunable properties such as thermal stability, bandgap energy and emission monochromaticity. Heavy organometallic complexes have been incorporated into organic polymers to facilitate intersystem crossing and thus harness triplet-state emissions. Unity internal quantum efficiency can be achieved with this material design. In order to minimise phase separation and aggregation of the triplet emitters, a strategy that involves attaching the emitters covalently onto the polymer backbone has been successfully employed in the fabrication of PLED with improved device efficiency. There are two types of metal-containing polymers: one is that the metal complexes are conjugatively linked onto the polymer backbone, and the other one is that the metal complexes are attached onto the polymer main chain as pendant. Electrophosphorescent conjugated polymers containing Ir(III)[21–27] or Ru(II)[28–31] metal centres have been extensively studied and have demonstrated impressive efficiency for OLED applications. Other than Ir(III), Pt(II) and Ru(II) complexes, some Re(I) complexes have been reported as highly efficient yellow emitters.[32] Several studies on phosphorescent OLEDs with Re(I) metal complexes as emissive dopant have been reported.[32–35]

8.4.2 Rigid-Rod Re(I)-Containing Conjugated Polymers

Until now, OLEDs comprised Re(I) units either blended with a polymer host or covalently linked onto the polymer backbone generally exhibited poor performance. Fundamental understanding for the photophysics and electronic processes in the devices with Re(I) compounds as active materials is essential. Schanze and coworkers have studied the photophysical and EL properties of a series of d^6 transition-metal complexes containing poly/oligo(aryleneethynylene). The origin of the EL feature is either from ^{3}MLCT or a $^3\pi-\pi$ excited state.[13,36,37]

Chan *et al.* have demonstrated several luminescent rigid-rod conjugated polymers based on Re(I) dipyridophenazine complexes **4.1a – f** (Scheme 8.5). The polymers have been synthesised by the palladium-catalysed coupling reaction using divinylbenzene, functionalised dipyridophenazine and aromatic dibromides as the monomers. From the photoluminescence spectra, an energy-transfer process between the conjugated backbone and the metal complexes is proposed. The electroluminescence (EL) of the polymers originated from the emission due to the π^*-π or π^*-d transitions.[9]

Another series of π-conjugated copolymers incorporated with Re(2,2'-bipyridine)(CO)$_3$Cl **4.2a** – **c** were reported by Zhang *et al.* (Scheme 8.6). The polymers were synthesised by Suzuki cross-coupling reaction of functionalised dihexylfluorene and 2,2'-bipyridine in different ratios, followed by complexation with rhenium(I) pentacarbonyl chloride. The absorbance of the MLCT transition of the (Re) dπ to (bpy) π^* increased with the content of bipyridine in the polymerisation. The polymers displayed a yellowish-orange emission with peak maximum at $\sim$590 – 600 nm. The difference in the normalised fluorescence decay curves of the higher-energy $\pi-\pi^*$ emission showed that there is an energy flow from the polymer backbone $\pi-\pi^*$ to the MLCT state of localised Re(I) complex that occurs *via* the resonant energy transfer, but not simply by reabsorption of photons from the polymer backbone.[38] Electroluminescent properties were examined with device structure ITO/PEDOT:PSS/4.2/LiF/Al. The EL peak maxima of **4.2b** and **4.2a** are located at $\sim$475 and $\sim$600 nm, respectively. Electron–hole recombination in Re(I) complexes and the transfer of triplet excitons from the $\pi-\pi^*$ state of the polymer backbone to the Re(I) complexes are evidenced by the increased in MLCT emission intensity in an EL device compared to the PL spectrum. The EL efficiency was far from satisfactory and was only <0.1 cd/A. The low efficiency of this system is conceivably related to charge-carrier separation of localised electron–hole pairs in the Re(I) complexes. A photovoltaic response was observed in these copolymers and suggested that the transfer of the

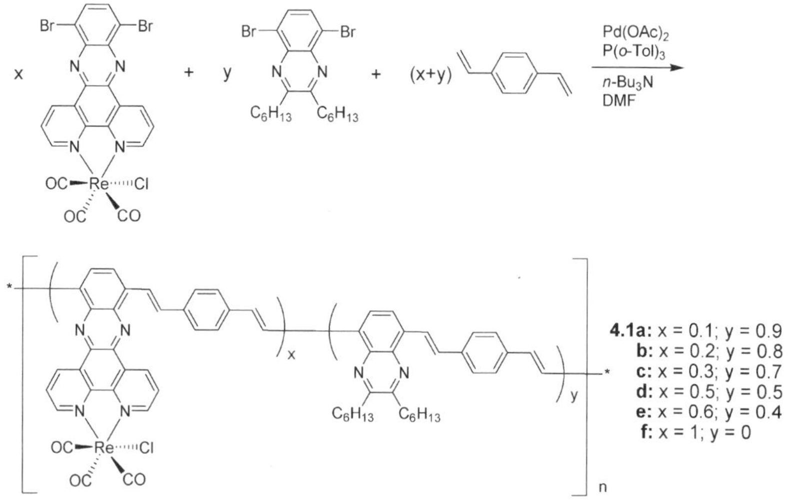

Scheme 8.5

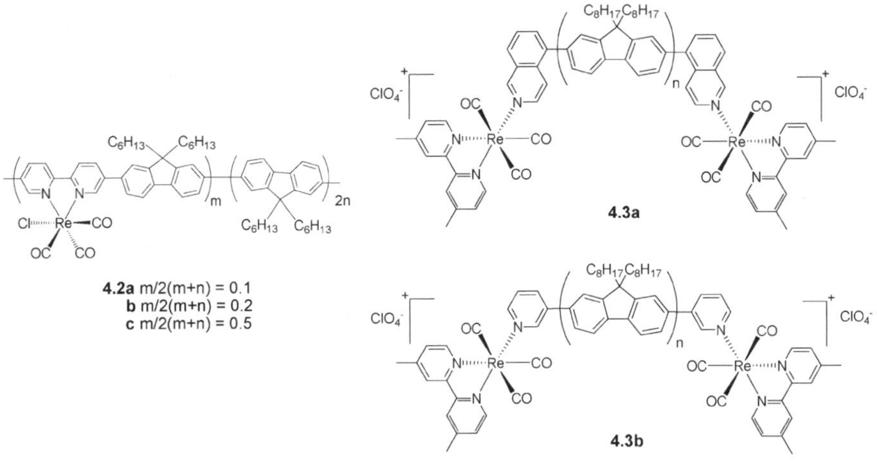

4.2a m/2(m+n) = 0.1
b m/2(m+n) = 0.2
c m/2(m+n) = 0.5

4.3a

4.3b

Scheme 8.6

electrons within the system is efficient and can potentially be used in photovoltaics.

A new method for incorporating organometallic complexes onto conjugated polymers to make phosphorescent polymers was reported by Lee and Hsu.[39] Quinoline **4.3a** or pyridine **4.3b** was used to end-cap and control the molecular weight of the polyfluorene and also for the complexation with $Re(CO)_5Cl$ after polymerisation (Scheme 8.6). The coordination of Re(I) complexes onto the end group was identified by FTIR and NMR spectroscopies. The HOMO level energy has been increased in the complexed polymers and thus a lower barrier was achieved for hole injection. Bluish-white EL emission was obtained from the polymers **4.3a** and **4.3b** with device configuration ITO/PEDOT/**4.3a** or **4.3b**/Ca/Ag.

8.4.3 Re(I)-Containing Dendrimers

Another class of macromolecule, dendrimer, was also utilised as the conducting host for the Re(I) complex. Unlike most of the conjugated polymers, dendrimers exhibit well-defined molecular structure. The control of charge-transport and emission properties of the active materials becomes possible by controlling the generation, the number or the types of dendrons. Burn and coworkers have reported the synthesis of a series of (1,10-phenanthroline)rhenium(I) carbonyl complexes with one (**4.4a**), two (**4.4b – d**) or three (**4.4e**) dendrons attached to the ligand (Scheme 8.7).[40] Six dendrimers were synthesised with the dendron containing biphenyl units and 2-ethylhexyloxy as solubilising groups. Based on the device configuration ITO/neat dendrimer film/Ca/Al, the external quantum efficiency and power efficiency of the best device performance based on one of these dendrimers at 100 cd/m^2 and 12.8 V are 0.8 cd/A and 0.2 lm/W, respectively. The low efficiency is due to the imbalance of charge injection with electrons injected more readily than holes.

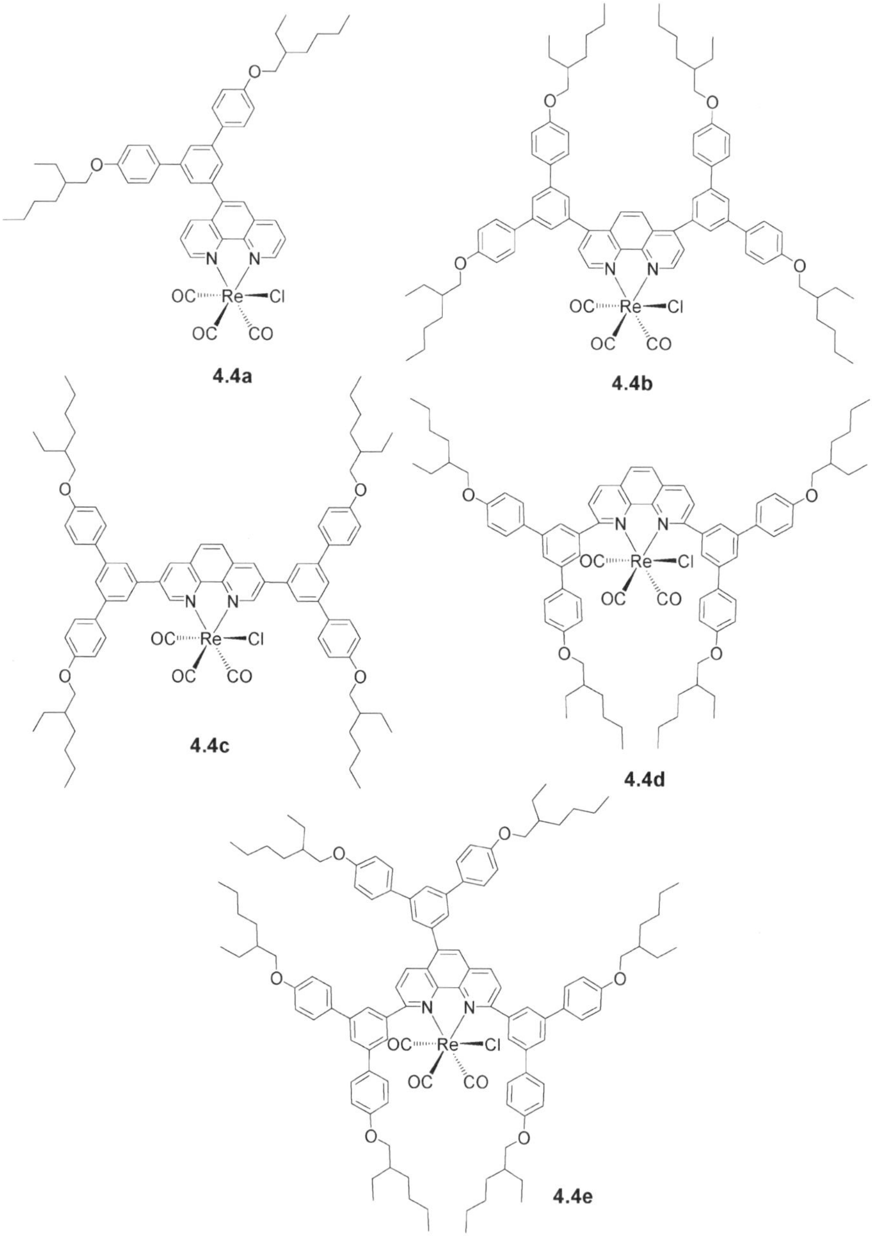

4.4a

4.4b

4.4c

4.4d

4.4e

Scheme 8.7

8.5 Photosensitisers for Photovoltaic Applications

The properties for the photovoltaic material are essentially opposite to that of
the OLED material. Materials with high extinction coefficient, broad

Table 8.1   Photovoltaic performance of rhenium(I) containing macromolecules under 100 mW cm^{-2} AM 1.5 simulated solar light illumination.

Device configuration	I_{sc} (mA cm^{-2})	V_{oc} (V)	FF	PCE (%)	Ref.
ITO/**5.7a**/C60/Al	0.022	0.16	0.14	0.002	54
ITO/(**5.8**/PTEBS)$_{80}$/Al	0.027	1.2	0.19	0.006	57
ITO/PEDOT:PSS/**5.9a**/PTCDI/Al	0.037	0.93	0.21	0.045	58
ITO/PEDOT:PSS/**5.9b**/PTCDI/Al	0.427	0.86	0.21	0.061	58
ITO/PEDOT:PSS/**5.9c**/PTCDI/Al	0.102	0.76	0.23	0.012	58
ITO/PEDOT:PSS/**5.10**:PCBM (1:2)/Al	0.08	0.92	0.23	0.02	60
ITO/PEDOT:PSS/**5.11**:PCBM (1:4)/Al	0.14	0.66	0.27	0.03	61
ITO/PEDOT:PSS/**5.12**:PCBM (1:4)/Al	0.11	0.48	0.25	0.01	61

absorption throughout the visible and near-infrared region, nonemissive excited state and short excited state lifetime for efficient charge separation are desired for efficient photovoltaic cells. Metallopolymers have drawn unabated attention in photovoltaic applications. Wong and coworkers have demonstrated impressive power-conversion efficiency with solution processable Pt(II) polyynes in polymer solar cells.[41–45] Ruthenium(II) complexes have been the most extensively used organometallic compounds for photovoltaic applications because of their excellent photo-/electrochemical stability, high extinction coefficient in the visible region and large versatility and flexibility with six coordination sites on the Ru(II) metal centre for ligand chelation. Also, the intrinsic high absorptivity of the MLCT transition of Ru(II) polypyridyl made it a suitable candidate as a photovoltaic material. Photovoltaic performances based on the polymers with Ru(II) polypyridyl either on the main chain[46,47] or as pendant[48] have also been investigated. On the contrary, the well-known emissive Re(I)-containing polymers reported in the literature do not fulfill the requirements, as they possess absorption at the higher energy of the visible range with relatively lower extinction coefficient compared to that of the Ru(II) complexes. Nevertheless, previous studies have shown that some Re(I)-complex-containing polymers, which possess both electron donor and acceptor conjugatively linked onto the main chain, may act as bipolar charge carriers.[9,29] Photovoltaic performances of Re(I)-containing macromolecules are summarised in Table 8.1. The power-conversion efficiency (PCE) is in the range of $10^{-2} – 10^{-3}$%. Research on the use of Re(I) compounds for photovoltaic applications is full of challenges. Further material development and device optimisation are needed in order to have significant improvement on the device efficiency.

8.5.1   Re(I)-Containing Nonconjugated Polymers

Most attempts reported in the literature use Re(I) complexes as photosensitisers in polymers. The photosensitising properties of Re(I)-containing polyimides **5.1a – f** were reported by Lam *et al.* in 2000 (Scheme 8.8).[49] The polymers were synthesised by the reaction of chlorotricarbonyl[1,4-bis(4-aminophenyl)-1,4-diazabutadiene]rhenium(I) [Re(CO)$_3$Cl(DAB)] with various dianhydrides in NMP followed by *in situ* imidisation using acetic anhydride as the dehydrating agent and pyridine as the catalyst.

The Re(I)-containing polyimides possessed an intense absorption peak at *ca.* 310 nm with the leading edge down to 600 nm. The photocurrent response was measured as a function of applied electric field at different wavelengths. A photocurrent of 2 nA was detected when the polyimide was excited at 514 nm and this value is one order of magnitude higher than Ru(II)-containing polyimides in the previous study, although the polymer itself was not a good charge-transport agent.[50] Polyurethanes generally composed of both hard and soft segments[51] and the film-forming and mechanical properties can be varied by tuning the ratio of these segments. The soft ether linkage segment may accommodate metal ions and can be used as a metal ion probe after the polymer functionalised with Re(I) diimine complexes, in which the MLCT excited states are highly sensitive to the surrounding environment.

Polyurethanes **5.2a – c** were synthesised by mixing 4,4′-methylenebis(phenyl isocyanate), Re(I) diimine monomer and PEG300 in distilled DMF at room temperature (Scheme 8.9). The chain extension was done by the addition of 1,4-butanediol and heated to 60 °C overnight.[52] Aggregation of the metal complexes was observed and 50 – 100 nm nanodomains were shown in the TEM photos. The intensity of the MLCT absorption decreased with the addition of lithium ions, or europium ions. The results showed that the polymers have sensing capability on particular metal ions. On the other hand, the photoconductivity of the polymer **5.2c** was enhanced two orders of magnitude after the addition of charge-transport materials. The photocurrent responses relied on the amount of doped charge transport agent. This result indicated that the photoconduction was limited by the charge-transport process within the system.

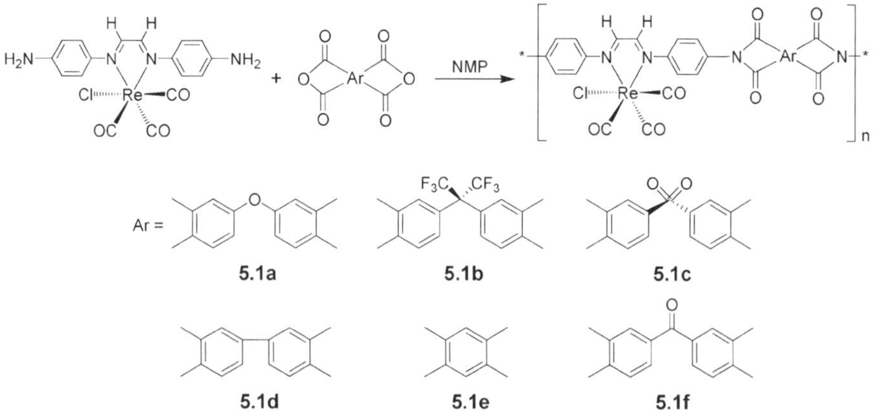

Scheme 8.8

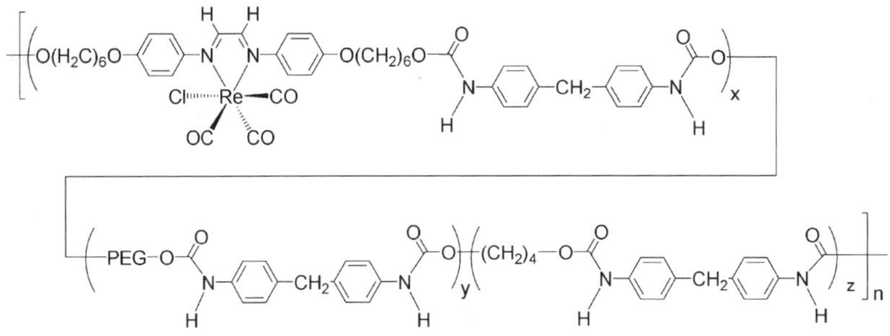

5.2a x:y:z = 0.1:1:1.9
b x:y:z = 0.3:1:1.7
c x:y:z = 0.5:1:1.5

Scheme 8.9

8.5.2 Re(I)-Containing Block Copolymers

Similar investigation has been carried out on the random or block copolymers **5.3a – c, 5.4a – c, 5.5a, b, 5.6a, b,** of ionic Re(I)-complex-containing poly(methylmethacrylate) (PMMA), which was synthesised by atom-transfer radical polymerisation (ATRP) with copper(I) bromide, 1,1,4,7,7-pentamethyl-diethylenetriamine (PMDETA) and methyl 2-bromopropinoate as catalyst and initiator (Scheme 8.10).[10] Photoconductivity was measured with the active polymer layer sandwiched between ITO anode and aluminium cathode and the values were approximately $10^{-12} - 10^{-13}$ Ω^{-1} cm^{-1}. Phase separation was observed between the charged Re(I) complexes and the neutral MMA molecules due to incompatibility. The Re(I) complexes aggregated to form microdomains, which affected the photosensitisation and the subsequent

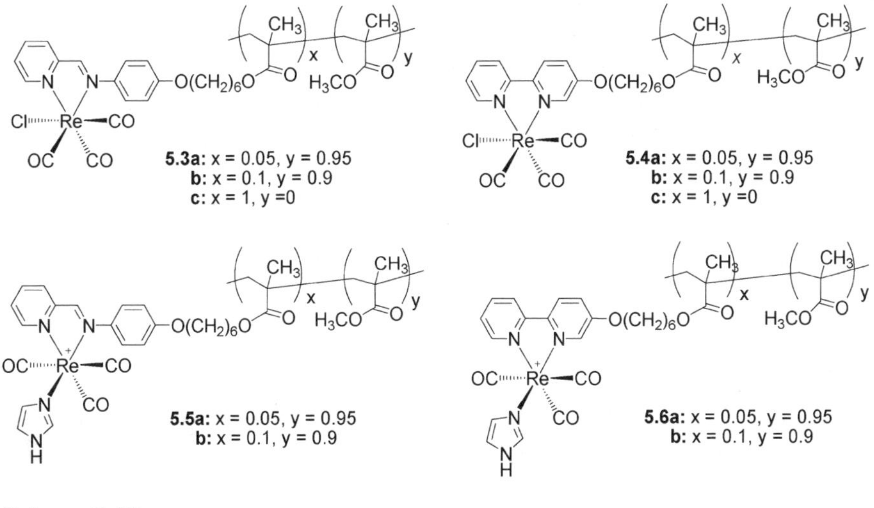

5.3a: x = 0.05, y = 0.95
b: x = 0.1, y = 0.9
c: x = 1, y = 0

5.4a: x = 0.05, y = 0.95
b: x = 0.1, y = 0.9
c: x = 1, y = 0

5.5a: x = 0.05, y = 0.95
b: x = 0.1, y = 0.9

5.6a: x = 0.05, y = 0.95
b: x = 0.1, y = 0.9

Scheme 8.10

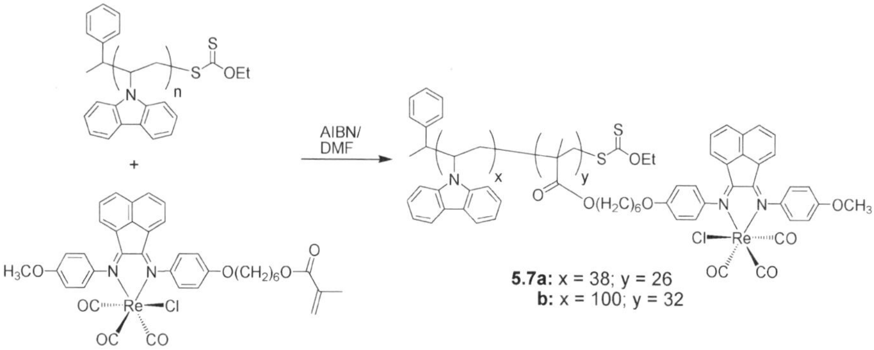

Scheme 8.11

charge separation. Nevertheless, the photosensitivity of the resulting materials was thus strongly enhanced by the Re(I) complexes in the polymer.

The first report on the synthesis of multifunctional poly(vinylcarbazole) (PVK) block copolymer tethered with Re(I) complexes (**5.7a, b**) by reverse addition-fragmentation chain transfer (RAFT) polymerisation[53] was carried out by Tam *et al.*[54] Substituted Re(DIAN) monomer was synthesised by acylation of the corresponding hydroxy precursor with methacryloyl chloride as shown in Scheme 8.11.

Polymers with well-defined block sizes of the Re(I) complex block and the PVK block were prepared *via* RAFT polymerisation using azoisobutyronitrile (AIBN), *O*-ethyl-*S*-(1-phenylethyl)dithiocarbonate as the chain-transfer agent[55] and PVK macroinitiator respectively. Polymers **5.7a** and **5.7b** possessed average molecular weights (M_n) of 29 600 and 46 600, respectively. AFM images of the polymer **5.7a** cast onto silicon wafer showed elongated spherical/ellipsoidal domains with average size approximately 50 – 80 nm. It was envisaged that the Re(I) sensitisers would generate excitons upon photoexcitation and the PVK block would function as a hole-transporting material after the generation of charge carriers by exciton dissociation. The PCE of the bilayer heterojunction photovoltaic device, ITO/**5a**/C$_{60}$/Al, was 1.9 $\times$ 10^{-3}%. Photocurrent in the incident photo-to-electron conversion efficiency (IPCE) spectrum spread beyond 600 nm and the response was extended over the leading edge of the absorption spectrum. The result indeed proved the role of the rhenium complex in photosensitisation. The poor PCE was probably due to inefficient harvesting of incident photons and a high recombination rate.

8.5.3 Re(I) Hyperbranched Polymers

Hyperbranched polymer is one type of dendritic polymers and its irregularity displays the major difference from the highly ordered dendrimer. The synthesis of Re(I) hyperbranched polymers was reported in numerous studies.[56,57] The synthetic route is shown in the following reaction scheme. The key linkage of the hyperbranched polymers contains a stilbazole and two diimine ligands and is synthesised *via* Heck coupling. Subsequent complexation reaction with Re(CO)$_5$Cl in DMF at ~50 °C afforded functionalised Re(I) complexes in good yield. Hyperbranched polymers **5.8** were synthesised in a one-step reaction using silver hexafluorophosphate to extract the chloride ligand on the complex, which then allows the coordination of the stilbazole moiety onto the Re(I) metal centre (Scheme 8.12). The radius of gyration of the polymer was studied by laser light scattering photometry. The resulting polymers are highly charged and can function as polyelectrolyte for layer-by-layer (LBL) deposition. Multilayer thin-film devices of Re(I) hyperbranched polymer and poly[2-(3-thienyl)ethoxy-4-butylsulfonate] (PTEBS) were fabricated by the LBL method. The two polyelectrolye layers do not exhibit distinct stratified layers and form an interpenetrating network instead. This thin-film morphology was supported by the observed peaks and valleys, which correspond to materials of different nature in the phase-contrast AFM images. Devices with configuration ITO/(5.8/PTEBS)$_{80}$/Al were deposited by 80 bilayers of polyelectrolytes. The Re(I) hyperbranched polymer and the PTEBS functioned as photosensitiser and hole-transport material, respectively. The best device performance is of the order of 0.006%. Although the device efficiencies were relatively low, this deposition

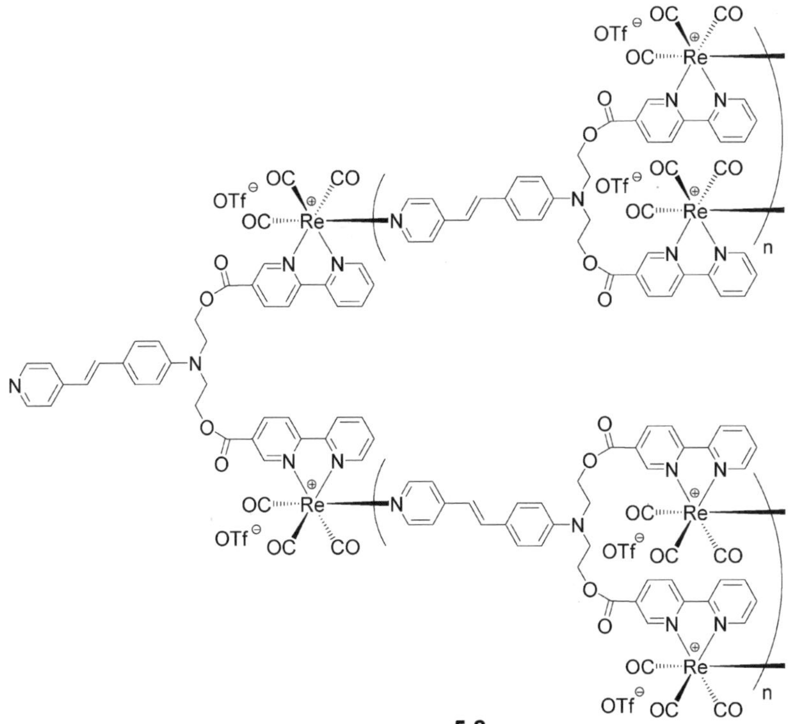

5.8

Scheme 8.12

method provides a simple and versatile approach to fabricate photovoltaic devices *via* a solution-processing method.

8.5.4 Re(I)-Containing Conjugated Polymers

Apart from nonconjugated polymers, well-known conducting polymers were also adapted as the covalently bonded host for the Re(I) sensitisers. Poly(phenylenevinylene) conjugatively linked with Re(DIAN) complexes **5.9** was synthesised by conventional Heck coupling reaction of the diiodo-Re(DIAN), 1,4-dihexoxy-2,5-diiodobenzene, 1,4-divinylbenzene in different ratios in the presence of 10 equivalent of lithium chloride.[58] Replacement of chloride ligand of the Re(I) complex by the iodide ion, which was generated *in situ* during the palladium-catalysed reaction, was observed. Lithium chloride was added to suppress the displacement reaction during the polymerisation. Similar to most of the reported Re(I)-containing polymers, both MLCT [$d\pi$ (Re) to π^* (DIAN)] and the intense intraligand $\pi-\pi^*$ transition transitions were observed, the absorption wavelength peaked at *ca.* 520 nm and 370 nm, respectively, in this system. The intensity of these two bands became comparable when the ratio of the metal content on the main chain to 1,4-dihexoxybenzene was greater than 3:7. Trilayered photovoltaic devices based on polymer **5.9a** – **c** with PEDOT:PSS as the hole-transporting layer and 3,4,9,10-perylenetetracarboxylic diimide (PTCDI) as the electron-transporting layer in a device structure ITO/PEDOT:PSS/**5.9**/PTCDI/Al were fabricated (Scheme 8.13). The overall PCE was of the order of 10^{-2}% with the fill factor

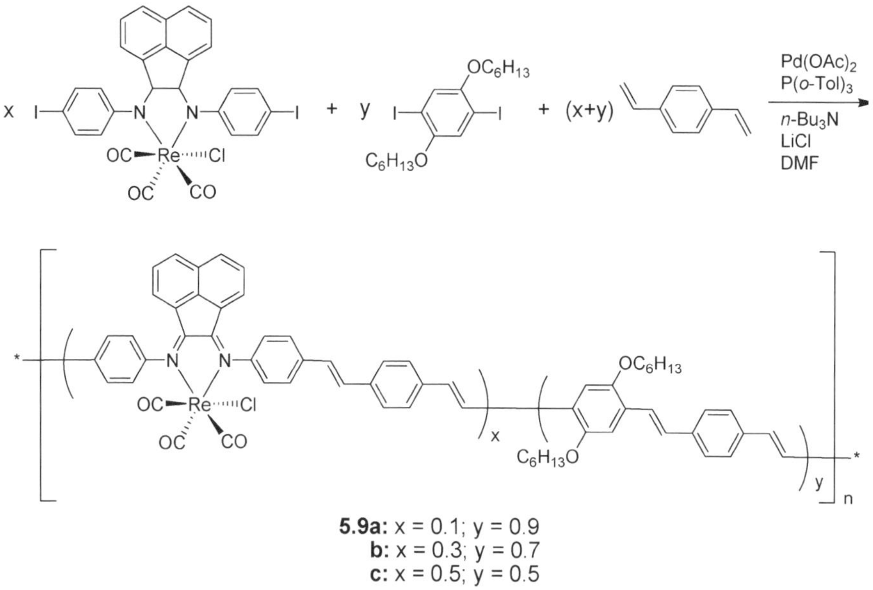

5.9a: x = 0.1; y = 0.9
b: x = 0.3; y = 0.7
c: x = 0.5; y = 0.5

Scheme 8.13

(FF) around 0.2 and open-circuit voltage (V_{oc}) ranging from 0.76 – 0.93 V. The low FF revealed that the charge transport within the system was slow and the recombination rate was high.[59]

Further study on low-bandgap Re(I)-complex-containing polymers with an intramolecular charge transfer (ICT) unit as ligand for photovoltaic applications were carried out by Mak *et al.* The synthetic routes towards the low band-gap polymers with Re (I) complex tethered[60] (**5.10**) or conjugatively linked[61] (**5.11** and **5.12**) onto the polymer main chain are presented below (Scheme 8.14).

Demetallation was observed by the colour change of the reaction mixture from dark green to purple when using complexed Re(I) monomer in the present of base during polymerisation. Therefore, polymerisation was carried out prior to the complexation. Polymerisation of functionalised ligand with fluorenyl diboronic ester *via* Suzuki cross-coupling afforded a purple fibrous polymer. Dark green polymers **5.11** and **5.12** were synthesised by the polymerisation of dibromo-moiety of ICT ligand with 2,7-di(4,4,5,5-tetramethyl-1,3,2-dioxaboralane)-9,9-diethylhexylfluorene *via* Suzuki cross-coupling and with 2,6-di(tributyltin)-4,4-dioctylcyclopenta[2,1-b:3,4-b′]dithiophene (bis(stannyl) (bis(stannyl)-CPDT) *via* Stille coupling, respectively. The nitrogen atoms on the pyridyl ring and the pyrazine core of the ligand can function as the coordination sites for metal ions. Similar to polymer **5.10**, polymerisation was carried out prior to the complexation of the ICT ligand with Re(CO)$_5$Cl. The molecular weight ranged from 6200 to 81 500 g mol^{-1} determined by GPC with a polydispersity index < 2.6. All polymers are readily soluble in common organic solvents such as chloroform 1,1,2,2-tetrachloroethane, chlorobenzene and THF. The absorption spectra of

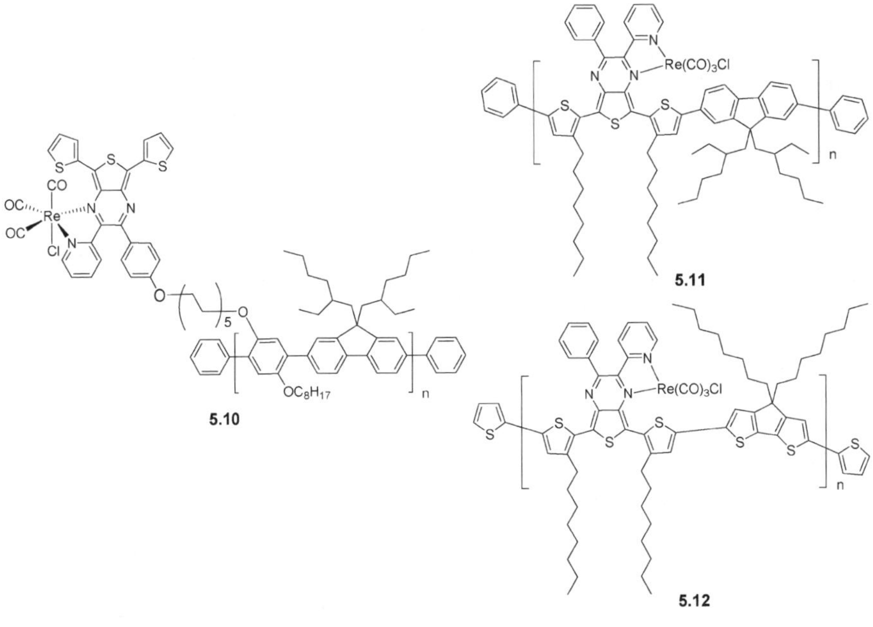

Scheme 8.14

the Re(I) tethered polymers showed a characteristic ITC band, which exhibited a significant red-shift from 555 nm to 633 nm in $CHCl_3$ upon metal chelation. This energy shift is attributed to the increase in electron delocalisation from the thienopyrazine unit to the pyridine unit, which is forced to be coplanar with the thienopyrazine with metal coordination. This is consistent with the electron-density plot from the DFT calculation. The singlet ground-state DFT calculation has a good agreement with the observation in the optical absorption spectrum.[60]

Bulk heterojunction solar cells in a device configuration of ITO/PEDOT:PSS/**5.10**:PCBM/Al based on **5.10** and [6,6]-phenyl-C_{61}-butyic acid methyl ester (PCBM) were fabricated and the photovoltaic performances were examined under simulated AM 1.5 solar illumination (100 mW cm^{-2}) and in the dark. Under AM 1.5 illumination, the PCE of this type of material was around 0.03%. The PCE is determined by several factors such as absorption coefficient of the active material for exciton formation, donor/acceptor interfaces for efficient exciton dissociation and the charge carrier mobility of the active material for charge transport to the corresponding electrodes. The low PCE may be due to the relatively low absorptivity of the charge-transfer transition in the visible range and the inadequate light harvesting limits the charge generation and hence lowers the power-conversion efficiency. Also, the surface morphology may account for the poor charge transport within the active layer. Since the exciton diffusion length of organic materials is generally around 10 nm, a good interpenetrating network of the donor and acceptor is essential to ensure charge separation can take place at the interfaces. In organic materials, electrons and holes can only be generated after the diffusion of excitons to the donor/acceptor interface. However, the diffusion length of an exciton is short, and only those excitons generated within this diffusion limit can be dissociated. This indirect charge-generation process in organic materials results in low efficiency. Therefore, precise control in the morphology of the active layer is of utmost importance to the overall performance of the solar cells. The disordered nature of the organic materials affects the charge-transport process. Charge transport in organic materials proceeds by a hopping mechanism rather than through bandlike transport.[62] This results in lower charge-carrier mobility and a higher chance for charge recombination. In addition, the mismatch of absorption spectra limits the fraction of photons to be absorbed by the organic photosensitiser. Better understanding of the photophysics of the Re(I) complex in macromolecular materials is essential and modification of the polymerisable ligand is critical in order to retain the absorption intensity after metal coordination.

8.6　Biological Applications of Bioconjugates and Polymers Incorporated with Re(I) Complexes

The intriguing and tunable spectroscopic and photophysical properties of Re(I) complexes have also been utilised in biological sensing and labelling. For example, some Re(I) bioconjugates were found to be exceptional biological probes by either covalent or noncovalent bonding to the target biomolecules such as DNA and proteins.[63–65] In addition, radioactive Re isotopes have been identified as a new class of potential antitumour pharmaceuticals.[66,67] These Re isotopes might be incorporated into certain biocompatible polymers for the delivery to tumour tissues.[68,69]

8.6.1   Covalent Bonding to Proteins

Luminescent Re(I) tricarbonyl polypyridine complexes functionalised with carboxylic acid group react with the amine groups of proteins to form an amide linkage upon the activation with *N*-hydroxysuccinimide. These complexes are good labels for proteins owing to their long emission lifetime (on the μs scale) and specific reactivity towards the amino acid residues. For example, the isonicotinic acid complex $[Re(Me_2Ph_2\text{-}phen)(CO)_3(py\text{ -}COOH)]^+$ (**6.1**) has been used to label the protein human serum albumin (HSA).[70] The iodoacetamide (IAA) complex $[Re(phen\text{-}IAA)(CO)_3Cl]$ (**6.2**) also targets on HAS (Scheme 8.15).[71] The iodide group of the complex undergoes substitution with the thiol group of the cysteine residues of HSA.

These Re-HSA bioconjugates exhibit intense and long-lived ^{3}MLCT emission that is assigned to the $[d\pi(Re) \rightarrow \pi^*(N^\wedge N)]$ transition. In addition, the biological activity and structural conformation of HSA are preserved despite the incorporation of Re(I) complexes. For example, the recognitions of both Re-HSA and HSA are achieved by anti-HSA,[72] indicating that these Re(I) complexes are suitable for labelling biomolecules.

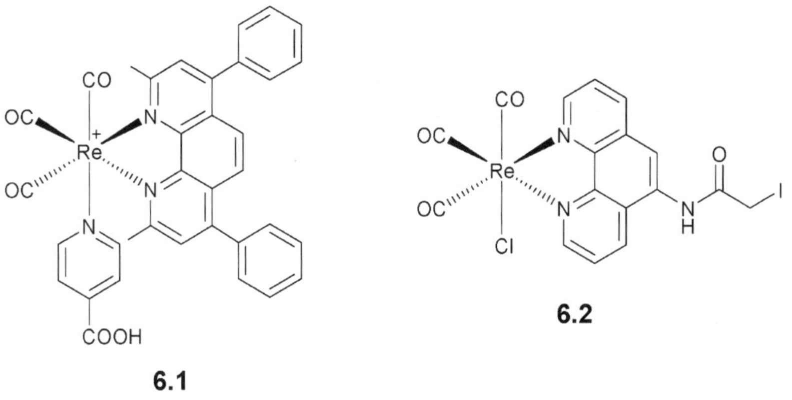

Scheme 8.15

8.6.2   Noncovalent Binding to Biomolecules

8.6.2.1   DNA Sensing by Intercalative Re(I) Complexes

Luminescent Re(I) tricarbonyl polypyridine complexes with a pendant planar aromatic unit such as anthracene and pyrene have proven to be effective sensors for double-stranded DNA. For example, in the complex $[\text{Re(bpy)(CO)}_3\text{(py-spacer-anthracene)}]^+$, the anthracene and pyridine are linked together by a spacer (**6.3**) (Scheme 8.16).[73]

The flexible spacer allows the hydrophobic anthracene unit to stay close to the Re(I) centre in an aqueous environment. Therefore, the ^{3}MLCT emission of the complex is quenched significantly by a Dexter exchange mechanism to the triplet state of anthracene. After the addition of DNA molecules to the solution, the anthracene unit intercalates into the base pairs of DNA and thus the spatial separation between the pendant quencher (anthracene) and the Re(I) centre is greatly increased. Consequently, the energy-transfer process is inhibited and the ^{3}MLCT emission of the complex is restored. The λ_{em} and the binding constant were determined to be 580 nm and $4.6 \pm 0.5 \times 10^5$ M^{-1}, respectively. The ^{3}MLCT emission enhancement upon intercalation into DNA suggests that the Re(I) complex is a promising sensor for this biomolecule.

8.6.2.2 *Avidin Sensing by Re(I) Biotin Complexes*

Avidin sensing techniques are important in various biological applications such as antibodies development and protein detection. The four biotin binding sites of tetrameric avidin allow the recognition of biotinylated fluorophores. The emission of most organic fluorescent probes is quenched upon binding to avidin due to resonance energy transfer (RET). However, the application of these biotinylated fluorophores is limited by self-quenching.[65] Luminescent biotinylated Re(I) complexes have been introduced as another class of avidin probes owing to their characteristic photophysical properties. The emission quantum yields (Φ_{em}) of hydrophobic Re(I) amido-dipyridoquinoxaline biotin complexes (**6.4a** and **6.4b**, see Scheme 8.17) have been shown to vary significantly with solvent polarity (from ~ 0.2 in CH$_2$Cl$_2$ to ~ 0.002 in aqueous buffer). This phenomenon is also attributed to the hydrogen bonding between water and the amido-dipyridoquinoxaline ligand.[74]

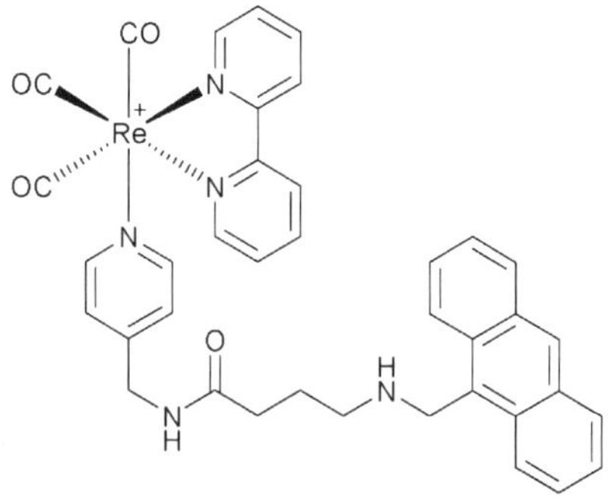

6.3

Scheme 8.16

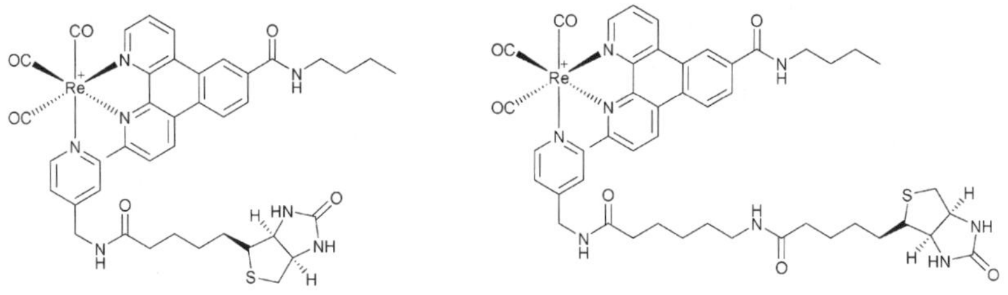

6.4a

6.4b

Scheme 8.17

After binding to avidin, complexes **6.4a** and **6.4b** demonstrate 8.1-fold and 3.1-fold increments in emission intensities respectively. The emission intensities increase because the complexes are in close proximity to the lipophilic and rigid environment of avidin. The less-efficient avidin-induced emission of complex **6.4b** suggest that the increased chain length of the spacer-arm separates the rhenium luminophore further away from the avidin core, hence the luminophore becomes more exposed to the aqueous media.

8.6.3 Re(I)-Containing Macromolecules for Antitumour Therapeutics

8.6.3.1 *Re(I)–Folate Bioconjugate*

Folate receptors (FR) on the cell surface facilitate the cellular uptake of folic acid (vitamin B_9) and also the folate derivatives. Since the FR is overexpressed in several types of tumours, targeting the FR has been considered as an ideal pathway to deliver pharmaceuticals into the tumour cells. The cellular uptake, imaging and cytotoxicity of Re(I)–folate bioconjugate [Re(N^N^N-B$_9$)(CO)$_3$]$^+$ (**6.5**) in A2780/AD ovarian cancer cells and Chinese hamster ovary (CHO) cells (non-FR expressing negative control) have been investigated (Scheme 8.18).[75] This Re(I)–folate bioconjugate contains a polyethylene glycol (PEG, M_r ~2000) unit as a spacer.

In vitro cell imaging by confocal microscopy has suggested that the cellular uptake of **6.5** in A2780/AD cancer cells was efficient (45 min after incubation) while no internalisation in CHO cells was observed. In addition, the

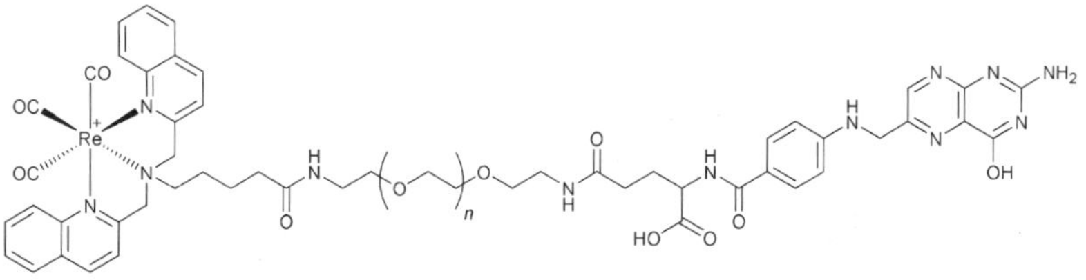

6.5

Scheme 8.18

cytotoxicity of **6.5** towards A2780/AD cancer cells is much higher than that of cisplatin, while the cytotoxicity of **6.5** towards CHO cells is much lower than that of cisplatin. The selective uptake folic acid derivatives into FR overexpressing cancer cells suggests that this Re(I)–folate bioconjugate is a potential candidate for antitumour treatment when radioactive [186/188]Re is used.

8.6.3.2 *Re(I)-Containing Poly(amidoamine)*

Poly(amidoamine)s (PAA) have been extensively studied for applications in medicinal chemistry because they are biodegradable, biocompatible and versatile for specific functions. They tend to remain in the circulatory system after intravenous injection but will be concentrated in tumour tissues by the enhanced permeation and retention (EPR) effect.[69] An amphoteric PAA functionalised with a thiol pendant (ISA23SH$_{10\%}$) has been reacted with [Re(CO)$_3$(H$_2$O)$_3$]$^+$ to form a Re(I)-containing PAA (**6.6**) (Scheme 8.19).[76] This polymer is soluble and stable in water, with the S and N atoms of the cysteamine moiety chelating to the metal centre, whereas the amphoteric nature is maintained by the free carboxylate groups.

6.6 does not show a hemolytic effect towards Hela cells up to a concentration of 5 mg/mL while the cytotoxic activity is not observed up to 100 ng/mL after 48 h of incubation. The low toxicity and tumour-targeting properties suggest that ISA23SH$_{10\%}$ is capable of delivering radioactive [186/188]Re for therapeutic purpose.

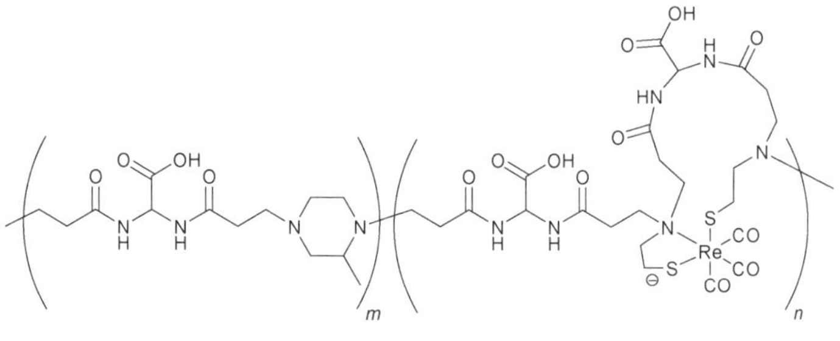

6.6 *m/n* = 9

Scheme 8.19

8.7 Concluding Remarks

The photophysical properties and characteristics of Re(I) polymers have been reviewed. Numerous potential applications of the Re(I) macromolecules in the areas of (1) nanostructures and nanopatterning; (2) triplet emitters for OLEDs; (3) photosensitising materials for photovoltaic devices and (4) luminescent probes in biological systems have been discussed. Owing to the simple coordination chemistry of these compounds, the chemical and photophysical properties can be easily tuned for desired functionality with tailor-made ligands *via* facile synthesis. Although the device efficiency of OLEDs and OPVs based on these materials is still far from satisfactory, substantial improvements and developments on Re(I)-containing macromolecules are believed to be achievable by good molecular design and well-controlled material synthesis.

Acknowledgements

The work described in this chapter was substantially supported by the University Grants Committee of the Hong Kong Special Administrative Region, China (AoE/P-03/08). Partial financial support from the Strategic Research Theme, University Development Fund and Small Project Fund administrated by the University of Hong Kong is also acknowledged.

References

1. D. R. Striplin and G. A. Crosby, *Coord. Chem. Rev.*, 2001, **211**, 163.
2. J. V. Caspar, B. P. Sullivan and T. J. Meyer, *Inorg. Chem.*, 1984, **23**, 2104.
3. J. V. Caspar and T. J. Meyer, *J. Phys. Chem.*, 1983, **87**, 952.
4. M. S. Wrighton and D. L. Morse, *J. Am. Chem. Soc.*, 1974, **96**, 998.
5. S. M. Fredericks, J. C. Luong and M. S. Wrighton, *J. Am. Chem. Soc.*, 1979, **101**, 7415.
6. D. R. Striplin and G. A. Crosby, *Coord. Chem. Rev.*, 2001, **211**, 163.
7. L. Sacksteder, M. Lee, J. N. Demas and B. A. DeGraff, *J. Am. Chem. Soc.*, 1993, **115**, 8230.
8. J. M. Villegas, S. R. Stoyanov, W. Huang and D. P. Rillema, *Inorg. Chem.*, 2005, **44**, 2297.
9. W. K. Chan, P. K. Ng, X. Gong and S. Hou, *J. Mater. Chem.*, 1999, **9**, 2103.
10. C. W. Tse, L. S. M. Lam, K. Y. Man, W. T. Wong and W. K. Chan, *J. Polym. Sci. A: Polym. Chem.*, 2005, **43**, 1292.
11. K. D. Ley, C. Ed Whittle, M. D. Bartberger and K. S. Schanze, *J. Am. Chem. Soc.*, 1997, **119**, 3423.
12. L. Sacksteder, A. P. Zipp, E. A. Brown, J. Streich, J. N. Demas and B. A. DeGraff, *Inorg. Chem.*, 1990, **29**, 4335.
13. K. A. Walters, K. D. Ley, C. S. P. Cavalaheiro, S. E. Miller, D. Gosztola, M. R. Wasielewski, A. P. Bussandri, H. Van Willigen and K. S. Schanze, *J. Am. Chem. Soc.*, 2001, **123**, 8329.
14. E. Wolcan and M. R. Féliz. *Photochem. Photobiol. Sci.*, 2003, **2**, 412.
15. L. L. B. Bracco, M. P. Juliarena, G. T. Ruiz, M. R. Féliz, G. J. Ferraudi and E. Wolcan, *J. Phys. Chem. B*, 2008, **112**, 11506.
16. S. Hou and W. K. Chan, *Macro. Rapid Commun.*, 1999, **20**, 440.

17. P. Xu, X. Ji, V. Abetz, S. Jiang and J. Shen, *J. Polym. Sci. B: Polym. Phys.*, 2008, **46**, 2047.
18. S. Hou, K. Y. K. Man and W. K. Chan, *Langmuir* 2003, **19**, 2485.
19. K. W. Cheng and W. K. Chan, *Langmuir*, 2005, **21**, 5247.
20. J.H. Burroughes, D. D. C. Bradley, A. R. Brown, R. N. Marks, K. Mackay, R. H. Friend, P. L. Burn and A. B. Holmes, *Nature*, 1990, **247**, 539.
21. A. J. Sandee, C. K. Williams, N. R. Evans, J. E. Davies, C. E. Boothby, A. Köhler, R. H. Friend and A. B. Holmes, *J. Am. Chem. Soc.*, 2004, **126**, 7041.
22. X. Chen, J.-L. Liao, Y. Liang, M. O. Ahmed, H.-E. Tseng and S.-A. Chen, *J. Am. Chem. Soc.*, 2003, **125**, 636.
23. N. R. Evans, D. L. Sudha, C. S. K. Mak, S. E. Watkins, S. I. Pascu, A. Köhler, R. H. Friend, C. K. Williams and A. B. Holmes, *J. Am. Chem. Soc.*, 2006, **128**, 6647.
24. C. Mei, J. Ding, B. Yao, Y. Cheng, Z. Xie, Y. Geng and L. Wang, *J. Polym. Sci. A: Polym. Chem.*, 2007, **45**, 1746.
25. H. Zhen, C. Jiang, W. Yang, J. Jiang, F. Huang and Y. Cao, *Chem. Eur. J.*, 2005, **11**, 5007.
26. K. M. Jung, K. H. Kim, J.-I. Jin, M. J. Cho and D. H. Choi, *J. Polym. Sci. Part A: Polym. Chem.*, 2008, **46**, 7517.
27. Y. Xu, R. Guan, J. Jiang, W. Yang, H. Zhen, J. Peng and Y. Cao, *J. Polym. Sci. A: Polym. Chem.*, 2008, **46**, 453.
28. C. T. Wong and W. K. Chan, *Adv. Mater.*, 1999, **11**, 455.
29. P. K. Ng, X. Gong, S. H. Chan, L. S. M. Lam and W. K. Chan, *Chem. Eur. J.*, 2001, **7**, 4358.
30. V. Duprez, M. Biancardo, H. Spanggaard and F. C. Krebs, *Macromolecules*, 2005, **38**, 10436.
31. P. D. Vellis, J. A. Mikroyannidis, C.-N. Lo and C.-S. Hsu, *J. Polym. Sci. A: Polym. Chem.*, 2008, **46**, 7702.
32. C. Liu, J. Li, B. Li, Z. Hong, F. Zhao, S. Liu and W. Li, *Appl. Phys. Lett.*, 2006, **89**, 243511.
33. S. Ranjan, S. Y. Lin, K.-C. Hwang, Y. Chi, W.-L. Ching and C.-S. Liu, *Inorg. Chem.*, 2003, **42**, 1248.
34. F. Li, M. Zhang, G. Cheng, J. Feng, Y. Zhao, Y. Ma, S. Liu and J. Shen, *Appl. Phys. Lett.*, 2004, **84**, 148.
35. N. J. Lundin, A. G. Blackman, K. C. Gordon and D. L. Officer, *Angew. Chem. Int. Ed.*, 2006, **45**, 2582.
36. K. D. Glusac, S. Jiang and K. S. Schanze, *Chem. Commun.*, 2002, 2504.
37. G. B. Cunningham, Y. Li, S. Liu and K. S. Schanze, *J. Phys. Chem. B*, 2003, **107**, 12569.88
38. M. Zhang, P. Lu, X. Wang, L. He, W. Zhang, B. Yang, L. Liu, L. Yang, M. Yang, Y. Ma, J. Feng, D. Wang and N. Tamai, *J. Phys. Chem. B.*, 2004, **104**, 13185.
39. P.-I. Lee and S. L.-C. Hsu, *J. Polym. Sci. A: Polym. Chem.*, 2007, **45**, 1492.

40. Y.-J. Pu, R. E. Harding, S. G. Stevenson, E. B. Namdas, C. Tedeschi, J. P. J. Markham, R. J. Rummings, P. L. Burn and I. D. W. Samuel, *J. Mater. Chem.*, 2007, **17**, 4255.

41. W.-Y. Wong, X. Z. Wang, Z. He, K.-K. Chan, A. B. Djurišić, K.-Y. Cheung, C.-T. Yip, A. M.-C. Ng, Y. Y. Xi, C. S. K. Mak and W.-K. Chan, *J. Am. Chem. Soc.*, 2007, **129**, 14372.

42. L. Liu, C.-L. Ho, W.-Y. Wong, K.-Y. Cheung, M.-K. Fung, W.-T. Lam, A. B. Djurišić and W. K. Chan, *Adv. Funct. Mater.*, 2008, **18**, 2824.

43. P.-T. Wu, T. Bull, F. S. Kim, C. K. Luscombe, S. A. Jenekhe, *Macromolecules* 2009, **42**, 671.

44. J. Mei, K. Ogawa, Y.-G. Kim, N. C. Heston, D. J. Arenas, Z. Nasrollahi, T. D. McCarley, D. B. Tanner, J. R. Reynolds and K. S. Schanze, *ACS Appl. Mater. Interfaces*, 2009, **1**, 150.

45. N. S. Baek, S. K. Hau, H.-L. Yip, O. Acton, K.-S. Chen and A. K.-Y. Jen, *Chem. Mater.*, 2008, **20**, 5734.

46. K. Y. K. Man, H. L. Wong, W. K. Chan, C. Y. Kwong and A. B. Djurišić, *Chem. Mater.*, 2004, **16**, 365.

47. K. Y. K. Man, H. L. Wong, W. K. Chan, A. B. Djurišić, E. Beach and S. Rozeveld, *Langmuir* 2006, **22**, 3368.

48. K. W. Cheng, C. S. K. Mak, W. K. Chan, A. M.-C. Ng and A. B. Djurišić, *J. Polym. Sci. A: Polym. Chem.*, 2008, **46**, 1305.

49. L. S. M. Lam, S. H. Chan and W. K. Chan, *Macromol. Rapid Commun.*, 2000, **21**, 1081.

50. W. Y. Ng, X. Gong and W. K. Chan, *Chem. Mater.*, 1999, **11**, 1165.

51. *Szycher's Handbook of Polyurethanes*, M. Szycher ed. CRC Press, New York, 1999.

52. C. S. Hui, L. S. M. Lam, C. Yin, W. K. Chan and A. B. Djurišić, *J. Polym. Sci. A: Polym. Chem.*, 2003, **41**, 1708.

53. C. Bamer-Kowollik, *Handbook of RAFT Polymerization*, Wiley-VCH, Weinheim, 2008.

54. W. Y. Tam, C. S. K. Mak, A. M. C. Ng and A. B. Djurišić, *Macromol. Rapid. Commun.*, 2009, **30**, 622.

55. H. Mori, H. Ookuma, S. Nakano and T. Endo, *Macromol. Chem. Phys.*, 2006, **207**, 1005.

56. C. W. Tse, K. W. Cheng, W. K. Chan and A. B. Djurišić, *Macromol. Rapid. Commun.*, 2004, **25**, 1335.

57. C. W. Tse, K. Y. K. Man, K. W. Cheng, C. S. K. Mak, W. K. Chan, C. T. Yip, Z. T. Liu and A. B. Djurišić, *Chem. Eur. J.*, 2007, **13**, 328.

58. W. K. Chan, C. S. Hui, K. Y. K. Man, K. W. Cheng, H. L. Wong, N. Zhu and A. B. Djurišić, *Coord. Chem. Rev.*, 2005, **249**, 1351.

59. J. Nelson, J. Kirkpatrick and P. Ravirajan, *Phys. Rev. B.*, 2004, **69**, 035337.

60. C. S. K. Mak, Q. Y. Leung, C. H. Li and W. K. Chan, *J. Polym. Sci. A: Polym. Chem.*, 2010, **48**, 2311.

61. C. S. K. Mak, W. K. Cheung, Q. Y. Leung and W. K. Chan, *Macromol. Rapid Commun.*, 2010, **31**, 875.

62. M. Zhang, P. Lu, X. Wang, L. He, H. Xia, W. Zhang, B. Yang, L. Liu, L. Yang, M. Yang, Y. Ma, J. Feng, D. Wang and N. Tamai, *J. Phys. Chem. B*, 2004, **108**, 13185.

63. K. K. W. Lo, M. W. Louie and K. Y. Zhang, *Coord. Chem. Rev.*, 2010, **254**, 2603.

64. K. K. W. Lo, K. Y. Zhang and S. P. Y. Li, *Eur. J. Inorg. Chem.*, 2011, 2011, 3551.

65. K. K. W. Lo, K. H. K. Tsang, K. S. Sze, C. K. Chung, T. K. M. Lee, K. Y. Zhang, W. K. Hui, C. K. Li, J. S. Y. Lau, D. C. M. Ng and N. Zhu, *Coord. Chem. Rev.*, 2007, **251**, 2292.

66. S. Jurisson, D. Berning, W. Jia and D. Ma, *Chem. Rev.*, 1993, **93**, 1137.

67. E. John, M. L. Thakur, J. DeFulvio, M. R. McDevitt and I. Damjanov, *J. Nucl. Med.*, 1993, **34**, 260.

68. C. Wu, M. W. Brechbiel, R. W. Kozak, O. A. Gansow, *Bioorg. Med. Chem. Lett.*, 1994, **4**, 449.

69. S. Richardson, P. Ferruti and R. Duncan, *J. Drug Target.*, 1999, **6**, 391.

70. X. Q. Guo, F. N. Castellano, L. Li, H. Szmacinski, J. R. Lakowicz and J. Sipior, *Anal. Biochem.*, 1997, **254**, 179.

71. J. D. Dattelbaum, O. O. Abugo and J. R. Lakowicz, *Bioconjugate Chem.*, 2000, **11**, 533.

72. X. Q. Guo, F. N. Castellano, L. Li and J. R. Lakowicz, *Anal. Chem.*, 1998, **70**, 632.

73. N. B. Thornton and K. S. Schanze, *Inorg. Chem.*, 1993, **32**, 4994.

74. K. K. W. Lo, K. H. K. Tsang and K. S. Sze, *Inorg. Chem.*, 2006, **45**, 1714.

75. N. Viola-Villegas, A. E. Rabideau, J. Cesnavicious, J. Zubieta and R. P. Doyle, *ChemMedChem*, 2008, **3**, 1387.

76. D. Donghi, D. Maggioni, G. D'Alfonso, F. Amigoni, E. Ranucci, P. Ferruti, A. Manfredi, F. Fenili, A. Bisazza and R. Cavalli, *Biomacromolecules*, 2009, **10**, 3273.

Recent Developments in Metal-Containing Complexes with Azo Chromophore Functionalities

A. S. ABD-EL-AZIZ*[1,2] AND E. A. STROHM[2]

[1] Department of Chemistry, University of Prince Edward Island, 550 University Avenue, Charlottetown, Prince Edward Island, C1A 4P3, Canada;
[2] Department of Chemistry, University of British Columbia, Okanagan Campus, 3333 University Way, Kelowna, British Columbia, V1V 1V7, Canada
*E-mail: abdelaziz@upei.ca

9.1 Introduction

There has been tremendous interest in azo dyes in light of their photochemical properties and their industrial applications.[1] Studies of metal-containing complexes with photochromic moieties ranging from small molecules to macromolecules have dramatically increased.[1,2] There are numerous reviews and books that illustrate the developments of coordination and organometallic compounds containing photochromic moieties.[1–28] In particular, those that contain azo dyes have attracted considerable interest due to their potential applications in the areas of photomemory and photoswitching devices,[3,4,29] nonlinear optical materials,[30–32] and optical information storage.[1,28,33] Moreover, compounds functionalised with azobenzene moieties exhibit unique photophysical[34,35] and photochromic properties.[1] Thus, they are well known for their ability to undergo *trans–cis* isomerisation of the N=N double bond in response to photoirradiation.[10] Under UV light, irradiation of the π–π* band

RSC Polymer Chemistry Series No. 2
Molecular Design and Applications of Photofunctional Polymers and Materials
Edited by Wai-Yeung Wong and Alaa S Abd-El-Aziz
© The Royal Society of Chemistry 2012
Published by the Royal Society of Chemistry, www.rsc.org

induces *trans*-to-*cis* isomerisation and the reverse photoinduced *cis*-to-*trans* isomerisation can occur when the n–π* band is irradiated with visible light or when heat is applied.[11] The incorporation of metal complexes into azo dyes can provide attractive versatile materials due to the combination of the *trans–cis* isomerisation characteristic of the azo moiety with the unique magnetic, optical, and redox properties of the metal species.[7,10,11,20] As a result, the inclusion of a metal centre with organic azo chromophores alters the physical and chemical properties of these molecular materials.[23]

There are a number of classes of metal complexes containing azo functionalities ranging from simple complexes to macromolecules that have been developed over the years.[11,20] Very recently, Espinet and coworkers[36] synthesised organometallic photochromic molecules (**1, 2**) that incorporate gold(I) and azobispyridine moieties. UV-vis studies indicated that these compounds photoisomerise in solution, but do not exhibit luminescence.

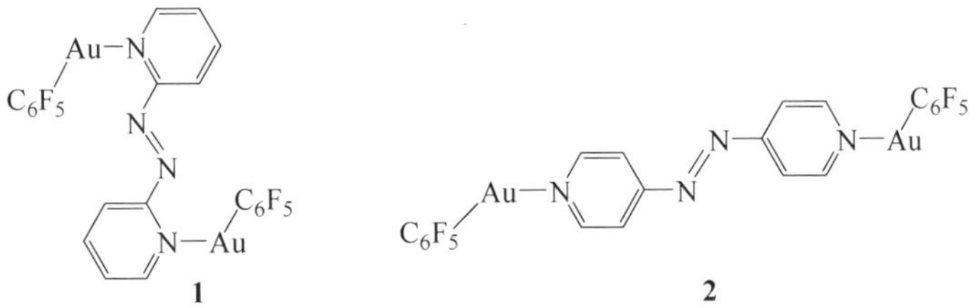

1

2

A photoresponsive, dendritic azo-conjugated palladium(II) coordination compound was recently reported by Park *et al.*[37] They were the first to report reversible photoisomerisation of a metallocomplex where the azobenzene moiety is exclusively bonded by a σ-bond between the azo nitrogen (-N=N-) and metal centre. The azobenzene moieties in Pd(II) dendrimers **3–5** all exhibited photoinduced *trans*-to-*cis* isomerisations *via* 350 nm irradiation. Additionally, the reverse *cis*-to-*trans* isomerisations occurred in the dark at room temperature (2–3 h), which is a photochromic behaviour that is consistent with the previously reported studies of organic azobenzene-containing dendrimers.[38,39]

The combination of an azo chromophore with a transition-metal complex enables these compounds to exhibit photochromic and redox properties, making them attractive materials that can be employed for use in a number of applications such as optical and electronic devices.[17,20] This chapter will discuss selective examples of azo-conjugated organometallic and coordination compounds ranging from small molecules to macromolecules.

9.2 Azo-Bridged Ferrocene Complexes

The incorporation of ferrocene moieties into various compounds are of particular interest due to their high thermal stability and tunable redox

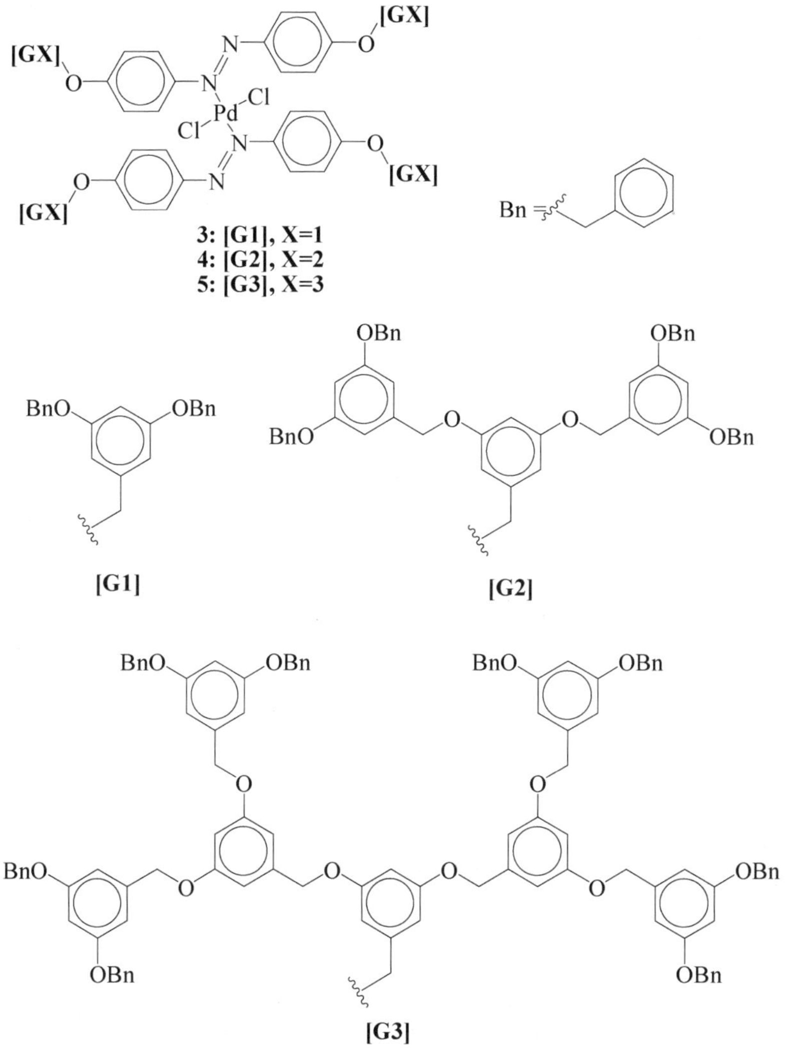

3: [G1], X=1
4: [G2], X=2
5: [G3], X=3

[G1]

[G2]

[G3]

properties. In 1997, Manners and coworkers reported the synthesis and thermal properties of liquid-crystalline poly(ferrocenylsilanes) functionalised with azobenzene mesogens.[40,41] The side-chain poly(ferrocenylsilanes) was prepared in two steps, as shown in Scheme 9.1. First, a ferrocenophane precursor, **6**, was used to synthesise the previously reported[42] polymer precursor, **7**, *via* ring-opening polymerisation (ROP) in the presence of heat. Next, the prepared mesogens containing azobenzene acrylate moieties **8** and **9** were reacted with poly(ferrocenylsilane) **7** *via* hydrosilylation to afford liquid-crystalline poly(ferrocenylsilanes) **10** and **11**. Approximately 80% of the silane units were substituted with the azobenzene mesogens for each polymer.

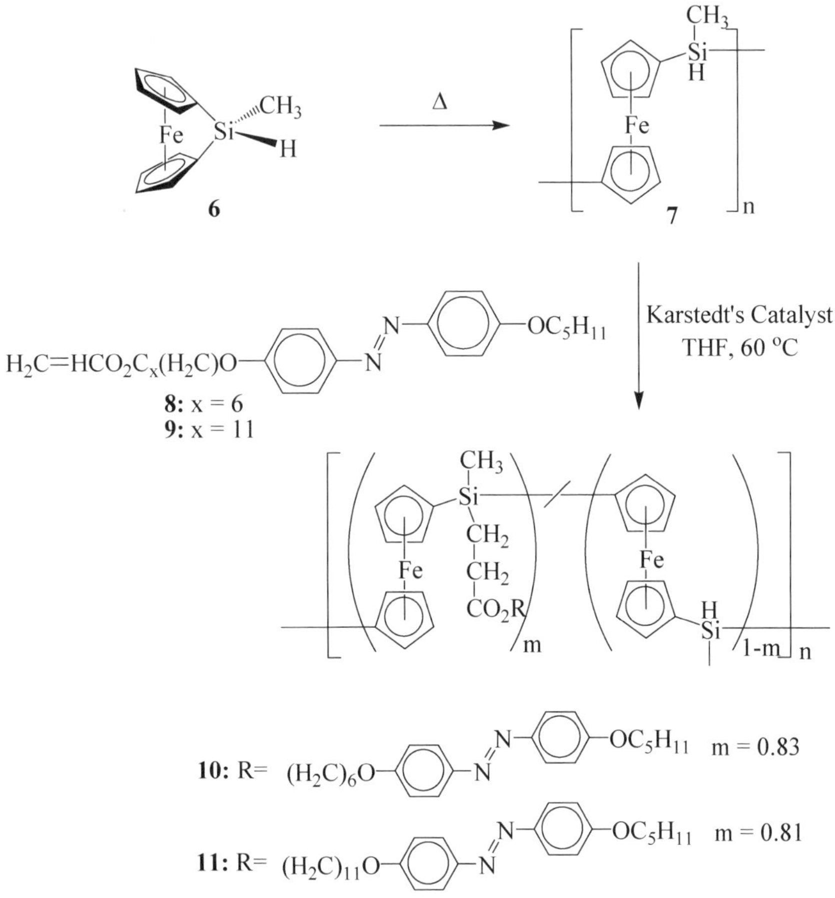

Scheme 9.1 Synthesis of póly(ferrocenylsilanes) **10** and **11**.

Thermal studies showed glass-transition temperatures between 30 and 35 °C and melting points between 36 and 56 °C. Compared to the azobenzene-functionalised mesogens, the polymers melted at considerably lower temperatures. After melting, a well-defined texture was observed in the nematic phase *via* polarised optical microscopy. The nematic phase remained stabilised when the polymers were heated to 250 °C, after which point thermal decomposition took place. Furthermore, functionalisation of these novel rod-like thermotropic side-chain liquid-crystalline poly(ferrocenylsilanes) can be tailored with the notion that their physical properties will vary accordingly.

Several poly(ferrocenylacetylenes) with conjugated spacers prepared *via* living metathesis polymerisation were reported by Schottenberger and coworkers in 1998.[43] Scheme 9.2 illustrates a slightly varied[44] synthetic pathway for the formation of one of the reported monomers, arylazoferrocene-substituted acetylene, **16**. Low-temperature recombination of lithioferrocene (**12**) and brominated diazonium salt **13** produced brominated azoferrocene **14**,

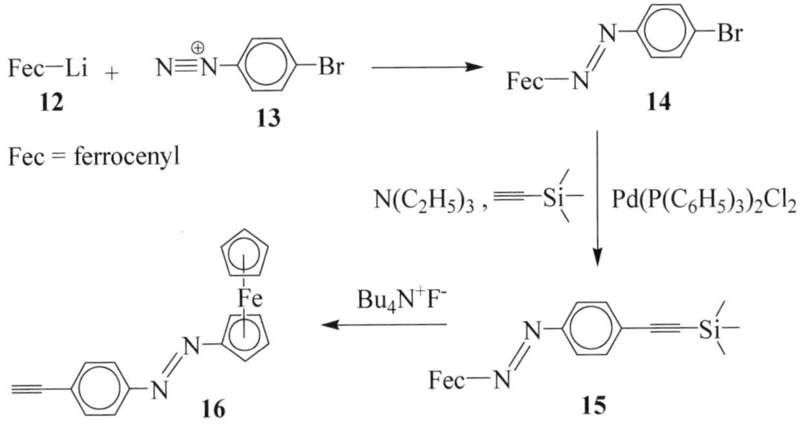

Scheme 9.2   Synthesis of monomeric azoferrocene acetylene **16**.

followed by Sonogashira–Hagihara coupling[45] to yield a trimethylsilyl-protected alkyne precursor **15**, which was subsequently deprotected to afford arylazoferrocene acetylene **16**. The resulting monomer **16** was found to be most stable in the (*E*)-configuration and exhibited near planarity. As a result, the incorporation of the diazene spacer allowed monomer **16** to exhibit resistance to photochemical isomerisation, a property that is desirable for nonlinear optical (NLO) conjugates.[46] The highly conjugated, linear diazene side group was easily incorporated into the conjugated polymer *via* living metathesis polymerisation. This was achieved specifically when the initiators, Mo(*N*-2,6-*i*-Pr$_2$C$_6$H$_3$)(CHCMe$_2$Ph)(OCMe-(CF$_3$)$_2$)$_2$ or Mo(*N*-2,6-Me$_2$C$_6$H$_3$)-(CHCMe$_2$Ph)(OCMe(CF$_3$)$_2$)$_2$ were selected and reacted with acetylene monomer **16** *via* β-insertion.

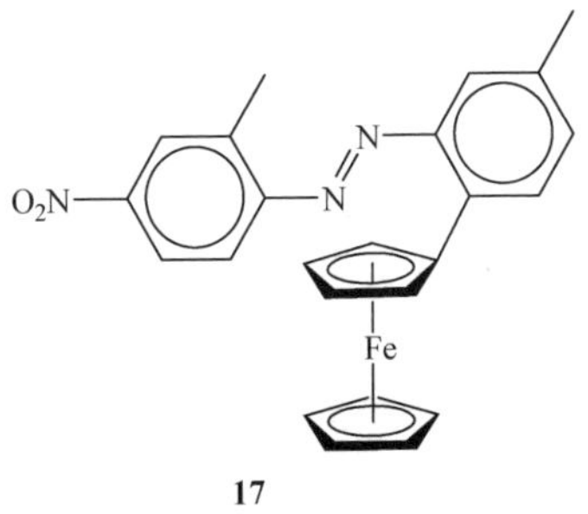

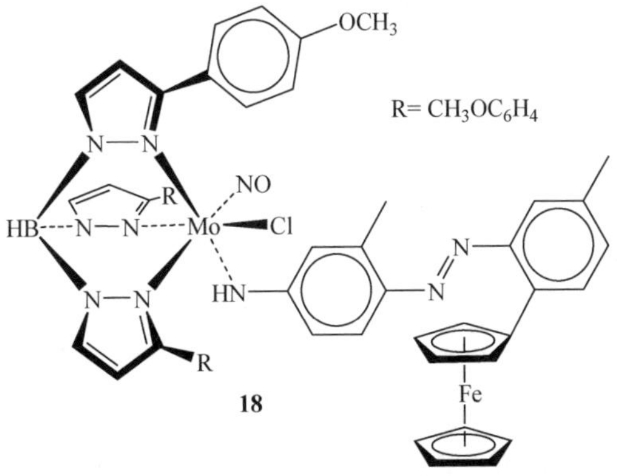

'Bent' ferrocenyl derivatives with azobenzene groups (complexes **17** and **18**) were synthesised and characterised, and their second-order NLO properties were investigated by Campo *et al.*[31] From the aspect of the diazo group, the ferrocene moiety was attached at the *ortho* position of the aryl group, resulting in the disruption of 'molecular linearity' at the bridging charge-transfer axis between the donor and acceptor groups. X-ray diffraction confirmed that the ferrocenyl complexes **17** and **18** are noncentrosymmetric crystalline structures. When these complexes were compared to the similar 'linear' compound that had the ferrocenyl group in the *para* position[47] of the aryl group, it was found that the second-order NLO response was significantly reduced as a direct result of molecular 'bending'. However, the study showed that for the *ortho* position complexes, the off-diagonal components of the tensor are strongly maintained.

Nishihara and coworkers[48] reported the synthesis of ferrocene complexes with bridging azo groups (**21–23**), from the reaction of lithioferrocenes **19** and **20**, using a modified route that was previously reported by Nesmeyanov *et al.*[49] (Scheme 9.3). Furthermore, this study was the first to report the crystal structure of azoferrocene **21** and confirm that the molecule resides in the *trans* confirmation. Electrochemical studies revealed that azoferrocene **21** exhibits two 1-electron oxidation waves through the construction of a mixed-valence cation. Additionally, nonsymmetrical azoferrocene **22** exhibits a reversible three-step 1-electron oxidation in which the oxidation pathway is assumed to have two mixed-valence states. Furthermore, symmetrical azoferrocene **23** undergoes reversible 2-electron and 1-electron oxidation waves, which implies that as a result of the azo group, the terminal ferrocene moiety contains nearly all of the positive charge. Consequently, the different structures of azoferrocenes **21**, **22**, and **23** considerably influence the electronic interaction between the ferrocene moieties. Moreover, it was found that the redox potentials and intervalence-transfer band characteristics of azoferrocenes **21**$^+$, **22**$^+$, and **23**$^{2+}$ are influenced by solvation effects. An azo-bridged ferrocene polymer **25** was also synthesised by reacting dilithioferrocene complex **24** with nitrous oxide under high-pressure conditions (Scheme 9.4). The degree of

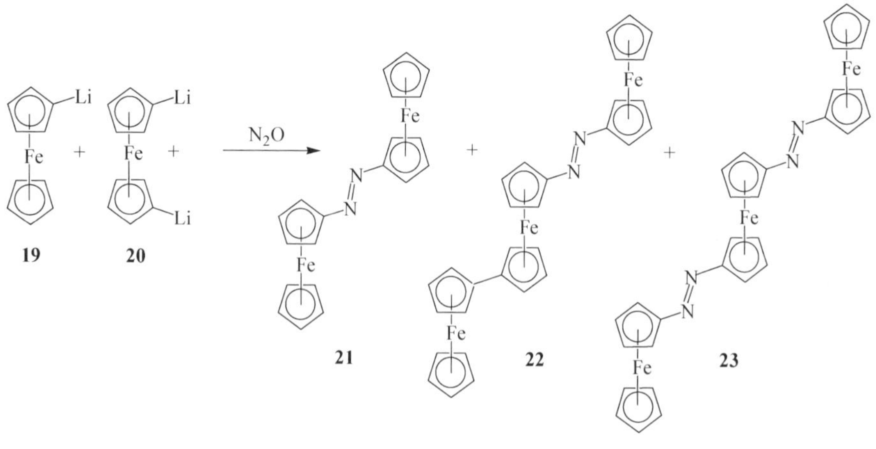

Scheme 9.3   Synthesis of azoferrocene complexes **21–23**.

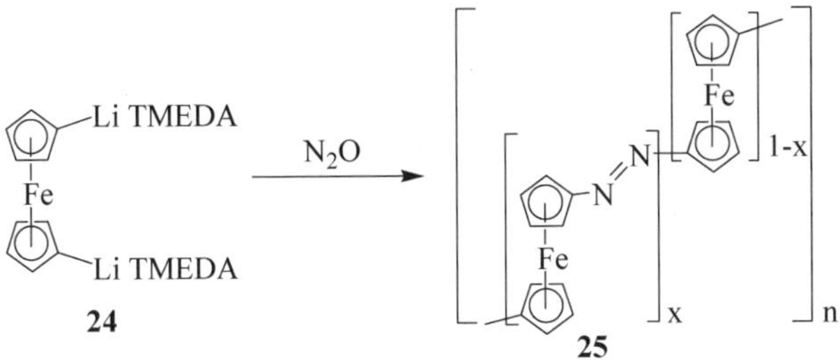

Scheme 9.4   Synthesis of poly(azoferrocene) **25**.

substitution for the azo-bridged ferrocene groups was *ca.* 60% and *ca.* 40% for the directly attached ferrocene units. In addition, polymer **25** had an average molecular weight of $M_w = 87\,000$ and $M_n = 13\,000$. UV-vis analysis revealed a metal-to-ligand charge transfer (MLCT) band at 535 nm and compared to the ferrocene monomers, the π–π^* transition shifted to a longer wavelength.

Photoisomerisation of azoferrocene **26** from the *trans* to *cis* form of the azo group *via* π-π^* transition under UV light of 365 nm and MLCT under UV light of 546 nm was reported by Nishihara and coworkers.[13,50] However, it was found that the reverse reaction from the *cis* to *trans* configuration is photochemically irreversible.[51] Electrochemical studies indicated that the *trans* isomer is thermodynamically stable in the mixed-valence state as a result of the reversible 1-electron oxidation waves at $E^{0\prime} = 0.29$ and 0.50 V *vs.* Ag/Ag$^+$. Conversely, the mixed-valence state of the *cis* form is not as thermodynamically stable as the *trans* form.

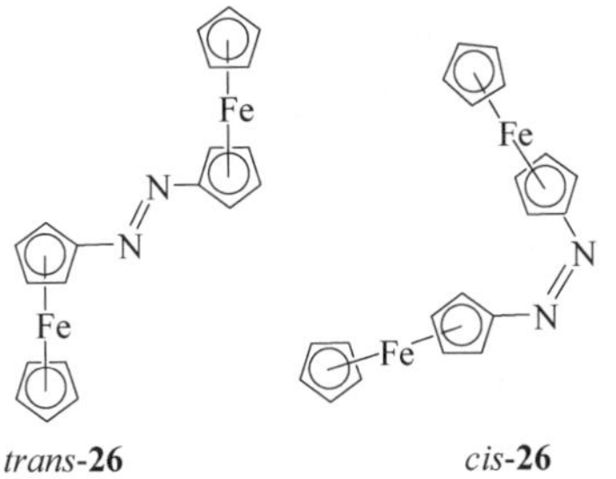

trans-**26**   *cis*-**26**

Since azoferrocene **26** does not undergo reversible photoisomerisation under visible light, Nishihara and coworkers[51] studied *meta*-azoferrocenylazobenzene **27** and discovered that the reversible oxidation and reduction of the iron centre, coupled with the azobenzene moiety, exhibits reversible isomerisation when irradiated with a monochromatic green (546 nm) light (Scheme 9.5).[10,11] In the reduced state, Fe(II), photoconversion from the *trans* to *cis* isomer occurs upon irradiation *via* the MLCT excited state.[21] Conversely, in the oxidised state, Fe(III), a *cis*-to-*trans* photoisomerisation occurs upon irradiation *via* the loss of the MLCT character. Consequently, the ability of 3-

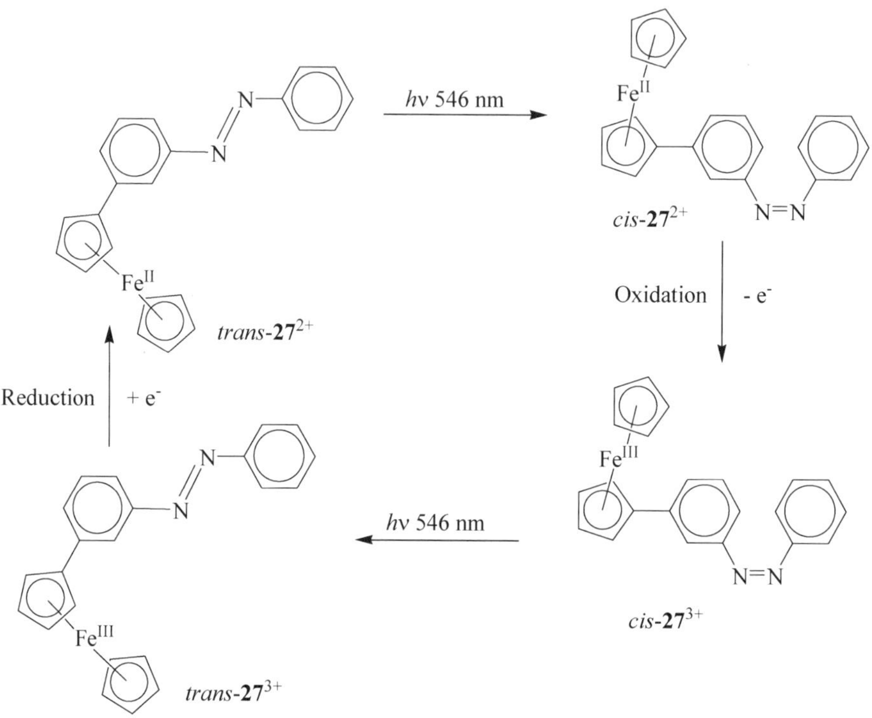

hv 546 nm

Fe^II

cis-27^2+   N=N

trans-27^2+   Fe^II

Oxidation   - e^-

Reduction   + e^-

Fe^III

hv 546 nm   N=N

cis-27^3+

Fe^III   trans-27^3+

Scheme 9.5   Reversible structural conversion of **27** using 546 nm irradiation and redox reaction.

ferrocenylazobenzene **27** to undergo redox-conjugated reversible photoconversion between *cis* and *trans* isomers by single 546 nm irradiation allows such complexes to be used as molecular switches in application to light-driven molecular devices.[21] Furthermore, degradation of photochromic molecules may be avoided if optical systems containing a monochromatic light source rather than a double light source are utilised.[51]

The discovery of reversible photoisomerisation of 3-ferrocenylazobenzene with a single green light and reversible redox reactions led Nishihara and coworkers[52] to investigate the isomerisation of azobenzene derivatives functionalised with ferrocene at various substitution points. The study shows that the conversion between *cis* and *trans* isomers can be regulated by photon-, electron-, and proton-catalysed reactions, depending on the redox state of the iron centre and its proximity to the arylazo group.

Similarly, Nishihara and coworkers[53] prepared a 3-ferrocenyl-4'-carboxylazobenzene complex, **28**, to make a single light-controllable azobenzene monolayer system *via* electrochemical oxidation. The monochromatic reversible photoisomerisation system of complex **28** was accomplished by combining 546 nm light with electrochemical reactions (Scheme 9.6). In this study, the authors elegantly demonstrated the photoswitching behaviour of

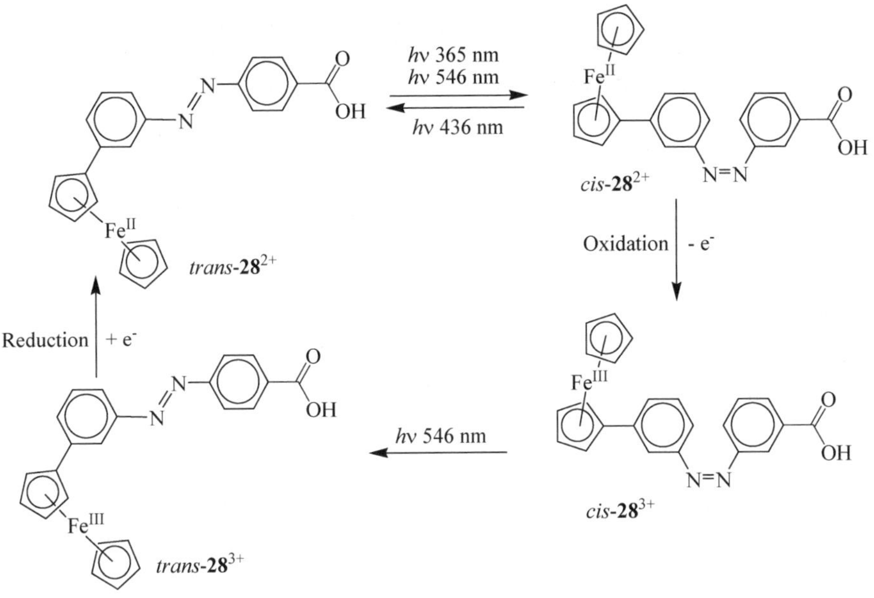

Scheme 9.6 Photoswitching behaviour of 3-ferrocenyl-4′-carboxylazobenzene complex, **28**.

this photochromic system and showed promising application in high-density optical data storage devices.

A ferrocene-based chiral molecular machine containing an azobenzene moiety (**29**) was reported by Aida and coworkers.[54] They discovered that complex **29** serves as a "light-driven molecular scissor",[54] in which the ferrocene moiety is used as the pivotal point that can produce an open–close motion, two phenyl groups are used as the blade units for this motion, and two phenylene moieties are used as the handle parts strapped by an azobenzene unit that can undergo photoisomerisation. Reversible *trans–cis* photoisomerisation of the azo moiety *via* UV and visible light irradiation allowed reversible opening and closing motions of the phenyl groups through the redox-active ferrocene pivot (Scheme 9.7).[55] Aida and coworkers[56,57] elegantly

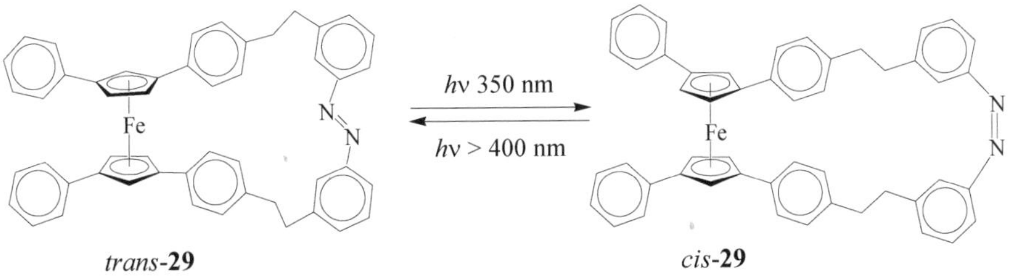

Scheme 9.7 Scissor-like motion of complex **29** induced *via* photoisomerisation.

took advantage of this "scissor-like" photoisomerisation behaviour of complex **29** to bind and structurally twist guest molecules. Two zinc porphyrin groups were incorporated at each cyclopentadienyl ring of the ferrocene complex, which behaved as the host molecule, to afford a guest-binding site. A stable host–guest system (**31**) was formed when a bidentate ligand (**30**) was introduced as a rotary guest molecule. Interestingly, when UV and visible light were used to irradiate host–guest complex **31**, photoisomerisation of the azobenzene moiety induced conformational changes in the zinc–porphyrin moieties, causing the guest molecule (**30**) to undergo a twisting motion (Scheme 9.8).

Dendrimers (9mer, 27mer and 81mer) functionalised with terminal 3-ferrocenylazobenzene moieties have been synthesised and investigated by Daniel *et al.*[58] Dendrimer (9mer) **32** is illustrated below as an example. These dendrimers were shown to exhibit reversible photoisomerisation and photochromism under 365, 436, and 546 nm irradiation. When the azobenzene moieties undergo *cis–trans* isomerism, the size and structure of the dendrimer alters.[24] Furthermore, upon photoisomerisation, electrochemical studies indicate that as the dendrimer generations increase, the diffusion coefficient

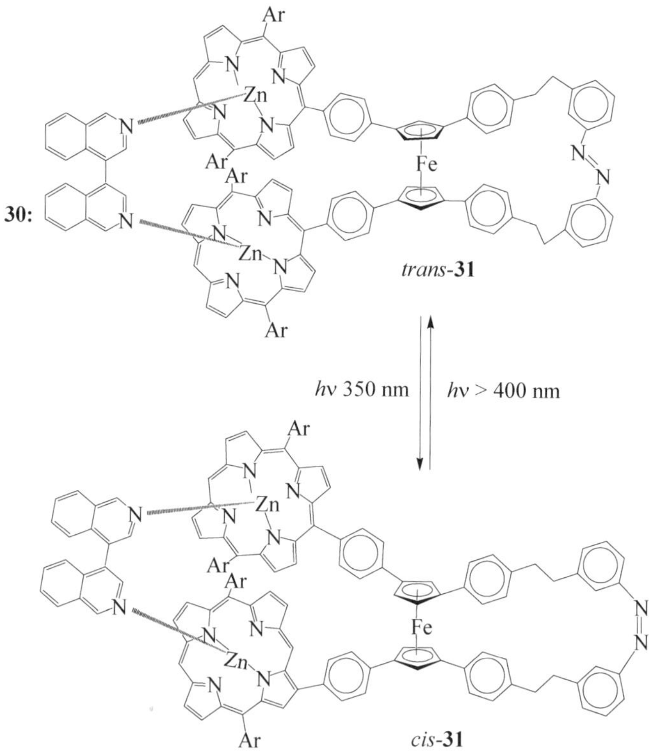

Scheme 9.8   Photoisomerisation of host–guest complex **31**.

decreases. Conversely, the diffusion coefficient increases for the smaller generation dendrimers and varies in size depending on *cis–trans* isomerisation.

Sakano *et al.*[59] reported the synthesis of 2-aza-[3]-ferrocenophanes functionalised with azobenzenes **33–35** *via* platinum- and ruthenium-catalysed reactions. UV-vis studies revealed absorption maxima at 413, 442, and 426 nm for complexes **33**, **34**, and **35**, respectively, as a result of the π–π^* transition of the *trans* isomer. Incomplete photoisomerisation from the *trans* to the *cis* isomer occurred when irradiated with 420 nm light. Thermal isomerisation from the *cis* to the *trans* isomer occurred when the azaferrocenophane derivatives were in the photostationary state. When the iron-coordinated centre undergoes a 1-electron oxidation, conversion between the *cis–trans* isomers occurs at a faster rate compared to isomerisation in the absence of electrochemical oxidation.

Azo dye-containing polymers **36** with both neutral and cationic organoiron moieties were first prepared by the Abd-El-Aziz and coworkers[60] in 2004. The incorporation of the cationic cyclopentadienyliron complex into the macro-

32

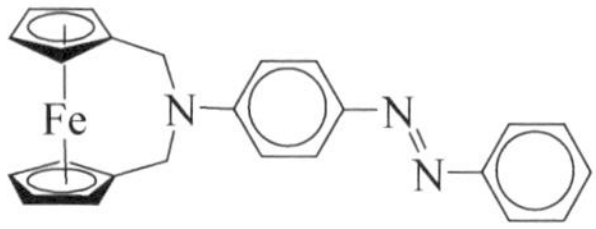

33

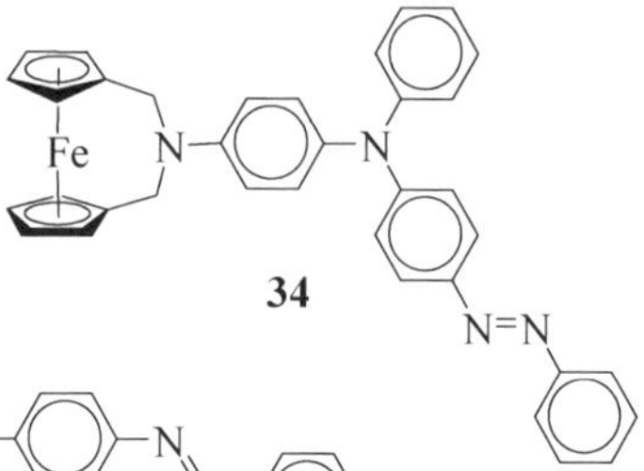

34

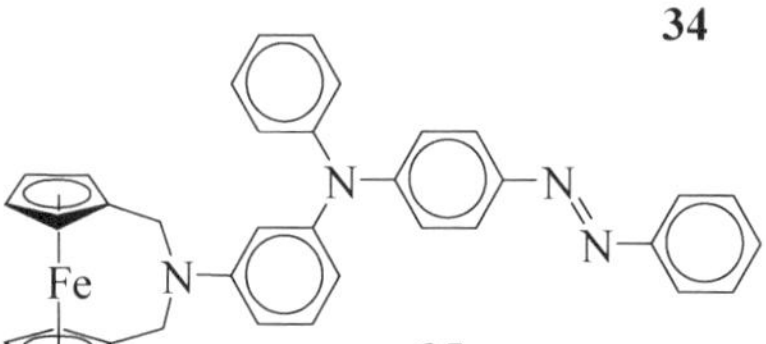

35

molecules enhanced their solubility in polar organic solvents. Electrochemical studies showed that both the neutral and cationic iron moieties are redox-active. The cationic cyclopentadienyliron iron centre was reduced at $E_{1/2} = -1.42$ V *vs.* Ag/Ag$^+$ and neutral ferrocene was oxidised at $E_{1/2} = 0.89$ V. Furthermore, UV-vis studies revealed that these polymers exhibit wavelength maxima near 419 nm. Upon the addition of hydrochloric acid, a bathochromic shift occurred near 530 nm. Photolysis was used to remove the cationic cyclopentadienyliron moieties to afford the analogous neutral polyferrocenes functionalised with azo chromophores.[23]

Similar to the neutral ferrocene-based polymers containing azo chromophores,[60] the preparation of polyester ferrocenes functionalised with arylazo dyes (**37**, **38**) along with chemical and physical properties were investigated by Akhter *et al.*[61] Thermal analysis *via* differential scanning calorimetry indicated that the polymers exhibit glass-transition temperatures between 60 and 85 °C. UV-vis studies revealed that the polymers exhibit wavelength maxima between 354 and 405 nm.

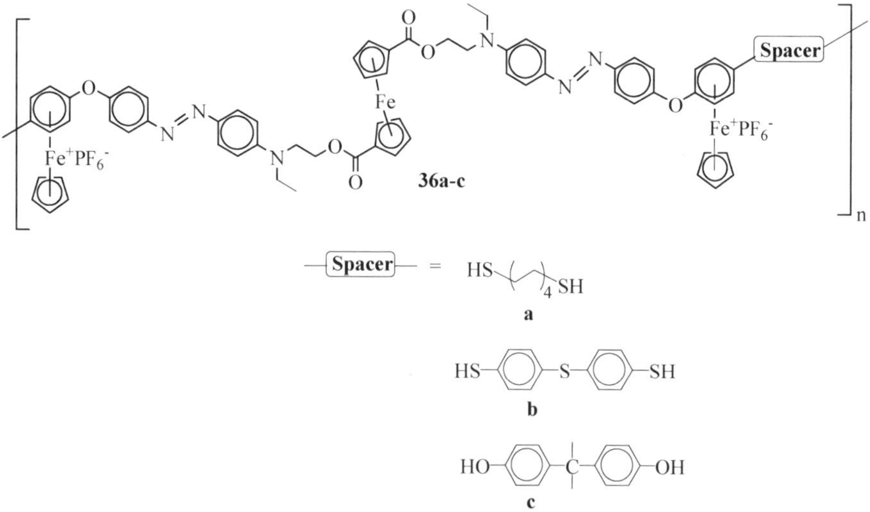

36a-c

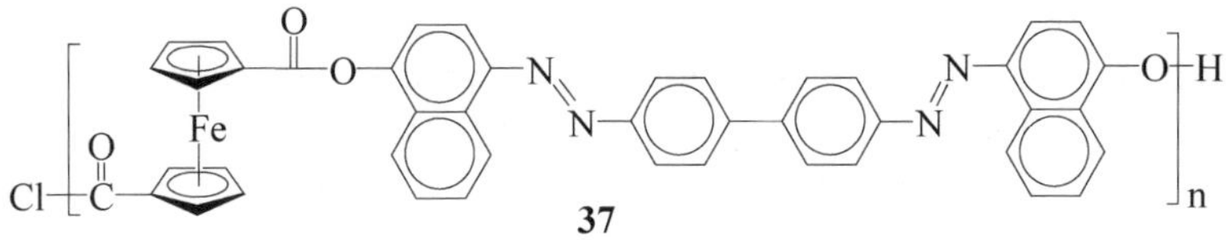

37

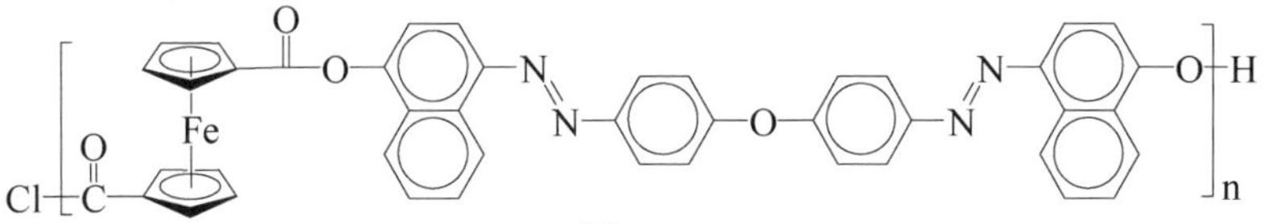

38

9.3 Cationic Iron-Coordinated Macromolecules

The incorporation of cationic cyclopentadienyliron moieties can enhance solubility and facilitate nucleophilic aromatic substitution and addition reactions, due to the intense electron-withdrawing ability of the iron centre.[62] Abd-El-Aziz *et al.*[63,64] reported the synthesis of cationic cyclopentadienyliron-complexed polyaromatic ethers and thioethers containing azo dyes (**41**). Synthesis took place *via* condensation reactions between assorted S- and O-containing dinucleophiles (**40**) and iron-coordinated arylazo dye monomers **39** (Scheme 9.9). As a result of the azo functionalities, UV-vis analysis revealed

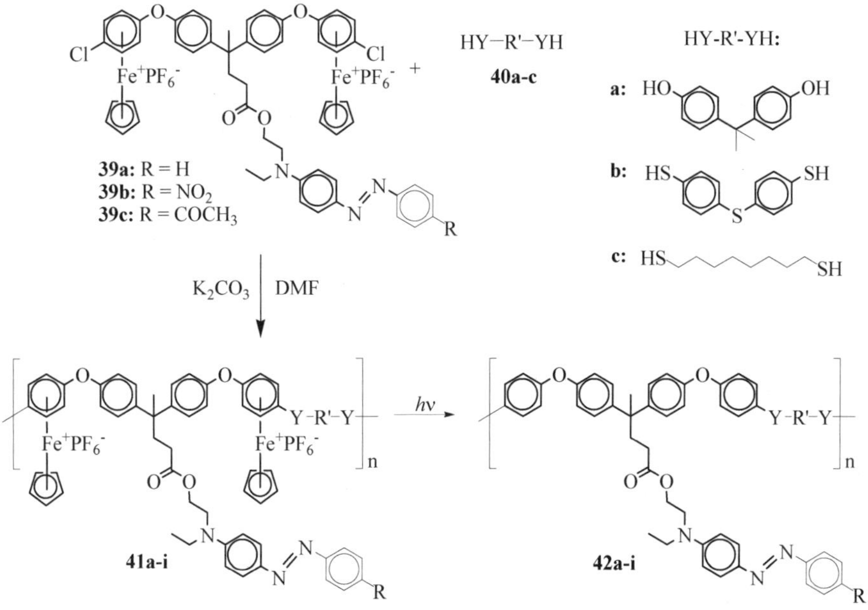

39a: R = H
39b: R = NO$_2$
39c: R = COCH$_3$

HY–R'–YH

40a-c

HY-R'-YH:

a: HO— —OH

b: HS— —SH

c: HS— —SH

K$_2$CO$_3$ | DMF

41a-i

hv

42a-i

Scheme 9.9 Synthesis of organoiron-based polyethers and polythioethers functionalised with arylazo dyes (**41**) and photolytic cleavage of organic analogues.

that these vibrantly coloured macromolecules exhibit wavelength maxima between 417 and 489 nm. Additionally, photolysis was used to isolate the corresponding organic polymer analogues (**42**), and the non-metallic azo-based polymers were photobleached in the presence of hydrogen peroxide.[8] Thermal analysis revealed increased glass-transition temperatures for the organometallic polymers compared to their organic analogues.

The synthesis of an organoiron-based calix[4]arene dendrimer functionalised with azo dyes on the upper rim (**43**) has been recently investigated.[65] The dendrimer's first generation was prepared by nucleophilic aromatic substitution of a calix[4]arene complexed with four cationic iron centres and an azo dye containing a terminal phenolic group. UV-vis studies indicated that wavelengths of maximum absorbance occurred between 430 and 456 nm and that bathochromic shifts occurred between 513 and 535 nm when subjected to acidic solutions. Furthermore, the azo dye-containing metallocalix[4]arene exhibited reversible colour changes when exposed to both acid and base. Under neutral conditions, the colour of the dendrimer was orange; upon the addition of an acid or base the colour changed to purple or yellow,

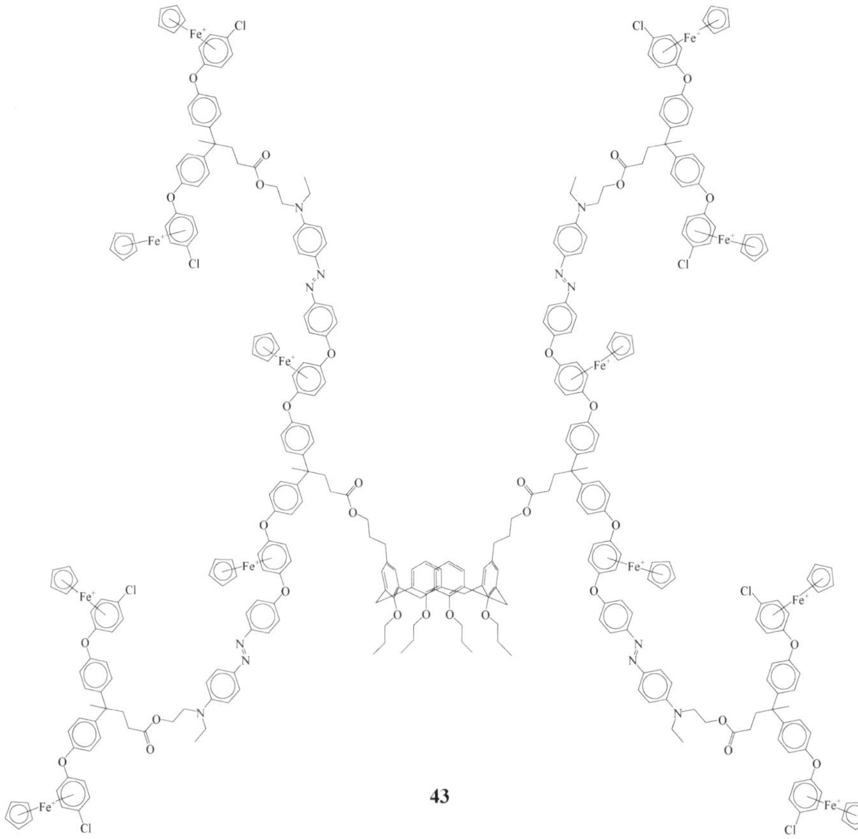

43

respectively. A reversible reduction of the cationic iron-coordinated complex was observed at $E_{1/2} = -1.49$ V according to cyclic voltammetric studies.

Star-shaped organoiron macromolecules functionalised with azo chromophores have been synthesised.[66] Convergent and divergent methods were employed by reacting a variety of azo dyes and iron-coordinated branches with multifunctional cores. As an example, organoiron oligomer **44** displayed a maximum wavelength at 427 nm and when exposed to an acidic media, the wavelength maximum shifted to 558 nm. Additionally, electrochemical studies showed a reversible reduction of the cationic organoiron complex at $E_{1/2} = -1.53$ V.

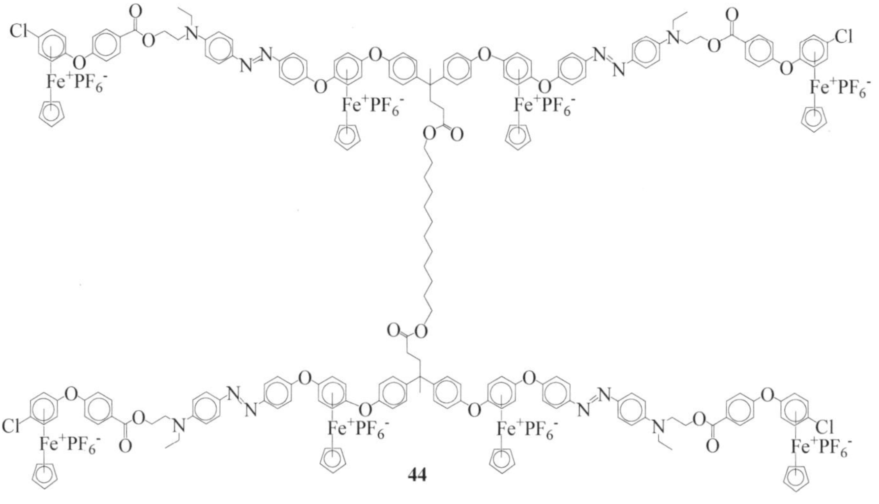

44

Cationic organoiron norbornene-based polymers containing arylazo dyes have been prepared[67] *via* ring-opening metathesis polymerisation (ROMP) of compound **45**, (polymer **46** as an example, see Scheme 9.10). Gel permeation chromatography (GPC) was used to determine the molecular weights of the polymers. However, due to the interactions between the cationic iron-coordinated complexes and the GPC columns, the molecular weights of the organoiron polymers were unable to be measured. Consequently, the molecular weights of the corresponding organic polynorbornenes isolated *via* photolytic demetallation were measured instead. These demetallated polymers were shown to have weight average molecular weights between 9200 and 21 800. Analysis of the UV-vis absorption spectra of the highly coloured cationic organoiron polynorbornenes revealed that these polymers exhibit wavelength maxima between 420 and 430 nm and display bathochromic shifts between 510 and 520 nm upon the addition of hydrochloric acid. Furthermore, electrochemical studies indicated that these polymers undergo reversible reduction between -1.2 and -1.4 V *vs.* Ag/Ag$^+$.

Organoiron polynorbornenes functionalised with benzothiazole azo chromophores (**47** and **48**) have also been prepared by Abd-El-Aziz *et al.*[68] The polymers were synthesised *via* ROMP using Grubb's catalyst. UV-vis

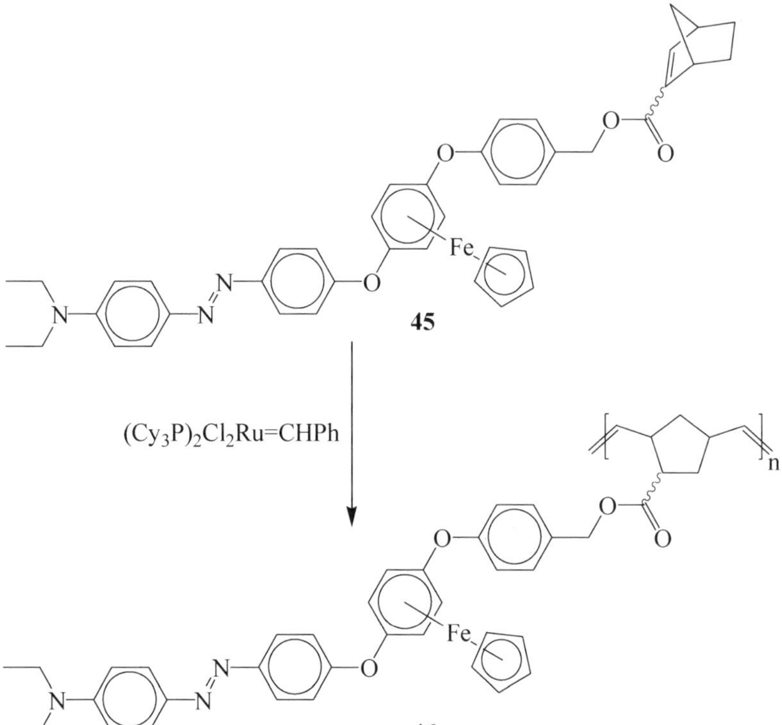

45

$(Cy_3P)_2Cl_2Ru=CHPh$

46

Scheme 9.10 Synthesis of organoiron-based polynorbornene with azo chromophores, **46**, as an example polymer.

absorption studies showed that they possess wavelength maxima (λ_{max}) between 423 and 520 nm. The difference in λ_{max} was attributed to the incorporation of various substituents on the heteroarylazo moieties and to the

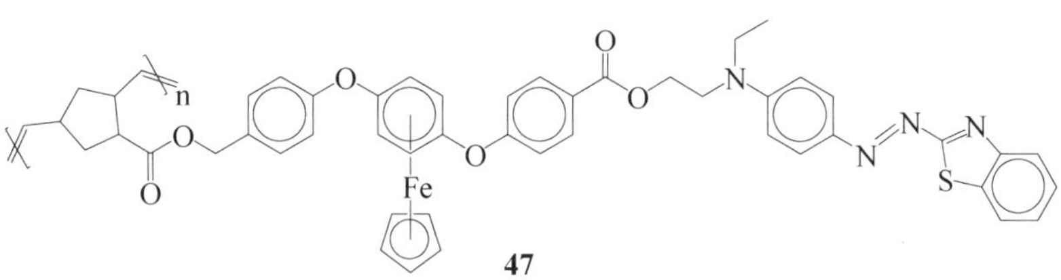

47

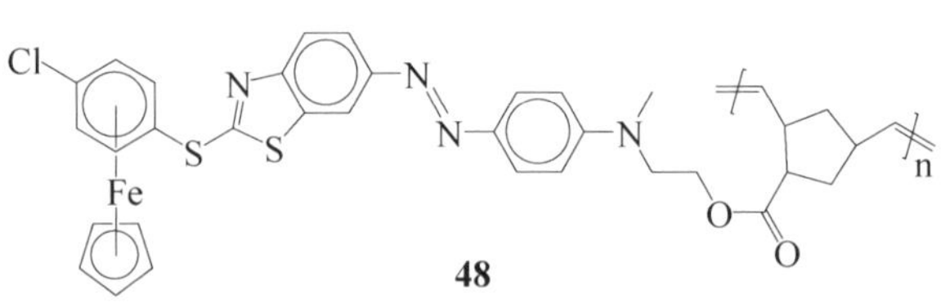

48

difference in conjugation.[68] Cyclic voltammetry was used to examine the redox properties of the organoiron polymers. The results showed that the cationic iron centres undergo a reduction at -1.08 V. Thermogravimetric analysis indicated that decomposition of the iron centre occurred between 225 and 231 °C and the polymer main chains degraded between 400 and 450 °C.

Photolytic discolouration of cationic organoiron polymers containing azo chromophores (**49**) has been investigated.[69] The synthesis of the monomers was achieved *via* metal-mediated nucleophilic aromatic substitution and 2,2′-azobis(2-methylpropionitrile) (AIBN) was used as a radical initiator for polymerisation.[70] Discolouration of the polymers was achieved *via* irradiation with 300 nm UV light, as a result of the decomplexation of the organoiron unit and its interaction with the azo dye.[69] In order for the complete loss of colour, the polymer concentration must be 0.025 M or lower.

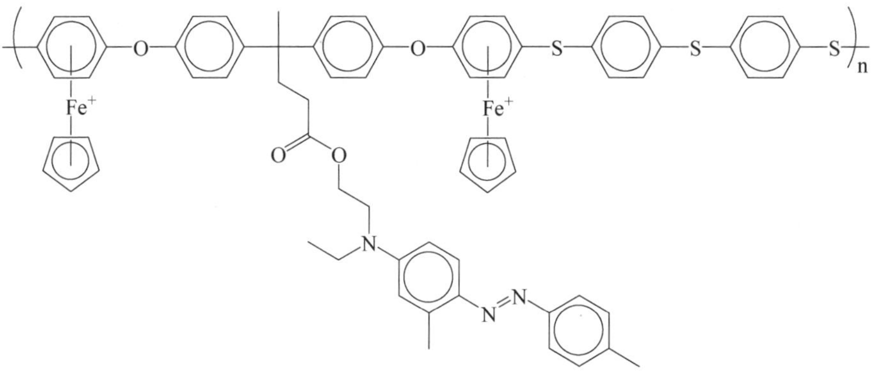

49

9.4 Pyridine and Bipyridine Derivatives

Studies have shown that pyridines and pyridine-containing compounds function as suitable monodentate ligands in coordination complexes.[71,72] Wong *et al.*[72] have reported the synthesis and characterisation of a series of homo-dinuclear tungsten and osmium complexes containing 4,4′-azopyridine and 4-phenylazopyridine ligands. Complexes **50** and **51** displayed below are shown as examples investigated in the study. The spectroscopic results indicated that upon irradiation (wavelength $>$ 400 nm) of the W $\rightarrow$ L (π^*) MLCT state of **51** in solution, the absorption energy was highly dependent on

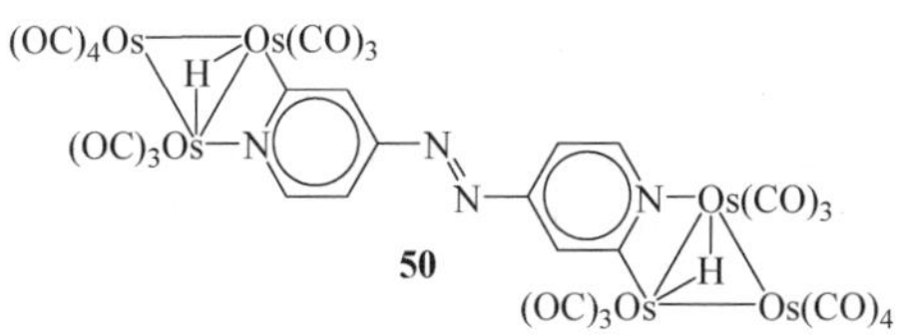

50

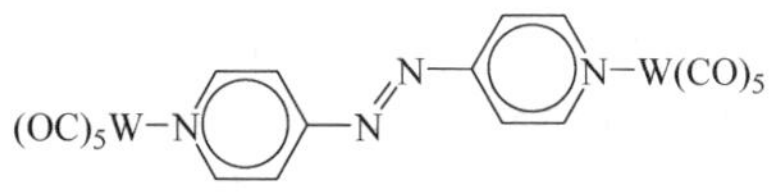

51

the different types of organic solvents used. However, complex **50** did not exhibit any solvatochromism in organic solvents. The electrochemical studies revealed that there was no electronic interaction between the two metal centres; however, it was observed that there was an improved interaction between the metal and the 4,4′azopyridine ligand in **51** compared to its corresponding monomeric complex. This finding was based on a large shift to the more positive reduction potential that was observed for dimeric complex **51**.

The incorporation of 2,2′- or 4,4′-azobis(pyridine) ligands into transition-metal complexes has been shown to reversibly isomerise under the influence of external stimuli and thus there has been considerable interest in their use as molecular switches in nanoscopic devices.[17,21] Very recently, the synthesis and spectroscopic behaviour of a 4,4′-azobis(pyridine)-containing tricarbonyl(2,2′-bipyridine) rhenium complex (**52**) was reported by Pourrieux *et al.*[73] At room temperature, Re(I) complex **52** does not luminesce; however, 2-electron reduction in acetonitrile or *trans*-to-*cis* photoisomerisation in dichloromethane enables the complex to exhibit luminescence. The emission spectrum of complex **52** at 77 K results from the rhenium to bipyridine MLCT excited state.

52

The synthesis and photoconversion between *trans–cis* isomers of a cobalt complex containing bipyridine ligands functionalised with azobenzene moieties, **53**, have been reported.[74] When cobalt(III) in complex **53** was reduced to cobalt(II) and irradiated with 366 nm light, the azobenzene moiety was converted from the *trans* to the *cis* isomer. Reversible isomerisation from the *cis*-to-*trans* form occurred when Co(II) was oxidised to Co(III) and irradiated with 366 nm light. Electrochemical studies showed that Co(III) undergoes a reversible reduction at −0.15 V and Co(II) partly undergoes a reversible reduction at −1.27 V *vs.* ferrocenium/ferrocene. UV-vis studies revealed that the azobenzene moiety exhibited wavelength maxima at 345, 360, and 374 nm as a result of its π–π^* transitions. Furthermore, the study showed

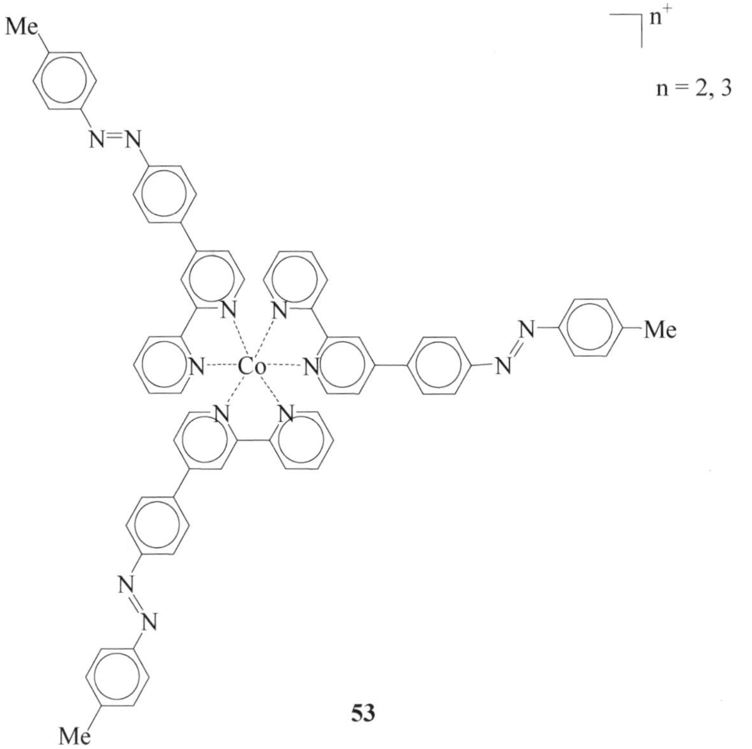

53

that the π–π^* transition decreased and the n–π^* transition increased when the complex was irradiated with 366 nm light.

Photoisomerisation studies were also carried out for tris(bipyridine)cobalt complexes containing six azobenzene groups (complex **54** is shown below as an example).[75] The results showed that the percentage of *trans*-to-*cis* conversion is dependent upon the oxidation state of the cobalt centre. This conclusion was based on the fact that the azo moieties from the Co(II) complexes yielded close to 50% conversion from the *trans* to the *cis* isomer, while only 10% conversion to the *cis* isomer was observed for the Co(III) complexes.

Nishihara and coworkers[76,77] reported the synthesis and investigated the *trans*–*cis* photoisomerisation and coordination behaviour of copper-coordinated cyclic molecules containing arylazo and bipyridine functionalities (**55**, **56**). The authors discovered that reversible photoinduced *trans*–*cis* isomerisation of the azobenzene moieties *via* UV and blue light induced a reversible ligand exchange reaction, which resulted in a redox reaction of the coordinated copper centre. The conformational change of the copper complexes is an example of mechanical motion of molecular machines.[77] The authors describe this cyclic mechanism as a photoelectric conversion system. Furthermore, this system is easily related to natural visual transduction since the application of

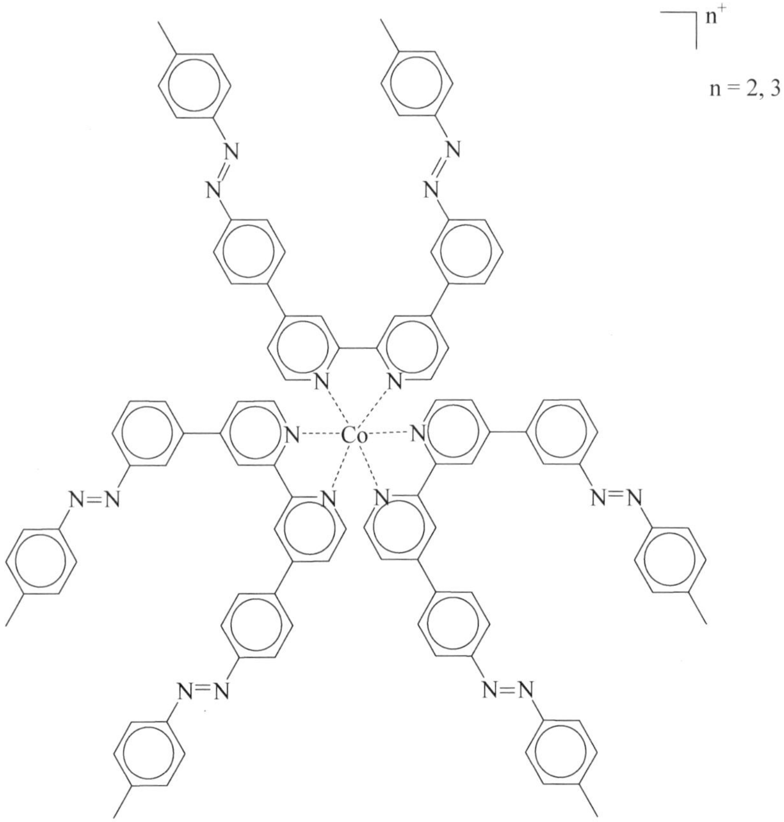

n = 2, 3

54

UV/blue light in this system produced a positive/negative current and a change in the potential of the system.[17]

A series of mono- and dinuclear azobis(2,2′-bipyridine) and azobis(terpyridine) derivatives coordinated with ruthenium have been prepared and characterised and their photochemical and electrical properties have been studied.[78–82] An azobis(2,2′-bipyridine) mononuclear complex (**57**) and a ruthenium azobis(terpyridine) dinuclear complex (**58**) are illustrated below as example compounds. It was found that bipyridine-based mononuclear Ru complexes exhibit reversible photoisomerisation under UV light.[79] However, the corresponding dinuclear complex does not isomerise reversibly. Moreover, the authors indicate that the low-energy MLCT excited state of the mononuclear complex sensitises the photoisomerisation exhibited by the azo group.[79] All of the azo-bridged Ru complexes with bipyridine ligands exhibit redox activity at the ruthenium centre and at both the substituted and non-substituted bipyridine ligands. Furthermore, electrochemical studies showed that the reduction potentials of the azo moieties were more positive compared to the bipyridine ligands. The reduction potential of the azo group on

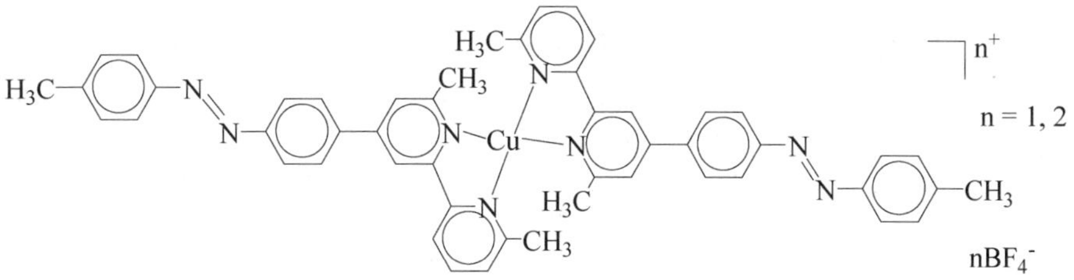

55

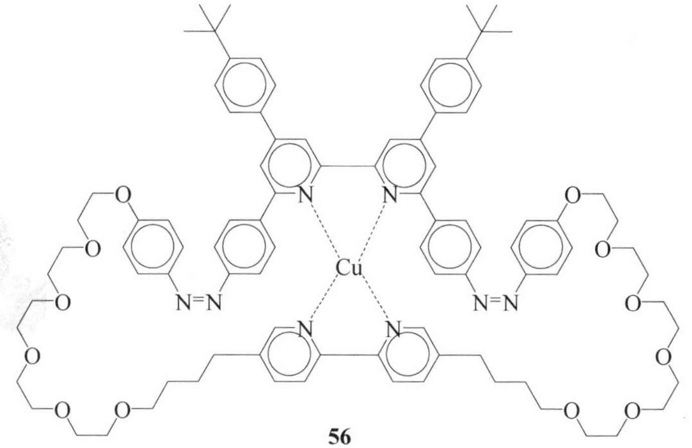

56

azobis(terpyridine) dinuclear ruthenium complexes was also more positive than that of the terminal terpyridine ligands.[82] In addition, both the azo and terpyridine ligands exhibited absorption bands as a result of the MLCT transitions.

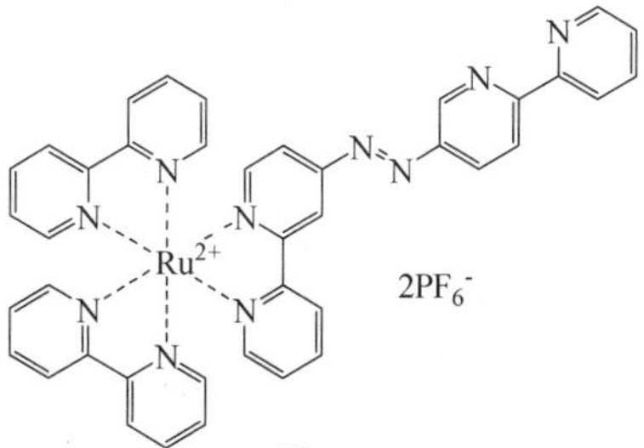

57

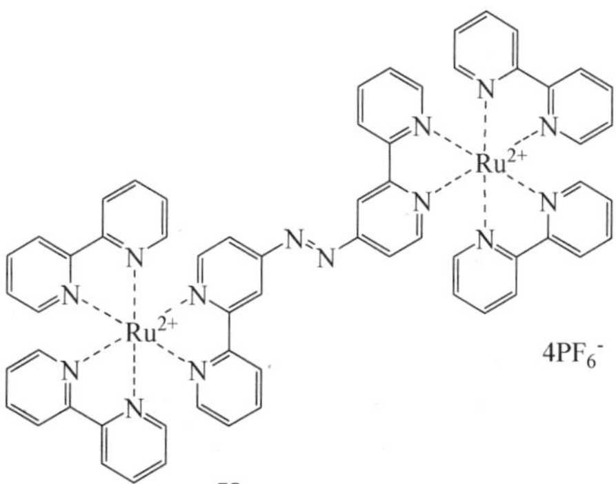

58

9.5   Terpyridine Derivatives

Rhodium-based azobis(terpyridine) complexes (**59** is shown as an example) have been reported by Nishihara and coworkers.[83] Photochemical properties were investigated and showed substantial *trans*-to-*cis* photoisomerisation that was found to be strongly influenced by their counterions and solvents. However, *cis*-to-*trans* photoisomerisation *via* irradiation at 430 nm was not observed. Additionally, thermal conversion from the *cis* isomer to the *trans* isomer occurred more slowly compared to the corresponding non-substituted ligands.

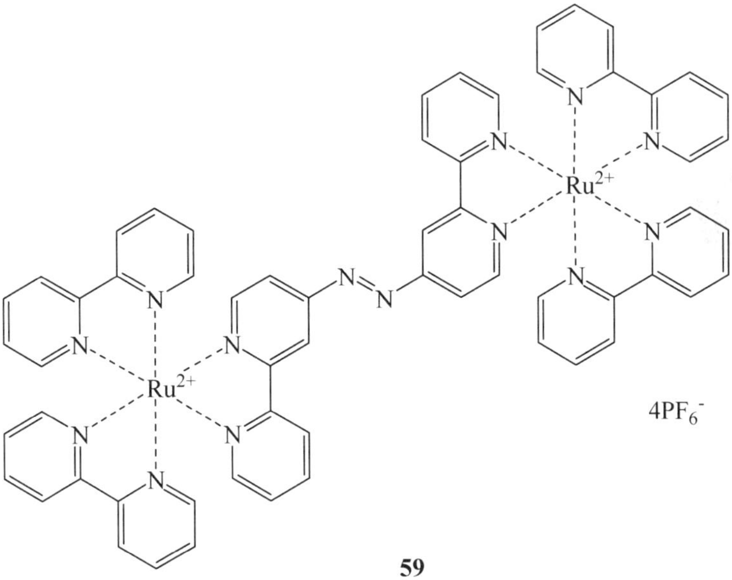

59

Mono- and disubstituted bis(terpyridine) ruthenium(II) and rhodium(III) complexes functionalised with azobenzenes (**60** is shown as an example) were prepared and characterised, and their photoisomerisation behaviour was studied.[84] The *trans* form of mono-substituted ruthenium complex was illuminated with 366 nm light and exhibited only 20% structural conversion to its *cis* form. On the other hand, when the *trans* isomer of the monosubstituted rhodium complex was irradiated with light, nearly complete structural conversion to its *cis* isomer was observed. When photoisomerisation was induced *via* UV light irradiation for the disubstituted Ru(II) and Rh(III) complexes, the ruthenium complex did not exhibit any *trans*-to-*cis* structural photoconversion, whereas the rhodium complex did. The authors attribute these differences in photochromic behaviour to the different coordinated metal centres, whose energy-transfer pathways to the MLCT states are different for each metal complex.

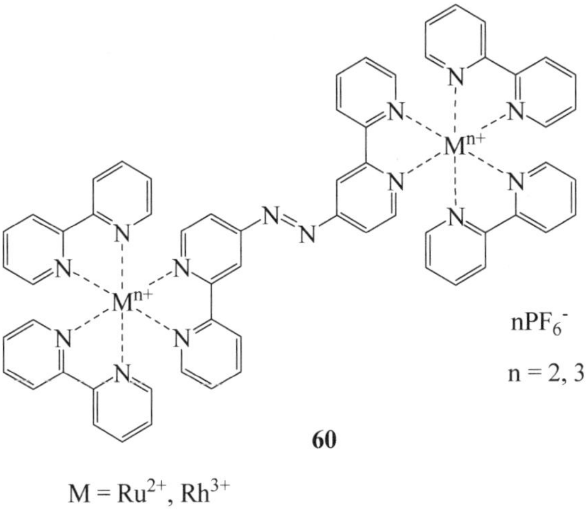

60

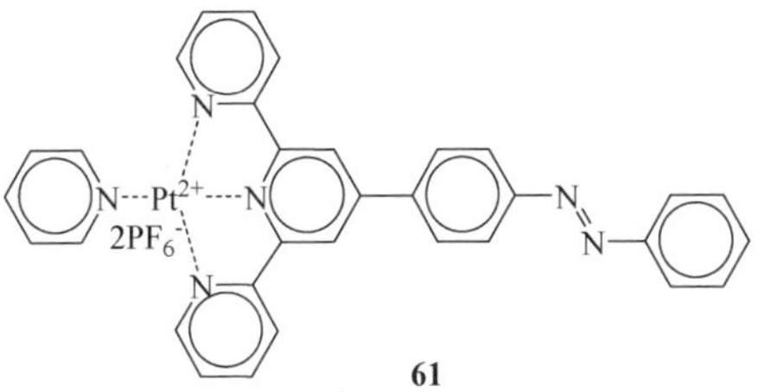

nPF$_6^-$

n = 2, 3

M = Ru^{2+}, Rh^{3+}

Platinum(II) has also been used to coordinate azobenzene-containing terpyridine complexes (**61**, **62**).[85] Under 366 nm irradiation, the azobenzene moieties exhibited *trans*-to-*cis* photoconversion. Furthermore, the azo groups reversibly isomerised from the *cis* to the *trans* form when irradiated with visible light greater than 430 nm in the presence of heat. Moreover, as a result of the *trans*-to-*cis* isomerisation *via* excitation with 600 nm light, a change in the emission spectra of the Pt(II) complexes was observed. The *cis* isomer was found to have an emission intensity that was four times greater compared to the *trans* isomer, which showed no emission upon excitation. Strong photoluminescence in the visible region was observed in solid or liquid solution as a result of the intraligand π–π* and MLCT state. Thus, Yukata *et al.*[85] demonstrated that these Pt(II) terpyridine complexes containing azobenzene moieties exhibit photoluminescence switching *via* photoinduced *trans–cis* isomerisation.

61

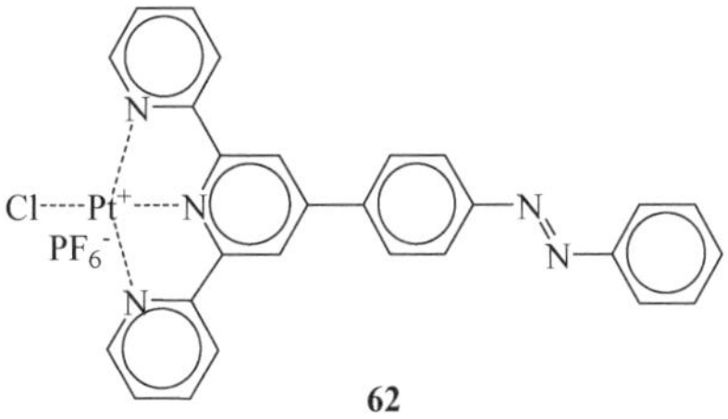

62

9.6 Metalladithiolene Complexes

Metalladithiolene complexes have attractive properties that include magnetism, conductivity, and redox activity, making them interesting compounds to study. Novel azo-containing Ni-, Pd-, and Pt-dithiolato complexes (**63–65**) that exhibit reversible *trans–cis* photoconversion and protonation behaviour were reported by Nishihara and coworkers.[86,87] The authors suggested the proton response with respect to the azo moiety occurred as a result of the strong electron-donating properties of the metalladithiolene. Furthermore, they discovered that this protonation behaviour led to protonation-catalysed *cis-to-trans* conversions. When the transition from metalladithiolene π to azobenzene π* band was irradiated with 405 nm light, this π–π* transition band decreased, which made it evident that *trans-to-cis* isomerisation had occurred. On the other hand, *cis-to-trans* isomerisation occurred *via* irradiation at 360 nm for complex **63** and 310 nm for complexes **64** and **65**. Interestingly, when a small quantity of acid was added to the *cis* isomer complexes in solution, *cis-to-trans* isomerisation occurred instantaneously.

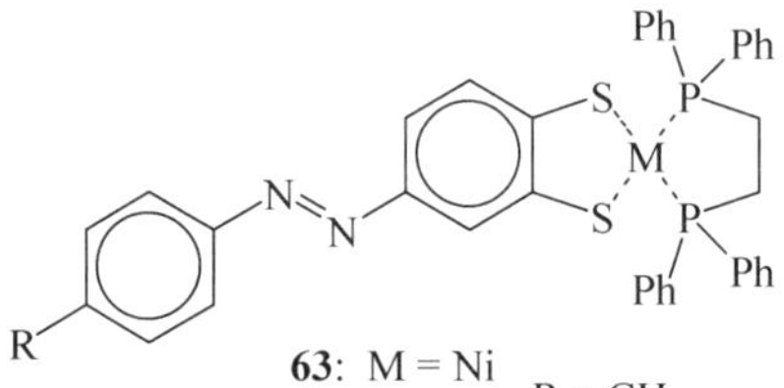

63: M = Ni
64: M = Pd R = CH₃
65: M = Pt

Photoisomerisation behaviour of Pt(II) complexes containing an azobenzene on the metalladithiolene side (**66**) and an azobenzene on the bipyridine side (**67** and **68**) was studied by Nishihara and coworkers.[88,89] These Pt(II) complexes had significant photoconversion efficiencies and were thermally stable in the *cis* isomers. Furthermore, photocontrolled, tristable dithiolato-bipyridine platinum complexes containing two different azobenzene groups, (*trans*, *trans*)-**69** and (*trans*, *trans*)-**70**, were also investigated. The results indicated

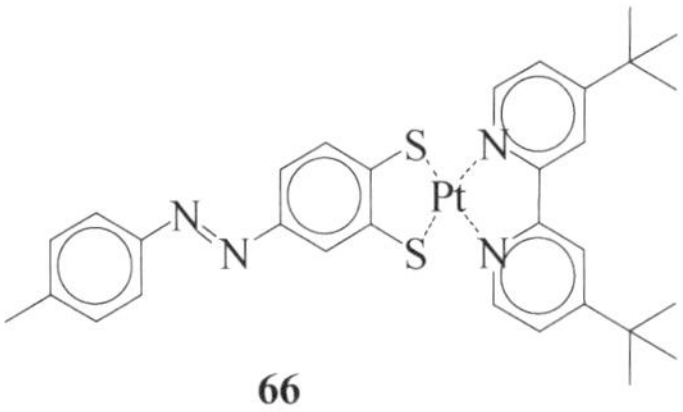

66

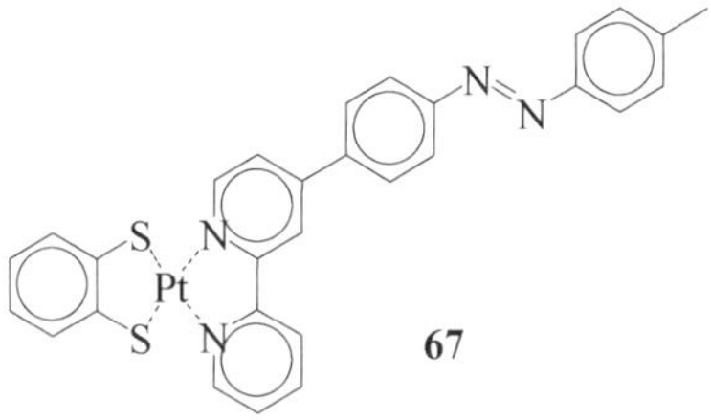

67

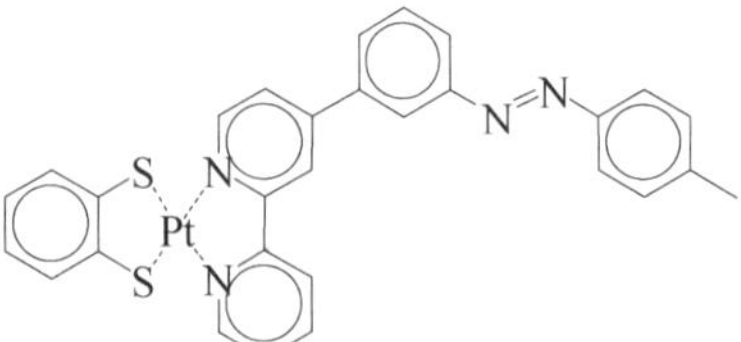

68

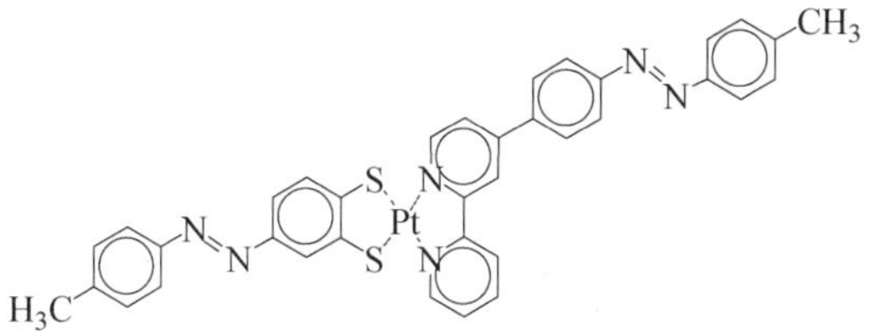

(*trans*, *trans*)-**69**

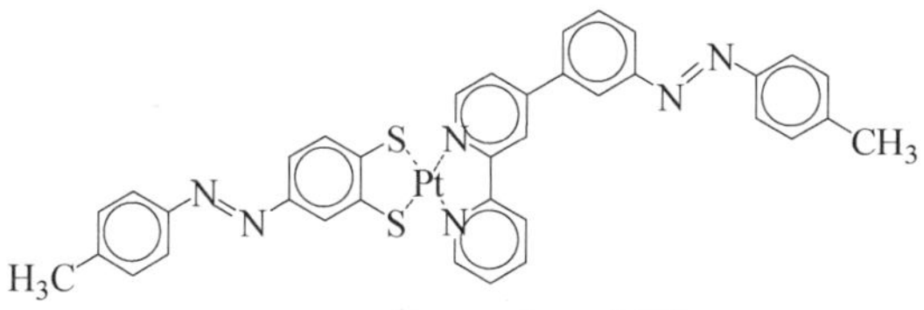

(*trans*, *trans*)-**70**

that these complexes had highly extended photoresponses to the long-wavelength region. Furthermore, the appearance of strong absorption bands in the UV region was observed for all Pt(II) complexes as a result of the π–π* transition state of the azobenzene and bipyridine moieties. Different wavelength excitations for (*trans,trans*)-**69** resulted in selective isomerisations of the azobenzene moieties (Scheme 9.11).[2,88,89] It was found that the

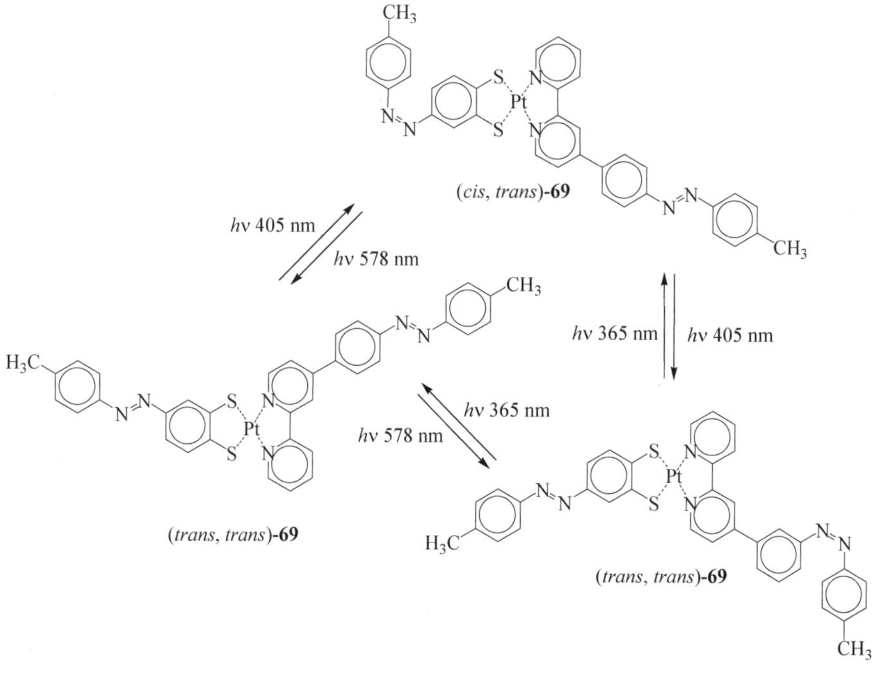

Scheme 9.11 Photoisomerisation behaviour of Pt(II) complex **69**.

trans,trans-, *trans,cis-*, and the *cis,trans*-forms can all be reversibly switched under various light irradiations.

9.7 Alkynyl-Based Metal Complexes

Photoisomerisation behaviour of a tetranuclear macrocyclic Au(II) alkynyl complex **71** (Scheme 9.12) was investigated by Tang *et al.*[90] The study demonstrated that *trans*-to-*cis* photoisomerisation occurred when irradiated with UV light into the π–π^* transition at 360 nm, resulting in the complete formation of the *trans,cis*-state. On the other hand, visible-light irradiation at 486 nm induced the reverse, *cis*-to-*trans* conversion. Furthermore, the authors discovered that photoisomerisation can be controlled by the addition or removal of Ag$^+$ ions, which demonstrates that this system can be employed as a "dual-input lockable molecular logic photoswitch".[90] This was observed when Ag$^+$ ions were added to the *trans,trans*-state, causing the alkyne moieties to trap the Ag$^+$ ions, which led to the inability to photoisomerise to the *trans,cis*-state. Conversely, this locked state can be converted back into an unlocked state by removing the Ag$^+$ ions *via* the addition of chloride anions.

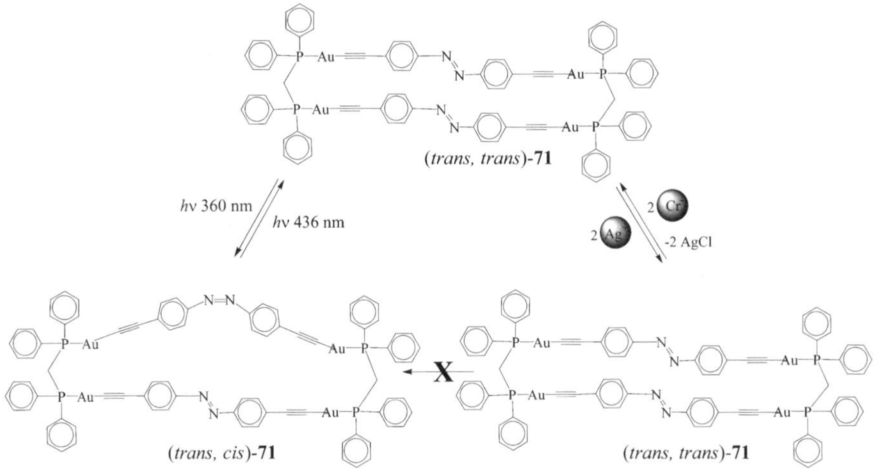

Scheme 9.12   Photoresponsive behaviour of Au complex **71** during light irradiation.

The synthesis and characterisation of alkynyl ruthenium complexes containing azobenzene moieties (**72** and **73**) for NLO has been reported by McDonagh *et al.*[30] Structural studies confirmed that the azobenzene moieties reside in the *trans* conformation. Electrochemical studies were also investigated and revealed that complexes **72** and **73** exhibited similar oxidation waves with potentials of 0.62 V and 0.66 V, respectively, due to the oxidation of Ru(II) to Ru(III). The authors reported that the NLO studies showed an increase in both the observed and two-level corrected nonlinearities for Ru complex **73** compared to Ru complex **72**.

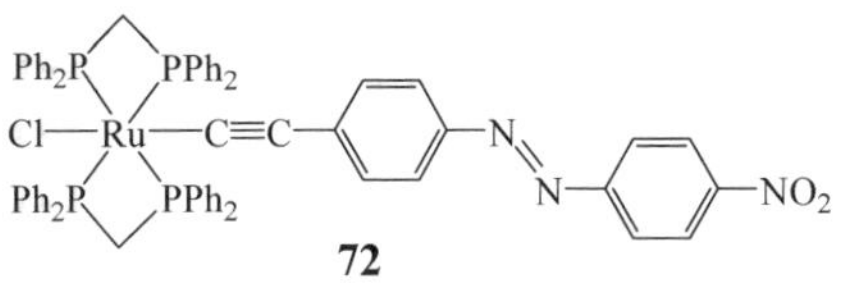

72

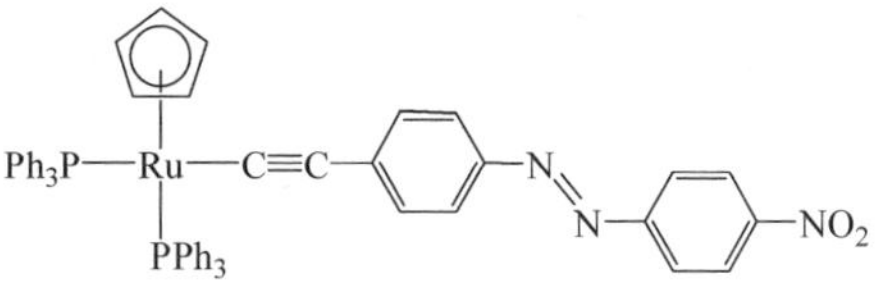

73

Similar azobenzene-containing ruthenium acetylide complexes (**74–76**) exhibiting second- and third-order NLO properties were prepared and surface relief grating (SRG) studies were investigated by Fillaut and coworkers.[33,91,92] The results indicated that the incorporation of a Ru(II) acetylide unit into the same conjugated system as the azobenzene moieties caused them to undergo *trans-cis-trans* photoconversion cycles. It was also discovered that *cis*-to-*trans* thermal isomerisation for the different Ru complexes increases in rate as the electronic richness increases.[33] Furthermore, these photoresponsive ruthenium acetylides were made into uniform thin films *via* spin coating and formed surface relief gratings on polymethyl methacrylate (PMMA) thin films using a short pulse (16 ps, 532 nm) laser irradiation. These SRGs were formed immediately due to the picosecond pulsed laser that resulted in a faster grating inscription process with minimal thermal effects than when using continuous laser irradiation. Furthermore, the authors concluded that the dynamics of the SRG formation depends on the polarisation states and on the light intensity of writing beams.[92]

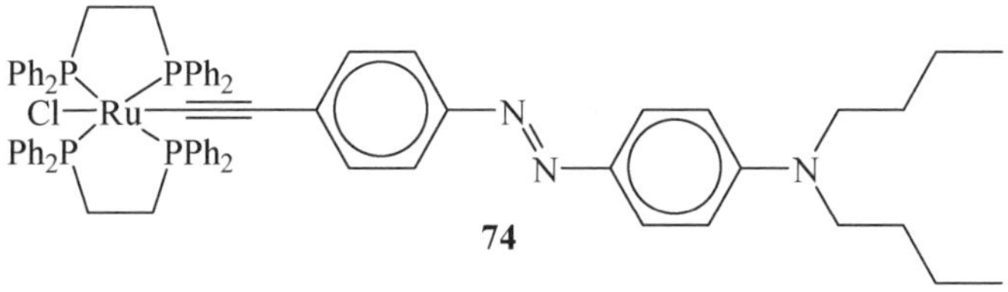

74

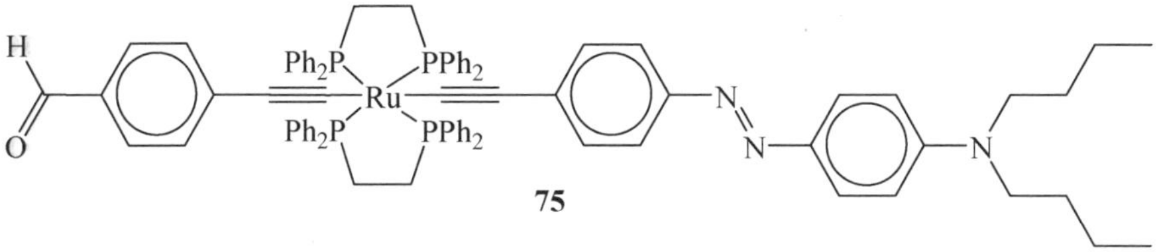

75

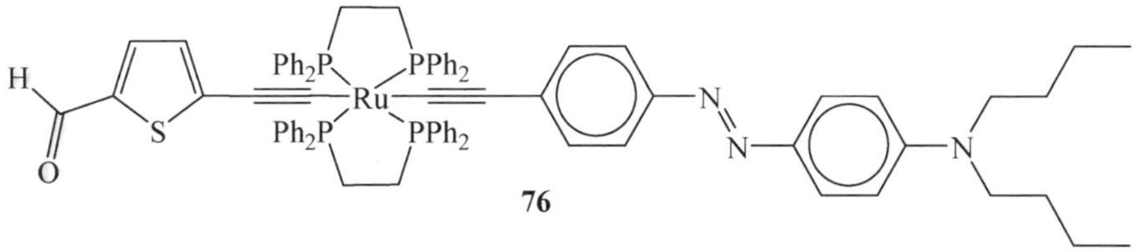

76

9.8 Methacrylate-Based Polymers

Photo- and electroactive polyazopyridines that are axially coordinated to zinc- and cobalt-containing porphyrins **79** were prepared by reacting polymer **77** with complex **78** and characterised by Zhao and coworkers[93,94] (Scheme 9.13). Thermogravimetric analysis indicated that axial coordination of the metal with the nitrogen on the pyridine moiety increased the rigidity of the polymers, leading to higher glass-transition temperatures and thermal stability. According to the UV-vis analysis, redshifts were observed in the wavelength maxima of the metalloporphyrin units, which further confirmed axial

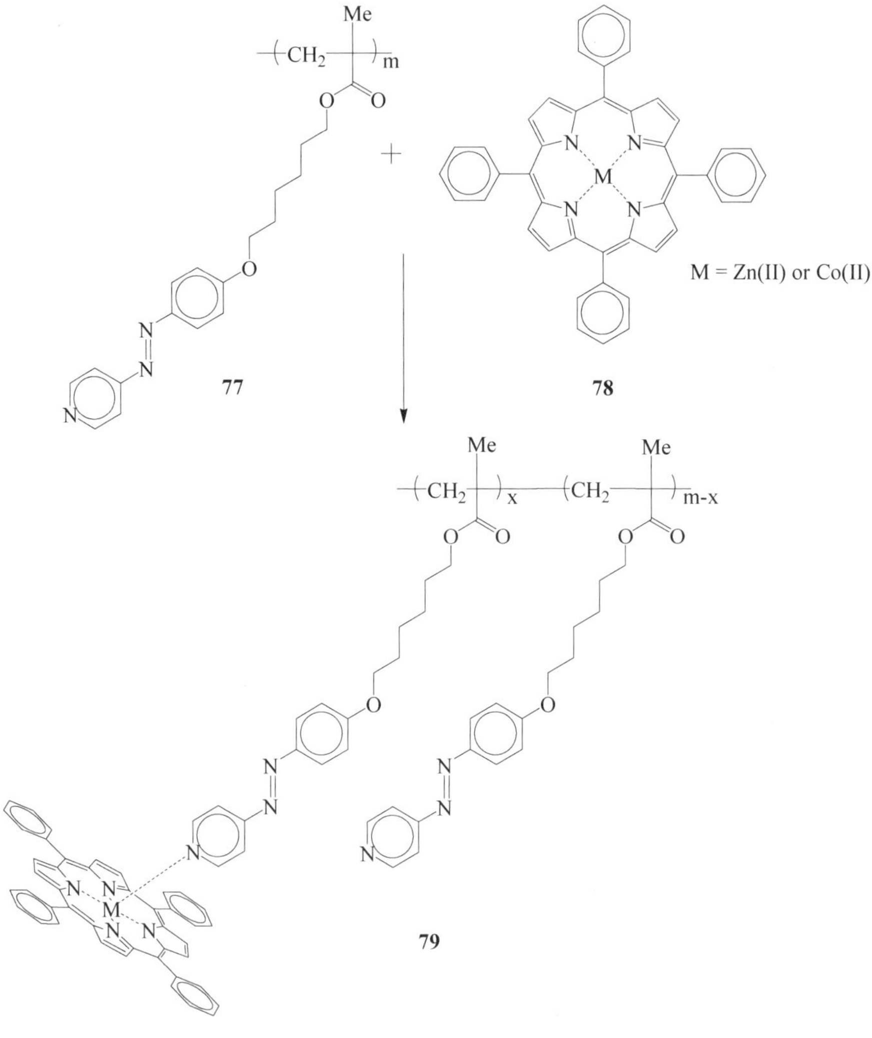

Scheme 9.13 Synthesis of polyazopyridines axially coordinated to zinc- and cobalt-containing porphyrins **79**.

coordination.[95] Furthermore, upon irradiation with visible and UV light, the polymers exhibited reversible *trans–cis* isomerism; however, the degree of photoisomerism that occurred for the azopyridine moieties was very low. Electrochemical studies indicated that in the solid state, the polymers exhibited irreversible oxidation–reduction processes at the metal centre of the porphyrin unit.

Savchenko and coworkers[96–98] reported the synthesis of cobalt-coordinated polymers containing azo chromophores (polymer **80** is shown as an example). Cobalt acetate was used to coordinate with the azobenzene monomers containing terminal methacryloyl groups that were subsequently used for polymerisation in the presence of the radical initiator, AIBN. The optical

80

properties in the presence of an electric field of the metallo-azopolymer thin films were investigated. When the polymer films were irradiated with linearly polarised light, it caused a reorientation of the azobenzene moieties *via cis–trans* photoisomerisation and optical anisotropy was induced. Furthermore, the photoinduced dipole moments of the azo groups were re-oriented by the electric field, causing an observed electro-optical effect. Moreover, the authors indicated that a rise in the dipole moments occurred as a result of the incorporation of electron donor/acceptor groups, which also reduced the effect of the cobalt(II) centre on the optical properties of the polymers under the influence of the applied electric field. In addition to the reported syntheses and electro-optical characteristics of these polymers, it was found that these polymeric films can be used for polarisation holographic recording due to their optically active properties.[97,98]

9.9 Conclusion

Azo dye-based molecules are a very important class of compounds that find applications in diverse areas such as nonlinear optics[30–32] and photoswitching devices.[32] One of the most significant and recognised properties of azo chromophores is the photochemical *trans–cis* isomerisation.[35,99] A number of examples of compounds that combine photochromic azo moieties with metal complexes have been described. These systems exhibit unique photophysical and chemical properties that are not found in similar organic photophores. It has been shown that such complexes undergo reversible *trans–cis* isomerisations upon UV light irradiation or by thermal application. Furthermore, due to the incorporation of the metal centre, many complexes were found to be redox-active. As a result of their unique behaviours, these photochromic materials can be used in a wide variety of applications and continue to be part of an area of research that is rapidly growing.

Acknowledgements

The support provided by the Natural Science and Engineering Research Council of Canada (NSERC), the University of British Columbia Okanagan and the University of Prince Edward Island is gratefully acknowledged.

References

1. V. Guerchais and B. H. Le, *Top. Organomet. Chem.*, 2010, **28**, 171.
2. C.-C. Ko and V. W.-W. Yam, *J. Mater. Chem.*, 2010, **20**, 2063.
3. M. Irie, *Chem. Rev.*, 2000, **100**, 1683.
4. Y. Kawata and S. Kawata, *Chem. Rev.*, 2000, **100**, 1777.
5. N. Tamai and H. Miyasaka, *Chem. Rev.*, 2000, **100**, 1875.
6. A. S. Abd-El-Aziz, *Coord. Chem. Rev.*, 2002, **233–234**, 177.
7. M. Kurihara and H. Nishihara, *Coord. Chem. Rev.*, 2002, **226**, 125.
8. A. S. Abd-El-Aziz and E. K. Todd, *Coord. Chem. Rev.*, 2003, **246**, 3.
9. A. S. Abd-El-Aziz, C. E. Carraher, Jr., C. U. Pittman, Jr., J. E. Sheats, M. Zeldin and Editors, *Macromolecules Containing Metal and Metal-Like Elements, Volume 2: Organoiron Polymers*, John Wiley & Sons, Inc., 2004.
10. H. Nishihara, *Bull. Chem. Soc. Jpn.*, 2004, **77**, 407.
11. H. Nishihara, *Coord. Chem. Rev.*, 2005, **249**, 1468.
12. V. W.-W. Yam and K. M.-C. Wong, *Top. Curr. Chem.*, 2005, **257**, 1.
13. K. Osakada, T. Sakano, M. Horie and Y. Suzaki, *Coord. Chem. Rev.*, 2006, **250**, 1012.
14. W. Y. Wong and C. L. Ho, *Coord. Chem. Rev.*, 2006, **250**, 2627.
15. A. S. Abd-El-Aziz, I. Manners and Editors, *Frontiers in Transition Metal-Containing Polymers*, John Wiley & Sons, Inc., 2007.
16. Y. Cao, J. X. Jiang and W. Yang, *J. Inorg. Organomet. Polym. Mater.*, 2007, **17**, 37.
17. S. Kume and H. Nishihara, *Struct Bond*, 2007, **123**, 79.
18. G. R. Whittell and I. Manners, *Adv. Mater.*, 2007, **19**, 3439.
19. I. Eryazici, C. N. Moorefield and G. R. Newkome, *Chem. Rev.*, 2008, **108**, 1834.
20. S. Kume and H. Nishihara, *Dalton Trans.*, 2008, 3260.
21. K. Szaciłowski, *Chem. Rev.*, 2008, **108**, 3481.
22. V. W.-W. Yam and E. C.-C. Cheng, *Chem. Soc. Rev.*, 2008, **37**, 1806.
23. A. S. Abd-El-Aziz, P. O. Shipman, B. N. Boden and W. S. McNeil, *Prog. Polym. Sci.*, 2010, **35**, 714.

24. D. Astruc, E. Boisselier and C. Ornela, *Chem. Rev.*, 2010, **110**, 1857.

25. N. C. Fletcher and M. C. Lagunas, *Top. Organomet. Chem.*, 2010, **28**, 143.

26. Y. Hasegawa, T. Nakagawa and T. Kawai, *Coord. Chem. Rev.*, 2010, **254**, 2643.

27. B. I. Kharisov, M. P. Elizondo, V. M. Jimenez-Perez, O. V. Kharissova, M. B. Najera and N. Perez, *J. Coord. Chem.*, 2010, **63**, 1.

28. W. Kaim, *Coord. Chem. Rev.*, 2001, **219–221**, 463.

29. T. Ikeda and O. Tsutsumi, *Science*, 1995, **268**, 1873.

30. A. M. McDonagh, N. T. Lucas, M. P. Cifuentes, M. G. Humphrey, S. Houbrechts and A. Persoons, *J. Org. Chem.*, 2000, **605**, 184.

31. J. A. Campo, M. Cano, J. V. Heras, C. López-Garabito, E. Pinilla, R. Torres, G. Rojo and F. Aguilló-López, *J. Mater. Chem.*, 1999, **9**, 899.

32. S. Xie, A. Natansohn and P. Rochon, *Chem. Mater.*, 1993, **5**, 403.

33. K. N. Gherab, R. Gatri, Z. Hank, B. Dick, R.-J. Kutta, R. Winter, J. Luc, B. Sahraoui and J.-L. Fillaut, *J. Mater. Chem.*, 2010, **20**, 2858.

34. R. L. M. Allen, *Colour Chemistry*, Appleton-Century-Crofts, New York, 1971.

35. H. Zollinger, *Azo and Diazo Chemistry; Aliphatic and Aromatic Compounds*, Interscience Publishers, New York, 1961.

36. M. Bardaji, M. Barrio and P. Espinet, *Dalton Trans.*, 2011, **40**, 2570.

37. S. Park, O. N. Kadkin, J.-G. Tae and M.-G. Choi, *Inorg. Chim. Acta*, 2008, **361**, 3063.

38. D. M. Junge and D. V. McGrath, *Chem. Commun.*, 1997, 857.

39. T. Aida, D.-L. Jiang, E. Yashima and Y. Okamoto, *Thin Solid Films*, 1998, **331**, 254.

40. X.-H. Liu, D. W. Bruce and I. Manners, *Chem. Commun.*, 1997, 289.

41. X. H. Liu, D. W. Bruce and I. Manners, *J. Organomet. Chem.*, 1997, **548**, 49.

42. D. Foucher, R. Ziembinski, R. Peterson, J. Pudelski, M. Edwards, Y.-Z. Ni, J. Massey, C. R. Jaeger, G. I. Vaneso and I. Manners, *Macromolecules*, 1994, **27**, 3992.

43. M. R. Buchmeiser, N. Schuler, G. Kaltenhauser, K.-H. Ongania, I. lagoja, I. Wurst and H. Schottenberger, *Macromolecules*, 1998, **31**, 3175.

44. K. H. Schündehütte, *Diarylazoverbindungen*, Georg Thieme Verlag, Stuttgart, 1965.

45. K. Sonogashira, *Comprehensive Organic Synthesis*, Pergamon Press, Oxford, U.K., 1990.

46. L. R. Dalton, A. W. Harper, R. Ghosn, W. H. Steier, M. Ziari, H. Fettermann, Y. Shi, R. V. Mustacich, A. K.-Y. Jen and K. J. Shea, *J. Chem. Mater.*, 1995, **7**, 1060.

47. C. López-Garabito, J. A. Campo, J. V. Heras, M. Cano, G. Rojo and F. Agulló-López, *J. Phys. Chem. B*, 1998, **102**, 10698.

48. M. Kurosawa, T. Nankawa, T. Matsuda, K. Kubo, M. Kurihara and H. Nishihara, *Inorg. Chem.*, 1999, **38**, 5113.

49. A. N. Nesmeyanov, E. G. Perevalova and T. V. Nikitina, *Dokl. Akad. Nauk SSSR*, 1961, **138**, 1118.
50. M. Kurihara, T. Matsuda, A. Hirooka, T. Yutaka and H. Nishihara, *J. Am. Chem. Soc.*, 2000, **122**, 12373.
51. M. Kurihara, A. Hirooka, S. Kume, M. Sugimoto and H. Nishihara, *J. Am. Chem. Soc.*, 2002, **124**, 8800.
52. A. Sakamoto, A. Hirooka, K. Namiki, M. Kurihara, M. Murata, M. Sugimoto and H. Nishihara, *Inorg. Chem.*, 2005, **44**, 7547.
53. K. Namiki, A. Sakamoto, M. Murata, S. Kume and H. Nishihara, *Chem. Commun.*, 2007, 4650.
54. T. Muraoka, K. Kinbara, Y. Kobayashi and T. Aida, *J. Am. Chem. Soc.*, 2003, **125**, 5612.
55. T. Muraoka, K. Kinbara and T. Aida, *Chem. Commun.*, 2007, 1441.
56. T. Muraoka, K. Kinbara and T. Aida, *Nature*, 2006, **440**, 512.
57. K. Kinbara, T. Muraoka and T. Aida, *Org. Biomol. Chem.*, 2008, **6**, 1871.
58. M.-C. Daniel, A. Sakamoto, J. Ruiz, D. Astruc and H. Nishihara, *Chem. Lett.*, 2006, **35**, 38.
59. T. Sakano, M. Horie, K. Osakada and H. Nakao, *Eur. J. Inorg. Chem.*, 2005, 644.
60. A. S. Abd-El-Aziz, R. M. Okasha, P. O. Shipman and T. H. Afifi, *Macromol. Rapid Commun.*, 2004, **25**, 1497.
61. Z. Akhter, M. S. Khan and M. A. Bashir, *J. Inorg. Organomet. Polym.*, 2004, **14**, 253.
62. A. S. Abd-El-Aziz and I. Manners, *J. Inorg. Organomet. Polym. Mater.*, 2005, **15**, 157.
63. A. S. Abd-El-Aziz, T. H. Afifi, W. R. Budakowski, K. J. Friesen and E. K. Todd, *Macromolecules*, 2002, **35**, 8929.
64. A. S. Abd-El-Aziz, N. M. Pereira, W. Boraie, E. K. Todd, T. H. Afifi, W. R. Budakowski and K. J. Friesen, *J. Inorg. Organomet. Polym. Mater.*, 2006, **15**, 497.
65. A. S. Abd-El-Aziz, P. O. Shipman and P. R. Shipley, *Macromol. Rapid Commun.*, 2010, **31**, 459.
66. A. S. Abd-El-Aziz, E. A. Strohm, M. Ding, R. M. Okasha, T. H. Afifi, S. Sezgin and P. R. Shipley, *J. Inorg. Organomet. Polym.*, 2010, **20**, 592.
67. A. S. Abd-El-Aziz, R. M. Okasha, T. H. Afifi and E. K. Todd, *Macromol. Chem. Phys.*, 2003, **204**, 555.
68. A. S. Abd-El-Aziz, R. M. Okasha and T. H. Afifi, *J. Inorg. Organomet. Polym.*, 2004, **14**, 269.
69. A. S. Abd-El-Aziz, B. Elmayergi, B. Asher, T. H. Afifi and K. J. Friesen, *Inorg. Chim. Acta*, 2006, **359**, 3007.
70. A. S. Abd-El-Aziz, C. R. de Denus and H. M. Hutton, *Can. J. Chem.*, 1995, **73**, 289.
71. F. A. Cotton, G. Wilkinson, C. A. Murillo and M. Bochmann, *Advanced Inorganic Chemistry*, 6th edn, Wiley, Chichester, 1999.

72. W. Y. Wong, S. H. Cheung, S. M. Lee and S. Y. Leung, *J. Organomet. Chem.*, 2000, **596**, 36.

73. G. Pourrieux, F. Fagalde, I. Romero, X. Fontrodona, T. Parella and N. E. Katz, *Inorg. Chem.*, 2010, **49**, 4084.

74. S. Kume, M. Kurihara and H. Nishihara, *Chem. Commun.*, 2001, 1656.

75. K. Yamaguchi, S. Kume, K. Namiki, M. Murata, N. Tamai and H. Nishihara, *Inorg. Chem.*, 2005, **44**, 9056.

76. S. Kume, M. Murata, T. Ozeki and H. Nishihara, *J. Am. Chem. Soc.*, 2005, **127**, 490.

77. S. Umeki, S. Kume and H. Nishihara, *Inorg. Chem.*, 2011, **50**, 4925.

78. J. Otsuki, K. Sato, M. Tsujino, N. Okuda, K. Araki and M. Seno, *Chem. Lett.*, 1996, 847.

79. J. Otsuki, N. Omokawa, K. Yoshiba, I. Yoshikawa, T. Akasaka, T. Suenobu, T. Takido, K. Araki and S. Fukuzumi, *Inorg. Chem.*, 2003, **42**, 3057.

80. J. Otsuki, M. Tsujino, T. Iizaki, K. Araki, M. Seno, K. Takatera and T. Watanabe, *J. Am. Chem. Soc.*, 1997, **119**, 7895.

81. R.-A. Fallahpour and M. Neuburger, *Helv. Chim. Acta*, 2001, **84**, 715.

82. T. Akasaka, J. Otsuki and K. Araki, *Chem. Eur. J.*, 2002, **8**, 130.

83. T. Yutaka, M. Kurihara, K. Kubo and H. Nishihara, *Inorg. Chem.*, 2000, **39**, 3438.

84. T. Yutaka, I. Mori, M. Kurihara, J. Mizutani, K. Kubo, S. Furusho, K. Matsumura, N. Tamai and H. Nishihara, *Inorg. Chem.*, 2001, **40**, 4986.

85. T. Yutaka, I. Mori, M. Kurihara, J. Mizutani, N. Tamai, T. Kawai, M. Irie and H. Nishihara, *Inorg. Chem.*, 2002, **41**, 7143.

86. M. Nihei, M. Kurihara, J. Mizutani and H. Nishihara, *Chem. Lett.*, 2001, 852.

87. N. Masayuki, M. Kurihara, J. Mizutani and H. Nishihara, *J. Am. Chem. Soc.*, 2003, **125**, 2964.

88. R. Sakamoto, M. Murata, S. Kume, H. Sampei, M. Sugimoto and H. Nishihara, *Chem. Commun.*, 2005, 1215.

89. R. Sakamoto, S. Kume, M. Sugimoto and H. Nishihara, *Chem. Eur. J.*, 2009, **15**, 1429.

90. H.-S. Tang, Z. Nianyong and V. W.-W. Yam, *Organometallics*, 2007, **26**, 22.

91. J. Luc, J. Niziol, M. Sniechowski, B. Sahraoui, J.-L. Fillaut and O. Krupka, *Mol. Cryst. Liq. Cryst.*, 2008, **485**, 990.

92. J. Luc, K. Bouchouit, R. Czaplicki, J. L. Fillaut and B. Sahraoui, *Opt. Exp.*, 2008, **16**, 15633.

93. Q. Bo, A. Yavrian, T. Galstian and Y. Zhao, *Macromolecules*, 2005, **38**, 3079.

94. S. Dahmane, A. Lasia and Y. Zhao, *Macromol. Chem. Phys.*, 2006, **207**, 1485.

95. F. D'Souza, Y.-Y. Hsieh and G. R. Deviprasad, *Inorg. Chem.*, 1996, **35**, 5747.

96. I. Savchenko, N. Davidenko, I. Davidenko, A. Popenaka and V. Syromyatnikov, *Mol. Cryst. Liq. Cryst.*, 2007, **468**, 203.

97. N. A. Davidenko, I. A. Savchenko, I. I. Davidenko, A. N. Popenaka, A. N. Shumelyuk and V. A. Bedarev, *Tech. Phys.*, 2007, **52**, 451.

98. N. A. Davidenko, I. I. Davidenko, A. N. Popenaka, I. A. Savchenko and A. N. Shumelyuk, *J. Appl. Spectrosc.*, 2007, **74**, 926.

99. A. S. Abd-El-Aziz, P. O. Shipman, E. G. Neeland, T. C. Corkery, S. Mohammed, P. D. Harvey, H. M. Mohamed, A. H. Bedair, A. M. El-Agrody, P. M. Aguiar and S. Kroeker, *Macromol. Chem. Phys.*, 2008, **209**, 84.

Hyperbranched Acetylenic Polymers from Metal-Free and Regioselective Polycyclotrimerization of Arylene Bipropiolates: Synthesis, Characterization, and Photonic Properties

CATHY K. W. JIM[1], ANJUN QIN[2], JACKY W. Y. LAM[1] AND BEN ZHONG TANG*[1,2]

[1] The Hong Kong University of Science and Technology, Department of Chemistry, Clear Water Bay, Kowloon, Hong Kong; [2] Zhejiang University, Department of Polymer Science and Engineering, Key Laboratory of Macromolecular Synthesis and Functionalization of the Ministry of Education of China, Hangzhou 310027, China
*E-mail: tangbenz@ust.hk

10.1 Introduction

The study of hyperbranched polymers is a "young" but "hot" area of research. Hyperbranched polymers are expected to show size-, shape-, branch-, and surface-related properties,[1] which may enable them to find an array of technological applications as nanoscale catalysts, chemical sensors, molecular antennae, supramolecular assemblies, micelle mimics, drug-delivery carriers,

RSC Polymer Chemistry Series No. 2
Molecular Design and Applications of Photofunctional Polymers and Materials
Edited by Wai-Yeung Wong and Alaa S Abd-El-Aziz

Published by the Royal Society of Chemistry, www.rsc.org

immunodiagnostic probes, and so forth.[2] Various hyperbranched polymers have been prepared by different synthetic strategies.[3] The most commonly used methods have been the condensation polymerizations of AB_n-type monomers, with A and B being mutually reactive functional groups and $n \geq 2$. However, the preparations of such multifunctional monomers often require nontrivial synthetic efforts.[4] The monomers are difficult to keep and handle and easy to self-oligomerize during storage. The polymerization reactions are often initiated by *in situ* deprotection under harsh conditions and the incomplete deprotection results in the formation of imperfect polymers with low molecular weights (MWs) and degrees of branching (DBs).

Acetylene cyclotrimerization is a century-old reaction for effective transformation of triple bonds to benzene rings. Polycyclotrimerizations of diyne molecules are anticipated to result in the formation of hyperbranched polyarylenes.[5] This simple A_2-type polycyclotrimerization approach will circumvent the synthetic difficulties encountered in the AB_n-type condensation polymerizations and produce stable polymers consisting of robust aromatic rings. This possibility, however, has not been actively explored, because alkyne cycloadditions can easily run out of control to yield crosslinked gels.[5,6]

We have embarked on a research program on the development of the polycyclotrimerizations of diynes (**I**) initiated by transition-metal catalysts into a useful synthetic protocol for the construction of hyperbranched polyarylenes (**PI**; Scheme 10.1).[7] While crosslinking or gelation was involved in the diyne

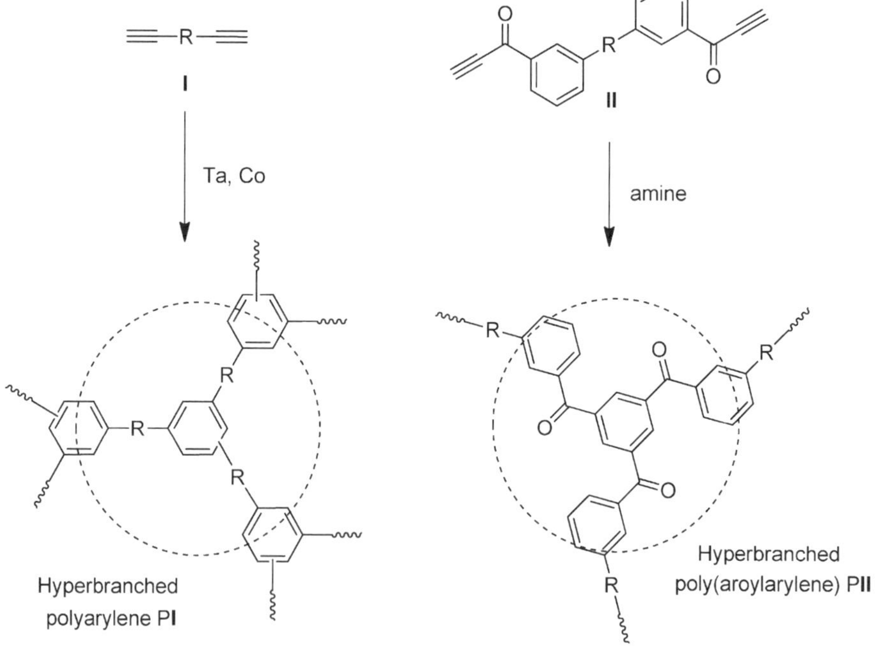

Scheme 10.1 [Copyright (2009) American Chemical Society]

polycyclotrimerization reactions, we succeeded in the preparation of hyperbranched polyarylenes with excellent solubility through optimization of polymerization conditions. The polymers have been found to exhibit a variety of unique properties.[8] They are, for example, highly luminescent with fluorescence quantum yields up to unity, nonlinear-optically active when photoexcited by laser pulses, thermally very stable (T_d up to 500 °C), and readily graphitized in high yields upon pyrolysis.

The tantalum-catalyzed diyne polycyclotrimerization, however, has some drawbacks: the catalysts are completely intolerant of polar functional groups and polymerization reactions therefore must be conducted under stringently moisture- and oxygen-free conditions. The polycyclotrimerization reactions proceed very rapidly, making the process control very difficult. Although the cobalt-based catalysts can polycyclotrimerize diyne monomers carrying certain functional groups, the resultant polymers generally have lower MWs and poorer optical and photonic properties than those prepared from the tantalum catalysts due to the presence of the catalyst residues in the polymer structures after precipitation. Moreover, both Ta and Co catalysts produce hyperbranched polymers consisting of regiorandom structural units of 1,2,4- and 1,3,5-trisubstituted benzene isomers.

We have recently discovered that the polycyclotrimerizations of aroylacetylenes (**II**) catalyzed by secondary amines (*e.g.*, piperidine) produce hyperbranched poly(1,3,5-triaroylarylene)s (**PII**) with high DBs in high yields (Scheme 10.1).[9] The polycyclotrimerization is tolerant to polar functional groups and is strictly regioselective, furnishing polymers comprising of sole structural units of 1,3,5-trisubstituted benzene regioisomers. The aroylacetylene monomers (**II**), however, are difficult to prepare. It takes many synthetic steps to prepare the monomers and the reactions involve the use of toxic heavy-metal oxidants such as MnO_2 and CrO_3.[9] It would be nice if the polycyclotrimerization can be extended or applicable to the "simple" diyne monomers that can be readily prepared from commercially available starting materials by one-step reactions in one-pot procedures in an environmentally benign fashion.

Careful examination of the molecular structure of monomer **II** reveals that this polymerization works for electron-deficient diynes, whose acetylene triple bonds are linked with electron-withdrawing groups. If the carbonyl linkage between the triple bond and the aromatic ring in the aroylacetylene can be replaced by an ester group, it will make the monomer synthesis much easier. Acetylenecarboxylic acid or propiolic acid is a commercially available reagent and can be readily esterified with arylene diol to give bipropiolate (Scheme 10.2). If the bipropiolate monomer can be polymerized, it will pave the way to facile and economic syntheses of functional hyperbranched polymers. However, a propiolate derivative is less electron deficient than an aroylacetylene because an ester group is less electron-withdrawing than a carbonyl group, which makes it uncertain whether or under what conditions an arylene bipropiolate will undergo polycyclotrimerization reaction.

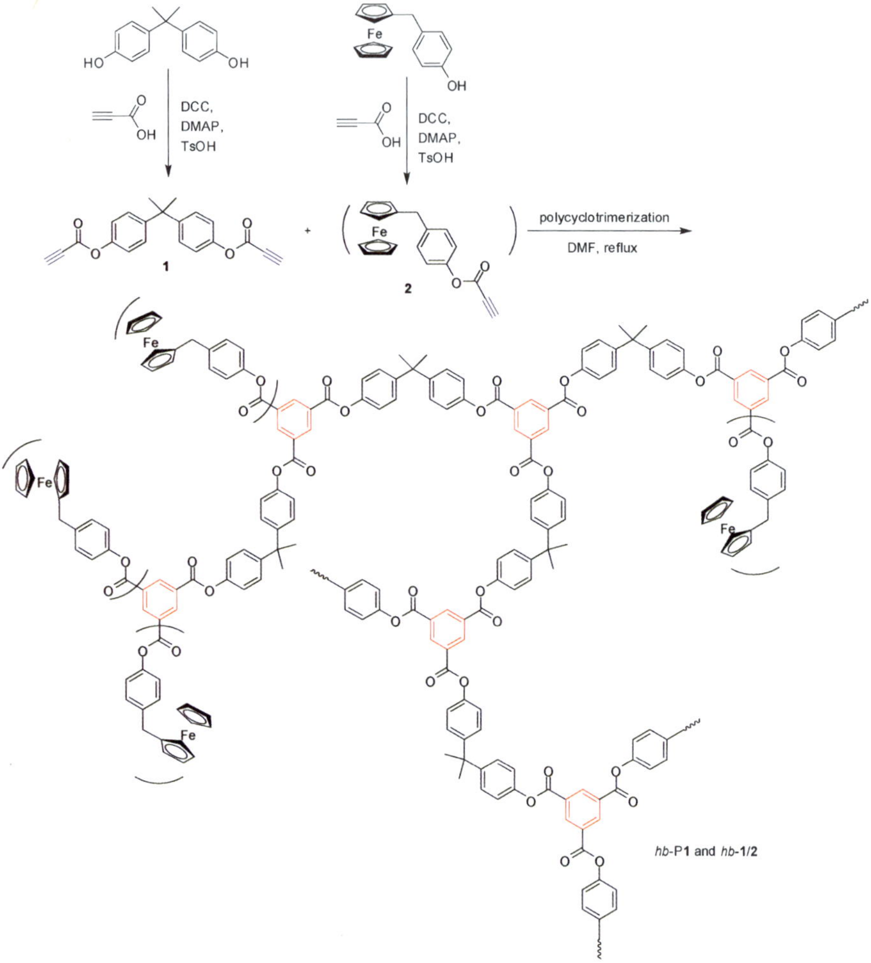

Scheme 10.2

In this work, we explore the utility of alkyne polycyclotrimerization of arylene bipropiolate as a useful tool for the preparation of hyperbranched polymers and their organometallic counterparts. In this chapter, we show that bipropiolate monomers can be effectively polycyclotrimerized in refluxed N,N-dimethylformamide (DMF) without adding any external catalysts, producing processable, regioregular, hyperbranched poly[1,3,5-tri(aroycarbonyl)phenylene]s or *hb*-PTACPs with high DB value in high yields.[10] Addition of ferrocene-containing propiolate during the polymerization results in the generation of *hb*-PTACPs with numerous ferrocene units on the surface. Structurally, the *hb*-PTACPs are polyesters but synthetically they are very difficult to access by the conventional polycondensation reactions of $A_2 + B_3$ monomers, with A_2 and B_3 being diol and triacid, respectively. Monomers of trisubstituted benzene derivatives (B_3) are generally difficult to prepare. In particular, those that carry three electron-withdrawing ester groups at the 1,3,5-positions are difficult to make due to the well-known deactivating effect of the ester group. To make hyperbranched polymers with high MWs and DBs by the ($A_2 + B_3$)-type polycondensation reaction, strict stoichiometric balance is theoretically required but practically difficult to meet. Moreover, the $A_2 + B_3$ polycondensations are equilibrium reactions with small MW compounds such as water generated as byproducts. The polycyclotrimerization of the A_2 monomers is free of all these problems and is thus of great value in terms of synthetic methodology.

10.2   Experimental

10.2.1   Materials

Tetrahydrofuran (THF, Labscan), toluene (BDH), and 1,4-dioxane (Aldrich) were distilled in an atmosphere of dry nitrogen from sodium benzophenone ketyl immediately prior to use. Dichloromethane (DCM) was distilled under nitrogen over calcium hydride. DMF was stirred with calcium hydride overnight, distilled under reduced pressure and kept under dry nitrogen. Other solvents such as dimethylsulfoxide (DMSO) and triethylamine were purified using standard procedures. Propiolic acid, bisphenol A (4,4′-isopropylidenediphenol), 1,3-dicyclohexylcarbodiimde (DCC), 4-dimethylaminopyridine (DMAP), *p*-toluenesulfonic acid monohydrate (TsOH), and all other chemicals were purchased from Aldrich and used as received without further purification. (4-Hydroxybenzyl)ferrocene (**2**) was prepared according to the literarture.[11]

10.2.2   Instrumentation

Relative number- ($M_{n,r}$) and weight-average ($M_{w,r}$) molecular weights and polydispersity indices (PDI or $M_{w,r}/M_{n,r}$) of the polymers were estimated by a Waters Associates gel permeation chromatography (GPC) system equipped with RI and UV detectors. THF was used as the eluent at a flow rate of 1.0 mL. A set of monodisperse linear polystyrenes was used as standards for MW calibration. Absolute weight-average molecular weights ($M_{w,a}$) of the polymers were measured by a commercial laser light scattering (LLS) spectrometer (ALV/DLS/SLS-5022F) equipped with a multi-τ-digital time correlator (ALV5000) and a cylindrical 22 mW He-Ne laser (λ = 632.8 nm, uniphase) as a light source. The RI increment (dn/dc) was determined to be 0.292 mL/g in THF at 25 °C on an Optilab DSP (Digital Signal Processing) refractometer (Wyatt Technology; λ = 632.8 nm, $c \leqq$ 1.0 mg/mL).

IR spectra were recorded on a Perkin-Elmer 16 PC FT-IR spectrophotometer. ^{1}H and ^{13}C NMR spectra were measured on a Bruker ARX 300 NMR spectrometer using CDCl$_3$, DMSO-d_6, or DCM-d_2 as deuterated solvents. Light transmission spectra were measured on a Milton Roy

Spectronic 3000 array spectrophotometer. MALDI-TOF spectra were recorded on a GCT Premier CAB048 mass spectrometer operating in a chemical ionization mode (CI) with methane as carrier gas. Elemental analyses were conducted with an Elementary Vario EL analyzer. Thermogravimetric analysis (TGA) measurements were carried out under nitrogen or in air on a Perkin-Elmer TGA 7 analyzer at a heating rate of 10 °C/min. RI values were measured on a Gaertner L116C ellipsometric thin-film thickness measurement system using 1 mW He-Ne laser beam (λ = 632.8 nm) as light source or determined on a J A Woollam variable-angle ellipsometry system with a wavelength tunability from 300 to 1700 nm. To fit the acquired Ψ and Δ curves with the data obtained from the 3-layer optical model consisting of crystalline silicon substrate, 2 nm SiO_2 layer and a uniform polymer film, the Levenberg–Marquardt regression algorithm was employed. The Cauchy dispersion law was applied to describe the polymer layer from visible to IR spectral region.

10.2.3 Monomer Preparation

Acetylene bipropiolate **1** was prepared by esterification of arylene diol with propiolic acid in the presence of DCC, DMAP, and TsOH (*cf.*, Scheme 10.2). The detailed experimental procedure for the synthesis of monomer **1** is given below as an example. In a 500 mL round-bottom flask were dissolved 3.26 g (14 mmol) of bisphenol A, 8.86 g (43 mmol) of DCC, 0.70 g (5.8 mmol) of DMAP, and 1.08 g (5 mmol) of TsOH in 240 mL of dry DCM/THF (3:1 v/v). The solution was cooled to 0 °C with an ice-water bath, into which 2.0 g (28.6 mmol) of propiolic acid (**4**) dissolved in 20 mL of DCM/THF (3:1 v/v) was added under stirring *via* a dropping funnel. The reaction mixture was stirred overnight. After filtering out the solid, the solution was concentrated by a rotary evaporator. The crude product was purified by a silica gel column using chloroform/hexane (1:2 v/v) as eluent.

Characterization Data for 4,4′-Isopropylidenediphenyl Bipropiolate (1): White solid; yield 53.2% (2.52 g). IR (thin film), v (cm^{-1}): 3266, 2934, 2124, 1730, 1634. ^{1}H NMR (300 MHz, CDCl$_3$), δ (TMS, ppm): 7.26 (d, 2H), 7.05 (d, 2H), 3.06 (s, 1H), 1.67 (s, 3H). ^{13}C NMR (75 MHz, CDCl$_3$), δ (TMS, ppm): 151.0, 148.5, 147.7, 127.9, 120.6, 76.6, 74.3, 42.6, 30.8. HRMS (MALDI-TOF): m/z 333.1080 [(M + H)$^+$, calcd 333.3493]. Anal. Calcd for $C_{21}H_{16}O_4$: C, 75.89; H, 4.85. Found: C, 75.17; H, 5.01.

4-(Ferrocenylmethyl)phenyl propiolate (2): It was synthesized by a procedure similar to that of **1**. Red solid; yield 75.8%. IR (thin film), v (cm^{-1}): 3254, 3090, 3042, 2944, 2922, 2846, 2118, 1722, 1504, 1432, 1292, 1208, 1106, 1058, 1018. ^{1}H NMR (300 MHz, DMSO-d_6), δ (TMS, ppm): 7.36 (d, 2H), 7.22 (d, 2H), 4.95 (s, 1H), 4.25 (s, 7H), 4.17 (s, 2H), 3.76 (s, 2H). ^{13}C NMR (75 MHz, DMSO-d_6), δ (TMS, ppm): 150.9, 147.5, 140.4, 129.4, 121.2, 87.7, 81.4, 74.3, 68.5, 68.3, 67.3, 34.7. Anal. Calcd for $C_{20}H_{17}O_2Fe$: C, 69.79, H, 4.69. Found: C, 69.34, H, 4.80.

10.2.4 Polymer Synthesis

All the polymerization reactions were carried out under dry nitrogen using a standard Schlenk technique, unless otherwise specified. A typical procedure for the polymerization of **1** is given below as an example. In a 15 mL Schlenk tube with a three-way stopcock on the sidearm was placed 88.7 mg of **1** (0.267 mmol) under nitrogen in a glovebox. Distilled DMF (1.5 mL) was added to dissolve the monomer using a hypodermic syringe. After stirring under reflux for 24 h, the mixture was added dropwise to ∼300 mL of methanol through a cotton filter under stirring. The precipitate was allowed to stand overnight and then collected by filtration. The isolated polymer (*hb*-**P1**) was washed with methanol and dried under vacuum at room temperature to a constant weight.

Characterization Data for hb-P1: Brown powder; yield 68.4% (Table 10.2, run 2). $M_{w,r}$ 15 600; M_w/M_n 2.4 (GPC, polystyrene calibration); $M_{w,a}$ 832000 (LLS). IR (thin film), v (cm^{-1}): 2966, 2922, 2872, 1743, 1603, 1504. ^{1}H NMR (300 MHz, DCM-d_2), δ (TMS, ppm): 9.16, 7.91, 7.26, 7.17, 7.03, 6.69, 1.70, 1.67. ^{13}C NMR (75 MHz, DCM-d_2), δ (TMS, ppm): 157.2, 154.9, 149.3, 136.5, 132.0, 128.7, 121.8, 115.5, 43.2, 31.4.

Characterization data of hb-P1/2: Brown powder; yield 46.0%. M_w 39 400; M_w/M_n 6.1 (GPC, polystyrene calibration). IR (thin film), v (cm^{-1}): 3086, 3038, 2966, 2932, 2870, 2378, 1742, 1646, 1602, 1506, 1466, 1408, 1394, 1334, 1294, 1218, 1200, 1170, 1104, 1080, 1016. ^{1}H NMR (300 MHz, DMSO-d_6), δ (TMS, ppm): 9.31, 9.07, 7.37, 7.35, 7.11, 6.76, 4.26, 4.20, 3.77, 3.46, 1.77, 1.71. ^{13}C NMR (75 MHz, DMSO-d_6), δ (TMS, ppm): 171.6, 171.3, 171.0, 170.8, 170.7, 170.5, 170.2, 169.9, 163.0, 155.1, 148.1, 140.2, 136.2, 130.8, 129.3, 127.7, 121.3, 114.7, 68.5, 67.3, 34.7, 30.4.

10.2.5 Model Reaction

Triphenyl benzene-1,3,5-tricarboxylate (**6**) was prepared as a model compound by cyclotrimerization of phenyl propiolate **5** (Scheme 10.3). The experimental

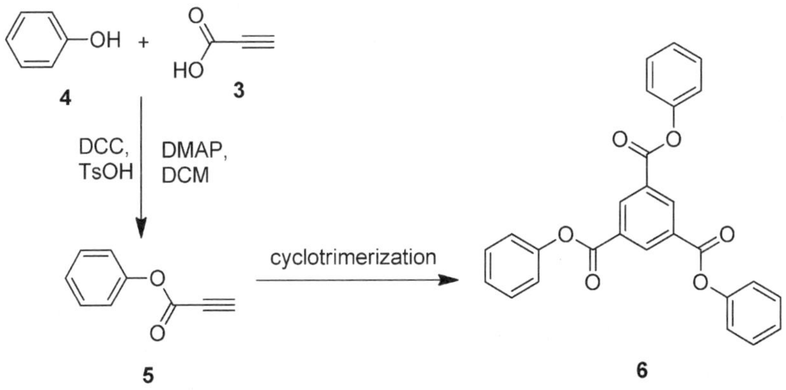

Scheme 10.3 [Copyright (2009) American Chemical Society]

procedures for the synthesis of **5** and its cycloaddition are similar to those described above for the synthesis of **1** and *hb*−**P1**.

Characterization data for **5**: Colorless oil; yield 68.3%. IR (thin film), v (cm^{-1}): 2931, 2854, 2117, 1732. ^{1}H NMR (300 MHz, CDCl$_3$), δ (TMS, ppm): 7.42 (t, 2H), 7.27 (t, 2H), 7.13 (d, 1H). ^{13}C NMR (75 MHz, CDCl$_3$), δ (TMS, ppm): 151.3, 121.6, 129.1, 125.5, 140.0, 76.8, 73.4.

Characterization data for **6**: White solid; yield 34.2% (Table 2.1, run 4). IR (thin film), v (cm^{-1}): 1748, 1589, 1487. ^{1}H NMR (300 MHz, DMSO-d_6), δ (TMS, ppm): 9.04 (s, 1H), 7.52 (t, 2H), 7.38 (m, 3H). ^{13}C NMR (75 MHz, DMSO-d_6), δ (TMS, ppm): 163.0, 150.4, 135.1, 130.9, 129.7, 126.4, 121.9.

10.3 Results and Discussion

10.3.1 Monoyne Model Reaction

Before studying the polycyclotrimerization of bipropiolates, we synthesized a monopropiolate carrying one triple bond (**5**) and utilized it as starting material for a model reaction (Scheme 10.3). According to our previous study, aroylacetylenes can be cyclotrimerized when refluxed in 1,4-dioxane using piperidine as a catalyst.[9b] We thus tried to cyclotrimerize **5** under the same conditions (Table 10.1, run 1). After 24 h the solvent was evaporated and the crude product was purified by silica gel column chromatography using chloroform/hexane (1:1 v/v) as eluent. Product isolation and structural characterization reveal that **6** is the sole product, confirming that **5** can undergo cyclotrimerization in the presence of the base in a regioselective fashion, although the conversion is not so efficient because the reaction conditions have not been optimized. Changing the solvent to toluene slightly lowered the yield, whereas when the reaction is carried out in a toluene/DMF mixture no product is isolated. The reaction conducted in DMF gives the best result (Table 10.1, run 4). Furthermore, triethylamine fails to initiate the cycloaddition reaction, probably due to its reaction with the acidic ethynyl proton of **5**.

Table 10.1 Cyclotrimerization of phenyl propiolate (**5**).

run	*solvent*	*catalyst*	*yield (%)*
1	1,4-dioxane	piperidine	25.0
2	toluene	piperidine	21.2
3	toluene/DMF^b		00.0
4	DMF		34.2
5	triethylamine^c		

^{a}Refluxed under nitrogen for 24 h; [**5**] = 0.178 M; [piperidine] = 5.7 mM. ^{b}Volume ratio = 1:1. ^{c}Reaction mixture turned black with heat dissipation immediately after triethylamine was added. Copyright (2009) American Chemical Society.

10.3.2 Diyne Polycyclotrimerization

After confirming that monopropiolate **5** can undergo regioselective cyclotrimerization, we utilized the reaction to synthesize new hyperbranched polymers. We prepared arylene bipropiolates **1** and **2** by esterification reactions of their corresponding phenol derivatives with propiolic acid in the presence of DCC, DMAP and TsOH (cf., Scheme 10.2). All of the monomers were fully characterized spectroscopically from which satisfactory analysis data corresponding to their molecular structures were obtained (see Experimental Section for details).

We attempted to cyclotrimerize **1** by piperidine in 1,4-dioxane but isolated no polymeric product. We tried the reaction in refluxed DMF and succeeded in the transformation of the diyne monomer into its polymer *hb*-**P1** (Table 10.2, run 1). The active catalytic species for the reaction is believed to be the trace amount of dimethylamine generated from the *in situ* decomposition of DMF at the high temperature.[12] We increased the monomer concentration by ~ 1.5-fold and obtained a soluble polymer in a higher yield ($\sim 68\%$). Further increasing the monomer concentration, however, promoted the formation of crosslinking product and hence decreased the yield of soluble polymer.

We followed the time course of the polymerization of monomer **1** in DMF. The yield was generally increased with time and reached its maximum value of $\sim 72\%$ at 24 h (Table 2.3). After polymerization for 18 h, the $M_{w,r}$ of the resultant polymer is high enough (14 200) for general-purpose applications. It should be pointed out that this relative value is probably considerably underestimated because of the hyperbranched nature of the polymer.[13] Our previous investigations reveal that the underestimation can be very large.[14] The absolute molecular weight ($M_{w,a}$) of *hb*-**P1** may be much higher than the relative value estimated from the GPC analysis ($M_{w,r}$). Indeed, analysis of *hb*-**P1** by a LLS spectrometer gives an $M_{w,a}$ value of 8.32×10^5 (*cf.*, Table 10.2, no. 2), which is 53-fold higher than its $M_{w,r}$ value.

Temperature exerts a strong influence on the polymerization reaction. Both the yield and molecular weight of the polymer are increased when the temperature is raised from 110 to 130 °C (Table 10.4). When the temperature is

Table 10.2 Effect of monomer concentration on the polymerization of monomer **1**.

run	*[1] (M)*	*yield (%)*	$M_{n,r}$[b]	$M_{w,r}$[b]	$M_{w,r}/M_{n,r}$[b]	$M_{w,a}$[c]
1	0.120	51.7	5700	14 200	2.5	
2	0.178	68.4	6500	15 600	2.4	832 000
3	0.267	44.8	5800	15 200	2.6	
4	0.534	37.5	5600	15 800	2.8	

[a]Carried out in refluxed DMF for 24 h under nitrogen. [b]Determined by GPC in THF on the basis of a linear polystyrene calibration. [c]Absolute value measured by LLS technique in THF. Copyright (2009) American Chemical Society.

Table 10.3 Time course for polymerization of monomer **1**.

run	time (h)	yield (%)	$M_{n,r}$[b]	$M_{w,r}$[b]	$M_{w,r}/M_{n,r}$[b]
1	6	19.7	4200	5400	1.3
2	12	22.1	4500	7200	1.6
3	18	55.8	6200	14 200	2.3
4	24	71.3	4800	13 800	2.9
5	36	71.2	4100	10 200	2.5

[a]Carried out in refluxed DMF under nitrogen; [**1**] = 0.178 M. [b]Determined by GPC in THF on the basis of a linear polystyrene calibration. Copyright (2009) American Chemical Society.

Table 10.4 Effect of temperature on polymerization of monomer **1**.

run	temp (°C)	yield (%)	$M_{n,r}$[b]	$M_{w,r}$[b]	$M_{w,r}/M_{n,r}$[b]
1	110	7.1	3600	5400	1.5
2	130	17.5	4200	6300	1.5
3	153[c]	68.4	4600	11 100	2.4

[a]Carried out in DMF for 24 h under nitrogen; [**1**] = 0.178 M. [b]Determined by GPC in THF on the basis of a linear polystyrene calibration. [c]Boiling point of DMF. Copyright (2009) American Chemical Society.

further increased to the boiling point of DMF, the polymerization results are improved significantly, with the polymer yield being ~ 4 times higher than that obtained at 130 °C.

The above investigations enable us to polymerize **1** and copolymerize **1** with **2** under optimal conditions. Table 10.5 summarizes the polymerization results. All the polycyclotrimerizations proceeded smoothly, giving *hb*-P**1** and *hb*-P**1/2** in satisfactory yields. Comparing the polymerization results of **1** obtained under nitrogen and in air, it is clear that oxygen and moisture exert little effect on the polymerization reaction (cf., Table 10.5, runs 1 and 2). This helps simplify the reaction procedures. As no transition-metal catalyst is used in the process, this metal-free polycyclotrimerization has the advantages of being less toxic, environmentally friendlier, and economically sounder.

Table 10.5 Polymerizations of bipropiolate monomers.

run	monomer	yield (%)	$M_{n,r}$[b]	$M_{w,r}$[b]	$M_{w,r}/M_{n,r}$[b]	$M_{w,a}$[c]
1[d]	**1**	68.4	6500	15 600	2.4	832 000
2[e]	**1**	72.3	4200	13 000	3.1	
3	**1** + **2**	46.0	6600	39 400	6.0	

[a]Carried out in refluxed DMF for 24 h under nitrogen; [**1**] = [**2**] = 0.178 M. [b]Determined by GPC in THF on the basis of a linear polystyrene calibration. [c]Absolute value measured by LLS technique in THF. [d]Data taken from Table 2.2, run 2. [e]Conducted in air.

10.3.3 Structural Characterization

In the model reaction, the cyclotrimerization of monoyne **5** gives **6** as the sole product. Diyne **1** thus must have been polycyclotrimerized in a 1,3,5-regioselective manner. To collect direct structural information, we characterized the polymers by spectroscopic methods. Examples of the IR spectra of **P1** and its monomer **1** are given in Figure 10.1. Diyne **1** shows absorption bands at 3266 and 2124 cm^{-1} due to $\equiv$C−H and C$\equiv$C stretching vibrations, respectively. All of these bands disappear in the spectrum of its polymer, indicating that the acetylene triple bonds have been converted to the benzene rings by the diyne polycyclotrimerization. Similarly, the spectrum of *hb*-**P1/2** shows no $\equiv$C−H and C$\equiv$C stretching vibrations of **1** and **2** but display absorption bands characteristic of monosubstituted ferrocene moiety at 1105 and 1000 cm^{-1}, confirming that the polycyclotrimerization reaction has proceeded as expected and has been unharmful to the Fc group.

The ^{1}H NMR spectra of *hb*-**P1** and its monomer **1** as well as model compound **6** are shown in Figure 10.2. The acetylene proton of **1** resonates at δ 3.06, which completely disappears after the monomer is subjected to polycyclotrimerization reaction. By comparison with the spectra of monomer **1** and model compound **6**, the resonance peaks in the spectrum of polymer *hb*-**P1** can be readily assigned (*cf.*, Chart 10.1). The polymerization shifts the resonance of the phenyl protons *ortho* to the ester group in **1** at 7.05 to δ 7.17 in *hb*-**P1**, while the phenyl protons in the periphery of *hb*-**P1** resonate at δ 7.03. The new peak at δ 9.16 is assigned to the proton resonances of the benzene rings newly formed by the cyclotrimerization polymerization. End-capping of one triple bond in a terminal branch by two triple bonds in one diyne

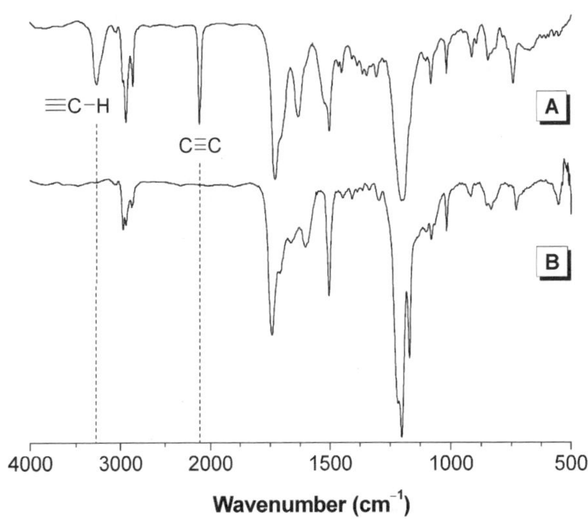

Figure 10.1 IR spectra of (A) monomer **1** and (B) its polymer *hb*-**P1** (sample taken from Table 10.2, run 2). Copyright (2009) American Chemical Society.

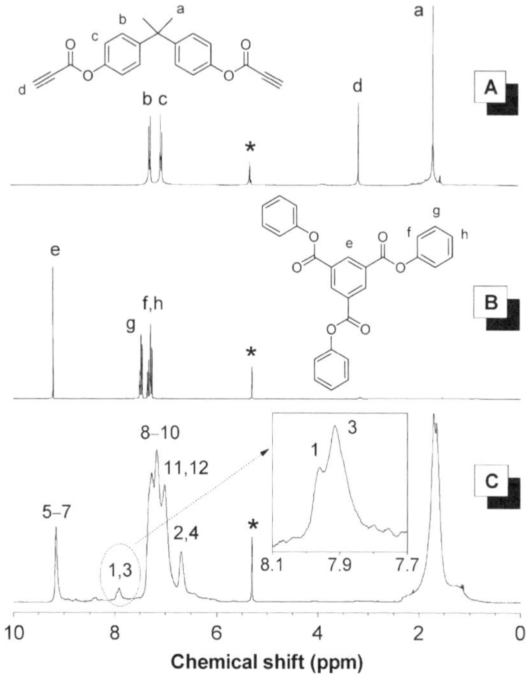

Figure 10.2 ^{1}H NMR spectra of (A) monomer **1**, (B) model compound **6** (sample taken from Table 10.1, run 4), and (C) polymer *hb*-**P1** (Table 10.2, run 2) in DCM-d_2. The solvent peaks are marked with asterisks. Copyright (2009) American Chemical Society.

monomer will generate a cyclophane ring, whose aromatic protons should resonate at δ ~9.3 and ~8.7. However, no such resonance peaks are observed in the spectrum of *hb*-**P1**. Instead, two peaks arising from the resonance of the olefinic unit formed by the alkyne hydroamination (Scheme 10.4) are detected at δ 7.91 and 6.69 (Figure 10.2C, peaks 1 and 3). The spectrum of *hb*-**P1/2** also exhibits no acetylene proton resonances of **1** and **2**. New resonance peaks associated with the proton resonances of the benzene rings of the terminal and dendritic units are observed at δ 9.29 and 9.06. No absorption peaks associated with the linear unit are observed. Thus, it is difficult for us to calculate the DB of *hb*-**P1/2**. The protons of the cyclopentadienyl ring of the ferrocene moiety resonate at δ 4.23. From its integral, the iron content of *hb*-**P1/2** is calculated to be 5.82%.

The ^{13}C spectrum of *hb*-**P1** shows no resonance peaks of the acetylenic carbons of **1** at δ 76.6 and 74.3 (Figure 10.3). New peaks corresponding to the absorptions of the triphenoxycarbonylphenyl and olefinic carbons are observed at δ 136.5, 132.0, and 115.5 due to the conversion of the acetylene triple bonds of **1** to the benzene rings and double bonds of *hb*-**P1**. Similar observations are also found in *hb*-**P1/2**.

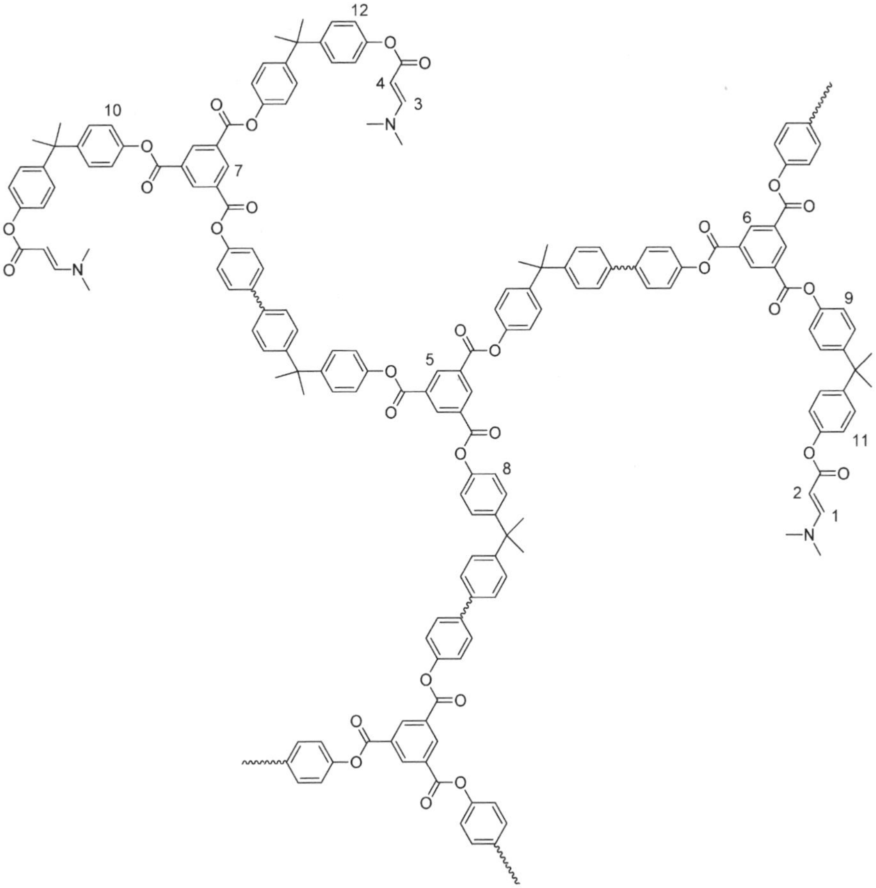

Chart 10.1   [Copyright (2009) American Chemical Society]

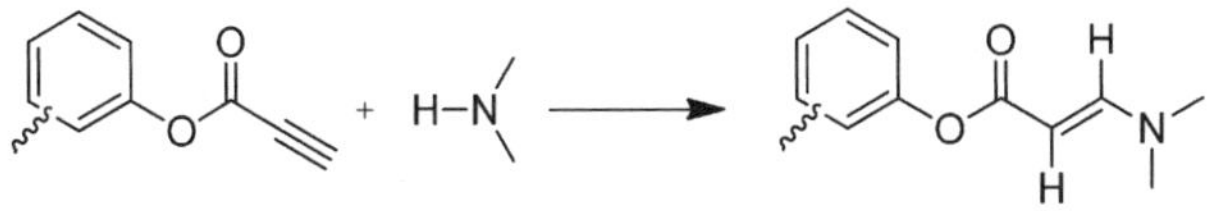

Scheme 10.4   [Copyright (2009) American Chemical Society]

10.3.4   Degree of Branching

As shown in Chart 10.2, there exist three structural components in *hb*-P1:
dendritic (D), linear (L) and terminal (T) units. Comparing the ^{1}H NMR
spectrum of *hb*-P1 with those of its monomer and the model compound, the
following relationships between the contents or fractions (*f*) of the structural
units can be established.

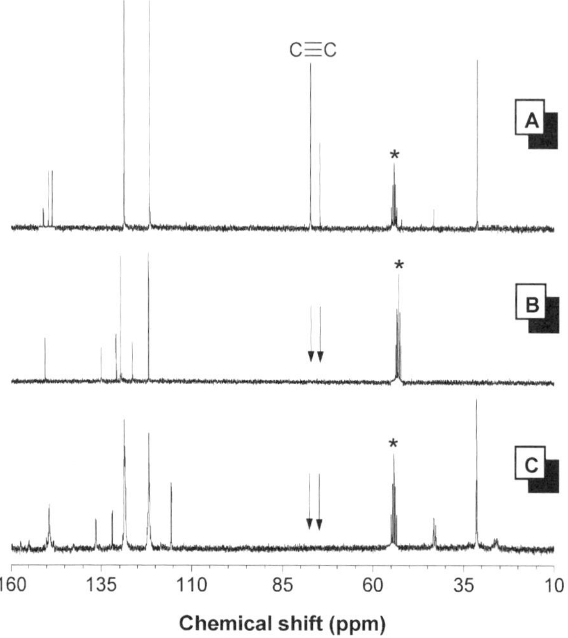

Chemical shift (ppm)

Figure 10.3 ^{13}C NMR of (A) monomer **1**, (B) model compound **6** (sample from Table 10.1, run 4), and (C) polymer *hb*-**P1** (Table 10.2, run 2) in DCM-d_2. The solvent peaks are marked with asterisks. Copyright (2009) American Chemical Society.

$$\frac{3f_{\text{L}} + 3f_{\text{T}} + 3f_{\text{D}}}{f_{\text{L}} + 2f_{\text{T}}} = \frac{A_{5-7}}{A_{1,3}} \tag{10.1}$$

$$f_{\text{D}} + f_{\text{L}} + f_{\text{T}} = 1 \tag{10.2}$$

where A_{5-7} and $A_{1,3}$ represent the integrals of the areas of resonance peaks (5−7) and (1,3), respectively, as labeled in Chart 10.1 and panel C of Figure 10.2. The values can be determined from the ^{1}H NMR spectral data, from which the following equation is deduced:

$$\frac{3f_{\text{L}} + 3f_{\text{T}} + 3f_{\text{D}}}{f_{\text{L}} + 2f_{\text{T}}} = \frac{1}{0.362} \tag{10.3}$$

Combining eqns (10.10) and (10.3) gives eqn (10.4):

$$f_{\text{L}} + 2f_{\text{T}} = 1.086 \tag{10.4}$$

The peak at δ 7.91 corresponds to the resonance of the olefinic proton capped by the dimethylamino group. Magnification of the peak manifests that it is actually a doublet and the amount of the olefinic bond in the L unit is $\sim$8-

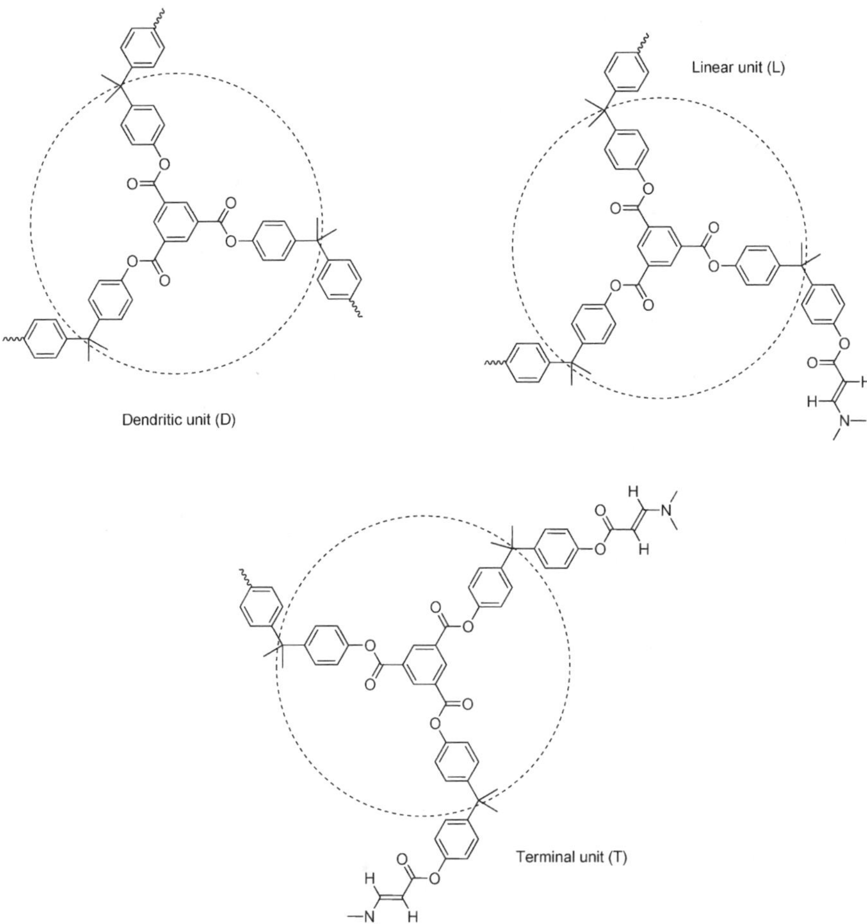

Chart 10.2 [Copyright (2009) American Chemical Society]

fold lower than that in the T unit. Because there are two such protons in the T unit, but only one in the L unit eqn (10.5) thus holds:

$$\frac{f_{\text{L}}}{2f_{\text{T}}} = \frac{1}{8} \tag{10.5}$$

From the above equations, f_{L} is calculated to be:

$$f_{\text{L}} = 0.121 \tag{6}$$

According to definition, DB is expressed as:[15]

$$\text{DB} = \frac{f_{\text{D}} + f_{\text{T}}}{f_{\text{D}} + f_{\text{L}} + f_{\text{T}}} \tag{10.7}$$

Incorporating eqns (10.2) and (10.6) into eqn (10.7) gives the DB value of *hb*-**P1**:

$$DB = 1 - f_L = 0.879 \qquad (10.8)$$

This value is much higher than those of the "conventional" hyperbranched polymer (commonly DB ~ 0.5),[1] which further confirms the hyperbranched structure of the polymer.

10.3.5 Solubility and Stability

All the hyperbranched polymers are completely soluble in common organic solvents, such as toluene, dichloromethane, chloroform, THF and dioxane, and can be readily fabricated into tough solid films by spin-coating or solution-casting processes. They also exhibit high thermal stability. As can be seen from Figure 10.4, the temperatures for their 5% weight loss or the degradation temperatures (T_d) are near or higher than 300 °C under nitrogen, indicative of their strong resistance to thermolysis and oxidation at high temperatures.

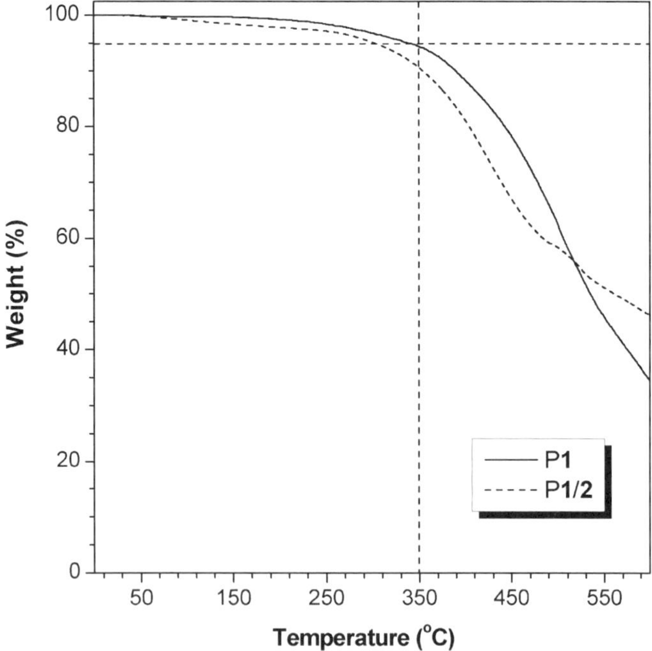

Figure 10.4　TGA thermograms of *hb*−**P1** and *hb*−**P1/2** recorded under nitrogen at a heating rate of 10 °C/min.

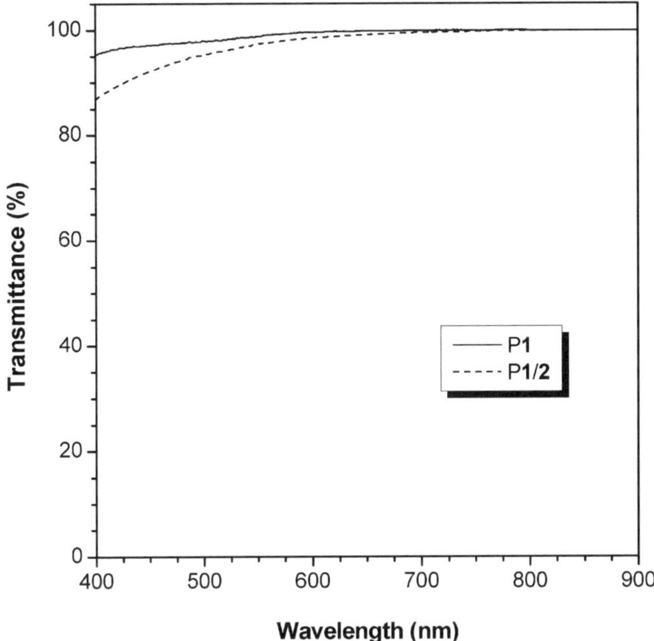

Figure 10.5 Light transmission spectra of THF solutions of *hb*−**P1** and *hb*−**P1/2**. Polymer concentration: 6.4 µg mL.

10.3.6 Optical Transparency

Polymers with high optical clarity are promising candidate materials for photonic applications.[16] As shown in Figure 10.5, both *hb*-**P1** and *hb*-**P1/2** absorb little light in the visible spectral region and allow all light at wavelengths longer than 600 nm to transmit through. This excellent optical transparency is due to their molecular structures. Polyesters often show high optical clarity, with poly(methyl methacrylate) (PMMA), poly(ethylene terephthalate) (PET) and polycarbonate (PC) being the best-known and widely used "organic glasses". The ester groups weaken the electronic communications between the aromatic rings in the polymer structure and decrease their extents of electronic conjugations, thereby enhancing their optical transparency.

10.3.7 Light Refractivity

hb-**P1** and *hb*-**P1/2** are comprised of polarizable aromatic rings, ester groups, and metallic species and they may thus show high refractive indices. Indeed, as can be seen from Figure 10.6, *hb*-**P1** displays high RI values (n = 1.6255−1.6181) in a wide wavelength region (400−1700 nm). The RI spectrum is almost flat: the RI value changes little over a wavelength span as wide as

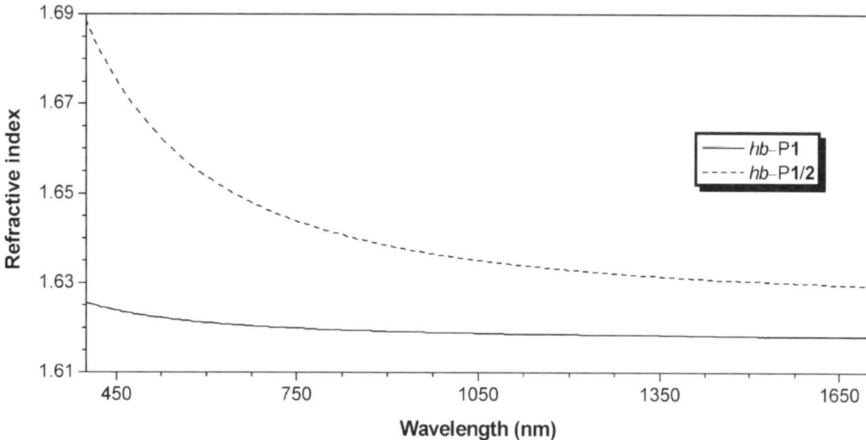

Figure 10.6 Wavelength dependence of refractive index of thin films of *hb*–**P1** and *hb*–**P1/2**.

1300 nm, which is very unusual and truly remarkable (*vide post*). Due the presence of ferrocene units in the polymer structure, *hb*-**P1/2** shows even igher RI values ($n = 1.6885$–1.6295) at the same wavelength region. The RIs of the polymers are much higher than those of the commercially important optical plastics (*e.g.*, $n \sim 1.49$ for PMMA). No or little birefringence is detected, which implicative of the amorphous nature of their thin solid films.

10.3.8   Chromatic Dispersion

For a material to be useful for practical applications, its optical aberrations should be small. The Abbé number (v_D) of a material is a measure of the variation or dispersion in its RI value with wavelength, which is defined as:

$$v_\mathrm{D} = \frac{n_\mathrm{D} - 1}{n_\mathrm{D} - n_\mathrm{C}} \qquad (10.9)$$

where n_D, n_F and n_C are the RI values at wavelengths of Fraunhofer D, F and C spectral lines of 589.2, 486.1, and 656.3 nm, respectively. A modified Abbé number (v_D') has been proposed to evaluate the application potential of an optical material, using its RI values at the nonabsorbing wavelengths of 1064, 1319 and 1550 nm.[17] The first two wavelengths are chosen in view of the practical interest of commercial laser wavelengths (Nd:YAG), while the last one is the wavelength for telecommunication. The modified Abbé number is defined as:

$$v_D' = \frac{n_{1319} - 1}{n_{1064} - n_{1550}} \qquad (10.10)$$

where n_{1319}, n_{1064}, and n_{1550} are the RI values at 1319, 1064, and 1550 nm, respectively. The chromatic dispersion ($D^{()}$) is the constringence of the Abbé number ($v_D^{()}$):

$$D^{()} = \frac{1}{v_D^{()}} \tag{10.11}$$

The v_D and v_D' values for *hb*-**P1** are as high as 258.9 and 1030.7, corresponding to D and D' values as low as 3.86×10^{-3} and 0.97×10^{-3}, respectively (Table 10.6). The chromatic dispersions of the polymer are much lower than those of the commercially important "organic glasses" such as PC ($D = 29.7 \times 10^{-3}$) and PMMA ($D = 17.5 \times 10^{-3}$).[18] The D' value is also much lower than those of the poly(aroyltriazole)s recently synthesized by our research groups by the metal-free "click" polymerizations ($D' = 6.1 \times 10^{-3}$–3.66×10^{-3}).[19] Although the Abbé numbers of *hb*-**P1/2** are larger than those of *hb*-**P1**, its chromatic dispersions ($D = 29 \times 10^{-3}$, $D' = 7 \times 10^{-3}$) are still comparable to those of the commercial optical plastics[18] and the newly synthesized poly(aroyltriazole)s.[19]

Table 10.6 Refractive indices and chromatic dispersions of *hb*-**P1** and *hb*-**P1/2**.

polymer	n_{1550}	v_D	v_D'	D	D'
hb-**P1**	1.6182	258.9	1030.7	0.0039	0.00097
hb-**P1/2**	1.6302	34.10	134.4	0.029	0.007

[a] All data taken from Figure 10.6, unless otherwise specified. Abbreviations: n = refractive index (at 1550 nm), v_D = Abbé number (calculated from equation 9), v_D' = modified Abbé number (calculated from eqn (10.10)), and $D^{()}$ = chromatic dispersion (calculated from eqn (10.11)).

10.4 Conclusion

In this work, we have developed a facile synthetic route to regioregular hyperbranched polymers. The homo-polycyclotrimerization of arylene bipropiolate **1** and its copolymerization with ferrocene-containing propiolate **2** proceed smoothly in refluxed DMF in the absence of externally added catalysts, furnishing soluble *hb*-**P1** and *hb*−**P1/2** with high MWs and DBs in high yields. The monomers are easy to access from commercially available starting materials and the polyaddition is strictly 1,3,5-regioselective and highly tolerant to functional groups, air and moisture. The resultant polymers are completely soluble, film-forming, thermally stable, and optically transparent. The polymers show high light refractivities with unprecedented low chromatic aberrations. These unique attributes make the polymers promising photonic materials for high-tech applications.

Acknowledgements

The work reported in this paper was partially supported by the Research Grants Council of Hong Kong (604509, 601608, and HKUST2/CRF/10), the National Science Foundation of China (20974028), the Ministry of Science and Technology of China (2009CB623605), and the Nissan Chemical Industries, Ltd. We thank Drs. Borong Shi and Fanny L. Y. Shek of our University, Profs. Chi Wu and To Ngai of the Chinese University of Hong Kong, and Prof. Yuping Dong of Beijing Institute of Technology for their technical assistance and helpful discussions. B.Z.T. thanks the support from the Cao Guangbiao Foundation of Zhejiang University.

References

1. (a) C. Gao and D. Yan, *Prog. Polym. Sci.*, 2004, **29**, 183. (b) D. A. Tomalia and J. M. Frechet, *J. Polym. Sci. Part A Polym. Chem.*, 2002, **40**, 2719. (c) M. Jikei and M. Kakimoto, *Prog. Polym. Sci.*, 2001, **26**, 1233. (d) Y. H. Kim, *J. Polym. Sci., Part A Polym. Chem.*, 1998, **36**, 1685. (e) T. E. Patten and K. Matyjaszewski, *Adv. Mater.*, 1998, **10**, 901. (f) M. Häussler, Q. Sun, K. Xu, J. W. Y. Lam, H. Dong and B. Z. Tang, *J. Inorg. Organomet. Polym. Mater.*, 2005, **15**, 67. (g) Z. Xie, H. Peng, J. W. Y. Lam, J. Chen, Y. Zheng, C. Qiu, H. S. Kwok and B. Z. Tang, *Macromol. Symp.*, 2003, **195**, 179. (h) J. W. Y. Lam, J. Luo, H. Peng, Z. Xie, K. Xu, Y. Dong, L. Cheng, C. Qiu, H. S. Kwok and B. Z. Tang, *Chin. J. Polym. Sci.*, 2001, **19**, 585. (i) M. Häussler, R. Zheng, J. W. Y. Lam, H. Tong, H. Dong and B. Z. Tang, *J. Phys. Chem. B*, 2004, **108**, 10645. (j) Q. Sun, K. Xu, H. Peng, R. Zheng, M. Häußler and B. Z. Tang, *Macromolecules*, 2003, **36**, 2309. (k) Q. Sun, J. W. Y. Lam, K. Xu, H. Xu, J. A. K. Cha, P. C. L. Wong, G. Wen, X. Zhang, X. Jing, F. Wang and B. Z. Tang, *Chem. Mater.*, 2000, **12**, 2617.
2. (a) B. Voit, *J. Polym. Sci. Part A, Polym. Chem.*, 2005, **43**, 2679. (b) C. R. Yates and W. Hayes, *Eur. Polym. J.*, 2004, **40**, 1257.
3. (a) J. M. J. Frechet, M. Henmi, I. Gitsov, S. Aoshima, M. R. Leduc and R. B. Grubbs, *Science*, 1995, **269**, 1080. (b) C. J. Hawker, J. M. J. Frechet, R. B. Grubbs and J. Dao, *J. Am. Chem. Soc.*, 1995, **117**, 10763. (c) S. G. Gaynor, S. Edelman and K. Matyjaszewski, *Macromolecules*, 1996, **29**, 1079. (d) A. Dworak, W. Walach and B. Trzebicka, *Macromol. Chem. Phys.*, 1995, **196**, 1963. (e) M. Suzuki, S. Yobbshida, K. Shiraga and T. Saegusa, *Macromolecules*, 1998, **31**, 1716. (f) H. Magnusson, E. Malmstrom, A. Hult, *Macromol. Rapid Commun.*, 1999, **20**, 453. (g) A. Sunder, R. Hanselmann, H. Frey and R. Muelhaupt, *Macromolecules*, 1999, **32**, 4240.
4. (a) P. F. W. Simon, W. Radke and A. H. E. Muller, *Macromol. Rapid Commun.*, 1997, **18**, 865. (b) K. Sakamoto, T. Aimiya and M. Kira, *Chem. Lett.*, 1997, 1245. (c) K. Matyjaszewski, S. G. Gaynor, A. Kulfan and M.

Podwika, *Macromolecules*, 1997, **30**, 5192. (d) D. Yan, Z. Zhou and A. H. E. Mueller, *Macromolecules*, 1999, **32**, 245. (e) D. Baskaran, *Polymer*, 2003, **44**, 2213. (f) C. Cheng, K. L. Wooley and E. Khoshdel, *J. Polym. Sci. Part A Polym. Chem.*, 2005, **43**, 4754.

5. (a) B. Z. Tang, *Macromol. Chem. Phys.*, 2008, **209**, 1303. (b) M. Häussler, A. Qin and B. Z. Tang, *Polymer*, 2007, **48**, 6181. (c) M. Häussler and B. Z. Tang, *Adv. Polym. Sci.*, 2007, **209**, 1.

6. (a) Y. H. Kim, *J. Polym. Sci. Part A Polym. Chem.*, 1998, **36**, 1685. (b) A. Hult, M. Johansson and E. Malmstrom, *Adv. Polym. Sci.*, 1999, **143**, 1. (c) C. J. Hawker, *Curr. Opin. Coll. Interf. Sci.*, 1999, **4**, 117. (d) K. Inoue, *Prog. Polym. Sci.*, 2000, **25**, 453. (e) B. I. J. Voit, *Polym. Sci. Polym. Chem.*, 2000, **38**, 2505. (f) S. M. Grayson and J. M. J. Frechet, *Chem. Rev.*, 2001, **101**, 3819. (g) D. A. Tomalia and J. M. J. Frechet, *J. Polym. Sci. Part A Polym. Chem.*, 2002, **40**, 2719.

7. (a) M. Häußler and B. Z. Tang, *Acc. Chem. Res.*, 2005, **38**, 745. (b) J. W. Y. Lam and B. Z. Tang, *J. Polym. Sci. Part A Polym. Chem.*, 2003, **41**, 2607. (c) K. K. L. Cheuk, B. Li and B. Z. Tang, *Curr. Trends Polym. Sci.*, 2002, **7**, 41. (d) J. Liu, R. Zheng, Y. Tang, M. Häußler, J. W. Y. Lam, A. Qin, M. Ye, Y. Hong, P. Gao and B. Z. Tang, *Macromolecules*, 2007, **40**, 7473. (e) M. Häußler, J. W. Y. Lam, A. Qin, K. K. C. Tse, M. K. S. Li, J. Liu, C. K. W. Jim, P. Gao and B. Z. Tang, *Chem. Commun.*, 2007, 2584. (f) M. Häußler, J. Liu, R. Zheng, J. W. Y. Lam, A. Qin and B. Z. Tang, *Macromolecules*, 2007, **40**, 1914. (g) R. Zheng, M. Häußler, H. Dong, J. W. Y. Lam and B. Z. Tang, *Macromolecules*, 2006, **39**, 7973. (h) Z. Li, J. W. Y. Lam, Y. Q. Dong, Y. P. Dong, H. H. Y. Sung, I. D. Williams and B. Z. Tang, *Macromolecules*, 2006, **39**, 6458. (i) Z. Li, A. Qin, J. W. Y. Lam, Y. Q. Dong, Y. Dong, C. Ye, I. D. Williams and B. Z. Tang, *Macromolecules*, 2006, **39**, 1436.

8. (a) M. Häussler, J. W. Y. Lam, R. Zheng, H. Peng and B. Z. Tang, *C. R. Chim.*, 2003, **6**, 833. (b) J. W. Y. Lam, J. Chen, C. C. W. Law, H. Peng, Z. Xie, K. K. L. Cheuk, H. S. Kwok and B. Z. Tang, *Macromol. Symp.*, 2003, **196**, 289. (c) J. W. Y. Lam, H. Peng, M. Häußler, R. Zheng and B. Z. Tang, *Mol. Cryst. Liq. Cryst.*, 2004, **415**, 43. (d) H. Peng, Y. Dong, D. Jia and B. Z. Tang, *Chin. Sci. Bull.*, 2004, **49**, 2637. (e) H. Peng, H. Dong, Y. Dong, D. Jia and B. Z. Tang, *Chin. J. Polym. Sci.*, 2004, **22**, 501. (f) H. Peng, J. W. Y. Lam and B. Z. Tang, *Macromol. Rapid Commun.*, 2005, **26**, 673. (g) H. Peng, R. Zheng, H. Dong, D. Jia and B. Z. Tang, *Chin. J. Polym. Sci.*, 2005, **23**, 1.

9. (a) A. Qin, J. W. Y. Lam, H. Dong, W. Lu, C. K. W. Jim, Y. Q. Dong, M. Häussler, H. H. Y. Sung, I. D. Williams, G. K. L. Wong and B. Z. Tang, *Macromolecules*, 2007, **40**, 4879. (b) H. Dong, R. Zheng, J. W. Y. Lam, M. Häussler and B. Z. Tang, *Macromolecules*, 2005, **38**, 6382. (c) B. Z. Tang, H. C. Dong and A. J. Qin, US Patent Pub No. US2006/ 0247410 A1, 2006.

10. B. Z. Tang, C. K. W. Jim, A. Qin, M. Häussler and J. W. Y. Lam, US Patent Pub. No. US60/933,884, 2007.

11. D. Plażuk, A. Vessières, F. L. Bideau, G. Jaouen and J. Zakrewski, *Tetrahedron Lett.*, 2004, **45**, 5425.

12. (a) K. K. Balasubramanian, S. Selvaraj and P. S. Venkataramani, *Synthesis*, 1980, 29. (b) K. Matsuda, N. Nakamura, K. Takahashi, K. Inoue, N. Koga and H. Iwamura, *J. Am. Chem. Soc.*, 1995, **117**, 5550. (c) F. C. Pigge, F. Ghasedi and N. P. Rath, *J. Org. Chem.*, 2002, **67**, 4547.

13. (a) Z. Muchtar, M. Schappacher and A. Deffieux, *Macromolecules*, 2001, **34**, 7595. (b) K. E. Uhrich, C. J. Hawker, J. M. J. Frechet and S. R. Turner, *Macromolecules*, 1992, **25**, 4583.

14. (a) R. H. Zheng, H. C. Dong, H. Peng, J. W. Y. Lam and B. Z. Tang, *Macromolecules*, 2004, **37**, 5196. (b) H. Peng, L. Cheng, J. Luo, K. Xu, Q. Sun, Y. Dong, F. Salhi, P. P. S. Lee, J. Chen and B. Z. Tang, *Macromolecules*, 2002, **35**, 5349.

15. (a) C. J. Hawker, R. Lee and J. M. J. Frechet, *J. Am. Chem. Soc.*, 1991, **113**, 4583. (b) H. Frey and D. Hoelter, *Acta Polym.*, 1999, **50**, 67.

16. (a) E. Otsuka, K. Kurumada, A. Suzuki, S. Matsuzawa and K. Takeuchi, *J. Sol-gel Sci. Tech.*, 2008, **46**, 71. (b) J. B. Chu, S. M. Huang, H. B. Zhu, X. B. Xu, Z. Sun, Y. W. Chen and F. Q. Huang, *J. Non-Cryst. Solids*, 2008, **354**, 5480.

17. (a) C. J. Yang and S. A. Jenekhe, *Chem. Mater.*, 1995, **7**, 1276. (b) C. J. Yang and S. A. Jenekhe, *Chem. Mater.*, 1994, **6**, 196.

18. (a) J. C. Seferis in *Polymer Handbook*, 3rd edn, J. Brandrup and E. H. Immergut, Wiley, New York, 1989; p. VI/451–461. (b) N. J. Mills, in *Concise Encyclopedia of Polymer Science and Engineering*; J. I. Kroschwitz, Wiley, New York, 1990; p. 683–687.

19. A. Qin, L. Tang, J. W. Y. Lam, C. K. W. Jim, Y. Yu, H. Zhao, J. Sun and B. Z. Tang, *Adv. Funct. Mater.*, 2009, **19**, 1891.

Organic Dyes for Dye-Sensitized Solar Cells

ZHIJUN NING[1,2] AND HE TIAN*[1]

[1] Key Lab for Advanced Materials and Institute of Fine Chemicals, East China University of Science and Technology, 130 Meilong Road, Shanghai, 200237, People's Republic of China; [2] Department of Theoretical Chemistry, School of Biotechnology, Royal Institute of Technology, S-106 91 Stockholm, Sweden
*E-mail: tianhe@ecust.edu.cn

11.1 Introduction

Exploiting clean and sustainable energy resources has become an urgent need worldwide. Solar energy is regarded as one of the most ideal energy resources.[1] Huge efforts have been invested to develop high efficiency solar energy conversion technologies. The emerging photovoltaic industry has been growing rapidly in recent years. However, the price of solar power from the state-of-the-art inorganic silicon technology significantly exceeds that from the electrical grid, which prohibits the large-scale application. There has been a continuous effort in searching for affordable organic-materials-based solar energy technology, among which dye-sensitized solar cells (DSCs) demonstrate thus far the highest energy-conversion efficiency and are regarded as the most prospective candidate.[2]

Among all the constituent components in DSCs (Fig. 11.1), the sensitizer that is charged with the task of the light absorption and electron injection is generally regarded as the most crucial one that determines its overall efficiency. Since the first report of DSCs by Grätzel and coworkers, Ru complex dyes are generally considered as the ideal sensitizers for DSCs.[3] Although a few metal-free dyes

RSC Polymer Chemistry Series No. 2
Molecular Design and Applications of Photofunctional Polymers and Materials
Edited by Wai-Yeung Wong and Alaa S Abd-El-Aziz
© The Royal Society of Chemistry 2012
Published by the Royal Society of Chemistry, www.rsc.org

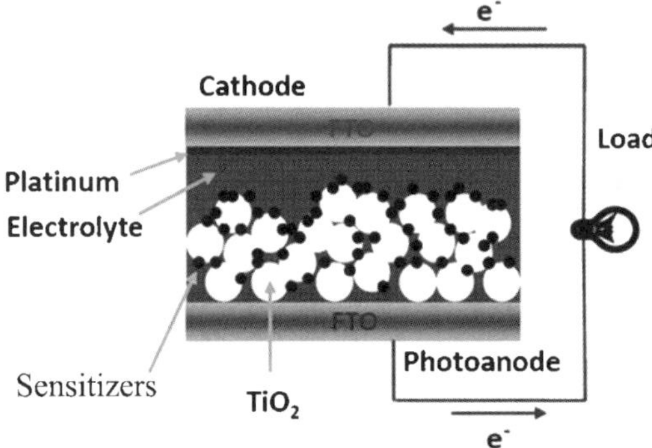

Figure 11.1   Schematic structure diagram of DSCs.

were reported intermittently in the early stage of the DSCs development, they are regarded as futureless at this time due to their low efficiencies. Nevertheless, at the beginning of the 21st century, the efficiencies of organic sensitizers began to leap upwards, and their efficiencies became comparable to the traditional noble metal complexes nowadays.[4] In summary, the development of organic dyes can be classified into three main stages. In the first phase, the development of organic sensitizers focused on dyes with large-π-aromatic structures such as nature chlorophyll and derivatives of perylene pigments. The next stage of the development of organic dyes can be dated back to the introduction of cyanine dyes. Since then, a great deal of effort was invested to develop sensitizers with donor (D)-conjugate linker (π)-acceptor (A) framework. The third stage of the development can be the modification of the D-π-A structured sensitizers. At this stage, novel kinds of sensitizer structures such as D-D-π-A and D-π-A-A, novel sensitizer configurations including starburst-shape and cone-shape, and the decoration of the conjugation framework such as the introduction of alkyl chains were developed.

With the improvement of the molecular design of the organic sensitizers, the conversion efficiencies of DSCs based on them increased step by step. In the following parts, the three stages of the development of organic dyes will be presented in sequence: (1) the development of dyes with large-π-aromatic structure; (2) the construction of D-π-A structured sensitizers; (3) the modification of D-π-A structured sensitizers. In the end, conclusions and prospects will be presented.

11.2   The First Stage of Development: Dyes with Large π-Conjugated Structure

11.2.1   Porphyrin Dyes

At the beginning of the development of organic dyes, there was no guideline for the molecular design of the sensitizers, the main issues considered were the

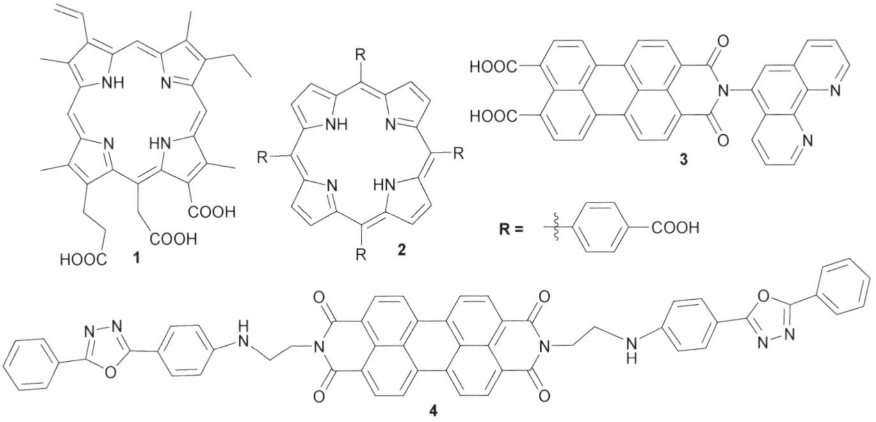

Scheme 11.1

absorption spectra (to absorb as much sunlight as possible) and the anchor units (to be adsorbed on TiO_2) of the sensitizers. The first metal-free organic sensitizer was chhorophyll dye. Inspired by photosynthesis, chhorophyll dye **1** (Scheme 11.1) was used for DSCs.[5] These chhorophyll dyes were firstly separated from the natural plant, and were successfully adsorbed on TiO_2 surface *via* the carboxyl groups. The porphyrin dyes show broad absorption spectra in most of the visible-light region. The highest incident photon-to-current conversion efficiency (IPCE) was 55%, and the highest energy-conversion efficiency was about 3%. To enhance the dye-adsorption capability, Cherian and Wamser synthesized a novel kind of porphyrin dye with four carboxyl groups,[6] sensitizer tetra(4-carboxyphenyl)porphyrin (TCPP) **2** (Scheme 11.1) that can adsorbs strongly onto TiO_2 and serve as an efficient photosensitizer with highest conversion efficiency of 3.5% under low intensity of irradiation.

11.2.2 Perylene Dyes

Perylene derivatives, a classic pigment material were introduced for DSCs due to their excellent stability under long-time irradiation. Gregg and coworkers firstly employed perylene dyes for DSCs.[7] Carboxyl substituents were implanted on the dye to make them absorbable on TiO_2. One example is the tetracarboxylic acid form of the commercially available dye perylene-3,4,9,-10-tetracarboxylic dianhydride, which adsorbs strongly to the surface of SnO_2 film (works similarly as TiO_2). The novel compound perylene-3,4-dicarboxylic acid-9,10-(5-phenanthroline)-carboximide **3** (Scheme 11.1) on SnO_2 yielded a short-circuit photocurrent (J_{sc}) of 3.26 mA/cm^2, an open-circuit photovoltage (V_{oc}) of 0.45 V, and an overall cell efficiency of 0.89%. To broaden the absorption spectra of the perylene dyes, Tian *et al.* introduced oxadiazole and naphthaldicarboximide on the perylene skeleton (**4** in Scheme 11.1).[8] A wide spectral photoresponse (from 310 to 700 nm) and a maximum IPCE of 24% were achieved.

11.2.3   Section Summary

At this stage, we have mainly discussed the development of organic dyes before 2000. Although both porphyrin and perylene dyes demonstrate broad absorption spectra and efficient photon to current conversion efficiency, their overall energy-conversion efficiencies were very low. The reason was, in this stage, no other characteristics of the sensitizers were considered except for the absorption spectra of the sensitizers and the connection of carboxyl group to enable it to be adsorbed on the TiO_2. With the development of the molecular design of the organic dyes, the efficiencies of porphyrin and perylene dyes were significantly improved, which will be discussed in Section 11.4.

11.3   The Second Stage of Development: Dyes with D-π-A Structure

11.3.1   Cyanine Dyes

Cyanine dyes were firstly introduced for DSCs by Tian and coworkers. DSC based on SnO_2 photoanode and sensitizer **5** (Scheme 11.2) shows the highest IPCE value of 46%.[9] Huang's group introduced hemi-cyanine dyes as sensitizers for DSCs.[10] Sensitizer **6** (Scheme 11.2) displayed a broad IPCE spectrum with a maximum of 34%, and a thin-layer DSCs based on **6** produced an overall efficiency of about 2.0%. Sensitizer with benzothiazolium salt (**7** in Scheme 11.2) as electron acceptor shows the highest IPCE over 90%.[11] Tian and coworkers using phenyl substituted indole iodide salt as acceptor group, the absorption spectrum of sensitizer **8** (Scheme 11.2) was effectively broadened, and energy-conversion efficiency of 4.9% was achieved.[12] It was speculated that the D-π-A configuration of the hemi-cyanine and cyanine sensitizers facilitated the electron separation and the electron injection into

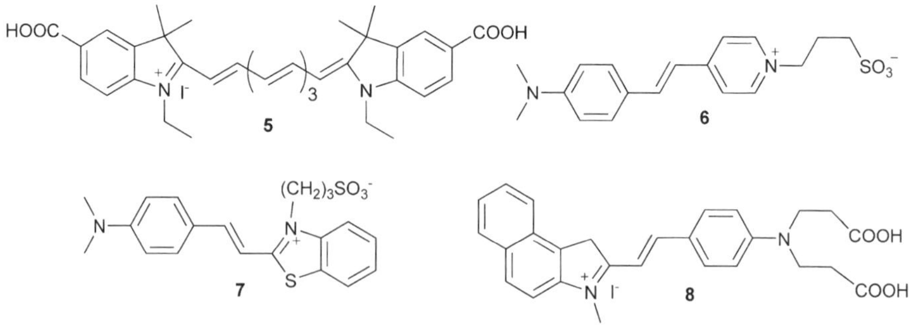

Scheme 11.2

TiO_2, leading to the enhanced efficiency of the cells. To increase the photon-absorption ability of cyanine dyes in the short-wavelength region, Tian and coworkers introduced the naphthalimide group on the cyanine sensitizer, and the highest efficiency was improved to 5.0%.[13] The cyanine dyes of the highest efficiency so far were synthesized by Tian and coworkers *via* the introduction of triphenylamine unit on the cyanine dye skeleton and an efficiency of 7.0% was obtained.[14] Since the introduction of cyanine dyes, the D-π-A structure has become widely employed for the molecular design of organic dyes.

11.3.2 Coumarin Dyes

Coumarin dye is one kind of traditional laser dyes due to its strong luminescence character. The rigid planar structure gives the coumarin unit a strong electron-donating character. Hara and coworkers firstly reported the usage of coumarin sensitizers for DSCs.[15] Sensitizer **9** (Scheme 11.3) with coumarin as the donor unit and cyanoacrylic acid as acceptor showed an overall conversion efficiency of 5.6%, with a J_{sc} of 13.8 mA cm^2, an V_{oc} of 0.63 V, and a fill factor (ff) of 0.63 under standard AM 1.5 irradiation. It is worth noting that in this sensitizer, cyanoacrylic acid was explored as an electron acceptor and anchor group for the first time. This electron acceptor considerably enhanced the electron overlap between the LUMO of the sensitizer and TiO_2, which is proved now to be important for the electron injection. Cyanoacrylic acid was widely adopted as an electron acceptor for organic sensitizers.

Following the first report of the coumarin sensitizers, much work was done to optimize their performance.[16] Considering the narrow absorption spectra of the coumarin cyanoacrylic acid dye **9**, some new sensitizers with longer linkers *via* the insertion of additional conjugate units such as thiophene or double bond were developed.[17] By extending the linker length, the absorption spectra were broadened and the overall energy-conversion efficiency was improved.

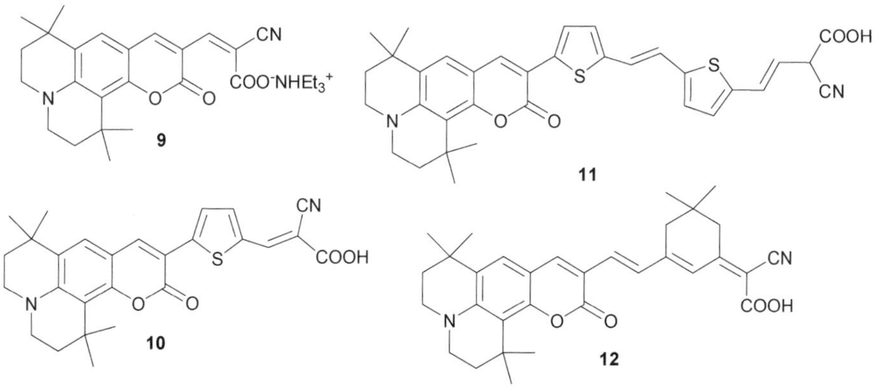

Scheme 11.3

Sensitizer **10** (Scheme 11.3) exhibited an overall conversion efficiency of 6.1%.[16] However, too long a conjugate linker results in serious molecular aggregation such as in sensitizer **11** (Scheme 11.3), and thus reduced conversion efficiency.[18] In order to prevent aggregation, coadsorbates such as deoxycholic acid (DCA) were used in the adsorption process of the sensitizer that effectively reduced the aggregation and improved the efficiency, and the highest efficiency was improved to 8.2%. Wang *et al.* introduced isophorone as the linker to suppress the molecular aggregation. Sensitizer **12** (Scheme 11.3) produced an overall energy-conversion efficiency of 6.7%.[19] Up to now, coumarin dye is one kind of sensitizers that shows the highest conversion efficiency.

11.3.3   Indoline Dyes

Indoline with a similar structure to coumarin is another kind of excellent electron-donor group. Horiuchi and coworkers first reported this type of organic dye for DSCs.[20] Dye **13** (Scheme 11.4) with indoline as electron donor and rhodanine as electron acceptor showed power-conversion efficiency of 6.1%. To further broaden the absorption spectra, an additional rhodanine unit was introduced on **13**. Sensitizer **14** (Scheme 11.4) with co-rhodanine as electron acceptor showed a prominent efficiency of 8.0%.[21] Based on a ZnO photoanode, DSC based on sensitizer **15** (Scheme 11.4) with thiophene as linker and cyanoacrylic acid as acceptor showed conversion efficiency of 3.8%.[22] Tian and coworkers reported a new kind of indoline dye **16** (Scheme 11.4) with isophorone as conjugate linker, which showed even higher efficiency than Ru complex sensitizer in DSCs with a thin TiO_2 film.[23] The high molar extinction coefficient of indoline dyes makes it advantageous for solid-state

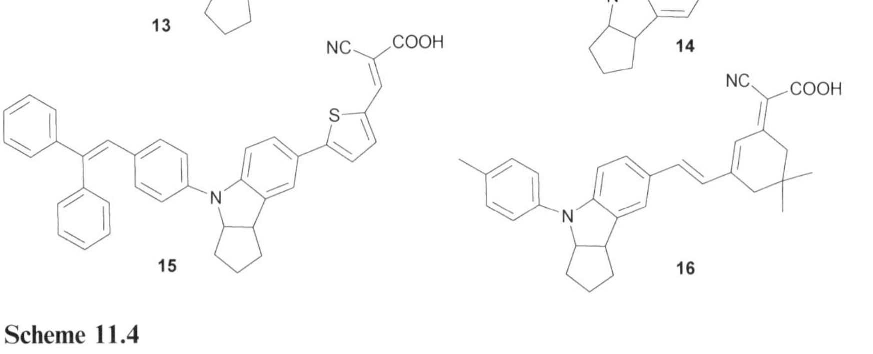

Scheme 11.4

DSCs, which require the usage of thin TiO_2 film. Solid-state DSCs based on **14** achieved efficiency over 4.0%.[24] To this day, the indoline unit is still one of the best donors for organic sensitizers.

11.3.4 Triarylamine Dyes

Triarylamine is another kind of widely used electron donor for organic sensitizers. Compared with the electron donors coumarin or indoline mentioned above, the electron-donating capability is weaker, while its special nonplanar spiral-like character can reduce intermolecular aggregation effectively.[4c] Yanagida and coworkers first explored triphenylamine as electron donor for organic dyes.[25] Sensitizer **17** (Scheme 11.5), employing triphenylamine as donor, and cyanoacrylic acid as acceptor, exhibited efficiency of 3.3%. To expand the absorption spectra, a thiophene unit was introduced as linker in sensitizer **18** (Scheme 11.5), which showed power-conversion efficiency of 5.1%,[26] and the sensitizer with phenyl as linker exhibited a conversion efficiency of 5.3%.[27] Tian and coworkers investigated sensitizers with isophorone as a bridging unit. [28] The DSC based on **19** (Scheme 11.5) showed an overall conversion efficiency of 5.7%. Chen and coworkers introduced rhodanine as electron acceptor for triphenylamine dyes. [29] Sun and coworkers

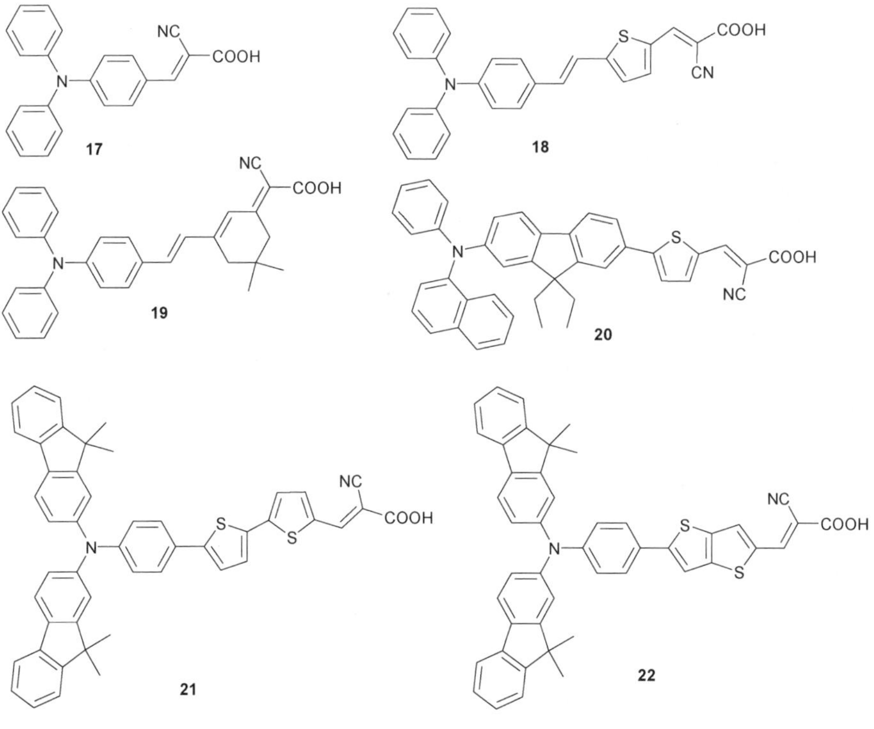

Scheme 11.5

exploited triphenylamine sensitizers with thiophene as linker and rhodanine as acceptor, achieving an energy-conversion efficiency of 2.3%.[30] Sensitizers with fluorene and thiophene as conjugate bridge were reported by Lin and coworkers.[31] The power-conversion efficiency of dye **20** (Scheme 11.5) was close to the Ru dye N719.

Ko and coworkers introduced a novel kind of triarylamine donor group consisting of two fluorenes groups and one phenyl unit. The DSC based on dye **21** (Scheme 11.5) exhibited a power-conversion efficiency of 8.0%.[32] With ionic liquid-based electrolytes, sensitizer **21** illustrated a conversion efficiency of 5.8%, and the performance can be essentially sustained under light soaking at 60 °C for 1200 h. Sensitizer **22** (Scheme 11.5) with the same donor and thienothiophene as linker was developed by Wang and coworkers.[33] With a solvent-free ionic liquid electrolyte, the conversion efficiency of sensitizer **22** reached 7.0% and the performance was highly stable under long-time solar irradiation. For the solid-state DSCs, dye **22** gave an impressive efficiency of 4.8%. Generally, dyes with single triarylamine as electron donor have much lower HOMO energy edge than iodine electrolyte, which is unfavorable for the extension of absorption spectra.[4c] The strategy to broaden the absorption spectra of the triarylamine-based sensitizers will be discussed in Section 11.4. However, triarylamine sensitizers show the highest efficiency among all organic dyes at the present.

11.3.5 Heteroanthracene Dyes

Sun and coworkers developed a series of organic dyes based on phenothiazine. Phenothiazine, showing strong electron-donating ability, is another kind of good electron donor for organic sensitizers. The simple sensitizer **23** (Scheme 11.6) with phenothiazine as electron donor and cyanoacrylic acid as electron acceptor achieved an efficiency of 5.5%.[34] Based on phenothiazine dyes, they reported a novel kind of dye **24** (Scheme 11.6) with phenoxazine as electron donor, which showed better electron-donating ability. Dye **24** with phenox-

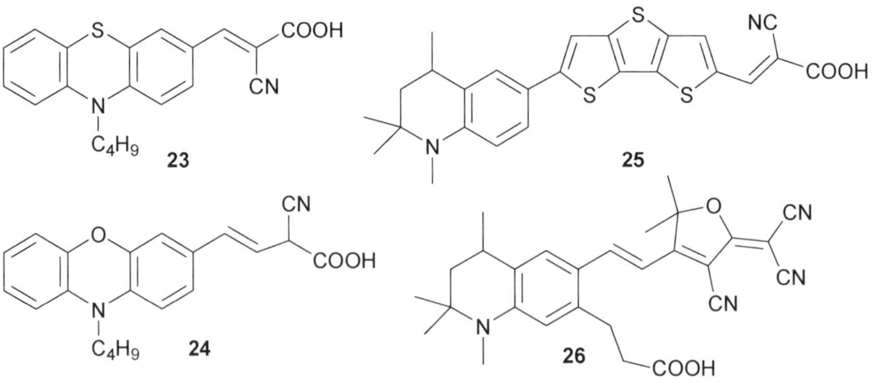

Scheme 11.6

azine as electron donor and cyanoacrylic acid as electron acceptor showed a conversion efficiency of 6.2%.[35] To broaden the absorption spectra of the sensitizers, co-rhodanine was introduced as electron acceptor, DSCs based on which showed a broad IPCE spectrum over the whole visible range extending into the near-IR region.[36] The efficiencies of phenoxazine dyes were enhanced further *via* the connection of triphenylamine on the sensitizer, which will be discussed in Section 11.4. Tian and coworkers developed several new phenothiazine dyes, and the highest efficiency is 4.4%.[37] Overall, heteroanthracene dyes exhibit good performance.

11.3.6   Tetrahydroquinoline Dyes

Tetrahydroquinoline donor is another kind of efficient electron donor for organic dyes. Sun and coworkers developed a series of tetrahydroquinoline based dyes with different conjugate linker, and their performance were compared.[38] A sensitizer with two thiophene units as the linker showed the highest conversion efficiency of 4.53%, higher than the sensitizer with one double bond and one thiophene unit. The low performance of double-bond-containing dyes was speculated to be due to the existence of intramolecular vibration.[39] It is worth noting that they first introduced dithieno[3,2-*b*;2′,3′-*d*]thienyl as conjugate linker in sensitizer **25** (Scheme 11.6), which showed a conversion efficiency of 2.9%. Moreover, they developed a new type of panchromatic dye **26** (Scheme 11.6) based on tetrahydroquinoline donor with strong absorption in the near-infrared (NIR) region. Dye **26**-based DSCs gave a maximum IPCE value of 86% at 660 nm and 3.7% of efficiency, which was one of the highest efficiencies reported so far for the D-π-A panchromatic organic dyes.[40]

11.3.7   Carbazole Dyes

Carbazole is another kind of widely investigated electron-donor unit for organic dyes. In general, the electron-donating capability is weaker than the donors listed above. Sensitizer **27** (Scheme 11.7) with carbabole as electron donor and three thiophene units as conjugation linker showed a conversion efficiency of 4.8%.[41] Ko and coworkers studied a series of carbazole sensitizers with a fluorene unit substituted on the carbazole unit. Sensitizer **28** (Scheme 11.7) with two thiophene units as linker and cyanoacylic acid as electron acceptor displayed a conversion efficiency of 5.2%.[42] Harima and coworkers developed a new kind of carbazole derivatives based sensitizers such as dye **29** and **30** (Scheme 11.7).[43] Sensitizer **29** showed a much higher efficiency than **30**. It was speculated that for **29**, the electron-acceptor part is close to the TiO_2 surface, which facilitates the electron injection. However, because of the planar structure of the carbazole dyes, the efficiencies of the DSCs based on these were normally low due to the serious aggregation on the TiO_2 surface. Their

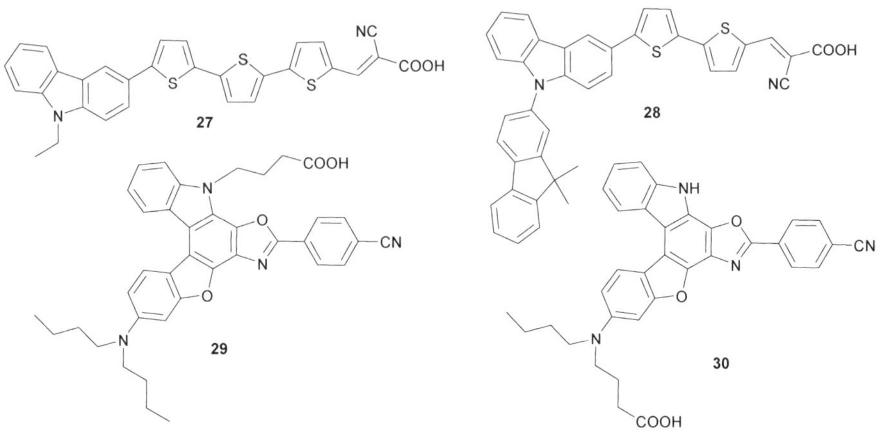

Scheme 11.7

efficiencies were significantly improved by suppressing the molecular aggregation, which will be discussed in Section 11.4.

11.3.8   Boron-dipyrromethene Dyes

Apart from electron donors, some new electron acceptors are also introduced. Akkaya and coworkers introduced the boron-dipyrromethene (BODIPY) in organic dyes, which is a strong electron-acceptor unit. Dye **31** (Scheme 11.8) show a NIR absorption spectrum (600–800 nm), and a DSC based on it achieved 1.7% of power-conversion efficiency and a panchromatic IPCE spectrum.[44] The development of panchromatic sensitizer is highly important for the efficient utilization of the whole solar irradiation spectrum. To improve

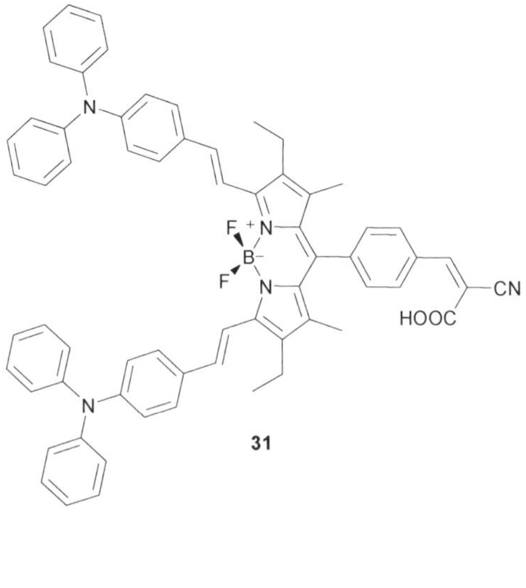

Scheme 11.8

the stability of these dyes, ethynyl units with side chains were adopted instead of the traditional fluorine atoms on boron.[45] The low efficiency of this kind of sensitizer might be due to the low electron overlap between the sensitizer LUMO and TiO_2, and the low LUMO energy level edge of the sensitizer.

11.3.9 Squarine Dyes

Squaraine dyes are another kind of panchromatic sensitizer that exhibit sharp and intense absorption bands in the visible and NIR regions. A number of squaraine dyes have been developed for DSCs. Das and coworkers found that DSCs employing unsymmetrical squaraines as sensitizers produced much higher efficiencies than sensitizers with symmetrical structure.[46] Dye **32** (Scheme 11.9) with unsymmetrical structure showed a higher efficiency value, 2.1%, than that of the symmetrical dye **33** (Scheme 11.9) that achieved 1.0% efficiency. To extend the absorption spectra, large conjugate units were adopted. A DSC based on sensitizer **34** (Scheme 11.9) achieved conversion efficiency was 1.6%.[47] Triphenylamine and thiophene were included in squaraines sensitizers (**35** in Scheme 11.9) by Wu and coworkers, and the highest efficiency is 2.6%. Similar as BODIPY dyes, squaraine sensitizers exhibit low efficiency as well.[48] It was speculated that the electron overlap between the sensitizer LUMO and TiO_2 needs to be enhanced.

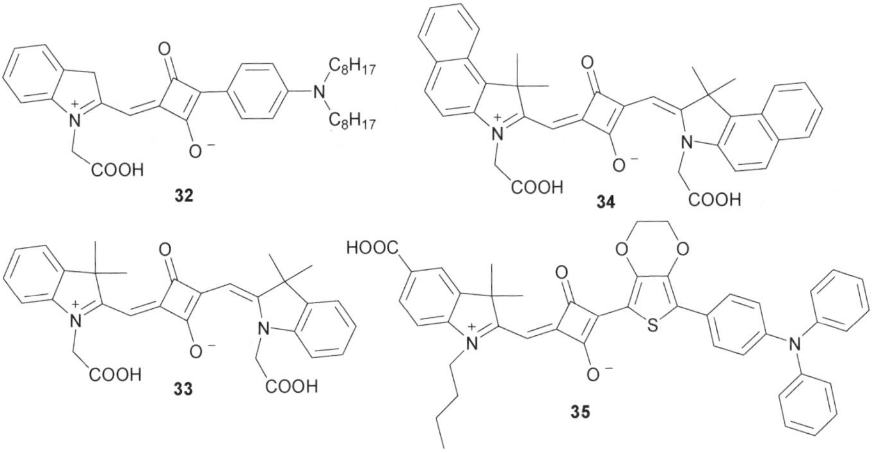

Scheme 11.9

11.3.10 Section Summary

At this stage, the development of organic dyes concentrated on the design of dyes with new electron donors, acceptors, as well as conjugate linker. Limited by the length of this chapter, some kinds of donors and acceptors are not introduced.[2a,4b,49] With the introduction of various kinds of electron donors or acceptors, the efficiency of organic dyes was significantly improved. However, their efficiencies are still lower than the traditional Ru complexes. The reasons including that organic dyes usually demonstrate lower V_{oc}, and there is a lack of panchromatic organic sensitizers that show high conversion efficiency.[4a] In the next section, we will introduce the strategies to further improve the efficiency of organic dyes based on the molecular modification of D-π-A sensitizers.

11.4 The Third Stage of Development: Molecular Modification of D-π-A Structural Dyes

11.4.1 D-D-π-A or D-π(D)-A Configuration

In the second stage of the development of organic sensitizers, much effort was made to broaden the absorption spectra of the sensitizers by inserting more conjugate linkers. However, the efficiency improvement was not salient and sometimes the conversion efficiency became even lower compared with the original sensitizers. The low efficiency was ascribed to the aggravation of sensitizer aggregation as the molecular length was extended. Generally, sensitizers with two conjugation units (such as thiophene, phenyl, pyrrole, thienothiophene, dithieno[3,2-*b*;2′,3′-*d*]thienyl) show high efficiency. Further insertion of conjugate units usually results in a decrease of efficiency.[4a]

Tian's group presented a novel strategy to broaden the absorption spectra of the D-π-A structured sensitizers by the development of sensitizers with D-D-π-A structure. Because of the nonplanar structure of the triphenylamine, the introduction of extra donor units on its outside was able to prevent the aggravation of the aggregation due to the extension of the molecular length. By the connection of carbazole units on the outside of the triphenylamine donor unit, (**36** in Scheme 11.10) the absorption spectra of the cells based on sensitizer **36** were broadened and the efficiencies were enhanced compared with the single triphenylamine donor.[28,50] Sun and coworkers introduced alkoxyl groups on the outside of the triphenylamine donor group, sensitizer **37** (Scheme 11.10) showed redshifted absorption spectrum and enhanced efficiency.[51] It is also effective to add the alkoxy groups to the conjugation linker. Chou and coworkers developed a kind of 3, 4-ethylenedioxythiophene bridged sensitizer **38** (Scheme 11.10) with additional electron donor alkoxy group substituted on thiophene linker.[52] The absorption spectrum was effectively extended and solar cell based on **38** showed power-conversion efficiency of 7.3%, compared to 5.2% for sensitizer without alkoxyl-substituent. With alkoxyl substituted triarylamine as electron donor, 3,4-ethylenedioxythiophene and thieno-thiophene bridged sensitizer **39** (Scheme 11.10) was synthesized by Wang and coworkers.[53] The absorption spectra were expanded further by the simultaneous substitution of alkoxy donor units on both the linker and triarylamine. Sensitizer **39** showed a conversion efficiency

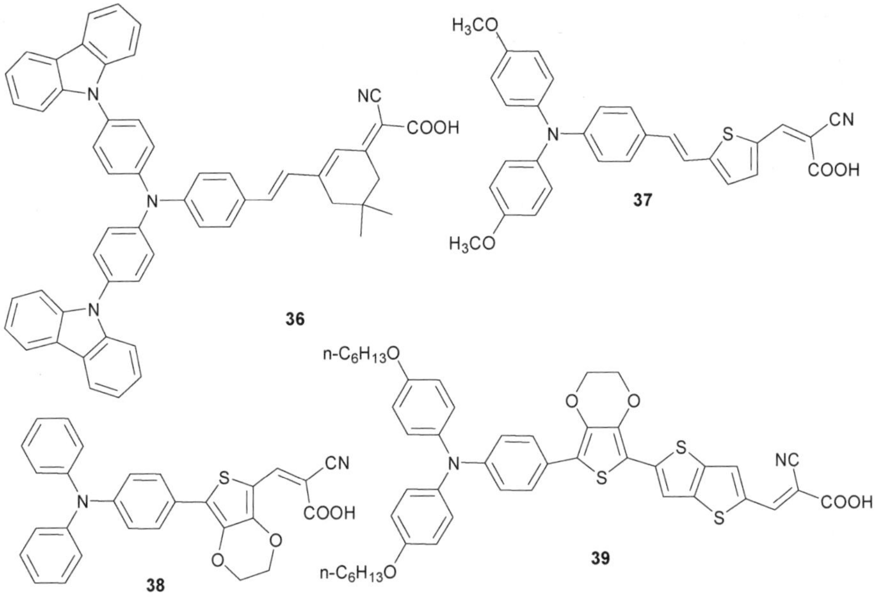

Scheme 11.10

of 9.8%. The adoption of D-D-π-A structure is especially useful for triarylamine sensitizers due to their low HOMO energy level, as discussed in Section 11.4. The addition of extra donor unit can up-lift the HOMO energy edge and reduce the energy waste. In addition to triarylamine sensitizers, the formation of D-D-π-A structure is also applicable for other sensitizers such as carbazole and phenoxazine derivatives. Sun and coworkers synthesized

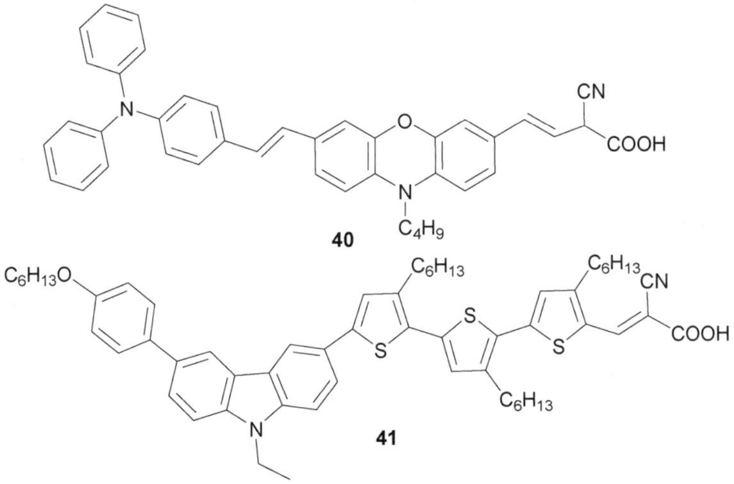

Scheme 11.11

sensitizer **40** (Scheme 11.11) with the connection of additional triphenylamine unit, which showed a significantly enhanced efficiency of 7.7% compared with 8.0% of N719.[35] Hara and coworkers developed a new kind of carbazole dye **41** (Scheme 11.11) with additional alkoxyl donor unit on the carbazole group, the efficiency was effectively improved as well.[54] However, it should be noted that the addition of too strong electron donors on the donor unit of the sensitizer will make the HOMO energy level edge of the sensitizer higher than the redox couple, affecting the sensitizer regeneration and reducing the efficiency.[28]

11.4.2 D-A-π-A and D-π-A-A Configuration

Apart from the addition of an electron-donor unit, some efforts were made to broaden the absorption spectra of the sensitizers *via* the addition of electron acceptor. Tian and coworkers added electron acceptor benzothiadiazole on the conjugate linker of the indoline sensitizer (D-A-π-A structure), which effectively extended the absorption spectra.[55] The efficiency of sensitizer **42** (Scheme 11.12) was as high as 8.7%. Wang and coworkers introduced a cyano group on the linker, the absorption spectrum was significantly broadened and the absorption molar extinction coefficient was enhanced.[56] Sensitizer **43** (Scheme 11.12) showed a higher efficiency than N719 in DSCs with thin TiO_2 film. However, Lin and coworkers observed a decline of the efficiency with the introduction of cyano group on the linker.[57] The different results with the same kind of configuration design might be derived from the different solar-cell fabrication procedures of different research groups. One possible drawback for the addition of electron acceptor in the linker is the reduction of the overlap of the sensitizer LUMO and TiO_2, which might degrade the electron-injection efficiency. Some efforts were made to shift the electron acceptor that was added in the linker close to the anchor unit to improve the electron-injection efficiency. Lin and coworkers reported a new kind of sensitizers with electron acceptor thiazole as linker, which effectively increased the efficiency.[58] Sensitizer **44** (Scheme 11.12) showed conversion efficiency of 6.9% compared

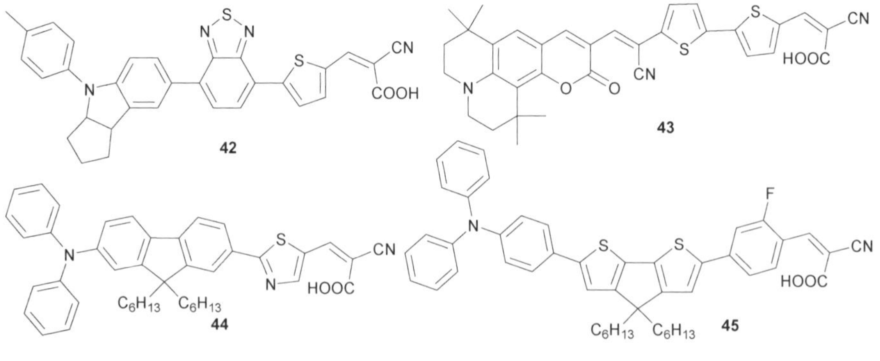

Scheme 11.12

with 7.2% of Ru sensitizer N719. Chou and coworkers investigated sensitizers with fluorine-substituted phenyl as acceptor and linker, which effectively broadened the absorption spectra and increased the efficiency as well.[59] Sensitizer **45** (Scheme 11.12) showed a conversion efficiency of 4.9% for solid-state DSCs. However, similar to the addition of the electron-donor group, the addition of a too strong electron acceptor will lead to a significantly downshifted LUMO energy level edge of the sensitizer, resulting in low electron-injection efficiency and energy-conversion efficiency.[27]

11.4.3 The Introduction of Alkyl Chains

Most organic sensitizers are constituted of donors, linkers, and acceptors, which are usually in the form of a rigid-rod configuration (Fig. 11.2). It can be imagined that the rod-shaped dyes facilitate the formation of a radial arrangement of sensitizer on TiO_2 surface.[4a] As shown in Fig. 11.2, there would be some interspaces formed between these rod-shaped sensitizers on the ball-like TiO_2 nanoparticles. Furthermore, the serious aggregation of the planar rod-shape sensitizer resulted in larger interspaces among the clustered dyes. The interspaces between sensitizers can lead to enhanced charge recombination between TiO_2 and redox couple. Charge recombination can cause the decrease of V_{oc} and the overall efficiency of the cells. In order to reduce the interspaces between sensitizers and cut off the charge recombination channel, flexible alky chains were introduced on the rigid skeleton of the sensitizer to fill in the intermolecular spaces and reduce the close $\pi-\pi$ aggregation between the dyes (Fig. 11.2).

The following are some examples of the effect of the alkyl chains. By linking long alkyl chains to the π-conjugated segment of the carbazole-based dye, Hara and coworkers found that the charge recombination was clearly reduced

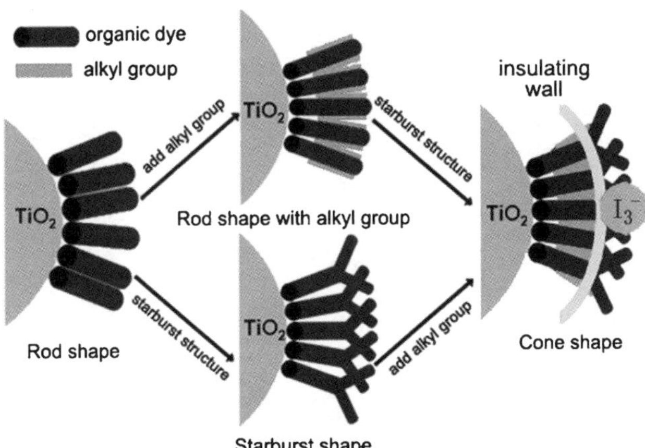

Figure 11.2 Schematic of sensitizers with different configuration adsorbed on TiO_2.

and V_{oc} was improved for a DSC based on sensitizer **46** (Scheme 11.13).[60] Chou and coworkers observed improved V_{oc} by introducing two 2-(2-methoxyethoxy)ethoxy to the linker thiophene unit (**47**) (Scheme 11.13).[52] Chen and coworkers introduced 3,4-propylenedioxythiophene units into triphenylamine-based dyes.[61] Sensitizer **48** (Scheme 11.13) shows a V_{oc} of 0.8 V, higher than the Ru dye N3. Not only on the conjugate linker, but the connection of alkyl chain on the electron donor can reduce charge recombination and enhance V_{oc} as well. Sensitizer **49** (Scheme 11.13) with alkyl chains substituted on triphenylamine donor part significantly enhanced the V_{oc} and the energy-conversion efficiency.[62] Furthermore, it was reported that this dye is quite stable under long-time irradiation. Alkyl chains were added both on donor and linker units in sensitizer **50** (Scheme 11.13),[63] which exhibited an overall conversion efficiency over 10%. For indoline-based sensitizers, dye **51** (Scheme 11.13) with a long alkyl chain effectively suppressed the charge-recombination processes and improved the energy-conversion performance, showing the highest efficiency of 9.4%.[64] Recently, Li and coworkers introduced multialkyl chains on the perylene sensitizers **52** (Scheme 11.13) with clearly improved V_{oc} and efficiency.[65] Long alkyl chains were also found to be useful for other organic dyes such as bisindoylmaleimide (**53**) (Scheme 11.13), [66] and benzothiazole merocyanine (**54**)[67](Scheme 11.13) sensitizers to reduce charge recombination. However, it should be noticed that too many alkyl chains may reduce sensitizer adsorption on the TiO_2 surface.[60]

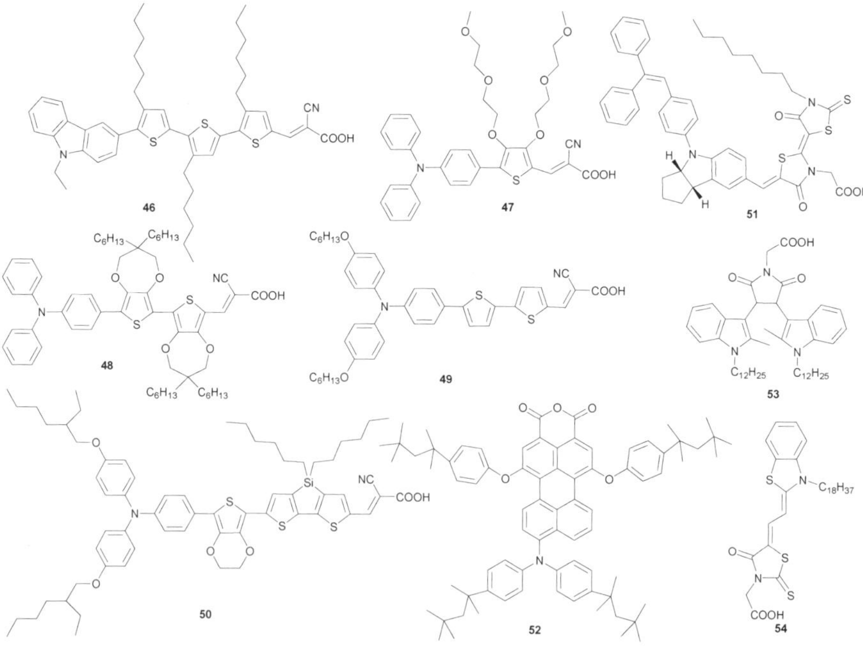

Scheme 11.13

11.4.4 Starburst-Shaped Sensitizer Structure

To reduce the interspaces between sensitizers, Ning and Tian designed and synthesized a novel kind of starburst-structured sensitizer.[4c] As shown in Fig. 11.2, unlike the radial arrangement of rod-shaped sensitizer on TiO_2, the starburst-shaped sensitizer is able to form a more compact sensitizer layer on the TiO_2 surface. Sensitizer **55** (Scheme 11.14) based on starburst triarylamine showed a V_{oc} value higher than sensitizes based on simple triphenylamine.[4c,50] Almost at the same time, Sun's and Lin's groups observed similar results respectively. The V_{oc} of starburst-triarylamine-based sensitizers **56** and **57** (Scheme 11.14) were enhanced compared with the single triarylamine counterpart.[51,68] The starburst triarylamine sensitizer provides a long lifetime of injected electrons in TiO_2, which was proved by transient V_{oc} attenuation experiment. Although sensitizer **56** employed two branches, the total sensitizer adsorption didn't decrease substantially compared with the single-branch counterpart, indicating that the sensitizer layer thus formed was more compact than the rod-like structured sensitizer.[68] **57**-sensitized solar cells yielded an overall energy-conversion efficiency of 7.2%, which was comparable to the performance of classic Ru dye N719. Xue and coworkers synthesized a novel kind of starburst sensitizer **58** (Scheme 11.14) based on truexe, which showed

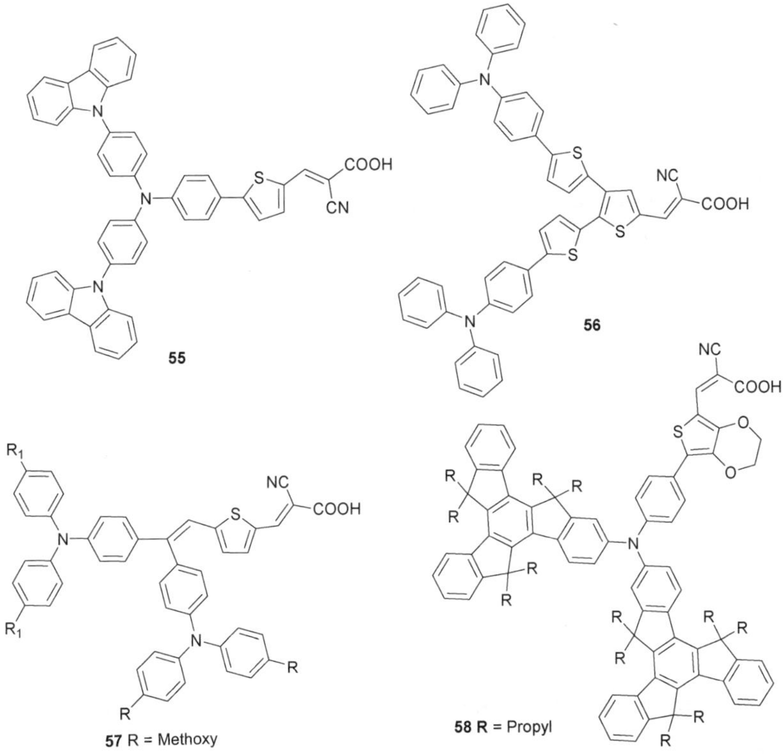

Scheme 11.14

quite high V_{oc} and conversion efficiency as well.[69] Furthermore, Tian and coworkers reported that solar cells based on starburst sensitizers showed much better stability under long-time irradiation than rod-shaped sensitizers, which is important for the future practical application of DSCs. However, not all branched sensitizers lead to enhanced V_{oc}. Similar to rod-shaped sensitizers, a too-long π-conjugation linker results in the decrease of V_{oc} for starburst-shape sensitizers as well. Bäuerle and coworkers developed a branched sensitizer, which showed lower photovoltage than its linear counterpart, implying that there might be other factors associated with the sensitizer configuration that affect the charge recombination. One possible factor is the complexation between sensitizer and redox couple.[70] However, the detailed mechanism is not clear at this time.

11.4.5 Cone-Shaped Sensitizer Structure

Although the starburst-type sensitizer effectively enhanced the V_{oc}, the charge recombination still exists and affects the efficiency of the cells. It was speculated that there would be small interspaces between the starburst sensitizers (Fig. 11.2). In order to further reduce the charge recombination, Tian's group developed the cone-shaped sensitizer **59** (Scheme 11.15) *via* the connection of multialkyl chains on the starburst sensitizer.[71] Sensitizer **59** showed considerably enhanced photovoltage compared with the sensitizer without the alkyl chains. This idea was further proved by the work of Yang and coworkers[72] Sensitizer **60** (Scheme 11.15) showed a V_{oc} of 0.76 V, much higher than the 0.70 V of N719. One drawback of the starburst or cone-shaped sensitizers is that the sensitizer adsorption amount might be reduced due to their big molecular size. However, Grätzel and coworkers reported a novel kind of similar cone-shaped sensitizer **61** (Scheme 11.15), which observed record conversion efficiency of 11%.[73]

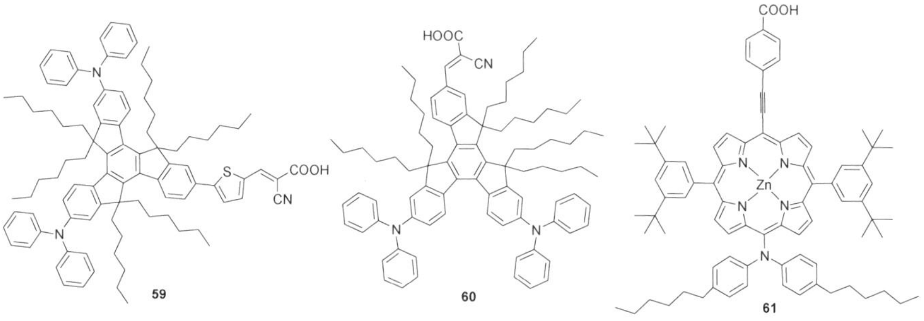

Scheme 11.15

11.4.6 Increasing the Electron-Injection Efficiency of Dyes

Apart from charge recombination, another critical issue of organic dyes that determines the conversion efficiency is the electron-injection efficiency.[4a] An important factor that affects the electron-injection efficiency is the overlap between sensitizer LUMO and TiO_2.[4a] Durrant and coworkers investigated the electron-injection efficiencies of porphyrin sensitizers with conjugated and nonconjugated linker between anchor group and the D-π-A framework.[74] The performance of the sensitizer with a conjugate linker was significantly higher than the one with a nonconjugate linker. This is generally observed in the higher efficiency of the sensitizer with cyanoacrylic acid as acceptor than that with rhodanine as acceptor. Another example is perylene dyes. With the usage of acid anhydride on the perylene derivative skeleton as anchor group (sensitizer **52**), the electron-injection efficiency was significantly improved.[75] However, although there is no conjugation between the anchor group and the D-π-A framework, the sensitizer with co-rhodanine as electron acceptor showed quite high energy-conversion efficiency, indicating it is not necessary to keep the conjugation between the anchor unit and the D-π-A framework.[21] The main factor that decides the electron injection might be the distance between the sensitizer LUMO and TiO_2 surface. Compared with Ru dyes such as N719 or black dye, most organic dyes possess only one anchor group, and the molecular length is usually longer, which determined that the electron overlap between sensitizer and TiO_2 is less efficient than Ru dyes.

Another important factor that affects the electron-injection efficiency is the LUMO energy level edge (E_{LUMO}) of the sensitizer. The higher-energy edge of the LUMO of the sensitizer favored high electron-injection efficiency.[4a] On the other hand, high E_{LUMO} corresponds to deeper electron injection in TiO_2, which can reduce charge recombination. With the adoption of magnesium doped TiO_2, high E_{LUMO} sensitizer triphenylamine cyanoacrylic acid showed a high V_{oc} of 1.0 V, much larger than N719 (0.75 V).[76] While for some panchromatic sensitizers with low LUMO energy edge, the V_{oc} is usually low.[9] Therefore, to improve the electron-injection efficiency, it is beneficial to design sensitizers with high E_{LUMO}. However, the high E_{LUMO} limited the absorption spectra of the sensitizers, which will suppress the efficiency improvement on the other hand. It is worth noting that the energy edges of the HOMO and LUMO of sensitizers can be adjusted *via* molecular modification. Generally, compared with the D-π-A structure, a sensitizer with stronger electron donor or D-D-π-A structure can bathochromically shift the absorption spectra by uplifting the HOMO energy edge,[4a] while the increase of the electron-accepting capability of the sensitizer and adoption of D-π-A-A structure can downshift the LUMO level.[4a]

Another factor that affects the electron-injection efficiency is the non-radiative decay of the sensitizer, which results in serious energy loss. Tian and coworkers investigated the relationship between the emission quantum yield and the electron-injection efficiency of sensitizers.[77] It was found that the electron-injection efficiency was consistent with the luminescence quantum yield of the sensitizer. Since less nonradiative decay guarantees a high luminescence quantum yield, to enhance the electron-injection efficiency, it is thus important to reduce the nonradiative decay that arises mainly from the molecular vibration. The ethylene linkage is susceptible to isomerization upon irradiation, which leads to serious energy loss in the form of vibration. For sensitizers with several ethylene units, the efficiencies are generally low.[4c] Hence, in the sensitizer design, the ethylene linkage is not a good candidate for the conjugate linker. On the other hand, to reduce the molecular vibration, it could be useful to form a compact sensitizer layer on the TiO_2 surface.

11.4.7 The Design of Aggregation-Resistant Dyes

The dye aggregates formed on the semiconductor surface are considered to be one of the major factors accounting for the low photovoltaic performance of DSCs.[4] Aggregation may lead to intermolecular quenching or molecules residing in the system not functionally attached to the TiO_2 surface and thus acting as filters, which will decrease the electron-injection efficiency greatly. In addition, the sensitizer regeneration by the redox couple can also be influenced by molecular aggregation.

To reduce molecular aggregation, one important approach is to develop sensitizers with nonplanar structure. Hence, it is critical to develop donors or linkers with aggregation-resistant character. As we discussed above, triarylamine was a good electron-donor unit that can prevent the molecular aggregation effectively. Furthermore, triarylamine can be connected on other dyes to enhance their aggregation-resistant character. Yeh and coworkers developed a series of aggregation-resistant porphyrin sensitizers such as **62** (Scheme 11.16) by connecting diphenylamine group on the porphyrin unit, which significantly enhanced the efficiency of porphyrin dyes.[78] Li *et al.*

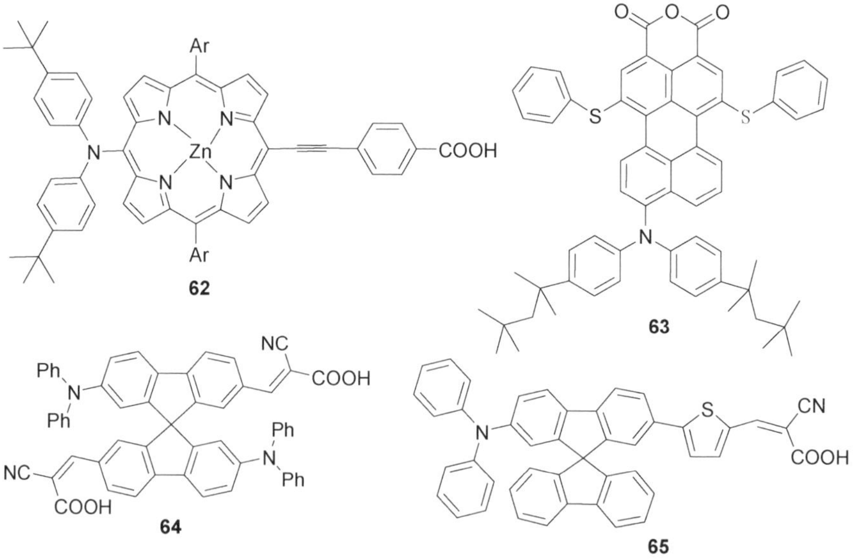

Scheme 11.16

synthesized a novel perylene-based sensitizer **63** (Scheme 11.16) with nonplanar diphenylamine unit on the rear part.[75] Sensitizer **63** achieved the highest efficiency among all perylene dyes. For sensitizers with planar structure susceptible to serious aggregation such as porphyrin and perylene, the introduction of additional nonplanar groups can enhance the efficiency effectively by rendering the dyes with the aggregation-resistant ability. In addition to the donor unit, it is also useful to introduce a conjugate linker with aggregation-resistant character. Spirobifluorene was introduced as conjugate linker for sensitizers **64** and **65** (Scheme 11.16), which significantly reduced the molecular aggregation.[79] However, it should be noted that the aggregation, especially J-type aggregation can broaden the absorption spectra of the sensitizers on TiO_2 film compared with that in solution, which can broaden the IPCE spectra. While for aggregation-resistant sensitizer, the absorption spectra of the sensitizer in solution and on TiO_2 film are almost the same.

11.4.8   Section Summary

At this stage, the efficiencies of organic dyes were significantly improved even to a level comparable with the traditional Ru complexes. One critical issue that contributes to the efficiency improvement is the suppression of the charge recombination. Moreover, the stability of organic dyes was enhanced significantly by forming a more compact sensitizer layer on the TiO_2 surface (starburst and cone-shaped sensitizer configurations and the connection of long alkyl chains). Tian's group did much work in the molecular modification of organic dyes in this stage.[4a] Work in this field was also done by other groups such as L. Sun, P. Wang, J. Ko, K. Hara, and J. Lin. However, it still needs to be noted that the electron-injection efficiency of organic dyes is still lower than Ru dyes, which needs to be enhanced further in the future.

11.5   Conclusions and Prospects

As can be seen above, the development of organic dyes mainly went through three stages. With the development of organic dyes, more and more characters that affect the DSCs performance were found. In summary, the general requirements for sensitizers are shown in Fig. 11.3: (1) A sensitizer should contain the adsorption group such as carboxyl to enable it to be adsorbed on TiO_2; (2) It should possess proper energy edges, including a more positive HOMO energy edge than the redox couple and a more negative LUMO edge than the TiO_2; (3) A sensitizer can form a compact layer on the TiO_2 surface to suppress the charge recombination; (4) It should possess strong molar extinction efficiency and broad absorption spectra in the solar irradiation region; (5) Good electron overlap with TiO_2 to fulfill fast electron injection into TiO_2. (6) Prominent stability under long-time irradiation. Nowadays, the efficiencies of organic dyes are comparable to Ru dyes. In addition, organic dyes show excellent stability under long-time irradiation. In addition to the

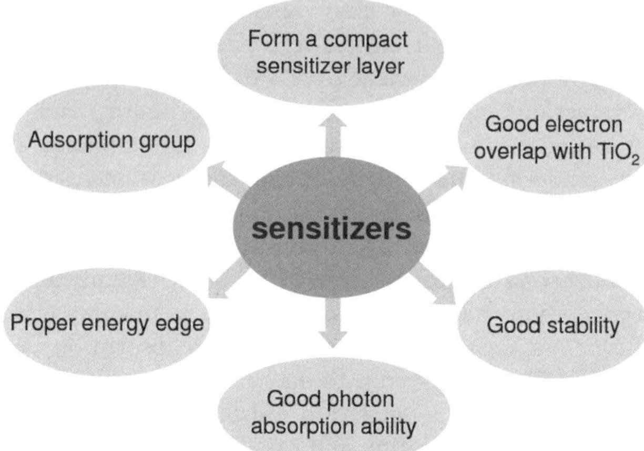

Figure 11.3 General requirements for high-performance sensitizers.

very low cost of organic dyes compared with Ru complexes, they are promising to be explored for the future industrialization of DSCs. However, the performance of the dyes still needs to be enhanced. Novel organic dyes with new donors or acceptors still need to be developed, meanwhile, the molecular modification of the structure and configuration of the sensitizers still need to be optimized.

Acknowledgements

The authors thank the NSFC/China, National Basic Research 973 Program (2011CB808400) and the Scientific Committee of Shanghai for financial support. Thanks also go to all coworkers working on organic solar cells in our laboratory.

References

1. N. S. Lewis, *Science*, 2007, **315**, 798.
2. (a) A. Hagfeldt, G. Boschloo, L. Sun, L. Kloo and H. Pettersson, *Chem. Rev.*, 2010, **110**, 6595. (b) M. Grätzel. *Acc. Chem. Res.*, 2009, **42**, 1788; (c) D. Li, D. Qin, M. Deng, Y. Luo and Q. Meng, *Energy Environ. Sci.*, 2009, **2**, 283.
3. (a) B. O'Regan, and M. Grätzel, *Nature*, 1991, **353**, 737. (b) Md. K. Nazeeruddin, A. Kay, I. Rodicio, R. Humpbry-Baker, E. Müller, P. Liska, N. Vlachopoulos and M. Grätzel. *J. Am. Chem. Soc.*, 1993, **115**, 6382. (c) Y. Chiba, A. Islam, Y. Watanabe, R. Komiya, N. Koide and L. Han, *Jpn. J. Appl. Phys., Part 2*. 2006, **45**, L638.
4. (a) Z. Ning, Y. Fu and H. Tian, *Energy Environ. Sci.*, 2010, **3**, 1170. (b) A. Mishra, M. K. R. Fischer and P. Bäuerle, *Angew. Chem. Int. Ed.*, 2009, **48**, 2474. (c) Z. Ning and H. Tian, *Chem. Commun.*, 2009, 5483.

5. (a) K. Kalyanasundaram, N. Vlachopoulos, V. Krishnan, A. Monnier and M. Grätzel, *J. Phys. Chem.*, 1987, **91**, 2342. (b) N. Vlachopoulos, P. Liska, A. J. McEvoy and M. Grätzel, *Surf. Sci.*, 1987, **189–190**, 823. (c) R. Dabestani, A. J. Bard, A. Campion, M. A. Fox, T. E. Mallouk, S. E. Webber and J. M. White, *J. Phys. Chem.*, 1988, **92**, 1872. (d) G. K. Boschloo and A. Goossens, *J. Phys. Chem.*, 1996, **100**, 19489.

6. S. Cherian and C. C. Wamser, *J. Phys. Chem. B* 2000, **104**, 3624.

7. (a) S. Ferrere, A. Zaban and B. A. Gregg, *J. Phys. Chem. B*, 1997, **101**, 4490. (b) S. Ferrere and B. A. Gregg, *New J. Chem.*, 2002, **26**, 1155.

8. H. Tian, P. Liu, W. Zhu, E. Gao, D. Wu and S. Cai, *J. Mater. Chem.*, 2000, **10**, 2708.

9. (a) L. Zhang, E. Gao, M. Yang, X. Qiao, Y. Hao, S. Cai, F. Meng and H. Tian, *Acta Physico-Chim. Sin.*, 1999, **4**, 293. (b) L. Zhang, M. Yang, E. Gao, X. Qiao, Y. Hao, Y. Wang, S. Cai, F. Meng and H. Tian, *Chem. J. Chinese U.*, 2000, **10**, 1543. (c) H. Tian and F. Meng, *Organic Photovoltaics: Mechanisms, Materials, and Devices*; CRC Press, London, 2005.

10. Z. Wang, F. Li, C. Huang, L. Wang, M. Wei, L. Jin and N. Li, *J. Phys. Chem. B* 2000, **104**, 9676.

11. Z. Wang, F. Li and C. Huang, *Chem. Commun.*, 2000, 2063.

12. Q. Yao, F. Meng, F. Li, H. Tian and C. Huang, *J. Mater. Chem.*, 2003, **13**, 1048.

13. W. Zhan, W. Wu, J. Hua, Y. Jing, F. Meng and H. Tian, *Tetrahed. Lett.*, 2007, **48**, 2461.

14. X. Ma, J. Hua, W. Wu, Y. Jin, F. Meng, W. Zhan and H. Tian, *Tetrahedron*, 2008, **64**, 345.

15. K. Hara, K. Sayama, Y. Ohga, A. Shinpo, S. Suga and H. Arakawa, *Chem. Commun.*, 2001, 569.

16. (a) K. Hara, Z. Wang, T. Sato, A. Furube, R. Katoh, H. Sugihara, Y. Dan-oh, C. Kasada, A. Shinpo and S. Suga, *J. Phys. Chem. B*, 2005, **109**, 15476, (b) K. Hara, T. Sato, R. Katoh, A. Furube, Y. Ohga, A. Shinpo, S. Suga, K. Sayama, H. Sugihara and H. Arakawa, *J. Phys. Chem. B*, 2003, **107**, 597.

17. K. Hara, Y. Dan-oh, C. Kasada, Y. Ohga, A. Shinpo, S. Suga, K. Sayama and H. Arakawa, *Langmuir*, 2004, **20**, 4205.

18. Z. Wang, Y. Cui, Y. Dan-oh, C. Kasada, A. Shinpo and K. Hara, *J. Phys. Chem. C*, 2007, **111**, 7224.

19. Z. Wang, K. Hara, Y. Dan-oh, C. Kasada, A. Shinpo, S. Suga, H. Arakawa and H. Sugihara, *J. Phys. Chem. B*, 2005, **109**, 3907.

20. T. Horiuchi, H. Miura and S. Uchida, *Chem. Commun.*, 2003, 3036.

21. T. Horiuchi, H. Miura, K. Sumioka and S. Uchida, *J. Am. Chem. Soc.*, 2004, **126**, 12218.

22. T. Dentani, Y. Kubota, K. Funabiki, J. Jin, T. Yoshida, H. Minoura, H. Miura and M. Matsui, *New J. Chem.*, 2009, **33**, 93.

23. B. Liu, W. Zhu, Q. Zhang, W. Wu, M. Xu, Z. Ning, Y. Xie and H. Tian, *Chem. Commun.*, 2009, 1766.

24. S. Ito, S. M. Zakeeruddin, R. Humphry-Baker, P. Liska, R. Charvet, P. Comte, M. K. Nazeeruddin, P. Pechy, M. Takata, H. Miura, S. Uchida and M. Grätzel, *Adv. Mater.*, 2006, **18**, 1202.

25. T. Kitamura, M. Ikeda, K. Shigaki, T. Inoue, N. A. Anderson, X. Ai, T. Lian and S. Yanagida, *Chem. Mater.*, 2004, **16**, 1806.

26. D. P. Hagberg, T. Edvinsson, T. Marinado, G. Boschloo, A. Hagfeldt and L. Sun, *Chem. Commun.*, 2006, 2245.

27. H. N. Tian, X. Yang, R. Chen, R. Zhang, A. Hagfeldt and L. Sun, *J. Phys. Chem. C*, 2008, **112**, 11023.

28. Z. Ning, Q. Zhang, W. Wu, H. Pei, B. Liu and H. Tian, *J. Org. Chem.*, 2008, **73**, 3791.

29. M. Liang, W. Xu, F. Cai, P. Chen, B. Peng, J. Chen and Z. Li, *J. Phys. Chem. C*, 2007, **111**, 4465.

30. T. Marinado, D. P. Hagberg, M. Hedlund, T. Edvinsson, E. M. J. Johansson, G. Boschloo, H. Rensmo, T. Brinck, L. C. Sun and A. Hagfeldt, *Phys. Chem. Chem. Phys.* 2009, **11**, 133.

31. K. R. Justin Thomas, J. T. Lin, Y. C. Hsu and K. C. Ho, *Chem. Commun.*, 2005, 4098.

32. (a) S. Kim, J. K. Lee, S. O. Kang, J. Ko, J. H. Yum, S. Fantacci, F. De Angelis, D. Di Censo, M. K. Nazeeruddin and M. Grätzel, *J. Am. Chem. Soc.*, 2006, **128**, 16701. (b) H. Choi, C. Baik, S. Kang, J. Ko, M. Kang, Md. K. Nazeeruddin and M. Grätzel, *Angew. Chem., Int. Ed.*, 2008, **47**, 327.

33. (a) M. Wang, M. Xu, D. Shi, R. Li, F. Gao, G. Zhang, Z. Yi, R. Humphry-Baker, P. Wang, S. M. Zakeeruddin and M. Grätzel, *Adv. Mater.*, 2008, **20**, 4460. (b) H. Qin, S. Wenger, M. Xu, F. Gao, X. Jing, P. Wang, S. M. Zakeeruddin and M. Grätzel, *J. Am. Chem. Soc.*, 2008, **130**, 9202.

34. H. N. Tian, X. Yang, R. Chen, Y. Pan, L. Li, A. Hagfeldt and L. Sun, *Chem. Commun.*, 2007, 3741.

35. H. N. Tian, X. Yang, J. Cong, R. Chen, J. Liu, Y. Hao, A. Hagfeldt and L. Sun, *Chem. Commun.*, 2009, 6288.

36. H. N. Tian, X. Yang, R. Chen, A. Hagfeldt and L. Sun, *Energy Environ. Sci.*, 2009, **2**, 674.

37. W. Wu, J. Yang, J. Hua, J. Tang, L. Zhang, Y. Long and H. Tian, *J. Mater. Chem.*, 2010, **20**, 1772.

38. R. Chen, X. Yang, H. N. Tian, X. Wang, A. Hagfeldt and L. Sun, *Chem. Mater.*, 2007, **19**, 4007

39. R. Chen, X. Yang, H. N. Tian and L. Sun, *J. Photochem. Photobiol., A*, 2007, **189**, 295.

40. Y. Hao, X. Yang, J. Cong, H. N. Tian, A. Hagfeldt and L. Sun. *Chem. Commun.*, 2009, 4031.

41. N. Koumura, Z. Wang, S. Mori, M. Miyashita, E. Suzuki and K. Hara, *J. Am. Chem. Soc.*, 2008, **130**, 4202.

42. D. Kim, J. K. Lee, S. O. Kang and J. Ko, *Tetrahedron*, 2007, **63**, 1913.

43. (a) Y. Ooyama, Y. Shimada, Y. Kagawa, I. Imae and Y. Harima, *Org. Biomol. Chem.*, 2007, 2046. (b) Y. Ooyama, A. Ishii, Y. Kagawa, I. Imae and Y. Harima, *New J. Chem.*, 2007, **31**, 2076.

44. S. Erten-Ela, M. D. Yilmaz, B. Icli, Y. Dede, S. Icli and E. U. Akkaya, *Org. Lett.*, 2008, **10**, 3299.

45. D. Kumaresan, R. Thummel, T. Bura, G. Ulrich and R. Ziessel, *Chem. Eur. J.*, 2009, **15**, 6335.

46. S. Alex, U. Santhosh and S. Das, *J. Photochem. Photobiol. A*, 2005, **172**, 63.

47. A. Burke, S. Ito, H. Snaith, U. Bach, J. Kwiatkowski and M. Grätzel, *Nano Lett.*, 2008, **8**, 977.

48. J. Li, C. Chen, C. Lee, S. Chen, T. Lin, H. Tsai, K. Ho and C. Wu, *Org. Lett.*, 2010, **12**, 5454.

49. (a) K. Hara, T. Sato, R. Katoh, A. Furube, T. Yoshihara, M. Murai, M. Kurashige, S. Ito, A. Shinpo, S. Suga and H. Arakawa, *Adv. Funct. Mater.*, 2005, **15**, 246. (b) S. Tan, J. Zhai, H. Fang, T. Jiu, J. Ge, Y. Li, L. Jiang and D. Zhu, *Chem. Eur. J.*, 2005, **11**, 6272. (c) K. Tanaka, K. Takimiya, T. Otsubo, K. Kawabuchi, S. Kajihara and Y. Harima, *Chem. Lett.*, 2006, **35**, 592.

50. J. Tang, J. Hua, W. Wu, J. Li, Z. Jin, Y. Long and H. Tian, *Energy Environ. Sci.*, 2010, **3**, 1736.

51. D. P. Hagberg, J.-H. Yum, H. Lee, F. De Angelis, T. Marinado, K. M. Karlsson, R. Humphry-Baker, L. Sun, A. Hagfeldt, M. Grätzel and M. K. Nazeeruddin, *J. Am. Chem. Soc.*, 2008, **130**, 6259.

52. W. Liu, I.-C. Wu, C.-H. Lai, C.-H. Lai, P.-T. Chou, Y.-T. Li, C.-L. Chen, Y.-Y. Hsu and Y. Chi, *Chem. Commun.*, 2008, 5152.

53. G. Zhang, H. Bala, Y. Cheng, D. Shi, X. Lv, Q. Yu and P. Wang, *Chem. Commun.*, 2009, 2198.

54. N. Koumura, Z. Wang, M. Miyashita, Y. Uemura, H. Sekiguchi, Y. Cui, A. Mori, S. Mori and K. Hara, *J. Mater. Chem.*, 2009, **19**, 4829.

55. W. Zhu, Y. Wu, S. Wang, W. Li, X. Li, J. Chen, Z. Wang and H. Tian, *Adv. Funct. Mater.*, 2011, **21**, 756.

56. (a) Z.-S. Wang, Y. Cui, K. Hara, Y. Dan-oh, C. Kasada and A. Shinpo, *Adv. Mater.*, 2007, **19**, 1138. (b) Z. Wang, Y. Cui, Y. Dan-oh, C. Kasada, A. Shinpo and K. Hara, *J. Phys. Chem. C*, 2008, **112**, 17011.

57. S. Huang, Y. Hsu, Y. Yen, H. Chou, J. T. Lin, C. Chang, C. Hsu, C. Tsai and D. Yin, *J. Phys. Chem. C*, 2008, **112**, 19739.

58. C. Chen, Y. Hsu, Hs. Chou, K. R. J. Thomas, J. T. Lin and C. Hsu, *Chem. Eur. J.*, 2010, **16**, 3184.

59. D. Chen, Y. Hsu, H. Hsu, B. Chen, Y. Lee, H. Fu, M. Chung, S. Liu, H. Chen, Y. Chi and P. Chou, *Chem. Commun.*, 2010, 5256.

60. Z.-S. Wang, N. Koumura, Y. Cui, M. Takahashi, H. Sekiguchi, A. Mori, T. Kubo, A. Furube and K. Hara, *Chem. Mater.*, 2008, **20**, 3993.

61. Y. Liang, B. Peng, J. Liang, Z. Tao and J. Chen, *Org. Lett.*, 2010, **12**, 1204.

62. J. Yum, D. P. Hagberg, S. Moon, K. Martin Karlsson, T. Marinado, L. Sun, A. Hagfeldt, Md. K. Nazeeruddin and M. Grätzel, *Angew. Chem. Int. Ed.*, 2009, **48**, 1576.

63. G. Zhang, H. Bala, Y. Cheng, D. Shi, X. Lv, Q. Yu and P. Wang, *Chem. Commun.*, 2009, 2198.

64. S. Ito, H. Miura, S. Uchida, M. Takata, K. Sumioka, P. Liska, P. Comte, P. Péchy and M. Grätzel, *Chem. Commun.*, 2008, 5194.

65. T. Edvinsson, C. Li, N. Pschirer, J. Scholneboom, F. Eickemeyer, R. Sens, G. Boschloo, A. Herrmann, K. Müllen and A. Hagfeldt, *J. Phys. Chem. C*, 2007, **111**, 15137.

66. Q. Zhang, Z. Ning, H. Pei and W. Wu, *Front. Chem. China*, 2009, **4**, 269.

67. K. Sayama, S. Tsukagoshi, K. Hara, Y. Ohga, A. Shinpou, Y. Abe, S. Suga and H. Arakawa, *J. Phys. Chem. B*, 2002, **106**, 1363.

68. K. R. Justin Thomas, Y. Hsu, J. T. Lin, K. Lee, K. Ho, C. Lai, Y. Cheng and P. Chou, *Chem. Mater.*, 2008, **20**, 1830.

69. M. Lu, M. Liang, H. Han, Z. Sun and S. Xue, *J. Phys. Chem. C*, 2011, **115**, 274.

70. M. K. R. Fischer, S. Wenger, M. Wang, A. Mishra, S. M. Zakeeruddin, M. Grätzel and P. Bäuerle, *Chem. Mater.*, 2010, **22**, 1836.

71. Z. Ning, Q. Zhang, H. Pei, J. Luan, C. Lu, Y. Cui and H. Tian, *J. Phys. Chem. C*, 2009, **113**, 10307.

72. S. Lin, Y. Hsu, J. Lin, C. Lin and J. Yang, *J. Org. Chem.*, 2010, **75**, 7877.

73. T. Bessho, S. M. Zakeeruddin, C. Yeh, E. W. Diau and M. Grätzel, *Angew. Chem. Int. Ed.*, 2010, **49**, 6646.

74. T. Dos Santos, A. Morandeira, S. Koops, A. J. Mozer, G. Tsekouras, Y. Dong, P. Wagner, G. Wallace, J. C. Earles, K. C. Gordon, D. Officer and J. R. Durrant, *J. Phys. Chem. C*, 2010, **114**, 3276.

75. (a) C. Li, J. Yum, S. Moon, A. Herrmann, F. Eickemeyer, N. G. Pschirer, P. Erk, J. Schöneboom, K. Müllen, M. Grätzel and M. K. Nazeeruddin, *ChemSusChem*, 2008, **1**, 615. (b) H. Imahori, T. Umeyama and S. Ito, 2009, **42**, 1809.

76. S. Iwamoto, Y. Sazanami, M. Inoue, T. Inoue, T. Hoshi, K. Shigaki, M. Kaneko and A. Maenosono, *ChemSusChem*, 2008, **1**, 401.

77. (a) Z. Ning, Q. Zhang, W. Wu and H. Tian, *J. Organomet. Chem.*, 2009, **694**, 2705. (b) S. Qu, W. Wu, J. Hua, C. Kong, Y. Long and H. Tian, *J. Phys. Chem. C* 2010, **114**, 1343.

78. C. Lee, H. Lu, C. Lan, Y. Huang, Y. Liang, W. Yen, Y. Liu, Y. Lin, E. W. Diau and C. Yeh, *Chem. Eur. J.*, 2009, **15**, 1403.

79. (a) D. Heredia, J. Natera, M. Gervaldo, L. Otero, F. Fungo, C. Lin and K. Wong, *Org. Lett.*, 2010, **12**, 12. (b) N. Ch, H. Choi, D. Kim, K. Song, M. Kang, S. O. Kang and J. Ko, *Tetrahedron*, 2009, **65**, 6236.

Conjugated Polymer Nanoparticles: Applications in Optoelectronics, Bioimaging and Biosensing

YUQIONG LI[1], KAI LI[2] AND BIN LIU*[1,2]

[1] Institute of Materials Research and Engineering, 3 Research Link, Singapore 117602; [2] Department of Chemical and Biomolecular Engineering, 4 Engineering Drive 4, National University of Singapore, Singapore 117576
*E-mail: cheliub@nus.edu.sg

12.1 Introduction

Conjugated polymers (CPs) are macromolecules with polymeric backbones composed of double or triple bonds of carbon linked by single bonds. This highly delocalized π-conjugated feature gives rise to the semiconductor characteristic of conjugated polymers. The electro- and photoluminescent characteristics of conjugated polymers can be fine tuned to satisfy specific requirements through molecular design. Up to now, CPs have become viable materials of choice for the construction of optoelectronic devices such as photovoltaic cells (PVs), organic light-emitting diodes (OLEDs), light-emitting electrochemical cells (LECs) and organic field effect transistors (OFETs).[1] In these devices, CPs serve as active layers, charge-injection layers or buffer layers for surface modification. The dimensions of these CP phase domains should be carefully controlled to achieve optimal device performance.[2] However, precise

RSC Polymer Chemistry Series No. 2
Molecular Design and Applications of Photofunctional Polymers and Materials
Edited by Wai-Yeung Wong and Alaa S Abd-El-Aziz
© The Royal Society of Chemistry 2012
Published by the Royal Society of Chemistry, www.rsc.org

control of CP microstructure and morphology of the bulk polymeric materials remain a challenge.

In addition to the construction of optoelectronic devices, CPs have also found promising applications in biosensing and cellular imaging. This generally requires CPs to have water solubility that is often achieved through sophisticated synthetic procedures. Recently, conjugated polymer nanoparticles (CPNs) in submicrometre dimensions with distinguished characteristics have emerged as a promising platform to overcome the limitations of bulk and discrete materials in optoelectronic and bioimaging applications. In this chapter, we summarize different strategies for CPN preparation. As the performance of CPNs in applications is highly dependent on their properties, the electronic conductivity, as well as electro- and photoluminescent properties of CPNs are further discussed. Finally, recent advances in applications of CPNs in optoelectronics, cellular imaging and biosensors are reviewed.

12.2 Preparation of Conjugated Polymer Nanoparticle (CPNs)

12.2.1 Direct Synthesis of CPNs from Monomers

Direct polymerization of various monomers affords a broad scope of nanoparticles in terms of size control and particle chemical structure.[3] During the past decades, nanoparticle dispersions of various types of conjugated polymers such as polyaniline, poly(p-phenylenevinylene) (PPV) and poly(phenyleneethylene) (PPE) have been explored through direct synthesis from the monomers.[4]

Polyaniline has been actively studied due to its environmental stability and ease of preparation.[5] To produce polyaniline colloids, Stejskal and Irina oxidized aniline hydrochloride with ammonium peroxy-disulfate in the presence of a stabilizer, poly(N-vinylpyrrolidone) (PVP). The polymerization of aniline was conducted at 20 °C in aqueous solution containing ammonium peroxy-disulfate. The reaction was finished in several minutes to yield CPNs with size of 241 ± 50 nm.[6] More recently, the stabilizer- and surfactant-free polyaniline nanoparticle dispersions have also been prepared by oxidative polymerization of boronic acid-substituted aniline in the presence of fluoride in 0.1 M phosphoric acid.[7] The formed polyaniline (**P1**) nanoparticles' size ranged between 25–50 nm and showed good colloidal stability due to the formation of a boron–phosphate complex (Chart 12.1).

A recent piece of work done by Kwan's group involves the synthesis of cyanovinylene-backboned PPV derivative (cv-PPV, **P2**) CPN as *in vivo* nanoprobes (Chart 12.1).[8] Surfactant-stabilized near-infrared (NIR) cv-PPV CPNs were directly synthesized from 2,5-bis(octyloxy)terephthalaldehyde and *p*-xylylene dicyanide *via in situ* colloidal polymerization in aqueous phase without using any organic solvents. The hydrophobic monomers were first dissolved in a neat liquid surfactant (Tween 80) and then micellized in water to

P1

P2

allow base-catalyzed Knoevenagel polymerization to occur in the nanoscopic nonpolar interior of the micelles. The obtained CPNs particle sizes ranged between 5–50 nm with an intense emission tail above 700 nm in water. It should be noted that the resulting fluorescent color of the CPN could be readily tuned throughout the broad spectral window by varying the aromatic structure of the monomers.

Poly(arylene ethynylene) (PAE) is one type of the most important CPs developed for biological imaging so far. Sonogashira[9] and Glaser[3] coupling polymerizations are the two key methods used in step-growth preparation of PAE NPs. A standard $(Ph_3P)_2PdCl_2/CuI$-catalyzed Sonogashira coupling with dibromo- and diethynyl-substituted benzenes and fluorenes in aqueous miniemulsion can lead to 60 – 120 nm size PAEs of molecular weights around 4×10^4 to 2×10^5 g mole^{-1} with solid contents up to 15 wt% through terminal alkynes and aryl or vinyl halides reactions. Incorporation of diethynyl pyrrolo-pyrrole or diethynyl-fluorenone in the miniemulsion polymerization can further tune the emission color from blue to orange due to intermolecular and intramolecular energy transfers to acceptors. Glaser coupling as another step-growth reaction yields PAEs of molecular weights around 10^4 to 10^5 g mole^{-1} with size less than 30 nm. In addition, Mecking and coworkers reported nanoparticles with tunable emission wavelengths synthesized from a mixture of monomers.[3] With incorporation of 0.1–2 mole% of perylene or 2–9 mole% of fluorenone dyes during polymerization, the emission color of obtained nanoparticles can be fine tuned to cover the spectrum range from bluish-green to red. A few reported issues limiting high conversion yield of these A-B-type polymerization reactions include the importance of strict stoichiometric balancing for obtaining the right molecular weights, stability of the polymer colloids in suspension, and reaction catalyst stability in the reaction medium.

12.2.2 Synthesis of CPNs *via* Precipitation

Precipitation, also known as reprecipitation or nanoprecipitation, involves a dilute hydrophobic CP solution in a good organic solvent, *e.g.* tetrahydrofuran (THF), added rapidly into an excess volume of a poor solvent, *e.g.* water, that is miscible with the good solvent.[10] When the polymer solution in organic solvent is injected into water, rapid mixing of the good and poor solvent causes a sudden decrease of polymer solubility. As a result, the polymer chains tend to avoid contact with water and form spherical hydrophobic aggregates (see

Figure 12.1). Sonication is always used to facilitate the formation of nanoparticles during mixing. Upon removal of the organic solvent *via* evaporation, the CPNs in water dispersion can be obtained for further studies. This method generally does not involve the use of excess surfactants or stabilizers and can be widely applied to the preparation of various hydrophobic CPNs that have good solubility in organic solvents. Moreover, the sizes of CPNs can be controlled through tuning the molecular weight and concentration of polymers used for nanoparticle preparation.

Moon *et al.* reported the synthesis of nanoparticles from PPE derivatives (**P3**) through phase inversion precipitation (Chart 12.2).[11] The polymers were dissolved in dimethyl sulfoxide (DMSO) and subsequently injected into SSPE (saline, sodium phosphate, EDTA) buffer to yield CPNs with a mean size of 400–500 nm determined by dynamic light scattering and 500–800 nm determined by transmission electron microscope (TEM). The large particle size should be attributed to the incorporation of bulky pentiptycene groups in the polymer backbones, which facilitates π–π stacking during nanoparticle formation. The authors suggested that the nature and density of the hydrophilic side groups affected particle formation. The side chains with protonated amine and short nonionic diethylene oxide moieties could stabilize the particle surface and prevent them from aggregating and precipitating. In 2007, the same group reported poly(phenylene-ethylene) (PPE) (**P4**)-based nanoparticles with a size of 97 nm through modification of polymer-backbone and side-chain structures.[12] Sequential ultrafiltration with acetic acid, ethylenediaminetetraacetic acid (EDTA), and water were used for the fabrication of stable nanometre-sized particles. The presence of acetic acid reduced PPE aggregation in the CPNs by generating repulsive forces.

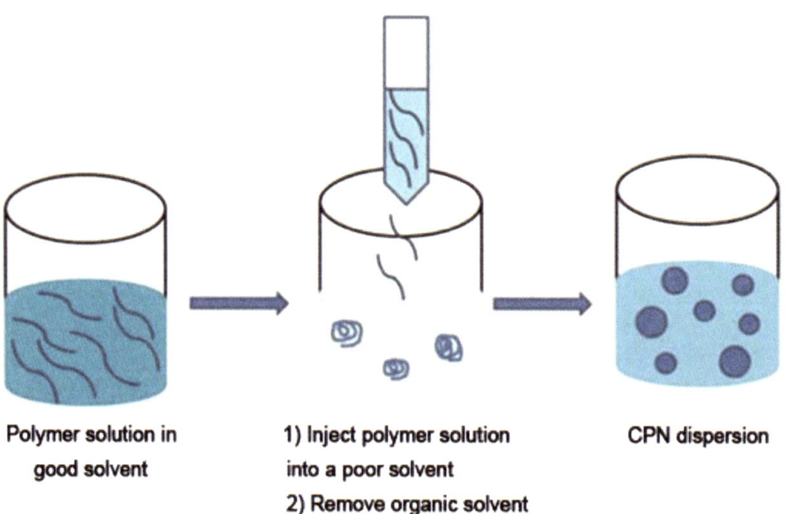

Figure 12.1   Precipitation method for CPN preparation (Copyright 2010 The Royal Society of Chemistry).

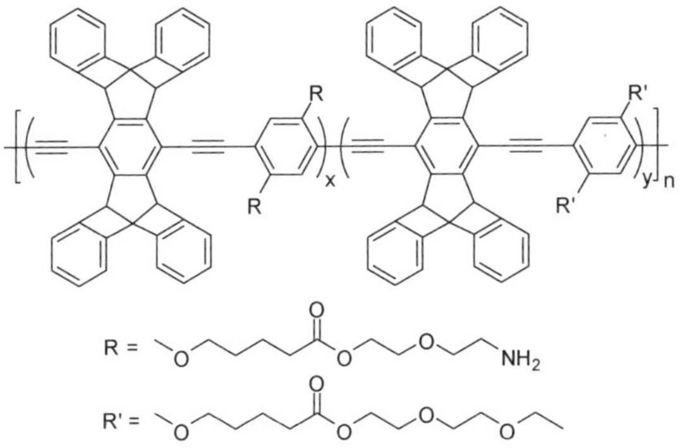

R =

R' =

P3

R =

R' =

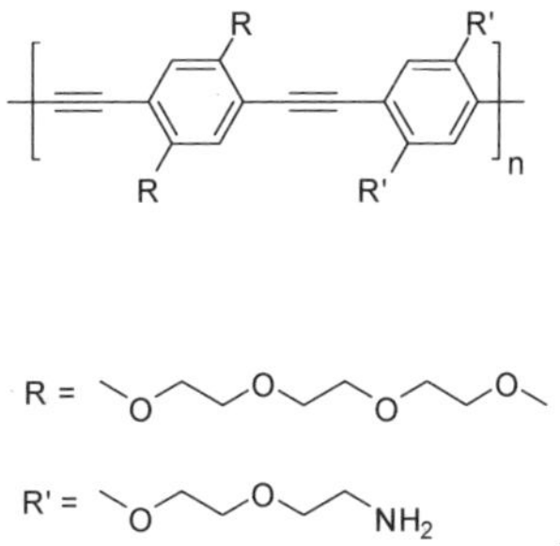

P4

Wu *et al.* also reported a general strategy to prepare CPNs through precipitation.[10] In this study, the authors used three different CPs, including polyfluorene (PF, **P5**), poly(fluorenyldivinylene) derivative (PFPV, **P6**) and poly(phenylenevinylene) PPV derivative (MEH-PPV, **P7**) (see Chart 12.3). The polymers were dissolved in THF with very low concentration and quickly added to an excess amount of water. The obtained CPNs had small particle sizes in the range of 7–14 nm, depending on the polymer species. The CPN dispersions were transparent and stable for weeks without aggregation or decomposition. Besides the bare CPNs without surface functional groups, the authors further developed a strategy to fabricate carboxyl-functionalized CPNs with size of 10 ± 3 nm through precipitation of a mixture of poly(fluorene-*alt*-benzothiadiazole) (PFBT, **P8**) and an amphiphilic copolymer, polystyrene-polyethylene glycol-COOH (PS-PEG-COOH).[13] The surface-functionalized CPNs allowed further conjugation with biomolecules through ethyl (dimethylaminopropyl) carbodiimide (EDC) mediated coupling reaction to yield specific imaging probes.

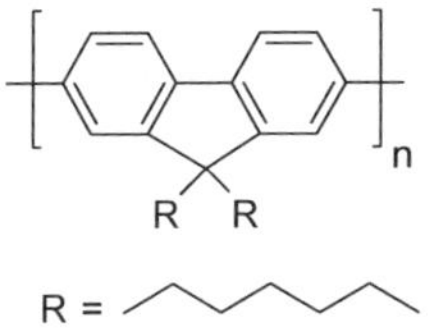

R =

P5

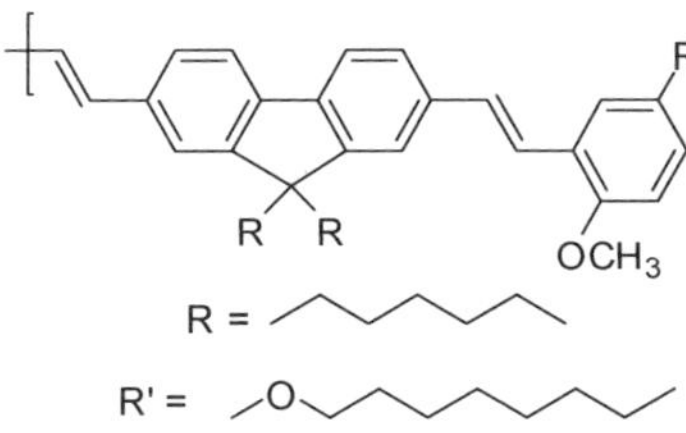

R =

R' =

P6

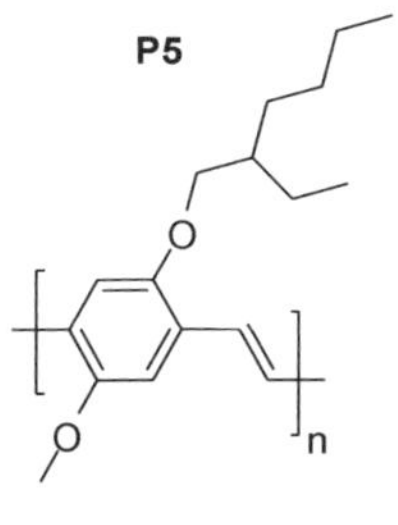

P7

R =

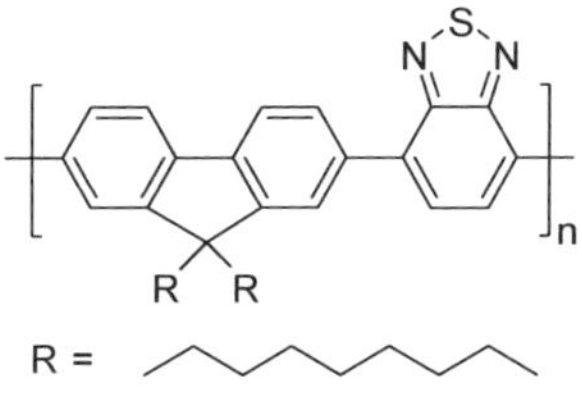

P8

Polyethylene glycol (PEG)-lipid was also employed to coprecipitate with **P8** to yield CPNs with different surface functional groups.[14] The PEG groups were hydrophilic and extended into aqueous medium, while the lipid end interacted with hydrophobic CP. CPNs with various functional groups including carboxyl and biotin were synthesized. These PEG-lipid-coated CPNs could be further concentrated by ultrafiltration up to 625 ppm and their apparent hydrodynamic size of 60 $\pm$ 2 nm was indistinguishable from the size of original unmodified particles (59 $\pm$ 2 nm), indicating no aggregation as a result of condensation. The excellent stability of CPNs in water could be attributed to surface charge contributed by the lipid end groups and PEG on CPN surface.

12.2.3   Synthesis of CPNs *via* Emulsion

Emulsion of CPs is an efficient method to impart surface functionality and solubility to the CPNs. In a typical emulsion process, the CP in a good hydrophobic solvent is mixed with an aqueous phase containing surfactant, followed by sonication to disperse the organic phase into the aqueous phase to form an emulsion. The CPNs in water suspension can be collected after evaporation of the organic solvent. When a polymer encapsulation matrix is introduced into the CP solution, the hydrophobic matrix segments can interact with CP chains to form a stable emulsion in the presence of an emulsifier. By properly choosing the encapsulation matrix or surfactant with desired functional terminal groups, surface-amendable CPNs can be easily synthesized (see Figure 12.2).

Howes *et al.* reported the synthesis of PPV CPNs through an emulsion route.[15] The PPV derivative (BEHP-PPV, **P9**, Chart 12.4) was first dissolved in dichloromethane (DCM) and mixed with an aqueous solution of PEG, followed by sonication. Varying the polymer concentration controlled particle size and an average size of 13 nm with a standard deviation of 5.4 nm was achieved for a 20 ppm BEHP-PPV solution. The same group further reported the synthesis of quantum-dot-sized CPNs in the range of 2 to 5 nm through a

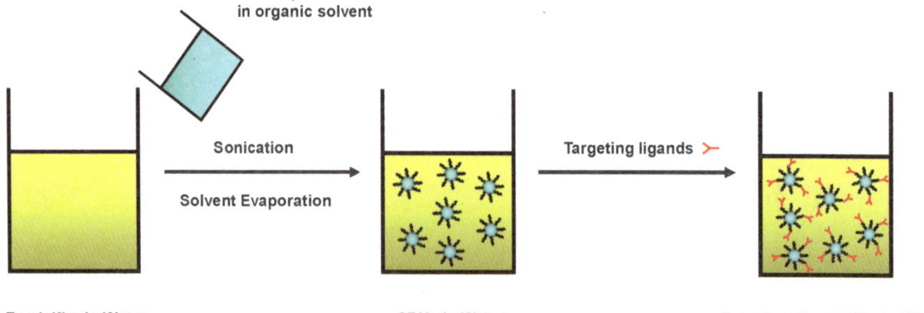

Figure 12.2   Preparation of CPNs using emulsion method (Copyright 2011 Wiley-VCH Verlag Gmbh & Co. KGaA, Weinheim).

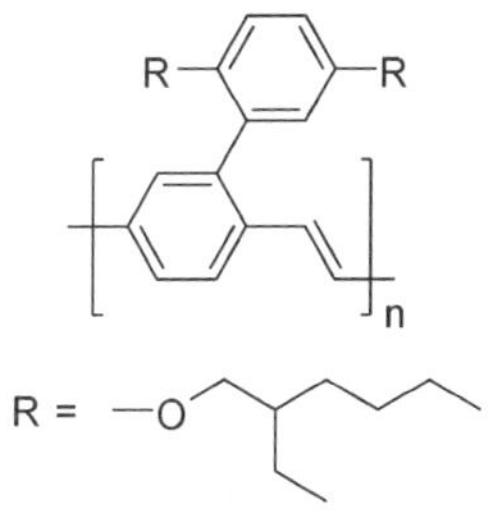

P9

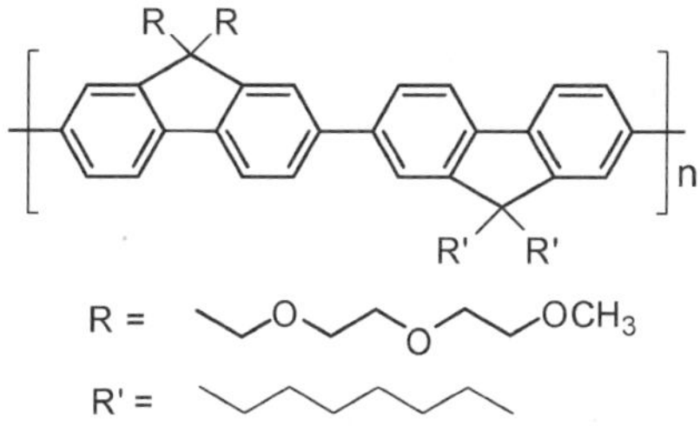

P10

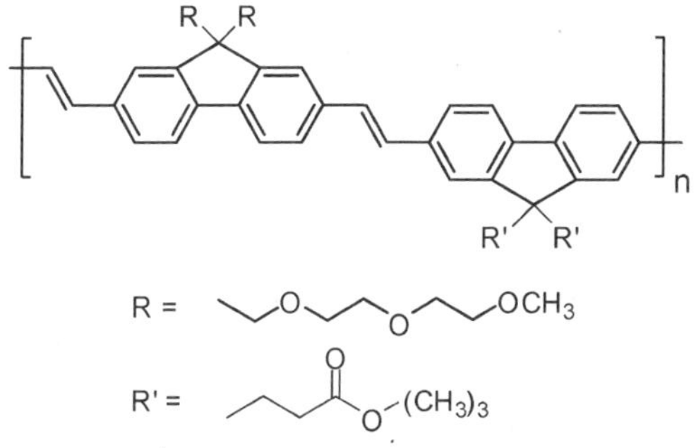

P11

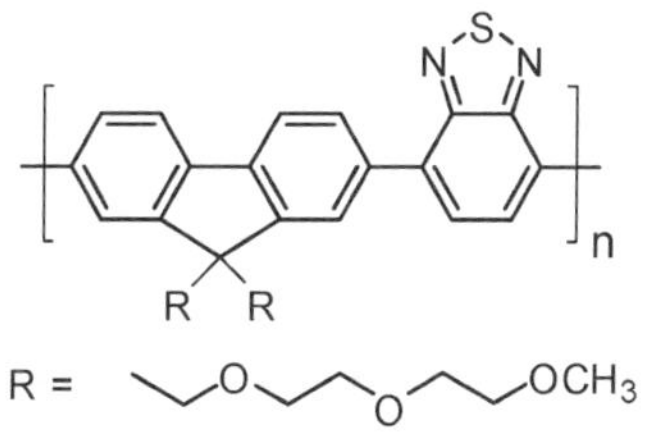

P12

miniemulsion using PEG as emulsifier.[16] Higher concentration of PEG was found to yield a larger amount of CPNs, indicating its important role in nanoparticle formulation. The authors also proved that the CPNs can be capped by PEG-dithiol to further facilitate modification of the fluorescent nanoparticles.

Liu's group[17] implemented the poly(DL-lactide-*co*-glycolide) (PLGA) encapsulation method to fabricate CPNs as a generic strategy towards a series of multicolor CPs (**P7**, **P10**, **P11** and **P12**) covering the emission range from 400 to 700 nm. Preparation of CP-loaded PLGA particles was achieved through a modified solvent extraction/evaporation and single emulsion method by sonication of the CP/PLGA solution in DCM and water containing poly(vinyl alcohol) (PVA) as the emulsifier. The volume-averaged hydrodynamic diameter of CPNs in water ranged between 240 to 270 nm. The resulting surface –COOH groups gave rise to a negative zeta-potential of −35 mV and good colloidal stability. The surface carboxylic acid groups further facilitated the conjugation reaction with aminated folic acid *via* *N*-(3-dimethylaminopropyl)-*N*'-ethylcarbodiimide hydrochloride (EDAC) mediated coupling reaction.

12.2.4 Self-Assembled CPNs

CPNs can self-assemble through both electrostatic and hydrophobic interactions. In addition to fabricating CPNs from hydrophobic CPs, water-soluble

CPs with charged side chains can be mixed with oppositely charged biomolecules or polymers to self-assemble into electrostatic-interaction-induced complexes in water.

Liu's group reported self-assembled CPNs consisting of a complex hybrid between anionic conjugated polyelectrolyte and cationic Arg-Gly-Asp (RGD) terminated peptide by electrostatic interaction (see Figure 12.3).[18] The backbone of the water-soluble conjugated polyelectrolyte (**P13**) contained polyfluorenyldivinylene (PFV) incorporated with 5 mol% of benzothiadiazole (BT), allowing for efficient energy transfer from PFV segments to the BT units upon CPN formation. Laser light scattering studies showed that the average sizes of the complexes ranged between 112 to 209 nm in water dispersions. In addition, the fluorescence of the corresponding CPNs is also tunable from blue to orange-red simply by changing the ratio of the peptide to polymer **P13**. Likewise, Wang and coworkers prepared similar electrostatically assembled CPN systems of positively charged poly(fluorenylene phenylene) (PFP, **P14**) and negatively charged poly(L-glutamic acid) conjugated with anticancer drug molecules, and achieved final particle size of around 50 nm.[19]

Another strategy to synthesize self-assembled CPNs is to introduce a hydrophilic side chain (*e.g.* PEG) to the polymer backbone to yield amphiphilic CPs. Upon transferring into an aqueous phase, the amphiphilic polymers tend to self-assemble into nanoparticles with the hydrophobic CP backbone as the core and the hydrophilic side chains as the shell. Liu's group has developed a novel synthetic strategy to form conjugated polyelectrolytes grafted with PEG side chains *via* alkyne-azide click chemistry.[20] After attaching PEG side chains to poly(fluorenyldivinylene-*alt*-benzothiadiazole) (PFV-BT, **P15**, see Chart 12.5), the PFVBT-*g*-PEG self-assembled into core–shell nanoparticles with a size of approximately 130 nm in water. Similarly, Wang's group used lipid modification to prepare amphiphilic PFPL (**P16**) CPNs.[21] The lipid used was dihexadecyl 2-(4-(2.5-dioxopyrrolidin-1-yl)-4)oxo-butanamido) pentanedioate (lipid-NHS). The formed PFPL CPN has two

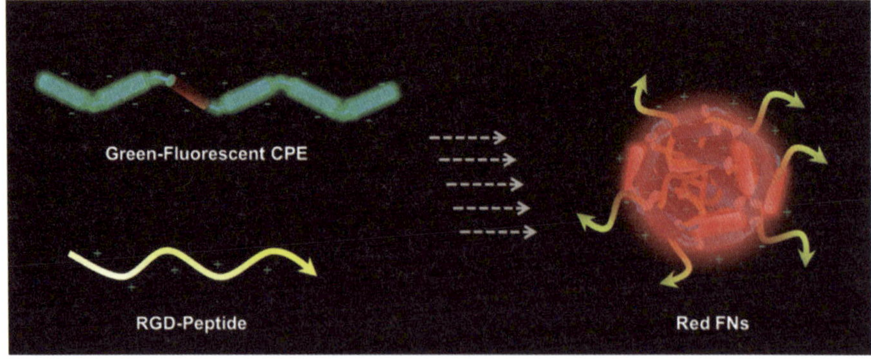

Figure 12.3   Illustration of preparation of CP/peptide nanoparticle (Copyright 2010 American Chemical Society).

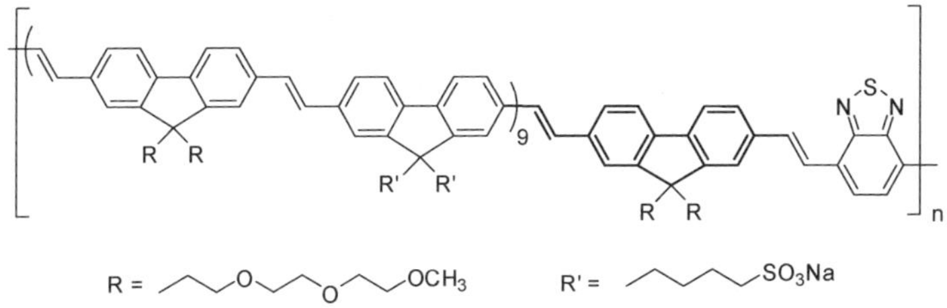

R = (structure with OCH$_3$)   R' = (structure with SO$_3$Na)

P13

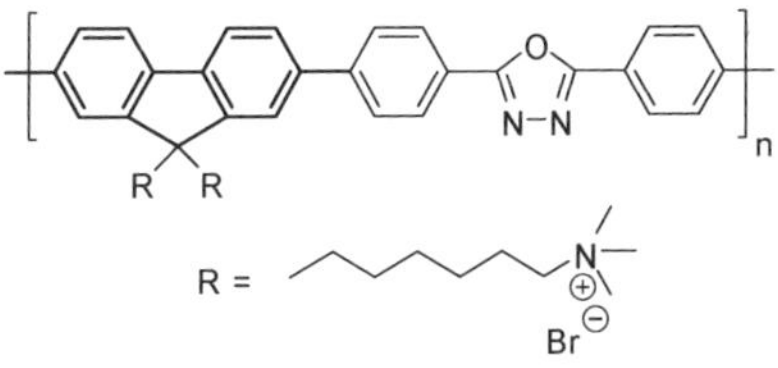

P14

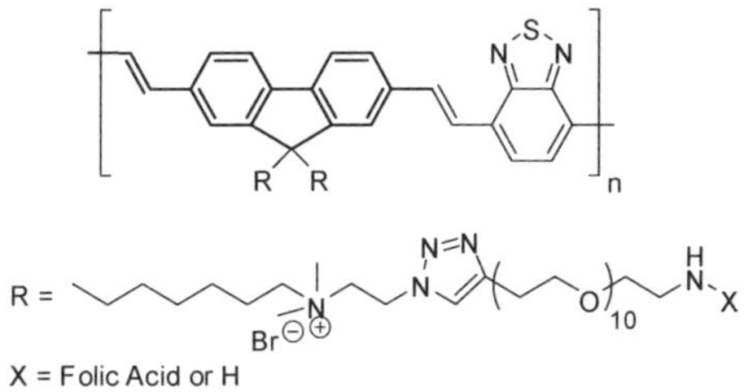

R = (structure with N$^+$, Br$^-$)

P15

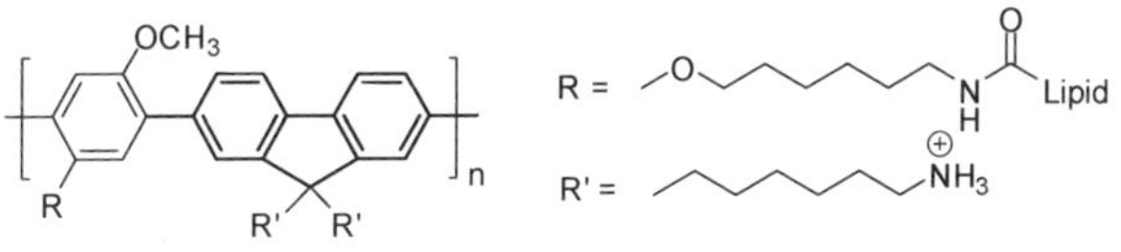

X = Folic Acid or H

P16

R = (structure with Lipid)

R' = (structure with NH$_3^+$)

structural features. First, the lipid is biocompatible and can provide protective layers to CP backbones. Secondly, the ammonium pendant groups can interface with genes that offer PFPL as a vector for gene delivery.

12.2.5 Unimolecular CPNs

Unimolecular nanoparticles of polyhedral oligomeric silsesquioxanes (POSS)-containing conjugated oligoelectrolytes (COE) are part of another category of organic NPs. Liu's group successfully demonstrated the synthesis of unimolecular COE-POSS nanospheres as light-harvesting materials using a bottom-up strategy.[22] The POSS, comprising of single-molecular nanospheres were synthesized from 2-(7-bromo-9,9-bis(6-*N*,*N*,*N*-trimethylammonium)hex-ylfluorenyl)-9,9-bis(6-*N*,*N*,*N*-trimethylammonium) hexylfluorene tetrabromide and commercially available octavinyl POSS *via* Heck coupling reaction. The average size of the obtained COE-POSS was around 3.6 nm, determined from TEM. This result was consistent with the single-molecule size calculated from molecular simulation. Moreover, the steric hindrance of the incorporated rigid POSS effectively prevented the COE arms from close packing and subsequent fluorescence quenching, resulting in a high quantum yield of 80% in aqueous solution. The small size of COE-POSS allowed them to possess whole-cell permeability across a nuclear pore complex to light up the cell nuclei. In a later study, two-photon absorbing (TPA) COE-POSS with fluorene-benzothiadia-zole (BT) monomer as the arm was synthesized to achieve nuclear imaging with high absorption action cross section.[23] The unimolecular CPNs also afford a supreme particle size around 3 nm in diameter with an efficient TPA cross section to facilitate two-photon excited cell nucleus imaging.

Hyperbranched conjugated polyelectrolytes (HCPE) represent another class of unimolecular CPNs that have three-dimensional structures and optical properties. Liu's group successfully synthesized water-soluble fluorescent HCPE (**P17**) with unique double-layered architecture *via* the combination of alkyne polycyclotrimerization and alkyne-azide "click" reaction (see Figure 12.4). The hydrophobic conjugated core with alkyne groups as the end-capper was synthesized from a diyne monomer, 9,9'-bis(6-bromohexyl)-2,7-diethynylfluorene. After quaternization to yield its water-soluble counter-part, azide-functionalized PEG was attached through click reaction to yield **P17**. The obtained HCPE consisted of a rigid conjugated polymer core and hydrophilic PEG shell. Furthermore, **P17** intrinsically formed single-molecular core–shell nanospheres with an average diameter of around 10.7 nm and a narrow size distribution, as determined by TEM. The HCPE nanospheres showed high quantum yields in buffer (30%) and good live-cell membrane permeability. The same group also reported the synthesis of another HCPE (**P18**) from 4-(9,9'-bis(6-bromohexyl)-7-ethynylfluorenyl)-7-ethynylbenzothia-diazole monomer (Chart 12.6), which has surface carboxyl groups for bioconjugation.[24] In water, **P18** had a small size of 32 nm in diameter.

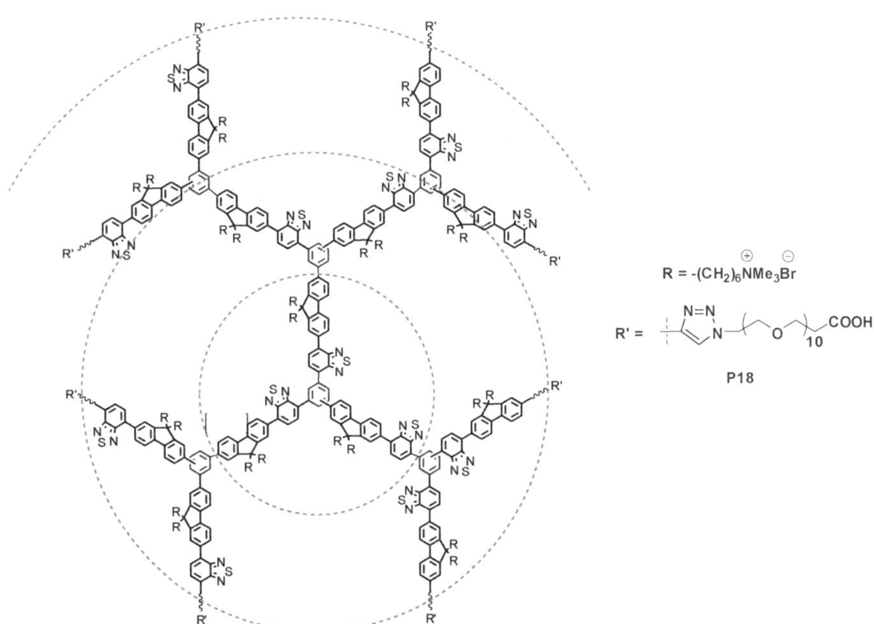

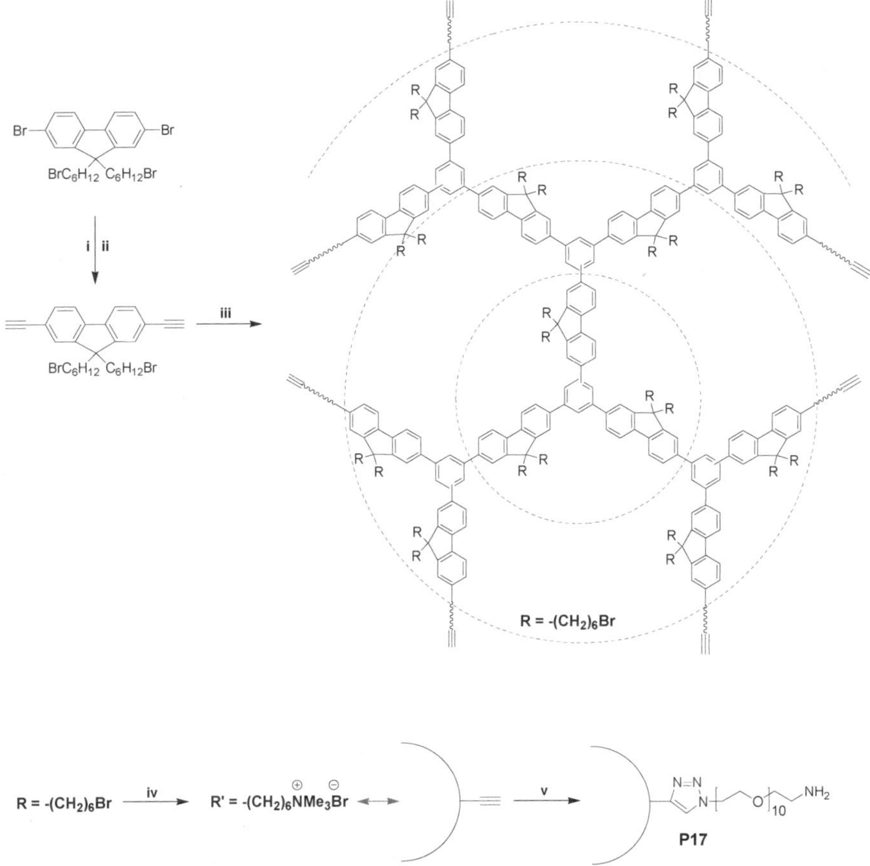

Figure 12.4 Synthetic entry to P17 (Copyright 2009 American Chemical Society).

12.3 Properties of CPNs

12.3.1 Electrical Conductivity

Electrical conductivity of CPNs is of high importance to determine their performance in optoelectronic applications. Generally, conductivity is measured in bulk samples through two typical methods, Van Der Pauw direct current[25] and four-point probe.[26] The samples used in measurements are generally prepared by compressing the dried CPNs to a pellet or spin casting the nanoparticle dispersions to obtain films on substrates.

Various parameters that significantly affect CPN conductivity include the presence of surfactants and other stabilizers, particle morphology and size, crystallinity, as well as molecular weight. Previous studies have shown a decrease in conductivity with an increase in the percentages of insulating surfactants or steric stabilizers for CPNs.[27] In contrast, incorporation of anionic surfactants in polypyrrole and polyaniline nanoparticles was found to enhance their nascent conductivity.[28] This can be attributed to the possibility that the anionic surfactants serve as counterions for cationic sites in CP backbone. Polymerization in miniemulsion has proven to yield CPNs with high crystallinity that is believed to benefit electrical conductivity.[29,30] Polyaniline nanoparticles synthesized with cationic surfactants at low temperatures yielded a conductivity of 85 Siemens/cm in bulk sample. The high conductivity may be attributed to the highly ordered structure of polymer chains as a result of reduced temperature and very small particle size.

12.3.2 Photoluminescent Properties

The photoluminescent properties of CPs are highly dependent on their morphology and chain configurations. As compared to CPs in a good solvent, the emission maxima of CPNs are generally redshifted. This may be attributed to the increased inter- and intrachain interactions favoring energy transfer to low energy and nonfluorescent aggregates upon nanoparticle formation.[31] In contrast, the absorption spectra of CPNs can exhibit either blue or redshifts as compared to the corresponding polymer solutions, depending on the preparation conditions.[32] When CPNs form rapidly (*e.g.* fabrication from precipitation), the formation of relaxed and ordered configuration is impeded, thus always leading to blueshifted absorption maxima. However, CPNs prepared over a more extended period (*e.g.* fabrication from emulsion or self-assembly) always show redshifted absorption maxima because of the formation of highly ordered structures.

The size of CPNs also has significant impact on optical properties. Masuhara and coworkers investigated the size-dependent optical properties of nanoparticles in the range of 40 to 400 nm, prepared from rapid precipitation of a polythiophene derivative (P3DDUT, **P19**, Chart 12.7).[33] The redshifts of about 30 nm and 50 nm in the absorption and emission spectra of the obtained nanoparticles were observed upon increasing the CPN sizes from 40 to 400 nm, respectively. The authors also found that annealing of nanoparticles prepared from rapid precipitation can lead to redshifted absorption maxima, thus indicating the rearrangement of distorted polymer chains to more ordered structures during the annealing process. For instance, the absorption maxima of 90 nm and 145 nm nanoparticles were redshifted by approximately 20 nm after annealing.

When a precipitation method is employed, the sizes of CPNs are highly dependent on the polymer precursor concentrations in solution. Huyal *et al.*

P19

analyzed the size-dependent optical properties of **P5**-based CPNs in suspension and in thin films by drop casting.[34] Nanoparticles with small sizes (5–30 nm) and large sizes (5–70 nm) were obtained by varying the **P5** concentration in solution. The results indicated that the peak of electronic transition of CPN film prepared from small particles was 430 nm, while that prepared from large particles was 429 nm, and thus redshifted in comparison to that in solution (418 nm). The redshifts of emission in CPNs were attributed to the initial collapse of the hydrophobic polymer chains into compact nanostructures. Additionally, **P5** CPNs with 5 to 30 nm sizes, exhibited a higher fluorescence quantum yield (68%) than nanoparticles with larger size (70 nm, 43%), which were higher than that of the bulk thin film of **P5** (23%).

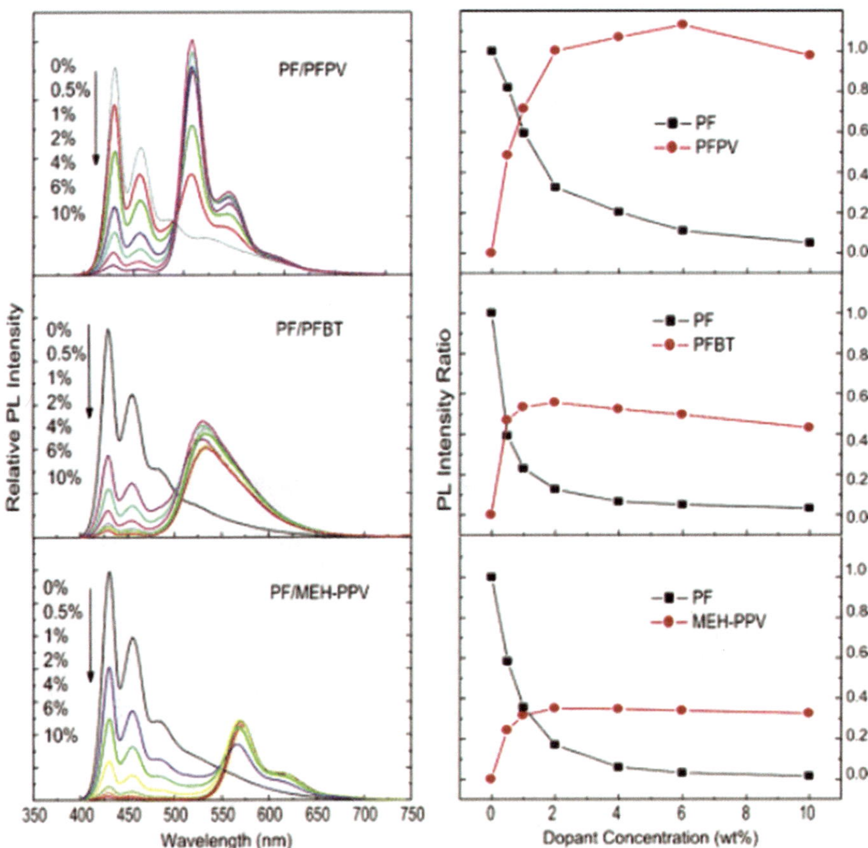

Figure 12.5 (Left) Concentration-dependent fluorescence spectra of polymer blend nanoparticles under 375 nm excitation. (Right) Fluorescence intensity change of P5 host and dopant polymers as a function of dopant concentration in blend nanoparticles. All fluorescence emission intensities were normalized to the 430 nm emission of pure PF nanoparticles (Copyright 2009 American Chemical Society).

Fluorescence resonance energy transfer (FRET) has also been successfully applied in the formation of CPNs to fine tune the emission wavelengths.[35] In 2006, McNeill and coworkers reported the synthesis of CPNs from the blends of different CPs with various bandgaps *via* a precipitation method.[10] A blue-emitting **P5** was used as the energy donor (host), while green-, yellow- and red-emitting **P6**, **P7** and **P8** were used as the energy acceptors (dopants), respectively. Analyses by the Stern–Volmer relation and the nanoparticle energy-transfer model revealed that highly efficient energy transfer occurred from the host to the dopant. Furthermore, the fluorescence spectra changed on changing the weight ratios between host and dopant (see Figure 12.5). The emission intensity of **P5** decreased with increasing dopant concentrations. In the FRET-based CPNs, the emission intensity of **P6**, **P7** and **P8** increased to maxima at 6-wt%, 2-wt% and 2-wt% doping percentage, respectively. These FRET-based CPNs exhibit different emission colors upon simultaneous excitation with a single laser, which holds great potentials for parallel identification of different analytes.

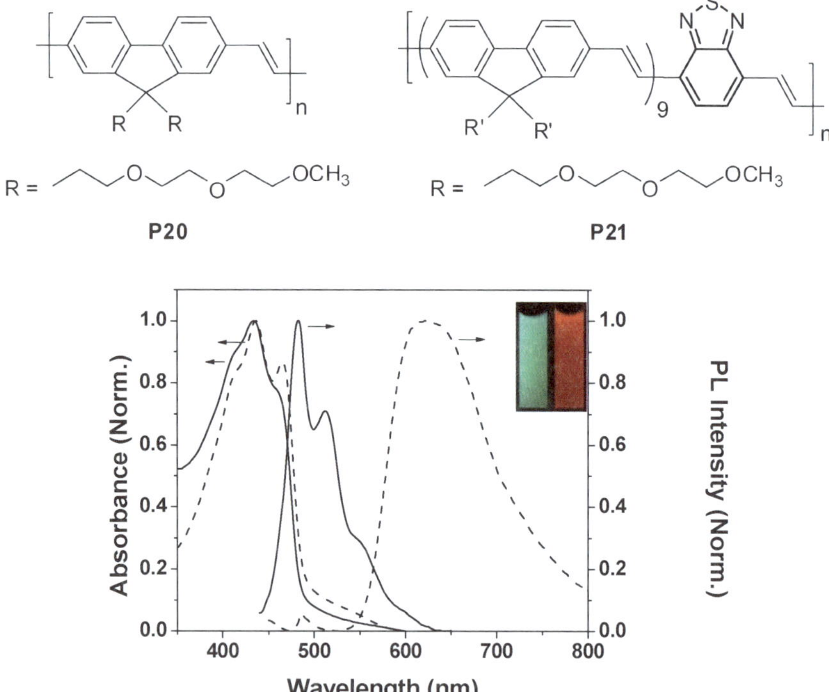

Figure 12.6   UV-visible absorption and PL spectra of surface functionalized PFV CPN (solid) and PFVBT CPN (dashed) suspensions in water at (λ_{ex} = 405 nm). The inset shows the photographs of PFV CPN (left) and PFVBT CPN (right) suspensions under UV excitation at 365 nm (Copyright 2011 American Chemical Society).

On the basis that three-dimensional interchain energy transfer could be more efficient than one-dimensional intrachain transfer,[36] great efforts have been made to synthesize multicolor conjugated polymers with energy donor–acceptor backbones. These polymers exhibit aggregation-enhanced acceptor emission with large apparent Stokes shifts upon excitation of the donor.[37,38] Liu and coworkers synthesized PFV (**P20**) and its derivative (PFVBT, **P21**) containing 10-mol% 2,1,3-benzothiadiazole (BT) (Chart 12.8).[39] Upon nanoparticle formation, efficient interchain or intrachain energy transfer from fluorenylvinylene moieties to BT units occurred to induce significantly redshifted emission for PFVBT CPNs, as compared to PFV CPNs. Meanwhile the UV-visible absorption maxima of PFV CPN and PFVBT CPN in aqueous suspensions are 434 and 436 nm, respectively (see Figure 12.6). The two CPNs possessed similar absorption but different emission properties with small fluorescence signal overlap.[40]

12.4 Applications of CPNs

12.4.1 Optoelectronic Devices

12.4.1.1 Electronic Devices

Efficient light generation has been achieved in inorganic semiconductors with direct bandgaps; however, these are not feasible economically for large-area display. Organic CPs have the main advantage over nonpolymeric semiconductors in their ease of processing to form useful structures. The studies of CPNs for LED fabrication are at a relatively early stage, compared to traditional fabrication directly using CP. Piok *et al.* reported the construction of LED devices using the single layer of methyl-substituted ladder-type poly(p-phenylene) (m-LPPP) nanoparticles.[40] The CPN-based LED was shown to be an improved version of previous devices. They were fabricated using thin films obtained directly from CP, due to the enhanced electron injection from a metallic cathode, resulting from *in situ* formation of a 'stalactite'-type nanostructured aluminum cathode. The same group also reported the fabrication of LED and LECs in planar surface geometry by ink-jet printing.[41] The anode and cathode were made of gold with an interelectrode spacing of 1 µm. The surface LED prepared from MEH-PPV dispersion by ink-jet printing, revealed extremely high current and light emission onset voltages (50–60 V). However, the surface LECs fabricated from their blends with PEG and lithium triflate showed distinctly improved performance with onset voltages slightly above 3 V.

CPNs with average diameters of 50 nm have been used in solar cells.[42] A hole-accepting polymer, poly(9,9-dioctylfluorene-2,7-diyl-*co*-bis-N,N'-(4-butylphenyl)-bis-N,N'-phenyl-1,4-phenylenediamine) (PFB, **P22**, Chart 12.9), and an electron-accepting polymer, **P8**, were used in this study. Different types of CPNs were synthesized using either pure polymer or the mixture of **P22** and

P22

P8. The relationship between external quantum efficiency and layer composition of the CPN-based photovoltaic device was studied. The incident photon to converted electron efficiency (IPCE) of layers prepared from blends of CPNs containing **P22** and **P8** was symmetrical with respect to the compositions. The highest external quantum efficiency of these devices was approximately 4% with a **P22**:**P8** weight ratio of 1:2.

12.4.1.2   *Opals and Photonics*

Other more specialized applications of CPNs in optoelectronics include opals and photonics.[2] Opals are highly regular arrays of monodispersed spherical particles that diffract certain wavelengths of light while permitting other wavelengths to be transmitted. These photonic crystals have inspired interest in new ways to harness and manipulate light waves. Most of the opal structures have been fabricated *via* self-assembly of colloidal silica or polystyrene microspheres. The interstitial voids of the opal template can then be filled with CPNs, followed by selective dissolution and removal of the silica or polystyrene nanocomposite, leaving behind a CPN inverse opal structure.[43] This approach has been successfully used in the preparation of PPV inverse opal films. Alternatively, PEDOT-silica and PEDOT-polystyrene composites have also been prepared by both Han and Foulger[44] and Kelly *et al.*[45] *via* self-assembly into opaline structures. In both cases, the resulting films display photonic bandgaps, electroconductivity and electrochromic switching properties, which are all attributed to the CPN constituent.

12.4.2   Cellular Imaging

Fluorescent tags have been widely implemented as effective visual and sensing tools for cellular and tissue imaging both *in vitro* and *in vivo*. These nanosized photoluminescent reporters have proven to be versatile for various functions including probing subcellular organelles with molecular specificity and targeting surface receptors to decipher cellular physiology. These functions are vital for the understanding of biological processes involved in cells. Traditional types of fluorophores that have emerged in the past decade include small organic dyes, dye-doped nanoparticles and colloidal inorganic semiconductor quantum dots (QDs). Recently, CPNs that are known to possess high fluorescence, good photostability and benign chemical constitution to live cells, have emerged into a novel class of promising alternatives to traditional fluorescent probes.

12.4.2.1 Targeted Cellular Imaging

The initial effort to apply CPNs for cellular imaging was accomplished by Moon *et al.*, who used PPE (**P4**) CPNs with good cell membrane permeability and photostability to achieve long-term live-cell imaging.[12] With subsequent development, many research groups have devised cellular labeling protocols with different CPNs. To achieve targeted cellular imaging, the CPNs are always functionalized with targeting moieties that have high binding affinity to the receptors in target cells. Christensen and coworkers reported targeted delivery of bioconjugated PEG-lipid **P8** CPNs to CD16/32 receptors on the surface of J774A.1 mouse macrophage cells *via* biotin-streptavidin linkage.[14]

Liu's group reported POSS-CP loaded PLGA nanoparticles with tratuzumab (Herceptin) functionalization for HER2-positive cancer cell detection.[46] They further demonstrated that simultaneous discrimination of different live cancer cells under single-wavelength excitation was possible using **P20**- and **P21**-based nanoparticles tagged with antihuman epidermal growth factor receptor 2 (HER2) affibody or RGD peptides to differentiate SKBR-3 breast cancer cells (HER2 overexpression) from HT-29 colon cancer cells (integrin receptor overexpression).[39] On the other hand, Wu *et al.* reported a strategy to fabricate amphiphilic PS-PEG-COOH encapsulated CPNs with surface carboxyl groups for the visualization of cell surface markers (EpCAM) in targeted cancer cell membranes after conjugation with antibodies.[13] The surface functional groups allowed further conjugation with streptavidin or immunoglobulin G (IgG) to yield CPN bioconjugates. The obtained bioconjugates can effectively label cell surface markers (EpCAM) in MCF-7 breast cancer cells with good specificity (see Figure 12.7).

Besides targeting cell membranes, Li *et al.* also reported the evidenced intracellular target detection, filamentous actin imaging in live Hela cells with phalloidin-functionalized HCPE (**P18**-phalloidin).[24] As shown in Figure 12.8, the cell periphery displayed brighter fluorescence than other region of the cells, which indicated specific interactions between HCPE-phalloidin and F-actin concentrated more along the cell periphery than in the cytoplasm of Hela cells.

12.4.2.2 Two-Photon Fluorescence Imaging

Several attempts have been made to explore CPNs in two-photon excited fluorescence (TPEF) imaging. TPEF represents a noninvasive diagnostic procedure for the detection of malignancy in deep tissue organs with reduced tissue photo damage. Moon and coworkers proposed and reported the superior TPEF characteristics of **P4** based CPNs and used them for imaging of

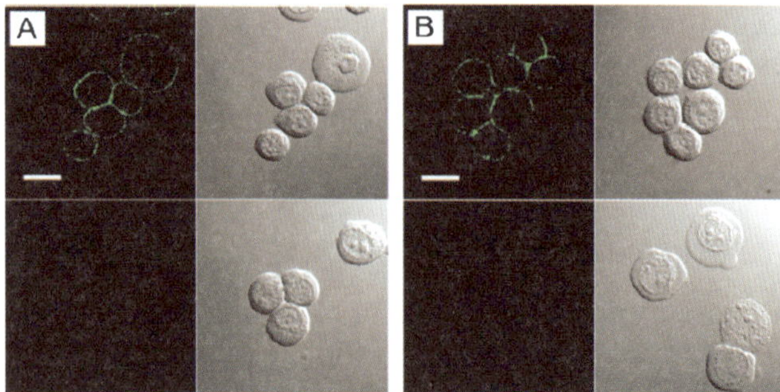

Figure 12.7   Fluorescence imaging of cell surface marker (EpCAM) in human breast cancer cells labeled with CPN bioconjugates. (A) Confocal images of live MCF-7 cells incubated sequentially with anti-EpCAM primary antibodies and CPN-IgG conjugates (top), and CPN-IgG conjugates only (bottom). (B) Confocal images of live MCF-7 cells incubated sequentially with anti-EpCAM primary antibodies, biotinylated goat-anti-mouse IgG secondary antibodies, and Pdot-streptavidin conjugates (top), and without secondary antibodies (bottom). Scale bar represents 20 μm (Copyright 2010 American Chemical Society).

endothelial cells in a tissue culture system. These CPNs exhibited extremely large two-photon cross sections ranging between 1000 and 11 000 GM with a maximum at about 730 nm for a typical particle hydrodynamic diameter of 8 nm and supreme photostability comparable to QDs over 3 months storage at room temperature. Furthermore, the hydrophilicity and nontoxicity of functionalized amine-containing PPE CPNs allowed for long-term monitoring of angiogenesis by endothelial cells, supporting their great potential in advanced biodistribution and histological analyses.[47]

Pecher *et al.* also studied the properties of multicolor PAE nanoparticles through copolymerization with diethynyl pyrrolo-pyrrole or diethynyl

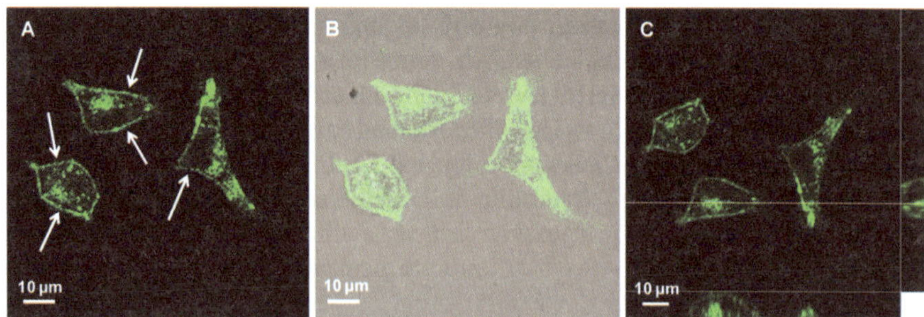

Figure 12.8   Confocal image (A), fluorescence/transmission overlapped image (B) and 3D sectional confocal image (C) of the Hela cells after incubation with 1 μg/mL HCPE-phalloidin in culture medium for 2 h at 37 °C (λ_{ex} = 405 nm, 1 mW laser power) (Copyright 2011 American Chemical Society).

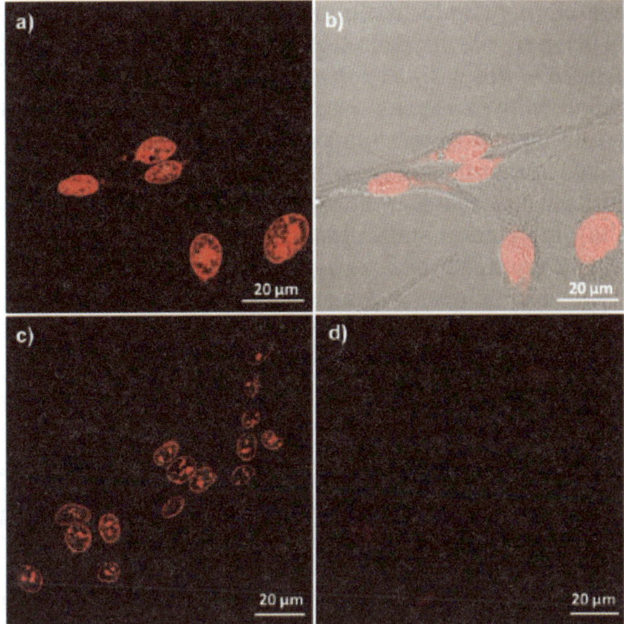

Figure 12.9   (a) OPEF and (b) OPEF/transmission overlapped images of MCF-cells stained with 1 μm COE-POSS. The signals were collected above 560 nm upon excitation at 488 nm. TPEF images of MCF-7 cells incubated with 1 μm COE-POSS (c) or SG (d) for 2 h. The signals are collected above 560 nm upon excitation at 760 nm (Copyright 2010 Wiley-VCH).

fluorenone *via* a direct miniemulsion polymerization method.[9] The multicolor CPNs showed high two-photon cross sections (10^6 to 10^7 GM) and promising potential in two-photon imaging. Likewise, Liu and coworkers studied positively charged COE-POSS of 3 nm size in TPEF application, which has a TPA cross section of 126 GM at 760 nm. The high TPA cross section allowed clear nuclei imaging using two-photon excited fluorescence (OPEF) microscopy with better resolution compared to one-photon excited fluorescence (TPEF) microscopy (see Figure 12.9).[23] The result also proved superior of COE-POSS to conventional commercial dye (SG) in TPEF imaging of cell nuclei.

12.4.2.3   *Magnetic-Fluorescent Nanoparticles for Bimodal Imaging*

Magnetic-fluorescent CP nanoparticles (MF-CPNs) combine the magnetic and optical properties of magnetic particles and CPNs. A wealth of applications emerging from these bimodal nanoparticles can be easily envisaged, including enhanced contrast in magnetic resonance imaging (MRI), simultaneous cell tracking and sorting, and localized delivery by manipulation with an external magnetic field. Howes *et al.* have synthesized such multifunctional particles by encapsulation of hydrophobic CPs and iron oxide nanoparticles in phospholipid micelles.[48] Superparamagnetic iron oxide nanoparticles and CPs were simultaneously encapsulated in PEG-phospholipid micelles by solvent evaporation, forming MF-CPNs of 100–400 nm sizes. The magnetic properties were studied and their application in cellular imaging was demonstrated after incubation with SHSY-5Y cells. These MF-CPNs responded to an external magnetic field while maintaining fluorescence.

12.4.2.4   Near-Infrared Fluorescence Imaging

At present, near-infrared (NIR) dyes such as cyanine and phthalocyanine suffer from several performance limitations comprising of dim fluorescence, self-aggregation, deficiency in surface chemical reactivity for simple modification, and small Strokes shift causing crosstalk between incident excitation light and the emitted fluorescence signals. By encapsulation of a NIR dye 775 into **P8** matrix, the obtained NIR fluorescent CPN emitted at 777 nm with about four times brighter photoluminescent intensity and a narrower emission peak compared to a water-soluble NIR QD that emits at 800 nm (ITK Qdot800).[49] By virtue of the excellent light-harvesting capability of PFBT and the efficient energy transfer from **P8** to NIR775, the NIR CPN dots display enhanced fluorescence signal beneficial for deep tissue NIR imaging. Similarly, Liu and coworkers attempted molecular brush constructs consisting of far-red/near infrared fluorescent conjugated polyelectrolyte **P15** grafted with dense PEG chains, which opened up a unique interrogation window for *in vivo* bioimaging with NIR probes to achieve minimal interferential absorption and deep tissue penetration.[20]

12.4.2.5   Fluorescent Tracking of Drug Release

Simultaneous delivery of disease therapeutics such as anticancer drug moieties and plasmid genes show promise as another major aspect of CPN applications. Wang's group[19] prepared an electrostatic assembly of cationic **P14** CPNs with anionic poly (L-glutamic acid) conjugated with doxorubicin (Dox). The CPN-drug composite displayed fluorescence quenching by Dox *via* electron transfer mechanism. After cellular uptake and hydrolysis of the poly (L-glutamic acid) to release Dox, the fluorescence of **P14** was restored and signified both cell targeting and therapeutic drug delivery. In addition, Liu's group effectively dispatched (PFVBT)-grafted-PEG-COOH (**P23**) nanoparticles with a cisplatin (Pt) anticancer component for concurrent imaging of HepG2 cells and *in vivo* drug distribution in nude mice upon intravenous administration (see Chart 12.10 and Figure 12.10).[50] *In vitro* drug release studies indicated that about 30% of the total loaded cisplatin were released in the initial 6 h. In the period following, the complex continuously released the drug up to 66% in 5 days.

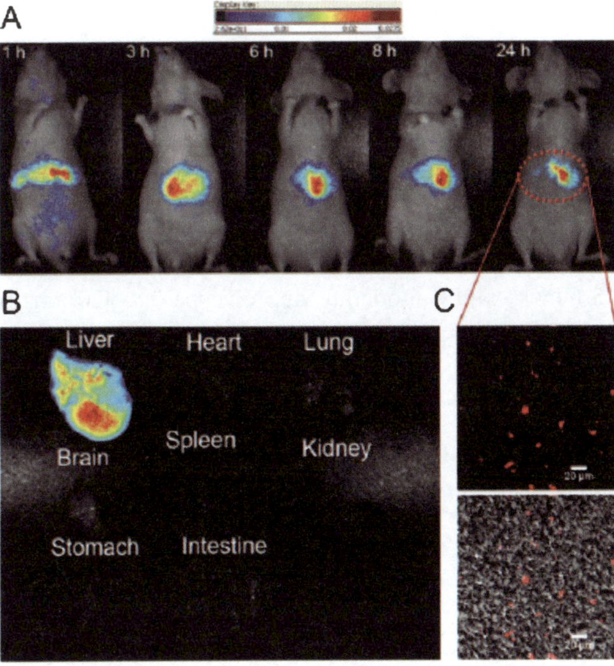

P23

The *in vivo* biodistribution and excretion profile of the **P23**-Pt was studied, indicating the CPNs were accumulated in the liver at 24 h postinjection. The results suggested that **P23**-Pt showed great promises in imaging and chemotherapy of liver cancer.

Figure 12.10 (A) *In vivo* noninvasive fluorescence imaging of a nude mouse after intravenous injection of P23-Pt nanoparticles. (B) Fluorescence imaging of various organs at 24 h postadministration. (C) Confocal images of liver slices at 24 h postadministration. The red fluorescence (upper) from the P23-Pt nanoparticles distributed in the liver tissue is overlaid with the optical image (lower) (Copyright 2011 The Royal Society of Chemistry).

12.4.3   Biosensors

Relying on the FRET platform, Chiu's group designed a CPN-based ratiometric pH probe by coupling a pH-sensitive dye, fluorescein, with a pH-insensitive PPE (**P24**, Chart 12.11).[51] This approach offered a rapid and robust sensor for pH determination. Upon excitation at a single wavelength, the CPN resulted in two emission peaks, one was pH sensitive from fluorescein (greenish yellow) and the other was pH insensitive from **P24** (blue) as an internal reference. The linear range of the designed pH probe was between 5.0 and 8.0, which is suitable for most cellular studies. Intracellular pH in Hela cells was tested following CPN-fluorescein uptake by endocytosis, thus demonstrating its utility in intracellular pH-sensing assays.

Another assay from the same group was designed based on Cu^{2+}- and Fe^{2+}-induced fluorescence quenching of PSMA encapsulated **P8** nanoparticles and their reversible fluorescence restoration by EDTA.[52] The loss of fluorescence from CPNs was attributed to the chelating interactions between the PSMA-COOH groups on the surface of CPNs with Cu^{2+} and Fe^{2+} in solution. Supplementation with EDTA as a copper-chelating agent passivated the metallic ions, which recovered the CPN fluorescence.

In addition to pH and ionic strength measurement, the concentration of molecular oxygen is also a key metabolite in aerobic biological systems and tissue hypoxia related to various tissue pathologies such as retinal diseases and cancer. McNeill and coworkers explored platinum (II) octaethylporphine (PtOEP) as the oxygen sensitive dye, with nanoparticles consisting of polyfluorene derivatives **P5** and **P25** as light-harvesting host materials. An energy transfer efficiency of 87% was determined for 10-wt% of PtOEP-doped **P25** nanoparticles. Similar to the fluorescein-PPE pH ratiometric probe, the PtOEP- **P25** CPNs exhibit moderate fluorescence (*ca.* 420 nm) from the **P25** host and oxygen-dependent phosphorescence (*ca.* 650 nm) from the PtOEP dopant. Oxygen saturation quenches the red fluorescence from the dopant molecule, leaving behind residual blue emission from the host **P25** backbone.[53]

R =

P24   P25

12.5   Conclusions and Future Perspectives

CPNs have emerged as a promising platform to overcome the limitations of bulk and discrete materials in optoelectronic and bioimaging applications. One of the very successful examples is the commercialized PEDOT-PSS CPNs, which have shown high conductivity and have been widely used as hole-transporting materials in various electronics. In addition, the high quantum yield, large photon absorption cross section, and low cytotoxicity of CPNs make them ideal fluorescent probes for biosensor and bioimaging applications. When coupled with therapeutic entities, these photoluminescent probes could also serve as drug carriers for *in vivo* imaging and chemotherapy. Although various strategies have been developed for the synthesis of CPNs, precise control of the nanoparticle size, nanostructure and morphology remains a challenge. Future efforts should be made toward the synthesis of well-controlled CPNs that will allow more detailed studies of the relationship between their size, structure and properties. As the electrical and optical properties of CPNs are largely dependent on the CP structures, rapid technological and scientific progresses in polymer chemistry and materials design will continue to provide fundamental guidelines and new research opportunities for the development of the next generation of CPNs for various applications.

Acknowledgements

The authors are grateful to the Institute of Materials Research and Engineering (IMRE/11-1C0213), National University of Singapore (R279-000-301-646), National Research Foundation (R279-000-323-281), Ministry of Education (R279-000-255-112), and Ministry of Defense (R279-000-301-232 and R279-000-340-232) for financial support.

References

1. C. R. McNeill and N. C. Greenham, *Adv. Mater.*, 2009, **21**, 3840.
2. T. L. Kelly and M. O. Wolf, *Chem. Soc. Rev.*, 2010, **29**, 1526.
3. C. M. Baier, J. Huber and S. Mecking, *J. Am. Chem. Soc.*, 2009, **131**, 14267.
4. J. Pecher and S. Mecking, *Macromolecules*, 2007, **40**, 7733.
5. R. B. Kaner and G. MacDiarmid, *Sci. Am.*, 1998, **258**, 106.
6. J. Stejskal and S. Irina, *Pure Appl. Chem.*, 2005, **77**, 815.
7. B. A. Deore and M. S. Freund, *Macromolecules*, 2009, **42**, 164.
8. S. Kim, C.-K. Lim, J. Na, Y.-D. Lee, K. M. Kim, K. Choi, J. F. Leary and I. C. Kwan, *Chem. Commun.*, 2010, **46**, 1617.
9. J. Pecher, J. Huber, M. Winterhalder, A. Zumbusch and S. Mecking, *Biomacromolecules*, 2010, **11**, 2776.
10. C. Wu, H. Peng, Y. Jiang and J. McNeill, *J. Phys. Chem. B*, 2006, **110**, 14148.
11. J. H. Moon, R. Deans, E. Krueger and L. F. Hancock, *Chem. Commun.*, 2003, 104.

12. J. H. Moon, W. McDaniel, P. MacLean and L. F. Hancock, *Angew. Chem. Int. Ed.*, 2007, **46**, 8223.

13. C. Wu, T. Schneider, M. Leigler, J. Yu, P. G. Schiro, D. R. Burnham, J. D. McNeill and D. T. Chiu, *J. Am. Chem. Soc.*, 2010, **132**, 15410.

14. P. K. Kandel, L. P. Fernando, P. C. Ackroyd and K. A. Christensen, *Nanoscale*, 2011, **3**, 1037.

15. P. Howes, R. Thorogate, M. Green, S. Jickells and B. Daniel, *Chem. Commun.*, 2009, 2490..

16. Z. Hashim, P. Howes and M. Green, *J. Mater. Chem.*, 2011, **21**, 1797.

17. K. Li, J. Pan, S. S. Feng, A. W. Wu, K. Y. Pu, Y. Liu and B. Liu, *Adv. Funct. Mater.*, 2009, **19**, 3535.

18. B. Liu, K. Y. Pu and K. Li, *Chem. Mater.*, 2010, **22**, 6736.

19. X. Feng, F. Lu, L. Liu, H. Tang, C. Xing, Q. Yang and S. Wang, *ACS Appl. Mater. Inter.*, 2010, **8**, 2429.

20. K. Y. Pu, K. Li and B. Liu, *Adv. Funct. Mater.*, 2010, **20**, 1.

21. X. Feng, Y. Tang, X. Duan, L. Liu and S. Wang, *J. Mater. Chem.*, 2010, **20**, 1312.

22. K. Y. Pu, K. Li and B. Liu, *Adv. Mater.*, 2010, **22**, 643.

23. K. Y. Pu, K. Li, X. Zhang and B. Liu, *Adv. Mater.*, 2010, **22**, 4186.

24. K. Li, K. Y. Pu, L. Cai and B. Liu, *Chem. Mater.*, 2011, **23**, 2113.

25. L. T. Van der Pauw, *Philips Research Report*, 1958, **13**, 1.

26. F. M. Smits, Bell System Technology Journal, 1958, 711.

27. R. B. Bjorklund and B. Liedberg, *J. Chem. Soc. Chem. Commun.*, 1986, 1293..

28. M. G. Han, S. K. Cho, S. G. Oh and S. S. Im, *Synth. Met.*, 2002, **126**, 53.

29. F. Yan and G. Xue, *J. Mater. Chem.*, 1999, **9**, 3035.

30. W. Zheng, M. Angelopoulos, A. J. Epstein and A. G. MacDiarmid, *Macromolecules*, 1997, **30**, 2953.

31. C. Szymanski, C. Wu, J. Hooper, M. A. Salazar, A. Perdomo, A. Dukes and J. McNeill, *J. Phys. Chem. B*, 2005, **109**, 8543

32. F. Wang, M. Y. Han, K. Y. Mya, Y. Wang and Y.-H. Lai, *J. Am. Chem. Soc.*, 2005, **127**, 10350.

33. N. Kurokawa, H. Yoshikawa, N. Hirota, K. Hyodo and H. Masuhara, *Chem. Phys. Chem.*, 2004, **5**, 1609.

34. I. O. Huyal, T. Ozel, D. Tuncel and V. Hilmi, *Opt. Exp.*, 2008, **26**, 13391.

35. J. I. Lee, I. N. Kang, D. H. Hwang, H. K. Shim, S. C. Jeong and D. Kim, *Chem. Mater.*, 1996, **8**, 1925.

36. B. J. Schwartz, *Ann. Rev. Phys. Chem.*, 2003, **54**, 141.

37. D. Yu, Y. Zhang and B. Liu, *Macromolecules*, 2008, **41**, 4003.

38. J. Shi, L. Cai, K.-Y. Pu and B. Liu, *Chem. Asian J.*, 2010, **5**, 301.

39. K. Li, R. Zhan, S. S. Feng and B. Liu, *Anal. Chem.*, 2011, **83**, 2125.

40. T. Piok, S. Gamerith, C. Gadermaier, H. Plank, F. P. Wenzl, S. Patil, R. Montenegro, T. Kietzke, D. Nehrer, U. Scherf, K. Landfester and E. J. W. List, *Adv. Mater.*, 2003, **15**, 800.

41. G. Mauthner, K. Landfester, A. Kock, H. Bruckl, M. Kast, C. Stepper and E. J. W. List, *Org. Electron.*, 2008, **9**, 164.
42. T. Kietzke, D. Nehrer, K. Landfester, R. Montenegro, R. Guntner and U. Scherf, *Nature Mater.*, 2003, **2**, 408.
43. M. Deutsch, Y. A. Vlasov and D. J. Norris, *Adv. Mater.*, 2000, **12**, 1176.
44. M. G .Han and S. H. Foulger, *Chem. Commun.*, 2004, 2154..
45. T. L. Kelly, Y. Yamada, S. P. Y. Che, K. Yano and M. O. Wolf, *Adv. Mater.*, 2008, **20**, 2616.
46. K. Li, Y. Liu, K. Y. Pu, S. S. Feng, R. Zhan and B. Liu, *Adv. Funct. Mater.*, 2011, **21**, 287.
47. N. Rahim, W. McDaniel, K. Bardon, S. Srinivasan, V. Vickerman, P. T. C. So and J. H. Moon, *Adv. Mater.*, 2009, **21**, 1
48. P. Howes, M. Green, J. Levitt, K. Suhling and M. Hughes, *J. Am. Chem. Soc.*, 2010, **132**, 3989.
49. Y. Jin, F. Ye, M. Zeigler, C. Wu and D. T. Chiu, *ACS Nano*, 2011, **5**, 1468.
50. D. Ding, K. Li, Z. Zhu, K. Y. Pu, Y. Hu, X. Jiang and B. Liu, *Nanoscale*, 2011, **3**, 1926.
51. Y. H. Chan, C. Wu, F. Ye, Y. Jin, P. B. Smith and D. T. Chiu, *Anal. Chem.*, 2011, **83**, 1448.
52. Y. H. Chan, Y. Jin, C. Wu and D. T. Chiu, *Chem. Commun.*, 2011, **47**, 2820.
53. Z. Tian, J. Yu, C. Wu, C. Szymanski and J. McNeill, *Nanoscale*, 2010, **2**, 1999.

White Light-Emitting Polymers and Devices

BIN ZHANG, WEI YANG* AND HONGBIN WU*

Institute of Polymer Optoelectronic Materials and Devices, State Key Laboratory of Luminescent Materials and Devices, South China University of Technology, Guangzhou 510640, People's Republic of China
*E-mail: pswyang@scut.edu.cn or hbwu@scut.edu.cn

13.1 Introduction

White polymer light-emitting devices (WPLEDs)[1,2] have attracted intense attention due to their potential applications as backlights for liquid-crystal displays as well as lighting sources. Despite the relatively lower efficiency when compared to its thermal vacuum deposition counterparts,[3,4] WPLEDs based on solution processing technology have particular advantages, such as low-cost manufacturing and easy processibility over large-area size by spin coating, ink-jet printing or roll-to-roll printing technology. In general, to achieve white emission devices, various approaches have been reported, including multilayer structures capable of sequential energy transfer that are fabricated by consecutive evaporations of red, green, and blue (RGB) light-emitting compounds,[3] multiple component emissive layers containing an appropriate ratio of RGB phosphorescent or fluorescent dopants,[2,5,6] polymer blends containing RGB-emitting species,[7,8] charge-transfer exciplexes or excimers broad emission[9] and single-component layers that utilize a polymer with broad emission covering the entire visible spectrum.[10,11]

RSC Polymer Chemistry Series No. 2
Molecular Design and Applications of Photofunctional Polymers and Materials
Edited by Wai-Yeung Wong and Alaa S Abd-El-Aziz

Published by the Royal Society of Chemistry, www.rsc.org

13.2   Single-Component White Light-Emitting Polymers

White emission from conjugated polymer can be obtained *via* the introduction of different chromophores (either fluorescent and/or phosphorescent) into host conjugated polymers with a binary chromophores type (blue-orange) or a

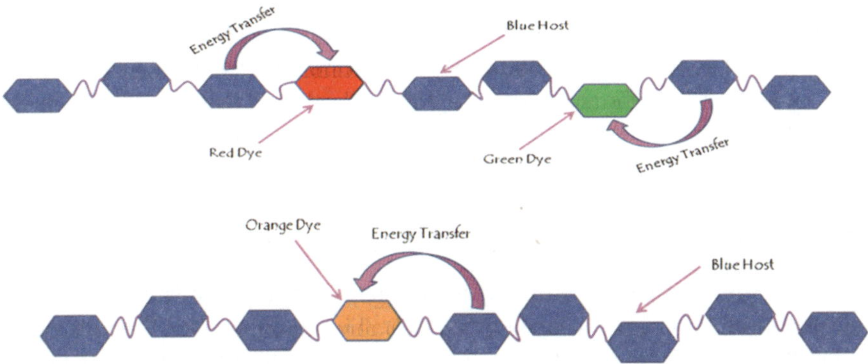

Figure 13.1   Schematic structure of WLEPs with chromophores incorporated into polymer main chain.

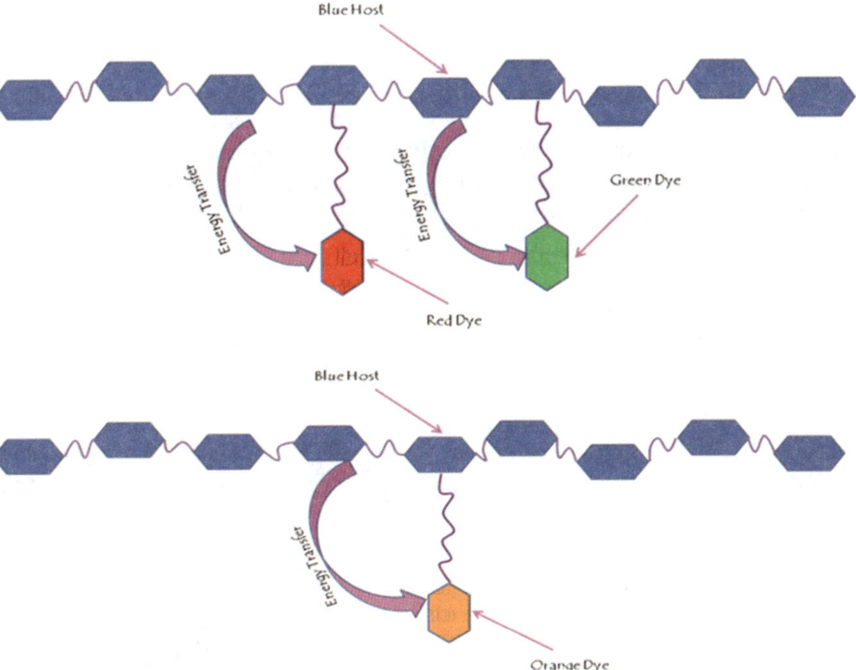

Figure 13.2   Schematic structure of WLEPs with chromophores incorporated into polymer side chain.

triple chromophores type (red-green-blue). Generally, there are two route methods available to realize white emitting polymers, which refer to (1) incorporating chromophores into polymer main chains, and (2) anchoring chromophores into polymer side chains (see Figures 13.1 and 13.2, respectively). The resulting single-component white light-emitting polymers (WLEPs) can offer several unique advantages, such as fine chemical tuning of the emission color, minor color shift upon changing of the driving voltage or current density, and easy device fabrication over large-area size. In addition, WLEPs can effectively prevent the phase segregation that is usually observed in dopant/host blend systems. Therefore, the scenario of incorporating chromophores in polymer main chains and side chains shows a feasible and efficient way to achieve the WPLEDs.

13.2.1 Fluorescent White Light-Emitting Polymers

Organic and inorganic fluorescent dyes have been widely utilized in laser devices and GaN-based light-emitting diodes (LEDs), ascribing to good thermal and optical stability, high photoluminescent efficiency, and flexible structure tunability. Similarly, one can use these organic fluorescent chromophores with appropriate structure modification and incorporate them into polymer chain to obtain white emission. In order to achieve a white emission, it is desirable to contain multiple emission bands covering the entire visible range. As shown in Figures 13.1 and 13.2, the most straightforward strategy to realize WLEPs is based on complementary blue-orange binary colors and red-green-blue triple colors, in which the addition of blue emission from polymer backbone and long-wavelength emission (orange or green/red) due to the energy transfer from the blue one is required. At the same time, incorporating a small amount of orange or green and red chromophores to the blue-emitting polymer chain (both in polymer main chain and side chain) is also a feasible approach to obtain WLEPs. Practically, construction of a donor-acceptor-donor (**D-A-D**) type narrow-bandgap moiety have been demonstrated as an effective approach to acquire long-wavelength emission (low-bandgap) fluorescent dyes, in which internal charge transfer (ICT) can lead to decreased bandgap. Of which, donors and acceptors can be selected

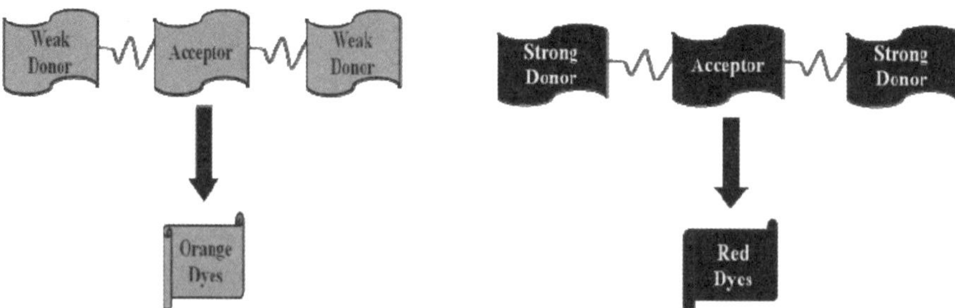

Figure 13.3   Constructive strategies of fluorescent dyes by D-A-D architecture.

from a wide range of candidates, for instance, 2,1,3-benzothiadiazole (BT) and quinoxaline are good electron acceptors, while thiophene and triphenylamine (TPA) are novel electron donors; if a weak donor (TPA) is attached to couple with a strong acceptor (BT), an orange emitting fluorescent dye can be obtained. On the other side, if a strong donor (thiophene) is chosen, a red fluorescent dye will be achieved (Figure 13.3). So, one can choose different donors and acceptors for synthesizing different fluorescent dyes.

13.2.1.1 Fluorescent Chromophores Incorporating into Polymer Main Chains

Wang and coworkers designed and synthesized blue-orange binary WLEPs that utilized polyfluorene backbones as the blue emission sites and BT-TPA derivative as the orange emitting species. Through differentiating the concentration of BT-TPA unit in the polymer main chain, two WLEPs **P1** and **P2** were obtained (Figure 13.4).[12] It was found that **P1** showed better WPLED device performance when compared with **P2,** with a lower turn-on voltage of 3.5 V, a luminance efficiency (LE) of 8.99 cd A^{-1}, and a power efficiency (PE) of 5.75 lm W^{-1}, while the displayed CIE coordinates of (0.34, 0.35) are very close to the ideal white emission color with CIE of (0.33, 0.33). **P2** presented a LE of 5.79 cd A^{-1}, and a PE of 2.5 lm W^{-1} with the CIE coordinates of (0.41, 0.41). The observed high efficiency and more pure white emission in **P1** can be understood since the blue emission in **P1** is more efficient than that in **P2**. Recently, Xie and coworker investigated the effects of thermal annealing[13] and solvent process[14] on the performance of **P2** and found that either thermal annealing or exposure to toluene can improve the efficiency and color purity dramatically. For instance, after thermal annealing, the LE and CIE coordinates were improved from 8.3 cd A^{-1} and (0.45, 0.46) to 10.7 cd A^{-1} and (0.39, 0.39), respectively. On the other hand, solvent treatment by toluene also enhanced the LE and CIE coordinates to 12.1 cd A^{-1} and (0.39, 0.39), respectively. The improvements in LE and color purity in both cases can be attributed to the presence of crystalline phase of polyfluorene that acts as efficient, self-dopant blue emitter, which boosts the blue emission in the polymer. Furthermore, these post-treatments can result in more balanced charge-carrier transport in the device as indicated by the experimental electron and hole mobilities. In addition, Li *et al.* reported the synthesis of a *p*-phenylenevinylene (PPV) derived orange dye combined with TPA groups and a binary-color WLEP **P3** (Figure 13.4) based on a polyfluorene backbone.[15] This polymer showed a LE of 0.03 cd A^{-1} and CIE coordinates of (0.30, 0.40), with broad electroluminescent spectra covering the entire visible spectra.

As compared with the linear WLEPs based on blue-orange colors as mentioned above, newly emerging star-shaped WLEPs consisting of a functional core and several supportive functional shells (arms or branches) *via* covalent bonding have received intense research attention. In principle, if an appropriate amount of orange dye is incorporated in the core position and supportive blue emitting polymers are used as functional shells, white emission can be realized *via* this versatile method through the energy-transfer mechanism (Figure 13.5). It is important to note that there exist some unique merits in these star-like WLEPs, such as inhibited dye aggregation, suppressed intermolecular interaction, and much reduced concentration quenching, all could potentially lead to improved efficiency and color purity. Wang's group also conducted research to design and synthesize this kind of star-shaped WLEPs. Recently, his group developed three kinds of orange cores, **Core 1**, **Core 2** and **Core 3**, and utilized the polyfluorene backbone as blue emission arms owing to their high PL and EL efficiencies in the solid state. Preliminarily, the **Core 1** was used as the orange core and star-like **P4** was synthesized through the A_4+AB Suzuki polycondensation. When the content of the **Core 1** was fixed at 0.03%, the resulting **P4** showed two emission bands peaking at around 420 nm/440 nm and 562 nm. The former was assigned to the emission from polymeric fluorene arms, while the latter was ascribed to the orange **Core 1** emission as a result of energy transfer. The turn-on voltage, LE, PE and CIE coordinates are 3.5 V, 7.06 cd A^{-1}, 4.43 lm W^{-1} and (0.35, 0.39), respectively.[16] Recently, Wang's group exploited other two orange cores, **Core 2** and **Core 3**, to synthesize two star-shaped WLEPs **P5** and **P6** (Figure 13.4).[17] When the **Core 2** was chemically doped with 0.01% in **P5**, white emission with CIE coordinates of (0.34, 0.31) can be obtained, and the LE, PE and EQE were found to be 6.10 cd A^{-1}, 3.83 lm W^{-1} and 2.74%, respectively. On the other hand, in order to obtain optimized device performance in **P6,** the doping concentration of **Core 3** is 0.02%. After thermal annealing at 120 °C for 30 min, the LE, EQE and CIE coordinates of the optimized device improved from 14.98 cd A^{-1}, 4.96% and (0.42, 0.46) to 18.01 cd A^{-1}, 6.36% and (0.33, 0.35), respectively The improvements can be ascribed to the combination of the increment of crystallized α-phase polyfluorene content that served as a self-dopant in the emitting layer, together with the optimized charge-transporting properties.

In order to acquire higher CRI for high color quality lighting sources, WPLEDs based on the RGB three-color system is more preferred than those based on the blue–orange binary color system. Luo *et al.* developed a WLEP **P7** with polyfluorene (blue), BT (green) and 4,7-bis(2-thienyl)-2,1,3-benzothia-dizole (DBT, red) as the RGB emissive components through a partial energy transfer from a wide-bandgap unit to a narrow-bandgap species.[18] The device based on **P7** displayed three distinguishable emission peaks, which were located at 452 nm, 524 nm and 616 nm, respectively. The device based on the copolymer containing 0.18% of green chromophore and 0.10% of red dye exhibited a maximal LE of 6.20 cd A^{-1} and a peak EQE of 3.84%, with a pure white emission color and CIE coordinates of (0.35, 0.34). In addition, quinoxaline derivatives are regarded as a kind of good electron-deficient moiety with novel green emissive properties. Lee *et al.* synthesized another RGB WLEPs **P8** that utilizes polyfluorene as blue host and 2,3-bis(4-methyloxyphenyl)quinoxaline as green dye and 5.8-bis(N,N-diphenylamino)-2,3-bis(4-methyloxyphenyl)quinoxaline as red chromophore.[19] As the blue, green and red contents are fixed at 99.45%, 0.5% and 0.05%, respectively, the

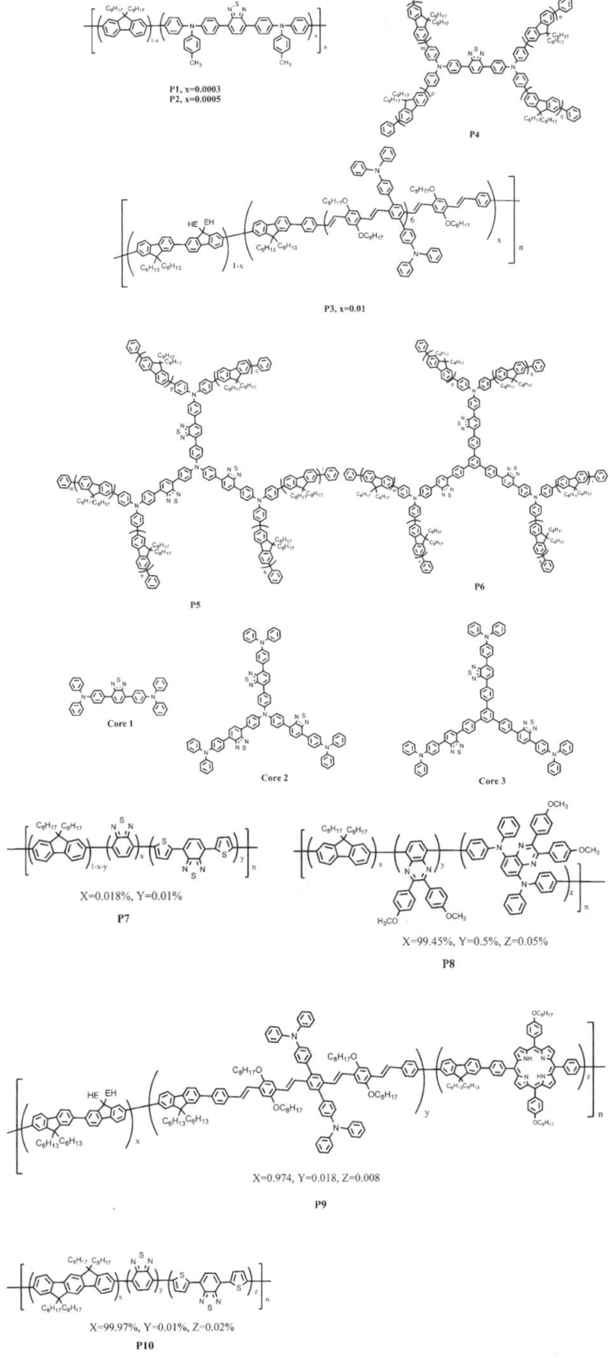

Figure 13.4 Structures of single-component WLEPs by incorporating fluorescent chromophores in polymer main chain.

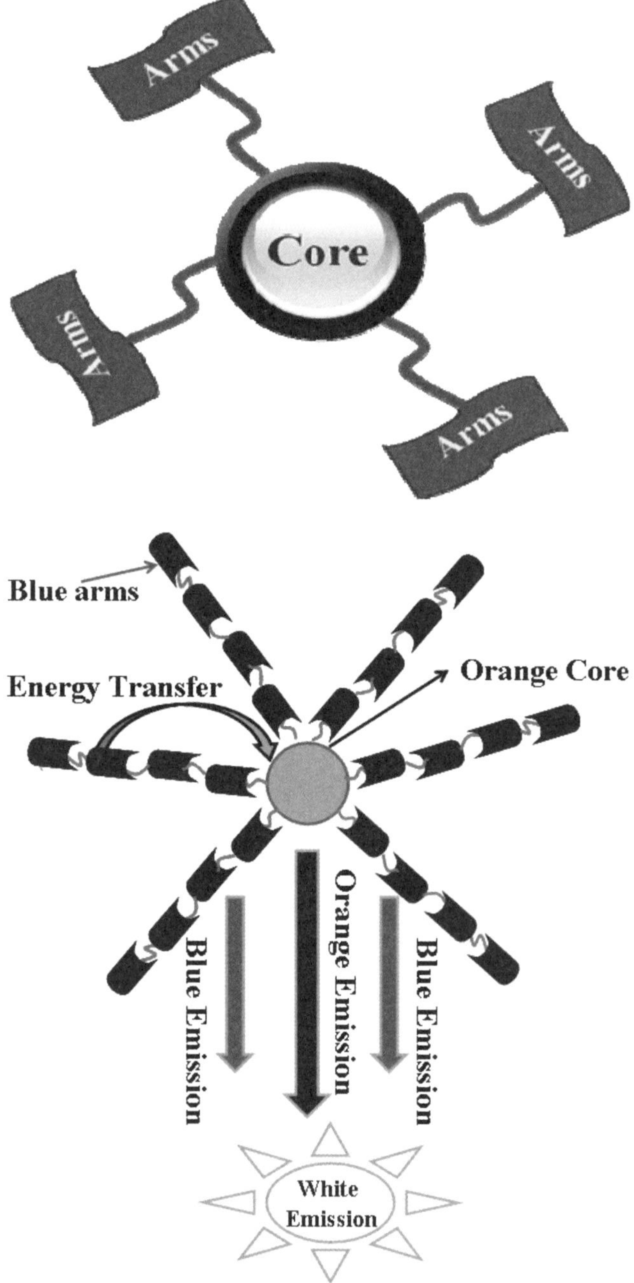

Figure 13.5 Working mechanism of star-shaped WLEPs.

obtained polymer **P8** showed the best electroluminescent performance with a maximal brightness of 12 300 cd m^{-2}, a maximal LE of 2.02 cd A^{-1} and CIE coordinates of (0.33, 0.40), respectively.

Since porphyrin derivatives exhibit saturated red emission with very narrow half-peak height (full width at half-maximum), thus have been regarded as one of the most important red dopants for OLEDs. Li *et al.* synthesized a new WLEPs **P9** in which polyfluorene plays the role of host, while PPV derivatives and porphyrins were incorporated as green chromophore and red emissive species, respectively.[20] **P9** exhibited moderate electroluminescent property with a LE of 0.08 cd A^{-1} and CIE coordinates of (0.29, 0.30). Although efficient white emitting polymers could be obtained by using deep blue-emitting polyfluorene as host, however this system suffers from low luminous efficiency since human eyes have an extremely weak response to deep blue emission. In contrast, the emission of polyindenofluorenes (PIFs) can be as sensitive as two to three times higher than that of polyfluoene, therefore it can be used for efficient WLEPs. Jeong *et al.* reported a WLEPs **P10** based on using PIFs as host and blue emitter, while BT and DBT were incorporated as green and red chromophores, respectively.[21] As expected, the device based on **P10** presented three distinct emission regions, of which the blue emission peaked at 444 nm and 471 nm originated from the indenofluorene moiety, while green emission peaked at 534 nm from the BT species and the red one at 615 nm from the DBT unit, respectively, although the observed LE was only 0.36 cd A^{-1}, with CIE coordinates of (0.34, 0.32).

13.2.1.2 Fluorescent Chromophores Incorporating into Polymer Side Chains

Besides incorporating organic dyes (orange or green and red) into the backbone of blue-emitting host (*e.g.* polyfluorene), white emitting polymers can be obtained by grafting a small amount of organic dyes to the side chain of the host with an alkyl spacer. Based on this strategy, Wang's group designed and synthesized two WLEPs **P11** and **P12** with two orange dyes **Dye 1** and **Dye 2** (Figure 13.6) noncovalently attached at the side of PFs.[22] Both polymers showed excellent electroluminescent performance with the peak LE, PE, maximum brightness and CIE coordinates of 10.66 cd A^{-1}, 6.88 lm W^{-1}, 21240 cd m^{-2} and (0.30, 0.40) for **P11**, and 4.29 cd A^{-1}, 2.67 lm W^{-1}, 13870 cd m^{-2} and (0.33, 0.28) for **P12**, respectively. For comparison, they also synthesized control WLEPs with the **Dye 1** and **Dye 2** incorporating into the PF main chain. It was noted that **P11/P12** exhibited a lower efficiency of 7.30 cd A^{-1} with CIE of (0.29, 0.37), verifying that the side-chain-type WLEPs can be more efficient than the corresponding main-chain-type WLEPs, which can be attributed to the absence of change in the electronic structure of the polymer host of the side-chain-type WLEPs.

Based on the previous study, Wang and coworkers synthesized another orange side-chain pendant type dye polymers based on **Dye 3** (Figure 13.6), in which a highly efficient sky-blue emitting dopant dimethylamino naphthalimide (DMAN) was incorporated into the polyfluorene side chain as the blue emitting species. As can be clearly seen, the replacement of polyfluorene by DMAN-containing polyfluorene significantly enhanced the efficiency of the obtained white emitting polymer (**P14**, Figure 13.6), as indicated by a maximal LE of 12.8 cd A^{-1}, a PE of 8.5 lm W^{-1} and CIE coordinates of (0.31, 0.36), which is higher than that of devices based on polyfluorene without DMAN dye (**P13**) (with a LE of 9.3 cd A^{-1}). Similarly, Coya *et al.* reported white electroluminescence with a LE of 22.62 cd A^{-1} and CIE coordinates of (0.30, 0.42) from a new WLEP **P15** based on 1,8-naphthalimide-containing orange dye.[23]

To achieve high color quality lighting sources, a high color rendering index (CRI) is needed. Wang's group reported two WLEPs **P16**[24] and **P17**,[25] in which **P16** employed PF backbone as blue host and another two side-chain-anchored dyes as the green and red chromophores, while **P17** utilized three blue (DMAN dye), green (**Dye 4**) and red (**Dye 5**) as the side-chain pendant dyes. The device based on **P16** exhibited a LE of 7.3 cd A^{-1}, a PE of 4.2 lm W^{-1} and CIE coordinates of (0.31, 0.32). Device based on the obtained copolymer **P17** showed very broad EL emission covering the entire visible range from 400 nm to 700 nm with four emission bands, which can be ascribed to the emission of the RGB dyes and the deep blue emitting host. As a result of the broad emission, the determined CRI was found to be between 88 and 93. Furthermore, they showed a high LE of 8.6 cd A^{-1}, a PE of 5.4 lm W^{-1}, with CIE coordinates of (0.33, 0.36).

13.2.2 Phosphorescent White Light-Emitting Polymers

Despite the success in fluorescent dyes based white light-emitting polymers, only the singlet excitons in the electroluminescent devices are utilized, while the major fraction of the excitons, the remaining triplet excitons, are wasted because radiative decay from triplet excited states is spin-forbidden by the spin selection rules in the fluorescent WLEPs. Therefore, if triplet excitons can be effectively utilized, it will allow for a conversion of up to 100% of injected charges that can be harvested into emitted photons, leading to a theoretical internal quantum efficiency (IQE) of 100%.[26,27] Given this guideline, the introduction of heavy-metal-based organometallic compounds into WLEPs either at main chains or side chains is becoming more and more attractive.

13.2.2.1 *Phosphorescent Chromophores Incorporating into Polymer Main Chains*

One of the applicable ways to obtain phosphorescent WLEPs is to incorporate phosphorescent chromophors into the main chain of polymer. Zhen *et al.* utilized benzothiophene-pyridine organic ligands to synthesize a red-emitting iridium complex, iridium(III)bis(2-(2'-benzo[4,5-α]thienyl)-pyridinato-N,C$^{3'}$)-2,2,6,6-tetramethyl-3,5-heptanedione [(btp)$_2$Ir(tmd)], and synthesized a WLEPs **P18** by incorporating small amount of green emitting benzothiadiazole (BT) and a small amount of above-red emitting iridium complex [(btp)$_2$Ir(tmd)] into the PF backbone.[28] By carefully adjusting the contents

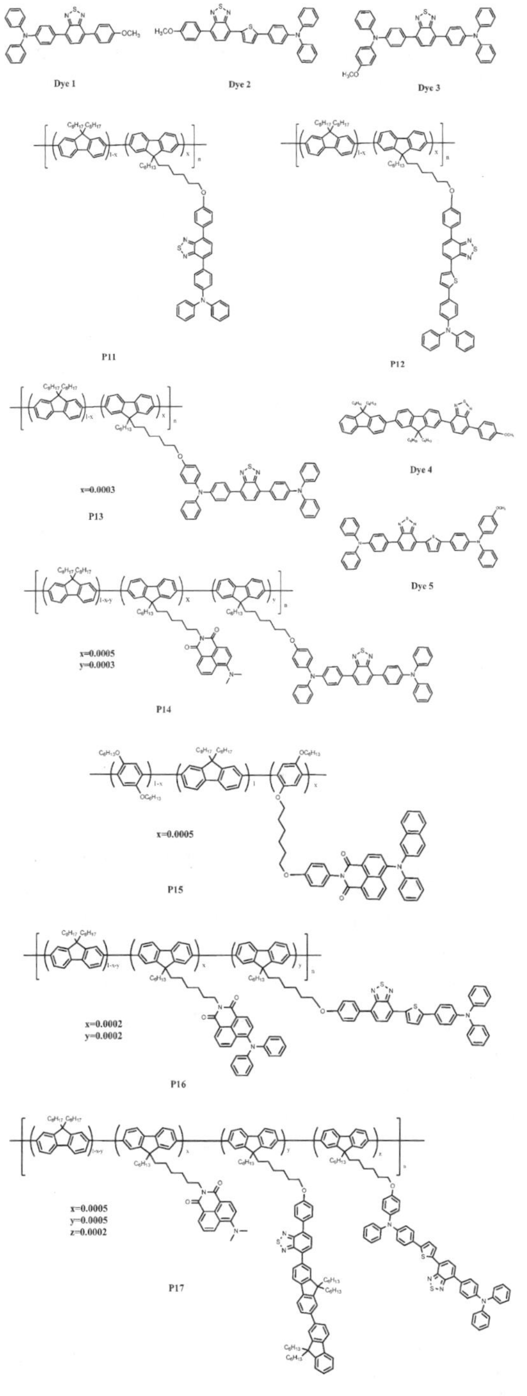

Figure 13.6 Structures of single-component WLEPs by incorporating fluorescent chromophores in polymer side chain.

of BT and [(btp)$_2$Ir(tmd)] in the polymer, the EL spectrum from the WLEPs can be adjusted to achieve white emission. The best device properties based on **P18** showed an EQE of 3.7% and a LE of 3.9 cd A^{-1} at the current density of 1.6 mA cm^{-2}, with CIE coordinates of (0.33, 0.34). A maximal luminance of 4180 cd m^{-2} was achieved at a current density of 268 mA cm^{-2} with CIE coordinates of (0.31, 0.32). Obviously, the device based on **P18** exhibited a stable white emission upon varied applied voltages, which is an important property for the application of solid-state lighting.

Since the emission of the phosphorescent dye (btp)$_2$Ir(tmd) in **P18** peaks at 650 nm where the response of human eyes is very weak and thus leads to the decrease in luminous efficiency, it is therefore rational to expect that luminous efficiency can be readily improved if less-saturated red emitters is incorporated into the polyfluorene backbone. For this reason, Chen *et al.* developed a new organic ligand, 2-naphthalene pyridine and synthesized a novel red emissive iridium complex, iridium(III) bis(2-(-naphthalene)pyridine- $C^{2'},N$)-2,2,6,6-tetramethyl-3,5-heptanedione [(1-npy)$_2$Ir(tmd)], which emits at 625 nm, where the response of human eyes is more sensitive than that on a saturated red emitter. By optimizing the components of the fluorescent green dye BT and the phosphorescent red chromophore (1-npy)$_2$Ir(tmd) in PF backbone, a WLEPs **P19** was synthesized (Figure 13.7).[29] The best device performance showed a LE of 5.3 cd A^{-1} and an EQE of 2.7%, with CIE coordinates of (0.34, 0.36). It is important to mention that as a result of broad and balanced emission in the visible range, the resulting devices show very good color quality, with high CRIs of 84–89. Fortunately, the device based on **P19** also presented a very stable white emission under different applied voltages, which makes it good candidate for high-quality lighting sources.

Besides WLEPs based on RGB phosphorescent dyes, polymers from complementary colors also attracted intense research interest due to the easy control over the monomer feed ratios in this system. Recently, Park *et al.* developed a new WLEPs in which a new red phosphorescent dye, bis(2-benzothiazole-2-yl-*N*-ethylcarbazole)iridium-1,3-bisphenyl-1,3-propanedione [(bec)$_2$Irdbm)] was incorporated into the main chain of 3,6-carbazole (**P20**, Figure 13.7).[30] Despite the fact that the green emitting segment is absent, **P20** showed three distinctive emissive peaks at 415, 495 and 651 nm, respectively. Unfortunately, **P20** only displayed a low EL efficiency of 0.05 cd A^{-1}, with CIE coordinates of (0.31, 0.32), which can be attributed to triplet energy quenching due to the energy transfer between fluorene or carbazole and iridium complex.

13.2.2.2 *Phosphorescent Chromophores Incorporating into Polymer Side Chains*

Although incorporation of triplet organic dyes into the conjugated main chain provides a straightforward method to realize WLEPs, the incorporation of dyes in the backbone of a polymer may alter the electronic properties of the main chain, and may ultimately deteriorate the performance of the obtained

WLEPs. Therefore, an alternative approach of covalently incorporating triplet dyes to the side chains of the host polymer is becoming an attractive focus through the energy transfer from the host triplet energy level to the pendants' one or the direct excitation of pendant triplet dyes *via* the charge trapping mechanism. Jiang *et al.* developed a side-chain WLEP **(P21)** by appending a triplet red emitter, iridium(III) [bis(2-phenylquinoline-N,C2'))-14-tri-fluoro-11,13-tetradecyldiketone)] to the alkyl chain, together with BT as a green emitter and polyfluorene as host (Figure 13.7).[31] By adjusting the contents of BT and/or iridium complex, the resulting EL spectra composed of fluorescent blue, green and phosphorescent red emission can be readily tuned to achieve white emission. A maximal LE of 6.1 cd A^{-1} was recorded at a current density of 2.2 mA cm^{-2} with a maximum luminance of 10 110 cd m^{-2} and CIE coordinates of (0.32, 0.44), respectively. The obtained white emission from the copolymers was stable at all of the applied voltages studied, and the LE only declined slightly with increase of current density.

13.3 Progress on the Structure of White Polymer Light-Emitting Devices

13.3.1 White Polymer Light-Emitting Devices Based on Single Active Layer Configuration

Generally, WPLEDs are fabricated in a thin-film stacked-type configuration, namely, the light-emitting layer is sandwiched between anode and cathode. On the basis of this, additional functional layers that are capable of facilitating charge injection and transport are usually utilized to optimize the electroluminescent properties. For instance, PEDOT:PSS and LiF or CsF are commonly used at the anode and cathode side for charge-carrier injection.

Despite the advantages of the WPLEDs based on single-component WLEPs, it is not easy to precisely control the doping concentration of each organic dye during the polymerization and reduce the batch-to-batch deviation, particularly in mass production. On the other hand, physically blending based on a host–guest system can provide an alternative approach towards white organic light-emitting devices, which can offer easy fabrication, excellent reproducibility and high efficiency. The method involves the blending of several dyes (guests) with different bandgaps and forming a uniform distribution of the dyes into the polymer matrix (see Figure 13.8).

Blending of fluorescent polymers is a straightforward method to realize WPLEDs. Shin *et al.* reported WPLEDs through a partial energy transfer by employing a new PFO-based blue polymer with a triphenylamine end-capped group PFO-TPA as the blue host, and MEH-PPV as the orange-red emissive guest.[32] Aiming at a higher efficiency, they utilized 2,2',2''-(1,3,5-benzinetriyl)-tris(1-phenyl-1-H-benzimidazole) (TPBi) as an electron-transporting and hole-blocking layer to improve the EL efficiency. With a device structure of ITO/PEDOT:PSS/PFO-TPA:MEH-PPV/TPBi/LiF/Al and a careful control of the

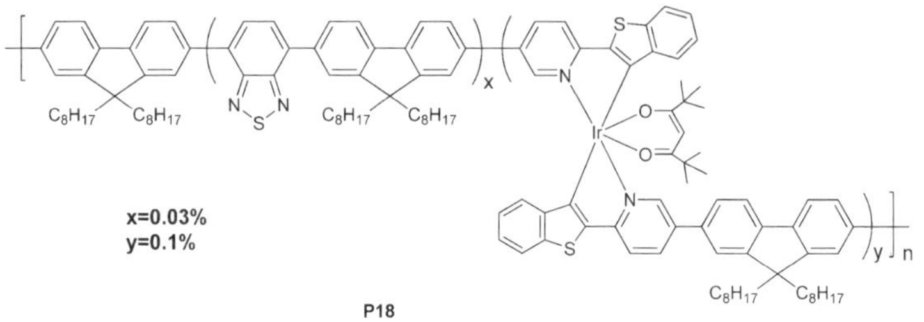

Figure 13.7   Structure of single-component phosphorescent WLEPs.

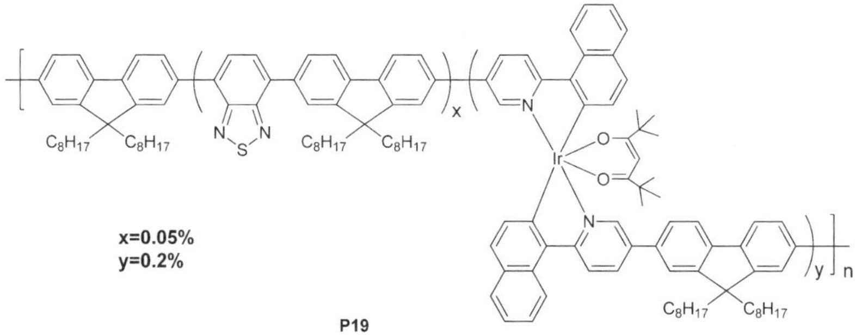

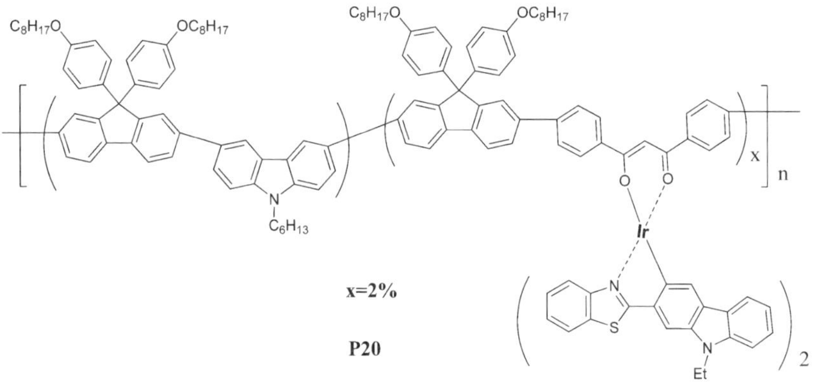

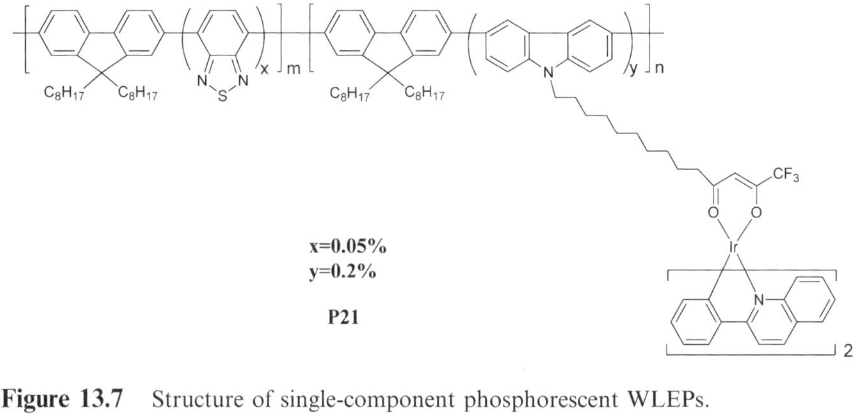

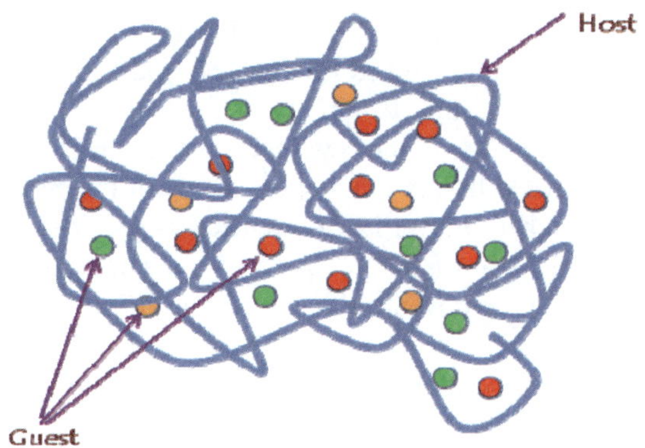

Figure 13.8   Device construction of blending system to realize WPLEDs.

blend concentration of MEH-PPV (*i.e.* 1.0wt%), the obtained devices exhibited a maximal LE of 7.8 cd A^{-1}, with CIE coordinates of (0.36, 0.33). Compared with the control device without TPBi, the optimized device showed a 7.5 times of improvement in efficiency, due to the excellent electron-transporting and hole-blocking properties of TPBi, which eventually confine the exciton recombination zone within the PFO-TPA:MEH-PPV layer and can effectively suppress the energy loss due to nonradiative recombination.

To further improve the overall performance of WPLEDs, one of the most effective approaches is to fabricate all phosphorescent dyes blended into a polymer host, which could allow conversion of up to 100% of injected charges into emitted photons, leading to an IQE of unity. Given these considerations, Wu *et al.* reported a series of single emission layer WPLEDs by double doping of blue-emitting iridium(III) bis(2-(4,6-difluorophenyl)-pyridinato-N, C^2)picolinate (FIrpic) and home-made efficient yellow iridium complexes (**C1**) (see Figure 13.9) with an appropriate ratio into poly(*N*-vinylcarbazole) (PVK) host with the presence of electron-transport material 1,3-bis[(4-*tert*-butylphenyl)-1,3,4-oxadiazolyl]phenylene (OXD-7).[33] When the blending ratio was PVK (63 wt%): OXD-7(27 wt%):FIrpic (10 wt%) and FIrpic to **C1** ratio was 20:1 under a device configuration of ITO/PEDOT:PSS (40 nm)/emitting layer (70–80 nm)/Ba (4 nm)/ Al (100 nm), the resulting WPLEDs have a peak PE of 20.3 lm W^{-1} for forward viewing at 6.0 V, and a peak LE of 42.9 cd A^{-1} for forward viewing at 1.5 mA cm^{-2} (7.2 V). At a typical luminance for solid-state lighting (*i.e.* 1000 cd m^{-2}), the PE is still retained at 16.8 lm W^{-1}, while the LE is as high as 41.7 cd A^{-1}, with CIE coordinates of (0.395, 0.452) at a current density of 12 mA cm^{-2}. In order to avoid the trade-off between efficiency and color quality, Zou *et al.* reported highly efficient all phosphorescent dyes WPLEDs based on a four-color system (blue-green-yellow-red).[34] The authors utilized FIrpic as blue phosphorescent emitter, *fac*-tris[2-(4-tolyl)pyridinato-N, C^2]iridium(III) Ir(mppy)$_3$ as green emitter, a newly synthesized **C2** as yellow emitter and a saturated red iridium(III)-based dendrimer **C3** as red emitter (Figure 13.9). When the blending ratio was fixed at FIrpic:Ir(mppy)$_3$:**C2**:**C3** = 70:1:1:1, the best device showed a peak PE of 37.4 lm

W^{-1} at 4.8 V and a peak LE of 60.1 cd A^{-1} at 0.1 mA cm^{-2} and 5.7 V. At a practical luminance of 100 cd m^{-2} and 1000 cd m^{-2}, the PE and LE are still maintained at 30.7 lm W^{-1}, 58.6 cd A^{-1} and 20.7 lm W^{-1}, 53.5 cd A^{-1}, respectively, thanks partially to the intrinsic nature of the new yellow emitter in overcoming the triplet–triplet annihilation caused by the interactions between the emitting molecules. It should be noted that when the WPLEDs are used as a solid-state lighting source, all the photons emitted out of the device in all directions should be taken into account for determining their overall efficiency. Thus, the fabricated WPLEDs can exhibit a total PE of ∼51 lm W^{-1}, which make the WPLEDs as efficient as fluorescent tubes that show a typical PE of 40–70 lm W^{-1}, significantly exceeding that of the incandescent light bulbs (with a typical PE between 12–17 lm W^{-1}). This is the first report that total PE of 50 lm W^{-1} was realized in WPLEDs without any light-extraction technology adopted. Such propitious progress minimizes the efficiency gap from the vacuum-deposited small-molecule devices and renders WPLEDs as competitive as the complicated multilayer WOLEDs currently in use for illumination purpose. This work definitely paves the road to low-cost and large-area WPLEDs for future solid-state lighting.

Recently, Liang *et al.* synthesized a new yellow phosphorescent emitter **C4**, which carries 3-phenylisoquinoline as the cyclometalated ligand (Figure 13.9).[35] By doping this complex **C4** into the blue emitting polyfluorene, efficient WPLEDs with CIE coordinates of (0.34, 0.31) have been obtained. Besides this study, blue emitting polyfluorene has been widely used as host and deep blue emitter by many other groups owing to its excellent photoluminescent and electroluminescent properties. For example, Lee *et al.* synthesized a new host polymer **P22** derived from polyfluorene, in which a fluorescent yellow chromophore was end-capped *via* the Yamamoto coupling reaction in the polymer backbone.[36] By carefully adjusting the content of the yellow emitting chromophore in the main chain and the doping concentration of a phosphorescent dye [bis(2-[2'-benzothienyl]-pyridinato-N,$C^{3'}$)]iridium(acetylacetonate) **C5** (see Figure 13.9), white emission was achieved *via* a device architecture of ITO/PEDOT:PSS P8000 (30 nm)/PVK (20 nm)/emitting layer (60–80 nm)/Ca (10 nm)/Ag (150 nm). A maximum LE of 0.75 cd A^{-1} with CIE coordinates of (0.32, 0.34) was obtained at an applied voltage of 4 V.

In order to avoid phase separation between host and dopant in a blending system, which has been identified as the major reason responsible for the degradation of performance in this type of WPLEDs, it is important to improve the miscibility of the blending system. Lee *et al.* developed a new blending system in which a commercially available PFO-based fluorescent blue polymer (Blue J) and green emitter (Green 1304) were employed, doped with a red phosphorescent dye tris(1-phenylisoquinoline)iridium(III) Ir(piq)$_3$.[37] By controlling the concentration of dyes and the phase morphology of the blended film, nearly "pure" white light was realized with CIE coordinates of (0.32, 0.35), and a maximal LE reached 8.22 cd A^{-1}. It is interesting to note that the devices offer good color stability upon change of applied voltage, which can be ascribed to the good homogeneity and miscibility in this three-component system.

The development of host materials is as important as that of the dopants. Although PVK has proven to be a good host material in blending systems due to its excellent hole transporting properties and its high triplet energy level, the use of this nonconjugated polymer also leads to a high operational voltage and reduction in power efficiency. On the other hand, 3,6-carbazole-based conjugated polymers show promising potential as host materials for PLEDs. Cheng *et al.* synthesized a novel polymer host **P23** by combining a 3.6-carbazole unit with δ-Si derivative.[38] It was found that **P23** displayed a wide bandgap of 3.26 eV and a high triplet energy level of 2.67 eV, which can ensure efficient energy transfer from host to guest. When blending with FIrpic, Ir(mppy)$_3$ and Ir(piq)$_3$ with a doping concentration of 8 wt%, 0.8 wt%, 0.4 wt%, respectively, the recorded peak LE of the resulting device was 8.70 cd A^{-1}, corresponding to an EQE of 4.26%. As a result of simultaneous red, green and blue emission covering the entire visible spectrum, high CRI of 82 was obtained, representing good color quality for lighting applications.

13.3.2 White Polymer Light-Emitting Devices Based on Multilayer Stacked Configuration

Besides the aforementioned single-component WLEPs, blends of semiconducting polymers or dyes in a host matrix, WPLEDs can be obtained from the state-of-art device fabrication engineering, namely, multilayer stacked technique. This technique has been verified as an effective approach for RGB monochromatic emitting PLEDs, where each layer plays specific functions such as charge injection and transport, exciton confinement, *etc.* Naturally, this strategy can also be applied in the fabrication of WPLEDs.

It is well known that electrochemical polymerization (**EP**) is a feasible method to achieve organic optoelectronic materials due to its easy manipulation. Gu *et al.* reported a novel method towards WPLEDs based on EP strategies, in which WLEPs were obtained by crosslinking or layer-by-layer deposition. In the first step, three fluorescent fluorene-based dyes containing peripheral carbazole groups were synthesized (TCPC for blue emission, TCBzC for green emission, TCNzC for red emission, see Figure 13.10 for the chemical structures), owing to the high electrochemical activity of carbazole.[39] By adjusting the contents of RGB dyes, crosslinked WLEPs were obtained *via* EP in a mixed dichloromethane solution. The triple-doped polymer WPLEDs exhibited three individual emissive bands peaking at 438, 512 and 574 nm, respectively. When the contents of TCBzC and TCNzC with respect to the TCPC were controlled at 2 wt% and 1 wt%, respectively, the resulting WPLEDs showed stable white emission with a peak LE of 6.7 cd A^{-1}, with CIE coordinates of (0.33, 0.35). As a result of balanced full-color emission covering the entire visible spectra, the device offers a high CRI of 92. Therefore, this technique represents a novel multilayer electrochemical polymerization method for the fabrication of multilayer, high-efficiency and color-stable WPLEDs from three fluorene-based dyes with an emission-adjustable unit and

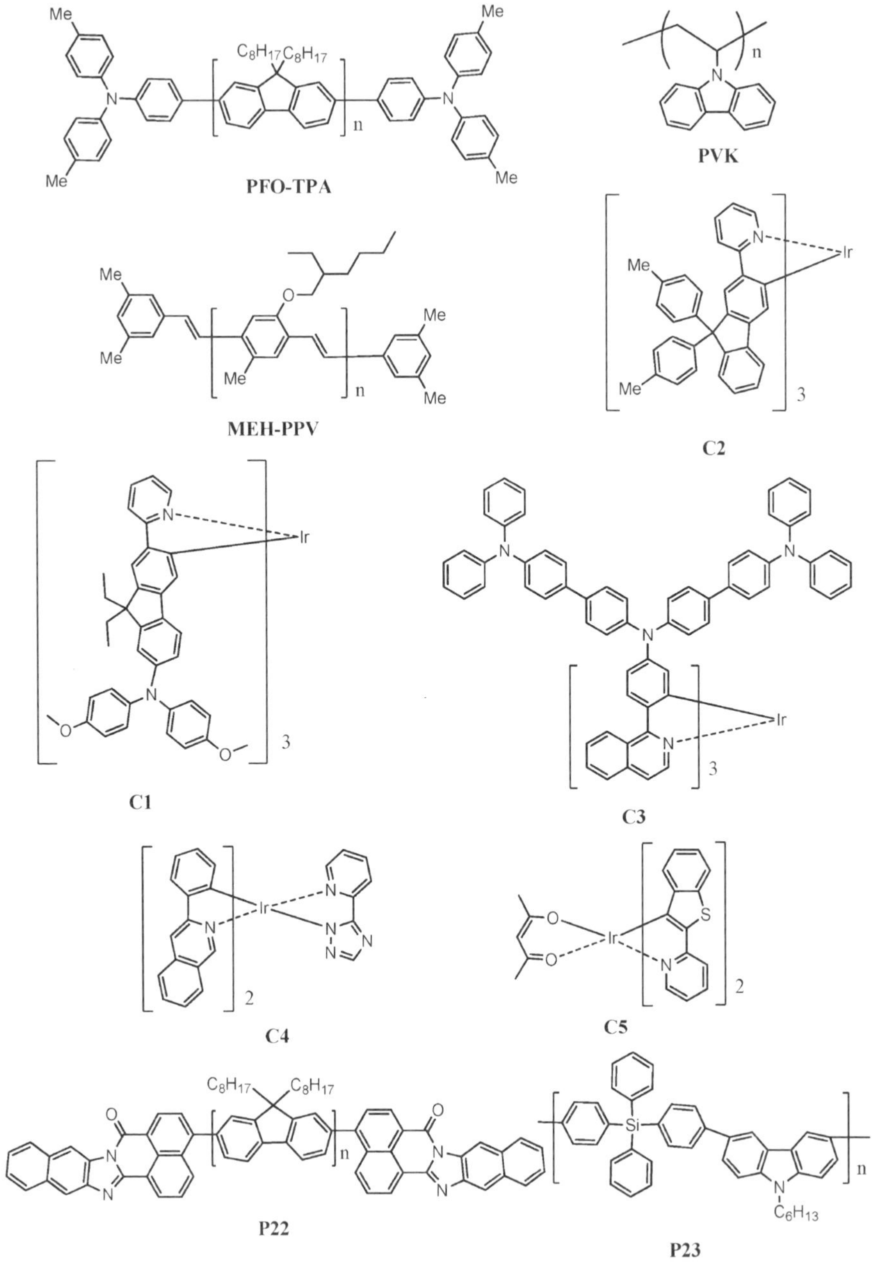

Figure 13.9 Structures of molecules and polymers for blending-type WPLEDs.

peripheral carbazole groups (TCPC, TCBzC, TCNzC).[40] During the fabrication of the device, a red emissive layer from TCNzC was deposited on the ITO substrate at first, then green and blue emitters based on TCBzC and TCPC were sequentially electrochemically oxidized and crosslinked onto the foregoing layer. The method can provide precise control over the electroluminescent property of the polymer films deposited, and allow flexible deposition sequence and position, therefore can be used for the fabrication of WPLEDs with complicated structure. Through controlling the thickness of each layer by different electrochemically scan rates and cycles, the authors achieved white emission from this multilayer device structure, in which each layer emitted RGB colors individually. The device exhibited a maximal LE of 5.5 cd A^{-1}, a power efficiency of 1.8 lm W^{-1}, with CIE coordinates of (0.35, 0.35) and a CRI of 93. Conclusively, the proposed EP method presents a feasible and versatile strategy to realize WPLEDs either scientifically or industrially.

Besides the above-mentioned chemically assisted method, physical stacking provides another attractive strategy to realize white emission devices. Recently, Joo *et al.* developed a stamp-transfer printing method to fabricate multilayer stacked WPLEDs.[41] First, a blue emissive layer based on PF was spin coated on a PEDOT:PSS coated ITO glass. Then, an individual yellow emitting layer was fabricated as follows (Figure 13.11): Yellow-green emitter with a commercial

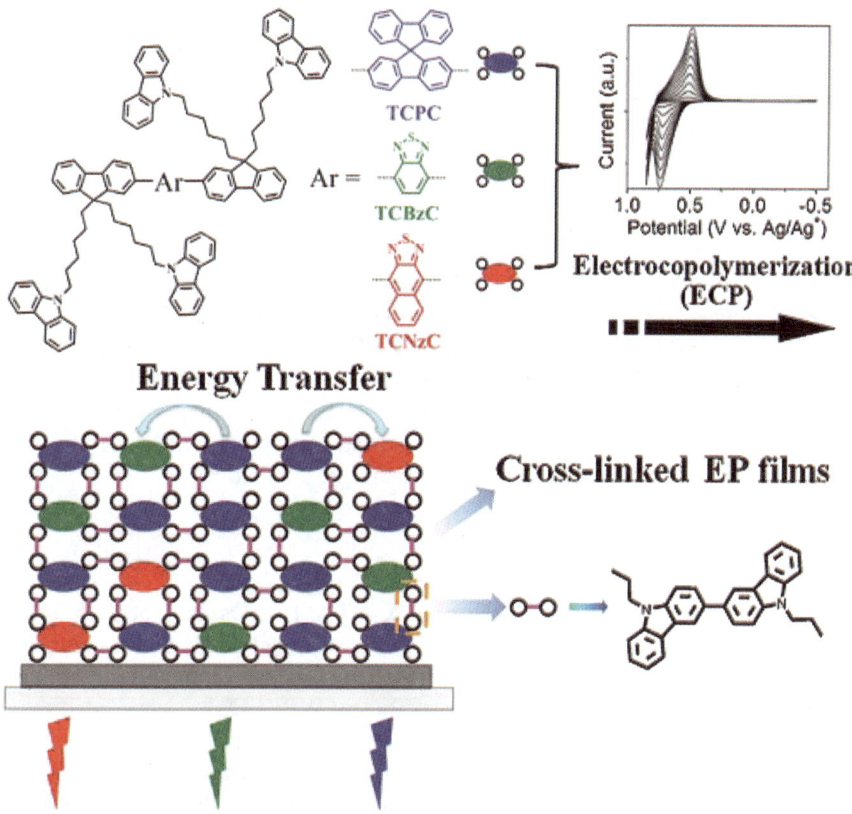

Figure 13.10   Molecular structures for electrochemical polymerization.

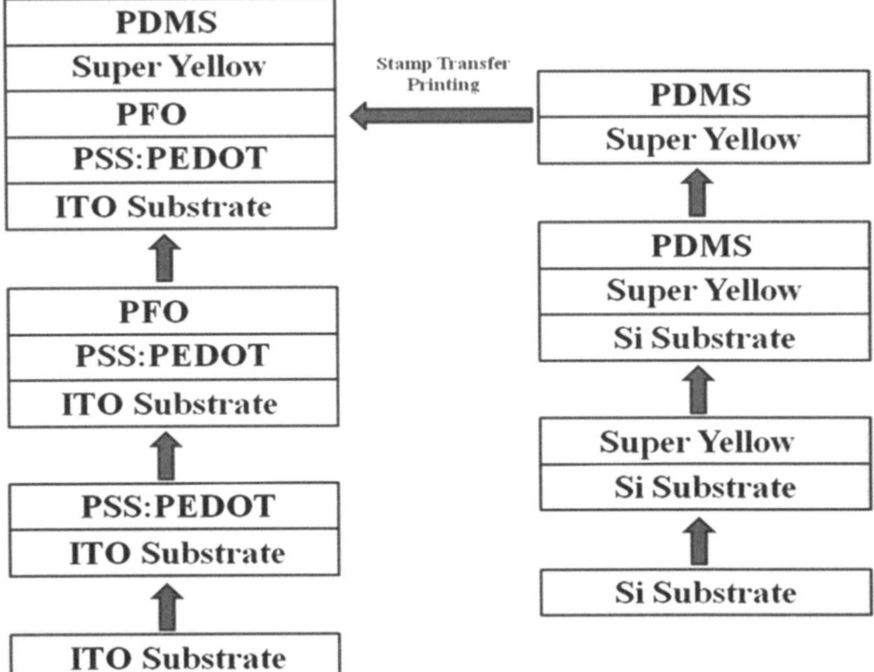

Figure 13.11   Fabrication procedures for stamp-transfer printing technique.

name of Super Yellow was spin coated from solution onto silicon substrate, and then a polydimethylsilozane (PDMS) stamp was casted onto the yellow emitting layer, followed by a stamp-transfer printing through pressing the PDMS stamp on the yellow polymer film at 120 °C for 5 s. When the thickness of the blue and yellow layers are both fixed at 40 nm, the device exhibited a maximal LE of 3.98 cd A^{-1} and white emission colors with CIE coordinates of (0.33, 0.41). Fortunately, this device had good efficiency sustainability with a LE of 3.27 cd A^{-1}, while it was operated under a luminance of 1000 cd m^{-2}.

13.4   Conclusions

In the past few years, the performance of WPLEDs has been improved significantly due to the progress of synthesis of novel single-component white emitting polymers, the novel device structure and the application of triplet emitters in WPLEDs. To date, the most efficient WPLEDs have exceeded that of incandescent light bulbs, and may find practical applications for displays, backlighting for liquid-crystal displays and lighting sources. However, even for the most efficient WPLEDs nowadays, their efficiency is still too low and further improvement is urgently needed. We note that all devices in this chapter are not fully optimized and there is still much room for improvement. Under the current rate, we anticipate the total power efficiency of more and more WPLEDs can reach 50 lm W^{-1} as more novel materials synthesis, optimized device structure and optical design efforts will be carried out in the coming few years.

Acknowledgements

The authors are grateful to the financial supports from the National Basic Research Program of China (No. 2009CB623602) , the National Natural Science Foundation of China (No. 21074038, 60906032, and 61177022), and National High Technology Research and Development Program 863 (No.2011AA03A110).

References

1. H. B. Wu, L. Ying, W. Yang and Y. Cao, *Chem. Soc. Rev.*, 2009, **38**, 3391.
2. H. B. Wu, J. H. Zou, F. Liu, L. Wang, A. Mikhailovsky, G. C. Bazan, W. Yang and Y. Cao, *Adv. Mater.*, 2008, **20**, 696.
3. Y. Sun, N. C. Giebink, H. Kanno, B. Ma, M. E. Thompson and S. R. Forrest, *Nature*, 2006, **440**, 908.
4. C.-L. Ho, W.-Y. Wong, Q. Wang, D.-G. Ma, L.-X. Wang and Z. Lin, *Adv. Funct. Mater.*, 2008, **18**, 928.
5. Y. H. Xu, J. B. Peng, J. X. Jiang, W. Xu, W. Yang and Y. Cao, *Appl. Phys. Lett.*, 2006, **87**, 193502.
6. D. An, J. H. Zou, H. B. Wu, J. B. Peng, W. Yang and Y. Cao, *Org. Electron.*, 2009, **10**, 299.
7. P. I. Shih, Y. H. Tseng, F. I. Wu, A. K. Dixit and C. F. Shu, *Adv. Funct. Mater.*, 2006, **16**, 1582.
8. J. H. Zou, J. Liu, H. B. Wu, J. B. Peng, W. Yang and Y. Cao, *Org. Electron.*, 2009, **10**, 843.
9. Q. J. Sun, B. H. Fan, Z. A. Tan, C. H. Yang, Y. F. Li and Y. Yang, *Appl. Phys. Lett.*, 2006, **88**, 163510.
10. G. L. Tu, C. Y. Mei, Q. G. Zhou, Y. X. Cheng, Y. H. Geng, L.X. Wang, D. G. Ma and X. B. Jing and F. S. Wang, *Adv. Func. Mater.*, 2006, **16**, 101.
11. J. X. Jiang, Y. H. Xu, W. Yang, R. Guan, Z. Q. Liu, H. Y. Zhen and Y. Cao, *Adv. Mater.*, 2006, **18**, 1769.
12. J. Liu, Q. G. Zhou, Y. X. Cheng, Y. H. Geng, L. X. Wang, D. G. Ma, X. B. Jing and F. S. Wang, *Adv. Funct. Mater.*, 2006, **16**, 957.
13. X. D. Niu, H. B. Zhang, Z. Y. Xie, Y. Cheng and L. Wang, *Appl. Phys. Lett.*, 2010, **96**, 073303.
14. B. Fraboni, R. DiPietro, A. Cavallini, P. Cosseddu, A. Bonfiglio, J.-O. Vogel, J. P. Rabe, and N. Koch, *Org. Electron.*, 2010, **11**, 273.
15. H. Li, L. Wang, B. Zhao et al., *J. Polym. Sci. Part A: Polym. Chem.* 2009, **47**, 296.
16. J. Liu, Y. X. Cheng, Z. Y. Xie, Y. H. Geng, L. X. Wang, X. B. Jing and F. S. Wang, *Adv. Mater.*, 2008, **20**, 1357.
17. L. Chen, P. C. Li, Y. X. Cheng, Z. Xie, L. Wang, X. Jing and F. Wang, *Adv. Mater.*, 2011, **23**, 2986.

18. J. Luo, X. Li, Q. Hou, J. B. Peng, W. Yang and Y. Cao, *Adv. Mater.*, 2007, **19**, 1113.
19. P. I. Lee, S. L. C. Hsu and P. Y. Lin, *Macromolecules*, 2010, **43**, 8051.
20. H. Li, N. Xiang, L. Wang, B. Zhao, P. Shen, J. Lu and S. Tan, *J. Polym. Sci. A: Polym. Chem.*, 2009, **47**, 5291.
21. E. Jeong, S. H. Kim, I. H. Jung, Y. Xia, K. Lee, H. Suh, H.-K. Shim and H. Y. Woo, *J. Polym. Sci. A: Polym. Chem.*, 2009, **47**, 3467.
22. J. Liu, X. Guo, L. J. Bu, Z. Y. Xie, Y. X. Cheng, Y. H. Geng, L. X. Wang, X. B. Jing and F. S. Wang, *Adv. Funct. Mater.*, 2007, **17**, 1917.
23. C. Coya, R. Blanco, R. Juárez, R. Gómez, R. Martínez, A. de Andrés, Á. L. Álvarez, C. Zaldo, M. M. Ramos, A. de la Peña, C. Seoane and J. Segura, *Eur. Polym. J.*, 2010, **46**, 1778.
24. J. Liu, Z. Y. Xie, Y. X. Cheng, Y. H. Geng, L. X. Wang, X. B. Jing and F. S. Wang, *Adv. Mater.*, 2007, **19**, 531.
25. J. Liu, L. Chen, S. Y. Shao, Z. Y. Xie, Y. X. Cheng, Y. H. Geng, L. X. Wang, X. B. Jing and F. S. Wang, *Adv. Mater.*, 2007, **19**, 4224.
26. M. A. Baldo, D. F. O'Brien, Y. You, A. Shoustikov, S. Sibley, M. E. Thompson and S. R. Forrest, *Nature*, 1998, **395**, 151.
27. C. Adachi, M. A. Baldo, M. E. Thompson and S. R. Forrest, *J. Appl. Phys.*, 2001, **90**, 5048.
28. H. Y. Zhen, W. Xu, W. Yang, Q. L. Chen, Y. H. Xu, J. X. Jiang, J. B. Peng and Y. Cao, *Macromol. Rapid Commun.*, 2006, **27**, 2095.
29. Q. L. Chen, N. L. Liu, L. Ying, W. Yang, H. B. Wu, W. Xu and Y. Cao, *Polymer*, 2009, **50**, 1430.
30. M.-J. Park, J. Kwak, J. Lee, I. H. Jung, H. Kong, C. Lee, D.-H. Hwang and H.-K. Shim, *Macromolecules*, 2010, **43**, 1379.
31. J. X. Jiang, Y. H. Xu, W. Yang, R. Guan, Z. Q. Liu, H. Y. Zhen and Y. Cao, *Adv. Mater.*, 2006, **18**, 1769.
32. S. B. Shin, S. C. Gong, H. M. Lee, J. G. Jang, M. S. Gong, S. O. Ryu, J. Y. Lee, Y. C. Chang, H. J. Chang, *Thin Solid Films*, 2009, **517**, 4143.
33. H. B. Wu, G. J. Zhou, J. H. Zou, C-L . Ho, W-Y. Wong, W. Yang, J. B. Peng and Y. Cao, *Adv. Mater.*, 2009, **21**, 4181.
34. J. H. Zou, H. Wu, C.-S. Lam, C. D. Wang, J. Zhu, C. M. Zhong, S. J. Hu, C.-L. Ho, G.-J. Zhou, H. B. Wu, W. C. H. Choy, J. B. Peng, Y. Cao and W.-Y. Wong, *Adv. Mater.*, 2011, **23**, 2976.
35. B. Liang, Y. H. Xu, Z. Chen, J. Peng and Y. Cao, *Synth. Met.*, 2009, **159**, 1876.
36. H. K. Lee, T. H. Kim, J. H. Park, J.-K. Kim and O. O. Park, *Org. Electron.*, 2011, **12**, 891.
37. G. Cheng, T. Fei, Y. Zhao, Y. Ma and S. Liu Org. *Electron.*, 2010, **11**, 498.
38. C. Gu, T. Fei, Y. Lv, T. Feng, S. Xue, D. Lu and Y. Ma, *Adv. Mater.*, 2010, **22**, 2702.
39. C. Gu, T. Fei, L. Yao, Y. Lv, D. Lu and Y. Ma, *Adv. Mater.*, 2011, **23**, 527.
40. C. W. Joo, S. O. Jeon, K. S. Yook and J. Y. Lee, *J. Phys D: Appl. Phys.*, 2009, **42**, 105115.

Strategies Towards Enhancing Charge Collection in Polymer Photovoltaic Devices

ZHI-YUAN XIE

State Key Laboratory of Polymer Physics and Chemistry, Changchun Institute of Applied Chemistry, Chinese Academy of Sciences, Changchun 130022, China
E-mail: xiezy_n@ciac.jl.cn

14.1   Introduction

Harvesting energy directly from sunlight using photovoltaic (PV) technology is being increasingly recognized as an essential component of future global energy production. The finite supply of fossil fuel sources and the detrimental long-term effects of CO_2 and other emissions into the atmosphere underscore the urgency of developing renewable-energy sources. PV technology is considered as one of the most effective solutions to the growing energy challenge.[1,2] Current PV production is dominated by single-crystal and polycrystalline silicon PV modules, which dominates over 90% of the market. One essential issue to the development of PV technologies is the reduction of the manufacturing cost. Organic semiconductors are a remarkable class of materials with present and potential applications in various optoelectronic devices.[3–6] In principle, devices using organic materials should be cheaper and simpler to manufacture than the corresponding ones using inorganic semiconductors. In 1986, Tang demonstrated the first efficient organic PV cell with a stacked bilayer of two organic materials, including a phthalocyanine

RSC Polymer Chemistry Series No. 2
Molecular Design and Applications of Photofunctional Polymers and Materials
Edited by Wai-Yeung Wong and Alaa S Abd-El-Aziz

Published by the Royal Society of Chemistry, www.rsc.org

derivative as electron-donating semiconductor and a perylene derivative as electron-accepting semiconductor, which were sandwiched between a transparent conducting oxide anode and a metal cathode.[7] The offset between the lowest unoccupied molecular orbital (LUMO) level of the donor and the LUMO level of the acceptor provides the driving force for efficient exciton dissociation. However, due to a short exciton diffusion length in organic semiconductors, only the excitons generated close to the interfaces dissociate into free charges and contribute to the photocurrent, resulting in limited power-conversion efficiency.[8,9] In 1995, a novel concept of the so-called bulk heterojunction (BHJ) was introduced,[10,11] accounting for the short exciton diffusion length in disordered organic semiconductors, as well as the required thickness for sufficient light absorption. In the past decade, polymer solar cells based on BHJ structures have been developed widely.[12–16] Although the BHJ structure can realize efficient light absorption and exciton dissociation, the charge collection in BHJ polymer solar cells plays an important role in determining the final performance of the cells.

In this chapter, in addition to a brief introduction to the physics of polymer PVs, approaches towards efficiently harvesting photogenerated carriers in polymer PVs are discussed, including the morphology control of polymer-fullerene derivative blends to realize balanced hole/electron transport in the active layer, the electrode modification to restrain the photogenerated carrier from recombination and enhance carrier extraction, and the tandem structure to improve the optical absorption and carrier collection.

14.2 Working Principles of Organic Photovoltaic Devices

The PV characteristics of a solar cell can be described by a current density-voltage (J-V) curve, which involves the parameters of the open-circuit voltage (V_{OC}), the short-circuit current (J_{SC}), the fill factor (FF) and the power-conversion efficiency (PCE), as shown in Figure 14.1.

The PCE of a solar cell that is the ratio of the maximum generated electrical power to the power of incident light is defined as:

$$\text{PCE} = \frac{P_{\text{out, max}}}{P_{\text{in}}} = \frac{V_{\text{max}} \times J_{\text{max}}}{P_{\text{in}}} = \frac{V_{\text{OC}} \times J_{\text{SC}} \times \text{FF}}{P_{\text{in}}} \tag{14.1}$$

where $P_{\text{out, max}}$ is the maximum electrical power generated by the solar cell and P_{in} the power of incident light. Obviously, in order for high PCE, we must obtain large V_{OC}, J_{SC} and FF simultaneously.

The J_{SC} is defined as the current under illumination at zero applied bias. To achieve large J_{SC}, more photons should be absorbed by the photoactive layer. This requires the photoactive materials, especially the conjugated polymers in BHJ solar cells, possessing high absorption coefficients and broad coverage to the solar spectrum. In polymer BHJ PV cells, the photogenerated excitons

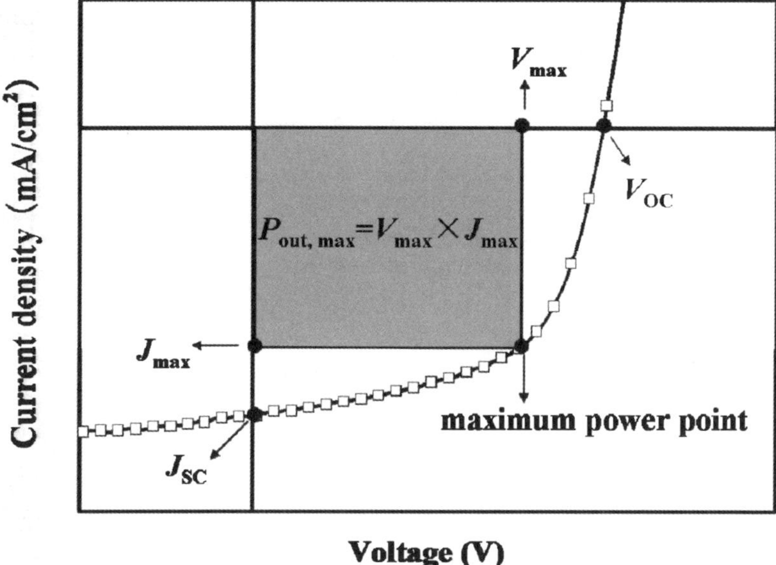

Figure 14.1 *J–V* curve of a solar cell under illumination.

should be separated into free electrons and holes and collected by the respective electrodes. Therefore, the J_{SC} of the polymer BHJ PV cells correlates with the BHJ morphology[13,17] and the electrode interfaces[18,19,20] in addition to the intrinsic properties of active materials.

The V_{OC} is defined as the applied external bias at which the photocurrent of the solar cell is zero. V_{OC} is a measure of the built-in potential, which is essential for charge separation and extraction in organic solar cells.[21] In polymer BHJ PV cells, V_{OC} scales with the energy difference between the highest occupied molecular orbital (HOMO) of the donor materials and the LUMO of the acceptor materials.[22,23] With respect to polymer-fullerene-based solar cells, there was a linear correlation between the V_{OC} and the oxidation potential of the conjugated polymer and reduction potential of the fullerene derivatives.[23] As a result, in order to maximize the V_{OC}, one should optimize the energy levels of the active materials, *i.e.* raising the LUMO of the electron acceptor and lowering the HOMO of the electron donor. In addition, the morphology of the active layers[24,25] and the interaction of electrode interfaces[26] could also affect the V_{OC}. The FF is defined as the quotient of the $P_{out, max}$ to the product of J_{SC} and V_{OC}:

$$FF = \frac{P_{out, max}}{V_{OC} \times J_{SC}} = \frac{V_{max} \times J_{max}}{V_{OC} \times J_{SC}} \tag{14.2}$$

Generally speaking, the FF correlates with the series resistance (R_S) and the shunt resistance (R_{SH}) of the solar cell. The R_S involves both the bulk resistance of organic semiconductors and the interfacial resistance at the

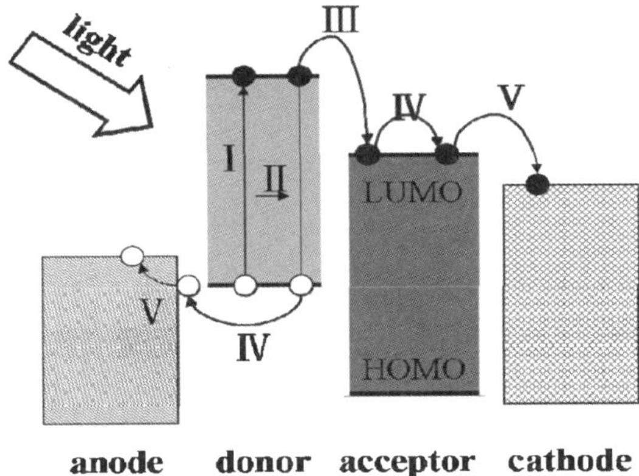

Figure 14.2   Illustration of the optoelectronic conversion processes in organic solar cells

interfaces between the active layer and electrodes, while the R_{SH} represents the total leakage current through the device. In order for a solar cell to achieve large FF and PCE, R_S has to be minimized and R_{SH} has to be maximized.

The optoelectronic conversion process of a polymer solar cell can be illustrated in Figure 14.2. There are several steps: (I) light absorption and exciton generation, (II) exciton diffusion, (III) exciton dissociation at the donor/acceptor interfaces, (IV) charge-carrier transport and (V) charge-carrier extraction at the respective electrodes.

14.2.1   Light Absorption and Exciton Generation

The active layers of polymer solar cells absorb the photons with energy beyond the absorption edge of the materials, and the excited electron–hole pairs (called excitons) are generated. Since the absorption coefficients of organic semiconductor materials are very high, the active layers of only a few hundred nanometres are thick enough to harvest most of the incident photons. However, a large bandgap of some conjugated polymers may cause a large mismatch between the absorption of active layers and the solar spectrum. Taking the poly(3-hexylthiophene) (P3HT, E_g =1.9 eV) as an example, it is only capable of absorbing the solar light in a wavelength range of 350–650 nm and about 46% of the available solar photons.[27] One of the most common approaches to broaden the spectral coverage is to utilize low-bandgap conjugated polymers as electron donors.[14,15,28]

14.2.2   Exciton Diffusion and Dissociation

The photogenerated excitons in organic semiconductors possess high binding energy. A major difference between organic solar cells and inorganic solar cells resides in the nature of the primary photoexcited state.[29] In inorganic semiconductors, the absorption of photons leads directly to the creation of free electrons and holes, which can be collected at the respective electrodes. However, in organic semiconductors with relatively low dielectric constants, the absorption of photons induces electron–hole pairs with binding energies of 50–200 mV. The internal electric field is not strong enough to overcome this binding energy. Therefore, an external driving force is needed, which is usually provided by the energy offset of the LUMO level between electron donor and acceptor. As a result, the exciton dissociation occurs only at the donor/acceptor interface. Because of the poor exciton diffusion length in organic semiconductor materials, the excitons are generated far away from the interface recombine and do not contribute to the photocurrent. To overcome this drawback, the concept of BHJ is proposed,[10,11] where electron donor and acceptor are blended together to obtain a bicontinuous donor–acceptor interpenetrating network for realizing efficient exciton dissociation.

14.2.3 Charge-Carrier Transport and Extraction

The free electrons and holes generated after exciton dissociation are transported through the acceptor and donor domains, and then collected at the respective electrodes. Photogenerated charge carriers should migrate to the collecting electrodes within their lifetime, and otherwise they would experience recombination causing photocurrent losses. Therefore, in addition to broad absorption of the active materials and efficient exciton dissociation, efficient charge extraction to the respective electrodes is another key issue to determine the photocurrent and the final PCE. The efficiency of charge collection correlates not only with the mobilities of the active materials and the BHJ morphology, but also with the electrode-active layer interfaces and device configuration. Below, we will discuss the strategies towards improving the charge-collection efficiency for polymer BHJ PV cells.

14.3 Optimization of the Film Morphology for Efficient Charge Collection

In polymer PV cells, the morphology of donor–acceptor blend films must be well controlled in order to realize efficient charge-carrier collection and high PCE. There are two fundamentals: (1) formation of a homogeneous BHJ morphology between the donor and acceptor materials with large interfacial areas for efficient charge separation; (2) well-connected individual pathways of the donors and acceptors for efficient hole and electron collection. The donor–acceptor BHJ morphologies can be manipulated through two main ways including solution-processing conditions and postprocessing treatments.

Processing conditions include the ratio of materials,[15] the selection of solvent[12] and the use of mixed solvents.[30,31] For the poly[*N*-9''-hepta- decanyl-2,7-carbazole-*alt*-5,5-(4 ',7'-di-2-thienyl-2',1',3'-benzothiadiazole)] (PCDTBT): [6,6]-phenyl C_{70}-butyric acid methyl ester ($PC_{70}BM$) blend system, a weight ratio of 1:4 was found to form a better film morphology to achieve efficient charge transport and collection.[15] A proper solvent can also induce the formation of BHJ film with nanoscale donor–acceptor phase separation and lead to enhancement of PCE, such as the poly[2-methoxy-5-(3,7-dimethyloctyloxy)]-1,4-phenylenevinylene) (MDMO-PPV):[6,6]-phenyl C_{60}-butyric acid methyl ester ($PC_{60}BM$) system processed with chlorobenzene solution.[32] In recent years, the use of processing additives, such as 8-diiodooctane (DIO), was implemented to construct the BHJ film with nanoscale donor–acceptor phase separation and increased hole mobility, boosting the PCE of the cells.[33,34] The processing additives with a higher boiling point may selectively dissolve one of the active materials in the blends and induce the donor–acceptor phase separation during the slow drying process.[30] Postprocessing treatments include the thermal annealing,[13] solvent annealing,[17] and even microwave annealing.[35] As a donor–acceptor blend film deposited from a solution is thermodynamically unstable, thermal annealing can provide energy for the blend film to equilibrate towards a favourable BHJ configuration.[36] "Solvent annealing" was also developed for the P3HT:PCBM system to construct BHJ film with balanced charge transport to boost the PCE.[17]

Taking the P3HT:PCBM system as an example, we have addressed how the BHJ morphology with a balanced charge transport in the P3HT and PCBM domains influences the charge collection and final PV performance.[37] The P3HT:PCBM (1:0.8 in weight ratio) blend film deposited from chlorobenzene (CB) was subjected to thermal annealing (150 °C, 1 min), 1,2-dichlorobenzene (ODCB) solvent vapor treatment (30 min) or both. The UV-visible absorption

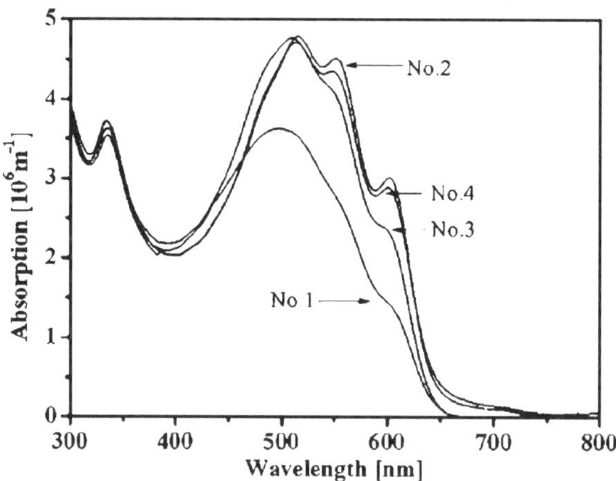

Figure 14.3   UV-vis absorption spectra for P3HT:PCBM blend films with different treatments: pristine (No. 1), DCB vapor treated (No. 2), annealed (No.3), and DCB vapor treated and annealed (No. 4). Reprinted with permission from Y. Zhao, Z. Xie, Y. Qu, Y. Geng and L. Wang, *Applied Physics Letters*, 2007, 90, 043504. Copyright 2007, American Institute of Physics.

of P3HT:PCBM films with different treatments is shown in Figure 14.3. For the pristine film (No. 1), the P3HT absorption peak was at 500 nm with a shoulder at 600 nm. In the annealed film (No. 3), the absorption increased largely and the three vibronic absorption shoulders became clear, indicating the formation of ordered P3HT domains. The absorption of the ODCB vapor-treated films (Nos. 2 and 4) became much stronger in the red region compared with the annealed film, attributing to an enhanced conjugation length and the more ordered structure of P3HT. This indicated that the ODCB vapor treatment was a more effective way to realize P3HT self-organization. Since the solvent molecules could penetrate into the film and increase the space between polymer chains, the chains became mobile and self-organization could occur to form the more ordered P3HT domains.

The so-called single-charge-carrier devices were fabricated to investigate the effect of different treatment on the hole and electron transport in P3HT and PCBM domains. It was found that the ODCB vapor-treated P3HT:PCBM blend film demonstrated more efficient hole transport than the thermally annealed film. The electron transport for the films with and without ODCB vapor treatment could be largely enhanced through thermal annealing, which indicated that thermal annealing could effectively activate PCBM molecules to diffuse and aggregate into clusters for better electron transport. By combining solvent vapor treatment and post-thermal annealing, efficient and balanced hole and electron transport were realized in the P3HT:PCBM blend film,

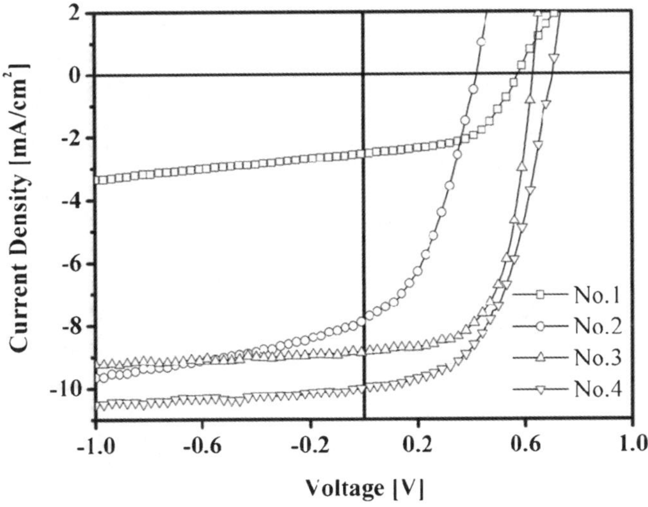

Figure 14.4   The illuminated *J–V* curves of the four solar cells under white light illumination with an intensity of 100 mWcm^{-2}: as produced (No. 1), ODCB vapor treated (No. 2), annealed (No. 3), and ODCB vapor treated and annealed (No. 4). Reprinted with permission from Y. Zhao, Z. Xie, Y. Qu, Y. Geng and L. Wang, *Applied Physics Letters*, 2007, 90, 043504. Copyright 2007, American Institute of Physics.

which provided efficient individual pathways for holes and electrons and improved the charge carrier collection. As shown in Figure 14.4, the solvent vapor-treated and thermally annealed device showed a largest J_{SC} among these devices, indicating the charge collection was improved by manipulating the BHJ film morphology.

The field-effect mobility of P3HT in thin-film transistors can be effectively improved *via* the highly interconnected P3HT nanofibrillar networks as the charge-transport conduits. We introduced P3HT nanofibrils in P3HT:PCBM composite films *via* P3HT preaggregation in solution by adding a small amount of acetone, aiming at improving the P3HT interconnectivity and the hole collection.[38] The transmission electron microscopy (TEM) images of the P3HT:PCBM (1:0.8 in weight ratio) composite films fabricated from CB solution containing various percentages of acetone are shown in Figure 14.5. Compared to the film fabricated from the neat CB solution, the P3HT:PCBM films showed obvious P3HT nanofibril networks with the addition of acetone. The P3HT nanofibrils were about 10–15 nm wide and 0.1–3 μm long. With an increase of the acetone content, the amount of P3HT nanofibrils in the composite film is increased. The X-ray diffraction (XRD) results of the P3HT:PCBM composite films showed that the intensity of the (100) peak of P3HT was enhanced with the increase of acetone content, indicating that more crystalline P3HT nanofibrils are formed. The hole transport in P3HT:PCBM composite films was gradually enhanced with the increase of the amount of

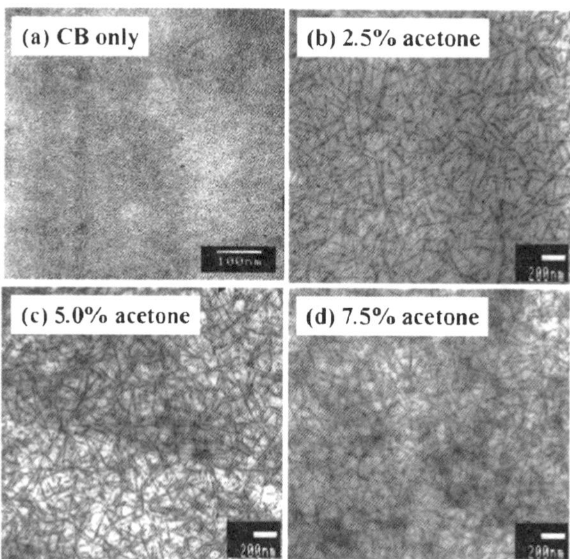

Figure 14.5   TEM images of the P3HT:PCBM composite films fabricated from CB solution containing various percentages of acetone: (a) none, (b) 2.2%, (c) 5%, (d) 7.5%. Reprinted with permission from Y. Zhao, S. Shao, Z. Xie, Y. Geng and L. Wang, *J. Phys. Chem. C*, 2009, 113, 17235. Copyright 2009 American Chemical Society.

crystalline P3HT nanofibrils.[38] The improvement might be attributed to the increase of P3HT crystallinity as well as the good connectivity of crystalline P3HT nanofibrils.

The maximum possible generation rate (G_{max}) for producing free carriers out of bound electron–hole pairs at the donor/acceptor interfaces can be calculated through eqn (14.3), where J_{sat} is the saturated photocurrent determined from the photocurrent (J_{ph}) at a high effective voltage ($(V_0-V) >$ 10 V), e the electric charge, and L the thickness of the active layer.

$$J_{sat} = eG_{max}L \qquad (14.3)$$

Figure 14.6 shows the value of G_{max} in P3HT:PCBM films as a function of the acetone content in CB solution. For the as-produced PV cells, the G_{max} increased with the increase of the added acetone percentage in CB solution of P3HT, indicating that the preaggregated P3HT nanofibrils in solution induced the PCBM demixing in P3HT:PCBM composite films to form individual pathways to efficiently collect free charges. High content of P3HT nanofabrils would decrease the G_{max} since the PCBM molecules were trapped among the crystalline P3HT nanofibrils. Therefore, the charge-collection efficiency was

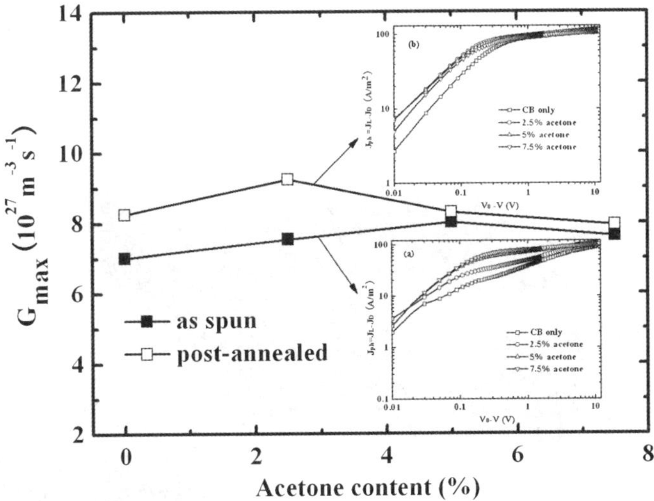

Figure 14.6 Calculated G_{max} of electron–hole pairs for the as-produced (filled squares) and postannealed (open squares) P3HT:PCBM BHJ PV cells with the P3HT:PCBM composite films fabricated from CB solution containing various percentages of acetone. The two inserted figures are photocurrent as a function of effective applied voltage (V_0-V) for the as-produced (a) and postannealed (b) P3HT:PCBM BHJ PV cells with the P3HT:PCBM composite films made from CB solution containing various percentages of acetone: 0% (open squares), 2.5% (open circles), 5% (open up triangles), and 7.5% (open down triangles). Reprinted with permission from Y. Zhao, S. Shao, Z. Xie, Y. Geng and L. Wang, *J. Phys. Chem. C*, 2009, 113, 17235. Copyright 2009 American Chemical Society.

decreased even when the hole transport was further enhanced. When the devices were annealed, the PCBM demixing into aggregates could easily occur resulting in a large enhancement of G_{max}. The small amount of P3HT nanofibrils (2.5% acetone) did not restrain PCBM molecules from demixing and G_{max} was further enhanced. However, the high content of P3HT nanofibrils (5% and 7.5% acetone) restrained PCBM from further demixing and the G_{max} was slightly improved.[38]

These results indicate that high contents of crystalline P3HT nanofibrils may restrain PCBM molecules from demixing with the P3HT component, which forms electron traps in the active layer, and hence reduce the charge-collection efficiency. Small contents of P3HT nanofibrils not only improve the demixing between P3HT and PCBM components, but also enhance the hole transport *via* crystalline P3HT nanofibrillar networks, resulting in efficient charge collection.[38]

As discussed above, the charge carrier collection correlates strongly with the morphology of the blend films in BHJ polymer solar cells. The two components of donor and acceptor should be separated into polymer donor and PCBM acceptor domains with high purity on an appropriate scale to favor the charge collection and at the same time not affect the exciton dissociation. The donor and acceptor pathways with high charge mobilities may improve the charge collection and the final PCE. However, it is difficult for the blend films to control a donor–acceptor separation without isolated islands or dead ends, which would act as charge trapping sites and increase the charge recombination. An ideal ordered BHJ structure is shown in Figure 14.7. The dimensions of both phases in the ideal structure should be controlled to ensure

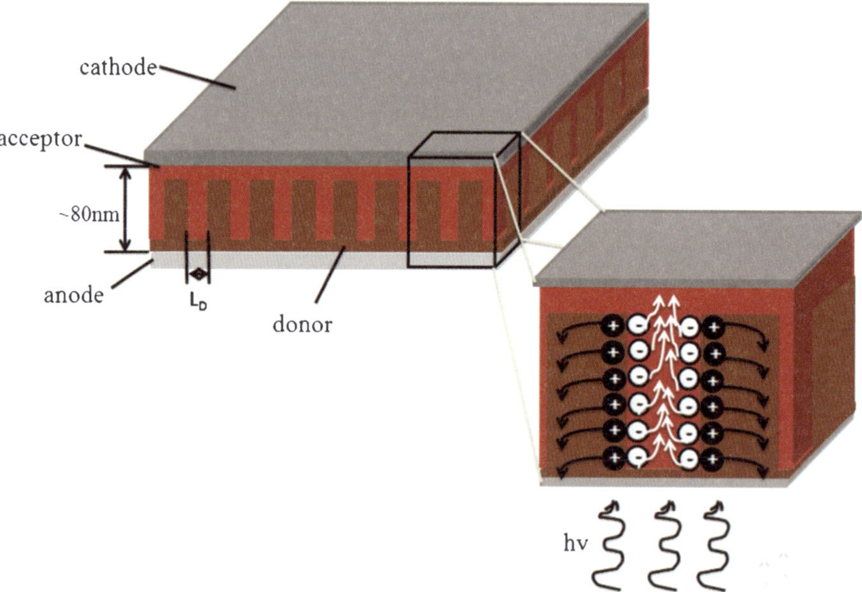

Figure 14.7   Schematic illustration of an ideal ordered BHJ PV cell.

that every spot in the film is within the exciton diffusion length of the interface between the two components. More importantly, there are no isolated islands and dead ends in the structure. After the excitons are dissociated, the electrons and holes have straight pathways to the electrodes. The ordered BHJ ensures that the carriers escape from the device as quickly as possible, which would reduce the charge recombination. The ordered BHJ structures are more difficult to fabricate than the disordered blends.

Designing new molecules with a structure of donor–acceptor can provide a way to fabricate such an ordered nanostructure film with donor–acceptor bicontinuous phase due to the intrinsically preferred composition-dependent nanophase separation.[39,40] Monodisperse conjugated oligomers are characterized by a uniform chemical structure as well as a tunable molecular length; therefore, they are ideal scaffolds for well-defined nanostructures with a length scale equal to or less than the exciton diffusion length of organic semiconductors. Following this concept, the D–A co-oligomers, F3T4-hP, F4T6-hP, and F5T8-hP with different molecular lengths were synthesized.[41] Figure 14.8 illustrates the chemical structures of these molecules. Oligo(fluorene-*alt*-bithiophene)s (OFbTs) are selected as electron-donor segments not only because they are typical p-type organic semiconductors but also because their liquid-crystalline property may provide an additional driving force for the formation of ordered films *via* post-treatment. Perylene diimide (PDI) and its derivatives are recognized as good electron-acceptor materials, and their strong π–π interactions qualify them as great building blocks for functional supramolecular architectures.[41]

Figure 14.9 shows the TEM images and the selected-area electron diffractions (SAED) patterns (inset) of F3T4-hP, F4T6-hP, and F5T8-hP thin films after thermal annealing or solvent-vapor annealing. After thermal annealing at 210 °C for 2 min, dark-bright stripes appeared in the films for all the three co-oligomers. However, the SAED patterns indicate that the order of the films is still relatively low. The film order of the co-oligomers can be significantly improved with CH_2Cl_2 vapor annealing. In Figures 14.9(d)–(f), the persistent lengths of the lamellae increase from 50–150 nm for thermal-annealed films to 100–400 nm for solvent-vapor annealed films, and the (002) and (003) diffraction rings in SAED become distinct. The periods of the lamellae are 5.6, 7.4, and 8.9 nm for F3T4-hP, F4T6-hP, and F5T8-hP, respectively.[41]

Donor moiety   Acceptor moiety

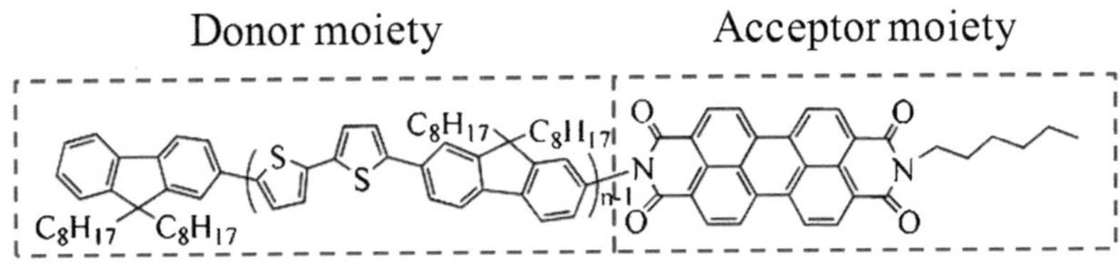

n=3,F3T4-hP; n=4, F4T6-hP; n=5, F5T8-hP

Figure 14.8   Chemical structures of F3T4-hp, F4T6-4h and F5T8-hp. Reprinted with permission from L. Bu, X. Guo, B. Yu, Y. Qu, Z. Xie, D. Yan, Y. Geng and F. Wang, *J. Am. Chem. Soc.*, 2009, **131**, 13242. Copyright 2009 American Chemical Society.

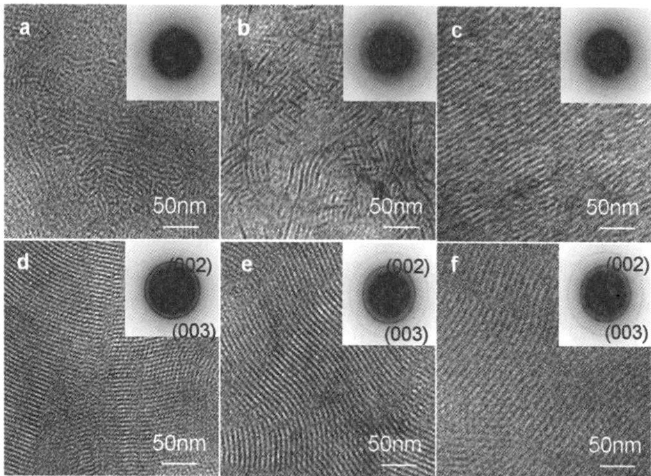

Figure 14.9 TEM images and SAED patterns (inset) of F3T4-hP (a,d), F4T6-hP (b,e), and F5T8-hP (c,f) thin films after thermal annealing (a–c) or solvent-vapor annealing (d–f). Reprinted with permission from L. Bu, X. Guo, B. Yu, Y. Qu, Z. Xie, D. Yan, Y. Geng and F. Wang, *J. Am. Chem. Soc.*, 2009, 131, 13242. Copyright 2009 American Chemical Society.

These results indicate that the co-oligomers have formed alternating D–A lamellar nanostructures with the period depending on the molecular length. As a result, single-molecular organic solar cells with external quantum efficiency (EQE) up to 46% at 410–500 nm and PCE up to 1.50% have been demonstrated. These results represent the best performance for single-molecular solar cells to date.

14.4 Electrode Modification Towards Enhancing Charge Collection

In BHJ polymer solar cells, the photogenerated charges must be collected at the respective electrodes to complete the power-conversion process. In addition to the active-layer morphology, the interfaces between the electrodes and the active layer are also crucial for efficient charge collection. The electrode modification can enhance not only the V_{OC}, but also the charge-collection efficiency. In this section, we will discuss the effect of the electrode modification on the charge collection.

14.4.1 Anode Modification Towards Enhancing Charge Collection

Conductive indium tin oxide (ITO) is widely used as an anode in polymer BHJ solar cells. The low work function of ITO (4.8 eV) does not match the HOMO level of most polymer donors used and results in energy loss during charge extraction. In addition, the direct contact of the ITO anode with the acceptor materials in BHJ may result in a large recombination of charges and a loss of photocurrent. A water-soluble poly(ethylenedioxythiophene):poly(styrenesulfonate) (PEDOT:PSS) anode buffer layer is commonly used in the BHJ polymer solar cells to enhance the PV performance due to its high work function (5.0–5.2 eV) and planarization to the ITO surface.[42] However, its electrical inhomogeneity limits the electron-blocking capability and its acidic nature has been demonstrated to etch ITO and degrade the device lifetime,[43–45] which has motivated the development of neutral anode buffer layers. The modification of ITO with silane-based self-assembled monolayers (SAMs) has been successfully employed to tune the work function of ITO. The work function of ITO was tuned to 5.16 eV by using SAMs with an electron-withdrawing CF_3 group.[46] Consequently, an ohmic contact between the active layer and ITO is realized and improves the hole collection. Transition-metal oxides such as vanadium oxide (V_2O_5),[47] molybdenum oxide (MoO_3),[48] nickel oxide (NiO),[49] and tungsten oxide (WO_3)[50] have also been employed as hole-transporting/electron-blocking buffers to improve the interfacial properties between ITO and the active layer in BHJ PV cells. These metal oxides with wide bandgaps possess high optical transparency in the visible region. The LUMO levels of these semiconducting oxides are sufficiently higher to block electron leakage to the anode. Moreover, the Fermi levels of the oxides align well with the HOMO of the conjugated polymers, forming an ohmic contact at the interface to enhance the hole extraction from polymer donor to ITO anode.

The above-mentioned anode modification approaches aim at increasing the effective work function of the anode or blocking electron transfer from the acceptor in the BHJ to the ITO anode. In the case of BHJ polymer solar cells, limited by low charge mobility of the semiconducting polymers, the active layers comprised of polymer-fullerene derivative blends should be thin enough to match the short hole drift length and restrain current loss from charge recombination. However, the thin active layers reduce the light-harvesting efficiency and hence hinder the further enhancement of PCE. How to increase the active-layer thickness and enhance optical absorption without compromising charge-carrier collection is a serious problem for BHJ polymer solar cells.

Recently, distinctive ITO nanorods were employed to serve as buried electrodes for BHJ polymer solar cells,[51] as shown in Figure 14.10. In this cell structure, the randomly oriented nanorod electrodes were deposited on an ITO-coated glass substrate using an oblique electron-beam evaporation method. The embedded nanoelectrodes allowed three-dimensional conducting pathways for low-mobility holes, offering a highly scaffolded cell architecture. The active-layer thickness is increased to 150–200 nm, which can efficiently absorb incident photons. As a result, the PCE of a P3HT:PCBM BHJ solar cell with ITO nanoelectrode is increased to about 3.4% and 4.4% under one-sun and five-sun illumination conditions, respectively, representing an enhancement factor of up to 10% and 36% compared to a conventional counterpart.[51]

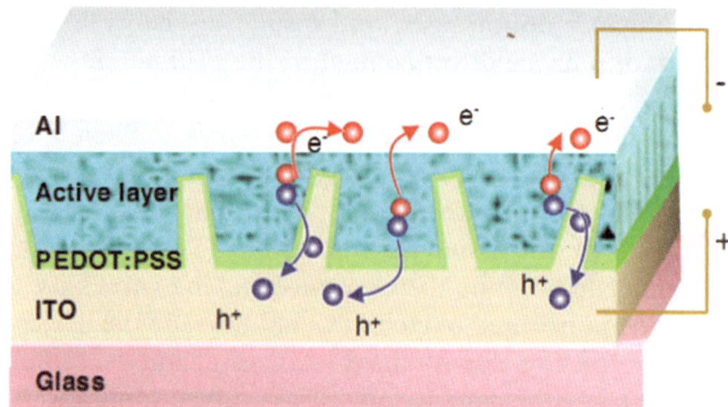

Figure 14.10 Device schematic with embedded nanoelectrodes. Reprinted with permission from P. C. Yu, C. H. Chang, M. S. Su, M. H. Hsu and K. H. Wei, *Appl. Phys. Lett.*, 2010, **96**, 153307. Copyright 2010, American Institute of Physics.

We developed a kind of BHJ polymer solar cell based on blends of poly[2-methoxy-5-(2'-ethylhexyloxy)-1,4-phenylene vinylene] (MEH-PPV) donor and n-type ZnO nanoparticle acceptor with high-mobility p-type Cu_2O nanocrystals as a hole-collection antenna to enhance the hole-collection efficiency.[52] The proposed device configuration of the PV cell with the Cu_2O nanocrystal as an anode buffer layer and the scanning electron microscopy (SEM) images of the Cu_2O nanocrystals on ITO glass substrate are shown in Figure 14.11. It can be seen that the Cu_2O nanocrystals were dispersed on the ITO surface and did not form a dense and continuous Cu_2O film. Spin coating the MEH-PPV:ZnO blend solution did not wash off the Cu_2O nanocrystals due to good adhesion of Cu_2O nanoparticles to the ITO surface after vacuum drying. The

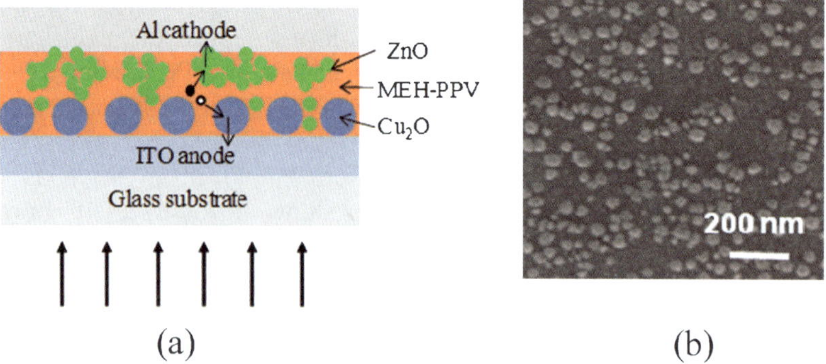

Figure 14.11 (a) The proposed PV cell configuration with Cu_2O nanocrystal as an anode buffer layer. (b) SEM images of Cu_2O nanocrystals on ITO substrate. Reprinted with permission from S. Shao, F. Liu, Z. Xie and L. Wang, *J. Phys. Chem. C.*, 2010, 114, 9161. Copyright 2010 American Chemical Society.

dispersed structure of Cu_2O nanocrystals on the ITO anode surface could increase the contact area between Cu_2O nanocrystals and MEH-PPV, and hence the hole collection was enhanced. In other words, the Cu_2O nanocrystals on the ITO anode surface percolated into the MEH-PPV:ZnO active layer to serve as a hole-collection antenna to efficiently collect holes. The HOMO and LUMO values of Cu_2O, MEH-PPV and ZnO are -3.3 and -5.4 eV, -3.0 and -5.3 eV, -4.2 and -7.6 eV, respectively. The work function of ITO, PEDOT:PSS and Al are -4.8, -5.0 and -4.3 eV, respectively. The Cu_2O nanocrystal has a HOMO level of -5.4 eV, close to the HOMO level of MEH-PPV (-5.3 eV), allowing the photon-generated holes on MEH-PPV to be transferred to Cu_2O.[52] It is noted that the HOMO level of Cu_2O is higher than that of PEDOT:PSS, which may render the PV cell a high V_{OC} by replacing Cu_2O for PEDOT:PSS.

The performance dependence of the PV cells with different anode structures on the active-layer thickness has been investigated. The variations of J_{SC} and PCE of the devices are shown in Figure 14.12 as a function of the active-layer thickness. The devices with different anode structures keep their V_{OC} values stable when the active-layer thickness is changed. For the devices with ITO only and ITO/PEDOT:PSS structure, the maximum J_{SC} and PCE are achieved at an active-layer thickness of 100 nm. On further increasing the active-layer thickness, the J_{SC} and PCE were decreased. The maximum J_{SC} and PCE were obtained at an active-layer thickness of 140 nm for the device with ITO/Cu_2O structure. The device with ITO/Cu_2O/MEH-PPV:ZnO (140 nm)/Al structure showed a J_{SC} of 3.92 mA cm^{-2}, a V_{OC} of 1.0 V, and a FF of 0.51, respectively, which gave an overall PCE over 2.0%, representing the highest value for a MEH-PPV:ZnO blend PV cell. The peak EQE of this device is about 53%,

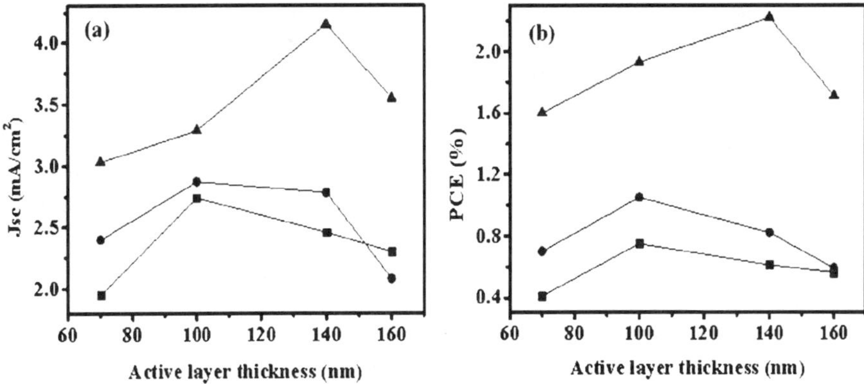

Figure 14.12   The variation of J_{SC} and PCE with the varied active-layer thickness for the MEH-PPV:ZnO (1:3.3) PV cells with different anode structures: ITO-only (solid squares), ITO/PEDOT:PSS (solid circles), and ITO/Cu_2O (solid triangles). Reprinted with permission from S. Shao, F. Liu, Z. Xie and L. Wang, *J. Phys. Chem. C.*, 2010, 114, 9161. Copyright 2010 American Chemical Society.

much higher than the values of 32% and 33% for the devices with ITO-only and ITO/PEDOT:PSS structure. These results highlight that the Cu_2O nanocrystal serves as a hole-collection antenna to enhance the charge-collection efficiency of the resulting PV cell.

14.4.2 Cathode Modification Towards Enhancing Charge Collection

In common polymer PV cells, the active layer is directly in contact with a thermally deposited cathode, such as Al. This kind of polymer PV cells suffers from some problems that significantly limit the charge-collection efficiency and final PCE. First, deposition of the cathode metal onto the BHJ active layer introduces defects (typically in the form of deep trap levels) resulting in increased leakage current and low R_{SH} and limits the enhancement of the PV cells.[53] Secondly, the incident optical electric field vanishes at the surface of the highly conducting electrode limiting the effective absorption of the active layer. Thirdly, the strong dipole effect at PCBM/cathode interface may cause an increase of the effective work function of metal cathode, resulting in a decrease of V_{OC}.[54] Hence, it is necessary to modify the BHJ active layer/cathode interface to enhance the performance of the polymer solar cells.[18,26,55,56] For example, the submonolayer LiF has been successfully utilized in polymer solar cells to modify the work function of the cathode and increase the V_{OC}.[18] However, LiF is thermally deposited in vacuum and is impossible to be used in printing techniques, which are the biggest advantage of polymer solar cells compared to their inorganic counterparts. Moreover, the submonolayer of LiF cannot effectively protect metal atoms from migrating into the active layer during the metal deposition process; the thicker LiF increases the R_S and reduces charge collection due to its insulating property.

We developed an efficient polymer solar cell based on P3HT:PCBM BHJ by introducing a conjugated polymer between the active layer and the cathode *via* spin coating from ethanol solution.[57] The charge-collection efficiency was greatly improved by using this kind of conjugated polymer cathode buffer layer. Unlike the thin insulating buffer layers such as PEO[56] or LiF,[18] the increased thickness of the semiconducting conjugated polymer buffer layer could more effectively protect metal atoms from penetrating into the active layer and hence the R_{SH} of the cell was increased. This kind of conjugated polymers was poly(9,9-bis(6'-diethoxylphosphorylhexyl)fluorene) (PF-EP), possessing good solubility in ethanol solution and its chemical structure is shown in the inset of Figure 14.13. Figure 14.13 shows the absorption spectra of pristine PF-EP film spin coated from ethanol solution, the pristine P3HT:PCBM blend film spin coated from CB solution, and the pristine P3HT:PCBM blend layer overlaid by a thin layer of PF-EP spin coated from ethanol solution. The absorption of P3HT:PCBM/PF-EP bilayer film exhibited a simple superposition of the individual P3HT:PCBM and PF-EP absorption. Since these films were measured under the same condition, the

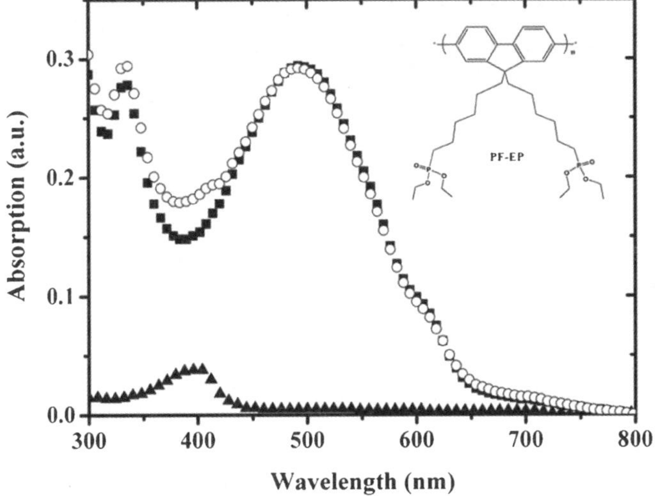

Figure 14.13 Absorption spectra of pristine PF-EP solid film (filled triangles), the pristine P3HT:PCBM blend film (filled squares), and the pristine P3HT:PCBM blend film overlaid by a thin layer of PF-EP *via* spin coating from ethanol solution (open circles). The inset shows the chemical structure of PF-EP. Reprinted from *Solar Energy Materials and Solar Cells*, 93, Y. Zhao, Z. Xie, C. Qin, Y. Qu, Y. Geng and L. Wang, Enhanced charge collection in polymer photovoltaic cells by using an ethanol-soluble conjugated polyfluorene as cathode buffer layer, 604–608, Copyright (2009), with permission from Elsevier.

almost identical absorption of P3HT:PCBM in both shape and intensities for the P3HT:PCBM and P3HT:PCBM/PF-EP films indicated that the PF-EP layer could be well deposited on the P3HT:PCBM film through spin coating from ethanol solution without destroying the morphology of the underlying P3HT:PCBM blend layer.

Figure 14.14 shows the dark and illuminated J–V curves of the P3HT:PCBM PV cells with Al or 5.0 nm PF-EP/Al cathode. It can be seen that the reverse current was reduced by two orders of magnitude due to the reduced leakage current by inserting 5.0 nm of PF-EP. This indicated that the PF-EP layer could restrain Al atoms from migrating into the P3HT:PCBM blend active layer and therefore the leakage current was reduced and R_{SH} increased. More importantly, post-thermal annealing of the PV cell reduced effectively the R_S from 10.19 to 6.32 Ω cm^2 and resulted in high charge collection with the J_{SC} increasing from 9.01 to 10.28 mA cm^{-2} due to the improved PF-EP/Al contact. These results show that the introduction of a conjugated PF-EP buffer layer can effectively improve charge collection and finally the PCE of the polymer solar cell. Its solution-processing property is also suitable for the future roll-to-roll printing techniques for large-area deposition.

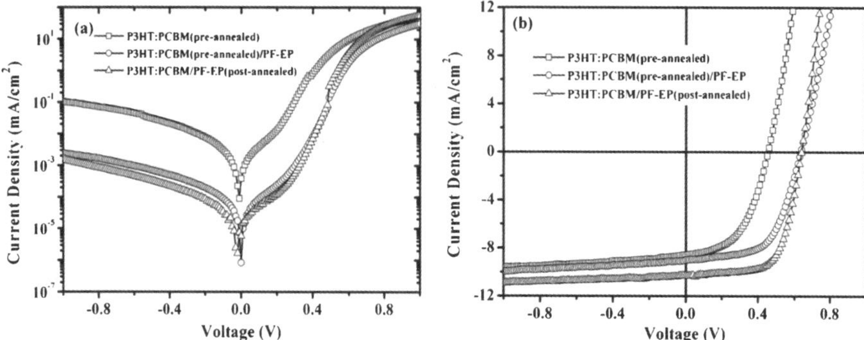

Figure 14.14   Dark (a) and illuminated (b) *J–V* curves of three kinds of P3HT:PCBM blend PV cells under 100 mW cm^{-2} white light illumination: P3HT:PCBM (annealed)/Al (open squares); P3HT:PCBM (annealed)/ 5.0 nm PF-EP/Al (open circles); P3HT:PCBM/5.0 nm PF-EP/Al (postannealed, open triangles). Reprinted from *Solar Energy Materials and Solar Cells*, 93, Y. Zhao, Z. Xie, C. Qin, Y. Qu, Y. Geng and L. Wang, Enhanced charge collection in polymer photo-voltaic cells by using an ethanol-soluble conjugated polyfluorene as cathode buffer layer, 604–608, Copyright (2009), with permission from Elsevier.

14.5 Parallel-Connected Tandem Polymer Solar Cells

In a single BHJ polymer solar cell, the optimized thickness of the active layer should be of the order of the carrier collection length, *i.e.* the distance that a free carrier can drift within the active layer prior to recombination with its opposite charge. However, limited by the low hole mobilities of conjugated polymer donors, the thickness of the BHJ active layer is usually as low as around 100 nm for realizing efficient charge collection. Although further increase of the BHJ layer thickness may enhance the incident light absorption, the decreased charge collection and increased R_S would result in a decreased PV performance. If the two single BHJ polymer solar cells with the optimized active-layer thickness are stacked in parallel connection together, the total absorption of the stacked cell can be greatly enhanced without compromising the carrier collection length.[58]

We fabricated a tandem polymer solar cell with the two subcells connected in parallel.[59] The device configuration is shown in Figure 14.15. The conjugated polymers of poly[2,6-(4,4-bis-(2-ethylhexyl)-4H-cyclopenta[2,1-*b*;3,4-*b'*]dithio-phene) -*alt*-4,7-(2,1,3-benzothiadiazole)] (PCPDTBT) and P3HT were selected to serve as electron donors in the two subcells, respectively. PCBM was used as the electron acceptor. The absorption of the PCPDTBT:PCBM (1:3) composite film covered from 400 to 900 nm with strong absorption located in the range of 650–800 nm. The absorption of the P3HT:PCBM (1:0.8) film covered the visible spectral range from 400 to 650 nm, just located at the valley of the PCPDTBT:PCBM absorption spectrum. The complementary absorption of

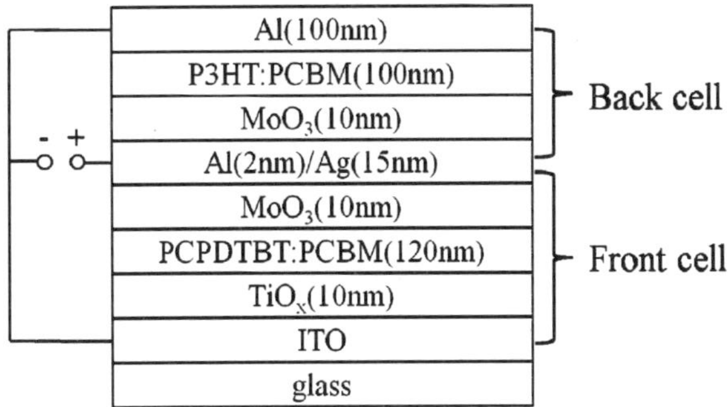

Figure 14.15   The device configuration of the parallel-connected tandem polymer solar cell. Reprinted from *Organic Electronics*, 10, X. Guo, F. Liu, W. Yue, Z. Xie, Y. Geng and L. Wang, Efficient tandem polymer photovoltaic cells with two subcells in parallel connection, 1174–1177, Copyright (2009), with permission from Elsevier.

the P3HT:PCBM (1:0.8) and PCPDTBT:PCBM (1:3) BHJ films enhanced the spectral coverage to the solar light. The ITO covered with 10-nm thick electron-transporting TiOx and the top Al cathode are used in the tandem cell to collect electrons. The middle structure of MoO_3 (10 nm)/Al (2 nm)/Ag (15 nm)/MoO_3 (10 nm) serves as a semitransparent anode to collect holes.

This kind of three-terminal device configuration allows for separate measurement of the two subcells and the tandem solar cell in one device. The front PCPDTBT:PCBM PV cell has a V_{OC} of 0.65 V, a J_{SC} of 6.49 mA cm^{-2} and a calculated FF of 0.34. Therefore, the overall PCE of this cell is *ca.* 1.45%. For the back P3HT:PCBM PV subcell, the V_{OC}, J_{SC} and FF are 0.60 V, 4.25 mA cm^{-2} and 0.49, respectively, leading to a PCE of 1.24%. The V_{OC} of the tandem polymer PV cell is 0.65 V, similar to each of the two subcells. The J_{SC} of the tandem cell reaches 11.32 mA cm^{-2}, close to the sum of the J_{SC}s of the two subcells, indicating that the two subcells are in parallel connection. The overall efficiency of the tandem polymer PV cell was *ca.* 3.10%. Interestingly, this value is even higher than the sum of those of the two subcells. This could be attributed to the reduced R_S when the two subcells are stacked in parallel connection. Figure 14.16 shows the spectra of the EQEs of the two subcells and the tandem cell. It can be seen that the subcells show the known spectral response, which are in agreement with their absorption spectra. The front PCPDTBT:PCBM cell covered from 350 to 800 nm showing two dominant EQE peaks with *ca.* 17% at 430 nm and 26% at 750 nm. The back P3HT:PCBM cell exhibited a EQE of 30% at a range of 450–620 nm. The EQE of the tandem cell was the superposition of the two subcells.

These results indicate that efficient carrier collection can be achieved by utilizing this kind of parallel-connected tandem polymer solar cell. For some conjugated polymer donors with low hole mobility, the BHJ active layer should be thin enough to realize efficient carrier collection with sacrificing of the incident-light absorption. At this time, a tandem cell with parallel stacking

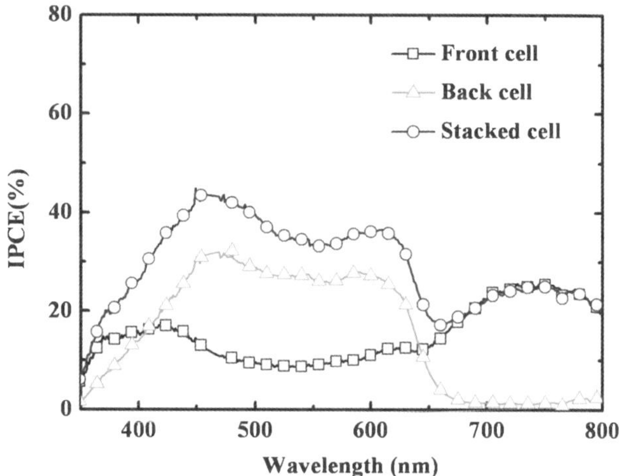

Figure 14.16 IPCE spectra of the front PCPDTBT:PCBM cell, the back P3HT:PCBM cell and the parallel-connected tandem polymer solar cell. Reprinted from *Organic Electronics*, 10, X. Guo, F. Liu, W. Yue, Z. Xie, Y. Geng and L. Wang, Efficient tandem polymer photovoltaic cells with two subcells in parallel connection, 1174–1177, Copyright (2009), with permission from Elsevier.

the two single BHJ cells with the optimized active-layer thickness can be fabricated to enhance the total absorption without compromising the carrier collection length. In addition, parallel-connected tandem cells do not require the match of J_{SC} in subcells, which is a prerequisite for series-connected tandem solar cells.[60,61] It allows for the subcells being optimized individually to achieve the highest J_{SC} since the overall J_{SC} of the stacked cell is the superposition of the two subcells.

14.6 Concluding Remarks

Efficient charge collection in BHJ polymer solar cells is an essential step for achieving high PCE. In general, both the morphology of active layers and the electrode interfaces play important roles in the charge-collection process. The active layer needs to form high-mobility bicontinuous interpenetrated networks to reduce the recombination of photogenerated holes and electrons and thus facilitate the charge-carrier transport and extraction. We take the P3HT:PCBM blend as an example to introduce several approaches to construct high-mobility bicontinuous P3HT:PCBM networks. In addition, efficient charge extraction can also benefit from the electrode interface modification. For example, p-type metal oxide semiconductor nanocrystals such as Cu_2O are tentatively developed as "hole-collection antenna" for improving the hole collection efficiency and the performance of the resulted PV cells is greatly enhanced. The interfaces of the top cathode/the active layer also influence the charge-collection efficiency. A kind of alcohol-soluble conjugated polymer is developed as the cathode buffer layer material successfully reducing the leakage current of the cell and enhancing the charge-collection efficiency. A parallel-connected stacked structure was made to enhance the charge-collection efficiency, which may be suitable for conjugated polymers with low hole mobilities. Thus, the absorption can be enhanced by increasing the total thickness of the active layer and the low charge collection in a single cell with a thick active layer can be avoided.

Acknowledgements

The author acknowledges the support from the National Natural Science Foundation of China (60977026, 20834005) and he also thanks Yingying Fu, Shuyan Shao, Jian Liu and Gang Fang for very helpful assistance in manuscript preparation.

References

1. J. H. Zhao, A. H. Wang, M. A. Green and F. Ferrazza, *Appl. Phys. Lett.*, 1998, **73**, 1991.
2. M. A. Green, K. Emery, D. L. King, S. Igari and W. Warta, *Prog. Photovolt.*, 2004, **12**, 365.
3. J. Roncali, *Chem. Rev.*, 1997, **97**, 173.
4. C. D. Dimitrakopoulos and P. R. L. Malenfant, *Adv. Mater.*, 2002, **14**, 99.
5. T. Tsuzuki, Y. Shirota, J. Rostalski and D. Meissner, *Sol. Energy Mater. Sol. Cells*, 2000, **61**, 1.
6. X. Yang and J. Loos, *Macromolecules*, 2007, **40**, 1353.
7. C. W. Tang, *Appl. Phys. Lett.*, 1986, **48**, 183.
8. J. G. Xue, S. Uchida, B. P. Rand and S. R. Forrest, *Appl. Phys. Lett.*, 2004, **84**, 3013.
9. M. Y. Chan, S. L. Lai, M. K. Fung, C. S. Lee and S. T. Lee, *Appl. Phys. Lett.*, 2007, **91**, 089902.
10. G. Yu, J. Gao, J. C. Hummelen, F. Wudl and A. J. Heeger, *Science*, 1995, **270**, 1789.
11. J .J. M. Halls, C. A. Walsh, N. C. Greenham, E. A. Marseglia, R. H. Friend, S. C. Moratti and A. B. Holmes, *Nature*, 1995, **376**, 498.
12. H. Hoppe, M. Niggemann, C. Winder, J. Kraut, R. Hiesgen, A. Hinsch, D. Meissner and N. S. Sariciftci, *Adv. Funct. Mater.*, 2004, **14**, 1005.
13. W. L. Ma, C. Y. Yang, X. Gong, K. Lee and A. J. Heeger, *Adv. Funct. Mater.*, 2005, **15**, 1617.
14. Y. Y. Liang, D. Q. Feng, Y. Wu, S. T. Tsai, G. Li, C. Ray and L. P. Yu, *J. Am. Chem. Soc.*, 2009, **131**, 7792.
15. S. H. Park, A. Roy, S. Beaupre, S. Cho, N. Coates, J.S. Moon, D. Moses, M. Leclerc, K. Lee and A. J. Heeger, *Nature Photon.*, 2009, **3**, 297.

16. Y. Y. Liang, Z. Xu, J. B. Xia, S. T. Tsai, Y. Wu, G. Li, C. Ray and L. P. Yu, *Adv. Mater.*, 2010, **22**, E135.
17. G. Li, V. Shrotriya, J. S. Huang, Y. Yao, T. Moriarty, K. Emery and Y. Yang, *Nature Mater.*, 2005, **4**, 864.
18. C. J. Brabec, S. E. Shaheen, C. Winder, N. S. Sariciftci and P. Denk, *Appl. Phys. Lett.*, 2002, **80**, 1288.
19. V. Shrotriya, G. Li, Y. Yao, C. W. Chu and Y. Yang, *Appl. Phys. Lett.*, 2006, **89**, 063505.
20. J. S. Huang, Z. Xu and Y. Yang, *Adv. Funct. Mater.*, 2007, **17**, 1966.
21. G. G. Malliaras, J. R. Salem, P. J. Brock and J. C. Scott, *J. Appl. Phys.*, 1998, **84**, 1583.
22. M. C. Scharber, D. Wuhlbacher, M. Koppe, P. Denk, C. Waldauf, A. J. Heeger and C. L. Brabec, *Adv. Mater.*, 2006, **18**, 789.
23. G. Dennler, M. C. Scharber, T. Ameri, P. Denk, K. Forberich, C. Waldauf and C.J. Brabec, *Adv. Mater.*, 2008, **20**, 579.
24. F. L. Zhang, K. G. Jespersen, C. Bjorstrom, M. Svensson, M. R. Andersson, V. Sundstrom, K. Magnusson, E. Moons, A. Yartsev and O. Inganas, *Adv. Funct. Mater.*, 2006, **16**, 667.
25. J. Liu, Y. J. Shi and Y. Yang, *Adv. Funct. Mater.*, 2001, **11**, 420.
26. Y. Zhao, Z. Y. Xie, Y. Qu, Y. H. Geng and L. X. Wang, *Synth. Met.*, 2008, **158**, 908.
27. B. C. Thompson and J. M. J. Frechet, *Angew. Chem., Int. Ed.*, 2008, **47**, 58.
28. J. Peet, J. Y. Kim, N. E. Coates, W. L. Ma, D. Moses, A. J. Heeger and G. C. Bazan, *Nature Mater.*, 2007, **6**, 497.
29. B.A. Gregg and M. C. Hanna, *J. Appl. Phys.*, 2003, **96**, 3605.
30. J. K. Lee, W. L. Ma, C.J. Brabec, J. Yuen, J.S. Moon, J. Y. Kim, K. Lee, G. C. Bazan and A. J. Heeger, *J. Am. Chem. Soc.*, 2008, **130**, 3619.
31. J. S. Moon, C. J. Takacs, S. Cho, R. C. Coffin, H. Kim, G. C. Bazan and A. J. Heeger, *Nano. Lett.*, 2010, **10**, 4005.
32. S. E. Shaheen, C. J. Brabec, N. S. Sariciftci, F. Padinger, T. Fromherz and J. C. Hummelen, *Appl. Phys. Lett.*, 2001, **78**, 841.
33. H. Y. Chen, J. H. Hou, S. Q. Zhang, Y. Y. Liang, G. W. Yang, Y. Yang, L. P. Yu, Y. Wu and G. Li, *Nature Photon.*, 2009, **3**, 649.
34. G. B. Zhang, Y. Y. Fu, Q. Zhang and Z. Y. Xie, *Chem. Commun.*, 2010, 4997..
35. H. Flügge, H. Schmidt, T. Riedl, S. Schmale, T. Rabe, J. Fahlbusch, M. Danilov, H. Spieker, J. Schöbel and W. Kowalsky, *Appl. Phys. Lett.*, 2010, **97**, 123306.
36. L. H. Nguyen, H. Hoppe, T. Erb, S. Günes, G. Gobsch and N. S. Sariciftci, *Adv. Funct. Mater.*, 2007, **17**, 1071.
37. Y. Zhao, Z. Y. Xie, Y. Qu, Y. H. Geng and L. X. Wang, *Appl. Phys. Lett.*, 2007, **90**, 043504.
38. Y. Zhao, S. Y. Shao, Z. Y. Xie, Y. H. Geng and L. X. Wang, *J. Phys. Chem. C*, 2009, **113**, 17235.

39. F. S. Bates and G. H. Fredrickson, *Annu. Rev. Phys. Chem.*, 1990, **41**, 525.
40. U. Scherf, A. Gutacker and N. Koenen, *Acc. Chem. Res.*, 2008, **41**, 1086.
41. L. Bu, X. Guo, B. Yu, Y. Qu, Z. Xie, D. Yan, Y. Geng and F. Wang, *J. Am. Chem. Soc.*, 2009, **131**, 13242.
42. F. L. Zhang, A. Gadisa, O. Inganäs, M. Svensson and M. R. Andersson, *Appl. Phys. Lett.*, 2004, **84**, 3906.
43. M. P. de Jong, L. J. van Ijzendoorn and M. J. A. de Voigt, *Appl. Phys. Lett.*, 2000, **77**, 2255.
44. T. P. Nguyen and S. A. de Vos, *Surf. Sci.*, 2004, **221**, 330.
45. K. W. Wong, H. L. Yip, Y. Luo, K. Y. Wong, W. M. Lau, K. H. Low, H. F. Chow, Z. Q. Gao, W. L. Yeung and C. C. Chang, *Appl. Phys. Lett.*, 2002, **80**, 2788.
46. J. S. Kim, J. H. Park, J. H. Lee, J. Jo, D. Y. Kim and K. Cho, *Appl. Phys. Lett.*, 2007, **91**, 112111.
47. V. Shrotriya, G. Li, Y. Yao, C. W. Chu and Y. Yang, *Appl. Phys. Lett.*, 2006, **88**, 073508.
48. F. M. Liu, S.Y. Shao, X.Y. Guo, Y. Zhao and Z.Y. Xie, *Sol. Energy. Mater. Sol. Cells*, 2010, **94**, 842.
49. M. D. Irwin, D. B. Buchholz, A. W. Hains, R. P. H. Chang and T. J. Marks, *PNAS*, 2008, **105**, 2783.
50. M. Y. Chan, C. S. Lee, S. L. Lai, M. K. Fung, F. L. Wong, H. Y. Sun, K. M. Lau and S. T. Lee, *J. Appl. Phys.*, 2006, **100**, 094506.
51. P. C. Yu, C. H. Chang, M. S. Su, M. H. Hsu and K. H. Wei, *Appl. Phys. Lett.*, 2010, **96**, 153307.
52. S. Y. Shao, F. M. Liu, Z. Y. Xie and L. X. Wang, *J. Phys. Chem. C.*, 2010, **114**, 9161.
53. K. Suemori, T. Miyata, M. Yokoyama and M. Hiramoto, *Appl. Phys. Lett.*, 2004, **85**, 6269.
54. C. J. Brabec, A. Cravino, D. Meissner, N. S. Sariciftci, T. Fromherz, M. T. Rispens, L. Sanchez and J.C. Hummelen, *Adv. Funct. Mater.*, 2001, **11**, 374.
55. J. Y. Kim, S. H. Kim, H. Lee, K. Lee, W. L. Ma, X. Gong and A. J. Heeger, *Adv. Mater.*, 2006, **18**, 572.
56. F. L. Zhang, M. Ceder and O. Inganäs, *Adv. Mater.*, 2007, **19**, 1835.
57. Y. Zhao, Z. Y. Xie, C. J. Qin, Y. Qu, Y. H. Geng and L. X. Wang, *Sol. Energy Mater. Sol. Cells*, 2009, **93**, 604.
58. S. Sista, Z. Hong, M.-H. Park, Z. Xu and Y. Yang, *Adv. Mater.*, 2010, **22**, E77.
59. X. Guo, F. Liu, W. Yue, Z. Xie, Y. Geng and L. Wang, *Org. Electron.*, 2009, **10**, 1174.
60. J. Y. Kim, K. Lee, N. E. Coates, D. Moses, T.-Q. Nguyen, M. Dante and A. J. Heeger, *Science*, 2007, **317**, 222.
61. X. Guo, F. Liu, B. Meng, Z. Xie and L. Wang, *Org. Electron.*, 2010, **11**, 1230.

Novel Fluorene-Based Functional "Click Polymers" for Dye-Sensitized Solar Cells

SUNG-HO JIN

Department of Chemistry Education & Interdisciplinary Program of Advanced Information and Display Materials, Pusan National University, Busan 609-735, Korea
E-mail: shjin@pusan.ac.kr

15.1 Introduction

The molecular structure of polymers is a key issue in modern polymer synthesis because of their potential applications in various optoelectronic devices, such as polymer light-emitting diodes (PLEDs),[1] organic lasers,[2] organic thin-film transistors (OTFTs),[3] nonlinear optical (NLO),[4] organic photovoltaic (OPV) devices,[5] and dye-sensitized solar cells (DSSCs).[6] DSSCs based on the sensitization of nanocrystalline TiO_2 by photoexcited dye molecules have been investigated extensively because of their lower cost and potential alternatives to traditional photovoltaic device. The power-conversion efficiencies (PCEs) of DSSCs based on liquid electrolytes using organic compounds, such as acetonitrile, propylene carbonate, and ethylene carbonate as a solvent and the iodide/triiodide (I^-/I_3^-) redox couple as the electrolyte have reached 10–11% under irradiation of AM 1.5G.[7,8] However, there are some problems with this type of liquid electrolyte junction cell, which include low long-term stability, which is caused by organic solvent evaporation and the leakage of

RSC Polymer Chemistry Series No. 2
Molecular Design and Applications of Photofunctional Polymers and Materials
Edited by Wai-Yeung Wong and Alaa S Abd-El-Aziz
© The Royal Society of Chemistry 2012
Published by the Royal Society of Chemistry, www.rsc.org

liquid electrolytes, high-temperature instability and difficulties in sealing the devices.[9] In order to overcome these problems, considerable effort has been made to replace liquid electrolytes with polymer or quasisolid-type charge transport materials.[10,11] Polymer electrolytes have some advantages compared with other types of charge-transport materials. These include high ionic conductivity, which is achieved by "trapping" a liquid electrolyte in polymer cages formed in a host matrix, good contact and excellent filling properties with the nanostructured electrode and counterelectrode. Therefore, polymer electrolytes have attracted considerable attention. Several types of polymer electrolytes based on different types of polymers are already used in DSSCs.[12]

The photophysical and device properties of those polymers depend on the molecular structure as well as the morphology of the material in the solid state. Recently, click chemistry, the Cu^I-catalyzed Huisgen's azide-alkyne [3+2] cycloaddition reaction, has attracted considerable attention.[13] This type of reaction is characterized by high yield, mild, and simple reaction conditions, oxygen and water tolerance, and the simple work-up of products. Moreover, it is highly chemoselective in the formation of the desired 1,4-disubstituted 1,2,3-triazole, even presence of functional groups. This methodology is applied widely in organic chemistry,[14] supramolecular chemistry,[15] drug discovery,[16] bioconjugations,[17] and materials science.[18] Polymer chemists have employed click chemistry to construct dendritic[19–25] and linear macromolecules.[26–28] The main obstacles in the synthesis of linear polymers *via* click chemistry are long reaction time and poor product solubility. However, there are only a few reports on the synthesis of π-conjugated polymers by 1,3-cycloadditions.[29]

We have been interested in using the click reaction to construct a new pathway for the synthesis of novel fluorene-based functional polymers that will be soluble in common organic solvents and easily spin coated to produce high-quality optical thin films. In our previous work, various types of Fréchet- and PAMAM-type dendrimers were synthesized using a convergent method employing click chemistry.[20–23] We have been interested to compare the three methods in click polymerization such as (1) Cu^I-catalyst 1,3-dipolar cycloaddition with alkylamine as ligand, (2) noncatalyst (using polar solvent at moderate temperature), and (3) end-capping Cu^I-catalyst 1,3-dipolar cycloaddition (azide end-capping). The current chapter focuses on the reaction method with an aim of improving the efficiency of the reaction condition and its utility in future research. We have also compared the electro-optical properties and photovoltaic performance of these three click polymerization methods.

We report the synthesis and characterization of a series of new click polymers (P1–P6), which consist of 9,9-dioctylfluorene as a monomer with various types of comonomers such as N-octylcarbazole, benzothiadiazole, and 9,9-dipropargylfluorene units. The new polymers were synthesized using "click chemistry", and used as a polymer matrix to trap liquid electrolyte to form a polymer electrolyte. The device was fabricated with the configuration of SnO_2:F/TiO_2/N3 Dye/polymer electrolyte/Pt. The effect of the click polymerization methods on the photovoltaic performance of the DSSCs was studied.

15.2 Experimental Techniques

15.2.1 Method (I) for the Synthesis of Click Polymers, P1–P4

2,7-Diazido-9,9-dioctyl-fluorene,[29,30] 9,9-dipropargylfluorene,[31] 4,7-diethynyl-benzothiadiazole,[32] 2,7-diethynyl-9,9-dioctylfluorene,[33,34] and N-octyl-3,6-diethynylcabazole,[35] were synthesized using a slight modification of the method reported in the literature. Diazide- and diethynyl-based monomers (1:1 equiv.) and sodium L-ascorbate (10 mol%) were dissolved in THF (2–3 mL) under N_2 flow in a flame-dried Schlenk flask and added to a mixture of triethylamine (TEA) (0.2–0.3 mL) as a ligand.[36] The flask was flushed with N_2 for 20–30 min. The mixture was frozen and evacuated three times, which was followed by the addition of $CuSO_4 \cdot 5H_2O$ (5 mol%) under a flow of N_2 gas. The mixture was stirred at 30–35 °C for 48 h. The THF was removed under vacuum and the mixture was dissolved in chloroform, washed with an aqueous NH_4OH solution followed by water. The organic layer was separated and the solvent was removed. The resulting polymer was precipitated into methanol. In the above procedure, the reaction will not proceed if TEA is not added, even after 10 days at room temperature. By adding TEA and increasing the reaction temperature, reasonable molecular weight polymers could be synthesized by click chemistry.

Polymer 1 (P1). [I+II], yellow solid. ^{1}H-NMR ($CDCl_3$ 500 MHz): δ (ppm) 8.39 (s), 8.35 (s), 8.03 (s), 7.95–7.91 (m), 7.86–7.79 (m), 7.78–7.73 (m), 7.09–7.07 (m), 7.03 (d), 2.19–2.10 (m), 2.05–1.99 (m), 1.27–1.10 (m), 0.95–0.78 (m), 0.72–0.66 (m). Anal Calcd for $(C_{62} H_{82} N_6)_n$: C, 81.71; H, 9.06, N, 9.22. Found: C, 80.18: H, 9.29; N, 9.26.

Polymer 2 (P2). [I+III], green-yellow solid. ^{1}H-NMR ($CDCl_3$ 300 MHz): δ (ppm) 8.87 (s), 7.91 (m), 7.83 (s) 7.75 (m), 7.03 (m), 2.11–2.04 (m), 1.26–1.11 (m), 0.85–0.66 (m). Anal Calcd for $(C_{39} H_{44} N_8 S)_n$: C, 71.31; H, 6.75, N, 17.06, S, 4.88. Found: C, 69.97; H, 6.91; N, 17.13; S, 5.08.

Polymer 3 (P3). [I+IV], light yellow-brownish solid. ^{1}H-NMR ($CDCl_3$ 300 MHz): δ (ppm) 8.79 (s), 8.65 (s) 8.39 (s), 8.35 (d), 8.12 (d), 7.96–7.71 (m), 7.61–7.49 (m), 7.38 (s), 4.35 (d), 3.10 (s), 2.18–2.06 (m), 1.92 (d), 1.36–1.11 (m), 0.88–0.77 (m). Anal Calcd for $(C_{53} H_{65} N_7)_n$: C, 79.56.; H, 8.19, N, 12.25. Found: C, 78.07: H, 8.39; N, 12.30.

Polymer 4 (P4). [I+V], brown-yellow solid. ^{1}H-NMR ($CDCl_3$ 300 MHz): δ (ppm) 7.66 (s), 7.61 (s), 7.38 (s), 7.26–7.25 (m), 7.25–7.24 (m), 1.91 (s), 1.26–1.02 (m), 0.81–0.77 (m) Anal Calcd for $(C_{46} H_{50} N_6)_n$: C, 80.43; H, 7.34, N, 12.23. Found: C, 78.92; H, 7.52; N, 12.28.

15.2.2 Method (II) for the Synthesis of Click Polymer, P5

Into a flame-dried Schlenk flask were placed diazide- and diethynyl-based monomers (1:1 equiv.) and injected mixture of DMF/toluene (1:1 by volume). The reaction mixture was stirred under N_2 gas at 100 °C for one day. After that the reaction mixture was dilute with chloroform and added dropwise of a 10:1

mixture (by volume) of hexane and chloroform through a cotton filter under stirring. The precipitates were allowed to stand overnight, collected by filtration, and dried under vacuum at room temperature. As no transition-metal catalyst is used in the process; this polymerization enjoys such advantages as being less toxic, environmentally friendly, and economically sounder. This helps simplify the reaction procedures and enhance the polymerization efficiency.

Polymer P5: Yellow solid. ^{1}H-NMR (CDCl$_3$, 300 MHz): δ (ppm) 8.35 (s), 7.98–7.94 (m), 7.89–7.84 (m), 7.76–7.74 (m), 7.66–7.58 (m), 7.48–7.43(m), 7.32(s), 7.02(s), 2.18(m), 2.01–1.82 (m), 1.25–1.08 (m), 0.81 (m), 0.66 (m). Anal Calcd for $(C_{62}H_{82}N_6)_n$: C, 81.71; H, 9.06, N, 9.22. Found: C, 80.18: H, 9.29; N, 9.26.

15.2.3 Method (III) for the Synthesis of Click Polymer, P6

Diazide- and diethynyl-based monomers (1:1 equiv.) and sodium L-ascorbate (10 mol%) were dissolved in THF (2–3 mL) under a flow of N_2 into a flame-dried Schlenk flask and added to the mixture of triethylamine (2–3 mL) as a ligand. The flask was flushed with a flow of N_2 for 20–30 min and the mixture was frozen and evacuated for three times and $CuSO_4 \cdot 5H_2O$ (5 mol%) was added under a flow of N_2 gas. The mixture was allowed to stir at 30–35 °C for 48 hr. A small amount of phenylacetylene was used as an azide end-capping material after checking the precipitation of that reaction mixture and stirring for more few minutes. This study demonstrated the potential of 1,3-dipolar cycloaddition reaction involving azides and alkynes for the preparation of conjugated polymer and meanwhile showed the necessity of controlling the polymerization conditions because of the high reactivity of the azide and alkyne functionalities. One possibility to gain control and inhibit autopoly-merization of the monomers is to put the functional groups in two separate monomers. Therefore, to control the autopolymerization, azide functional group of polymer end chain was end-capped by phenylacetylene. The THF was removed under vacuum and the mixture was dissolved in chloroform, washed with aqueous NH_4OH solution and then water. The organic layer was separated and the solvent removed. The resulting polymer was precipitated into the methanol.

Polymer P6: Yellow solid. ^{1}H-NMR (CDCl$_3$, 300 MHz): δ (ppm) 8.36 (s), 8.34 (s), 8.01 (s), 7.97–7.90 (m), 7.84–7.79 (m), 7.69–7.66 (m), 7.51–7.50 (m), 7.49 (s), 3.16 (s), 2.17–2.16 (m), 2.05–1.99 (m), 1.26–1.11 (m), 0.80–0.77 (m), 0.08 (m). Anal Calcd for $(C_{62}H_{82}N_6)_n$: C, 82.96; H, 8.75, N, 8.29. Found: C, 81.41: H, 8.97; N, 8.33.

15.2.4 Fabrication of DSSCs

The polymer electrolyte consisted of I_2 (95 mM), tetrabutylammonium iodide (TBAI) (70 mM), and 1-ethyl-3-methylimidazolium iodide (0.34 M) in a cosolvent of ethylene carbonate (EC) and propylene carbonate (PC) (0.4 mL, EC/PC=4/1 as weight ratio) with P1–P6 (40 mg) in an acetonitrile solution (0.4 mL). The DSSCs were fabricated using Ruthenium dye (N3 dye) as the photosensitizer and sandwiched between a TiO_2 thin film and a Pt counter-electrode as the two electrodes. The DSSC was fabricated using the following process; a volume of *ca.* 10 μl/cm^2 of the transparent pastes (Ti-Nanoxide HT) was spread on FTO glass using the doctor-blade method. The FTO glass spread TiO_2 nanoparticles were heated to *ca.* 100 °C for approximately 30 min and *ca.* 450 °C for approximately 30 min. The TiO_2 deposited electrode was then cooled from 100 °C to 60 °C at a controlled cooling rate (3 °C/min) to avoid cracking of the glass. A Pt counterelectrode was fabricated by spreading on FTO glass using the doctor-blade method. The FTO glass spread Pt catalyst T/SP was heated to approximately 100 °C for 10 min before firing at 400 °C for 30 min. The N3 dye photosensitizer was dissolved in absolute ethanol to a concentration of 20 mg per 100 mL of solution. The nanoporous TiO_2 film was dipped in this solution at room temperature for 24 h. The dye-sensitized TiO_2 electrode was then rinsed with absolute ethanol and dried in air. The solid-state electrolyte was cast onto the N3-dye-impregnated TiO_2, and dried at 60 °C for 2 h. The effective area of the DSSCs was 25 mm^2. The performances of photovoltaic devices were measured using a calibrated AM 1.5G solar simulator (Orel 300 W simulator, models 81150) with a light intensity of 100 mW/cm^2 adjusted using a standard PV reference cell (2 cm × 2 cm monocrystalline silicon solar cell, calibrated at NREL, Colorado, USA) and a computer-controlled Keithley 236 source measure unit.

The PCE (η) of a solar cell given by

$$\eta = P_{out}/P_{in} = (J_{sc} \times V_{oc}) \times FF/P_{in}$$
$$\text{with } FF = P_{max}/(J_{sc} \times V_{oc}) = (J_{max} \times V_{max})/(J_{sc} \times V_{oc})$$

where P_{out} is the output electrical power of the device under illumination, and P_{in} is the intensity of incident light (*e.g.*, in W/m^2 or mW/cm^2). V_{oc} is the open-circuit voltage, J_{sc} is the short-circuit current density, and the fill factor (FF) is calculated from the values of V_{oc}, J_{sc}, and the maximum power point, P_{max}. All fabrication steps and characterization measurements were carried out in an ambient environment without a protective atmosphere. While measuring the current density–voltage (J–V) curves for DSSCs, a black mask was used and only the effective area of the cell was exposed to light irradiation. The data reported in this paper was confirmed by making each device more than 5 times.

15.3 Results and Discussion

Scheme 15.1 shows the normal click polymerization routes for the diazide- and diethynyl-based monomers and their corresponding polymers. The molecular structures of the fluorene-based monomers with diazide and diethynyl units are published elsewhere.[29,30] However, monomers III, IV, and V were first used to synthesize the functional polymers using click chemistry. The desired 1,4-

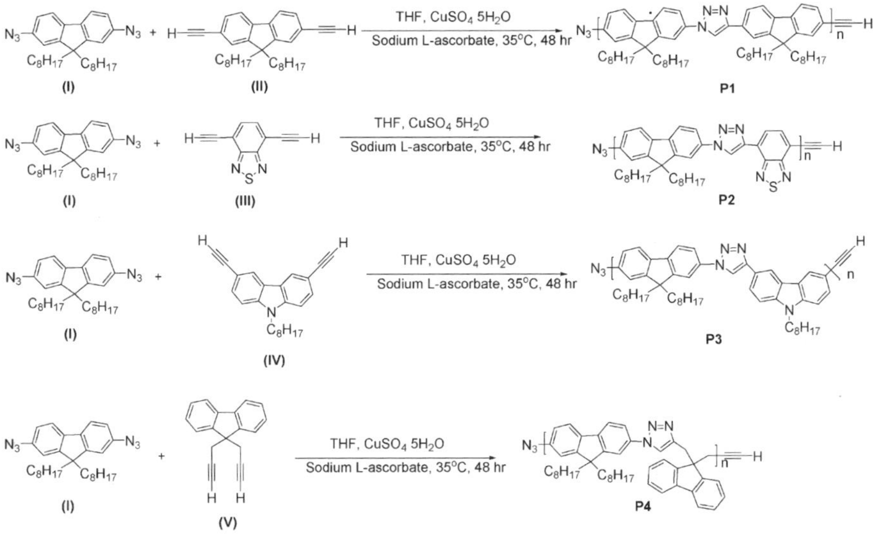

Scheme 15.1 Synthetic scheme for the synthesis of click polymers, P1–P4.

disubstituted 1,2,3-triazole ring units were introduced into the fluorene-based polymer backbone using 2,7-diazido-9,9-dioctylfluorene as the common monomer for click coupling with a different molecular structure, such as 2,7-diethynyl-9,9-dioctylfluorene (II), 4,7-diethynyl-benzothiadiazole (III), N-octyl-3,6-diethynylcabazole (IV), and 9,9-dipropargylfluorene (V) at 1:1 mol. equivalent ratio, using THF as solvent, sodium L-ascorbate (10 mol%) as the reducing agent, $CuSO_4 \cdot 5H_2O$ (5 mol%) as the catalyst and a small amount of TEA. The resulting mixture was stirred constantly at 35 °C for 48 h under a N_2 atmosphere. Polymers P1, P2, P3, and P4 were obtained from the above methodology using monomers II, III, IV and V, respectively. However, the reaction mixture became highly viscous after stirring for 48 h at 35 °C. The THF was removed under vacuum and the reaction mixture was dissolved in chloroform, washed with aqueous NH_4OH solution and then with water. The organic layer was separated and removed by evaporation under reduced pressure. The residue was precipitated with methanol. In order to improve the purity of the polymers and device performance, the precipitated polymers were further purified by multiple Soxhlet extraction with methanol, hexane and finally extracted with chloroform. From this process, a highly purified and narrow polydispersity of the polymers were obtained. The resulting polymers were completely soluble in various organic solvents such as chloroform, chlorobenzene, THF, toluene, xylene, *etc.* Table 15.1 summarizes the polymerization results, molecular weights and thermal characteristics of the polymers, P1–P4. The weight average molecular weight (M_w) and polydispersity of the polymers, P1–P4 ranged from $(8.1-33) \times 10^3$ and 1.38–2.41, respectively. The GPC results revealed these polymers to have a relatively narrow polydispersity index. These polymers had a better solubility in the

Table 15.1 Polymerization results and thermal properties of P1–P4.

Polymers	Yield (%)	M^{a}_{w} ($\times 10^{-3}$)	PDI^{a}	DSC	TGA^{b}
P1	92	16	1.92	115	339
P2	90	8.1	1.38	163	323
P3	89	8.6	1.92	148	335
P4	90	33	2.41	135	346

[a]Measured by GPC using polystyrene standards. [b]Measured at temperature of 5% weight loss for the polymers.

reaction system due to the long alkyl side chain. Therefore, the reagents can react with each other in a manner to afford a narrow molecular weight distribution.

The structure and thermal properties of the polymers, P1–P4 were identified by ^{1}H-NMR, infrared spectroscopy, elemental analysis, DSC, and TGA thermograms. The disappearance of the characteristics of the acetylenic proton peaks from the monomers at approximately 2.6–3.7 ppm in ^{1}H-NMR and 2100 cm^{-1} in infrared spectroscopy, and the appearance of the vinylic proton peaks in the 1,4-disubstituted 1,2,3-triazole ring units confirmed the polymerization reaction. The other peaks were consistent with the proposed chemical structure of the polymers. The thermal stability of the polymers, P1–P4 was determined by TGA under a N_2 atmosphere. The 1,4-disubstituted 1,2,3-triazole units were lost at approximately 350 °C followed by polymer decomposition at higher temperatures. The thermally induced phase transition properties of the polymers were also examined by DSC under a N_2 atmosphere. Most fluorene-based EL polymers do not show a distinct glass-transition temperature (T_g). However, the T_g of the polymers ranged from 115–163 °C. These values are higher than those of poly(9,9-dioctylfluorene)[37] and poly(9,9-dihexylfluorene).[38] It is evident that the incorporation of 1,4-disubstituted 1,2,3-triazole ring units in the main chain can increase the T_g of the resulting fluorene-based polymers. This is very important if these polymers are to be used as active materials for electronic applications such as PLEDs, OPVs, and DSSCs. The higher thermal stability of the polymers prevents the deformation and degradation of the active layer from the heat induced during the operation of the devices.

The absorption and PL data of the polymers, P1–P4 were measured in both the solution and film states. The UV-visible absorption spectra of the polymers in chloroform and in the thin films coated onto the quartz substrates were examined. In the solution state, P1–P4 showed absorption maxima at 350, 328, 337, and 323 nm, respectively, as shown in Figure 15.1(a). Among the polymers, P4 showed a significantly enhanced blueshift of its maximum peak in solution compared with the other polymers. This blueshift is consistent with the reduction of the π-conjugated systems induced by the introduction of 9,9-dipropargylfluorene moieties. P2 showed two UV-visible absorption peaks at approximately 328 and 434 nm, which were attributed to the introduction of fluorene and benzothiadiazole units along with the polymer backbone through the 1,4-disubstituted 1,2,3-triazole linkage. As shown in Figure 15.1(b), the UV-visible absorption spectra of film states were similar in solution with a similar maximum absorption wavelength. This indicates a similar conformation of the polymers in both states with tailing structures in the low-energy regions in front of steep main absorption band edges.

Figure 15.2 shows the PL spectra of the polymers, P1–P4, in a chloroform solution (a) and in the thin-film state (b). The PL spectra of the polymers in chloroform were similar and emitted a blue color between 370 and 406 nm, which can be explained using fluorene-moiety-induced emission bands. These blue bands have well-developed vibronic structures with approximately 130

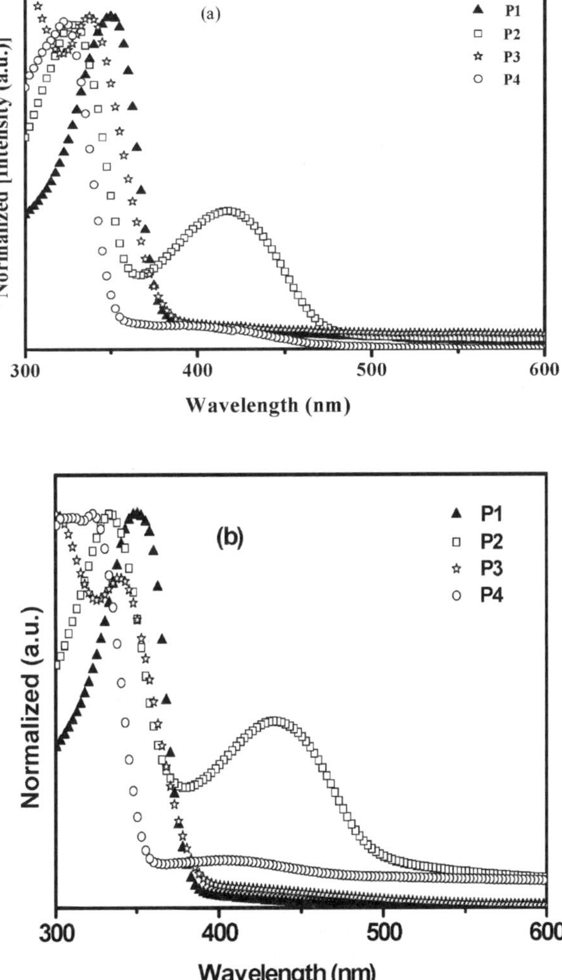

Figure 15.1 UV-visible absorption spectra in chloroform (a) and film (b) of P1–P4.

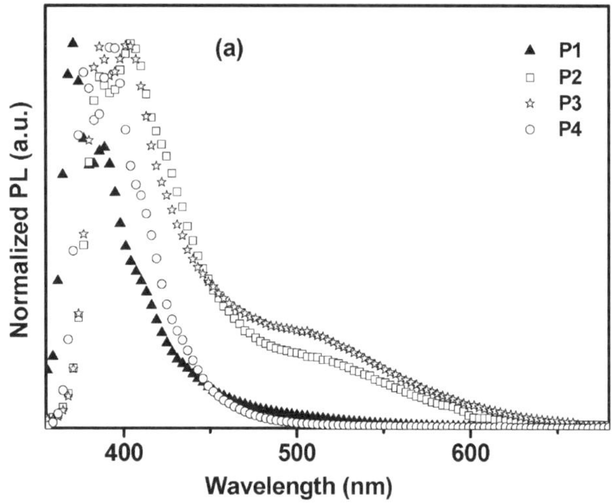

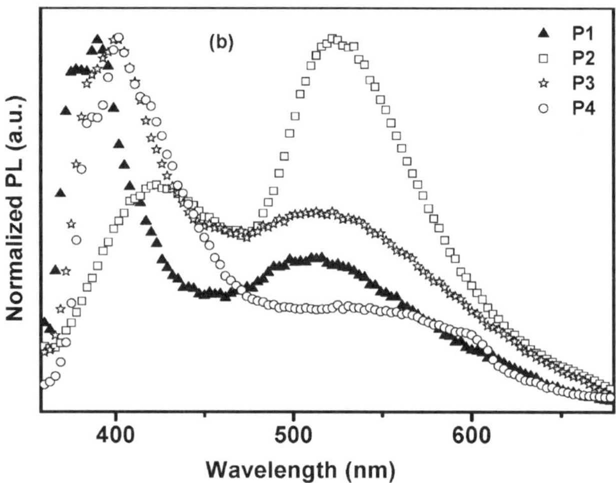

Figure 15.2   PL emission in solution (a) and film (b) of P1–P4.

meV vibronic bandgaps, which are typical of such conjugated systems due to their carbon-carbon double-bond stretching mode. These pronounced vibronic structures are common features of both the solution and film PL spectra except for the P2 film PL spectrum. In addition to these blue bands, P2 and P3 have pronounced energetically lower-lying structureless bands centered at approximately 525 nm. This can be explained by the introduction of benzothiadiazole and carbazole moieties with different molecular structures in the polymer chain. As shown in Figure 15.2(b), the PL spectra of the polymers in the film states are quite different compared with the solution states. The emission spectra of P1–P4 in the solid films were slightly redshifted by 15–40 nm, and had a significantly pronounced energetically lower-lying band centered at approximately 525 nm. This redshift can often be explained by the formation of an interchain excimer. Another possible explanation might be the strong reabsorption due to relatively steep absorption edge where there is also an emission band for the $S_0 \leftarrow S_1$ 0-0 transition. The most dramatic change was observed in the film PL spectrum of P2. P2 had a maximum emission peak at 521 nm with a residual blue band at 422 nm, which were attributed to the benzothiadiazole and fluorene units, respectively. This indicates that the electron-deficient benzothiadiazole unit dominates the luminescence properties in its solid state through an energy-transfer effect due to the perfectly overlapped blue emission band and absorption band centered at approximately 420 nm. In addition, this suggests that the benzothiadiazole containing polymer has the lowest energy level. In order to understand the pronounced low energy bands of the P1, P3 and P4 films as well as the P2 film PL spectra, the 3-dimensional exciton dynamics due to the disordered but closely packed solid states of the polymer chains while exciton dynamics are limited 1 dimensionally (1D) due to less interchain migration probability in dilute solutions.

Redox measurements were carried out using cyclic voltammetry (CV) to determine the electrochemical properties of the polymers, P1–P4 and to evaluate their HOMO and LUMO energy levels. The HOMO binding energies of the polymers with respect to the ferrocene/ferrocenium (4.8 eV) standard were approximately 5.23 eV for P1, 5.39 eV for P2, 5.35 for P3, and 4.70 eV for P4. From the onsets of the absorption spectra, the bandgaps of P1–P4 were calculated to be 3.25, 2.44, 3.14, and 3.46 eV, respectively. The LUMO energy levels were calculated from the bandgaps and HOMO energies. It was reported that the HOMO and LUMO energy levels of poly(9,9-dioctylfluorene) measured using an electrochemical method were 5.8 eV and 2.12 eV, respectively.[39] There is a significant difference in electrochemical behaviour between the poly(9,9-dioctylfluorene) and P1, which suggests that the electrochemical properties of the P1 had been altered through the introduction of a 1,4-disubstituted 1,2,3-triazole group between the fluorene units along with the polymer backbone *via* click chemistry. The LUMO energy level and bandgap of P2 were lower than the other polymers due to the introduction of electron-deficient benzothiadiazole units to the polymer backbone. The HOMO energy level was lower than those of the other polymers through linkage of the propargyl unit to the polymer backbone in P4, which suggests the easy injection of holes from the anode in the PLEDs.

Figure 15.3 shows the device configuration of DSSCs using click polymers as the polymer matrix. Figure 15.4 shows the current density–voltage (J–V) curves of a SnO_2:F/TiO_2/N3 Dye/polymer electrolyte/Pt device using P1–P4 as the polymer matrix for the polymer electrolyte under AM 1.5G illumination (100 mW/cm^2). Table 15.2 summarizes the photovoltaic properties of the DSSCs. The absorption characteristics and molar absorption coefficient of the

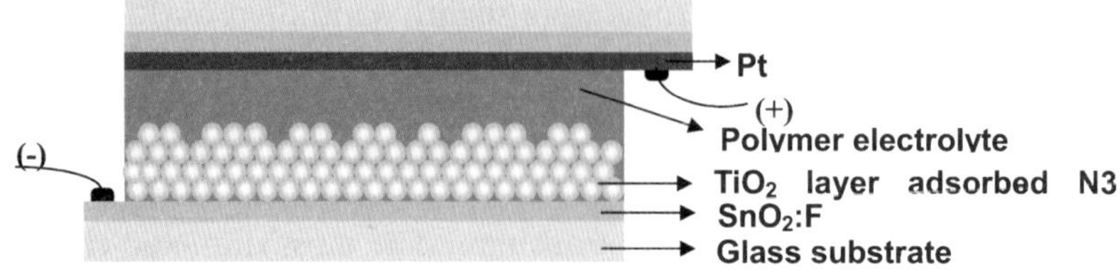

Figure 15.3   The cross-sectional structure of SnO_2:F/TiO_2/N3 Dye/polymer electrolyte/Pt device.

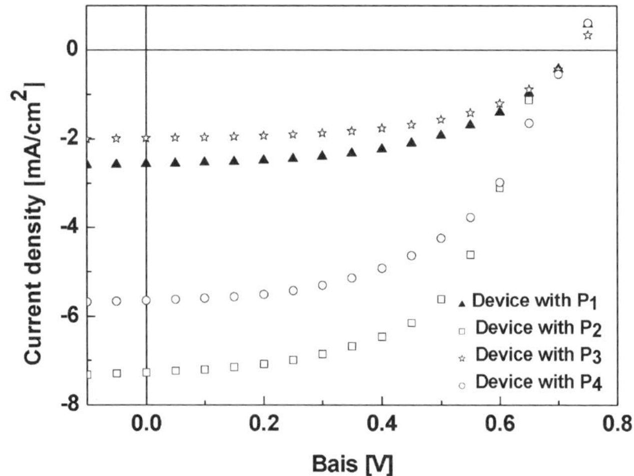

Figure 15.4   Photocurrent–voltage characteristics of the DSSCs fabricated with P1–P4 as a polymer matrix of polymer electrolyte under AM 1.5 sunlight illumination (100 mW/cm^2).

Table 15.2   Photovoltaic properties of the DSSCs made with P1–P4.

DSSCs with polymer	Molar absorption coefficient $(\varepsilon)^a$	V_{OC} (V)	J_{SC} (mA/cm^2)	FF	PCE (%)
P1	430 300	0.72	2.56	0.50	0.96
P2	496 000	0.68	7.27	0.57	2.80
P3	225 400	0.73	1.99	0.54	0.78
P4	485 600	0.72	5.65	0.52	2.12

^{a}Measured at a solution concentration 1.0×10^{-5} M.

dye or polymer play important roles in determining the performance of the DSSCs. Although dyes such as N719 and N3 showed good dye coverage on TiO_2 and were also reasonably stable, they still suffer from low molar absorption coefficients. The molar absorption coefficients of these polymers at the λ_{max} in solution ranged from 225 400 to 496 000 M^{-1} cm^{-1}. Relatively high molar absorption coefficients can result in high efficiency because of the extended π-electron delocalization. The values of V_{oc} and FF were similar, which might be due to the triazole linkage at the polymer chain with fluorene as the commonly repeating unit in all polymers. The DSSCs exhibited a photovoltaic performance with a PCE of 0.96, 2.80, 0.78, and 2.12% for P1, P2, P3, and P4, respectively. Among the four cells, the cell using P2 had the highest efficiency, which reached 2.80% (V_{oc}=0.68 V, J_{sc}=7.27 mA/cm^2, FF=0.57). The higher photovoltaic performance of P2 was attributed to the low molecular weight and higher molar absorption coefficient, which allowed the polymer electrolyte based on P2 to easily penetrate the dye-adsorbed nanocrystalline porous TiO$_2$ electrode.

Scheme 15.2 shows three different polymerization routes between the diazide- and diethynyl-based monomers. The desired 1,4-disubstituted 1,2,3-triazole ring units were introduced into the fluorene-based polymer backbone using 2,7-diazido-9,9-dioctylfluorene monomer by click coupling with 2,7-diethynyl-9,9-dioctylfluorene at 1:1 mole ratio using three different methods such as catalyst, noncatalyst and catalyst with end-capping. Polymers, P5 and P6 were obtained from the above three different methods, respectively. Table 15.3 summarizes the polymerization results, molecular weights, and thermal characteristics of the polymers, P1, P5, and P6. The weight average molecular weight (M_w) and polydispersity of the polymers ranged from $(16-33) \times 10^3$ and 2.1–2.4, respectively. The GPC results revealed these polymers to have a relative narrow polydispersity index. These polymers had a better solubility in the reaction system due to the long alkyl side chain. Therefore, the reagents can react with each other in a manner to afford a narrow molecular weight distribution.

The HOMO binding energies of the polymers with respect to the ferrocene/ferrocenium (4.8 eV) standard were approximately 5.67 eV for P65 and 5.62 eV for P6. From the onsets of the absorption spectra, the bandgaps of P5 and P6 were calculated to be 2.39 and 2.32 eV, respectively. The LUMO energy levels were calculated from the bandgaps and HOMO energies.

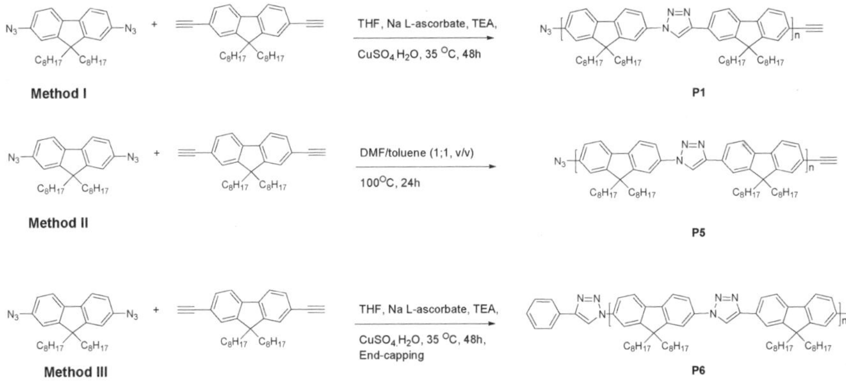

Scheme 15.2 Synthetic scheme for the synthesis of click polymers, P1, P5, and P6.

Table 15.3 Polymerization results and thermal properties of P5 and P6.

Polymers	Yield (%)	M^a_w ($\times 10^{-3}$)	PDI^a	DSC	TGA^b
P5	91	16	2.15	104	362
P6	94	33	2.41	97	346

[a]Measured by GPC using polystyrene standards. [b]Measured at temperature of 5% weight loss for the polymers.

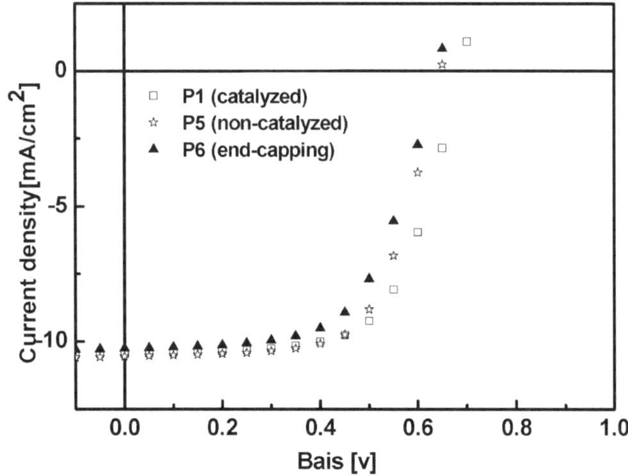

Figure 15.5 Photocurrent–voltage characteristics of the DSSCs fabricated with P1, P5, and P6 as a polymer matrix of polymer electrolyte under AM 1.5 sunlight illumination (100 mW/cm^2).

Figure 15.5 shows the J–V curves of a SnO_2:F/TiO_2/N719 Dye/polymer electrolyte/Pt device using polymer, P1, P5, and P6 as the polymer matrix for the polymer electrolyte under AM 1.5G illumination (100 mW/cm^2). Table 15.4 summarizes the photovoltaic properties of the DSSCs. The DSSCs exhibited photovoltaic performances with PCE of 4.62, 4.41, and 4.02% for P1, P5, and P6, respectively. P6 shows the lower J_{sc} and PCE than P1 and P5. Due to the higher molecular weight of P6 than P1 and P5, the decrease in the J_{sc} of DSSC with P6 electrolyte is mainly originates from the lowered I_3^- diffusion coefficients, which reduce the supply of I_3^- to the counterelectrode, retard the

Table 15.4 Photovoltaic properties of the DSSCs made with P1, P5 and P6.

DSSCs with polymer	V_{oc} (V)	J_{sc} (mA/cm^2)	FF (%)	PCE (%)	Area (mm^2)
P1	0.65	10.46	68.0	4.62	21.003
P5	0.60	10.56	69.6	4.41	20.001
P6	0.60	10.25	65.4	4.02	22.161

regeneration of dye and as a result lower the PCE. P1 shows the highest photovoltaic performance, which reached at 4.62% (V_{oc}: 0.65 V, J_{sc}: 10.46 mA/cm^2, FF: 68.0) under AM 1.5G illumination (100 mW/cm^2) and slightly higher V_{oc} than P5 and P6. The higher photovoltaic performance of P4 was attributed to the low molecular weight, which allowed the polymer electrolyte based on P5 to easily penetrate the dye-adsorbed nanocrystalline porous TiO$_2$ electrode.

15.4 Summary

A series of novel soluble fluorene-based functional polymers were synthesized by Cu^I-catalyzed 1,3-dipolar "click chemistry" between various types of diazides and diethynyl-based monomers. The click polymers, which were linked *via* 1,4-disubstituted 1,2,3-triazole units, exhibited good solubility and reasonable molecular weight. The PL spectra of the polymers showed blue emission. From the absorption, emission, and CV results, it is believed that these fluorene-based functional "click polymers" will be promising candidates as a polymer matrix in polymer electrolyte for DSSCs. We successfully employed the three different methods of click chemistry such as Cu^I-catalyzed, noncatalyzed and Cu^I-catalyzed with azide end-capping to synthesize fluorene-based polymers for the first time. The three different methods of 1,3-dipolar cycloaddition click reaction were employed between diazide- and diethynyl-based fluorene monomers to compare electro-optical as well as photovoltaic performance and to determine the suitable reaction method for conjugated click polymers. From the absorption, emission, CV results, and DSSCs performance, we were able to conclude that the Cu^I-catalyzed 1,3-dipolar cycloaddition click polymer exhibited the highest PCE of 4.62% and better electro-optical properties than both noncatalyzed and azide end-capping click polymers.

Acknowledgement

This work was supported by the National Research Foundation of Korea (NRF) grant funded by the Korea government (MEST) (No. 2011-0028320).

References

1. L. Akcelrud, *Prog. Polym. Sci.*, 2003, **28**, 875.
2. J. H. Schön, C. H. Kloc and B. Batlogg, *Science*, 2000, **288**, 2338.
3. A. Tsumura, H. Koezuka and T. Ando, *Appl. Phys. Lett.*, 1986, **49**, 1210.
4. Q. Zeng, L. Zhong'an, L. Zhen, Y. Cheng, Q. Jingui and B. Z. Tang *Macromolecules*, 2007, **40**, 5634.
5. L. Gang, S. Vishal, J. Huang, Y. Yan, M. Tom, E. Keith and Y. Yang, *Nature Mater.*, 2005, **4**, 864.
6. B. O'Regan and M. Grätzel, *Nature*, 1991, **353**, 737.

7. M. K. Nazeeruddin, P. Pechy and M. Grätzel, *J. Am. Chem. Soc.*, 2006, **123**, 1613.

8. M. K. Nazeeruddin, A. Kay, I. Rodicio and M. Grätzel, *J. Am. Chem. Soc.*, 2006, **115**, 6382.

9. G. Smestad, *Sol. Energy Mater. Sol. Cells*, 2003, **76**, 17.

10. B. O'Regan, F. Lenzmann, R. Muis, J. Wienke, *Chem. Mater.*, 2002, **14**, 5023.

11. F. Cao, G. Oskam and P. C. Searson, *J. Phys. Chem. B*, 1995, **991**, 17071.

12. P. Wang, S. M. Zakeeruddin, J. E. Moser and M. Grätzel, M. *Nature Mater.*, 2003, **2**, 402.

13. H. C. Kolb, M. G. Finn and K. B. Sharpless, *Angew. Chem. Int. Ed.*, 2001, **40**, 2004.

14. V. V. Rostovtsev, L. G. Green, V. V. Fokin and K. B. Sharpless, *Angew. Chem. Int. Ed.*, 2002, **41**, 2596.

15. J. D. Badjic, V. Balzani, A. Credi, J. N. Lowe, S. Silvi and J. F. Stoddart, *Chem. Eur. J.*, 2004, **10**, 1926.

16. R. Manetsch, A. Krasinski, Z. Radic, J. Raushel, P. Taylor and K. B. Sharpless, *J. Am. Chem. Soc.*, 2004, **126**, 12809.

17. Q. Wang, T. R. Chan, R. Hilgraf, V. V. Fokin, K. B. Sharpless and M. G. Finn, *J. Am. Chem. Soc.*, 2003, **125**, 3192.

18. D. A. Ossipov and J. Hilborn, *Macromolecules*, 2006, **39**, 1709.

19. M. Malkoch, K. Schleicher, E. Drockenmuller, C. J. Hawker, T. P. Russell, P. Wu and V. V. Fokin, *Macromolecules*, 2005, **38**, 3663.

20. J. W. Lee, B. K. Kim and S. H. Jin, *Bull. Korean Chem. Soc.*, 2005, **26**, 833.

21. J. W. Lee, B. K. Kim and S. H. Jin, *Bull. Korean Chem. Soc.*, 2005, **26**, 715.

22. J. W. Lee, B. K. Kim, J. H. Kim, W. S. Shi and S. H. Jin, *Bull. Korean Chem. Soc.*, 2005, **26**, 1790.

23. J. W. Lee, H. J. Kim, S. C. Han, W. S. Shin and S. H. Jin, *Macromolecules*, 2006, **39**, 2418.

24. M. J. Joralemon, R. K. O'Reilly, J. B. Matson, A. K. Nugent, C. J. Hawker and K. L. Wooley, *Macromolecules*, 2005, **38**, 5436.

25. E. Fermandez-Megia, J. Correa, I. Rodríguez-Meizoso and R Riguera, *Macromolecules*, 2006, **39**, 2113.

26. N. V. Tsarevsky, B. S. Sumerlin and K. Matyjaszewski, *Macromolecules*, 2005, **38**, 3558.

27. B. S. Sumerlin, N. V. Tsarevsky, G. Louche, R. Y. Lee and K. Matyjaszewski, *Macromolecules*, 2005, **38**, 7540.

28. H. F. Gao and K. Matyjaszewski, *Macromolecules*, 2006, **39**, 4960.

29. S. Bakbak, P. J. Leech, B. E. Carson, S. Saxena, W. P. King and U. H. F. Bunz, *Macromolecules*, 2006, **39**, 6793.

30. S. Nimura, O. Kikuchi, T. Ohana, A. Yabe, S. Kondo and M. Kaise, *J. Phys. Chem. A*, 1997, **101**, 2083.

31. Y. S. Gal, S. H. Jin, J. W. Park, W. C. Lee, H. S. Lee and S. Y. Kim, *J. Polym. Sci., Part A. Polym. Chem.*, 2001, **39**, 4101.

32. B. A. DaSilveira Neto, A. S. Ana Lopes, G. Ebeling, R. S. Goncalves, V. E. U. Costa, F. H. Quina and J. Dupont, *Tetrahedon*, 2005, **61**, 10975.
33. P. Anuragudom, S. S. Newaz, S. Phanichphant and T. R. Lee, *Macromolecules*, 2006, **39**, 3494.
34. B. Liu, W. L. Yu, J. Pei, S. Y. Liu, Y. H. Lai and W. Huang, *Macromolecules*, 2001, **34**, 7932.
35. Y. Takihana, M. Shiotsuki, F. Sanda, T. Masuda, *Macromolecules*, 2004, **37**, 7578.
36. Y. Zhu, Y. Huang, W. D. Meng, H. Li and F. L. Qing, *Polymer*, 2006, **47**, 6272.
37. J. Ding, M. Day, G. Robertson and J. Roovers, *Macromolecules*, 2002, **35**, 3474.
38. J. I. Franklin and L. T. Michael, *J. Mater. Chem.*, 2006, **16**, 83.
39. S. Janietz, D. D. C. Bradley, M. Grell, C. Giebeler, M. Inbasekaran, E. P. Woo, *Appl. Phys. Lett.*, 1998, **73**, 2453.

Subject Index

Reference to figures are given in *italic* type.

Abbé number, 368–9
acetylene bipropiolates, 356
σ-acetylide complexes, 59–60, 74–8, 86–7
  azo-, 342–3
  by metal
    gold, 68–71, 69–71, 75, 78, 260–1, 342
    iron, 91, 92–3
    manganese, 94
    mercury, 71–2
    nickel, 75–6
    osmium, 72–3
    palladium, 60–2, 75
    platinum, 60, 60–1, 75, 260, 269–74
    ruthenium, 72–3, 99–101
    silver, 71
    tungsten, 93, *94*
  colour tuning, 59–60
  conductance measurements, 96
  diamagnetic anisotropy, 77
  molecular conductance, 88
  nonlinear optical (NLO) properties, 74–5, 256–9, 343
  polynuclear, 90
  pyridyl, 269–74
  σ-bond formation, 87–8
  stepwise synthesis
    methane elimination, 91–2
    vinylidene formation, 90–1
  terminal binding groups, 88–9

acridine, 148
additive colour mixing, 2
albumins, 152, 157, 308
σ-alkynyl derivatives *see* σ-acetylide complexes
amine-reactive protein labels, 152–6
anthracene, 149–50
antitumour complexes, 311–12
Arthrobacter globiformis copper amine oxidase, 169–70
arylamine, 15–16
atom-transfer radical polymerisation (ATRP), 210–11, 303
avidin, 310–11
azo complexes, 199
  cobalt, 334–5, 344–5
  dyes, 317–18, 327–8
  ferrocene, 318–28
  iron, 329–33
  metalladithiolene, 339–42
    platinum, 340–2
  methacylate polymers, 344–5
  osmium, 333–4
  palladium(II), 318, 339–40
  pyridine and bipyridine, 333–7
  terpyridine, 338–9

benzothiadiazole, 386
benzothiazole, 4
benzyl azide, 47
biotin complexes, 144–5, 154–5, 164–9